Essentials of electromagnetics for engineering

Essentials of Electromagnetics for Engineering provides a clearly written introduction to the key physical and engineering principles of electromagnetics. Throughout the book, the author describes the intermediate steps in mathematical derivations that many other textbooks leave out.

The author begins by examining Coulomb's law and simple electrostatics, covering in depth the concepts of fields and potentials. He then progresses to magnetostatics and Maxwell's equations. This approach leads naturally to a discussion of electrodynamics and the treatment of wave propagation, waveguides, transmission lines, and antennas. At each stage, the author stresses the physical principles underlying the mathematical results.

A large number of exercises is provided; these include several in MATLAB® and MATHEMATICA® formats. The book contains a separate chapter on numerical methods in electromagnetics, and a broad range of worked examples to illustrate important concepts. It is suitable as a textbook for undergraduate students of engineering and applied physics taking introductory courses in electromagnetics.

David de Wolf is Professor of Electrical Engineering at Virginia Polytechnic Institute and State University. He has carried out pioneering work in the field of electromagnetic scattering and is the author of the book *Basics of Electron Optics*. He is a Fellow of the Optical Society of America and the IEEE.

ESSENTIALS OF ELECTROMAGNETICS FOR ENGINEERING

DAVID A. DE WOLF

Virginia Polytechnic Institute and State University

CAMBRIDGE
UNIVERSITY PRESS

PUBLISHED BY THE PRESS SYNDICATE OF THE UNIVERSITY OF CAMBRIDGE
The Pitt Building, Trumpington Street, Cambridge, United Kingdom

CAMBRIDGE UNIVERSITY PRESS
The Edinburgh Building, Cambridge CB2 2RU, UK
40 West 20th Street, New York, NY 10011-4211, USA
10 Stamford Road, Oakleigh, VIC 3166, Australia
Ruiz do Alarcón 13, 28014 Madrid, Spain
Dock House, The Waterfront, Cape Town 8001, South Africa

http://www.cambridge.org

First published 2001

Printed in the United Kingdom at the University Press, Cambridge

Typeface Times 10/14pt. System 3B2 6.03 [ADVENT]

A catalogue record for this book is available from the British Library

ISBN 0 521 66281 8 hardback

Contents

Selection of contents for a one-semester course

In some university curricula there is room only for a one-semester course in electromagnetics. The following selection from the contents of this text is suggested in that case. Not all of the material excerpted below is essential, but at least some from each of these sections is recommended.

1

Introduction

Why yet another intermediate text on electromagnetic field theory when there are so many to choose from, for upper-level electrical-engineering undergraduates? The present text grew from course notes developed during many years of teaching first a three-quarter, later a two-semester course on the subject to third-year undergraduates in electrical engineering at Virginia Polytechnic Institute and State University (Virginia Tech.). It was hard to agree upon a textbook that satisfied both students and faculty; in a period of roughly 12 years six changes of text occurred. One text was easy to read and short, and therefore appetizing to the students; its coverage of the topics, however, was deemed insufficient by the faculty. Another text gave excellent coverage of the desired material, but was considered hard to read by the students. A third text distinguished itself by extensive coverage of applications, but was lacking in explanation of the principles. All of the existing texts seem – at least to the present author – to fall short in explaining in clear English what is going on in the mathematics. While that seems a minor problem to many instructors, it is a severe problem for today's students, who seem to rely more and more upon software tools and are less prepared to deal with the mathematical concepts than students of a generation ago. Thus the present text is motivated strongly by the author's wish to communicate the *content* of a mathematical statement in physical and visual terms, as these – in his experience – seem to leave the strongest, and thus longest lasting, imprint in the minds of his audience. In contrast to many other texts, less space is devoted here to market-place applications (which can be dealt with more easily at a later stage, or which in any event are not really crucial to understanding). Many *examples*, however, replace discussion of the more elaborate applications; such examples were chosen so as to avoid the many nonessential details of real-world applications and yet to illustrate how the basic material applies to actual physical situations.

Another goal of this text is to provide the student with more detailed intermediate steps in leading from the formulation of a problem to the equations and from the equations to the physical quantities they describe. As there are always arguments about the amount of detail that is needed, many of the more detailed sections have been enclosed in shaded boxes to indicate that they are optional reading material, which can be omitted without losing the line of reasoning.

There appear to be several schools of thought about the organization of material. Many of the older texts follow a more or less chronological order, in which electrostatics, then magnetostatics, and finally electrodynamics are discussed, starting with Coulomb's law and leading to Maxwell's equations with intermediate steps. Some of the newer texts start from Maxwell's equations and discuss the consequences, often relegating electrostatics and magnetostatics to later chapters and even more often abbreviating their coverage. I prefer the former approach because – sad to say – the large majority of today's students are ill-prepared to make the mental leap into the unknown required by an *ab initio* treatment of electromagnetic field theory from Maxwell's equations. The advantage of a gentle slope in dealing with the increasing intricacies of the mathematics along the path from Coulomb's law towards the four Maxwell equations offsets, in my opinion, the relative 'disadvantage' of having to spend time on a formulation closer to the historical development. Besides, a number of numerical-integration methods are closely related to, say, Coulomb's law. The time spent in building up to Maxwell's equations is time well spent, from a didactic point of view!

This text adheres strictly to the rationalized-MKS–Giorgi system of units, as is common practice today. To adapt these to practical units such as microns, picofarads, and gauss where needed seems quite sufficient, in our opinion. No matter what system of units is used, some parameters have vast ranges (e.g. conductivities range over 20 orders of magnitude).

The material is organized as follows. The present chapter outlines the philosophy and structure of this text. Chapter 2 reintroduces the student to the essentials of vector calculus needed for understanding the material and it is probably useful to include this as part of a course in electromagnetism. The mathematics that our students seem to have most difficulty with – the use of gradient, divergence, and curl – has been relegated to those sections of the subject matter where they first become essential.

Chapter 3 introduces the electrostatic field via Coulomb's law and works its way towards the Gauss's-law integral formulation. Special attention is devoted to line integrals; these, in my experience, pose great difficulties for our students, which have not been eased by any of the texts used at Virginia Tech. This also is a natural place to introduce the concept of gradient and to tie it closely to its geometrical meaning, so that it can be used in Chapter 4 to link the electrostatic field to the electrostatic potential in a geometrically meaningful way. In Chapter 4 we also discuss the elementary electrostatic dipole, which is the starting point for understanding the behavior of electrostatic fields in dielectric materials.

Chapter 5 prepares the reader for Maxwell's equations for the electrostatic fields and introduces these at the end of the chapter. Here, the reader needs to know what the divergence and curl of a vector field are, and these concepts are accordingly treated in a fashion that makes them relevant. The Helmholtz theorem seems to me to be too advanced a concept to explain in great detail to a third-year engineering student, but it is possible to say something relevant about it without invoking all the detail. Consequently this chapter includes what is needed to understand why potentials are used to replace zero-curl fields (electrostatics) or zero-divergence fields (magnetostatics). General properties of the electric and magnetic field lines are derived here from Maxwell's equations.

Metallic conductors and dielectrics are introduced in Chapter 6 by considering unbound charges in the former and bound charges in the latter. The discussion of dielectrics moves from dipoles to polarization to permittivity in a relatively straightforward fashion. The discussion of boundary conditions is applied to various types of cavity in a dielectric. Finally, some comments on anisotropy in a dielectric (e.g. as in biaxial or uniaxial crystals) are included.

The concepts of work and energy in Chapter 7 lead naturally to that of energy storage and thus capacitance. Chapter 8 is devoted to reducing the $\operatorname{div} \mathbf{D} = \varrho_{\mathrm{v}}$ equation to tractable partial second-order differential equations for the potential such as the Poisson and Laplace equations, which have been studied exhaustively for over 200 years. Conditions for obtaining solutions unique to an experimental situation are given, and the simplest analytical methods are discussed. The separation-of-variables method is perhaps given more space than it merits in a text of this nature, but much of it can be skipped in even a two-semester course.

As analytical solutions of the differential equations discussed in Chapter 8 are seldom possible, Chapter 9 devotes attention to numerical solutions. At the time of writing (1992–99), it appears that the method of finite elements is the most flexible and useful numerical tool for solving, say, Laplaces's equation in a finite closed region. However, it is arguably too complicated to describe in a text of this nature, and I have decided to retain the older and much simpler method of finite-mesh relaxation to explain in more detail. Several MATHEMATICA® and MATLAB® programs are included to illustrate two-dimensional applications; both types of program can be modified without great difficulty.

As in some other texts, we choose at this stage to introduce the concept of current, in Chapter 10, and in particular to define a conduction current carefully, as it is not intuitive to third-year students that such a current is determined to have a steady *velocity*, rather than an acceleration, as the result of a force. Chapter 11 starts the treatment of magnetostatics. We have chosen a somewhat unusual way of introducing the Biôt–Savart law, which emphasizes the analogy with Coulomb's law; this analogy is well rooted in the special theory of relativity, itself beyond the realm of treatment here. The student is thus given to understand from the very start that electric and magnetic fields are closely related. As in electrostatics, a straightforward progression is made to Maxwell's laws, but now those for magnetostatics. Chapter 12 introduces the potentials by which the magnetic fields can be calculated and also the magnetic dipole, in preparation for the discussion of magnetic materials.

Chapter 13 is a corollary to Chapter 7 and it connects the circuit concept of inductance to that of magnetic stored energy. The forces derived from the stored energy in the fields are more complicated than is the case in electrostatics, and Chapter 14 applies them to find the torque on a current loop and thus to explain the effect of a magnetic field upon a magnetic dipole. The behavior of magnetic fields in magnetic materials, with emphasis upon ferromagnetic materials, forms the rest of the chapter. An introductory treatment of diamagnetic and paramagnetic materials is given; a fuller discussion would require quantum mechanics, which is also beyond the scope of this text.

At this stage, we are ready to extend Maxwell's equations to electrodynamics, and Chapter 15 does so through Faraday's law and Maxwell's displacement current successively. Close attention is devoted to Faraday's law, which is more subtle than seems to be realized in most engineering texts. Some illuminating examples from Feynman[1] are included to help make that point. Chapter 16 then deduces the wave equation from Maxwell's equations and discusses its consequences for a narrow-band signal in the form of idealized plane time-harmonic waves, which are treated at length in Chapter 17. Likewise, the concept of power flow or flux is introduced here in connection with the Poynting vector.

The behavior of electromagnetic waves in transition from one medium to another is introduced by regarding the incidence of plane time-harmonic waves upon an interface; this is done in Chapter 18. The various associated phenomena of reflection and refraction upon transmission are discussed with particular emphasis upon some refractive phenomena that occur in nature. Diffraction, however, is left to a short discussion in Chapter 21.

In contrast to many texts, we prefer to precede a discussion of transmission lines by one on waveguides in general, because transmission lines are, after all, only a special type of waveguide. After a systematic treatment of the principal types of waveguide in Chapter 19 (the treatment is by no means exhaustive), the student is then introduced in Chapter 20 to transmission lines. The treatment concentrates on the essentials and does not delve deeply into the many techniques of impedance matching, etc., that are usually the topic of follow-up microwave courses.

[1] R.P. Feynman *et al.*, *The Feynman Lectures On Physics*, Addison-Wesley, Reading MA, 1964.

Each of the above chapters has a set of problems at the end to exercise the student in the concepts discussed in the chapters. Highly recommended for auxiliary problems are the following texts: *Theory and Problems of Electromagnetics*, second edition, J.A. Edminster (Schaum's Outline Series, McGraw-Hill, New York, 1993) and *2000 Solved Problems in Electromagnetics*, S.A. Nasar (Schaum's Solved Problems Series, McGraw-Hill, New York, 1992).

Finally, in Chapter 21 we give some material on a number of topics such as antennas, diffraction theory, geometrical optics, etc. Each of these topics is usually treated extensively in special courses at either the undergraduate or graduate level, and it is probably unrealistic to expect greater coverage in a two-semester or three-quarter undergraduate course.

Appendix A.1 reviews a few essentials from calculus and complex numbers. It could be omitted from a taught course but is included here so that the student can brush up as needed. A number of other appendix sections give lists of physical constants, vector relationships and material constants.

As in all texts, the content is dictated largely by the curriculum requirements at a university, and it seems that the curriculum at Virginia Tech is somewhat more comprehensive than that at many similar institutions. Of course there is also an element of personal choice in deciding how much or how little coverage to give to each topic. I hope that my own choices, based on long experience at Virginia Tech, will adequately reflect the needs at other places also.

The level of preparedness that I expect students to have should include several semesters of calculus, with exposure to multiple integrals, some linear algebra, the essentials of complex numbers, and some elementary vector algebra. I do not expect much proficiency in the latter. I assume some high-school-attained ability in trigonometry, and some programming skill in computer languages. We seem to be in a transition towards higher-level computer languages such as MATHEMATICA®, so that I am loathe to prescribe any particular ones. As the ability to support such higher-level languages seems to depend upon choices made by individual departments or colleges, I have had to make some choices. For the present, I have incorporated MATHEMATICA® and/or MATLAB® programs as a minimum.

It remains for me to thank my colleagues I.M. Besieris, R.O. Claus, S.M. Riad, and W.L. Stutzman for reading sections of this text and providing me with feedback and suggestions. Several of my students have contributed to parts of the MATHEMATICA® and MATLAB® programs. I am grateful also to my students as well as other readers of this text for pointing out typographical and other errors. I hope readers will forward to me any other errors they find so that corrections can be incorporated in future printings. I am profoundly grateful to Dr Susan Parkinson of Cambridge University Press (CUP), who did a remarkable job of editing, and to whose critical acumen I can attribute the removal of dozens of inaccuracies that escaped my notice as well as that of many other readers. It also has been a great pleasure to work with Dr Phil Meyler, Dr Jo Clegg, and many others at CUP in preparing the manuscript for publication.

David A. de Wolf

Virginia Tech.

2

Some elements of vector analysis

Although the student is assumed to have taken a course in vector analysis before embarking upon a study of electromagnetics, we believe it would be useful to review briefly some of the most basic aspects of vectors and vector fields. Students need to be familiar with vectors because the significant quantities in electromagnetics are vectors or vector fields. Indeed, an important feature usually not emphasized in undergraduate treatments of vectors is the distinction between vectors and vector fields. This distinction is somewhat subtle, but it is of importance and an understanding of it can help the student of electromagnetics overcome difficulties often obscured by similarities in notation. It will be elaborated below. Another less routine feature we constantly encounter is vector operators; in fact Maxwell's equations are formulated in terms of them. Vector operations are associated with basic physical concepts, and an understanding of these will aid the student greatly. This association is also a good reason to postpone the treatment of the three main vector operators to later chapters where it fits in logically with their subsequent use in Maxwell's equations.

Finally, this chapter will deal with coordinate systems, and at the outset the reader will perceive that vectors are almost universally expressed in terms of a coordinate system. The magnitude and direction of a vector cannot depend upon a coordinate system, so that it does make sense to speak of vectors independently of coordinate systems, but even the earliest experiences with vectors normally employ a coordinate system to express the magnitude and direction of a vector. Various different coordinate systems are commonly used, and it is helpful to be familiar with the transformations from one to another. Treatment of these transformations is simplified appreciably by looking at projections of a three-dimensional vector upon certain well-chosen planes, as we will see.

2.1 Simple numerical vectors

As you may remember, a *scalar* is the name given to a simple numerical quantity. Temperature, weight, altitude, pressure, length, volume, etc., are all scalar quantities because each can be specified by an algebraic variable having a range of numerical values. A *vector* differs from a scalar because its specification includes a *direction* as well as a numerical value. Examples of vectors are velocity, force, momentum, acceleration, etc.

Let us spend a moment on the notation used in this book. Whereas we denote a scalar quantity with an italic lower- or upper-case letter, e.g. a or A, we will denote a vector by means of a boldface roman letter, e.g. **a** or **A** (this is done in handwritten text by adding an arrow to the scalar symbol, e.g. $\vec{a}$ or $\vec{A}$, or by underlining with a straight or wavy line). Some vectors have unit length, i.e. the magnitude of such a vector is 1. Because the vector can point in any direction, we still need to indicate such *unit vectors* by a symbol. We will place a *caret* above a boldface Roman symbol, e.g. $\hat{\mathbf{a}}$ or $\hat{\mathbf{A}}$, to indicate such unit vectors. Summarizing:

$$\begin{aligned} &a, A \text{ are scalars} \\ &\mathbf{a}, \mathbf{A} \text{ are vectors of arbitrary magnitude} \\ &\hat{\mathbf{a}}, \hat{\mathbf{A}} \text{ are vectors of unit magnitude} \end{aligned} \tag{2.1}$$

Any vector can be written as a product of a scalar (representing its magnitude) and a unit vector (representing its direction), e.g. $\mathbf{A} = A\hat{\mathbf{a}}$ or $\mathbf{A} = \hat{\mathbf{a}}A$. Sometimes the magnitude A is written more fully as $|\mathbf{A}|$. Thus, a unit vector can be defined as $\hat{\mathbf{a}} \equiv \mathbf{A}/A$ or $\hat{\mathbf{a}} \equiv \mathbf{A}/|\mathbf{A}|$.

Example 2.1

A vector represents a magnitude with a given direction, but where the vector is located might not be exactly specified. If so, you may 'pick up' the vector and displace it in any direction, as long as you do not change the magnitude or direction the vector is pointing. For example, the velocity vector of a car traveling due west at a speed of 50 mph represents both the direction (west) and the magnitude (50 mph) of the car's velocity (if we ignore the curvature of the Earth for distances of 50–100 km or so). You do not have to say where the car is; the same vector describes any car traveling due west in the region at this speed.

Vectors obey some almost self-evident rules of algebra. Multiplication of a vector **A** by a scalar quantity b produces a new vector in the same direction that is b times as long; i.e. the new vector is $b\mathbf{A}$ or, equivalently, $\mathbf{A}b$. The order in which b and **A** are written down is arbitrary, but it is customary to write $b\mathbf{A}$. The geometrical depiction of vector **A** is an arrow of length A pointing in the direction $\hat{\mathbf{A}}$.

Any two vectors **A** and **B** can be added together and the sum vector can be written as either $\mathbf{A} + \mathbf{B}$ or as $\mathbf{B} + \mathbf{A}$ (*commutative* rule of addition). The geometrical construction of the sum vector occurs by moving one vector (either **A** or **B**) without changing its direction until the base of that vector touches the tip of the other vector.

The sum vector then points from the base of one vector to the tip of the other vector; see Fig. 2.1.

In an equivalent method of addition, the parallelogram rule, one vector is moved until its base touches the base of the other vector. The sum vector is then the diagonal of a parallelogram formed from the ensuing pair of vectors; see Fig. 2.2. The other diagonal must then correspond either to $\mathbf{A} - \mathbf{B}$ or to $\mathbf{B} - \mathbf{A}$, depending in which of the two possible directions the arrow on this diagonal is made to point; see Fig. 2.3. In particular, $\mathbf{A} - \mathbf{B}$ is interpreted as $\mathbf{A} + (-\mathbf{B})$; see again Fig. 2.3. Later, when we discuss coordinate systems, we will learn how to calculate the sum or difference algebraically by adding vector components, but for now we restrict ourselves to the geometrical method.

The *distributive* rule also holds: $\mathbf{A} + (\mathbf{B} + \mathbf{C}) = (\mathbf{A} + \mathbf{B}) + \mathbf{C}$. In fact, the commutative and distributive rules simply confirm that the addition of vectors can be done in any order or sequence.

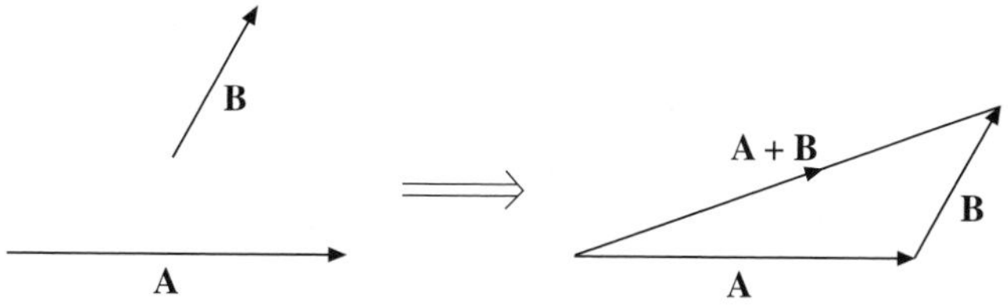

Fig. 2.1 Illustration of vector addition. The arrowhead may be placed at the centre of the arrow if more convenient.

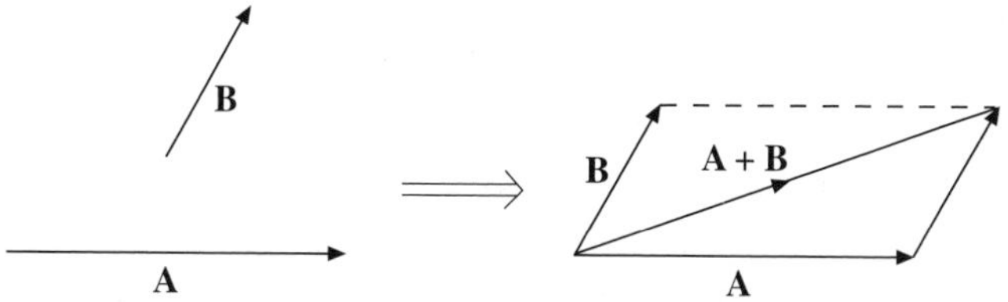

Fig. 2.2 Alternative vector addition: the parallelogram rule.

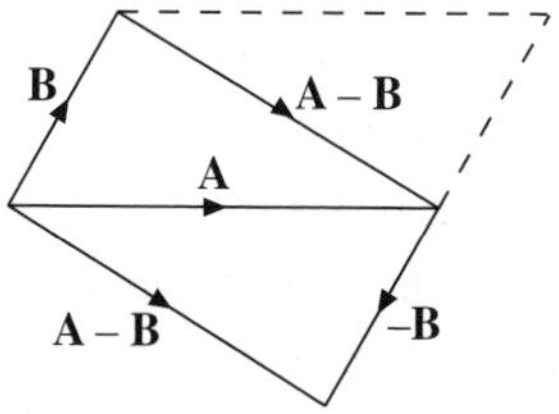

Fig. 2.3 Addition and subtraction of vectors.

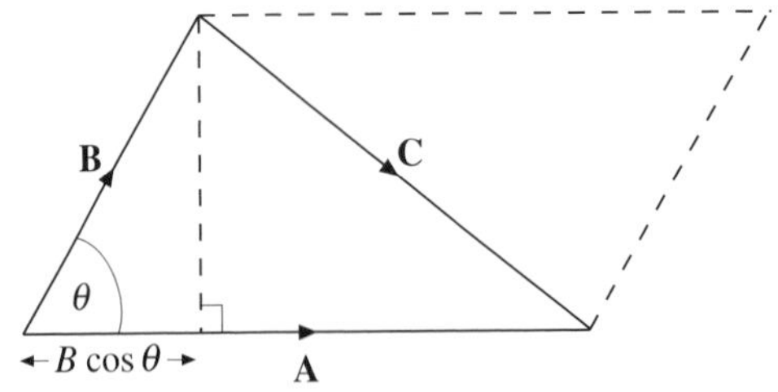

Fig. 2.4 Projection of **B** on **A** and the cosine rule.

Two 'products' involving the vectors **A** and **B** are of importance. The first, the *scalar* or *dot product*, is defined as

$$\mathbf{A} \cdot \mathbf{B} \equiv AB \cos \theta \tag{2.2}$$

One may think of the scalar product of the magnitude of one vector times the magnitude of the *projection* of the other vector upon the first, e.g. as shown in Fig. 2.4. It allows us to define the magnitude of a vector conveniently:

$$A = \sqrt{\mathbf{A} \cdot \mathbf{A}} \tag{2.3}$$

Other obvious consequences of (2.2) are $\mathbf{A}{\cdot}\mathbf{B} = \mathbf{B}{\cdot}\mathbf{A}$ and

$$\mathbf{A} \cdot (\mathbf{B} + \mathbf{C}) = \mathbf{A} \cdot \mathbf{B} + \mathbf{A} \cdot \mathbf{C}$$

The second of these is most easily proved using a coordinate system (see later). Another consequence is the *cosine law* of triangles. In this triangle of Fig. 2.4, vector $\mathbf{C} = \mathbf{A} - \mathbf{B}$ and therefore, using the above rules,

$$\begin{aligned} C^2 &= (\mathbf{A} - \mathbf{B}) \cdot (\mathbf{A} - \mathbf{B}) = A^2 + B^2 - 2\mathbf{A} \cdot \mathbf{B} \\ &= A^2 + B^2 - 2AB \cos \theta \end{aligned} \tag{2.4}$$

which proves the cosine law.

The other product of **A** and **B** is the *vector* or *cross product*, and we write it as

$$\mathbf{A} \times \mathbf{B} = \hat{\mathbf{n}} AB \sin \theta \tag{2.5}$$

It represents the *area* of the parallelogram spanned by vectors **A** and **B**, and the unit vector $\hat{\mathbf{n}}$ is perpendicular ($\perp$) to the plane in which both vectors lie, in such a way that its direction (up or down) is found by rotating **A** towards **B** through the *shortest* angle θ; $\hat{\mathbf{n}}$ is then in the direction in which a right-handed screw would move (see Fig. 2.5).

The cross product therefore has the property that

$$\mathbf{B} \times \mathbf{A} = -\mathbf{A} \times \mathbf{B}$$

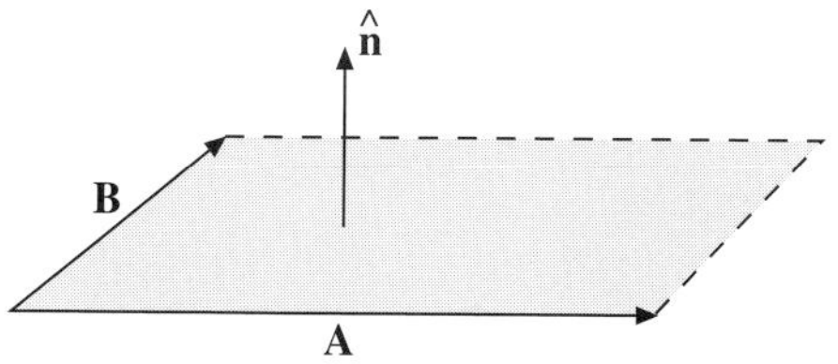

Fig. 2.5 Direction of the cross product $\mathbf{A}\times\mathbf{B}$.

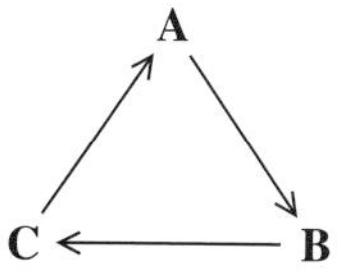

Fig. 2.6 Cyclical permutation.

which implies that the cross product of a vector with itself must be zero since only the zero vector can be its own negative. It also has an obvious distributive law:

$$\mathbf{A}\times(\mathbf{B}+\mathbf{C})=\mathbf{A}\times\mathbf{B}+\mathbf{A}\times\mathbf{C}$$

which – again – is most easily proved after we discuss coordinate systems.

Finally, there are only two product combinations of three vectors that make any sense. They are

$$\begin{aligned}\mathbf{A}\cdot(\mathbf{B}\times\mathbf{C})&=\mathbf{B}\cdot(\mathbf{C}\times\mathbf{A})=\mathbf{C}\cdot(\mathbf{A}\times\mathbf{B})\\ \mathbf{A}\times(\mathbf{B}\times\mathbf{C})&=\mathbf{B}(\mathbf{A}\cdot\mathbf{C})-\mathbf{C}(\mathbf{A}\cdot\mathbf{B})\end{aligned}\tag{2.6}$$

A proof is postponed to subsection 2.5.4 after we have introduced coordinate components. The first of these *triple products* is a scalar, which represents the *volume* of the parallelepiped formed by three arbitrary vectors **A**, **B**, and **C** when their bases meet at one point. Note the *cyclical permutation* of the three terms: we can find the others from the first by thinking in terms of an equilateral triangle, as in Fig. 2.6, and rotating clockwise by 120° to replace the old vertices by the new ones. Thus, one rotation would lead to the three substitutions $\mathbf{A}\to\mathbf{B}$, $\mathbf{B}\to\mathbf{C}$, $\mathbf{C}\to\mathbf{A}$. The second of these triple products is the *vector triple product*; the 'bac–cab' rule for it, given in (2.6), is shown to be true fairly easily with a coordinate system.

Example 2.2

An application of the vector triple product that we shall encounter often in electromagnetics is at the surface of a plane, where $\hat{\mathbf{n}}$ is a unit vector $\perp$ to that plane and **A** is an arbitrary vector:

$$\mathbf{A}=-\hat{\mathbf{n}}\times(\hat{\mathbf{n}}\times\mathbf{A})+\hat{\mathbf{n}}(\hat{\mathbf{n}}\cdot\mathbf{A})\tag{2.7}$$

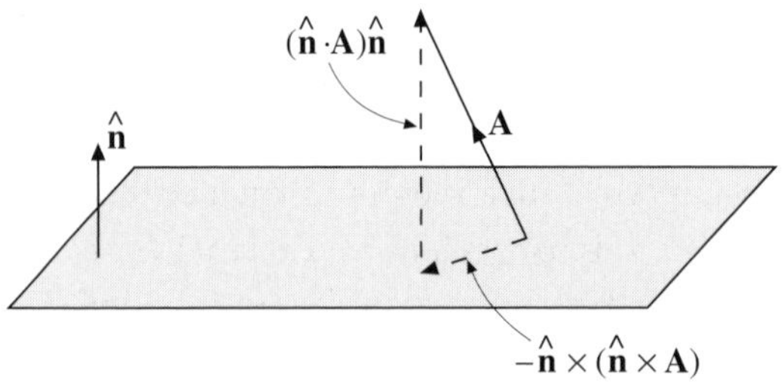

Fig. 2.7 Decomposition of a vector into normal and tangential components.

This follows from the second of Eqs. (2.6) with the substitutions $\mathbf{A} \rightarrow \hat{\mathbf{n}}$, $\mathbf{B} \rightarrow \hat{\mathbf{n}}$, and $\mathbf{C} \rightarrow \mathbf{A}$ and with the use of $\hat{\mathbf{n}} \cdot \hat{\mathbf{n}} = 1$. Finally we rearrange, to obtain (2.7). The meaning of the first term on the right-hand side becomes clear when it is realized that the second term, $\hat{\mathbf{n}}(\hat{\mathbf{n}} \cdot \mathbf{A})$, is the projection of vector **A** *perpendicular* to the plane: the first term, $-\hat{\mathbf{n}} \times (\hat{\mathbf{n}} \times \mathbf{A})$, must then be the projection of **A** *upon* the plane. Thus, the right-hand side of (2.7) represents, respectively, the *tangential* and *normal* components of **A** with respect to the plane (see Fig. 2.7).

Some texts incorrectly label $\hat{\mathbf{n}} \times \mathbf{A}$ as the tangential component, but this vector does not lie in the plane formed by **A** and its normal component, even though it has the correct magnitude of the tangential component!

2.2 Vector fields

The concept of a *vector field* is deceptively simple, but it requires careful attention by the reader encountering it for the first time. An intuitive model for a vector field is that of the wind velocity in the atmosphere. We have a three-dimensional volume of air, at each point of which there is motion at a certain velocity. We represent the magnitude and direction of this velocity by a *velocity vector*. The ensemble of such vectors over the three-dimensional continuum of points in the volume constitutes a vector field.

Each vector **A** in the vector field belongs to a point $\mathbf{r} = (x, y, z)$ in space. Here **r** indicates the *location* of the point; it is a vector representing the length and direction of the point from the origin. The three Cartesian coordinates x, y, z represent the projections of the vector **r** along three mutually perpendicular directions from the origin.

That **A** depends on position is indicated by the notation $\mathbf{A}(\mathbf{r})$.

We can combine a number of vector fields acting in the same volume; this means that we combine their representative vectors at each point. For example, $\mathbf{A}(\mathbf{r}) + \mathbf{B}(\mathbf{r})$ is a vector field that consists of vectors $\mathbf{A} + \mathbf{B}$ at each point **r** of the space in which it is defined. We also can perform certain elementary operations such as obtaining the vector component field perpendicular to the (flat) earth's surface. It is the field composed of all vectors $[\hat{\mathbf{n}} \cdot \mathbf{A}(\mathbf{r})]\hat{\mathbf{n}}$, if $\hat{\mathbf{n}}$ is the unit vector perpendicular to the earth's surface. Here, we may consider $\hat{\mathbf{n}}$ to be just a vector, not a vector field, and it can be combined with all the representative vectors of a vector field and still give something meaningful. (However, this particular $\hat{\mathbf{n}}$ is in a trivial sense itself a vector field: one in which all $\hat{\mathbf{n}}(\mathbf{r})$ are identical in magnitude and direction.)

Vector fields are important in electromagnetics because charged particles are influenced in their motion at every point of a combined electric and magnetic vector field.

2.3 Vector operators

Although engineering undergraduate calculus does not always stress it, *scalar operators* are familiar quantities. For example, the derivative of a function $f(x)$ can be written as $df(x)/dx$. If we write this as $\partial_x f(x)$, where $\partial_x \equiv d/dx$, then we observe that ∂_x is an operator that *works on* $f(x)$. Furthermore, the product of a scalar and an operator, e.g. $a\partial_x$, has obvious meaning since it symbolizes the operation of finding a derivative and multiplying this by the constant a, for any differentiable function $f(x)$.

When we have a three-dimensional function of the Cartesian coordinates x, y, z, we can define partial derivatives with respect to each of these three coordinates, $\partial_x f(x, y, z)$, $\partial_y f(x, y, z)$, and $\partial_z f(x, y, z)$, where ∂_y and ∂_z are the derivative operators with respect to y and z respectively. We can also define $\boldsymbol{\partial}_x \equiv \hat{\mathbf{x}}\partial_x$, where $\hat{\mathbf{x}}$ is a unit vector pointing in the x direction. This is a *vector operator*, and it assigns a vector in the direction $\hat{\mathbf{x}}$ with magnitude $df(x)/dx$ when it works upon $f(x)$. Such operators are very useful in electromagnetics because they help define the electric and magnetic fields, which we will come to know well shortly, in terms of other, scalar, fields that are easier to calculate. In fact, although at this time it seems as if we can define only three such vector operators (namely for the three Cartesian directions), we will extend this concept, when the time comes, to express the idea of taking derivatives in *any* direction, even though the three main axial directions have been fixed.

2.4 The three major coordinate systems

Having introduced the notion of labeling points by vectors $\mathbf{r}$ in order to be able to characterize a vector field $\mathbf{A}(\mathbf{r})$, it is logical that we now turn our attention to schemes for describing the location of any point with respect to some conveniently defined fixed *origin*. This is the purpose of a *coordinate system*, the best-known and simplest of which is the *Cartesian coordinate system*, mentioned above.

2.4.1 Cartesian coordinates

A general method of defining coordinates is by introducing three sets of (curved) surfaces in three-dimensional space, because any point is the intersection of three arbitrary surfaces (with some exceptions such as the case where two surfaces are parallel). While this can be done for Cartesian coordinates, we suggest a somewhat simpler procedure that amounts to the same thing. When a fixed point of reference (the origin) has been decided upon, we define three mutually orthogonal straight-line axes through it and label them the x, y, and z axes. Points in space are labeled by the distances of their projection upon each of these axes from the origin, i.e. the *coordinates* x, y, and z. Each coordinate can take values from $-\infty$ to $+\infty$. The origin O is by definition at the point characterized by $x = 0$, $y = 0$, and $z = 0$, or – more succinctly – by the point $(0, 0, 0)$. An arbitrary point P in space is then labeled by its three Cartesian coordinates, (x, y, z). Its distance from the origin is $r = \sqrt{x^2 + y^2 + z^2}$. Its projection upon the $z = 0$ plane is the point P_0, $(x, y, 0)$, the distance of which from the origin is $\rho = \sqrt{x^2 + y^2}$. The line OP makes a *polar* angle θ with the z axis, and the line OP_0 makes an *azimuthal* angle φ with the x axis; see Fig. 2.8.

Finally, the normal convention is to use *right-handed* coordinate systems, i.e. if we move the x axis towards the y axis in the $z = 0$ plane the ensuing rotation should correspond to that of a right-handed screw, which would then move in the $+z$ direction. The unit vectors $\hat{\mathbf{x}}$, $\hat{\mathbf{y}}$, and $\hat{\mathbf{z}}$ are true constants (their magnitude and direction do not depend upon location) and hence do not need arguments. (In other coordinate systems the unit vectors may vary from point to point.) The dot product of any two *different* unit vectors is zero, since they are mutually perpendicular and

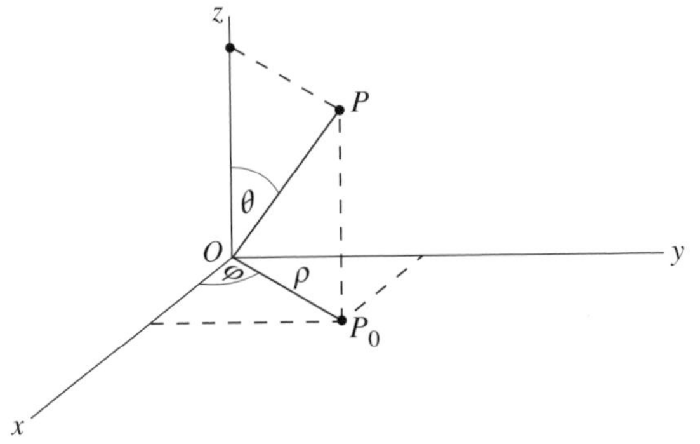

Fig. 2.8 Relation between Cartesian and cylindrical coordinate systems.

application of (2.2) would yield $\cos \pi/2 = 0$. Likewise, the cross product of any unit vector *with itself* must be zero, as (2.5) would yield $\sin 0 = 0$. The right-handedness of the coordinate system, and (2.5), yield

$$\hat{\mathbf{x}} = \hat{\mathbf{y}} \times \hat{\mathbf{z}}, \qquad \hat{\mathbf{y}} = \hat{\mathbf{z}} \times \hat{\mathbf{x}}, \qquad \hat{\mathbf{z}} = \hat{\mathbf{x}} \times \hat{\mathbf{y}} \tag{2.8}$$

2.4.2 Cylindrical coordinates

Another useful description, especially when there is a strong element of cylindrical symmetry in the situation to be described (e.g. a cylindrically shaped piece of metal), is the *cylindrical coordinate system*. Here, the z axis and coordinate are unchanged from the Cartesian system, and the idea is to assign the z direction to the rotational axis of symmetry or to whatever direction corresponds with the least change in properties of a scalar or vector field. Figure 2.9 discloses that the length ρ and angle φ are useful alternatives to x and y for describing the position of a point in a $z =$constant plane. All points with constant z lie in a plane $\perp$ the z axis and at height z above the xy plane. All points with constant ρ lie on a circular cylinder centered on the z axis with radius ρ. All points with constant φ lie on a plane through the z axis that makes an angle φ with the x axis. The intersection of three such surfaces defines a point. We assign a vector to the length and direction of OP_0, namely $\rho \equiv x\hat{\mathbf{x}} + y\hat{\mathbf{y}}$, which we also will write (as is commonly done) as $\rho = (x, y, 0)$, and we note from the figure that

$$\begin{aligned} x &= \rho \cos \varphi \qquad & \rho &= \sqrt{x^2 + y^2} \\ y &= \rho \sin \varphi \qquad & \varphi &= \arctan(y/x) \end{aligned} \tag{2.9}$$

It is useful to keep in mind that we regard φ as an independent variable and, while ρ is often also an independent variable, we should remember that a cylindrical surface (not necessarily with a *circular* cross section) is usefully described by defining $\rho(\varphi)$ for all φ between 0 and 2π. We need $0 < \rho < \infty, 0 < \varphi < 2\pi$, and $-\infty < z < \infty$ in order to describe every point in three-dimensional space.

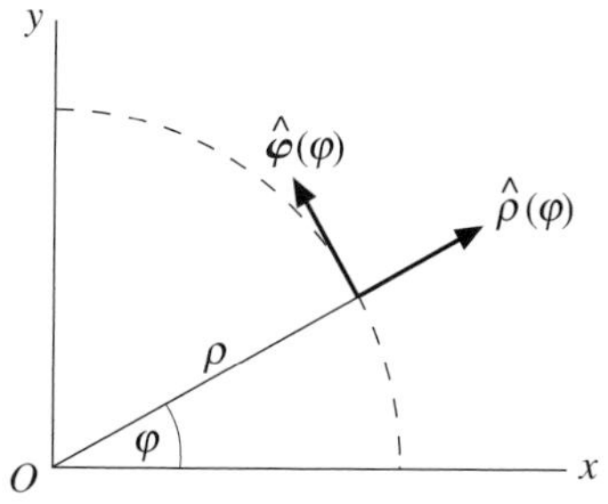

Fig. 2.9 Projection of the cylindrical coordinate system upon the $z = 0$ plane.

The unit vectors at a point in a $z =$ constant plane are $\hat{\boldsymbol{\rho}}(\varphi)$, $\hat{\boldsymbol{\varphi}}(\varphi)$, and $\hat{\mathbf{z}}$; see Fig. 2.9. Unlike the unit vectors of the Cartesian coordinate system, the first two unit vectors are functions of position, and specifically of the angle φ, i.e. their *direction* depends upon φ.

For the cylindrical unit vectors described here we have

$$\hat{\boldsymbol{\rho}} = \hat{\boldsymbol{\varphi}} \times \hat{\mathbf{z}} \qquad \hat{\boldsymbol{\varphi}} = \hat{\mathbf{z}} \times \hat{\boldsymbol{\rho}} \qquad \hat{\mathbf{z}} = \hat{\boldsymbol{\rho}} \times \hat{\boldsymbol{\varphi}} \tag{2.10}$$

2.4.3 Spherical coordinates

A third useful frame of reference is the *spherical coordinate system*, which exploits spherical symmetry best but can also be used to describe volumes bounded by surfaces that do not deviate too strongly from a spherical shape (see below). In this system, the location of a point P is given by its distance r to the origin, the *polar angle* θ that the vector $\overrightarrow{OP}$ makes with the z axis, and the *azimuthal angle* φ of the projection ($OP_0 = \rho$) of $\overrightarrow{OP}$ upon the $z = 0$ plane; see Fig. 2.10.

The three surfaces that now define a point are a sphere of radius r, a cone around the z axis with half-angle θ and with apex at the origin, and a plane through the z axis making an angle φ with the x axis. Obviously, φ has the same meaning as in the cylindrical coordinate system. We observe, from Fig. 2.10, that $z = r\cos\theta$ and $\rho = r\sin\theta$. We note that (2.9) still holds, hence

$$\begin{aligned} x &= r\sin\theta\cos\varphi \qquad & r &= \sqrt{x^2 + y^2 + z^2} \\ y &= r\sin\theta\sin\varphi \qquad & \theta &= \arctan(z/r) \\ z &= r\cos\theta \qquad & \varphi &= \arctan(y/x) \end{aligned} \tag{2.11}$$

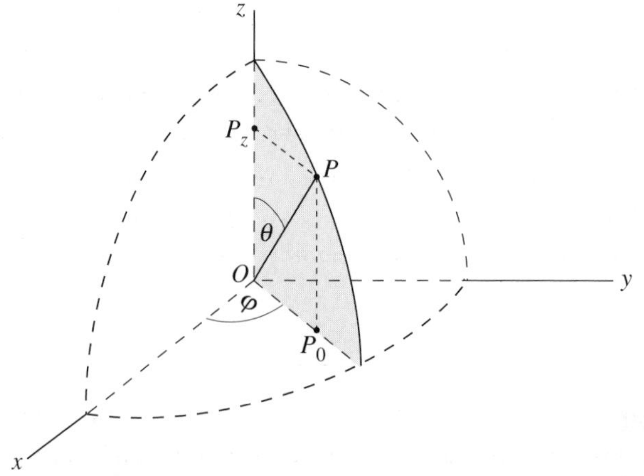

Fig. 2.10 The spherical coordinate system; $OP_0 = \rho$.

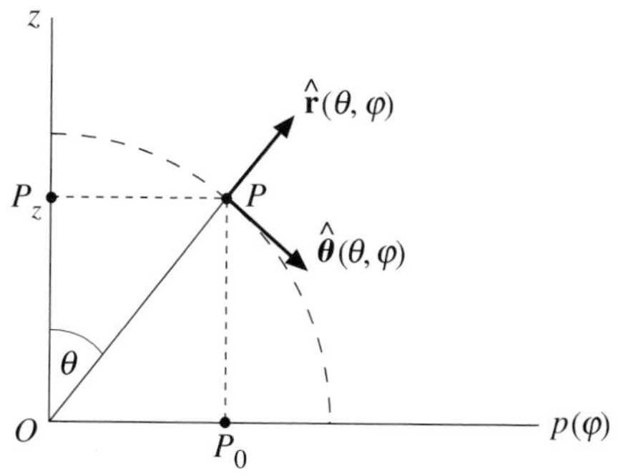

Fig. 2.11 Projection of the spherical coordinate system upon a plane where ϕ is constant.

In order to describe every point in three-dimensional space, we need $0 < r < \infty$, $0 < \theta < \pi$, and $0 < \varphi < 2\pi$. Figure 2.10 illustrates that these are the required bounds. Now, *two* of the unit vectors, $\hat{\mathbf{r}}(\theta, \varphi)$, $\hat{\boldsymbol{\theta}}(\theta, \varphi)$, are functions of *both* angles but $\hat{\boldsymbol{\varphi}}(\varphi)$ remains a function only of the azimuthal angle φ. This is illustrated in Fig. 2.11, which shows the plane containing OPP_0. The unit vector $\hat{\boldsymbol{\varphi}}(\varphi)$ at P is not shown; it would be straight *into* the paper. One can convince oneself of the φ-dependence of the unit vectors by 'taking hold' of the plane OPP_0 and swinging it around the z axis (thus varying the angle φ). Likewise, variation of the angle θ without changing angle φ (in Fig. 2.11) shows the dependence of $\hat{\boldsymbol{\rho}}$ and $\hat{\boldsymbol{\theta}}$ upon angle θ, and the fact that $\hat{\boldsymbol{\varphi}}$ is independent of θ. For spherical unit vectors we have

$$\hat{\mathbf{r}} = \hat{\boldsymbol{\theta}} \times \hat{\boldsymbol{\varphi}} \qquad \hat{\boldsymbol{\theta}} = \hat{\boldsymbol{\varphi}} \times \hat{\mathbf{r}} \qquad \hat{\boldsymbol{\varphi}} = \hat{\mathbf{r}} \times \hat{\boldsymbol{\theta}} \tag{2.12}$$

Note that Eqs. (2.8), (2.10), and (2.12) all exhibit the usefulness of cyclical permutation. One only needs to establish the first equation in each case; the second follows from the first and the third from the second through cyclical permutation of the coordinate variables in the order shown above.

A final comment follows regarding the use of all three coordinate systems. Remember, a coordinate system is there to serve the user, and one is free to choose the origin and the coordinate system as one wishes! For example, whenever we appear to have symmetry about a straight line, we would be wise to make that straight line the z axis and (in most cases) resort to a cylindrical coordinate system. Judicious choice of a coordinate system often makes a calculation much simpler.

2.5 Some vector identities

We now return to some of the statements made about vectors in the earlier sections of this chapter and use what we have learned so far about coordinate systems to verify those statements. Such verification also is a useful exercise in learning to use coordinate systems.

2.5.1 Sums and differences of vectors

The vectors **A** and **B** can be written in the Cartesian system as

$$\mathbf{A} = A_x\hat{\mathbf{x}} + A_y\hat{\mathbf{y}} + A_z\hat{\mathbf{z}} \qquad \text{and} \qquad \mathbf{B} = B_x\hat{\mathbf{x}} + B_y\hat{\mathbf{y}} + B_z\hat{\mathbf{z}}$$

The sum (top sign) or difference (bottom sign) vector **C** is therefore

$$\mathbf{C} = (A_x \pm B_x)\hat{\mathbf{x}} + (A_y \pm B_y)\hat{\mathbf{y}} + (A_z \pm B_z)\hat{\mathbf{z}} \tag{2.13}$$

If we redraw Fig. 2.1 and add a coordinate system, we obtain Fig. 2.12, which shows the result for addition. Subtraction goes in a similar fashion with Fig. 2.3.

2.5.2 Dot and cross products in the Cartesian system

By using the properties of the dot product of three mutually perpendicular unit vectors, namely that for $i, j = 1, 2,$ or 3 we have $\hat{\mathbf{x}}_i \cdot \hat{\mathbf{x}}_j = \delta_{ij}$ (δ_{ij} is the Kronecker delta: it is 1 when $j = i$ and 0 otherwise), we can see that

$$\begin{aligned}\mathbf{A}\cdot\mathbf{B} &= (A_x\hat{\mathbf{x}} + A_y\hat{\mathbf{y}} + A_z\hat{\mathbf{z}})\cdot(B_x\hat{\mathbf{x}} + B_y\hat{\mathbf{y}} + B_z\hat{\mathbf{z}})\\ &= A_xB_x + A_yB_y + A_zB_z \end{aligned}\tag{2.14}$$

Likewise, using (2.8) we can establish that

$$\begin{aligned}\mathbf{A}\times\mathbf{B} &= (A_x\hat{\mathbf{x}} + A_y\hat{\mathbf{y}} + A_z\hat{\mathbf{z}})\times(B_x\hat{\mathbf{x}} + B_y\hat{\mathbf{y}} + B_z\hat{\mathbf{z}})\\ &= (A_yB_z - A_zB_y)\hat{\mathbf{x}} + (A_zB_x - A_xB_z)\hat{\mathbf{y}} + (A_xB_y - A_yB_x)\hat{\mathbf{z}} \end{aligned}\tag{2.15a}$$

This is sometimes written as a determinant:

$$\mathbf{A}\times\mathbf{B} = \begin{vmatrix} \hat{\mathbf{x}} & \hat{\mathbf{y}} & \hat{\mathbf{z}} \\ A_x & A_y & A_z \\ B_x & B_y & B_z \end{vmatrix} \tag{2.15b}$$

Of course, one must be careful in using this latter form to remember that the unit vectors play a special role, and that they should either precede or follow the products of the quantities in the second and third rows when evaluating a determinant in the usual way.

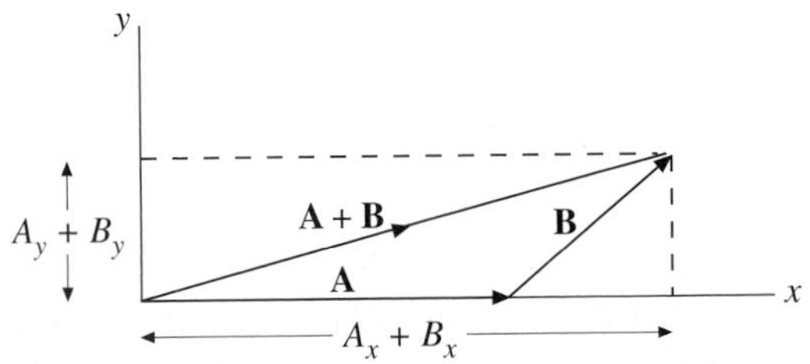

Fig. 2.12 Vector sum in a coordinate system.

2.5.3 Distributive law

Now we will consider $\mathbf{A} \times (\mathbf{B} + \mathbf{C})$. Let $\mathbf{D}$ be an abbreviation of the sum vector $\mathbf{B} + \mathbf{C}$. So we must calculate $(A_x\hat{\mathbf{x}} + A_y\hat{\mathbf{y}} + A_z\hat{\mathbf{z}}) \times (D_x\hat{\mathbf{x}} + D_y\hat{\mathbf{y}} + D_z\hat{\mathbf{z}})$. By applying (2.8) and the fact that $\hat{\mathbf{x}} \times \hat{\mathbf{x}} = 0$, etc., we find that the x component of this cross product is $A_yD_z - A_zD_y$. Now we replace the D's, so that this last expression becomes $A_y(B_z + C_z) - A_z(B_y + C_y)$, and reorganize this into $(A_yB_z - A_zB_y) + (A_yC_z - A_zC_y)$, which is just the x component of $\mathbf{A} \times \mathbf{B} + \mathbf{A} \times \mathbf{C}$. We continue to do the same thing for the other two components and have thus proved the distributive law written down soon after (2.5).

2.5.4 Triple vector products

Here, we will establish Eqs. (2.6). The left-hand side of the first is, by virtue of (2.14) and (2.15a),

$$A_x(B_yC_z - B_zC_y) + A_y(B_zC_x - B_xC_z) + A_z(B_xC_y - B_yC_x)$$

and we find the right-hand side by reorganizing these six terms around the components of $\mathbf{B}$. For example, take the fifth term first and the fourth term second, to obtain $B_x(C_yA_z - C_zA_y)$; this is the first term in the dot product of $\mathbf{B}$ with $\mathbf{C} \times \mathbf{A}$. A similar reorganization produces the remaining two terms.

Note that $|\mathbf{B} \times \mathbf{C}|$ is the *area* of the parallelogram spanned by vectors $\mathbf{B}$ and $\mathbf{C}$. The dot product of $\mathbf{A}$ with $\mathbf{B} \times \mathbf{C}$, by virtue of (2.2), is therefore equal to the product of $A\cos\theta$ with the area of the parallelogram. But $A\cos\theta$ is the projection of vector $\mathbf{A}$ *perpendicular* to the plane of that parallelogram, i.e. it is the height of the parallelepiped formed by the three vectors $\mathbf{A}$, $\mathbf{B}$, and $\mathbf{C}$ meeting jointly at their bases. Hence this dot product represents the volume of the parallelepiped.

The 'bac-cab' rule also is easily established in Cartesian coordinates. We do it as follows:

$$\begin{aligned}
[\mathbf{A} \times (\mathbf{B} \times \mathbf{C})]_x &= A_y(\mathbf{B} \times \mathbf{C})_z - A_z(\mathbf{B} \times \mathbf{C})_y \\
&= A_y(B_xC_y - B_yC_x) - A_z(B_zC_x - B_xC_z) \\
&= B_x(A_yC_y + A_zC_z) - C_x(A_yB_y + A_zB_z)
\end{aligned} \tag{2.16}$$

To the first of these terms we add $B_x(A_xC_x)$ and subtract it as $C_x(A_xB_x)$ – which is the same thing! – from the second term. It is then obvious by referring to (2.14) that we have obtained the x component of the right-hand side of the 'bac–cab' rule. We again obtain the other two components by cyclical permutation.

2.6 General orthogonal coordinate systems

In orthogonal coordinate systems, the unit vectors are mutually perpendicular at every point in space. Three of the most common orthogonal systems – the Cartesian, cylindrical and spherical systems – have been introduced in Section 2.4. There are, however, significant advantages to the student of having at least some understanding of *generalized* orthogonal systems, and the corresponding formulas can be most helpful in rapidly recalling, or re-deriving, line, surface, and volume elements for doing integrals that occur in electromagnetics.

We have seen that the three coordinate systems of Section 2.4 could each be defined in terms of three intersecting *surfaces*; on each surface *one* of the three coordinates is constant. In the Cartesian system, a $z =$ constant plane is perpendicular to the z axis everywhere, and only x and y vary in that plane. It is the same for the planes $y =$ constant and $x =$ constant, so that each point in space is defined by the intersection of three mutually orthogonal planes.

In the cylindrical coordinate system, circular cylinders around the z axis define $\rho =$ constant surfaces, and vertical planes fanning out from the z axis to $\rho = \infty$ define the $\varphi =$ constant surfaces.

In the spherical coordinate system, spheres around the origin define $r =$ constant surfaces; cones around the z axis (each with its apex at the origin) define $\theta =$ constant surfaces; and vertical planes through the z axis (as for the cylindrical system) define the $\varphi =$ constant surfaces. Figure 2.13(*a*)–(*c*) illustrates these.

As we have said, in each of these coordinate systems a point in space is defined uniquely by the intersection of three mutually orthogonal surfaces. It can be shown

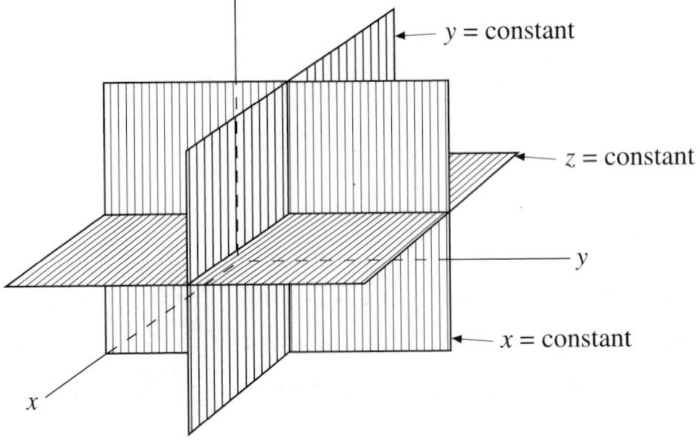

(*a*)

Fig. 2.13 Orthogonal coordinate systems: (*a*) Cartesian, (*b*) cylindrical, (*c*) spherical, (*d*) general.

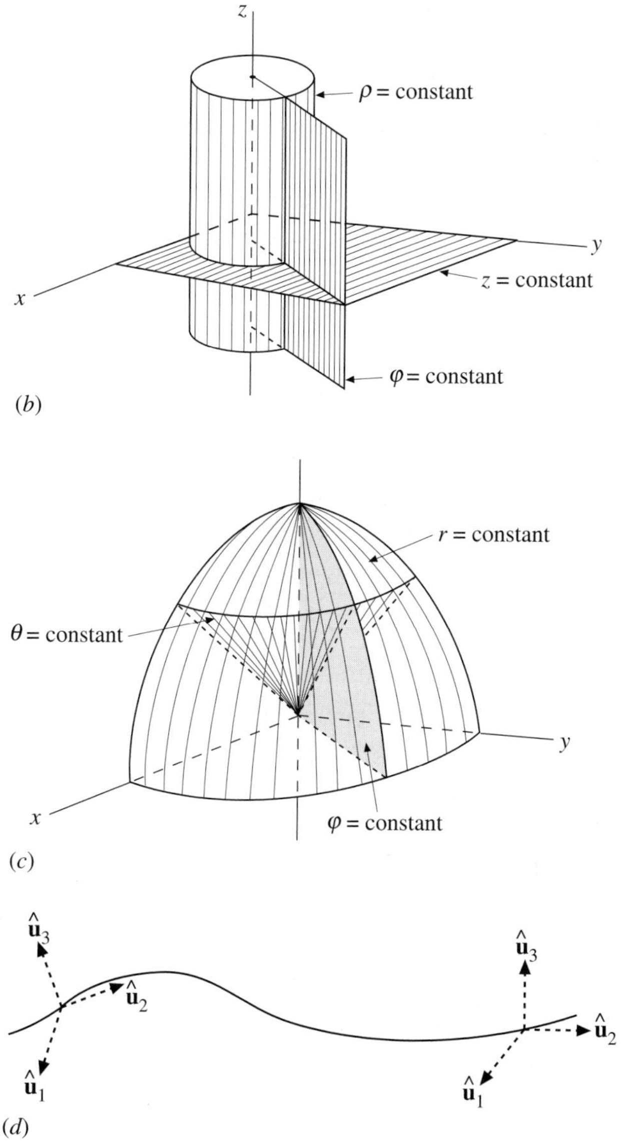

Fig. 2.13 (*contd*)

that a surface in space is defined in the Cartesian system by a single equation relating x, y, and z, which can always be written as a function $f(x, y, z) = 0$. For example, the three surfaces for the spherical coordinate system are defined by

$$
\begin{aligned}
&x^2 + y^2 + z^2 - r^2 = 0\\
&x^2 + y^2 - z^2 \tan^2\theta = 0\\
&y - x\tan\varphi = 0
\end{aligned}
\tag{2.17}
$$

in the understanding that here r, θ, φ are constants. The first equation describes spheres around the origin, the second cones around the z axis, and the third planes through the z axis.

Consider now a triple set of (possibly curved) general surfaces in Cartesian space, each defined by a function of x, y, and z:

$$\begin{aligned} u_1 &= f_1(x, y, z) \\ u_2 &= f_2(x, y, z) \\ u_3 &= f_3(x, y, z) \end{aligned} \tag{2.18}$$

Let us assume that the three surfaces intersect each other perpendicularly, i.e. *orthogonally*, at each point (x, y, z). The constants u_1, u_2 and u_3 represent three *curvilinear* coordinates, which can describe any point in the entire three-dimensional space if the three sets of surfaces cover it sufficiently completely. It should become clear that, unlike in the Cartesian coordinate system, we can no longer draw three fixed axes in space. In fact, that is why we need to introduce surfaces. Each set of u coordinates is uniquely defined by the x, y, z values at a point. (We shall later discuss *transformations* back and forth between the two systems.) We sometimes write $u_i(x, y, z)$, for $i = 1, 2, 3$, explicitly. At each point (u_1, u_2, u_3) we use the three local intersections of each pair of surfaces $f_i(x, y, z)$, $i = 1, 2, 3$, to define *local coordinate unit vectors* $\hat{\mathbf{u}}_1$, $\hat{\mathbf{u}}_2$, and $\hat{\mathbf{u}}_3$ along each of the intersections, see Fig. 2.13(d) where these unit vectors are shown at two different points. This is how we indicate direction in the absence of fixed coordinate axes (which we have only in the Cartesian system).

At any point in space, we now can describe a vector field $\mathbf{A}(\mathbf{r})$ by

$$\mathbf{A}(\mathbf{r}) = \sum_{i=1}^{3} \mathcal{A}_i(u_x, u_y, u_z)\hat{\mathbf{u}}_i \tag{2.19}$$

where the $\mathcal{A}_i$ are the components of $\mathbf{A}$ along the $\hat{\mathbf{u}}_i$ (i.e. they are the projections of the vector $\mathbf{A}$ along the directions defined by the $\hat{\mathbf{u}}_i$). The unit vectors $\hat{\mathbf{u}}_i$ as before obey the dot product rule $\hat{\mathbf{u}}_i \cdot \hat{\mathbf{u}}_j = \delta_{ij}$ and the cross product rule $\hat{\mathbf{u}}_i \times \hat{\mathbf{u}}_j = \hat{\mathbf{u}}_k$, where (i, j, k) is any of the three cyclic-permutation triplets $(1, 2, 3)$, $(2, 3, 1)$ or $(3, 1, 2)$. The dot product of two vectors can then be written as

$$\mathbf{A} \cdot \mathbf{B} = \mathcal{A}_1\mathcal{B}_1 + \mathcal{A}_2\mathcal{B}_2 + \mathcal{A}_3\mathcal{B}_3 \tag{2.20}$$

and the cross product (in determinant form) as

$$\mathbf{A} \times \mathbf{B} = \begin{vmatrix} \hat{\mathbf{u}}_1 & \hat{\mathbf{u}}_2 & \hat{\mathbf{u}}_3 \\ \mathcal{A}_1 & \mathcal{A}_2 & \mathcal{A}_3 \\ \mathcal{B}_1 & \mathcal{B}_2 & \mathcal{B}_3 \end{vmatrix} \tag{2.21}$$

As a final general comment, in Table 2.1 we make three identifications of the u_i for our three major coordinate systems.

Table 2.1. Coordinates u_1, u_2, u_3 in the three major coordinate systems

	Cartesian	Cylindrical	Spherical
u_1	x	ρ	r
u_2	y	φ	θ
u_3	z	z	φ

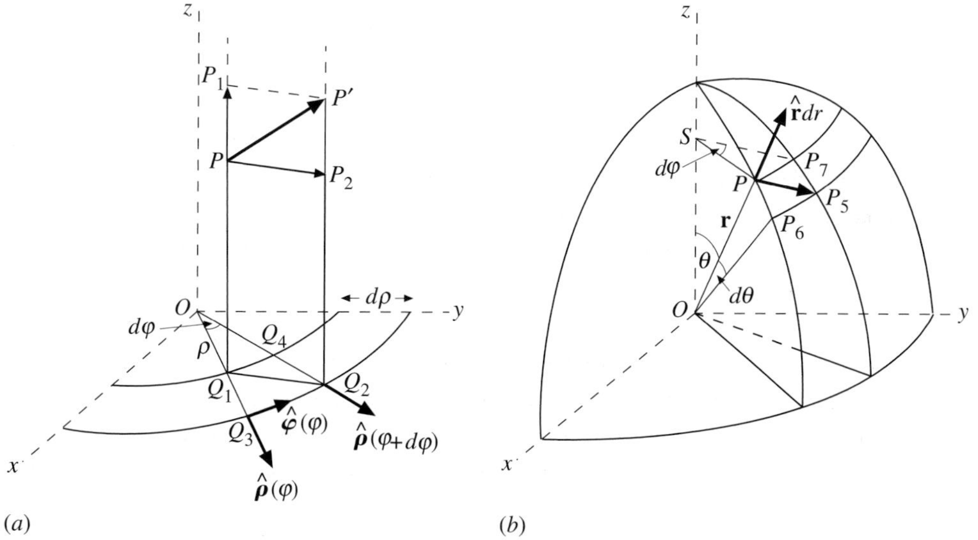

Fig. 2.14 A line element $d\boldsymbol{l} = \overrightarrow{PP'}$: (*a*) in the cylindrical coordinate system, (*b*) its components $\|$ and $\perp$ to the sphere $r =$ constant, in the spherical coordinate system.

2.6.1 Line elements

Consider two points P and P', respectively labeled $\mathbf{r} = (x, y, z)$ and $\mathbf{r}' = (x + dx, y + dy, z + dz)$. If these points are infinitesimally close to each other, then the distance vector $d\boldsymbol{l} = \hat{\mathbf{x}}\,dx + \hat{\mathbf{y}}\,dy + \hat{\mathbf{z}}\,dz$ is known as the *line element* vector in the direction given by the vector $\mathbf{r}' - \mathbf{r}$. Sometimes it will be written as $d\mathbf{r}$; there is no preference for either notation. Note that we interpret $\mathbf{r}$ as a vector from the origin to the point P, but the difference vector $\mathbf{r}' - \mathbf{r}$ is independent of the origin!

It is obvious that $d\boldsymbol{l}$ must be a length in dimension; and therefore each of its three components must also be a length. We immediately see that something is wrong if we simply extend the Cartesian definition to $d\boldsymbol{l} = \hat{\mathbf{u}}_1\,du_1 + \hat{\mathbf{u}}_2\,du_2 + \hat{\mathbf{u}}\,du_3$, because Table 2.1 shows clearly that some of the du_i, for the non-Cartesian systems, do not have the correct dimension. Let us see what we need to do.

Table 2.2. The coefficients h_i

System	Coordinate			Metric coefficient		
	u_1	u_2	u_3	h_1	h_2	h_3
Cartesian	x	y	z	1	1	1
cylindrical	ρ	φ	z	1	ρ	1
spherical	r	θ	φ	1	r	$r\sin\theta$

Consider the cylindrical coordinate system; see Fig. 2.14(a). The infinitesimal difference $\overrightarrow{PP'}$ is a line element vector $d\boldsymbol{l}$ that is the sum of a vertical piece $\overrightarrow{PP}_1 = \hat{\mathbf{z}}\,dz$ and a horizontal piece $\overrightarrow{PP}_2 = \overrightarrow{Q_1Q_2} = \overrightarrow{Q_1Q_3} + \overrightarrow{Q_1Q_4}$. The geometry of the $z = 0$ plane shows that $\overrightarrow{Q_1Q_3} = \hat{\boldsymbol{\rho}} d\rho$ and $\overrightarrow{Q_1Q_4} = \hat{\boldsymbol{\varphi}}\rho\,d\varphi$. Therefore we see that in cylindrical coordinates

$$d\boldsymbol{l} = \hat{\boldsymbol{\rho}}\,d\rho + \hat{\boldsymbol{\varphi}}\rho\,d\varphi + \hat{\mathbf{z}}\,dz \tag{2.22}$$

Now we express $d\boldsymbol{l}$ in spherical coordinates; see Fig. 2.14(b). For visual clarity we omit $d\boldsymbol{l}$ itself but show its components $\hat{\mathbf{r}}\,dr$ in the radial direction and $\overrightarrow{PP}_5$ in the tangential direction. We can reexpress $\overrightarrow{PP}_5$ as $\overrightarrow{PP}_6 + \overrightarrow{PP}_7$. Since $OP = r$, $PP_6 = r\,d\theta$ so that $\overrightarrow{PP}_6 = \hat{\boldsymbol{\theta}}r\,d\theta$. Likewise, $\overrightarrow{PP}_7 = PS\,d\varphi$ and $PS = r\sin\theta$ so that $\overrightarrow{PP}_7 = \hat{\boldsymbol{\varphi}}r\sin\theta\,d\varphi$. As a result, we find that in spherical coordinates

$$d\boldsymbol{l} = \hat{\mathbf{r}}\,dr + \hat{\boldsymbol{\theta}}r\,d\theta + \hat{\boldsymbol{\varphi}}r\sin\theta\,d\varphi \tag{2.23}$$

The unit vectors $\hat{\boldsymbol{\rho}}$, $\hat{\mathbf{r}}$, $\hat{\boldsymbol{\theta}}$, and $\hat{\boldsymbol{\varphi}}$ are shown separately in Figs. 2.17 and 2.18. It should be noted that each unit vector points in the direction of increase of the corresponding variable.

We deduce from Eqs. (2.22) and (2.23) that the line element in a general orthogonal coordinate system must be given by

$$d\boldsymbol{l} \equiv \sum_{i=1}^{3} d\boldsymbol{l}_i = \sum_{i=1}^{3} \hat{\mathbf{u}}_i\,dl_i = \sum_{i=1}^{3} h_i(u_1, u_2, u_3)\hat{\mathbf{u}}_i\,du_i \tag{2.24}$$

where the *metric coefficients* $h_i(u_1, u_2, u_3)$ are factors that can have a dimension (depending upon whatever type of coordinate u_i is). They are needed to make sure that the components dl_i all have the correct dimension of length, even if du_i itself is not a length.

It follows from (2.24) that $dl_i = h_i\,du_i$, so that we can identify h_i as $\partial l_i/\partial u_i$. If the length components dl_i are known as functions of u_1, u_2, and u_3, then the h_i can be calculated directly. That is what has been done in (2.22) and (2.23)! Table 2.2 enumerates the metric coefficients h_i for the three major coordinate systems.

The advantage of having this table at hand is that expressions for the line element in the three systems are then easily assembled from (2.24).

The material in this paragraph is meant for the reader who wishes for more detail, but it can be skipped in a first reading. The above metric coefficients play an important role in more advanced treatments of electromagnetics, and some brief comments may be helpful. The most general expression for a line element in three-dimensional space is in terms of the square of its length,

$$dl^2 = \sum_{i,j=1}^{3} g^{ij}\, du_i\, du_j$$

The coefficients g^{ij} are known as (*contravariant*) *metric coefficients*, and they have a *covariant* counterpart, both of which are useful in the tensor analysis needed in the formulation of relativistic physics. In nonrelativistic physics, that distinction is unnecessary and we shall not comment on it further. The coefficients are nonzero for $j \neq i$ when we have *nonorthogonal* coordinate systems. Off-diagonal coefficients do not exist for orthogonal coordinate systems, as can be seen by taking the dot product of $d\boldsymbol{l}$ with itself in (2.24). Thus, the g^{ii} are the *squares* of the $h_i(u_1, u_2, u_3)$, which we have called metric coefficients. We will work only with the metric coefficients h_i, as appears to be common practice in electrical-engineering undergraduate texts.

2.6.2 Surface and volume elements

Surface element vectors are defined as infinitesimal areas with a direction. The general definition for a surface element in the plane $\perp$ the unit vector $\hat{\mathbf{u}}_k$ is

$$d\mathbf{S}_k \equiv d\boldsymbol{l}_i \times d\boldsymbol{l}_j = \hat{\mathbf{u}}_k h_i(u_1, u_2, u_3) h_j(u_1, u_2, u_3)\, du_i\, du_j \tag{2.25}$$

where i, j, k are the three cyclical permutations of the integers 1, 2, and 3 (see Fig. 2.15).

Note that the definition has 'right-handedness' because we rotate $\hat{\mathbf{u}}_i$ towards $\hat{\mathbf{u}}_j$ through the shortest angle to find $d\mathbf{S}_k$ pointing in the $\hat{\mathbf{u}}_k$ direction. The Cartesian counterparts of (2.25) are obvious; one of them is $d\mathbf{S}_z = \hat{\mathbf{z}}\, dx\, dy$. The three cylindrical surface elements are easily obtained from (2.25) with the aid of Table 2.2:

$$\begin{aligned} d\mathbf{S}_\rho &= \hat{\boldsymbol{\rho}}\rho\, d\varphi\, dz \\ d\mathbf{S}_\varphi &= \hat{\boldsymbol{\varphi}}\, d\rho\, dz \\ d\mathbf{S}_z &= \hat{\mathbf{z}}\rho\, d\rho\, d\varphi \end{aligned} \tag{2.26}$$

and Fig. 2.16 should confirm these expressions.

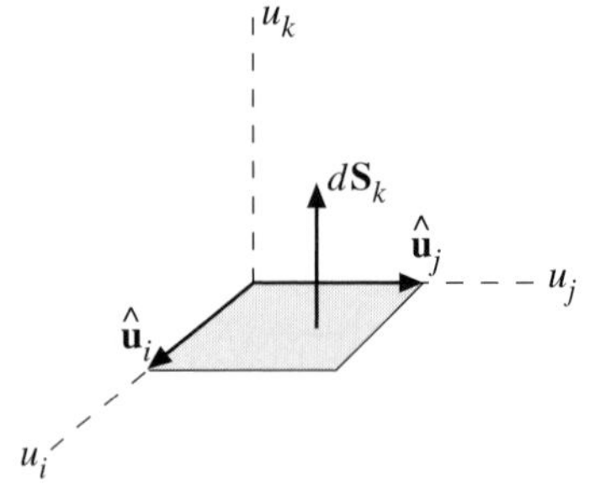

Fig. 2.15 A surface element.

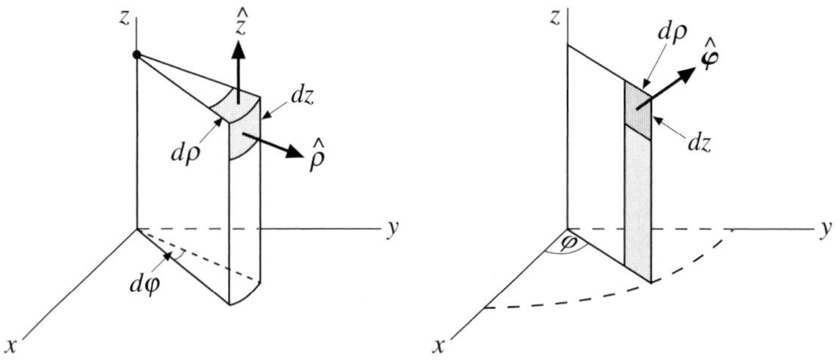

Fig. 2.16 Surface elements for the cylindrical coordinate system.

The three spherical surface elements are obtained likewise:

$$\begin{aligned} d\mathbf{S}_r &= \hat{\mathbf{r}} r^2 \sin\theta \, d\theta \, d\varphi \\ d\mathbf{S}_\theta &= \hat{\boldsymbol{\theta}} r \sin\theta \, dr \, d\varphi \\ d\mathbf{S}_\varphi &= \hat{\boldsymbol{\varphi}} r \, dr \, d\theta \end{aligned} \tag{2.27}$$

These surface elements are shown in Fig. 2.17, but we have omitted some of the unit vectors for greater clarity; they are perhaps most easily visualized by using the cross products $d\boldsymbol{l}_i \times d\boldsymbol{l}_j$ given in (2.25) (the various dl_i are indicated explicitly in the diagrams).

The *volume element* (there is only one in three dimensions) is $dv = d\mathbf{S}_k \cdot d\boldsymbol{l}_k = (d\boldsymbol{l}_i \times d\boldsymbol{l}_j) \cdot d\boldsymbol{l}_k$, which works out to give

$$dv = h_1(u_1, u_2, u_3) h_2(u_1, u_2, u_3) h_3(u_1, u_2, u_3) \, du_1 \, du_2 \, du_3 \tag{2.28}$$

For the three major coordinate systems, this becomes

$$\begin{aligned} dv &= dx \, dy \, dz && \text{(Cartesian coordinates)} \\ dv &= \rho \, d\rho \, d\varphi \, dz && \text{(cylindrical coordinates)} \\ dv &= r^2 \sin\theta \, dr \, d\theta \, d\varphi && \text{(spherical coordinates)} \end{aligned} \tag{2.29}$$

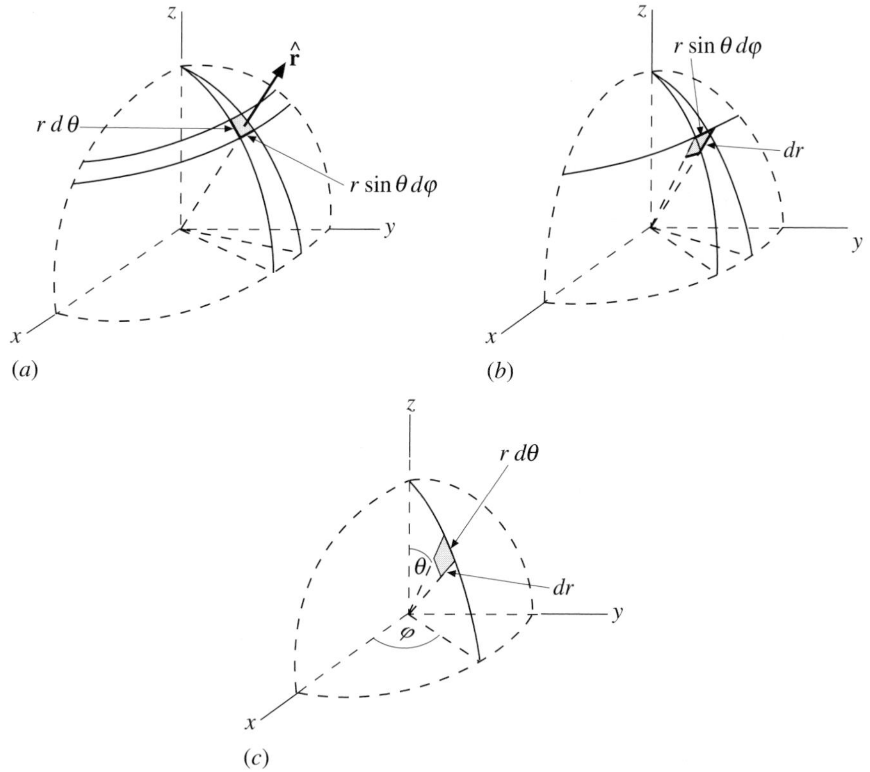

Fig. 2.17 Surface elements for the spherical coordinate system.

2.7 Transformations between coordinate systems

Rather than delving deeply into general transformations from one coordinate system into another, let us look into the more common special cases, such as transformations to and from the Cartesian coordinate system. Consider a vector field **A** as given by (2.19), i.e. its three components in the $\hat{\mathbf{u}}_i$ system are $\mathcal{A}_i(u_1, u_2, u_3)$. To transform to Cartesian coordinates (which we rename temporarily as $x_1 \equiv x$, $x_2 \equiv y, x_3 \equiv z$), we write

$$\mathbf{A} = A_1\hat{\mathbf{x}}_1 + A_2\hat{\mathbf{x}}_2 + A_3\hat{\mathbf{x}}_3$$

Upon dot-multiplying this by each $\hat{\mathbf{x}}_i$, we learn that

$$A_i = \hat{\mathbf{x}}_i \cdot \mathbf{A}$$

If, now, we insert (2.19) into this last dot product, we obtain a sum of scalar products:

$$A_i = \hat{\mathbf{x}}_i \cdot \mathbf{A} = \hat{\mathbf{x}}_i \cdot \sum_{j=1}^{3} \mathcal{A}_j \hat{\mathbf{u}}_j$$
$$= \sum_{j=1}^{3} \mathcal{A}_j \hat{\mathbf{x}}_i \cdot \hat{\mathbf{u}}_j \qquad (2.30)$$

Note that $\mathcal{A}_i$ is not the same as A_i. There are still two things left to do. The $\mathcal{A}_j$ are given as functions of the coordinates u_1, u_2, u_3, so first we must substitute for each u_i as a function of the three Cartesian coordinates wherever we find a dependence upon u_i in $\mathcal{A}_j$. In the second place, we need to know how each unit vector $\hat{\mathbf{u}}_j$ transforms into Cartesian unit vectors so that we can find the scalar $\hat{\mathbf{x}}_i \cdot \hat{\mathbf{u}}_j$.

An equivalent way of stating the above transformation into Cartesian coordinates is to say that $\mathbf{A} = \mathcal{A}_1\hat{\mathbf{u}}_1 + \mathcal{A}_2\hat{\mathbf{u}}_2 + \mathcal{A}_3\hat{\mathbf{u}}_3$, and to substitute

$$\hat{\mathbf{u}}_j = (\hat{\mathbf{u}}_j \cdot \hat{\mathbf{x}}_1)\hat{\mathbf{x}}_1 + (\hat{\mathbf{u}}_j \cdot \hat{\mathbf{x}}_2)\hat{\mathbf{x}}_2 + (\hat{\mathbf{u}}_j \cdot \hat{\mathbf{x}}_3)\hat{\mathbf{x}}_3 \qquad \text{for } j = 1, 2, 3 \qquad (2.31)$$

into this sum of weighted unit vectors. The reader can see, by rearranging the nine terms, that this leads to (2.30).

As an illustration, consider the following. Suppose that

$$\mathbf{A} = (r\cos\theta)\hat{\mathbf{r}} + (r\sin\theta)\hat{\boldsymbol{\theta}} + (\cos\varphi)\hat{\boldsymbol{\varphi}} \qquad (2.32a)$$

and we want to transform to Cartesian coordinates. From (2.11) it is seen that $r\cos\theta = z$, $r\sin\theta = \rho$ (here, an abbreviation for $\sqrt{x^2 + y^2}$), and $\cos\varphi = x/\rho$. Equation (2.32a) translates into

$$\mathbf{A} = z\hat{\mathbf{r}} + \rho\hat{\boldsymbol{\theta}} + (x/\rho)\hat{\boldsymbol{\varphi}}. \qquad (2.32b)$$

How do we transform the unit vectors of the spherical coordinate system? We need the dot products $\hat{\mathbf{u}}_j \cdot \hat{\mathbf{x}}_i$. We can obtain

$$\hat{\mathbf{x}}_i = \sum_{j=1}^{3} (\hat{\mathbf{x}}_i \cdot \hat{\mathbf{u}}_j)\hat{\mathbf{u}}_j \qquad (2.33)$$

in a very similar fashion to the way (2.31) was derived (by interchanging the roles of the u_i and the Cartesian coordinates); for the spherical coordinate system we get

$$\hat{\mathbf{x}}_i = (\hat{\mathbf{x}}_i \cdot \hat{\mathbf{r}})\hat{\mathbf{r}} + (\hat{\mathbf{x}}_i \cdot \hat{\boldsymbol{\theta}})\hat{\boldsymbol{\theta}} + (\hat{\mathbf{x}}_i \cdot \hat{\boldsymbol{\varphi}})\hat{\boldsymbol{\varphi}} \qquad (2.34)$$

In fact we can use either (2.31) or (2.34). Let us use the former, in connection with Fig. 2.18. Then part (*b*) of the figure gives us $\hat{\mathbf{r}} = \hat{\boldsymbol{\rho}}\sin\theta + \hat{\mathbf{z}}\cos\theta$ and part (*c*) $\hat{\boldsymbol{\rho}} = \hat{\mathbf{x}}\cos\hat{\boldsymbol{\varphi}} + \hat{\mathbf{y}}\sin\varphi$. Take the dot product of the first equation with $\hat{\boldsymbol{\rho}}$ to give

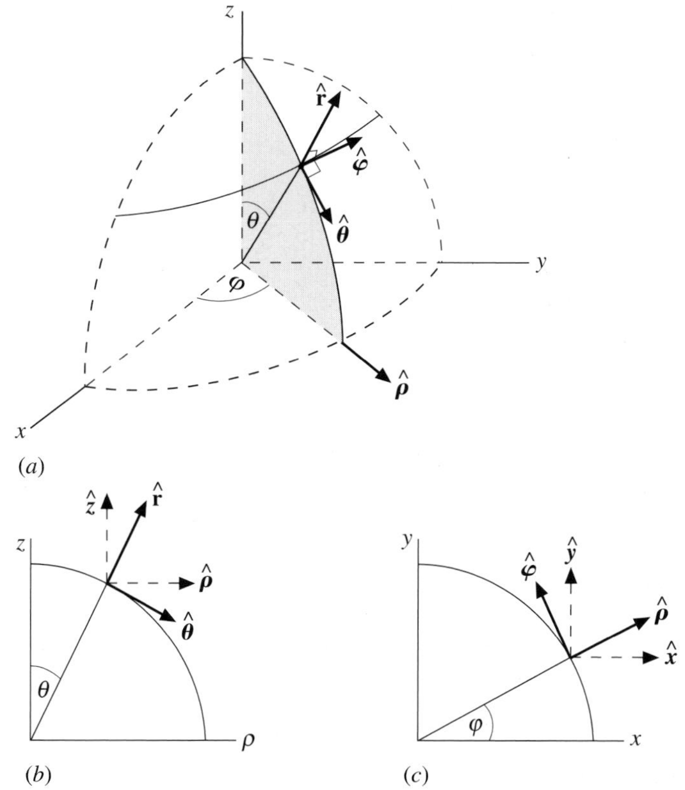

Fig. 2.18 (*a*) The transformation between spherical and Cartesian coordinate systems; (*b*) unit vectors in the plane φ = constant; (*c*) unit vectors in the plane z = constant.

$\hat{\boldsymbol{\rho}} \cdot \hat{\mathbf{r}} = \sin\theta$ and then with $\hat{\mathbf{z}}$ to give $\hat{\mathbf{z}} \cdot \hat{\mathbf{r}} = \cos\theta$. The other relationships are written down similarly. This establishes

$$\hat{\mathbf{r}} = \hat{\mathbf{x}} \sin\theta \cos\varphi + \hat{\mathbf{y}} \sin\theta \sin\varphi + \hat{\mathbf{z}} \cos\theta \tag{2.35a}$$

Figure 2.18(*b*) with (2.31) also shows that $\hat{\boldsymbol{\theta}} = \hat{\boldsymbol{\rho}} \cos\theta - \hat{\mathbf{z}} \sin\theta$. This, together with the just-established $\hat{\boldsymbol{\rho}} = \hat{\mathbf{x}} \cos\varphi + \hat{\mathbf{y}} \sin\varphi$, gives

$$\hat{\boldsymbol{\theta}} = \hat{\mathbf{x}} \cos\theta \cos\varphi + \hat{\mathbf{y}} \cos\theta \sin\varphi - \hat{\mathbf{z}} \sin\theta \tag{2.35b}$$

Finally, we also get from Fig. 2.18(*c*)

$$\hat{\boldsymbol{\varphi}} = -\hat{\mathbf{x}} \sin\varphi + \hat{\mathbf{y}} \cos\varphi \tag{2.35c}$$

We gather these three equations together for convenience:

$$\begin{aligned} \hat{\mathbf{r}} &= \hat{\mathbf{x}} \sin\theta \cos\varphi + \hat{\mathbf{y}} \sin\theta \sin\varphi + \hat{\mathbf{z}} \cos\theta \\ \hat{\boldsymbol{\theta}} &= \hat{\mathbf{x}} \cos\theta \cos\varphi + \hat{\mathbf{y}} \cos\theta \sin\varphi - \hat{\mathbf{z}} \sin\theta \\ \hat{\boldsymbol{\varphi}} &= -\hat{\mathbf{x}} \sin\varphi + \hat{\mathbf{y}} \cos\varphi \end{aligned} \tag{2.36a}$$

or, in the notation of matrix multiplication,

$$\begin{bmatrix} \hat{\mathbf{r}} \\ \hat{\boldsymbol{\theta}} \\ \hat{\boldsymbol{\varphi}} \end{bmatrix} = \begin{bmatrix} \sin\theta\cos\varphi & \sin\theta\sin\varphi & \cos\theta \\ \cos\theta\cos\varphi & \cos\theta\sin\varphi & -\sin\theta \\ -\sin\varphi & \cos\varphi & 0 \end{bmatrix} \begin{bmatrix} \hat{\mathbf{x}} \\ \hat{\mathbf{y}} \\ \hat{\mathbf{z}} \end{bmatrix} \tag{2.36b}$$

The reader can check quite easily that all three unit vectors are $\perp$ each other.

As an aside, we note that by comparing (2.36a, b) with (2.31) we can obtain explicit expressions for all the dot products $\hat{\mathbf{u}}_j \cdot \hat{\mathbf{x}}_i$. Substitution into (2.34) then yields the inverse transformation

$$\begin{aligned} \hat{\mathbf{x}} &= \hat{\mathbf{r}}\sin\theta\cos\varphi + \hat{\boldsymbol{\theta}}\cos\theta\cos\varphi - \hat{\boldsymbol{\varphi}}\sin\varphi \\ \hat{\mathbf{y}} &= \hat{\mathbf{r}}\sin\theta\sin\varphi + \hat{\boldsymbol{\theta}}\cos\theta\sin\varphi + \hat{\boldsymbol{\varphi}}\cos\varphi \\ \hat{\mathbf{z}} &= \hat{\mathbf{r}}\cos\theta - \hat{\boldsymbol{\theta}}\sin\theta \end{aligned} \tag{2.37}$$

We now return to our main purpose: transforming (2.32a) to Cartesian coordinates. Insertion of (2.36a) into (2.32b) finishes the transformation; we obtain

$$\begin{aligned} \mathbf{A} = \hat{\mathbf{x}}&[(z\sin\theta + \rho\cos\theta)\cos\varphi - (x/\rho)\sin\varphi] \\ &+ \hat{\mathbf{y}}[(z\sin\theta + \rho\cos\theta)\sin\varphi + (x/\rho)\cos\varphi] \\ &+ \hat{\mathbf{z}}[z\cos\theta + \rho\sin\theta] \end{aligned} \tag{2.38}$$

We note that in the course of arriving at (2.38) we have obtained the transformations of unit vectors back and forth, (2.36a), (2.37). Incidentally, once we have a transformation in one direction, the opposite direction can, of course, be obtained by inverting the matrix of the first transformation. Thus (2.37) can also be obtained directly from (2.36b) by finding the inverse of the 3×3 matrix, using the usual techniques for matrix inversion.

The transformation from the Cartesian into the spherical coordinate system is very similar. We take

$$\mathbf{A} = A_1\hat{\mathbf{x}}_1 + A_2\hat{\mathbf{x}}_2 + A_3\hat{\mathbf{x}}_3$$

(remember: the $\hat{\mathbf{x}}_i$ are alternative symbols for the three Cartesian unit vectors), and substitute (2.37) into this. Then we use (2.11) in the three A_i coefficients to transform them from the Cartesian to the spherical coordinate description.

The transformation of the Cartesian unit vectors from and into cylindrical coordinates is accomplished with diagrams similar to those of Fig. 2.18. In particular, Fig. 2.18(c) illustrates the transformation from x, y to ρ, φ. We see from it that both $\hat{\boldsymbol{\rho}}$ and $\hat{\boldsymbol{\varphi}}$ must be a linear superposition of vectors $\hat{\mathbf{x}}$ and $\hat{\mathbf{y}}$, e.g. $\hat{\boldsymbol{\rho}} = a\hat{\mathbf{x}} + b\hat{\mathbf{y}}$. Dot-multiply this first by $\hat{\mathbf{x}}$ and then $\hat{\mathbf{y}}$. Thus we find $a \equiv \hat{\mathbf{x}} \cdot \hat{\boldsymbol{\rho}} = \cos\varphi$ and $b \equiv \hat{\mathbf{y}} \cdot \hat{\boldsymbol{\rho}} = \sin\varphi$. Likewise, $\hat{\boldsymbol{\varphi}} = c\hat{\mathbf{x}} + d\hat{\mathbf{y}}$ with $c \equiv \hat{\mathbf{x}} \cdot \hat{\boldsymbol{\varphi}} = -\sin\varphi$ and $d \equiv \hat{\mathbf{y}} \cdot \hat{\boldsymbol{\varphi}} = \cos\varphi$. The inverse transformations can be obtained in similar fashion, or by calculating the inverse 2×2 matrix for the transformation from one to the other set of coordinates. The transformations and inverse transformation are given by

$$\begin{aligned}\hat{\mathbf{x}} &= \hat{\boldsymbol{\rho}} \cos\varphi - \hat{\boldsymbol{\varphi}} \sin\varphi \\ \hat{\mathbf{y}} &= \hat{\boldsymbol{\rho}} \sin\varphi + \hat{\boldsymbol{\varphi}} \cos\varphi \\ \hat{\mathbf{z}} &= \hat{\mathbf{z}}\end{aligned} \tag{2.39}$$

$$\begin{aligned}\hat{\boldsymbol{\rho}} &= \hat{\mathbf{x}} \cos\varphi + \hat{\mathbf{y}} \sin\varphi \\ \hat{\boldsymbol{\varphi}} &= -\hat{\mathbf{x}} \sin\varphi + \hat{\mathbf{y}} \cos\varphi \\ \hat{\mathbf{z}} &= \hat{\mathbf{z}}\end{aligned} \tag{2.40}$$

Problems

2.1. A vector **A** points 4 units northwards, and vector **B** points 3 units eastwards.
(a) Calculate the angle (in degrees) by which **A** + **B** is east of north.
(b) Repeat for **A** – **B**.

2.2. Calculate the length of the projection of the vector $\mathbf{A} = 3\hat{\mathbf{x}} - 2\hat{\mathbf{y}} + \hat{\mathbf{z}}$ on the vector $\mathbf{B} = -2\hat{\mathbf{x}} + 3\hat{\mathbf{y}} + 6\hat{\mathbf{z}}$. Also specify the unit vector in the direction of **B**.

2.3. On the surface of the Earth, the vector **A** points 2 units eastward, and vector **B** points 5 units southwest. Specify the length and direction of $\mathbf{A} \times \mathbf{B}$.

2.4. Explain briefly why it does not matter which way you measure the angle from **A** to **B** when you calculate the cross product $\mathbf{A} \times \mathbf{B}$.

2.5. What is the angle between the two vectors $\mathbf{A} = 2.5\hat{\mathbf{x}} - 3.6\hat{\mathbf{y}} + 1.2\hat{\mathbf{z}}$ and $\mathbf{B} = 4.1\hat{\mathbf{x}} + 0.6\hat{\mathbf{y}} - 0.3\hat{\mathbf{z}}$?

2.6. A plane is given by the equation $3x + 4y + 2z = 12$. Use the cross product of vectors to find a unit vector perpendicular to this plane. *Hint*: First find two vectors in the plane.

2.7. A vector **A** has its base at the origin and its tip at the point ($\rho = 2.3$, $\varphi = 60°$, $z = 3.7$). Calculate its magnitude.

2.8. Repeat the calculation of problem 7 when the tip of **A** is at the point ($r = 4$, $\theta = 30°$, $\varphi = 45°$).

2.9. Two points are characterized in cylindrical coordinates by $P = (5, \pi/6, 3)$ and $Q = (3, \pi/6, -2)$. Specify the magnitude and direction of the vector $\overrightarrow{QP}$.

2.10. Two points on a sphere are given in spherical coordinates by $P = (4, 0.578, 0.126)$ and $Q = (4, 0.998, 2.554)$ (with the angles in radians). Specify the magnitude and direction of the vector $\overrightarrow{QP}$.

2.11. Specify the vector $\mathbf{A} = -y\hat{\mathbf{x}} + x\hat{\mathbf{y}} + 2\hat{\mathbf{z}}$ in cylindrical coordinates.

2.12. Vector **A** has its base at the origin and its tip at the point Q $(r = 2.5, \theta = \pi/4, \varphi = \pi/6)$. Vector **B** has its base at Q and its tip at P $(\rho = 4.5, \varphi = \pi/6, z = 6)$. Specify the sum vector $\mathbf{A} + \mathbf{B}$.

2.13. Find the area of the cylindrical surface at $\rho = 3$ between the planes $z_1 = -1$ and $z_2 = 2.5$ and the planes $\varphi_1 = 30°$ and $\varphi_2 = 45°$.

2.14. (a) What percentage of the Earth's surface lies between the 42° latitude circle (parallel to the equator) and the equator?

(b) Calculate also the percentage of the volume of the Earth that is formed by intersection of parallel planes through this 42° latitude circle and the equator.

2.15. Calculate the volume that lies between the spheres with radii $r = 1$ and $r = 2$ and within a cone with half-angle $\theta = 60°$, the apex of which is at the center of the spheres.

2.16. The mass per unit volume of an inhomogeneous sphere of radius $r = 2\,\text{m}$ is given by $\rho_m(r, \theta, \varphi) = \rho_0 \sin 2\theta \cos^2 \varphi$, with $\rho_0 = 5\,\text{kg}\,\text{m}^{-3}$. Calculate the mass of the sphere.

2.17. The vector $\mathbf{A} = -z\hat{\boldsymbol{\rho}}(\varphi) + \rho(\varphi)\hat{\mathbf{z}}$. Express **A** in terms of spherical coordinates and unit vectors.

2.18. Express the following vectors in terms of the most suitable coordinate system:

(a) $\mathbf{A}_1 = (x\hat{\mathbf{x}} + y\hat{\mathbf{y}})/(x^4 + 2x^2y^2 + y^4)$

(b) $\mathbf{A}_2 = (x\hat{\mathbf{x}} + y\hat{\mathbf{y}} + z\hat{\mathbf{z}})/(x^4 + y^4 + z^4 + 2x^2y^2 + 2y^2z^2 + 2z^2x^2)^{1.5}$.

3

The electrostatic field

3.1 Introduction

There are so many different ways to approach the subject of electromagnetic theory, which is ultimately expressed concisely by Maxwell's laws, that there is no general agreement on how to do it. It is perhaps a matter of taste. In this text we prefer to start from the concept of the force upon a charged particle. There is a problem right here at the very start: What is 'charge?' It is the same problem that occurs in the teaching of mechanics, where some concepts, such as 'mass' or 'force', simply have to be taken as quantities that everyone understands. Somewhere, we must start from an assumption that there is such a thing as an intuitive feel for what 'charge' is. We shall have to assume that all readers have an intuitive understanding of certain basic quantities, under which we range *mass*, *force*, and *charge*.

Let us consider two very small bodies with masses m_1 and m_2; let us suppose that these bodies also carry *charge*, of which we shall say only that it is the source of electric forces, just as mass is the source of gravitational forces. The basic experimental fact is that two such point masses experience a force owing to their charge in the direction of the line between them. Experiments showed that both attraction and repulsion can happen and that there are *two* kinds of charge. We name these *positive* (+) and *negative* (−) charges. We say that a body has charge q on it, and by that we mean that its charge is q times as large as the charge of a reference body with *unit charge*; q may take positive or negative values.

In its simplest form, the basic experimental law, named after *Charles Augustin Coulomb* (1736–1806), states that *electrostatic force* F between two point charges q_P at point P and q_Q at point Q (with a distance R_{QP} between them) is

$$F \propto \frac{q_Q q_P}{R_{QP}^2} \tag{3.1}$$

This version of Coulomb's law gives only the magnitude of the force and not its direction (apart from a sign). We shall adopt the convention from here on that Q in general will refer to a source point (e.g. where there is a source charge) and P to a

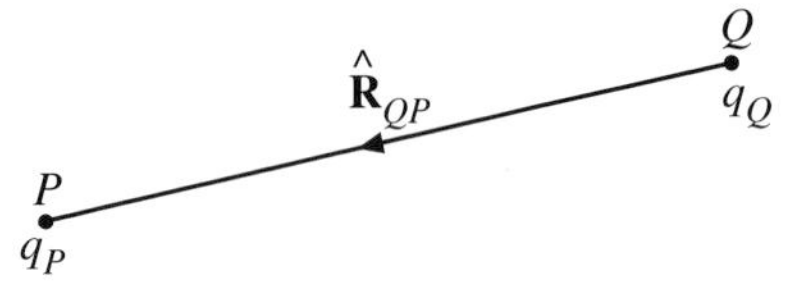

Fig. 3.1 Force between two electric charges.

point where the electrostatic force is observed. A vector form of Coulomb's law is more precise because it also indicates the direction as well as the magnitude of the electrostatic force: the force of q_Q on q_P is given by

$$\mathbf{F}_{QP} = \kappa_e \frac{q_Q q_P}{R_{QP}^2} \hat{\mathbf{R}}_{QP} \tag{3.2a}$$

The unit vector $\hat{\mathbf{R}}_{QP}$ points from Q to P, as shown in Fig. 3.1, so that the electrostatic force vector $\mathbf{F}_{QP}$ also points in that direction if the two charges *are of the same polarity* (i.e. are both + or both −).

We also note that the force of q_P upon q_Q is equal but opposite in direction: $\mathbf{F}_{PQ} = -\mathbf{F}_{QP}$. Clearly, two charges of the same polarity *repel* each other, whereas charges of opposite polarity *attract* each other. The constant $\kappa_e \approx 8.988 \times 10^9$ and it has the dimension Nm^2/C^2, if distances are measured in meters (m), charges in coulombs (C), and forces in newtons (N). The dimension of κ_e is also expressed as meters/farad (m/F), where $1\,\text{F} = 1\,\text{C}^2\,\text{J}^{-1}$. Of course, the numerical value of κ_e will change if different units are used for length, charge, and force.

Example 3.1

Suppose that point $P = (1, 1, 2)$ and point $Q = (0, 0, 1)$ in Cartesian coordinates and that charge $q_Q = 1.5\,\text{nC}$ ($1\,\text{nC} = 10^{-9}\,\text{C}$). Calculate the magnitude and direction of the force upon a unit charge at P. To do so, note that the vector $\mathbf{R}_{QP} = (1, 1, 2) - (0, 0, 1) = (1, 1, 1)$. The length of this vector is $\sqrt{1^2 + 1^2 + 1^2} = \sqrt{3}$. Therefore we find that

$$\mathbf{F}_{QP} = 8.988 \times 10^9 \frac{1.5 \times 10^{-9}}{(\sqrt{3})^2} \left[\frac{1}{\sqrt{3}} (\hat{\mathbf{x}} + \hat{\mathbf{y}} + \hat{\mathbf{z}}) \right] \approx 2.595(\hat{\mathbf{x}} + \hat{\mathbf{y}} + \hat{\mathbf{z}})\,\text{N}$$

The force is directed along one of the diagonals of a unit cube with a bottom rear corner at the origin. Its magnitude is close to 4.5 N. This example shows that the coulomb is an impractically large unit; forces with magnitudes close to unity are obtained when much smaller charges act upon each other.

In modern parlance, Coulomb's law is considered to be a special form of one of Maxwell's equations in which not κ_e but another constant, $\varepsilon_0 \approx 8.854 \times 10^{-12}\,\text{F/m}$, figures prominently. It turns out that

$$\kappa_e = \frac{1}{4\pi\varepsilon_0} \tag{3.2b}$$

hence the above-quoted numerical value. In textbook examples, one sometimes finds the approximation $\varepsilon_0 \approx 10^{-9}/36\pi \approx 8.842 \times 10^{-12}$ F/m, which yields $\kappa_e \approx 9 \times 10^9$ m/F, with an error of roughly 0.1%.

The inverse-square dependence upon the distance R_{QP} is a most important facet of (3.2). It is intimately tied to the fact that there can be no electrostatic force inside a spherical conducting surface with charge on it (see later). Early experiments were devised to test this as accurately as possible, and the inverse-square law has been shown to hold to within the highest experimental accuracy available. Of course, there is also an inverse-square gravitational force between the two small bodies due to their masses m_1 and m_2, but this force is extremely small compared with the electrostatic force (because the constant κ_m that occurs in this law is minuscule compared to (3.2b)). We will ignore the effect of mass in almost all parts of this course on electromagnetics. In fact, the concept of mass will play a role only where electrostatic forces give rise to the motion of charges.

3.2 The electrostatic field

It is immediately apparent from (3.2a, b) that the electrostatic force of a charge q_Q upon any charge q_P at point P is proportional in magnitude to q_P. As a result, it suffices to describe the force upon a unit charge at P. The actual force upon the charge at P then can be obtained by multiplying this by q_P. The *force upon a unit charge* is known as the *electrostatic field* and is denoted by the vector **E**. The **E** field is the first of several vector fields we shall encounter. According to Coulomb's law, (3.2a, b), the electrostatic field at a point P due to a point charge q_Q at point Q is

$$\mathbf{E}_P = \frac{q_Q}{4\pi\varepsilon_0} \frac{\mathbf{R}_{QP}}{R_{QP}^2} \tag{3.3a}$$

Thus, the force on a charge q_P at P is $\mathbf{F}_P = q_P\mathbf{E}_P$. Equation (3.3a) can be rewritten in the two forms:

$$\mathbf{E}_P = \frac{q_Q}{4\pi\varepsilon_0} \frac{\mathbf{R}_{QP}}{R_{QP}^3} = \frac{q_Q}{4\pi\varepsilon_0} \frac{\mathbf{r}_P - \mathbf{r}_Q}{|\mathbf{r}_P - \mathbf{r}_Q|^3} \tag{3.3b}$$

The first shows a *length vector* divided by the third power of a distance, which is equivalent to the *unit vector* divided by the square of a distance shown in (3.3a). In the second equation in (3.3b) the vector between the observation and charge points is written as a difference between two vectors with their base at the origin: $\boldsymbol{R}_{QP} = \mathbf{r}_P - \mathbf{r}_Q$. Sometimes we will use one notation, and at other times the other. Convenience will decide which.

The unit for the electrostatic field follows from its definition as the force per unit charge: it is newtons/coulomb (N/C). According to (3.3) this must be the same as coulombs/farad meter (C/F m). As mentioned above, 1 F = 1 coulomb/volt (C/V), so we see that newtons/coulomb may be expressed as volts/meter (V/m); this is the practical unit customarily used for the electric (or electrostatic) field.

This brings us to another point, which is more of an aside: the problem of action at a distance. It became obvious early in the modern history of electromagnetics that the forces described by (3.2a) or (3.3a, b) are not perceptibly changed when a high vacuum is produced in the space between the two points. Like the gravitational force, the electrostatic force appeared to cause *action at a distance*. The idea of action at a distance, without an intervening medium, was anathema to researchers in the late eighteenth and nineteenth centuries. The field concept brought welcome relief because it meant that a *virtual force field* was always in place at all points in space (even in a vacuum) once a charge source was placed somewhere. This virtual force was *felt* only when another charge was placed at a point of observation, but it was always there. In this way, the worrisome concept of action at a distance could be avoided. The field concept thus gained a philosophical importance. The irony of this is that the 20th-century formulation of quantum electrodynamics has returned us somewhat to the earlier position of action at a distance; actually, the quantum-electrodynamics formulation is somewhat intermediate between the two. The modern interpretation is that the electrostatic force is carried by the emission of electromagnetic particles without mass, called *photons*, which speed (at the velocity of light) towards the object charge and interact with it there. For the purposes of this text, however, we consider the electrostatic field merely to be the force upon a unit charge, and the philosophical implications will be cast aside.

The next item of importance is to consider the force or field due to more than one point charge. The joint force on a given charge is found to be merely the sum of the separate forces due to the other charges. This is known as the *principle of superposition*. It implies that the force of charge q_1 at point $\mathbf{r}_1$ upon charge q at $\mathbf{r}$ is uninfluenced by the presence of any other charge in space. Therefore, if we have N charges $q_1, q_2, \ldots, q_N$ at N points in space, the electrostatic field at $\mathbf{r}$ is

$$\mathbf{E}(\mathbf{r}) = \sum_{n=1}^{N} \frac{q_n}{4\pi\varepsilon_0} \frac{\mathbf{r} - \mathbf{r}_n}{|\mathbf{r} - \mathbf{r}_n|^3} \tag{3.4}$$

Here, the positions of the charges Q_n ($n = 1, \ldots, N$) are indicated by vectors $\mathbf{r}_n$. This formula exhibits another important property of electrostatic fields in a vacuum: the *linearity* of the fields with respect to charge. That is, if we double the amount of charge at all points the field will double also. This property will turn out later to be untrue in *dielectric* (nonconducting) media, because strong fields can pull extra charges off the atoms (or ionized atoms or molecules) of which the material consists, which introduces a nonlinear change in the fields. The property of superposition is strongly connected to the linearity of the fields, and both properties hold for sufficiently weak fields in materials.

3.3 Volume-, surface-, and line-charge densities

3.3.1 Volume-charge density

In materials, charges are carried by ionized atomic particles or electrons. It suffices here to state that electrons are either bound to the vicinity of the nucleus of a neutral or ionized atomic particle, or they have the capability of more or less free motion between the lattice points of a metallic material. The latter is the case for those electrons having energies that place them in a conduction band. At any rate, on any *laboratory* or *macroscopic* scale, there are huge numbers of charges per volume unit over which the macroscopic properties of the material do not vary appreciably. In a relatively dilute gas such as air at atmospheric temperature and pressure, there are of the order of 10^{19} molecules per cubic centimeter, so that there will be a very large number of charges per cubic centimeter even if only a small fraction of the atoms are ionized. It becomes apparent that on the macroscopic scale we do not see the individual charges and that the description entailed in (3.4) is too detailed. For that reason, a *fluid* or *continuous-medium* description will be introduced. Consider Fig. 3.2. We will consider the electric field $d\mathbf{E}$ at P due to a small volume element dv_Q at Q at some distance from P. Because dv_Q is a *macroscopically small but*

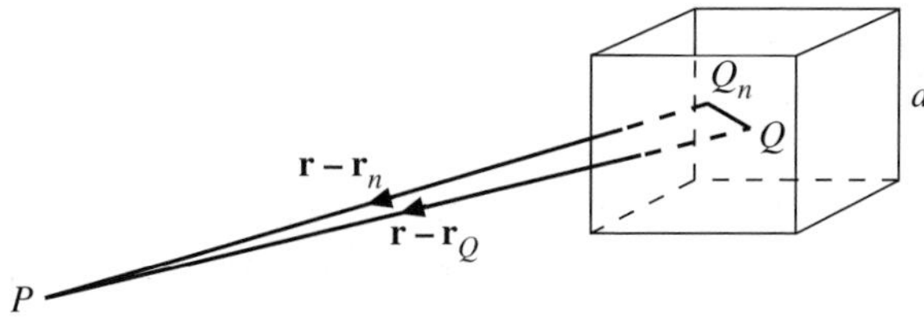

Fig. 3.2 Contribution to $\mathbf{E}(\mathbf{r})$ of a macroscopically small but microscopically large volume element dv_Q filled with point charges. Points P, Q, Q_n have position vectors $\mathbf{r}$, $\mathbf{r}_Q$, $\mathbf{r}_n$; point Q_n carries a charge q_n.

microscopically large volume element, it will contain very many charges, but all macroscopic properties (temperature, pressure, etc., and especially the *total* charge) will not vary perceptibly if we move the volume dv_Q in any direction by, say, half the length of a side.

We now wish to discuss briefly a mathematical approximation that will be very useful again and again. It is derived from a truncation of the *binomial expansion* of a polynomial. Let ϵ be a positive or negative number with $|\epsilon| < 1$. Let p be any real (or complex) number; in most applications $|p| \leq 5$ or so. Then

$$(1+\epsilon)^p = 1 + p\epsilon + \frac{1}{2!}p(p-1)\epsilon^2 + \frac{1}{3!}p(p-1)(p-2)\epsilon^3 + \cdots \tag{3.5}$$

We make use of a truncation after the second term:

$$(1+\epsilon)^p = 1 + p\epsilon + O(p\epsilon^2) \tag{3.6}$$

The last term of (3.6) refers to terms of the order of $p\epsilon^2$ or more. So we require not only $|\epsilon| < 1$ but also that $|p\epsilon| \ll 1$, in order that $1 + p\epsilon$ is a good approximation to $(1+\epsilon)^p$.

Now consider the vector-to-length ratio in (3.4). It can be simplified for the situation of Fig. 3.2 by using the *binomial theorem* (explained in more detail in the above optional paragraph). This theorem states that the power p of a number $1+\epsilon$ between 0 and 2 (i.e. for $|\epsilon| < 1$), namely $(1+\epsilon)^p$, is approximated by $1 + p\epsilon$ when $|p\epsilon| \ll 1$. We can use it here as follows. Let $\mathbf{r}$ and $\mathbf{r}_Q$ be the position vectors of P and Q and let $\mathbf{r}_n$ be the position vector of charge q_n at point Q_n in the volume element dv_Q. Then, we see that

$$\frac{\mathbf{r}-\mathbf{r}_n}{|\mathbf{r}-\mathbf{r}_n|^3} = \frac{(\mathbf{r}-\mathbf{r}_Q)-(\mathbf{r}_n-\mathbf{r}_Q)}{|(\mathbf{r}-\mathbf{r}_Q)-(\mathbf{r}_n-\mathbf{r}_Q)|^3} \approx \frac{\mathbf{r}-\mathbf{r}_Q}{|\mathbf{r}-\mathbf{r}_Q|^3} \tag{3.7}$$

where we have used the binomial theorem to write

$$\begin{aligned}|(\mathbf{r}-\mathbf{r}_Q)-(\mathbf{r}_n-\mathbf{r}_Q)|^3 &= |\mathbf{r}-\mathbf{r}_Q|^3\left|1-\frac{\mathbf{r}_n-\mathbf{r}_Q}{\mathbf{r}-\mathbf{r}_Q}\right|^3 \\ &\approx |\mathbf{r}-\mathbf{r}_Q|^3\left|1-3\frac{\mathbf{r}_n-\mathbf{r}_Q}{\mathbf{r}-\mathbf{r}_Q}\right| \\ &\approx |\mathbf{r}-\mathbf{r}_Q|^3\end{aligned}$$

because all magnitudes $\mathbf{r}_n - \mathbf{r}_Q$ are very small compared to the magnitude of $\mathbf{r}-\mathbf{r}_Q$. The approximation (3.7) renders (3.4) into an expression for the contribution

$d\mathbf{E}(\mathbf{r})$ to the field at P from the charge in the small volume element dv_Q:

$$d\mathbf{E}(\mathbf{r}) = \frac{1}{4\pi\varepsilon_0}\left(\sum_{n \in dv_Q} q_n\right)\frac{\mathbf{r} - \mathbf{r}_Q}{|\mathbf{r} - \mathbf{r}_Q|^3} \tag{3.8a}$$

The sum $\sum q_n$ of all charges in dv_Q represents a *macroscopic* charge, and therefore, as mentioned above, will not change appreciably when we move the volume element by, say, one diameter. For this reason we may also assume that the sum is proportional to the volume dv_Q; it can therefore be replaced by $\varrho_\mathrm{v}(\mathbf{r}_Q)\,dv_Q$, where $\varrho_\mathrm{v}(\mathbf{r}_Q)$ is the *volume charge density*, i.e. the charge per unit volume. Some texts define it as the limit of the ratio dq/dv (the charge to volume ratio) as $dv \to 0$, but that obviously fails as dv becomes so small that only a few particles are left in it. It is preferable to speak of a *macroscopically small but microscopically large* volume element, as defined above. So, with the above replacement, the increment in the electrostatic field at P due to the charge in dv_Q becomes

$$d\mathbf{E}(\mathbf{r}) = \frac{1}{4\pi\varepsilon_0}\varrho_\mathrm{v}(\mathbf{r}_Q)\frac{\mathbf{r} - \mathbf{r}_Q}{|\mathbf{r} - \mathbf{r}_Q|^3}\,dv_Q \tag{3.8b}$$

Now we integrate over the entire distribution of charge to obtain for the field at P

$$\mathbf{E}(\mathbf{r}) = \frac{1}{4\pi\varepsilon_0}\iiint_V dv_Q\,\varrho_\mathrm{v}(\mathbf{r}_Q)\frac{\mathbf{r} - \mathbf{r}_Q}{|\mathbf{r} - \mathbf{r}_Q|^3} \tag{3.9}$$

This is the counterpart of (3.4) for a *continuous distribution of charge* (which is what we usually have in macroscopic situations). Note that the dimension of the volume-charge density ϱ_v is $\mathrm{C/m^3}$.

It may be remarked that the approximation (3.8a) does *not* hold for the region immediately around **r**, and this is true. However, only in a relatively small fraction of the contributing volume is the denominator of the integrand expected to be so small that appreciable errors can be anticipated, and then one can argue that the forces on an infinitesimal spherical surface around P (which contains essentially uniform charge density) cancel each other. The error made in using the approximation close to P is negligible for these reasons.

3.3.2 Surface-charge density

Charges on metallic materials tend to congregate on the surface (the reasons will be discussed later). The concept of surface charge can be approached in the following manner. We can think of the charge on the surface as a volume-charge density times a 'pancake-like' volume element $dv_Q \equiv dS_Q\, dl$, where dl is the (very narrow) width of the volume element (see Fig. 3.3). Thus

$$\varrho_v(\mathbf{r}_Q)\, dv_Q \equiv \varrho_v(\mathbf{r}_Q)\, dl\, dS_Q \tag{3.10a}$$

For all practical purposes in cases where the charge is narrowly confined to the vicinity of a surface, we can regard the product $\varrho_v(\mathbf{r}_Q)\, dl$ as a *surface-charge density* $\varrho_s(\mathbf{r}_Q)$, which is measured in C/m^2. The integral in (3.9) is then replaced by

$$\mathbf{E}(\mathbf{r}) = \frac{1}{4\pi\varepsilon_0} \int\!\!\int dS_Q\, \varrho_s(\mathbf{r}_Q) \frac{\mathbf{r} - \mathbf{r}_Q}{|\mathbf{r} - \mathbf{r}_Q|^3} \tag{3.10b}$$

Note that the dimension of the right-hand side remains the same as in (3.9).

Example 3.1 Disk with uniform charge density

Consider a surface-charge density ϱ_s smeared uniformly over a disk of radius a, as shown in Fig. 3.4. Let us calculate the **E** field at a point P on the axis of the disk (along which we will set the z axis). Let P be characterized by $\mathbf{r} = (0, 0, z)$. Let us choose an arbitrary point Q on the disk, characterized by $\mathbf{r}_Q = (\rho, \varphi, 0)$.

The diagram shows that $\mathbf{r} = z\hat{\mathbf{z}}$ and $\mathbf{r}_Q = \rho\hat{\boldsymbol{\rho}}(\varphi)$. Thus $\mathbf{r} - \mathbf{r}_Q = z\hat{\mathbf{z}} - \rho\hat{\boldsymbol{\rho}}(\varphi)$ and $|\mathbf{r} - \mathbf{r}_Q| = \sqrt{z^2 + \rho^2}$. The surface element, in cylindrical coordinates,

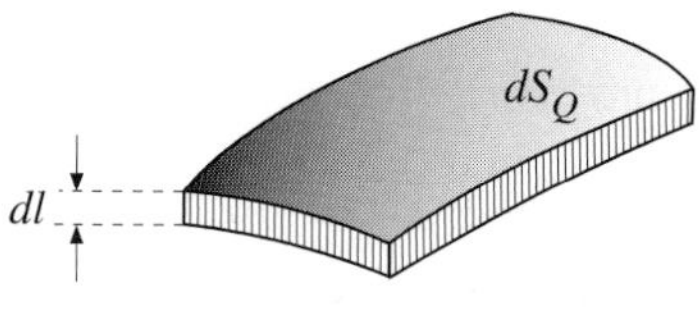

Fig. 3.3 Surface-charge density as a limit of volume charge in a thin layer.

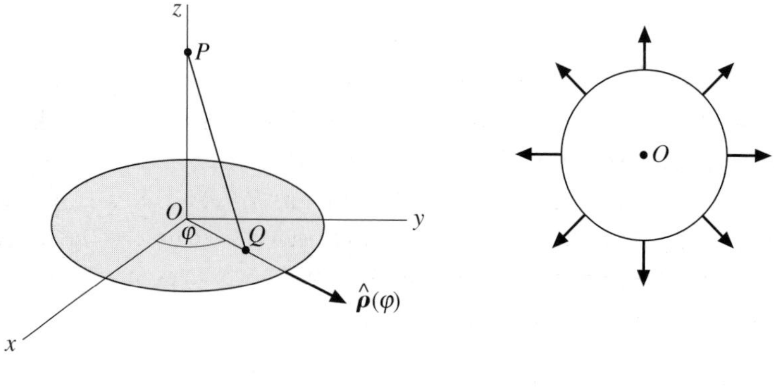

Fig. 3.4 Disk with uniform charge density.

is $dS_Q = \rho\, d\rho\, d\varphi$, as we can see from (2.26), so that we obtain from (3.10b)

$$\mathbf{E}(\mathbf{r}) = \frac{\varrho_s}{4\pi\varepsilon_0}\int_0^{2\pi} d\varphi \int_0^a d\rho\rho \frac{z\hat{\mathbf{z}} - \rho\hat{\boldsymbol{\rho}}(\varphi)}{(z^2+\rho^2)^{3/2}} \tag{3.11a}$$

From the paragraph after (2.34) we have that $\hat{\boldsymbol{\rho}}(\varphi) = \hat{\mathbf{x}}\cos\varphi + \hat{\mathbf{y}}\sin\varphi$. As this is the only φ-dependence in (3.11a), we can perform the $d\varphi$ integration from 0 to 2π over either $\sin\varphi$ or $\cos\varphi$, obtaining zero for the $\rho\hat{\boldsymbol{\rho}}(\varphi)$ part and 2π for the $z\hat{\mathbf{z}}$ part. That the $\rho\hat{\boldsymbol{\rho}}(\varphi)$ part gives zero is also easy to understand from symmetry considerations because, as we rotate $\hat{\boldsymbol{\rho}}(\varphi)$ with increasing φ, the only thing that changes in the integrand of the $\rho\hat{\boldsymbol{\rho}}(\varphi)$ part is the direction of $\hat{\boldsymbol{\rho}}(\varphi)$. We are adding up a circle of radially outward pointing unit vectors, as also shown in Fig. 3.4. Thus we have left

$$\mathbf{E}(\mathbf{r}) = \frac{2\pi\varrho_s}{4\pi\varepsilon_0}\int_0^a d\rho \frac{z\rho}{(z^2+\rho^2)^{3/2}}\hat{\mathbf{z}} \tag{3.11b}$$

which is a field in the $\hat{\mathbf{z}}$ direction, as might be expected from symmetry considerations; these rule out a $\hat{\boldsymbol{\rho}}$ component. The integral can be facilitated by transforming variable ρ to a new variable, $w = \rho/z$. Upon dividing numerator and denominator by z^3, we obtain

$$\mathbf{E}(\mathbf{r}) = \frac{\varrho_s}{2\varepsilon_0}\int_0^{a/z} dw \frac{w}{(1+w^2)^{3/2}}\hat{\mathbf{z}} \tag{3.11c}$$

This integral is elementary and can be either calculated (by transforming from w to a new variable $u = w^2$) or looked up in an integrals table. The result is

$$\mathbf{E}(\mathbf{r}) = \frac{\varrho_s}{2\varepsilon_0}\left(1 - \frac{1}{\sqrt{1+a^2/z^2}}\right)\hat{\mathbf{z}} \tag{3.11d}$$

Note that the **E** field always points *away* from the disk of charge if ϱ_s is positive; hence the **E** field below the disk (i.e. at $z < 0$) points *downwards*. This result is due to symmetry: when we put P on the axis, we simply defined that semi-infinite part of the axis to be the positive z axis. If we had placed P on the other half-axis, we also would have obtained (3.11d), but then $\hat{\mathbf{z}}$ would have been chosen (and would point) in the direction opposite to our initial choice. We obtain a result when $a/z \to \infty$ that we shall encounter later in the infinite-disk case:

$$E(\mathbf{r}) = \frac{\varrho_s}{2\varepsilon_0}\frac{z}{|z|}\hat{\mathbf{z}} \qquad \text{for } a/z \to \infty \tag{3.12a}$$

Note that we have a useful notation here that gives the correct direction for any z, because $z/|z|$ is $+1$ for $z > 0$ and -1 for $z < 0$. The truncated binomial expansion (3.6) can be applied to (3.11d) when the ratio $z/a \to \infty$ (so that $a/z \to 0$): we set $[1 + (a/z)^2]^{-1/2} \approx 1 - \frac{1}{2}(a/z)^2$, thus obtaining

$$E(\mathbf{r}) = \pm \frac{\varrho_s a^2}{4\varepsilon_0 z^2} \hat{\mathbf{z}} = \pm \frac{\varrho_s \pi a^2}{4\pi\varepsilon_0 z^2} \hat{\mathbf{z}} \qquad \text{for } z/a \to \pm\infty \tag{3.12b}$$

Because $\varrho_s \pi a^2$ is the total charge q on the disk, we observe that (3.12b) is Coulomb's law for a point charge $q = \varrho_s \pi a^2$ at distance z. The disk collapses essentially to a point at very large distances compared with the disk radius. The $\pm$ sign is needed to give $\mathbf{E}$ the correct direction when $z < 0$, and it follows from the fact that $\sqrt{z^2} = \pm|z|$, with the minus sign to be chosen whenever $z < 0$.

3.3.3 Line-charge density

Charges on thin wires form an example of *line charge*. In this case (see Fig. 3.5), we can consider the volume element $dv_Q = dS\, dl_Q$ to be long and cylindrical, dS being a small cross-sectional area such that $\varrho_v(\mathbf{r}_Q)\, dS$ may be considered to be a *line-charge density* $\varrho_\ell(\mathbf{r}_Q)$, which is measured in C/m. The field at a point P is then

$$\mathbf{E}(\mathbf{r}) = \frac{1}{4\pi\varepsilon_0} \int dl_Q\, \varrho_\ell(\mathbf{r}_Q) \frac{\mathbf{r} - \mathbf{r}_Q}{|\mathbf{r} - \mathbf{r}_Q|^3} \tag{3.13}$$

Example 3.2 Infinite straight line with uniform charge density

Now consider an infinite straight line (coinciding with the z axis) with uniform line-charge density ϱ_ℓ (see Fig. 3.6). Note the similarity of the diagram to Fig. 3.4. However, points P (observation) and Q (charge) are reversed in comparison to Fig. 3.4. The point P is $(\rho, \varphi, 0)$ in cylindrical coordinates. Let $\overrightarrow{OP} = \rho\hat{\boldsymbol{\rho}}(\varphi)$. Because $\overrightarrow{OQ} = z\hat{\mathbf{z}}$, we have $\mathbf{r} - \mathbf{r}_Q = \rho\hat{\boldsymbol{\rho}}(\varphi) - z\hat{\mathbf{z}}$, and so

$$\mathbf{E}(\mathbf{r}) = \frac{\varrho_\ell}{4\pi\varepsilon_0} \int_{-\infty}^{\infty} dz \frac{\rho\hat{\boldsymbol{\rho}}(\varphi) - z\hat{\mathbf{z}}}{(z^2 + \rho^2)^{3/2}} \tag{3.14a}$$

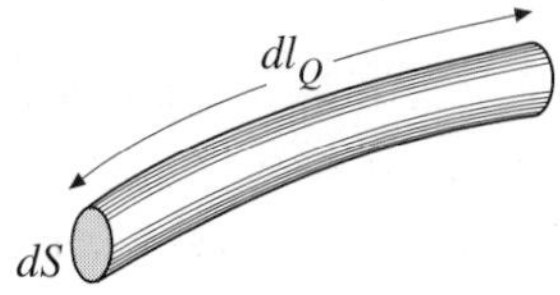

Fig. 3.5 Volume element for a short section of a thin wire.

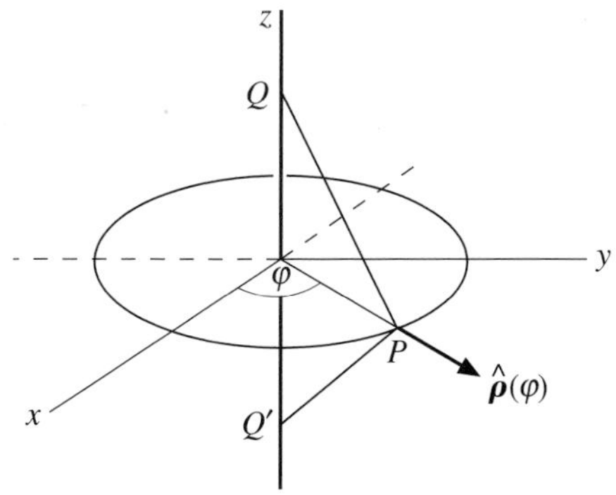

Fig. 3.6 Circularly symmetric locations around an infinite uniformly charged infinitesimally thin straight wire.

The $z\hat{\mathbf{z}}$ part integrates to zero because that part of the integrand is odd in z and the integral bounds are symmetric around zero. A more physical expression of the same fact is that for every point Q there is a mirror-image point Q' below the $z = 0$ plane for which the $\hat{\mathbf{z}}$ component of $\mathbf{r} - \mathbf{r}_{Q'}$ points in the $-\hat{\mathbf{z}}$ direction with the same magnitude. We define $w = z/\rho$ in the remaining part, to obtain

$$\mathbf{E}(\mathbf{r}) = \frac{\varrho_\ell}{4\pi\varepsilon_0\rho} \int_{-\infty}^{\infty} dw \frac{1}{(1+w^2)^{3/2}} \hat{\boldsymbol{\rho}}(\varphi) \tag{3.14b}$$

The integrand is even in w, and it is twice the same integral but with limits $w = 0$ and $w = \infty$, which can be found from an integrals table to equal unity. Thus we obtain

$$\mathbf{E}(\mathbf{r}) = \frac{\varrho_\ell}{2\pi\varepsilon_0\rho} \hat{\boldsymbol{\rho}}(\varphi) \tag{3.14c}$$

So, if $\varrho_\ell > 0$ we see that the electrostatic field points *away* from the line charge (and is $\perp$ to it). The field is symmetric around the line charge. The field falls off with an inverse-distance law as ρ is increased.

3.4 Gauss's law for the electric field in free space

When one tries to calculate from Coulomb's law (3.9) the electrostatic field produced by an arbitrarily shaped charge distribution, it becomes clear that the mathematics of doing the integrals is generally intractable. It then may seem that a numerical scheme of integration is needed, but the numerical *quadrature* of a three-dimensional integral is often a costly and time-consuming matter, even with relatively large computers. It is usually easier to solve numerically differential equations. However, we have not yet developed such equations. This is what we are now aiming to do.

The first major step in going beyond Coulomb's law is the formulation of *Gauss's law*. Considered on its own merits, Gauss's law has the advantage of helping to increase our insight into the nature of the electrostatic field. We now write down *Gauss's law in free space* and then explain what is conveyed by it:

$$\oint dS_Q \hat{\mathbf{n}}_Q \cdot \mathbf{E}_Q = q_{\text{encl}}/\varepsilon_0 \tag{3.15a}$$

A crucial item here is that q_{encl} represents the charge *inside* the volume v enclosed by the surface S (see Fig. 3.7). It specifically excludes all charges *outside* that volume! The unit vector $\hat{\mathbf{n}}_Q$ is $\perp$ to the surface at the position of surface element dS_Q, and it points *out of* v. $\mathbf{E}_Q$ is the field at dS_Q.

Note that (3.15a) is a surface integral, but we have written down only one integral sign (with a circle through it); the surface element dS_Q indicates clearly that a surface integral is involved. Henceforth, we will omit double- and triple-integral signs, as given in (3.9) and (3.10b), except in special cases where bounds are specified. The circle through the surface-integral sign indicates that integration is over a closed surface.

We have seen, when we encountered surface elements, that $d\mathbf{S}_Q \equiv \hat{\mathbf{n}}_Q dS_Q$, and therefore we often prefer to write (3.15a) as

$$\oint d\mathbf{S}_Q \cdot \mathbf{E}_Q = q_{\text{encl}}/\varepsilon_0, \tag{3.15b}$$

but there is no special preference for one notation over the other, and we shall use both. We emphasize that only the portion of the charge distribution *inside* the

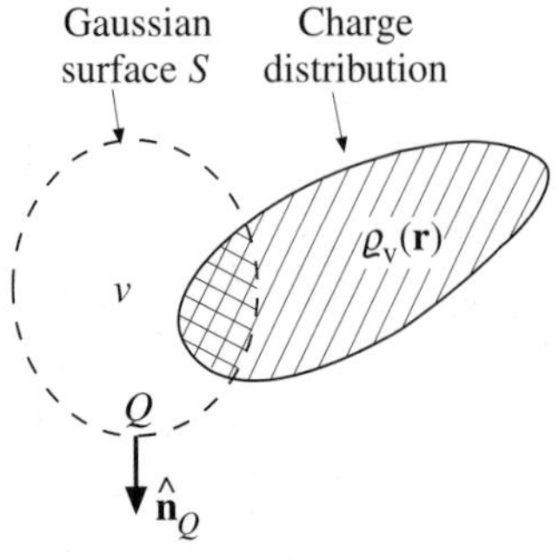

Fig. 3.7 Illustration of the charge contributing to Gauss's law with respect to a volume v enclosed by surface S.

surface S (see the cross-hatching in Fig. 3.7) counts towards q_{encl}, namely

$$q_{\text{encl}} = \int_v dv_P\, \varrho_v(\mathbf{r}_P) \tag{3.16}$$

The surface S, known as a *Gaussian surface*, can be a real material surface, or it can be any imaginary surface. But whatever surface it is, only the charge inside it must be included on the right-hand side of (3.15). The left-hand sides of (3.15a, b) give the total *electrostatic flux* out of the surface S. This name is in analogy with fluid velocity: if $\mathbf{E}$ were the velocity vector field of a fluid, then the left-hand sides of (3.15a, b) would express the total flow of that '$\mathbf{E}$-fluid' across (and out of) the closed surface S.

Gauss's law can be derived from Coulomb's law as follows (see Fig. 3.8). Consider a region of charge surrounded by a closed surface S. Inside the region is a point charge q at point Q. The electrostatic field at point P (on the surface S) is $\mathbf{E}_P = q\hat{\mathbf{R}}_{QP}/(4\pi\varepsilon_0 R_{QP}^2)$. Dot-multiply this by a small surface element $d\mathbf{S}_P = \hat{\mathbf{n}}_P dS_P$ to obtain the electrostatic flux through dS_P:

$$\mathbf{E}_P \cdot d\mathbf{S}_P = \frac{q}{\varepsilon_0} \frac{dS_P(\hat{\mathbf{n}}_P \cdot \hat{\mathbf{R}}_{QP})}{4\pi R_{QP}^2}$$

The quantity $dS_P(\hat{\mathbf{n}}_P \cdot \hat{\mathbf{R}}_{QP}) = dS_P \cos\psi$ represents the small piece of surface area $dS = R_{QP}^2\, d^2\Omega = R_{QP}^2 \sin\theta\, d\theta\, d\varphi$, which is part of a spherical surface with Q as center and R_{QP} as radius. The element $d^2\Omega = \sin\theta\, d\theta\, d\varphi$ is an element of solid angle, and is independent of the distance R_{QP}. So we can write

$$\mathbf{E}_P \cdot d\mathbf{S}_P = \frac{q}{\varepsilon_0} \frac{d^2\Omega_P}{4\pi}$$

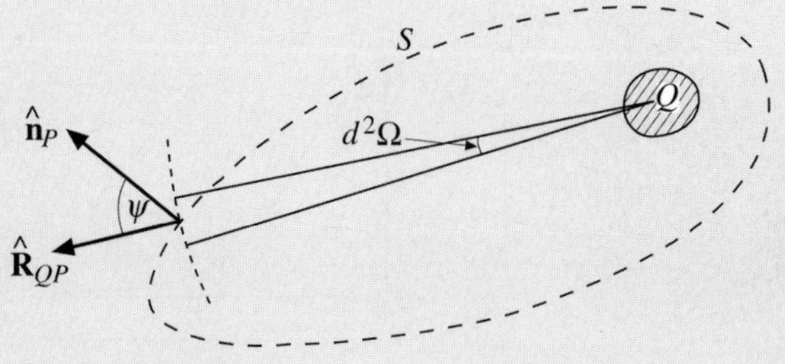

Fig. 3.8 The proof of Gauss's law.

By integrating over the whole surface S all possible values of $0 \le \theta \le \pi$ and $0 \le \varphi \le 2\pi$ are achieved, so that

$$\oint \mathbf{E}_P \cdot d\mathbf{S}_P = \frac{q}{4\pi\varepsilon_0} \oint d^2\Omega_P = \frac{q}{4\pi\varepsilon_0} \int_0^{2\pi} d\varphi \int_0^{\pi} d\theta \sin\theta = \frac{q}{\varepsilon_0}$$

Finally, the superposition of **E** fields for diverse point charges q_n inside S yields the same result, with q replaced by the sum of all q_n inside S, i.e. the total charge inside S. For a continuous charge distribution, we replace the infinitesimal charge q and Q by $\varrho_v(\mathbf{r}_Q)\, dv_Q$. This completes the derivation from Coulomb's law.

Gauss's law can be made plausible without recourse to the derivation outlined in the above optional paragraph, and we shall proceed in doing so by means of some simple examples.

Example 3.3 Spherically symmetric distribution of charge

As shown in Fig. 3.9, consider a spherical charge distribution $\varrho_v(r)$ around the origin O. We surround the origin at distance r by a spherical Gaussian surface. From (2.29) we observe that $dv = r^2\, dr \sin\theta\, d\theta\, d\varphi$, and from (2.27) that $d\mathbf{S} = \hat{\mathbf{r}} r^2 \sin\theta\, d\theta\, d\varphi$. The right-hand side of (3.15b) becomes

$$\frac{1}{\varepsilon_0} q_{\text{encl}} = \frac{1}{\varepsilon_0} \int_0^r dr' r'^2 \varrho_v(r') \int_0^{\pi} d\theta \sin\theta \int_0^{2\pi} d\varphi = \frac{1}{\varepsilon_0} 4\pi \int_0^r dr' r'^2 \varrho_v(r') \qquad (3.17a)$$

Note that we have placed $\varrho_v(r)$ *in front of* the angular integrals because it does not depend upon θ or φ. The two angular integrals give the factor 4π (this is a result we will use often!). Now consider the left-hand side of (3.15b). The spherical symmetry

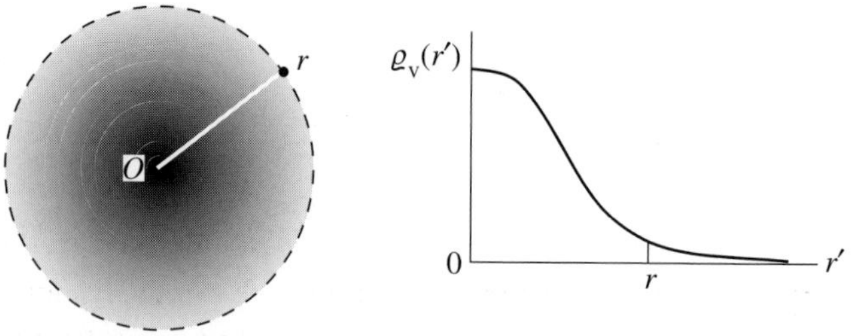

Fig. 3.9 A spherically symmetric distribution of charge.

forces us to conclude that $\mathbf{E}(r, \theta, \varphi)$ cannot depend upon θ or φ, hence we write it as $\mathbf{E}(r)$, a function of only the radial spherical coordinate. Similar symmetry arguments tell us that $\mathbf{E}(r)$ can only be in the *radial* direction, because there is no 'preferred' $\hat{\boldsymbol{\theta}}$ or $\hat{\boldsymbol{\varphi}}$ direction. Therefore, the left-hand side of (3.15b) is

$$r^2 \int_0^{\pi} d\theta \sin\theta \int_0^{2\pi} d\varphi \, \hat{\mathbf{r}} \cdot \mathbf{E}(r) = 4\pi r^2 E_r(r) \tag{3.17b}$$

By equating the left-hand side of (3.17a) to the right-hand side of (3.17b), we obtain

$$\mathbf{E}(r) = \frac{q_{\mathrm{encl}}}{4\pi\varepsilon_0 r^2} \hat{\mathbf{r}} \tag{3.18a}$$

which is Coulomb's law for the $\mathbf{E}$ field at distance r from a point charge q_{encl} at the origin. In other words, the symmetric charge distribution *inside* the spherical surface might just as well be lumped as a point charge at the center, as far as the $\mathbf{E}$ field is concerned. This is a typical consequence of inverse-square laws and therefore holds for the gravitational force, too. We can also substitute the right-hand side of (3.7a) into (3.18a) to obtain, for spherical symmetry:

$$\mathbf{E}(r) = \frac{1}{\varepsilon_0 r^2} \int_0^{r} dr' r'^2 \varrho_{\mathrm{v}}(r') \hat{\mathbf{r}} \tag{3.18b}$$

Suppose we have a charge distribution $\varrho_{\mathrm{v}}(r')$ that is uniform inside a sphere of radius r_0 and is zero outside it:

$$\begin{aligned} \varrho_{\mathrm{v}}(r') &= \varrho_{\mathrm{v}} \qquad \text{for } r' < r_0 \\ \varrho_{\mathrm{v}}(r') &= 0 \qquad \text{for } r' > r_0 \end{aligned} \tag{3.19a}$$

Now we apply (3.18b) carefully, both for values of r less than r_0 and for values of r greater than r_0. We then obtain

$$\begin{aligned} E_r(r) &= \frac{1}{\varepsilon_0 r^2} \int_0^{r} dr' r'^2 \varrho_{\mathrm{v}} = \frac{\varrho_{\mathrm{v}}}{3\varepsilon_0} r = \frac{(4\pi r^3/3)\varrho_{\mathrm{v}}}{4\pi\varepsilon_0 r^2} \qquad \text{for } r < r_0 \\ E_r(r) &= \frac{1}{\varepsilon_0 r^2} \int_0^{r_0} dr' r'^2 \varrho_{\mathrm{v}} = \frac{\varrho_{\mathrm{v}}}{3\varepsilon_0} \frac{r_0^3}{r^2} = \frac{v_0 \varrho_{\mathrm{v}}}{4\pi\varepsilon_0 r^2} \qquad \text{for } r > r_0 \end{aligned} \tag{3.19b}$$

Here v_0 is the volume occupied by the charge. Note the difference in the upper bounds of the integrals! We recognize Coulomb's law for a point charge ($q_{\mathrm{encl}} = \text{volume} \times \varrho_{\mathrm{v}}$) in these formulas; the reason is that they are merely special cases of (3.18a). Figure 3.10 depicts the r-dependence of the radial field E_r calculated

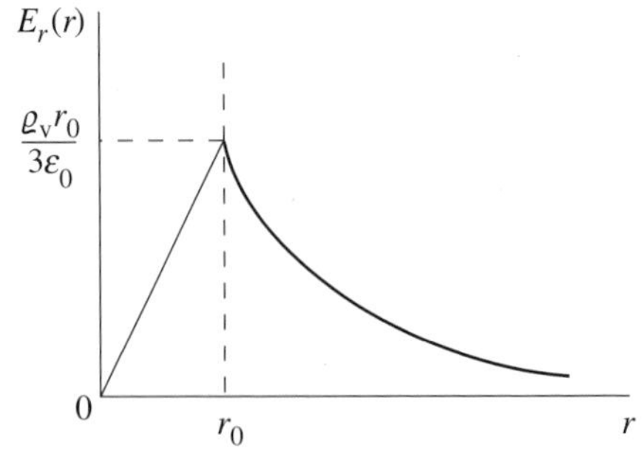

Fig. 3.10 $E(r)$ as a function of r for a sphere of uniform charge density; see Eq. (3.19b).

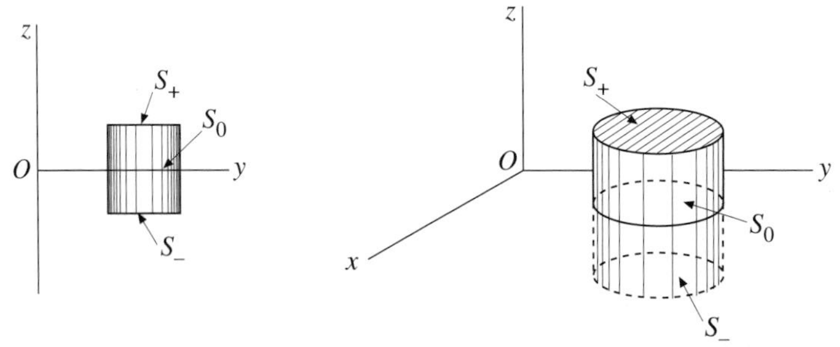

Fig. 3.11 Gaussian volume for an infinite plane with uniform surface-charge density.

in (3.19b). The E_r field builds up linearly from the central zero value to a maximum value $\varrho_v r_0/(3\varepsilon_0)$ at $r = r_0$; then it decays with the inverse of the distance to the center.

Example 3.4 Infinite planar uniform distribution of charge

Here, we assume that the surface-charge density ϱ_s is constant over the plane $z = 0$ and that there is no charge anywhere else. We show a Gaussian cylinder in Fig. 3.11 (not necessarily circular, but with its cylindrical or 'wrap-around' surface $\perp$ the $z = 0$ plane). The bottom surface S_- is at $-z$ and the top surface S_+ is at $+z$; surface S_0 is the intersection with the $z = 0$ plane. All three of these planar surfaces are parallel, have the same shape, and the same area S.

Owing to the symmetry, we expect $\mathbf{E}(\mathbf{r}) = E_z(z)\hat{\mathbf{z}}$ not to be a function of x or y, and not to have $\hat{\mathbf{x}}$ or $\hat{\mathbf{y}}$ components; furthermore, $\mathbf{E}(-z) = -\mathbf{E}(z)$. Application of Gauss's law (3.15b) to this situation then yields

$$\begin{aligned} &\mathbf{S}_+ \cdot \mathbf{E}(z) + \mathbf{S}_- \cdot \mathbf{E}(-z) = S_0\varrho_s/\varepsilon_0 \qquad \text{or} \\ &S[E_z(z) - E_z(-z)] = S\varrho_s/\varepsilon_0 \qquad \longrightarrow \qquad 2E_z(z) = \varrho_s/\varepsilon_0 \end{aligned} \tag{3.20a}$$

To understand the second line here, note that $\mathbf{S}_+ = +S\hat{\mathbf{z}}$ and $\mathbf{S}_- = -S\hat{\mathbf{z}}$ since surface vectors are measured outwards. As this calculation is in terms of $z > 0$, we conclude, as in (3.12a), that

$$\mathbf{E}(z) = \pm\frac{\varrho_s}{2\varepsilon_0}\hat{\mathbf{z}} = \frac{\varrho_s}{2\varepsilon_0}\frac{z}{|z|}\hat{\mathbf{z}} \tag{3.20b}$$

which shows that the field is a constant vector on either side of the charge distribution (pointing away from it if $\varrho_s > 0$). There is an interesting way to understand this independence of height above the charge distribution. On the one hand Coulomb's law tells us that the **E** field of a point charge must fall off as z^{-2}, but on the other hand the *area* of charge 'seen' inside the cone on $z = 0$ that has a given angle pointing downwards from the point of observation, increases as z^2. These two factors therefore cancel each other during integration of the Coulomb's-law field for a surface-charge distribution! Finally, (3.20b) can be used to calculate the field due to two or more parallel planes, each with constant surface-charge density, by means of the principle of superposition: the field at a point P is the sum of fields (3.20b), where each z is the distance to a plane.

Example 3.5 Uniform distribution of charge density inside an infinite circular cylinder

The case of an infinite straight wire with constant circular cross section (with radius ρ_0) and constant volume-charge distribution ϱ_v is handled by imagining a coaxial Gaussian cylinder with radius ρ that can be greater or less than ρ_0 around the z axis; see Fig. 3.12. The z axis is chosen as the central axis of the charge distribution. The symmetry dictates that in terms of cylindrical coordinates $\mathbf{E}(\rho, \varphi, z) = E_\rho(\rho)\hat{\boldsymbol{\rho}}(\varphi)$. Now there is no contribution to $d\mathbf{S}\cdot\mathbf{E}$ from the surfaces S_+ or S_- (because the vectors representing these surfaces are $\perp$ the **E** fields everywhere). The magnitude of E_ρ is constant on the 'wrap-around' surface with area $S_w = 2\pi\rho h$, for a length h of

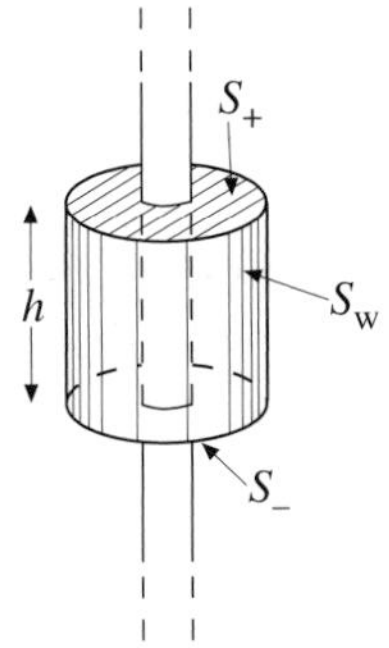

Fig. 3.12 Gaussian volume for a circularly cylindrical distribution of uniform charge. The wire has radius ρ_0 and the Gaussian cylinder is shown for $\rho > \rho_0$.

cylinder, and the directions of **E** and the surface-element vector $d\mathbf{S}$ are radial everywhere on that surface. We therefore find from Gauss's law, (3.15b),

$$
\begin{aligned}
2\pi\rho h E_\rho(\rho) &= \pi\rho^2 h\varrho_\mathrm{v}/\varepsilon_0 && \text{for } \rho < \rho_0 \\
2\pi\rho h E_\rho(\rho) &= \pi\rho_0^2 h\varrho_\mathrm{v}/\varepsilon_0 && \text{for } \rho > \rho_0
\end{aligned} \tag{3.21a}
$$

from which expressions it is readily seen that

$$
\begin{aligned}
\mathbf{E}(\rho,\varphi) &= \frac{\pi\rho^2\varrho_\mathrm{v}}{2\pi\varepsilon_0\rho}\hat{\boldsymbol{\rho}}(\varphi) \equiv \frac{\varrho_\ell(\rho)}{2\pi\varepsilon_0\rho}\hat{\boldsymbol{\rho}}(\varphi) && \text{for } \rho < \rho_0 \\
\mathbf{E}(\rho,\varphi) &= \frac{\pi\rho_0^2\varrho_\mathrm{v}}{2\pi\varepsilon_0\rho}\hat{\boldsymbol{\rho}}(\varphi) \equiv \frac{\varrho_\ell(\rho_0)}{2\pi\varepsilon_0\rho}\hat{\boldsymbol{\rho}}(\varphi) && \text{for } \rho > \rho_0
\end{aligned} \tag{3.21b}
$$

where $\varrho_\ell(\rho) \equiv \pi\rho^2\varrho_\mathrm{v}$ is an effective *line-charge density*, measured in C/m. Thus, the field inside the wire rises linearly with ρ to a maximum value $\varrho_\mathrm{v}\rho_0/2\varepsilon_0$, then it decays with the inverse of the distance to the central axis. Compare the second result to (3.14c), which was obtained directly from Coulomb's law. The present derivation certainly seems much easier to do. The dependence of E_ρ upon ρ is similar to that shown for E_r upon r in Fig. 3.10.

These examples clarify several important points. First, Gauss's law helps exploit the symmetry of a situation to simplify a surface integral into a product of an area times a constant **E** field, which greatly simplifies calculation of the latter. In the second place, this very same fact limits the usefulness of Gauss's law for calculating the electrostatic field of less symmetric situations: the **E** field then is not separable from the surface element in the surface integral. Nevertheless, Gauss's law helps give important insights into the nature of the *local* electrostatic field in connection with the construction of *field lines* (see below).

3.5 Electrostatic field lines

In Example 3.3, concerning a spherically symmetric distribution of charge, we observed that the **E** field always points in the $\hat{\mathbf{r}}(\theta,\varphi)$ direction. We can 'connect' all the arrows that represent $\mathbf{E}(r,\theta,\phi)$ for fixed θ and φ. Those arrows will all point in the same direction, but they have different lengths determined by the magnitude $E = |\mathbf{E}|$ at each r. We can draw a line through these arrows, and this line is a *field line*. Thus the field lines in Example 3.3 are in the radial direction. In Example 3.5, a cylindrical charge distribution, the field lines are in the $\hat{\boldsymbol{\rho}}(\varphi)$ direction, i.e. they point away from the z axis.

Field lines supply us with a visual picture of the direction of the forces from one point to the next. However, we need a more general definition of a field line (see Fig. 3.13):

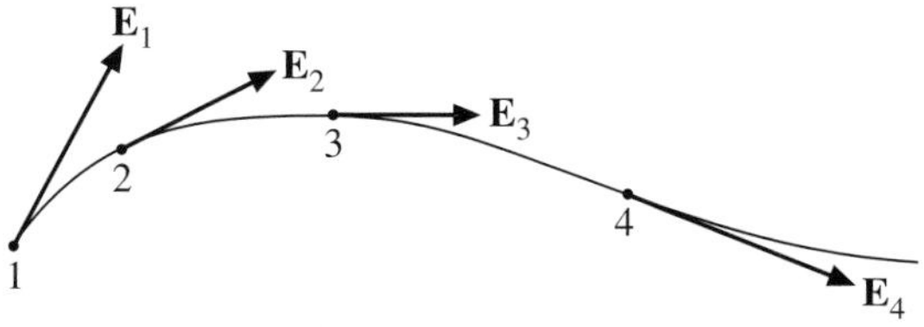

Fig. 3.13 The field line concept.

A field line is a line such that the field at any point on the line is tangential *to that line (in a space in which a vector field is defined).*

The usual convention is to put an arrow through a field line in the same direction as the tangential fields. An electrostatic field line therefore points from + to − charges, and we shall see later that field lines start and stop on charges. One consequence of great importance is that *field lines cannot cross each other*! For if they did, there would be two simultaneous tangential directions at the point of intersection, but there can only be one field direction at that point. The mathematical form of the definition, for any orthogonal coordinate system, is that at any location P on a field line, given that $\mathbf{E}$ is the field at P and that $d\boldsymbol{l}$ is the length element at P,

$$dl_i = \text{constant} \times E_i \qquad (i = 1, 2, 3) \tag{3.22}$$

The constant can change from point to point, but it must have the same value for all three components at any single point. For Example 3.3 we have spherical coordinates, in which case E_i is nonzero only for the radial component, which gives $d\boldsymbol{l} = d\boldsymbol{l}_1 = \hat{\mathbf{r}}\, dr$.

Example 3.6 Field lines for a dipole field

A dipole consists of two equal and opposite changes, separated by a fixed distance. For now, let us just take for granted that the $\mathbf{E}$ field of an infinitesimal dipole oriented along the z axis is proportional to $2\hat{\mathbf{r}}\cos\theta + \hat{\boldsymbol{\theta}}\sin\theta$. Therefore, in any plane where $\varphi = \text{constant}$, we see that the $\mathbf{E}$ field lines are given by

$$\begin{aligned} dl_r &= dr = \text{constant} \times 2\cos\theta \\ dl_\theta &= rd_\theta = \text{constant} \times \sin\theta \end{aligned} \tag{3.23}$$

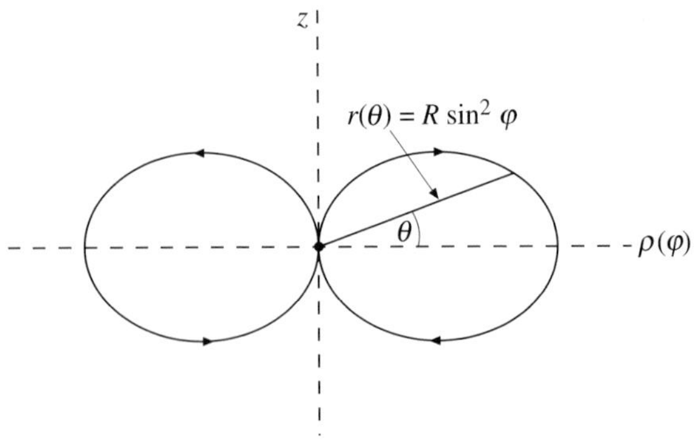

Fig. 3.14 Field lines for a small electrostatic dipole, oriented along the z axis.

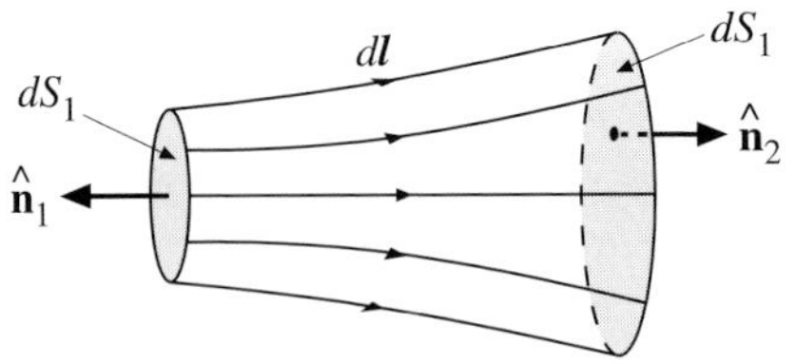

Fig. 3.15 A tube of field lines.

By division, we find $dr/d\theta = 2r\cos\theta/\sin\theta$, or $dr/r = 2d(\sin\theta)/\sin\theta$. Integration tells us that $\ln r = \ln\sin^2\theta + \text{constant}$, and exponentiation then gives $r(\theta) = R\sin^2\theta$, where we have renamed the constant as $\ln R$ for this particular field line. If we choose a particular value for R and then vary θ from 0 to 2π, we obtain Fig. 3.14.

The *spacing between field lines* gives information about the way the strength of the field is changing locally. Consider the infinitesimal bundle of field lines in Fig. 3.15; $d\mathbf{l}$ is along the field lines. If there is no charge in the tube, then Gauss's law (3.15b) predicts

$$E_1\, dS_1\hat{\mathbf{n}}_1 + E_2\, dS_2\hat{\mathbf{n}}_2 = 0 \qquad \longrightarrow \qquad E_2 = E_1\, dS_1/dS_2 \tag{3.24}$$

because $\hat{\mathbf{n}}_2 = -\hat{\mathbf{n}}_1$. We do not need integral signs because the surface elements are infinitesimal. Note that the 'wrap-around' surface of the bundle does not contribute, even though it is needed to make a closed surface, because $\mathbf{E} \perp d\mathbf{S}$ everywhere on that part of the total surface. We observe that $E_i \propto 1/dS_i$, so that if the surface element dS_2 increases, implying that the field lines are *diverging*, then the field becomes *weaker*. *Converging* field lines lead to *stronger* fields.

3.6 Line integrals in vector fields

We have considered volume and surface integrals. *Line integrals* are also very useful, but their calculation requires considerable care. A line integral is formally defined for a vector field $\mathbf{A}(\mathbf{r})$ as

$$\int_{P_\mathrm{i}}^{P_\mathrm{f}} d\boldsymbol{l} \cdot \mathbf{A} = \lim_{N\to\infty,\ d\boldsymbol{l}_n\to 0 \text{ for all } n} \sum_{n=1}^{N} d\boldsymbol{l}_n \cdot \mathbf{A}_n \tag{3.25}$$

See Fig. 3.16. The integral, in general, is dependent upon the path from P_i to P_f. Each infinitesimal line element $d\boldsymbol{l}_n$ is dot-multiplied by the field vector $\mathbf{A}_n$ at its location, and the result is summed over all line elements on the path between P_i and P_f. Let us discuss (3.25) in Cartesian coordinates for convenience although any other coordinate system is equally valid (and perhaps more useful in some cases!). Let the path be defined by a curve in three-dimensional space. Mathematically, a curve in a three-dimensional space can always be defined as the intersection of two surfaces, e.g.

$$f_1(x, y, z) = 0 \qquad \text{and} \qquad f_2(x, y, z) = 0 \tag{3.26a}$$

Each of these two functions defines a (possibly curved) surface in three-dimensional space; this can be seen by rewriting each of these equations as $z = F_i(x, y)$. Although it may be hard to do so in practice, in principle these latter two equations in three variables can then be used to write x and y as functions of z, i.e. $x = X(z)$ and $y = Y(z)$. By taking differentials of each function f_i $(i = 1, 2)$ we obtain from (3.26a)

$$\begin{aligned} df_1 = 0 \quad &\longrightarrow \quad (\partial f_1/\partial x)\, dx + (\partial f_1/\partial y)\, dy + (\partial f_1/\partial z)\, dz = 0 \\ df_2 = 0 \quad &\longrightarrow \quad (\partial f_2/\partial x)\, dx + (\partial f_2/\partial y)\, dy + (\partial f_2/\partial z)\, dz = 0 \end{aligned} \tag{3.26b}$$

As the partial derivatives should be known, because the two surfaces are specified, in (3.26b) we have *two* equations in the *three* unknowns dx, dy and dz, so that any two

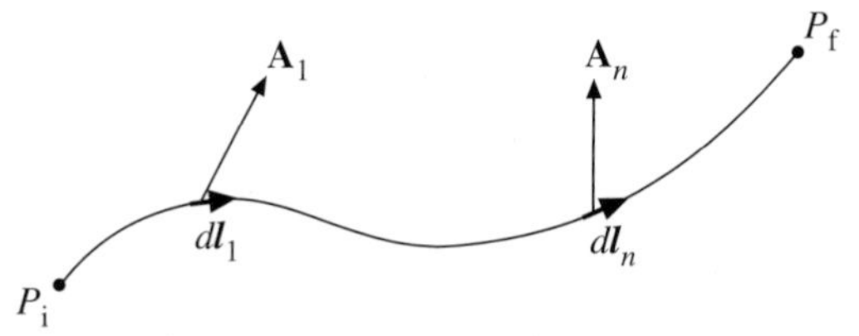

Fig. 3.16 A line integral of a vector field.

of these can be described in terms of the third: e.g. $dx = g_x(x, y, z)\,dz$ and $dy = g_y(x, y, z)\,dz$. Then dx and dy can be given solely as functions of z and dz by replacing x and y in g_x and g_y by respectively $X(z)$ and $Y(z)$, as described above for the functions f_i. We now rename g_x and g_y as follows:

$$\begin{aligned} G_x(z) &= g_x[X(z), Y(z), z] \\ G_y(z) &= g_y[X(z), Y(z), z] \end{aligned} \tag{3.27a}$$

Summarizing, we have thus obtained in principle:

$$\begin{aligned} x &= X(z) \qquad & dx &= G_x(z)\,dz \\ y &= Y(z) \qquad & dy &= G_y(z)\,dz \end{aligned} \tag{3.27b}$$

The line integral becomes

$$\begin{aligned} \int_{P_i}^{P_f} d\boldsymbol{l} \cdot \mathbf{A} &= \int_{P_i}^{P_f} [A_x(x, y, z)\,dx + A_y(x, y, z)\,dy + A_z(x, y, z)\,dz] \\ &= \int_{z_i}^{z_f} dz\{A_x[X(z), Y(z), z]G_x(z) + A_y[X(z), Y(z), z]G_y(z) \\ &\qquad + A_z[X(z), Y(z), z]\} \end{aligned} \tag{3.28}$$

It has thus has been reduced to a one-dimensional definite integral in a single variable, z, and the bounds (integration limits) are simply the z bounds of the points P_i and P_f. The z coordinate was singled out here for special attention, but the same procedure could have been followed for any one of three orthogonal coordinates, even in coordinates other than Cartesian. The performance of line integrals as above is a straightforward procedure in principle but can be a little tricky in practice. We will illustrate it with a example in which we derive two different line integrals between a given pair of points. We shall see that the second line integral breaks down naturally into sections where first one, then another coordinate is the only one that varies over the section.

Example 3.6 Line integrals in a vector field

Consider the vector field $\mathbf{A}(\mathbf{r}) = -y\hat{\mathbf{x}} + x\hat{\mathbf{y}}$ and integrate it over the parabolic path $y = x^2$ from $P_i = (0, 0, 0)$ to $P_f = (2, 4, 0)$; this is path I in Fig. 3.17. If we use (2.9) and (2.38) to transform to cylindrical coordinates we find

$$\mathbf{A}(\mathbf{r}) = -\rho \sin\varphi(\hat{\boldsymbol{\rho}} \cos\varphi - \hat{\boldsymbol{\varphi}} \sin\varphi) + \rho\cos\varphi(\hat{\boldsymbol{\rho}} \sin\varphi + \hat{\boldsymbol{\varphi}} \cos\varphi) = \rho\hat{\boldsymbol{\varphi}}$$

This vector field has field lines that consist of counterclockwise circles around the z axis. Let us, however, work in Cartesian coordinates and designate x as the variable

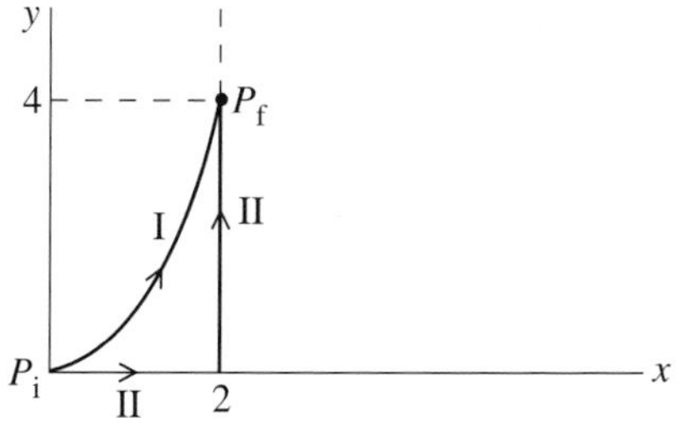

Fig. 3.17 Diagram for Example 3.6: two possible integration paths from P_i to P_f.

to transform all others into. Our work proceeds in steps from $\mathbf{A}(\mathbf{r}) = -y\hat{\mathbf{x}} + x\hat{\mathbf{y}}$ as follows:

(1) Evaluate the first integrand in (3.28):

$$A_x\,dx + A_y\,dy + A_z\,dz = -y\,dx + x\,dy + 0$$

(2) Along path I we have $y = x^2$, so in the above integrand replace y by x^2 everywhere, giving $-x^2\,dx + x\,dy$.

(3) From the equation $y = x^2$ for the path we obtain $dy = 2x\,dx$, so in the new form of the integrand replace dy by $2x\,dx$, giving $-x^2\,dx + 2x^2\,dx$, i.e. $x^2\,dx$.

(4) Now integrate this from $x_i = 0$ to $x_f = 2$:

$$\int_{P_i}^{P_f} d\mathbf{l}\cdot\mathbf{A} = \int_{P_i}^{P_f} (A_x\,dx + A_y\,dy) = \int_0^2 dx\,x^2 = \frac{8}{3} \tag{3.29a}$$

Let us now integrate between the same points P_i and P_f but over a *different* path, II, as illustrated in Fig. 3.17. We can retain step (1) above, as this does not depend upon the path. But for steps (2) and (3), there now are two straight-line pieces of path: first along the line $y = 0$ from the origin to $x = 2$, and then vertically up to the point P_f. So now we must look at

$$\int_{x=0}^{x=2} (-y\,dx + x\,dy)_{\mathrm{II}} + \int_{y=0}^{y=4} (-y\,dx + x\,dy)_{\mathrm{II}}$$

In the first integral, $y = 0$ and $dy = 0$, so it is entirely zero. In the second integral $dx = 0$, so that the first part does not contribute and all that is left is

$$\int_{y=0}^{y=4} (x\,dy)_{\mathrm{II}} = \int_{y=0}^{y=4} 2\,dy = 8 \tag{3.29b}$$

Thus for path II part of the integral is over x and part is over y.

A note of caution: suppose that path II were *reversed*, so that integration is from P_f to P_i. In this case *only* the bounds interchange in step (4), giving the negative of the previous answer. (It is tempting to think that we should also replace dx by $-dx$ (and dy by $-dy$), but that would be incorrect.) Note also that the two answers in (3.29a) and (3.29b) are *different*, of course, because the line integral is path dependent. Later, we will see that certain vector fields always give the *same* answer for any path between two points: such vector fields are called *conservative*.

3.7 The gradient of a scalar field

At this stage, it becomes important to introduce a generalization of the derivative $f'(x) \equiv df(x)/dx$: we need to consider the rate of change of a function $\Phi(\mathbf{r})$ in *any* direction. This is known as the *gradient* of $\Phi(\mathbf{r})$. We introduce it as follows. Consider a (curved) surface $\Phi(\mathbf{r}) = \Phi_0$ in space. This is a surface because Φ_0 is a constant so that, for example, $\Phi(x, y, z) = \Phi_0$ is an equation between the three Cartesian coordinates, which defines that surface. Now consider a second surface that almost coincides with the first: $\Phi(\mathbf{r}) = \Phi_1$ (i.e. Φ_1 is infinitesimally close in numerical value to Φ_0). Choose points P on one and Q on the other surface that differ by an infinitesimal line element $d\boldsymbol{l} \equiv d\mathbf{r}$. Consequently the function Φ at point P is $\Phi(\mathbf{r}) = \Phi_0$ and that at Q is $\Phi(\mathbf{r} + d\mathbf{r}) = \Phi_1$. The difference between these two functions is

$$\Phi_1 - \Phi_0 = \Phi(\mathbf{r} + d\mathbf{r}) - \Phi(\mathbf{r}) = dx\frac{\partial \Phi}{\partial x} + dy\frac{\partial \Phi}{\partial y} + dz\frac{\partial \Phi}{\partial z} + O(|d\mathbf{r}|^2) \tag{3.30}$$

In (3.30) terms of second and higher order in dx, dy, dz, or products of these, have been neglected. This is consistent with the choice that $\Phi_1 - \Phi_0$ must be infinitesimal. If we write (3.30) for general orthogonal coordinates, then it becomes

$$\begin{aligned}\Phi_1 - \Phi_0 = \Phi(\mathbf{r} + d\boldsymbol{l}) - \Phi(\mathbf{r}) &= dl_1\frac{\partial \Phi}{\partial l_1} + dl_2\frac{\partial \Phi}{\partial l_2} + dl_3\frac{\partial \Phi}{\partial l_3} + O(|d\boldsymbol{l}|^2), \\ &= dl_1\frac{\partial \Phi}{h_1 \partial u_1} + dl_2\frac{\partial \Phi}{h_2 \partial u_2} + dl_3\frac{\partial \Phi}{h_3 \partial u_3} + O(|d\boldsymbol{l}|^2)\end{aligned} \tag{3.31}$$

Here we have used (2.23) for the line elements in the denominator and we consider $\Phi(u_1, u_2, u_3)$ to be a function of the coordinates u_1, u_2, u_3. In the limit $dl \to 0$ we can write

$$\Phi(\mathbf{r} + d\boldsymbol{l}) - \Phi(\mathbf{r}) = dl_1\frac{\partial \Phi}{h_1 \partial u_1} + dl_2\frac{\partial \Phi}{h_2 \partial u_2} + dl_3\frac{\partial \Phi}{h_3 \partial u_3} \tag{3.32a}$$

The right-hand side of this can be considered as a dot product of $d\boldsymbol{l}$ with a vector, of which the components are $(\partial\Phi/h_1\partial u_1, \partial\Phi/h_2\partial u_2, \partial\Phi/h_3\partial u_3)$. This vector is the *gradient* of Φ, and we write it as $\nabla\Phi$ (the symbol ∇ is called *del* or *nabla*).

Equation (3.32a) for an infinitesimal increment in Φ then is written as follows:

$$\Phi(\mathbf{r}+d\boldsymbol{l})-\Phi(\mathbf{r})=d\boldsymbol{l}\cdot\nabla\Phi$$
$$\nabla\Phi\equiv\hat{\mathbf{u}}_1\frac{\partial\Phi}{h_1\,\partial u_1}+\hat{\mathbf{u}}_2\frac{\partial\Phi}{h_2\,\partial u_2}+\hat{\mathbf{u}}_3\frac{\partial\Phi}{h_3\,\partial u_3} \tag{3.32b}$$

Using Table 2.2 it is easy to find expressions for the gradient in the three major coordinate systems:

$$\begin{aligned} &\text{Cartesian} && \nabla\Phi\equiv\hat{\mathbf{x}}\frac{\partial\Phi}{\partial x}+\hat{\mathbf{y}}\frac{\partial\Phi}{\partial y}+\hat{\mathbf{z}}\frac{\partial\Phi}{\partial z}\\ &\text{cylindrical} && \nabla\Phi\equiv\hat{\boldsymbol{\rho}}\frac{\partial\Phi}{\partial\rho}+\hat{\boldsymbol{\varphi}}\frac{\partial\Phi}{\rho\,\partial\varphi}+\hat{\mathbf{z}}\frac{\partial\Phi}{\partial z}\\ &\text{spherical} && \nabla\Phi\equiv\hat{\mathbf{r}}\frac{\partial\Phi}{\partial r}+\hat{\boldsymbol{\theta}}\frac{\partial\Phi}{r\,\partial\theta}+\hat{\boldsymbol{\varphi}}\frac{\partial\Phi}{r\sin\theta\,\partial\varphi}\end{aligned} \tag{3.33}$$

Let us look at some aspects of the gradient. Consider a point Q that lies on the *same* surface as P, for which $\Phi(\mathbf{r}+d\boldsymbol{l})=\Phi_0$ also. Then (3.32b) predicts that $d\boldsymbol{l}\cdot\nabla\Phi=0$, which in turn implies that $\nabla\Phi$ is $\perp$ to this choice of $d\boldsymbol{l}$. Thus $\nabla\Phi$ must be $\perp$ to the constant-Φ surface on which P and Q lie. For the more general case, shown in Fig. 3.18, where Q is on a different surface, we can say that $d\boldsymbol{l}$ makes an angle ψ with the surface-normal unit vector $\hat{\mathbf{n}}$ of surface Φ_0, and so $d\boldsymbol{l}\cdot\nabla\Phi=|d\boldsymbol{l}|\,|\nabla\Phi|\cos\psi$. This dot product is clearly maximal when $\psi=0$, i.e. when $d\boldsymbol{l}$ is *parallel* to $\hat{\mathbf{n}}$, which implies from (3.32b) that $\Phi(\mathbf{r}+d\boldsymbol{l})-\Phi(\mathbf{r})$ is *maximal* for a displacement $\perp$ to the surface. Thus *the gradient is the derivative in the direction of maximal change*, which is the direction, $\hat{\mathbf{n}}$, of the normal to the surfaces of constant $\Phi(\mathbf{r})$.

From (3.32b) this can be expressed algebraically as

$$\nabla\Phi=\lim_{dl\to0}\left\{\frac{[\Phi(\mathbf{r}+d\boldsymbol{l})-\Phi(\mathbf{r})]_{\max}}{dl}\right\}\hat{\mathbf{n}}=\lim_{dl\to0}\left(\frac{d\Phi}{dl}\right)_{\max}\hat{\mathbf{n}} \tag{3.34}$$

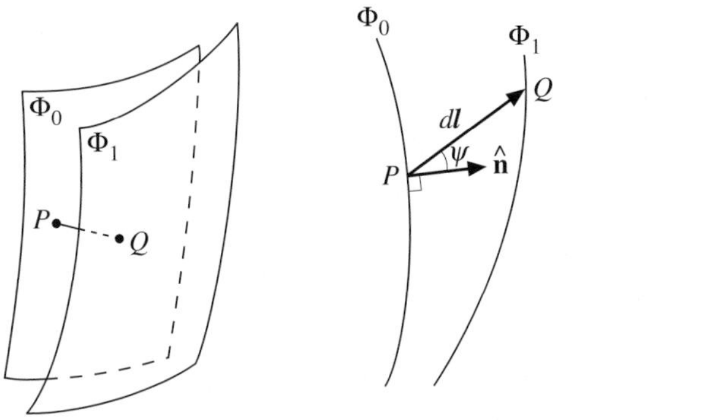

Fig. 3.18 Two equipotential planes with closely spaced Φ-values.

A symbolic meaning is attached to the operator ∇:

$$\nabla \equiv \hat{\mathbf{u}}_1 \frac{\partial}{h_1 \partial u_1} + \hat{\mathbf{u}}_2 \frac{\partial}{h_2 \partial u_2} + \hat{\mathbf{u}}_3 \frac{\partial}{h_3 \partial u_3} \tag{3.35}$$

∇ must always operate on a scalar to produce a gradient. The left-hand side of (3.34) is usually pronounced 'del phi'. Another consequence of (3.32b) is obtained by applying it successively to $N \to \infty$ infinitesimal pieces of line element $d\boldsymbol{l}_n$ ($1 \le n \le N$), strung together to form a nonzero curved line in space linking points P and Q:

$$\Phi(\mathbf{r}_Q) - \Phi(\mathbf{r}_P) = \int_P^Q d\boldsymbol{l} \cdot \nabla\Phi \tag{3.36}$$

This implies that the *closed-curve* integral $\oint d\boldsymbol{l} \cdot \nabla\Phi$ is *zero* for any scalar field $\Phi(\mathbf{r})$, because then the end point Q coincides with the initial point P. It also implies that the line integral from P to Q of a vector field that is a gradient is *independent* of the path taken from P to Q.

Example 3.7 Another line integral in a vector field

Consider the vector field $\mathbf{A}(\mathbf{r}) = y\hat{\mathbf{x}} + x\hat{\mathbf{y}}$ and integrate it over path I (see Fig. 3.19), i.e. from $P = (1, 0)$ to $(0, 0)$ to $Q = (0, 1)$, and then over the straight-line slant path from P to Q. We have $A_x = y$ and $A_y = x$, so for path I we obtain

$$\int d\boldsymbol{l} \cdot \mathbf{A} = \int_1^0 (dx\, y)_{\mathrm{I}} + \int_0^1 (dy\, x)_{\mathrm{I}} = 0$$

since in the first integral $y = 0$ and in the second $x = 0$. For path II we obtain

$$\int d\boldsymbol{l} \cdot \mathbf{A} = \int_1^0 dx(1 - x) + \int_0^1 dy(1 - y) = 0$$

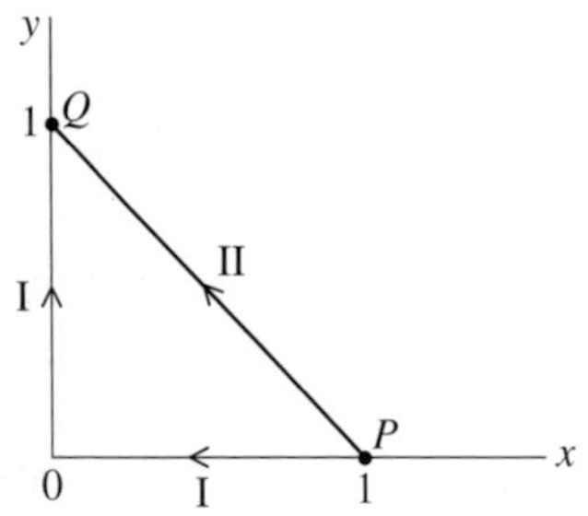

Fig. 3.19

because $x + y = 1$ on the slant path II. The reason that these line integrals are the same is that $\mathbf{A} = \nabla(xy)$. The line integral from one point to another over an integrand that is a gradient of a vector is independent of the path.

Thus, summarizing this section, the gradient of a scalar field $\Phi(\mathbf{r})$ is a vector field that, everywhere, is perpendicular to surfaces of constant $\Phi(\mathbf{r})$; the magnitude of the gradient is the derivative in that perpendicular direction (which is the direction of greatest change in $\Phi(\mathbf{r})$).

References

1. C.A. de Coulomb, Première mémoire sur l'électricité et magnétisme, *Histoire de l'Académie Royale des Sciences*, p. 569, 1785. A good historical summary of this as well as of many other matters in the present text is found in R.S. Elliott, *Electromagnetics: History, Theory, and Applications*, IEEE press series on Electromagnetic Waves, New York, 1993.
2. See for example any elementary text on solid-state physics. A well-known widely available text is C. Kittel, *Introduction to Solid State Physics*, second edition, J. Wiley & Sons, New York, 1956.

Problems

3.1. Charge $q_1 = 3.5\,\text{nC}$ is at location Q_1 ($\rho = 1$, $\varphi = \pi/6$, $z = -2$), and charge $q_2 = -2\,\text{nC}$ is at Q_2 ($\rho = 5$, $\varphi = -\pi/6$, $z = 6$). Calculate the magnitude and direction of the force of q_1 upon q_2.

3.2. Repeat the above calculation if Q_1 is at ($r = 1$, $\theta = \pi/6$, $\varphi = \pi/6$) and Q_2 is at ($r = 5$, $\theta = \pi/3$, $\varphi = -\pi/6$).

3.3. Two charges q_1 and q_2 are placed at locations z_1 and z_2, respectively, on the z axis.
(a) If q_1 and q_2 have the same polarity, determine the location where a negative charge q_0 can be placed so that it experiences no net force.
(b) If q_0 is constrained mechanically so that it can move only on the z axis, explain why its location is (electro)dynamically stable or unstable, depending upon the polarity of the other two charges.

3.4. An electrostatic field is produced by a charge $q = 4.427\,\text{pC}$ ($1\,\text{pC} = 10^{-12}\,\text{C}$) located at Q ($r = 3\,\text{m}$, $\theta = 30^\circ$, $\varphi = 45^\circ$). Find the electrostatic field at P ($r = 6\,\text{m}$, $\theta = 0$, $\varphi = 0$).

3.5. Show that the total force on a charge q at the origin O from a distribution of surface charges $\varrho_s(\theta, \varphi)$ placed on a surface with $r = \text{constant}$, $0 < \theta < \theta_0$, $0 < \varphi < \varphi_0$, around q is independent of the distance r for given θ_0, φ_0.

3.6. A uniform line charge ϱ_ℓ (in C/m) extends from $z = -L/2$ to $z = L/2$.
(a) Find the electrostatic field at an arbitrary location in the $z = 0$ plane.
(b) Show that Eq. (3.14c) ensues when $L \to \infty$.
(c) How large must L be before the answer to (a) differs by 5% from the $L \to \infty$ answer?

3.7. A plane contains the surface charge distribution $\varrho_s(\rho, 0) = \varrho_{s0} e^{-\rho^2/a^2}$, where ϱ_{s0} is a constant surface charge density in C/m^2.
(a) Express the electrostatic field $\mathbf{E}(0, 0, z)$ as an integral over a variable $w = \rho^2/z^2$.
(b) Find an approximate closed-form expression for $\mathbf{E}(0, 0, z)$ for $z \gg a$.

3.8. Two infinite uniform planar charge distributions are $\varrho_{s1} = 5 \times 10^{-11}$ C/m^2 at $z = 2$ m and $\varrho_{s2} = 4 \times 10^{-11}$ C/m^2 at $z = -2$ m. Calculate the electrostatic field vector
(a) at $z = 5$ m
(b) at $z = 0$ m
(c) at $z = -5$ m.

3.9. A circular disk of radius a with uniform charge ϱ_ℓ surrounds the z axis at $z = 0$. Find the electrostatic field vector $\mathbf{E}$ at $(0, 0, z)$ for both positive and negative z.

3.10. An infinite circular cylinder of radius $a = 5$ m contains within it a uniform charge with density $\varrho_v = 30$ pC/m^3 (1 pC $= 10^{-12}$ C).
(a) Use Gauss's law to find an expression for $\mathbf{E}$ in the region $\rho < a$.
(b) Do the same for $\mathbf{E}$ in the region $\rho > a$.
(c) Calculate the magnitude of the electric field $\mathbf{E}$ at the point $P = (5, 12, 0)$.
(d) Calculate the potential V_Q at point $Q = (6, 6, 0)$, given that $V_P = 0$ at $\varrho = a$.
All Cartesian coordinate distances are in meters.

3.11. Use Gauss's law to calculate the electrostatic field at an arbitrary point P for a given spherically symmetric volume-charge density $\varrho_v(r, \theta, \varphi) = Ar^2 e^{-(r/a)^5}$. For what value of r is the field maximal?

3.12. Repeat problem 3.11 for a cylindrically symmetric distribution of charge $\varrho_v(\rho, \varphi, z) = \rho e^{-(\rho/a)^3}$.

3.13. Calculate the charge inside a cube with sides of length $L = 2$ parallel to the Cartesian axes and symmetric around the origin, if the electrostatic field is given by

$$\mathbf{A} = (2\hat{\mathbf{x}} + \hat{\mathbf{y}})/(x^2 + y^2).$$

3.14. Repeat the calculation in problem 3.13 to find the charge between the original cube and one that is concentric (and parallel) with it but has sides of length $L = 2^{5/3}$. *Hint*: Solve problem 3.13 for arbitrary L first.

3.15. An electrostatic field is given by

$$E_r(r, \theta, \varphi) = a \cos\theta / r^2 \qquad E_\theta(r, \theta, \varphi) = a \sin\theta / r^2 \qquad E_\varphi(r, \theta, \varphi) = 0$$

Give equations for the field lines.

3.16. A vector field is given by $\mathbf{A} = 3x^2y\hat{\mathbf{x}} + x^3\hat{\mathbf{y}}$. Calculate the line integral from $P = (0, 1, 0)$ to $Q = (2, 5, 0)$

(a) along the path $(0, 1, 0) \to (2, 1, 0) \to (2, 5, 0)$,

(b) along the path $y = x^2 + 1$ (do this in at least two different ways).

(c) Show that this field is conservative (and therefore could be an electrostatic field).

(d) Regarding $\mathbf{A}$ as an electrostatic field, find the potential field $V(\mathbf{r})$.

3.17. A vector field is given by $\mathbf{A} = -y\hat{\mathbf{x}} + x\hat{\mathbf{y}}$. Calculate the line integral from $P\ (5, 0, 0)$ to $Q\ (0, 5, 0)$

(a) along the path $(5, 0, 0) \to (0, 0, 0) \to (0, 5, 0)$,

(b) along the quarter circle from P to Q.

3.18. A scalar field is given by $\Phi = ayz + bzx + cxy$. Find the gradient $\nabla\Phi$ in terms of the three constants a, b, c.

3.19. Calculate the gradients of the scalar fields

(a) $\Phi = x/(x^2 + y^2)$

(b) $\Phi = z/(x^2 + y^2 + z^2)$.

3.20. A plane is given by the equation $2x + 4y + \sqrt{5}z = 1$. Find the unit vector perpendicular to this plane that points away from the origin.

3.21. An electrostatic field is given as $\mathbf{E} = \hat{\mathbf{r}}(Ar/\varepsilon_0)$ for $r \leq a$ and $\mathbf{E} = \hat{\mathbf{r}}(Aa^3/\varepsilon_0 r^2)$ for $r > a$. Find the magnitude of the charge q producing this electrostatic field in free space.

4

The electrostatic potential

The electrostatic potential is a key parameter in the analysis of electromagnetic fields. It is first encountered by students of electrical engineering in lumped circuits as the *voltage difference* between two different points. Because the power is defined to be $p = Vi$ in simple current branches with current i flowing over voltage difference V, it would follow that V could be defined as the increment of power per unit change in current: $V = dp/di$. Because power is energy transferred per unit of time, and current is charge flow per unit of time, it is also possible to consider V as the change in energy per unit charge moved between the two points. This, in fact, is the approach from the point of view of electrostatic fields. This chapter will introduce it as such.

4.1 Definition

The concept of *potential* is tied closely to that of electrostatic *work*, i.e. to the work done owing to the presence of an electrostatic field. Assume an electrostatic field $\mathbf{E}(\mathbf{r})$ in space. If, using an external force $\mathbf{F}_{\text{ext}}$ to balance the field force $q\mathbf{E}$, we push a point charge q an infinitesimal distance $d\boldsymbol{l}$, we perform an amount of work $dw = \mathbf{F}_{\text{ext}} \cdot d\boldsymbol{l} = -q\mathbf{E} \cdot d\boldsymbol{l}$. If the particle is moved along a path from point P to point Q, then the work performed against the field is

$$w(P \to Q) = -q \int_P^Q d\boldsymbol{l} \cdot \mathbf{E} \tag{4.1}$$

Let us now consider a round trip. The work required to move the point charge from P to Q along one path and back to P along another (see Fig. 4.1) is zero, owing to conservation of energy, regardless of the shape of the round-trip path. As a consequence, $w(P \to Q)$ is *independent* of the path between P and Q, hence must depend only upon the end points: $w(P \to Q) = w(Q) - w(P) \equiv q(V_Q - V_P)$.

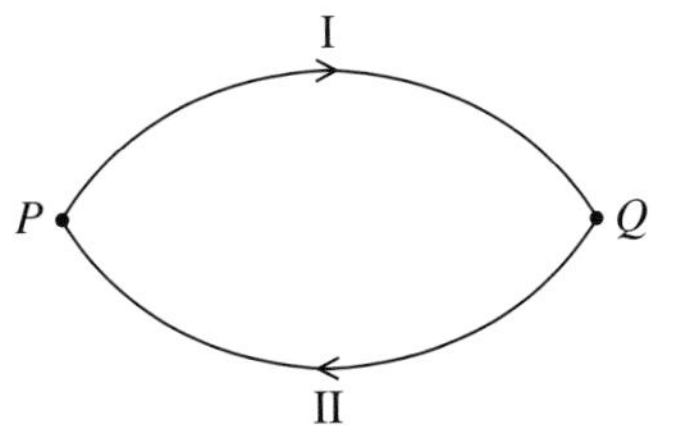

Fig. 4.1 A closed path in an electrostatic field.

Hence the work required to move a *unit* charge is

$$V_Q - V_P = -\int_P^Q d\boldsymbol{l} \cdot \mathbf{E} \tag{4.2}$$

and is known as the *potential difference* between Q and P. An explicit expression for V_P can be obtained as follows. From Eq. (3.8b) we observe that the electrostatic field at P can be written as

$$\mathbf{E}_P = \frac{1}{4\pi\varepsilon_0}\int dv_Q\, \varrho_{\mathrm{v}}(\mathbf{r}_Q)\frac{\hat{\mathbf{R}}_{QP}}{R_{QP}^2} \tag{4.3}$$

with $\mathbf{R}_{QP} \equiv \mathbf{r}_P - \mathbf{r}_Q$, $R_{QP} \equiv |\mathbf{r}_P - \mathbf{r}_Q|$, and $\hat{\mathbf{R}}_{QP} \equiv \mathbf{R}_{QP}/R_{QP}$ as the unit vector pointing from Q to P. A crucial step, that we will often use in future development of the material, is to note that

$$\frac{\hat{\mathbf{R}}_{QP}}{R_{QP}^2} = -\nabla_P\left(\frac{1}{R_{QP}}\right) \tag{4.4}$$

For those who wish a demonstration, (4.4) is easily proved by use of Cartesian coordinates. Consider the length of the difference vector $\mathbf{R}_{QP}$:

$$R_{QP} = [(x_P - x_Q)^2 + (y_P - y_Q)^2 + (z_P - z_Q)^2]^{1/2}$$

We take the $\hat{\mathbf{x}}$ component of the gradient, $\hat{\mathbf{x}}\partial/\partial x$, of the inverse of R_{QP}:

$$-\hat{\mathbf{x}}\frac{\partial}{\partial x_P}\left\{[(x_P - x_Q)^2 + (y_P - y_Q)^2 + (z_P - z_Q)^2]^{-1/2}\right\}$$
$$= \frac{1}{2}\frac{2(x_P - x_Q)}{R_{QP}^3}\hat{\mathbf{x}} = \frac{(x_P - x_Q)\hat{\mathbf{x}}}{R_{QP}^3}$$

The numerator of the right-hand side is just the $\hat{\mathbf{x}}$ component of $\mathbf{R}_{QP}$. The procedure is similar for the other two Cartesian components, and addition of all three results in R_{QP}/R_{QP}^3, which is identical to the left-hand side of (4.4) after cancellation in numerator and denominator of one length R_{QP}.

This allows (4.3) to be rewritten as

$$\mathbf{E}_P = -\frac{1}{4\pi\varepsilon_0}\int dv_Q\, \varrho_{\mathrm{v}}(\mathbf{r}_Q)\nabla_P\left(\frac{1}{R_{QP}}\right) \tag{4.5}$$

The gradient operator ∇_P works only on quantities that are a function of $\mathbf{r}_P$, and therefore it can be moved in front of the integral sign to obtain

$$\mathbf{E}_P \equiv -\nabla_P\left[\frac{1}{4\pi\varepsilon_0}\int dv_Q \frac{\varrho_{\mathrm{v}}(\mathbf{r}_Q)}{R_{QP}}\right] \tag{4.6}$$

If the quantity inside the square brackets is assigned the label $V_P \equiv V(\mathbf{r}_P)$, we obtain

$$\mathbf{E}_P = -(\nabla V)_P \qquad \text{with} \quad V_P \equiv \frac{1}{4\pi\varepsilon_0}\int dv_Q \frac{\varrho_{\mathrm{v}}(\mathbf{r}_Q)}{R_{QP}} \tag{4.7}$$

as required. When this is inserted into (4.1), we obtain under the integral sign $-d\boldsymbol{l}\cdot\nabla V$. That the left-hand side of (4.2) then follows can be seen by referring to (3.32b), which states that the infinitesimal product $d\boldsymbol{l}\cdot\nabla V$ is the difference of two infinitesimally separated values of V. A Riemann sum of infinitesimals then yields $V_Q - V_P$. Our assumption that energy is conserved in a round trip was justified.

Several important facts perhaps need emphasis here. First, although a *potential* V_P is defined at any point P by (4.7), this quantity has no direct physical meaning. Only the potential difference $V_Q - V_P$ is defined physically as the work in displacing a point charge from P to Q in the presence of an electrostatic field. An arbitrary constant could be added to the potential at each point without modifying the potential difference between two points. Second, *the potential is continuous everywhere*, because a jump in potential would imply an infinite gradient, i.e. an infinite electrostatic field, which is not possible (except infinitesimally close to a point charge, which is in itself an abstraction from reality!). Third, the potential is often simpler to calculate than the **E** field because (i) it is a scalar field and (ii) the defining integral has only an inverse-distance singularity rather than an inverse-square one as does **E** (occurring when point Q equals point P, which complicates the numerical evaluation of such integrals). Summarizing, we have now obtained

$$V(\mathbf{r}_P) \equiv \frac{1}{4\pi\varepsilon_0}\int dv_Q\, \varrho_{\mathrm{v}}(\mathbf{r}_Q)\frac{1}{R_{QP}} \tag{4.8a}$$

$$\mathbf{E}(\mathbf{r}_P) \equiv \frac{1}{4\pi\varepsilon_0}\int dv_Q\, \varrho_{\mathrm{v}}(\mathbf{r}_Q)\frac{\hat{\mathbf{R}}_{QP}}{R_{QP}^2} \tag{4.8b}$$

$$\mathbf{E}(\mathbf{r}) = -\nabla V(\mathbf{r}) \tag{4.8c}$$

The fields $\mathbf{E}(\mathbf{r})$ and $V(\mathbf{r})$ also can be written as surface or line integrals, depending upon whether the distribution of charge is on a surface or along a line.

Note that, as discussed above, the properties of the integral of a gradient imply in (4.2) that

$$\oint d\boldsymbol{l} \cdot \mathbf{E} = 0 \qquad \text{for any closed path } C \tag{4.9}$$

The left-hand side of (4.9) is known as the *circulation integral* of $\mathbf{E}$ over the path C. It is tied closely to another expanded-derivative operation on a vector field, the *curl of a field*, which will be treated in due course.

4.2 Potentials for elementary charge distributions

The potential $V_P \equiv V(\mathbf{r}_P)$ is calculated fairly easily for many elementary charge distributions from what has already been done in Section 4.1, using (4.8). Some examples follow.

Example 4.1 Potential due to a point charge

Equation (4.1) gives us the potential at P due to an extended charge distribution. If this is shrunk down to a *point charge at Q* we find that

$$V_P = \frac{q_Q}{4\pi\varepsilon_0 R_{QP}} + \text{constant} \tag{4.10}$$

The constant in (4.10) is usually chosen as zero, giving us, as $P \to \infty$, $V_\infty = 0$. Inspection of (4.10) shows that all points P on any spherical surface around point Q (as center) have the same potential value. Such surfaces on which the potential has a constant value are known as *equipotential surfaces*. They play a key role in defining the concept of *field lines* (see below) and also in discussing good conductors.

Example 4.2 Potential due to a uniform line-charge density on an infinite straight line

Let ϱ_ℓ be a uniform line-charge density on the z axis. In discussing the electrostatic field due to an infinite uniform line charge, (3.14c), we observed that the $\mathbf{E}$ field was always in the $\hat{\boldsymbol{\rho}}(\varphi)$ direction and that it was a function only of $1/\rho$. This implies, by performing the line integral $\int d\rho\,(1/\rho)$ in (3.14c), that the V field is a function of $\ln \rho$. In fact, we obtain for a point P at distance ρ from the z axis

$$V_P = -\frac{\varrho_\ell}{2\pi\varepsilon_0} \ln \rho + \text{constant} \tag{4.11a}$$

As a check, the negative gradient of (4.11a) is proportional to $\nabla \ln \rho = \hat{\boldsymbol{\rho}}/\rho$, thus yielding (3.14c). We observe that the constant is a bit harder to choose in this case, because $V_P \to \pm\infty$ as point P is moved to infinity ($\rho \to \infty$). This infinite potential is due to the effect of the *infinite* charge on the z axis (even though the charge density is finite), which suffices to overcome the inverse fall-off in the potential due to any point on the line of charge, even at an infinite distance. However, as stated earlier, all physical forces are related to potential differences, so that the work done in moving a unit charge from P_1 to P_2 is finite as long as the two points are a finite radial distance from each other:

$$V_{P_2} - V_{P_1} = -\frac{\varrho_\ell}{2\pi\varepsilon_0}\ln\frac{\rho_2}{\rho_1} \tag{4.11b}$$

The equipotential surfaces are circular cylinders around the z axis.

Example 4.3 Potential due to a uniform charge density on a circular disk

This can be obtained by referring to (3.11b), where the field due to a circular disk of charge with radius a was considered, and by writing that expression as a negative gradient. The second term in (3.11d) can be written as $|z|/\sqrt{z^2+a^2}$, and it can be seen that this is just the derivative of $\sqrt{z^2+a^2}$. It follows that the potential on the z axis (and *only* on the z axis) is given by

$$V(0,0,z) = \frac{\varrho_s}{2\varepsilon_0}\left(\sqrt{z^2+a^2} - |z|\right) \tag{4.12a}$$

except for an additive constant. The need for the absolute-value bars follows from the fact that the **E** field of (3.11) is *antisymmetric*, i.e. it is equal in magnitude but opposite in direction when mirrored (with respect to the $z = 0$ plane here). This result also can be derived directly from the surface-integral equivalent to the expression for the potential in (4.8):

$$\begin{aligned} V(\mathbf{r}_P) &\equiv \frac{1}{4\pi\varepsilon_0}\int dS_Q\,\varrho_s(\mathbf{r}_Q)\frac{1}{R_{QP}} = \frac{1}{4\pi\varepsilon_0}2\pi\varrho_s\int_0^a d\rho\,\rho\frac{1}{\sqrt{z^2+\rho^2}} \\ &= \frac{\varrho_s}{2\varepsilon_0}\left[\sqrt{z^2+\rho^2}\right]_{\rho=0}^{\rho=a} = \frac{\varrho_s}{2\varepsilon_0}\left(\sqrt{z^2+a^2} - |z|\right) \end{aligned} \tag{4.12b}$$

which is identical to (4.12a).

Example 4.4 Potential due to a uniform surface-charge density on an infinite plane

This is a special case of the previous example for the limit $a \to \infty$, and the result could be written down immediately, again except for an additive constant. However, as an exercise it will be derived in an alternative fashion. If ϱ_s is the uniform charge density (in C/m^2) on the $z = 0$ plane, then, as shown in (3.20b), the **E** vector always points away from the plane (for positive ϱ_s) but has a constant magnitude, namely

$E = \varrho_s/(2\varepsilon_0)$. If we choose point Q infinitesimally above the $z = 0$ plane and P at height z in the $z > 0$ half-space, then, using (4.2) and (3.20b),

$$V_P - V_Q = -\int_Q^P d\boldsymbol{l} \cdot \mathbf{E} = -\int_0^z dz\,\hat{\mathbf{z}} \cdot \mathbf{E} = -\frac{\varrho_s |z|}{2\varepsilon_0} \tag{4.12c}$$

We can choose $d\boldsymbol{l}$ as $dz\,\hat{\mathbf{z}}$ because the $\mathbf{E}$ field is in the $\hat{\mathbf{z}}$ direction, so that the dot product pulls out only the z component of the vector line element. The absolute bars around z again make this expression valid for P in the $z < 0$ half-plane too. The equipotential surfaces are planes $\|$ the $z = 0$ plane, and the potential decreases linearly *on either side* with distance from the charge plane.

4.3 Potential and field of an elementary dipole

So far, the effect of charges upon other charges has been discussed only when they are located in a vacuum (or free space). In order to understand the effect of fields inside a *dielectric* material it is necessary to understand the fields produced by an *electric dipole*, which, as mentioned in Example 3.6, is formed by a pair of charges of equal but opposite polarity. The connection with dielectrics will be explained later. At this time, it suffices to state that the fields produced at a distance by a dipole will be weak compared to those produced by each of the charges separately because (i) one field 'pulls' and the other 'pushes' a test charge,[1] and (ii) the fields go to zero as the separation between the two charges decreases to zero. However, neutral atomic or molecular charges in a dielectric separate under the influence of an imposed electric field, and thus many dipoles are formed over a large volume; the combined effect of those dipoles is not negligible, as will be shown later.

Consider the situation in Fig. 4.2, in which a charge $+q$ is at location $z = l/2$ and another charge $-q$ is at $z = l/2$. We will calculate the potential at a distant point P

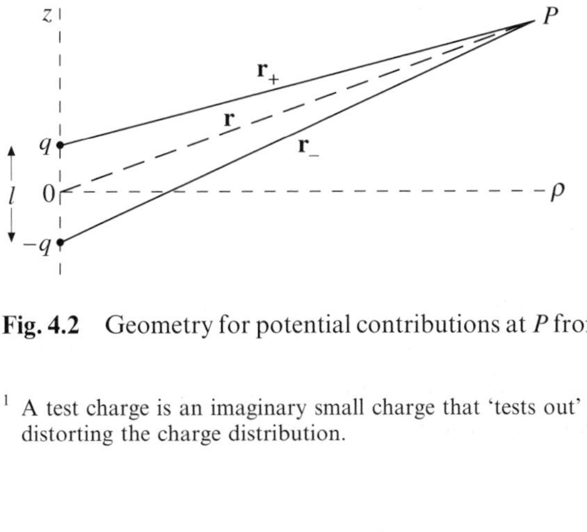

Fig. 4.2 Geometry for potential contributions at P from an electrostatic dipole.

1 A test charge is an imaginary small charge that 'tests out' the $\mathbf{E}$ field at points in a region without distorting the charge distribution.

at a vectorial distance **r** from the origin. The principle of superposition (see before (3.4)) then gives for the potential

$$V_P = \frac{q}{4\pi\varepsilon_0}\left(\frac{1}{r_+} - \frac{1}{r_-}\right) = \frac{q}{4\pi\varepsilon_0}\left(\frac{1}{|\mathbf{r} - l\hat{\mathbf{z}}/2|} - \frac{1}{|\mathbf{r} + l\hat{\mathbf{z}}/2|}\right) \tag{4.13}$$

Note that $\mathbf{r}_+$ and $\mathbf{r}_-$ are the vectors from the $+q$ and $-q$ charges to the point of observation P. Because we shall always assume that the lengths of these vectors are large compared with the separation distance l of the charges, we will be able to use the binomial approximation (3.6):

$$\frac{1}{|\mathbf{r} \mp l\hat{\mathbf{z}}/2|} = (r^2 \mp l\hat{\mathbf{z}}\cdot\mathbf{r} + l^2/4)^{-1/2} = \frac{1}{r}(1 \mp (l/r)\hat{\mathbf{z}}\cdot\mathbf{r} + (l/r)^2)^{-1/2} \tag{4.14a}$$

Referring to (3.6), we have $p = -\frac{1}{2}$ and $\epsilon = (l/r)\hat{\mathbf{z}}\cdot\hat{\mathbf{r}} + (l/r)^2$, but we may neglect the small $(l/r)^2$ term in ϵ. Hence it follows that

$$\frac{1}{|\mathbf{r} \mp l\hat{\mathbf{z}}/2|} = \frac{1}{r}\left[\left(1 \pm \frac{l}{2r}\hat{\mathbf{z}}\cdot\mathbf{r}\right) + O(l/r)^2\right] \approx \frac{1}{r}\left[1 \pm \hat{\mathbf{r}}\cdot\left(\frac{l}{2r}\right)\hat{\mathbf{z}}\right] \tag{4.14b}$$

where we have used $\mathbf{r}/r = \hat{\mathbf{r}}$, the unit vector pointing from the center of the dipole to P. If this is inserted into (4.13) we obtain up to (but not including) terms of $O(l/r)^2$:

$$V_P \approx \frac{q}{4\pi\varepsilon_0 r}\left[1 + \hat{\mathbf{r}}\cdot\left(\frac{l}{2r}\hat{\mathbf{z}}\right) - 1 + \hat{\mathbf{r}}\cdot\left(\frac{l}{2r}\hat{\mathbf{z}}\right)\right] = \frac{ql\hat{\mathbf{z}}\cdot\hat{\mathbf{r}}}{4\pi\varepsilon_0 r^2} \tag{4.15a}$$

If we identify a separation vector $\boldsymbol{l} \equiv l\hat{\mathbf{z}}$, we observe that the product ql occurs in the numerator of (4.15a). This product is known as the *electric dipole moment* and it is designated by $p = ql$. If the vectorial direction is included, we then obtain

$$V_P \approx \frac{\mathbf{p}\cdot\hat{\mathbf{r}}}{4\pi\varepsilon_0 r^2} = \frac{p\cos\theta}{4\pi\varepsilon_0 r^2} \qquad \mathbf{p} = ql\hat{\mathbf{z}} \tag{4.15b}$$

Of course this expression is only approximate, as we have neglected terms of order $(l/r)^2$ compared to unity, but there is one case in which it can be considered to be an exact expression. This is the case in which we allow $l \to 0$ and, simultaneously, $q \to \infty$ in such a fashion that the dipole moment $p = ql$ remains finite. For then, and then only, is l/r equal to zero for nonzero r, and there are no higher-order terms in l/r. This limiting case is known as the *infinitesimal electric dipole*, and it is the basis for discussing dielectric materials. Note that the dimension of the dipole moment is coulomb meters (C m). We immediately obtain the **E** field from (4.15b) by applying the negative gradient operator on $\cos\theta/r^2$. This is best done in spherical coordinates:

$$-\nabla\frac{\cos\theta}{r^2} = -\hat{\mathbf{r}}\frac{\partial}{\partial r}\left(\frac{\cos\theta}{r^2}\right) - \frac{1}{r}\hat{\boldsymbol{\theta}}\frac{\partial}{\partial\theta}\left(\frac{\cos\theta}{r^2}\right) = \frac{1}{r^3}(2\hat{\mathbf{r}}\cos\theta + \hat{\boldsymbol{\theta}}\sin\theta)$$

The result is

$$\mathbf{E}_P = \frac{p}{4\pi\varepsilon_0 r^3}(2\hat{\mathbf{r}}\cos\theta + \hat{\boldsymbol{\theta}}\sin\theta) \tag{4.16a}$$

but it is also useful to write the result in a coordinate-free notation by using $p\cos\theta = \mathbf{p}\cdot\mathbf{r}$, and also $\hat{\mathbf{z}} = \hat{\mathbf{r}}\cos\theta - \hat{\boldsymbol{\theta}}\sin\theta$ (see (2.34)). Using the latter,

$$2\hat{\mathbf{r}}\cos\theta + \hat{\boldsymbol{\theta}}\sin\theta = 3\hat{\mathbf{r}}\cos\theta - (\hat{\mathbf{r}}\cos\theta - \hat{\boldsymbol{\theta}}\sin\theta) = 3\hat{\mathbf{r}}\cos\theta - \hat{\mathbf{z}}$$

which, when inserted into (4.16a), gives

$$\mathbf{E}_P = \frac{1}{4\pi\varepsilon_0 r^3}[3(\mathbf{p}\cdot\hat{\mathbf{r}})\hat{\mathbf{r}} - \mathbf{p}] \tag{4.16b}$$

The results in (4.15) and (4.16) show that, as expected, both the V and $\mathbf{E}$ fields are cylindrically symmetric around the z axis and that both fields fall off by an extra factor $1/r$ compared with fields produced by a single point source (a *monopole*). We have already shown how to construct electrostatic field lines in Example 3.6, where we found that $r(\theta) \propto \sin^2\theta$. Location of the equipotential surfaces (which are cylindrically symmetric around the z axis) is also straightforward. For each equipotential surface, we have a constant $V_0 = p\cos\theta/(4\pi r^2)$; hence $r^2(\theta) = p\cos\theta/(4\pi\varepsilon_0 V_0)$ and therefore the intersection curves of the equipotential surfaces with planes through the z axis are given by $r(\theta) \propto \sqrt{\cos\theta}$. The equipotential lines

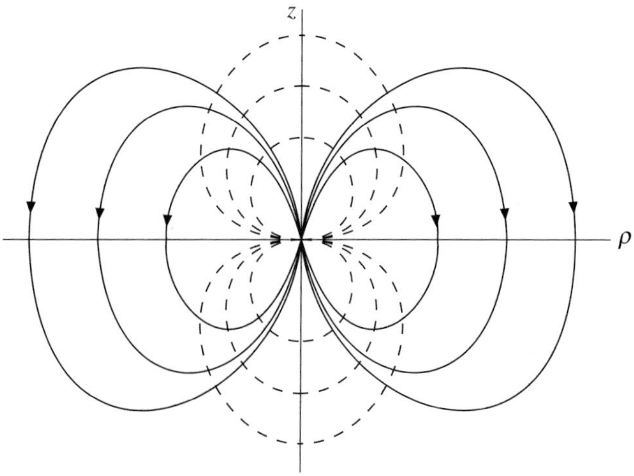

Fig. 4.3 Equipotential surfaces (broken) and field lines (solid) of an infinitesimal electrostatic dipole oriented vertically.

thus produced, and field lines, are shown in Fig. 4.3. Note that each field line is everywhere $\perp$ to the equipotential surfaces. In summary, we have

$$r(\theta) \propto \sqrt{\cos\theta} \qquad \text{equipotential curves}$$

$$r(\theta) \propto \sin^2\theta \qquad \text{field lines}$$

for an electric dipole oriented in the direction $+\mathbf{z}$. We offer two personal-computer programs by means of which diagrams similar to Fig. 4.3 for equipotentials can be produced. These programs can be used not only for dipoles but for almost any configuration (limited here to four charges), to produce equipotential plots. The first MATHEMATICA 3.0® program is given below, for a quadrupole with +1 and −1 charges placed in the corners of a square. A dipole is obtained by setting $q3 = 0$, $q4 = 0$. The program shows a contour plot of *equipotentials* $V(x, y) = V_0$, for various values of V_0, in the square region $0 < x < 50$, $0 < y < 50$. Changes are easily made to accommodate more charges, or other regions, or for specific equipotential values.

CHARGE.NB

```
q1 = +1;   (* Charge #1 *)
q2 = +1;   (* Charge #2 *)
q3 = −1;   (* Charge #3 *)
q4 = −1;   (* Charge #4 *)
(* Choose charges +1 or −1. This choice is for a quadrupole *)
(* Choose q1 = 1, q2 = −1, q3 = 0, q4 = 0 for a vertical dipole *)
r1 = {25.01, 30}; (* x and y location of charge #1 *)
r2 = {25.01, 20}; (* x and y location of charge #2 *)
r3 = {20, 25.01}; (* x and y location of charge #3 *)
r4 = {30, 25.01}; (* x and y location of charge #4 *)
m = 50;  (* x-size of grid *)
n = 50;  (* y-size of grid *)
r = {i,j};
R1 = r − r1;
R2 = r − r2;
R3 = r − r3;
R4 = r − r4;
R1a = (Part[R1, 1]^2 + Part[R1, 2]^2)^.5;
R2a = (Part[R2, 1]^2 + Part[R2, 2]^2)^.5;
R3a = (Part[R3, 1]^2 + Part[R3, 2]^2)^.5;
R4a = (Part[R4, 1]^2 + Part[R4, 2]^2)^.5;
V = N[q1/R1a + q2/R2a + q3/R3a + q4/R4a, 4]; (* Potential at r *)
Vmatrix = Table [V, {j, n}, {i, m}];
ListContourPlot[Vmatrix, ContourShading → False, ContourSmoothing → Automatic];
```

The following program is written in the student version of MATLAB 5.0®, and it restricts the region to $0 < x < 32$, $0 < y < 32$. To change from a dipole to a quadrupole situation, simply make nonzero the $q3$ and $q4$ charge values, as indicated in the comments.

CHARGE.M

```
clear;
for n = 1:32              % Initialize array V to zero
  for m = 1:32 V(n,m) = 0;
  end;
end
q1 = +100;                % q1 and q2 are dipole charges
q2 = -100;
q3 = 0;                   % For quadrupole set q3 = -100
q4 = 0;                   % For quadrupole set q4 = +100
r1 = [16.01 20];
r2 = [16.01 12];
r3 = [20.01 20];
r4 = [20.01 12];
    for n = 1:32
        for m = 1:32;
        R1 = [n m] - r1;
        R2 = [n m] - r2;
        R3 = [n m] - r3;
        R4 = [n m] - r4;
        V1(n,m) = q1/(R1(1)^2 + R1(2)^2)^0.5 + q2/(R2(1)^2 + R2(2)^2)^0.5;
        V2(n,m) = q3/(R3(1)^2 + R3(2)^2)^0.5 + q4/(R4(1)^2 + R4(2)^2)^0.5;
        V(n,m) = V1(n,m) + V2(n,m);
        end;
    end
v1 = -20:4:20;            % Generate range of voltages for contour plot
ran = [v1];
figure(1)                 % Generate two-dimensional contour plot
contour(V,ran);
C = contour(V,ran);
clabel (C,ran)
grid on
hold off
xlabel ('charges at x = 16,20, y = 12,20')
```

As in the MATHEMATICA program, it is relatively easy to make changes to accommodate different locations for the charges, etc. However, the student version of MATLAB (at the time of writing) does not allow a grid that is larger than 32×32.

Figure 4.4 shows a contour plot of equipotentials for CHARGE.NB, with opposite charges at (20, 25) and (30, 25). Figure 4.5 depicts a CHARGE.M contour plot for a situation similar to that of Fig. 4.4.

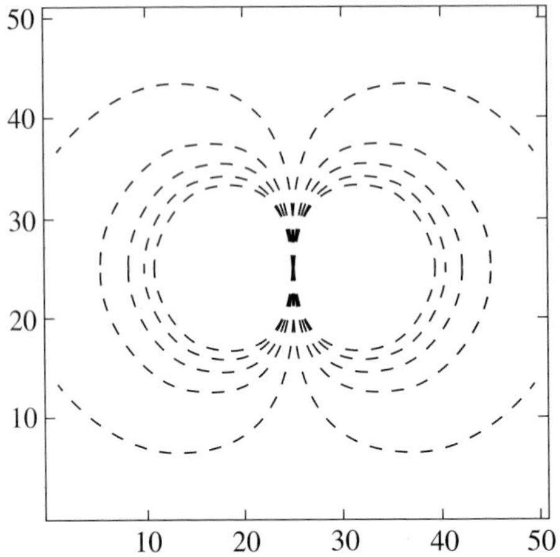

Fig. 4.4 Output figure of CHARGE.NB program, showing the equipotentials for a horizontally oriented dipole.

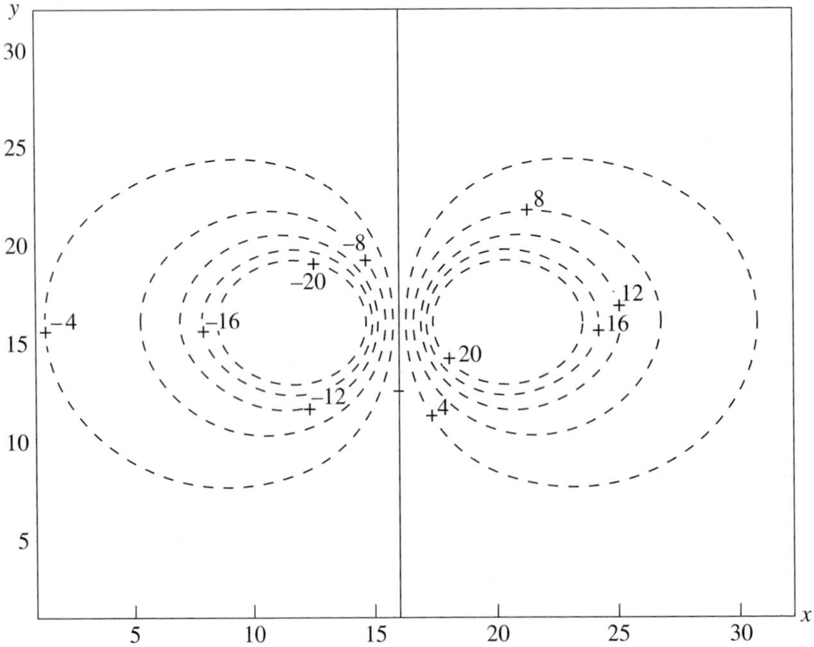

Fig. 4.5 Output figure of CHARGE.M program, showing the equipotentials for a horizontally oriented dipole. The charges are at $x = 12, 20$, $y = 16$.

Problems

4.1. When a particle with charge $q = 3.141 \times 10^{-12}$ C was placed at Q in a uniform electrostatic field, it was moved by the field from Q to P. The work done by the field was 2×10^{-10} J. What is the potential difference $V_P - V_Q$?

4.2. Calculate the work done in moving a particle with charge $q = 4 \times 10^{-11}$ C from $Q(0, 0, 0)$ to $P(1, 1, 1)$ against a uniform electrostatic field $\mathbf{E} = 2\hat{\mathbf{x}} + 3\hat{\mathbf{y}} + 4\hat{\mathbf{z}}$ in V/m. All lengths are in meters.

4.3. Calculate the work done in moving a unit charge from $Q(r = 1\,\text{m}, \theta = \pi/12, \varphi = 1.1\pi)$ to $P(r = 4\,\text{m}, \theta = -\pi/12, \varphi = -1.1\pi)$ against the electrostatic field produced by a point charge $q = 0.5 \times 10^{-11}$ C at the origin.

4.4. Show that no work is done in moving a test charge q from the origin to point $P = (3, 1, 4)$ in an electrostatic field $\mathbf{E} = 7\hat{\mathbf{x}} + 3\hat{\mathbf{y}} - 6\hat{\mathbf{z}}$ V/m.

4.5. A uniform line charge with density $\varrho_\ell = 3.5 \times 10^{-11}$ C/m coincides with the z axis. An equal but opposite-in-sign charge density exists on a parallel line at $x = 0$, $y = 5$ m.

(a) At what locations Q is the potential $V_Q = 0$?

(b) Calculate the potential at $P = (1\,\text{m}, 0.5\,\text{m}, 5\,\text{m})$.

4.6. Calculate the potential V_P at $P(x, y, 0)$ given that a uniform line charge ϱ_ℓ exists on the z axis for $-L/2 < z < L/2$. (See problem 3.9.)

4.7. An equipotential planar surface with $V = 6$ V is defined by the three points $P_1 = (2, 0, 0)$, $P_2 = (0, 3, 0)$, and $P_3 = (0, 0, 1)$.

(a) If $V = 0$ at the origin, and if all other equipotential planes are parallel to the above plane, specify the direction of the electrostatic field $\mathbf{E}$ at an arbitrary point on the planar surface given above.

(b) If the potential changes by 5 V/m in the direction of maximal change, what then is the electric field?

4.8. Use the solution to problem 4.5 to find the potential at the center of a square formed by four straight thin wires, each of length L and line-charge density ϱ_ℓ.

4.9. Consider a circular disk with radius r_0 and uniform surface-charge density ϱ_s. Show that the potential on the central axis through the disk reduces to that of a point charge when observed at a distance $R \gg r_0$. Also give the equivalent point charge in that case.

4.10. At the Earth's surface the electrostatic field is approximated by $\mathbf{E} = -50e^{-z/h}\hat{\mathbf{z}}$ (V/m), where z is a coordinate measuring height above the surface and $h \approx 8$ km. Find the potential difference between the Earth's surface and outer space.

4.11. The electric field in a planar diode is $\mathbf{E} = -\gamma z^{1/3}$. The potential at $z = 0$ is zero.

(a) What is the potential at the anode, which is at $z = d$?

(b) At what fraction of d is the potential equal to half the anode potential?

4.12. An infinitesimal dipole with moment vector $\mathbf{p} = p\hat{\mathbf{z}}$ is at the origin. At which locations does the electrostatic field have no component in the $\hat{\mathbf{z}}$ direction?

4.13. Another type of dipole distribution is formed by two infinite lines at $(0, 0, \pm d/2)$. One line has a uniform charge density ϱ_ℓ, the other $-\varrho_\ell$. Find an expression for V_P in the limit $d \to 0$, $\varrho_\ell \to \infty$, $\varrho_\ell d \equiv K =$ constant.

5

The transition towards Maxwell's equations for electrostatics

5.1 Introduction

So far, we have developed electrostatics from Coulomb's law towards two alternatives that appear to express somewhat different aspects of the electrostatic fields contained in Coulomb's law and are a half-way marker to the modern formulation in terms of Maxwell's equations. We have found

$$\oint d\mathbf{S} \cdot \mathbf{E} = q_{\text{encl}}/\varepsilon_0 \qquad \text{and} \qquad \oint d\boldsymbol{l} \cdot \mathbf{E} = 0 \tag{5.1}$$

The first is Gauss's law, (3.15b) (the total *flux* of the electrostatic field equals the enclosed charge divided by ε_0), and the second, (4.9), expresses the conservative nature of the electrostatic field (no net work is performed on or by the electrostatic field over a closed-contour path). While these two properties of the electrostatic field seem quite fundamental, they suffer from the liability of being *integrals over the field*, so that it is difficult – except in especially symmetric situations – to extract $\mathbf{E}(\mathbf{r})$ at any particular point in space. The purpose of working towards what are now called Maxwell's equations is to convert (5.1) into 'point form', i.e. into equations for $\mathbf{E}(\mathbf{r})$ at a single location. Two new derivative operators, *divergence* and *curl*, are needed; we will see that the definitions of these new operators have much to do with the physics of flow. These definitions will be discussed in Sections 5.2 and 5.3. We will need also some preparation, beyond the definitions of divergence and curl, on the Helmholtz theorem and this is given in Section 5.4. Readers are strongly advised to read this section in such a way as to be able to grasp the four main points which flow from the Helmholtz theorem (5.18) – see subsection 5.4.2.

5.2 The divergence of a vector field and Gauss's divergence theorem

Consider a vector field $\mathbf{A}(\mathbf{r})$, and its *flux* through a closed surface S, which is $\oint d\mathbf{S} \cdot \mathbf{A}$. As shown in Fig. 5.1, the vector $\mathbf{A}$ at a point on the surface can be decomposed into the sum of a normal component $\mathbf{A}_\perp = \hat{\mathbf{n}}(\hat{\mathbf{n}} \cdot \mathbf{A})$ and a tangential or parallel

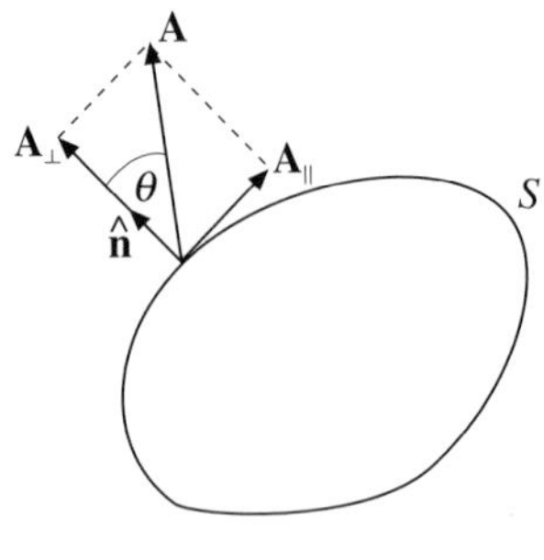

Fig. 5.1 Geometry for illustration of the divergence concept.

component $\mathbf{A}_{\parallel} = -\hat{\mathbf{n}} \times (\hat{\mathbf{n}} \times \mathbf{A})$; see also (2.7). The reader should check that the vector sum of these two is indeed $\mathbf{A}$. The flux $\oint d\mathbf{S} \cdot \mathbf{A}$ integrates the normal components over the entire closed surface S, and represents the total 'outflow' of $\mathbf{A}$ through the surface S.

The *divergence* of $\mathbf{A}$ is defined by the integral of $\mathbf{A}$ over a surface around $\mathbf{r}$ surrounding a vanishingly small volume δv:

$$\operatorname{div} \mathbf{A} \equiv \lim_{\delta v \to 0} \left(\frac{1}{\delta v} \oint d\mathbf{S} \cdot \mathbf{A} \right) \tag{5.2}$$

As a specific example, let us consider the *electric flux density* $\mathbf{D}$, which in free space equals $\varepsilon_0 \mathbf{E}$. The behavior of $\mathbf{D}$ in a medium will be discussed at length in Chapter 6. From Gauss's law, as restated in (5.1), it follows that

$$\oint d\mathbf{S} \cdot \mathbf{D} = q_{\text{encl}}$$

If $\mathbf{D}$ replaces $\mathbf{A}$ in (5.2) then the right-hand side of (5.2) becomes the limit of $q_{\text{encl}}/\delta v$ as both numerator and denominator tend to zero; this limit is the charge density ϱ_{v}. It follows that

$$\operatorname{div} \mathbf{D} = \varrho_{\text{v}}$$

and we interpret ϱ_{v} as the *scalar source field* of the vector field $\mathbf{D}$.

More generally, the divergence of a vector field thus seems to be connected intimately to the source of that vector field. In the definition (5.2) it represents *the total outflow (flux) away from point* $\mathbf{r}$ *per unit volume*, and it has the dimension of $\mathbf{A}$, divided by a length.

Differential expressions, using the three major coordinate systems, can be obtained for div $\mathbf{A}$, which is a *scalar field*.

In terms of the general orthogonal coordinate systems (u_1, u_2, u_3) we will now show that

$$\operatorname{div} \mathbf{A} = \frac{1}{h_1 h_2 h_3} \left[\frac{\partial}{\partial u_1}(h_2 h_3 A_1) + \frac{\partial}{\partial u_2}(h_3 h_1 A_2) + \frac{\partial}{\partial u_3}(h_1 h_2 A_3) \right] \tag{5.3}$$

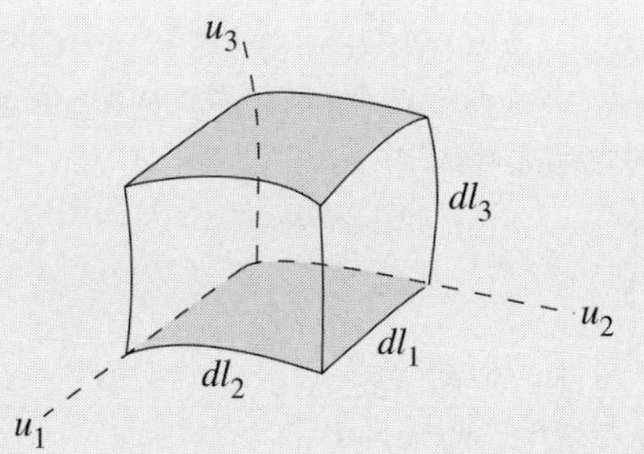

Fig. 5.2 An infinitesimal volume element in an orthogonal coordinate system. The shaded surface elements are $dS_3(u_1, u_2, u_3)$ and $dS_3(u_1, u_2, u_3 + du_3)$.

Consider an infinitesimal cube in the u_i coordinates at point $\mathbf{r}$; it has sides dl_1, dl_2, dl_3 (see Fig. 5.2).

Consider the two infinitesimal surface elements $dS_3(u_1, u_2, u_3)$ and $dS_3(u_1, u_2, u_3 + du_3)$. Their two contributions to $\oint d\mathbf{S} \cdot \mathbf{A}$ are

$$A_3(u_1, u_2, u_3 + du_3)\, dS_3(u_1, u_2, u_3 + du_3) - A_3(u_1, u_2, u_3)\, dS_3(u_1, u_2, u_3) \qquad (5.4)$$

The minus sign in the second term is due to the opposing directions of $\mathbf{A}_\perp$ and $\hat{\mathbf{n}} \equiv -\hat{\mathbf{u}}_3$ on the lower surface. Let $F(u_1, u_2, u_3) = A_3(u_1, u_2, u_3) dS_3(u_1, u_2, u_3)$ be a convenient abbreviation for the product of field times surface element, and note that $dS_3 \equiv dl_1\, dl_2 = h_1 h_2\, du_1\, du_2$. Then, from Taylor's theorem we obtain

$$F(u_1, u_2, u_3 + du_3) = F(u_1, u_2, u_3) + du_3 \left(\frac{\partial F}{\partial u_3}\right)_{u_1, u_2, u_3} + \cdots$$

as an approximation to the first term of (5.4) in terms of the contribution at (u_1, u_2, u_3). The difference (5.4) in the limit of infinitesimal du_3 becomes

$$du_3 \left(\frac{\partial F}{\partial u_3}\right)_{u_1, u_2, u_3} = du_3 \frac{\partial}{\partial u_3}(A_3 h_1 h_2\, du_1\, du_2) = du_1\, du_2\, du_3 \frac{\partial}{\partial u_3}(A_3 h_1 h_2)$$

It is only necessary to remember that the volume element is

$$dv = dl_1\, dl_2\, dl_3 = h_1 h_2 h_3\, du_1\, du_2\, du_3$$

to turn this into

$$du_3 \left(\frac{\partial F}{\partial u_3}\right)_{u_1, u_2, u_3} = \frac{dl_1\, dl_2\, dl_3}{h_1 h_2 h_3} \frac{\partial}{\partial u_3}(A_3 h_1 h_2) = \frac{dv}{h_1 h_2 h_3} \frac{\partial}{\partial u_3}(A_3 h_1 h_2) \qquad (5.5)$$

It has thus been established that (5.5) is equal to (5.4), which is that part of $\oint d\mathbf{S} \cdot \mathbf{A}$ due to the two infinitesimal surfaces $\perp \hat{\mathbf{u}}_3$. We get part of div $\mathbf{A}$ from (5.5) and (5.2) by dividing by dv, and taking the limit $dv \to 0$. That gives exactly the third term of (5.3). We obtain the remaining terms by considering the pairs of surfaces $\perp \hat{\mathbf{u}}_1$ and $\hat{\mathbf{u}}_2$. That completes the proof, except possibly for a number of mathematical details concerning the existence of derivatives of the components of **A** and the like.

Referring to Table 2.2, we can convert (5.3) easily to the three major coordinate systems:

$$\begin{aligned}
&\text{Cartesian} && \operatorname{div}\mathbf{A} = \frac{\partial A_x}{\partial x} + \frac{\partial A_y}{\partial y} + \frac{\partial A_z}{\partial z} \\
&\text{cylindrical} && \operatorname{div}\mathbf{A} = \frac{1}{\rho}\frac{\partial}{\partial \rho}(\rho A_\rho) + \frac{1}{\rho}\frac{\partial}{\partial \varphi}(A_\varphi) + \frac{\partial}{\partial z}(A_z) \qquad (5.6) \\
&\text{spherical} && \operatorname{div}\mathbf{A} = \frac{1}{r^2}\frac{\partial}{\partial r}(r^2 A_r) + \frac{1}{r\sin\theta}\left[\frac{\partial}{\partial \theta}(A_\theta \sin\theta) + \frac{\partial}{\partial \varphi}(A_\varphi)\right]
\end{aligned}$$

The Cartesian expression suggests that we can also write div A as a dot product of ∇ and **A**, i.e.

$$\operatorname{div}\mathbf{A} = \nabla \cdot \mathbf{A}$$

Using the expression for ∇ in cylindrical coordinates, which we can find from (3.33), we can write down

$$\left(\hat{\boldsymbol{\rho}}\frac{\partial}{\partial \rho} + \hat{\boldsymbol{\varphi}}\frac{1}{\rho}\frac{\partial}{\partial \varphi} + \hat{\mathbf{z}}\frac{\partial}{\partial z}\right) \cdot (A_\rho \hat{\boldsymbol{\rho}} + A_\varphi \hat{\boldsymbol{\varphi}} + A_z \hat{\mathbf{z}})$$

By using $\partial\hat{\boldsymbol{\rho}}/\partial\varphi = \hat{\boldsymbol{\varphi}}$ and $\partial\hat{\boldsymbol{\varphi}}/\partial\varphi = -\hat{\boldsymbol{\rho}}$ we then obtain the middle expression in (5.6), as required. A similar check can be applied to the spherical-coordinate case.

Let us return to Gauss's law. It involves $\oint d\mathbf{S} \cdot \mathbf{A}$ for the vector field **A** over a surface enclosing a noninfinitesimal volume (in contrast to (5.2)). *Gauss's divergence theorem* for any flux of a vector field taken over a noninfinitesimal surface states that

$$\oint d\mathbf{S} \cdot \mathbf{A} = \int dv(\nabla \cdot \mathbf{A}) \qquad (5.7)$$

This theorem follows from the definition (5.2) of the divergence, as follows. Consider the volume v to consist of infinitely many infinitesimal volume elements dv_i, as shown in Fig. 5.3. For element, dv_i, with surface area S_i, we may apply the definition (5.2) to write

$$\left(\oint d\mathbf{S} \cdot \mathbf{A}\right)_{S_i} = dv_i(\nabla \cdot \mathbf{A})_i$$

If we add up these terms for all the volume elements, then the right-hand side gives $\int dv(\nabla \cdot \mathbf{A})$ because the volume elements dv_i are infinitesimal. If we add up all the left-hand sides, then for each pair of adjoining volume elements dv_i, dv_j with an interface surface element dS – as shown in Fig. 5.3 – we observe that the vector surface elements of the interface are equal and *opposite in sign*. Their contributions

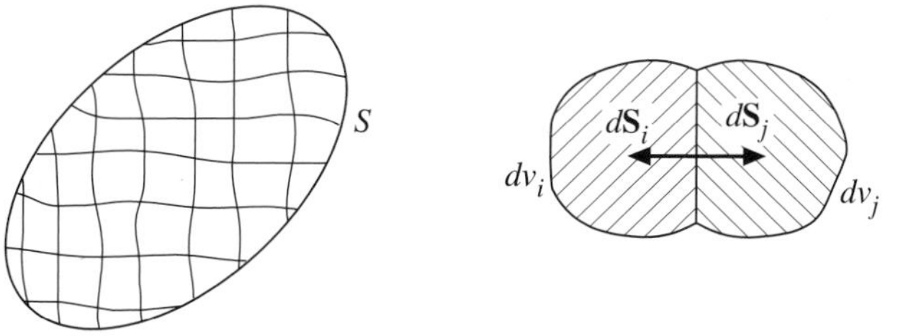

Fig. 5.3 Gauss's divergence theorem for a body of volume v, shown divided into small elements of volume dv_i. The two elements shown on the right have interface surface vectors $d\mathbf{S}_i$ and $d\mathbf{S}_j$ equal in magnitude to dS.

to the surface integral of $\mathbf{A}$ cancel two by two! Hence, only a surface integral over the *outside surface* S remains: $\oint d\mathbf{S}\cdot\mathbf{A}$. This establishes (5.7).

Example 5.1 Gauss's divergence theorem for a sphere around the origin in a field $\mathbf{A}(\mathbf{r}) = r^2\hat{\mathbf{r}}$

For a unit sphere with radius r, we have $\oint d\mathbf{S}\cdot\mathbf{A} = 4\pi r^4$, as $\mathbf{A}$ and $d\mathbf{S}$ both point in the $\hat{\mathbf{r}}$ direction and A is constant on a spherical surface. From the spherical-coordinate form of the divergence, we see that $\nabla\cdot\mathbf{A} = (1/r^2)\partial r^4/\partial r = 3r$. Owing to the angular symmetry the volume integral then becomes

$$\int dv(\nabla\cdot\mathbf{A}) = 4\pi\int_0^r dr_1\, r_1^2 \times 3r_1 = 4\pi r^4$$

Example 5.2 Gauss's divergence theorem for a sphere around a point charge

The application of Gauss's divergence theorem (5.7) sometimes requires caution. Consider for example the Coulomb field $\mathbf{E} = \kappa\hat{\mathbf{r}}/r^2$, where $\kappa \equiv q/(4\pi\varepsilon_0)$, in a sphere of radius R around a point charge q at the origin. The left-hand side of (5.7) becomes

$$\oint d\mathbf{S}\cdot\mathbf{E} = R^2\int_0^\pi d\theta\sin\theta\int_0^{2\pi} d\varphi\,\hat{\mathbf{r}}\cdot\hat{\mathbf{r}}\frac{\kappa}{R^2} = 4\pi\kappa \tag{5.8a}$$

but, if we instead use the right-hand side of (5.7) we obtain

$$\int dv\,\nabla\cdot\mathbf{E} = \int_0^R dr\,r^2\int_0^\pi d\theta\sin\theta\int_0^{2\pi} d\varphi\frac{1}{r^2}\frac{\partial}{\partial r}\left(r^2\frac{\kappa}{r^2}\right) = 0 \tag{5.8b}$$

It appears as if we obtain zero here, because $\partial\kappa/\partial r = 0$ and the $1/r^2$ singularity is canceled by the $r^2\,dr$ part of the volume element. Nevertheless, this is not correct when compared to (5.8a), because the right-hand side of (5.8b) must exclude the

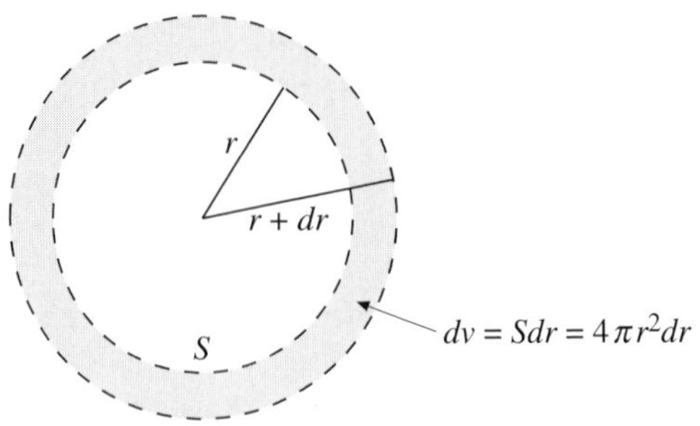

Fig. 5.4 Annular volume element for a spherically symmetric charge density.

origin, whereas (5.8a) represents a closed surface *around* the origin. The right-hand side of (5.8b) represents the so-called *principal value* of the volume integral

$$\lim_{\epsilon \to 0} \int_{\epsilon}^{R} dr\, r^2 \int_{0}^{\pi} d\theta \sin\theta \int_{0}^{2\pi} d\varphi \frac{1}{r^2} \frac{\partial}{\partial r} \left(r^2 \frac{\kappa}{r^2} \right) \tag{5.9}$$

The use of spherical coordinates in (5.8b), and hence in (5.9), implies that an infinitesimal sphere around the origin is excluded by the volume integral. The contribution of this sphere to $\oint d\mathbf{S} \cdot \mathbf{E}$, and therefore to $\int dv \nabla \cdot \mathbf{E}$, is $4\pi\kappa$ by virtue of (5.8a). Thus zero is indeed the outcome of (5.8b) when the infinitesimal sphere is excluded.

Question: why did the difficulty in Example 5.2 not occur in Example 5.1? Answer: it did, but the 'charge' of the field there, inside a sphere of radius r, by virtue of Gauss's law is $q(r) = 4\pi\varepsilon_0 r^4$, hence the 'charge' *density* must be $\varrho_v(r) = [q(r + dr) - q(r)]/dv = 12\pi\varepsilon_0 r^3 dr/4\pi r^2 dr = 3\varepsilon_0 r$. The volume element dv is illustrated in Fig. 5.4. So there is no charge at the center ($r = 0$), and negligible charge inside an infinitesimal sphere! The field in Example 5.1 is not due to a point charge (which is really a very special case).

5.3 The curl of a vector field; Stokes's theorem

Referring back to Fig. 5.1, we concentrate upon the component of $\mathbf{A}$ tangential to the surface: $\mathbf{A}_{\|} = -\hat{\mathbf{n}} \times (\hat{\mathbf{n}} \times \mathbf{A})$. This component represents the rotation of the vector $\mathbf{A}$ around the volume. If we intersect the volume by a plane, we obtain a curve. Let us assume it is a simply connected closed curve $C = \oint dl$ as in Fig. 5.1. The ith component of the *curl* of the vector field $\mathbf{A}$ is defined as the limit of a closed-contour line integral enclosing a vanishingly small surface element $\delta S_i = |\boldsymbol{dl}_j \times \boldsymbol{dl}_k|$, with i, j, k as the three right-handed permutations of 1, 2, 3:

$$(\text{curl}\,\mathbf{A})_i \equiv \lim_{\delta S \to 0} \left\{ \frac{1}{\delta S_i} \oint_{C_i} \boldsymbol{dl} \cdot \mathbf{A} \right\} \tag{5.10}$$

The sense of the contour C_i must be consonant with the right-hand rule, i.e. a right-handed screw rotating in the direction of $\boldsymbol{dl}$ on C_i should move in the direction of positive flux of **A** through $\mathbf{C}_i$. The curl is a *vector* with three components that depend upon the coordinate system used. Its dimension is (like div) that of **A** divided by length. It is sometimes referred to as the *rotation* of the vector field.

In an arbitrary orthogonal coordinate system, it is given by

$$\text{curl}\,\mathbf{A} = \sum_{k=1}^{3} \hat{\mathbf{u}}_k \frac{1}{h_i h_j} \left[\frac{\partial}{\partial u_i}(h_j A_j) - \frac{\partial}{\partial u_j}(h_i A_i) \right] \tag{5.11a}$$

Another way of writing (5.11a) is as a symbolic determinant,

$$\text{curl}\,\mathbf{A} = \frac{1}{h_1 h_2 h_3} \begin{vmatrix} h_1 \hat{\mathbf{u}}_1 & h_2 \hat{\mathbf{u}}_2 & h_3 \hat{\mathbf{u}}_3 \\ \partial/\partial u_1 & \partial/\partial u_2 & \partial/\partial u_3 \\ h_1 A_1 & h_2 A_2 & h_3 A_3 \end{vmatrix} \tag{5.11b}$$

in which case one must be careful to form triple products of the matrix components in the order 'first row–second row–third row'. With either form of curl **A**, the three main coordinate-system expressions are obtained easily, just by using Table 2.2.

Here follows a brief derivation of the above result. Consider an infinitesimal rectangle in a plane with $u_3 =$ constant, built out from the point $\mathbf{r} = (u_1, u_2, u_3)$; see Fig. 5.5. It will be used to derive the third component of (5.11). The four

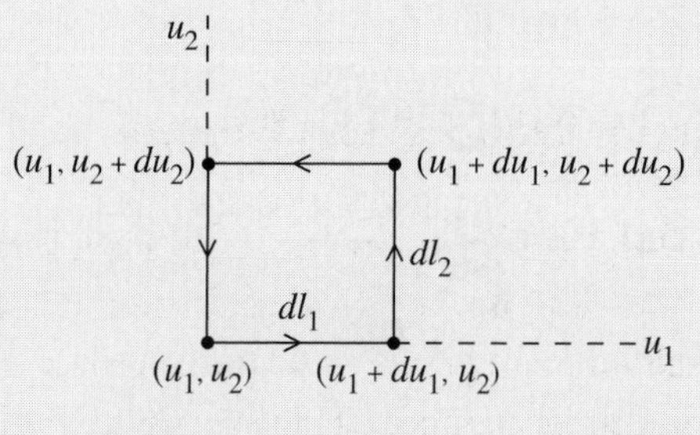

Fig. 5.5 An infinitesimal surface element in an orthogonal coordinate system.

contributions to the path integral of A around the rectangle, $\oint d\boldsymbol{l} \cdot \mathbf{A}$, are

$$dl_1(u_1, u_2)A_1(u_1, u_2) - dl_1(u_1, u_2 + du_2)A_3(u_1, u_2 + du_2)$$
$$- dl_2(u_1, u_2)A_2(u_1, u_2) + dl_2(u_1 + du_1, u_2)A_2(u_1 + du_1, u_2) \tag{5.12}$$

The minus signs in the second and third terms are due to the opposing directions of $\hat{\mathbf{u}}_i$ and $d\boldsymbol{l}_i$ on two sides of the rectangular contour of $\oint d\boldsymbol{l} \cdot \mathbf{A}$. Now use the expressions (2.23) for the line elements:

$$dl_1(u_1, u_2 + du_2) = h_1(u_1, u_2 + du_2)du_1$$

$$dl_2(u_1 + du_1, u_2) = h_2(u_1 + du_1, u_2)du_2$$

and so we obtain

$$\begin{aligned}
&\oint (d\boldsymbol{l} \cdot \mathbf{A})_{u_3=\text{constant}} \\
&= du_1[h_1(u_1, u_2)A_1(u_1, u_2) - h_1(u_1, u_2 + du_2)A_1(u_1, u_2 + du_2)] \\
&\quad - du_2[h_2(u_1, u_2)A_2(u_1, u_2) - h_2(u_1 + du_1, u_2)A_2(u_1 + du_1, u_2)] \\
&= -du_1 du_2 \frac{\partial}{\partial u_2}(h_1 A_1) + du_2 du_1 \frac{\partial}{\partial u_1}(h_2 A_2) + O(du)^3
\end{aligned} \tag{5.13}$$

We ignore the higher-order terms in du, and divide by $\delta S_3 = h_1 h_2\, du_1\, du_2$ to obtain for the infinitesimals:

$$\frac{1}{\delta S_3} \oint (d\boldsymbol{l} \cdot \mathbf{A})_{u_3=\text{constant}} = \frac{1}{h_1 h_2}\left[\frac{\partial}{\partial u_1}(h_2 A_2) - \frac{\partial}{\partial u_2}(h_1 A_1)\right] \tag{5.14}$$

We have thus found that the right-hand side of (5.14) is equal to the coefficient of $\hat{\mathbf{u}}_3$ in (5.11), which is the expression for curl $\mathbf{A}$. We obtain the remaining terms by considering rectangles in the surfaces $\perp \hat{\mathbf{u}}_1$ and $\perp \hat{\mathbf{u}}_2$. This completes the proof, except for a number of mathematical details concerning the existence of derivatives of the components of $\mathbf{A}$ and the like.

The rather lengthy expressions for curl $\mathbf{A}$ are as follows:

Cartesian $$\operatorname{curl}\mathbf{A} = \hat{\mathbf{x}}\left(\frac{\partial A_z}{\partial y} - \frac{\partial A_y}{\partial z}\right) + \hat{\mathbf{y}}\left(\frac{\partial A_x}{\partial z} - \frac{\partial A_z}{\partial x}\right) + \hat{\mathbf{z}}\left(\frac{\partial A_y}{\partial x} - \frac{\partial A_x}{\partial y}\right)$$

cylindrical $$\begin{aligned}\operatorname{curl}\mathbf{A} &= \hat{\boldsymbol{\rho}}\left(\frac{1}{\rho}\frac{\partial A_z}{\partial \varphi} - \frac{\partial A_\varphi}{\partial z}\right) + \hat{\boldsymbol{\varphi}}\left(\frac{\partial A_\rho}{\partial z} - \frac{\partial A_z}{\partial \rho}\right) \\ &\quad + \hat{\mathbf{z}}\frac{1}{\rho}\left(\frac{\partial(\rho A_\varphi)}{\partial \rho} - \frac{\partial A_\rho}{\partial \varphi}\right)\end{aligned}$$

spherical $$\begin{aligned}\operatorname{curl}\mathbf{A} &= \hat{\mathbf{r}}\frac{1}{r\sin\theta}\left(\frac{\partial(A_\varphi \sin\theta)}{\partial\theta} - \frac{\partial A_\theta}{\partial\varphi}\right) + \hat{\boldsymbol{\theta}}\frac{1}{r}\left(\frac{1}{\sin\theta}\frac{\partial A_\rho}{\partial\varphi} - \frac{\partial(rA_\varphi)}{\partial r}\right) \\ &\quad + \hat{\boldsymbol{\varphi}}\frac{1}{r}\left(\frac{\partial(rA_\theta)}{\partial r} - \frac{\partial A_r}{\partial\theta}\right)\end{aligned}$$

(5.15)

Just as we can write div $\mathbf{A} = \nabla \cdot \mathbf{A}$, we can write curl $\mathbf{A}$ as the cross product of ∇ and $\mathbf{A}$, i.e.

$$\text{curl}\,\mathbf{A} = \nabla \times \mathbf{A}$$

There is also a form of curl $\mathbf{A}$ for a noninfinitesimal area: it is known as *Stokes's theorem*, and it states that

$$\int_S d\mathbf{S} \cdot (\nabla \times \mathbf{A}) = \oint_C d\boldsymbol{l} \cdot \mathbf{A} \tag{5.16}$$

given that contour C encloses area S. It is proved in a fashion similar to Gauss's divergence theorem: we build a noninfinitesimal area S up out of infinitesimal areas δS_i, and the part of the contour of one δS_i shared with a contiguous neighbor gives zero contribution to $d\boldsymbol{l} \cdot \mathbf{A}$, as one $d\boldsymbol{l}$ is opposite in direction to that of the neighbor (and $\mathbf{A}$ is the same on that shared line element).

Example 5.3 Stokes's theorem for a circle around the z axis in a field $\mathbf{A}(\mathbf{r}) = -x\hat{\mathbf{y}} + y\hat{\mathbf{x}}$

This example is best done in cylindrical coordinates so that the vector field can be written as

$$\begin{aligned}\mathbf{A}(\mathbf{r}) &= \rho\cos\varphi[\hat{\boldsymbol{\rho}}(\varphi)\sin\varphi + \hat{\boldsymbol{\varphi}}(\varphi)\cos\varphi] - \rho\sin\varphi[\hat{\boldsymbol{\rho}}(\varphi)\cos\varphi - \hat{\boldsymbol{\varphi}}(\varphi)\sin\varphi] \\ &= \rho\hat{\boldsymbol{\varphi}}(\varphi)\end{aligned}$$

The contour element on the circle is $d\boldsymbol{l} = (\rho\, d\varphi)\hat{\boldsymbol{\varphi}}$ and, therefore,

$$\oint_C d\boldsymbol{l} \cdot \mathbf{A} = \rho \int_0^{2\pi} d\varphi\, \hat{\boldsymbol{\varphi}} \cdot \rho\hat{\boldsymbol{\varphi}} = 2\pi\rho^2 \tag{5.17a}$$

With the second of (5.15) we find $\nabla \times \mathbf{A} = 2\hat{\mathbf{z}}$, because $\mathbf{A}$ has only an A_φ component and this depends only on ρ. We have assumed counterclockwise rotation on C, in agreement with $d\boldsymbol{l} \propto \hat{\boldsymbol{\varphi}}(\varphi)$, so that the right-hand rule leads to a curl in the $+\hat{\mathbf{z}}$ direction. The surface element is $d\mathbf{S} = \hat{\mathbf{z}}\rho\, d\rho\, d\varphi$ so we have

$$\int_S d\mathbf{S} \cdot (\nabla \times \mathbf{A}) = \int_0^{2\pi} d\varphi \int_0^{\rho} d\rho_1 \rho_1 \hat{\mathbf{z}} \cdot 2\hat{\mathbf{z}} = 2\pi\rho^2 \tag{5.17b}$$

showing that Stokes's theorem is indeed satisfied.

5.4 Irrotational (conservative) and solenoidal fields

5.4.1 The Helmholtz theorem

Consider a region of space in which an arbitrary vector field $\mathbf{F}(\mathbf{r})$ defines a scalar divergence field $\sigma(\mathbf{r}) \equiv \nabla \cdot \mathbf{F}(\mathbf{r})$ and a vector curl field $\boldsymbol{J}(\mathbf{r}) \equiv \nabla \times \mathbf{F}(\mathbf{r})$, provided of course that the appropriate derivatives exist at all points of the space in which the fields are defined. Either of the *source* fields $\sigma(\mathbf{r})$ and $\mathbf{J}(\mathbf{r})$ may or may not be zero everywhere, and we shall see that those fields $\mathbf{F}(\mathbf{r})$ that have only one of these source fields associated with them are special. But, while we can see that a vector field $\mathbf{F}(\mathbf{r})$ defines corresponding source fields $\sigma(\mathbf{r})$ and $\mathbf{J}(\mathbf{r})$, *Helmholtz's theorem* tackles the more difficult question whether the existence of two *arbitrary* source fields $\sigma(\mathbf{r})$ and $\mathbf{J}(\mathbf{r})$ allows us to determine a field $\mathbf{F}(\mathbf{r})$ such that $\sigma(\mathbf{r}) \equiv \nabla \cdot \mathbf{F}(\mathbf{r})$ and $\mathbf{J}(\mathbf{r}) \equiv \nabla \times \mathbf{F}(\mathbf{r})$. That this is the case is less obvious, but it is of importance; compare the case of the electrostatic field $\mathbf{E}(\mathbf{r}) = -\nabla V(\mathbf{r})$, where the new field, $V(\mathbf{r})$, helps simplify matters considerably. The first part of Helmholtz's theorem states that, indeed, there *is* in general such a vector field $\mathbf{F}(\mathbf{r})$, and it is unique except possibly for a constant. There are some conditions that need to be met, such as knowledge of the values of $\mathbf{F}(\mathbf{r})$ and its normal derivative on the boundary of the region.

The theorem has an important second part. For if it *is* always possible, given suitable boundary-condition knowledge and existence of the needed derivatives, to find such an $\mathbf{F}(\mathbf{r})$ from a given pair $\sigma(\mathbf{r})$ and $\mathbf{J}(\mathbf{r})$, then Helmholtz's theorem also states that we can set[1]

$$\mathbf{F}(\mathbf{r}) = \mathbf{F}_{\mathrm{i}}(\mathbf{r}) + \mathbf{F}_{\mathrm{s}}(\mathbf{r}) \tag{5.18a}$$

with

$$\begin{array}{ll} \nabla \cdot \mathbf{F}_{\mathrm{i}}(\mathbf{r}) = \sigma_{\mathrm{s}}(\mathbf{r}) & \nabla \cdot \mathbf{F}_{\mathrm{s}}(\mathbf{r}) = 0 \\ \nabla \times \mathbf{F}_{\mathrm{i}}(\mathbf{r}) = 0 & \nabla \times \mathbf{F}_{\mathrm{s}}(\mathbf{r}) = \mathbf{J}_{\mathrm{s}}(\mathbf{r}) \end{array} \tag{5.18b}$$

The field $\mathbf{F}_{\mathrm{i}}$ has no curl and therefore is known as *irrotational*, whereas the field $\mathbf{F}_{\mathrm{s}}$ has no divergence and is known as *solenoidal*. The reasons for these names are clear: we have associated the curl with rotation and there is no rotation if the curl is zero everywhere; likewise, if the divergence is zero then the field lines can never end (we will elaborate later) and so they must be closed, hence are what is called 'solenoidal'. Thus, the theorem states that there is also a *unique decomposition of a vector field* $\mathbf{F}(\mathbf{r})$ *into an irrotational component* $\mathbf{F}_{\mathrm{i}}$ *and a solenoidal component* $\mathbf{F}_{\mathrm{s}}$ (except possibly for an additive constant). The theorem implies that $\mathbf{F}(\mathbf{r})$ *is determined uniquely by its divergence and its curl* (except possibly for an additive constant). The proof of the theorem is not trivial and it goes beyond the scope of this text.

5.4.2 Some vector theorems and some consequences of the Helmholtz theorem

There are several special properties associated with gradients, divergences, and curls that are easily seen to be true and that are very useful.

[1] In this subsection, the subscript 's' refers to 'solenoidal' not, as elsewhere, to 'surface'.

(1) *The curl of the gradient of any scalar field is zero*: $\nabla \times \nabla\Phi(\mathbf{r}) = 0$. (5.19)

To prove (5.19), consider e.g. the x component of this composite, $\partial^2\Phi/\partial y\partial x - \partial^2\Phi/\partial x\partial y$, which is zero. The same holds for the other Cartesian components.

(2) *The divergence of the curl of any vector field is zero*: $\nabla \cdot [\nabla \times \mathbf{A}(\mathbf{r})] = 0$. (5.20)

This too is easy to prove in Cartesian coordinates. The composite works out to be

$$(\partial/\partial x)(\partial A_z/\partial y - \partial A_y/\partial z) + (\partial/\partial y)(\partial A_x/\partial z - \partial A_z/\partial x) + (\partial/\partial z)(\partial A_y/\partial x - \partial A_x/\partial y) = 0$$

as is obvious upon rearranging the terms and remembering that $\partial^2 A_z/\partial x\partial y = \partial^2 A_z/\partial y\partial x$, and so on.

(3) *If* $\mathbf{F}(\mathbf{r})$ *has zero curl, then it has a scalar source field* $\sigma(\mathbf{r}) = \nabla \cdot \mathbf{F}$. (5.21)

Helmholtz's theorem predicts $\mathbf{F}(\mathbf{r}) = \mathbf{F}_i(\mathbf{r}) + \mathbf{F}_s(\mathbf{r})$, as in (5.18a), but if $\nabla \times \mathbf{F} = 0$ then certainly $\nabla \times \mathbf{F}_s = 0$ since we already know from (5.18b) that $\nabla \times \mathbf{F}_i = 0$. But if $\mathbf{F}_s$ has zero curl as well as, by definition, zero divergence, then it must itself be a constant (which we can take as zero); hence $\mathbf{F} = \mathbf{F}_i$, an *irrotational vector field.*

(4) *If* $\mathbf{F}(\mathbf{r})$ *has zero divergence, then it has a vector source field* $\mathbf{J}(\mathbf{r}) = \nabla \times \mathbf{F}$. (5.22)

Helmholtz's theorem predicts $\mathbf{F}(\mathbf{r}) = \mathbf{F}_i(\mathbf{r}) + \mathbf{F}_s(\mathbf{r})$, as in (5.18a), but if $\nabla \cdot \mathbf{F} = 0$ then certainly $\nabla \cdot \mathbf{F}_i = 0$, as we already know from (5.18b) that $\nabla \cdot \mathbf{F}_s = 0$. But if $\mathbf{F}_i$ has zero divergence as well as, by definition, zero curl, then it must itself be a constant (which we can take as zero). Hence $\mathbf{F} = \mathbf{F}_s$, *a solenoidal vector field.*

5.5 Maxwell's equations for electrostatic fields

We finally are able to return to the electrostatic field at the point where we left off, in Section 5.1, and draw some conclusions based upon the material in Sections 5.2–5.4.

In (5.1) we observed that the line integral of $d\mathbf{l} \cdot \mathbf{E}$ is zero for any closed contour. Stokes's theorem (5.16) then states that the surface integral of $\nabla \times \mathbf{E}$ is also zero for *any* surface enclosed by a closed contour. As the surface is arbitrary, it can be concluded that

$$\nabla \times \mathbf{E} = 0 \tag{5.23a}$$

The mathematical property (5.21) now tells us that the divergence of $\mathbf{E}$ is a source field, i.e. $\nabla \cdot \mathbf{E} = \sigma$. But, using Gauss's divergence theorem, it also follows that

$$\int dv\,\varrho_v/\varepsilon_0 \equiv q_{encl}/\varepsilon_0 = \oint d\mathbf{S}\cdot\mathbf{E} = \int dv(\nabla\cdot\mathbf{E}) \quad\longrightarrow\quad \nabla\cdot\mathbf{E} = \varrho_v/\varepsilon_0$$

because the volume is arbitrary; hence the mathematical 'source' σ of the **E** field is indeed related simply to the physical source ϱ_v. Using the free-space *electric flux density* $\mathbf{D} \equiv \varepsilon_0\mathbf{E}$, mentioned above in Section 5.2, the above relationship becomes

$$\nabla\cdot\mathbf{D} = \varrho_v \tag{5.23b}$$

D is also known as the *electric displacement vector*.

Obviously $\nabla\times\mathbf{D} = 0$, so that ϱ_v coincides with the mathematical 'source field' for **D**. Equations (5.23a, b) are Maxwell's equations for electrostatics, and they contain essentially everything discussed so far about electrostatic fields. Equation (5.23b) can be restated, again using Gauss's divergence theorem, to obtain Gauss's law in terms of **D**:

$$\oint d\mathbf{S}\cdot\mathbf{D} = \int dv\,\nabla\cdot\mathbf{D} = \int dv\varrho_v \equiv q_{encl} \tag{5.24}$$

Equations (5.23) are partial differential equations, from which the Maxwell integral equations (5.1) could be obtained by a procedure inverse to the one we have used. In fact, many textbooks do start from Eqs. (5.23) rather than pursuing the historical and perhaps easier-to-grasp path we have followed here.

The equation $\mathbf{D} \equiv \varepsilon_0\mathbf{E}$ is actually known as a *constitutive equation* because although in a vacuum it is a somewhat trivial relationship between the **E** and **D** fields, in an isotropic material the corresponding relationship $\mathbf{D} \equiv \varepsilon\mathbf{E}$ expresses an important property of the material: the extent to which dipoles are formed in it in the presence of an electric field.[1] With this modified definition of **D**, the two equations in (5.23) and also (5.24) are otherwise unchanged.

Equations (5.23) have a geometrical meaning that helps us understand the relationship between fields and their sources. Consider (5.23a) stating that the curl of **E** is zero. This tells us that *electrostatic field lines never can close upon themselves*, for, if that were possible, then the field line shown in Fig. 5.6 could have a contour as drawn along its path, and therefore **E** would be parallel to $+d\mathbf{l}$ (or $-d\mathbf{l}$) at every point of the contour. In this case the addition of all the $\mathbf{E}\cdot d\mathbf{l}$ infinitesimals would be either positive ($\mathbf{E} \parallel +d\mathbf{l}$) or negative ($\mathbf{E} \parallel -d\mathbf{l}$) but certainly not zero. And if $\oint d\mathbf{l}\cdot\mathbf{E} \neq 0$ then from (5.16) $\nabla\times\mathbf{E} \neq 0$, by virtue of the definition of a curl, which contradicts (5.23a).

Equation (5.23b), stating that divergence of **D** is the volume-charge density, tells us that *electrostatic field lines start and stop on charges*. The proof requires two small

[1] For nonisotropic media, the relationship between **D** and **E** is more complex, as we shall see later.

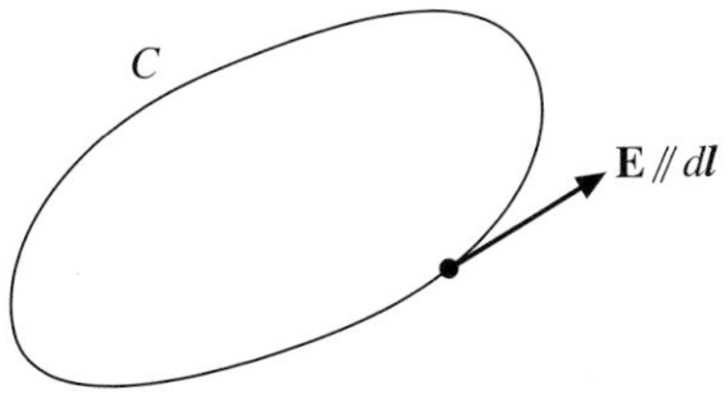

Fig. 5.6 A hypothetical closed electrostatic field line.

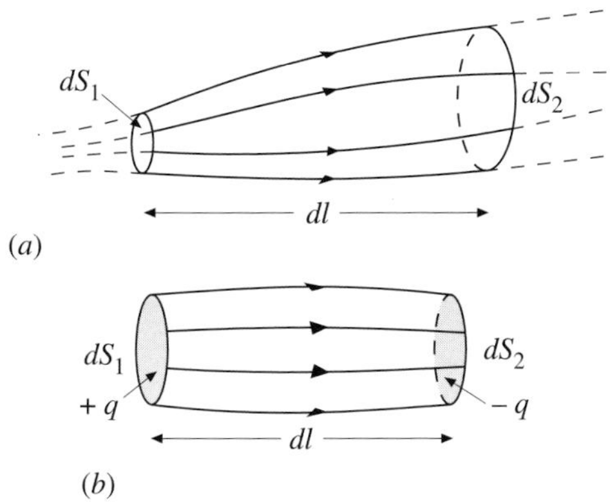

Fig. 5.7 A tube of **E** field lines: (*a*) in general; (*b*) from charge q to charge $-q$.

steps. The first makes use of what we have just proven: electrostatic field lines cannot close upon themselves, so that they indeed must start and stop at different places. The second step uses Fig. 5.7(*a*), which shows an infinitesimal bundle of electrostatic field lines between dS_1 and dS_2, which are at an infinitesimal distance dl. The assumption is that the field lines have traveled to dS_1 and the question is whether they stop there or whether they continue on to dS_2. Let us assume there is no charge inside the bundle. Then Gauss's law states that $\oint d\mathbf{S} \cdot \varepsilon_0 \mathbf{E} = q_{\text{encl}} = 0$. It follows from (5.24) that this statement is the same as saying that $\nabla \cdot \mathbf{D} = 0$ at all points in the bundle. We have already observed, in the discussion of the definition of field lines, that if $dS_1 \neq 0$ and $\oint d\mathbf{S} \cdot \varepsilon_0 \mathbf{E} = 0$, then $dS_2 \neq 0$ because field lines cannot cross or intersect each other. Hence the field lines continue on to dS_2 when $\nabla \cdot \mathbf{D} = 0$. There is only one other possibility in which the field lines really do start and stop as shown in the sketch of Fig. 5.7(*b*), and that is if there is a charge $q_1 = \varrho_s\, dS_1$ on surface dS_1 and a charge $q_2 = -q_1$ on surface dS_2 at some nonzero distance dl, because then it would still be true that $\oint d\mathbf{S} \cdot \varepsilon_0 \mathbf{E} = 0$, as the enclosed charge is $q_1 + q_2 = 0$. It is not possible that $q_2 \neq -q_1$. Thus, field lines start and stop on charges, and we observe that any bundle of field lines stops on a charge that is equal and opposite to the charge on which the bundle started, no matter how distorted the bundle has become!

Problems

5.1. Calculate the divergence of the vector field

$$\mathbf{A} = \sqrt{y^2+z^2}\,\hat{\mathbf{x}} + \sqrt{z^2+x^2}\,\hat{\mathbf{y}} + \sqrt{x^2+y^2}\,\hat{\mathbf{z}}.$$

5.2. Calculate the divergence of the vector field $\mathbf{A} = (\sin 2x)\hat{\mathbf{x}} - (2y\cos 2x)\hat{\mathbf{y}} + (xyz)\hat{\mathbf{z}}$.

5.3. For an infinite uniform line charge on the z axis, calculate the divergence of the electrostatic field

(a) at $\rho \neq 0$

(b) at $\rho = 0$.

5.4. Calculate the divergence of the vector field $\mathbf{A} = (5/r^{2.05})\hat{\mathbf{r}}$.

5.5. Calculate the divergence of the vector field

$$\mathbf{A} = [\sin(3\theta+4)/r^2]\hat{\mathbf{r}} + (5r\,\mathrm{cotan}\,\theta)\hat{\boldsymbol{\theta}} + (2r\sin\theta\sin\varphi)\hat{\boldsymbol{\varphi}}.$$

5.6. (a) Calculate the divergence of the vector field $\mathbf{A} = (2\cos\theta)\hat{\mathbf{r}} + (\sin\theta)\hat{\boldsymbol{\theta}}$ outside the origin.

(b) How could you have known in advance what the answer is?

5.6. For the vector field $\mathbf{A} = (x-3)\hat{\mathbf{x}} + (y-4)\hat{\mathbf{y}} + z^2\hat{\mathbf{z}}$, evaluate both sides of the divergence theorem (5.7) for the volume enclosed by a cube with sides of length 2 parallel to the coordinate axes and centered at the origin.

5.7. For the vector field $\mathbf{A} = (10\rho^3)\hat{\boldsymbol{\rho}}$, evaluate both sides of the divergence theorem (5.7) for the volume enclosed by $\rho = 2$ m, $z = 0$, and $z = 10$ m.

5.8. For the vector field $\mathbf{A} = (10\sin\theta)\hat{\mathbf{r}} + (2\cos\theta)\hat{\boldsymbol{\theta}}$, evaluate both sides of the divergence theorem (5.7) for the volume enclosed by the spherical surface $r = 2$ m.

5.9. Calculate the curl of the vector field $\mathbf{A} = y\hat{\mathbf{x}} + z\hat{\mathbf{y}} + x\hat{\mathbf{z}}$.

5.10. Given $\mathbf{A} = (ay+3z)\hat{\mathbf{x}} + (bz+4x)\hat{\mathbf{y}} + (cx+5y)\hat{\mathbf{z}} = 0$, find values of a, b, and c such that $\mathrm{curl}\,\mathbf{A} = 0$.

5.11. (a) Calculate the curl of $\mathbf{A} = \rho\hat{\boldsymbol{\varphi}}$.

(b) Check your answer by calculating the curl of $\mathbf{B} = -y\hat{\mathbf{x}} + x\hat{\mathbf{y}}$.

(c) Why are the answers the same?

5.12. Calculate the curl of the vector field $\mathbf{A} = a\hat{\boldsymbol{\varphi}}/\rho$ for $\rho \neq 0$. Use the integral definition to find the curl on the z axis.

5.13. Calculate the curl of the vector field $\mathbf{A} = (r\sin\theta)\hat{\boldsymbol{\varphi}} - (r\cos\theta)\hat{\mathbf{r}}$.

5.14. An equilateral triangle has its vertices at $(L/2, 0, 0)$, $(-L/2, 0, 0)$, $(0, L\sqrt{3}/2, 0)$. The path around this triangle is counterclockwise. Evaluate both sides of Stokes's theorem (5.16) for the vector field $\mathbf{A} = y\hat{\mathbf{x}} - x^2\hat{\mathbf{y}}$.

5.15. Given the vector field $\mathbf{F} = (\rho^2\sin^2\varphi)\hat{\boldsymbol{\rho}} + (\rho^2\cos^2\varphi)\hat{\boldsymbol{\varphi}} + (2z^2)\hat{\mathbf{z}}$, evaluate both sides of Stokes's theorem (5.16) for the path formed by the intersection of the cylinder $\rho = 2$ and the plane $z = 1$, and for the surface defined by $\rho = 2$, $1 < z < 3$, and $z = 3$, $0 \le \rho \le 2$. All lengths are in meters.

5.16. Can $\mathbf{F} = (6xy + z^3)\hat{\mathbf{x}} + (3x^2 - z)\hat{\mathbf{y}} + (3xz^2 - y)\hat{\mathbf{z}}$ be an electrostatic field? Prove your answer.

5.17. The vector field $\mathbf{F}(\mathbf{r})$ has zero curl and zero divergence inside and on a sphere of radius $R = 2\,\text{m}$. It is also given that $\mathbf{F} = 5\,\hat{\mathbf{r}}$ on the surface of the sphere. Why is this impossible?

6

Electrostatic fields in material media

6.1 Introduction

So far, we have discussed the properties of electrostatic fields *in vacuo* (that is to say, in free space), but in almost all applications and real-life situations the fields occur in *material media*. Even the air in the atmosphere is a material medium, and drastic electric phenomena such as lightning bolts that appear to take place in a vacuum really depend upon a chain of intermediate particles (disassociated molecules, atoms, and ions)! The properties of electric fields in material media require us to have some understanding of the underlying basic properties of media that can affect electromagnetism.

Gaseous materials consist of molecules that exhibit little or no structural order in space. It is possible for some or all of these molecules to be ionized, in which case the gas is known as a *plasma*. Liquid materials have higher densities than gases in general, and there is some structural order at short ranges of up to several tens of molecules, but no long-range order in general. Solid materials differ in that they mostly have even higher densities and they may exhibit longer-range order. Many solids can be considered to be more-or-less ordered lattices of atoms or molecules that exhibit a highly repetitive pattern.

An adequate, albeit primitive, picture of an atom, which dates from the Bohr model preceding modern quantum mechanics, is that of a tightly confined *nucleus* bearing a positive charge, surrounded by a cluster of much smaller negatively charged *electrons*. The positive charge exactly balances the sum of the negative electron charges. Modern quantum mechanics no longer regards the structure of an atom as analogous to that of a star (the nucleus) with a surrounding planetary system (the cluster of electrons), but that analogy is of some use here with one refinement: we must not try to localize the electrons exactly into orbits around the nucleus but instead should think of them as being spread around the nucleus in a *cloud-like* form concentrated in *shells* at certain distances from the nucleus. The distance of a shell from the nucleus is characterized by a *quantum number*, and higher quantum numbers are associated with greater distances from the nucleus. Electrons in shells with higher quantum numbers are less tightly 'bound' to the nucleus (i.e. have lower binding energy) because the lower-quantum-number electrons partly shield the positive charge of the nucleus.

The outermost electrons are largely responsible for the *chemical binding* of atoms into molecules. Molecules can be formed from two or more atoms (either atoms of the same *element* in the Periodic System, as in oxygen or nitrogen gas, or atoms of different elements) that are bound to each other by electromagnetic and/or quantum-mechanical forces. Chemistry distinguishes two basic types of atom in this respect: those that wish to add electrons to, and those that prefer to have electrons removed from, their outermost shell. The former readily become *negative ions* and the latter *positive ions*. The underlying reason is quantum mechanical and it has to do with the preferred-energy status of the various shells when they are *full*. (The maximum number is a function of the shell quantum number.) Some elements such as *sodium* form positive ions easily, because they have only one electron (or a few electrons in general) in their outermost shell, so that yielding it (or them) to bring the electron structure back to perfectly filled shells is energetically preferable. Other elements such as *chlorine* prefer the opposite; they have almost-filled outer shells and the addition of one (or more) electrons would bring them into the preferred-energy status. Thus combination of a sodium atom and a chlorine atom, for example, forms a very stable molecule because when they are close to each other the 'extra' outer electron of sodium is transferred to the outer shell of chlorine (which lacks exactly one electron); the resulting *table salt* molecule consists of two ions, each of which is closer to the 'ideal' state with perfectly filled outer shells than each atom would be if separated. This is an example of *ionic bonding*. Furthermore, there is another common type of bonding, *covalent bonding*, in which one or more electrons are *shared* between atoms in order to achieve full outer shells. An example is the water molecule, H_2O. In either case the electrostatic forces to the outside of the positive–negative ion pair or molecule are not zero. As a result, a whole sequence of pairs of ions can be held close to each other in some kind of regular lattice, or perhaps in a less ordered *amorphous* way, thus forming a *solid*. The shape and structure of the solid depends very much upon its constituents; we only have sketched here the simplest types of chemical binding and we have ignored almost entirely other quantum-mechanical binding forces.

The importance for electromagnetics lies in the *ability of the atomic electrons to be displaced under the influence of electric fields*. In almost all gases and liquids that ability is fairly limited. Most of these gases and liquids belong to the class of *dielectrics*, the most extreme of which are those with transferred rather than shared electrons (and strong electromagnetic binding forces) and consequently which are almost impervious to laboratory or practical electric fields: these are the *insulators*.

Conductors are an especially interesting and important class of materials for electromagnetics. These are materials, such as metals, in which a *conduction current* can flow. A metallic conductor consists of a lattice of positive ions of a particular group in the Periodic System that has some shells with high quantum numbers containing only one or two electrons. These shells are not necessarily the outermost, but even so they are well shielded by inner-electron shells, and the electrons in them are not tightly bound either by electromagnetic or by quantum-mechanical forces. Consequently, in a metal one or more of these electrons can roam almost freely through the lattice structure. An electrostatic field exerted upon these *conduction* or *free* electrons will attempt to accelerate the electrons ($\mathbf{F} \equiv -q\mathbf{E} = m\ddot{\mathbf{r}}$ by virtue of Newton's law of acceleration by forces), but collisions with the much larger ions, which randomize the electrons' movement, will occur so often that the accelerated motion turns into an average *drift velocity* $\mathbf{u}$ in the $-\mathbf{E}$ direction (see later). The drift velocity forms the conduction current, because, as we shall see, the current density $\mathbf{J} \propto \sum_i q\mathbf{u}_i$, where $\mathbf{u}_i$ is the velocity of the ith electron.

Semiconductors, for example germanium and silicon, are materials with electron-carrier behavior that is intermediate between insulators and conductors. Small currents can flow, but the number of charge carriers varies greatly, depending upon the energy available from outside in the form of heat or light. Furthermore, the pure substance can be 'doped' in a controlled way so as to *increase* the number of charge carriers, either the number of free electrons (n-doped) or the number of outer-shell vacancies, i.e. *holes* (p-doped). A hole jumps rapidly from one lattice site to another in the direction of the $\mathbf{E}$ field, thus constituting a positive quasi-charge-carrier.

6.2 Metallic conductors

Consider a macroscopic piece of a conducting material, say, a metal, and suppose that, somehow, we are able to place some excess charge inside the material. What will happen to it? It comes into a charge-balanced situation, so we conclude that it cannot be bound to any particular lattice sites; therefore if the excess charge is electronic, these extra electrons will be in the 'conduction band' of nearly free electrons. If the charges were positive or negative ions and therefore too big to roam through the metal, these would either attract or repel conduction electrons in their vicinity, setting off a cascade of charge adjustments throughout the entire material. This cascade would continue until there is some kind of equilibrium, and an adequate picture here is that the equilibrium is attained only when all excess charge (but not necessarily the original charge *carriers*) is spread over the surface of the material, so that the electrostatic energy of the excess of charge is minimized. We shall see that the time constant for this to occur in a metallic conductor is extremely short, perhaps about 10^{-17}–10^{-19} seconds. Thus, good conductors are characterized by the ability of charges to move freely and very rapidly (i.e. as conduction currents) even at very low field strengths.

This implies that in the equilibrium state – which occurs instantaneously from a macroscopic time point of view – there is *no net internal charge* inside the conductor,

$$\varrho_V = 0$$

There may well be a *surface-charge density* ϱ_s (and there surely is, if excess charge was introduced somewhere in or on the conductor, or if outside electric fields 'pull' electrons to a surface).

There are also no electromagnetic forces in the equilibrium state, otherwise the surface charges would still be jostling about. In other words there is *no macroscopic electrostatic field* in the interior of a good conductor:

$$\mathbf{E} = 0$$

The next point is also easily understood from the discussion about electrostatic potential in Section 4.1. Because the interior electrostatic field $\mathbf{E} = 0$, it follows that $-\int dl \cdot \mathbf{E} = 0$ for any path inside or on the surface of a conductor and thus from (4.2) that all potential differences between different points are zero! Consequently, *the entire conductor is at one potential value* and in particular it is true that

The surface of an isolated conductor is an equipotential surface. (6.1)

However, if the surface of a conductor is an equipotential, then the *gradient of the potential field* must be $\perp$ to the surface. But that gradient is $-\mathbf{E}$, so that there is no tangential component of the $\mathbf{E}$ field at the surface. That implies that $\mathbf{E}$ is in the normal direction $\hat{\mathbf{n}}$, as a consequence of which $\hat{\mathbf{n}} \times \mathbf{E} = 0$. Thus we have found for all surface locations $\mathbf{r}_s$ that

$$\hat{\mathbf{n}}(\mathbf{r}_s) \times \mathbf{E}(\mathbf{r}_s) = 0 \tag{6.2a}$$

We now have established that all field lines are perpendicular to a conductor's surface, as shown in Fig. 6.1.

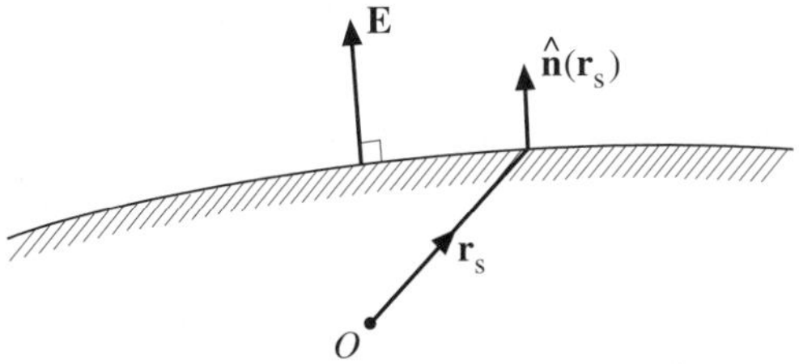

Fig. 6.1 The boundary condition for **E** at a vacuum–conductor interface.

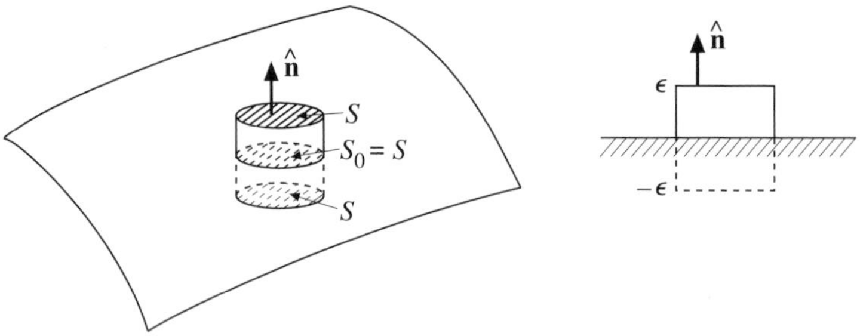

Fig. 6.2 A Gaussian pillbox at a vacuum–conductor interface.

The related and final important property at the surface of a conductor is that

$$\varrho_s(\mathbf{r}_s) = \varepsilon_0 \mathbf{E}(\mathbf{r}_s) = \mathbf{D}(\mathbf{r}_s) \tag{6.2b}$$

i.e. the **D** field at the surface is exactly equal to the surface-charge density. The proof is fairly easily given using a 'pillbox' argument. Suppose that the surface of a conductor is intersected by a cylindrical pillbox of infinitesimal semi-height ϵ, as shown in Fig. 6.2. The middle surface S_0 of the pillbox is at the conductor's surface. Of course, the two end surfaces and S_0 all have the same shape and area S. Now apply Gauss's law to all the (infinitesimal) outside surfaces of the pillbox. Within the conductor $\mathbf{E} = 0$; the field at the top surface of the pillbox equals the surface field **E** because ϵ is very small, and the 'wraparound' surface of the pillbox is perpendicular to **E**. Hence

$$\oint d\mathbf{S} \cdot \varepsilon_0 \mathbf{E} \equiv SE \qquad \text{and} \qquad q_{\text{encl}} = \oint dS\, \varrho_s = S\varrho_s \tag{6.3}$$

and so by Gauss's law $SE = \varepsilon_0 S \varrho_s$, which reduces to (6.2b).

6.3 Induced charge separation in conductors

What happens when we place a neutral piece of metallic conductor in a (uniform) electrostatic field? Before an equilibrium situation occurs, the *external* **E** field has penetrated the material and it works first on the conduction electrons, which will respond even to a very small field. These electrons will drift against the field lines (see Fig. 6.3) to the far surface. The electrons may not each themselves move all the way; they may cause a cascade of electron motion that leads to the same result.

Because charge neutrality must be maintained, and because electrons will move as much as possible against the field, it follows that an equal but opposite *positive* charge will occur on the right-hand surface (see Fig. 6.3), from which the electrons were repelled by the field.

Why do the electrons on the left-hand side not continue further? The reason is that the surface layer would become positive if electrons left the metal and

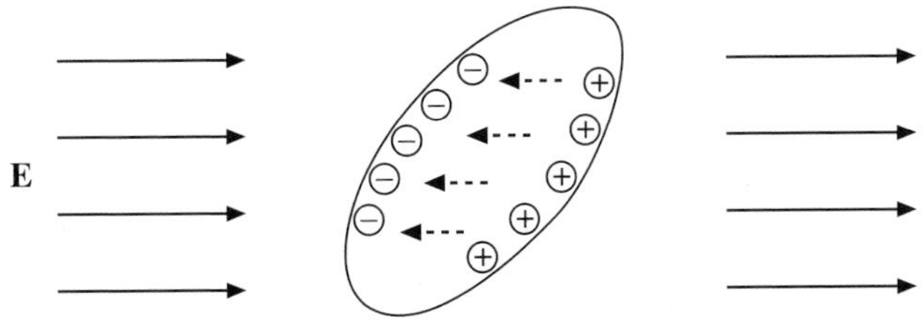

Fig. 6.3 Separation of charges in a conductor due to an outside **E** field.

so would rapidly produce a field pulling the electrons back. Such fields are usually much stronger than the exterior field. The *potential barrier* that thus arises would require a very high external field to force electrons to leak out of the conductor.

The negative and positive *surface charges* (see again Fig. 6.3) are responsible for an *internal* **E** field that has a direction opposite to the external field. This field grows in strength until it is equal and opposite to the external field, as a new equilibrium situation arises in which all charge carriers in the metal again experience no electrostatic force. The net result is a *separation of charge by induction.* Thus an electric field will induce a separation of charges on the conductor. By connecting either the 'near' or 'far' side by a wire to the ground, we connect this side to a huge reservoir of charge carriers (infinite for all practical purposes). Much or all of one of the two collections of separated charge will thus be neutralized, e.g. if in Fig. 6.3 the right-hand side is 'grounded' then electrons from the ground (which is a good conductor) will flow to that side of the conductor to neutralize it. The net positive charge left on the ground will dissipate by mutual repulsion. If the wire is disconnected, and then the exterior field is removed, the negative charge on the left-hand side of the conductor will redistribute itself over the conductor's surface. In some sense, the earth has become loaded with the extra positive charge, but this is such a minute fraction – with respect both to its total charge and also to the constant changes in charge due to the earth's atmosphere – that we can consider the earth to have remained neutral.

Example 6.1 A spherical shell of conductor surrounding a point charge

A point charge q at the origin (see Fig. 6.4) is surrounded by a concentric spherical shell of conductor with inner radius r_i and outer radius r_o. The point charge produces a field $\mathbf{E}(\mathbf{r}) = (q/4\pi\varepsilon_0 r^2)\hat{\mathbf{r}}$ in the absence of the shell. What do we find when the shell is there?

From the preceding, we may suspect that separation of charges is going to take place in the shell. Let us see how it goes. First apply Gauss's law to an imaginary spherical surface (a *Gaussian surface*) around the origin with radius $r < r_i$.

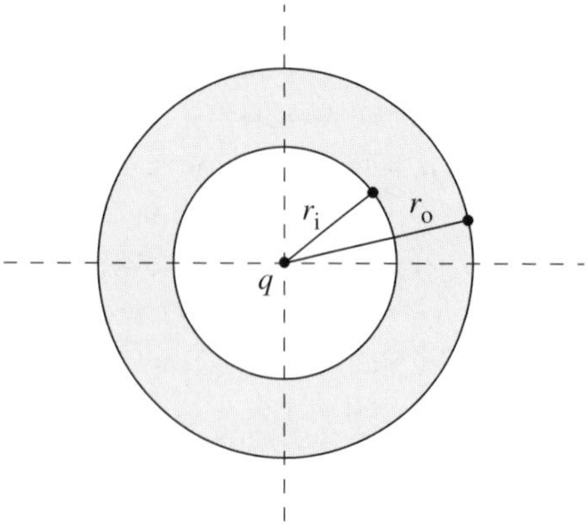

Fig. 6.4 Geometry for Example 6.1.

The enclosed charge can only be q, the point charge at the center. The symmetry of the situation tells us that $\mathbf{E}(r)$ has the same value everywhere on that surface, and points outward (if $q > 0$). Hence $\oint d\mathbf{S} \cdot \mathbf{E}$ must equal $4\pi r^2 E_r(r)$. This, in turn, is equal to q/ε_0, as (3.15b) confirms. Hence the field in the space enclosed within the shell is the same as it would have been in the absence of the shell.

Now consider a spherical Gaussian surface with r such that $r_i < r < r_o$, i.e. the surface lies *inside* the shell. We know there will be zero field in the equilibrium situation (which occurs instantaneously for all practical purposes). Thus $\oint d\mathbf{S} \cdot \mathbf{E} \equiv 0$, which tells us that $q_{\text{encl}} = 0$ by virtue of Gauss's law. However, q_{encl} certainly includes the point charge q, which is nonzero. So there must be a negative charge $-q$ somewhere, and the only possible place is on the near surface $r = r_i$ of the shell. It, of course, is the induced charge. As the shell is still neutral, the far surface at $r = r_o$ must carry the charge $+q$.

When the spherical Gaussian surface is placed outside the shell ($r > r_o$), then $q_{\text{encl}} = q - q + q = q$, and $\oint d\mathbf{S} \cdot \mathbf{E}$ must be equal to $4\pi r^2 E_r(r)$ again. The $\mathbf{E}$ field is again equal to what it would have been in the absence of the conducting shell. Summarizing,

$$\mathbf{E}(\mathbf{r}) = \frac{q}{4\pi\varepsilon_0 r^2}\hat{\mathbf{r}} \qquad \text{for } r < r_i \quad \text{and} \quad r > r_o \tag{6.4a}$$

$$\mathbf{E}(\mathbf{r}) = 0 \qquad \text{for } r_i < r < r_o \tag{6.4b}$$

The charge densities on the two spherical surfaces are found easily, since by symmetry the charge must be uniform on each:

$$\varrho_s(r_i) = -\frac{q}{4\pi r_i^2} \qquad \varrho_s(r_o) = +\frac{q}{4\pi r_o^2} \tag{6.5}$$

The potential V in each of the regions can be obtained using $\mathbf{E} = -\nabla V$. We then invoke the useful property (4.5) of the gradient of $1/r$, namely $\nabla(1/r) = -\hat{\mathbf{r}}/r^2$.

Starting with the outside region $r > r_o$, it follows that $V(r) = q/4\pi\varepsilon_0 r + V_1$ (V_1 is an integration constant). However, it was clear, in discussing potential values due to point charges (Example 4.1), that we can choose $V(\infty) = 0$, which results here in $V_1 = 0$. Thus, $V(r) = q/4\pi\varepsilon_0 r$ outside the shell. This must hold even at $r = r_o$, as the potential must be continuous everywhere. But the entire shell, being a metallic conductor, is an equipotential, hence $V(r) = q/4\pi\varepsilon_0 r_o$ on and in the shell. Inside the shell, where $r < r_i$, we again have $V(r) = q/4\pi\varepsilon_0 r + V_2$, with a second integration constant V_2. The potential must be continuous at $r = r_i$, hence $q/4\pi\varepsilon_0 r_i + V_2 = q/4\pi\varepsilon_0 r_o$, which allows us to conclude that $V_2 = (q/4\pi\varepsilon_0)(1/r_o - 1/r_i)$. Summarizing,

$$
\begin{aligned}
V(r) &= \frac{q}{4\pi\varepsilon_0}\left(\frac{1}{r} + \frac{1}{r_o} - \frac{1}{r_i}\right) && \text{for } r < r_i \\
V(r) &= \frac{q}{4\pi\varepsilon_0 r_o} && \text{for } r_i < r < r_o \\
V(r) &= \frac{q}{4\pi\varepsilon_0 r} && \text{for } r > r_o
\end{aligned}
\tag{6.6}
$$

Of course, we could have started at the inside instead, but we would have needed to carry one or more constants until we arrived at $r \to \infty$, at which point we would have determined the constants from the requirement that $V(\infty) = 0$.

Example 6.2 A circular cylindrical shell of conductor surrounding a line charge

This is done in an entirely analogous manner to the previous example. We will call the radii of the circular cylinder ρ_i and ρ_o, and we suppose that the z axis has uniform line-charge density ϱ_ℓ. The Gaussian surfaces are now cylinders around the z axis, and we have little trouble finding

$$
\begin{aligned}
\mathbf{E}(\mathbf{r}) &= \frac{\varrho_\ell}{2\pi\varepsilon_0 \rho}\hat{\boldsymbol{\rho}} && \text{for } \rho < \rho_i \quad \text{and} \quad \rho > \rho_o \\
\mathbf{E}(\mathbf{r}) &= 0 && \text{for } \rho_i < \rho < \rho_o
\end{aligned}
\tag{6.7}
$$

The major problem is in obtaining the potential because $1/\rho$ is the negative gradient of $-\ln \rho$, which becomes infinite in absolute value both at $\rho = 0$ and as $\rho \to \infty$. The potential at infinity is not well defined, as we have seen! Let us define the potential on the shell as V_{shell}. We may not know what it is, but this gives

$$
\begin{aligned}
V(\rho) &= V_{\text{shell}} - \frac{\varrho_\ell}{2\pi\varepsilon_0}\ln\left(\frac{\rho}{\rho_o}\right) && \text{for } \rho < \rho_i \\
V(\rho) &= V_{\text{shell}} && \text{for } \rho_i < \rho < \rho_o \\
V(\rho) &= V_{\text{shell}} - \frac{\varrho_\ell}{2\pi\varepsilon_0}\ln\left(\frac{\rho}{\rho_i}\right) && \text{for } \rho > \rho_o
\end{aligned}
\tag{6.8}
$$

These expressions were obtained in a simple fashion. The term $-(\varrho_\ell/2\pi\varepsilon_0)\ln\rho$ obviously gives the correct **E** fields (6.7) when its negative gradient is taken, and no further dependence upon ρ can then occur in $V(\rho)$. The remaining part of each expression simply results from the continuity of $V(\rho)$ at the boundaries $\rho = \rho_i$ and $\rho = \rho_o$. If V_{shell} is unknown but we do know $V(\rho)$ at some ρ, then (6.8) still can be used because we can determine V_{shell} from (6.8) when it is applied to the radial distance at which $V(\rho)$ is known.

6.4 Dielectric materials

6.4.1 Polarization

We have indicated briefly above that *dielectric materials* are characterized by the limited motion of charges of one polarity with respect to the other (in contrast to *conductors* in which conduction-band electrons are almost entirely free to respond to electrostatic forces from outside). However, in dielectrics with *polar molecules*, the charges are essentially stationary with respect to each other under the influence of weak fields. Such molecules, however, have a permanent dipole moment because asymmetry in shape places the geometrical center of the negative ions in a position that differs from that of the positive ions.

The water molecule, for example, consists of two positive hydrogen ions attached to a negative oxygen ion asymmetrically, as shown in Fig. 6.5. There is a permanent dipole as a result with dipole moment $p = qd \sim 10^{-30}$ C m, and these dipoles tend to *align* along an exterior **E** field. Perfect alignment will not occur as there are thermal and other forces that work against orientation along the field. However, there remains the tendency for dipoles to orient themselves along the field, and more strongly so as it increases.

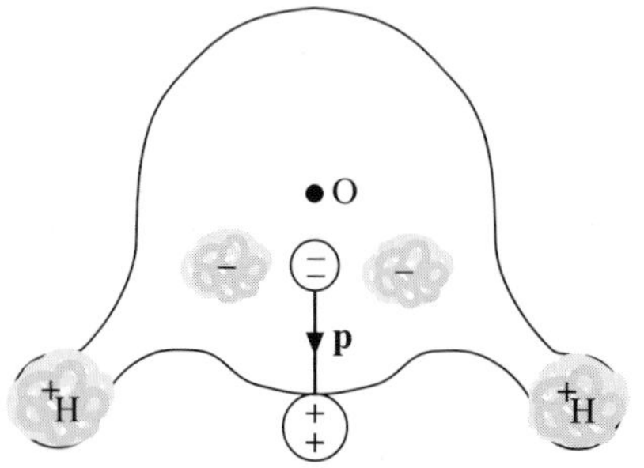

Fig. 6.5 A water molecule showing the centers of the charge distributions. The equivalent dipole **p** is shown, drawn from $-$ to $+$ according to convention.

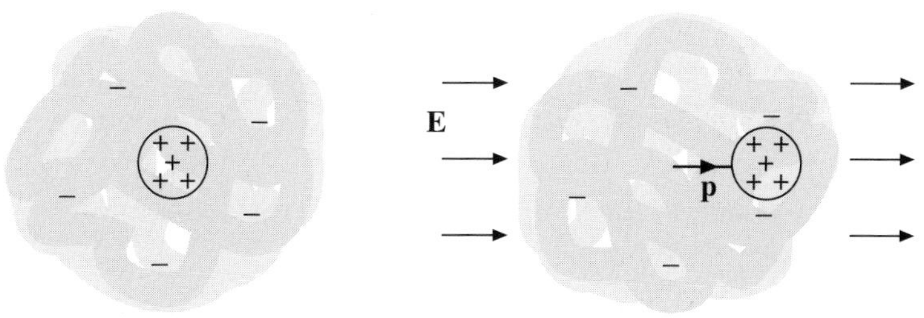

Fig. 6.6 Shifting of the charge center in a nonpolar molecule.

That the dipole moment in the water molecule will be of the order of 10^{-30} C m can be seen with a simple argument. The charge of an electron is $\sim 1.6 \times 10^{-19}$ C. The dipole charge will be perhaps twice or thrice as large. The radius of a hydrogen atom can be estimated from the Bohr radius, which is $\sim 5 \times 10^{-11}$ m. The oxygen atom will be somewhat larger but, as Fig. 6.5 shows, the distance between the negative-charge and positive-charge centers will be some fraction of the atom's radius. If that distance d is $\sim 10^{-11}$ m, then it becomes clear how the dipole moment value of $p = qd \sim 10^{-30}$ C m is arrived at.

Nonpolar molecules obtain dipole moments for a different reason. In the absence of a field, they have no dipole moment because the geometrical center of the negative ions, or of the electron cloud around a positive nucleus, coincides with that of the positive charges. When an **E** field is turned on, the electron cloud is shifted slightly with respect to the positive charges, as indicated in Fig. 6.6. A dipole moment is thereby created, which was not there in the absence of the field, and its vector is by definition in the same direction as the field **E**.

Thus, whether polar or nonpolar, molecule m has a dipole moment, say $\mathbf{p}_m$, in the presence of an imposed electrostatic field. We shall see shortly how that dipole moment depends upon the field strength E of the electrostatic field working upon it. The key point to be established is that such dipole moments cause new *polarization-charge densities* to appear within the volume and/or at the surface.

We now apply (4.15b) to calculate the potential $V_{\mathrm{d}}(\mathbf{r})$ at a point $\mathbf{r}$ in a dielectric containing electric dipoles. From (4.15b) the contribution $dV(\mathbf{r})$ at point $\mathbf{r}$ to the electrostatic potential from all the atomic or molecular dipoles in a small volume element dv' around point $\mathbf{r}$ is given by

$$dV(\mathbf{r}) = \sum_m \frac{\mathbf{p}_m \cdot (\mathbf{r} - \mathbf{r}_m)}{4\pi\varepsilon_0 |\mathbf{r} - \mathbf{r}_m|^3} \tag{6.9}$$

It is understood that only the dipole moments $\mathbf{p}_m$ at points $\mathbf{r}_m$ inside dv' are to be included in (6.9); see Fig. 6.7. If the volume element is infinitesimal compared to the distance $|\mathbf{r} - \mathbf{r}'|$, then all $|\mathbf{r} - \mathbf{r}_m|$ will be negligibly different from $|\mathbf{r} - \mathbf{r}'|$.

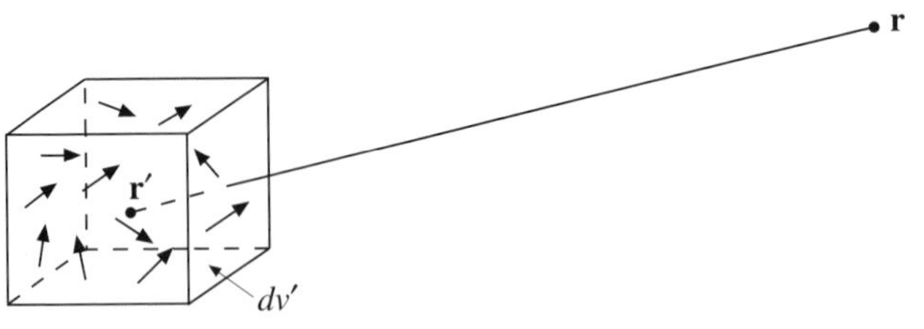

Fig. 6.7 Geometry for potential contributions at $\mathbf{r}$ from a collection of electrostatic dipoles in a volume element dv' at $\mathbf{r}'$.

In fact, we can be somewhat more precise after setting $\mathbf{r}-\mathbf{r}' \equiv \mathbf{R}$ and $\mathbf{r}_m - \mathbf{r}' \equiv \delta\mathbf{R}_m$, because then $|\mathbf{r}-\mathbf{r}_m| = |\mathbf{R}-\delta\mathbf{R}_m|$, so that

$$|\mathbf{r}-\mathbf{r}_m| = (R^2 - 2\mathbf{R}\cdot\delta\mathbf{R}_m + \delta\mathbf{R}_m^2)^{1/2} = R[1 - \hat{\mathbf{R}}\cdot(\delta\mathbf{R}_m/R) + \tfrac{1}{2}(\delta R_m/R)^2] \tag{6.10}$$

with $|\delta\mathbf{R}_m| \ll R$. Use of the binomial expansion (3.6) tells us that we may set $|\mathbf{r}-\mathbf{r}_m| \approx R \equiv |\mathbf{r}-\mathbf{r}'|$ with error terms $O(\delta R_m/R)$, both in the numerator and the denominator of (6.9).

As a consequence, (6.9) can be simplified to

$$dV(\mathbf{r}) = \left(\sum_m \mathbf{p}_m\right)\cdot\frac{(\mathbf{r}-\mathbf{r}')}{4\pi\varepsilon_0|\mathbf{r}-\mathbf{r}'|^3}. \tag{6.11}$$

Actually, the volume element dv' at $\mathbf{r}'$ is infinitesimal only in the sense that it is *macro*scopically small; it must be *micro*scopically large. That is, on the one hand it should be so small that all distances of dipoles $\mathbf{p}_m$ to the center $\mathbf{r}'$ of the volume element are negligible compared with $|\mathbf{r}-\mathbf{r}'|$ for all volume elements in the dielectric, except possibly those in a very small volume around the observation point $\mathbf{r}$. By 'very small' is meant that the contribution of those volume elements $dV(\mathbf{r})$ close to $\mathbf{r}$ is negligible. On the other hand, dv' should be sufficiently large that it contains very many dipoles. These conditions can be met in solid and liquid dielectrics in general, because there are of the order of 6×10^{23} molecules in a gram molecule of material (i.e. an amount of material equal in grams to the molecular weight of the substance). Hence if, say, the macroscopic properties of a dielectric material in the laboratory typically do not vary appreciably over distances of the order of a millimeter, then a cubic millimeter of that dielectric can be expected to contain 10^{17}–10^{20} dipoles and still be macroscopically small. The fact that it is microscopically large means that the sum $\sum_m \mathbf{p}_m$ changes very slowly and essentially smoothly as we move from one dv' to the next and that it is proportional to the volume dv'. Therefore

$$\sum_m \mathbf{p}_m \equiv \mathbf{P}(\mathbf{r}')\, dv' \tag{6.12}$$

where $\mathbf{P}(\mathbf{r}')$ is the dipole moment per unit volume at $\mathbf{r}'$, also known as the *polarization vector*. It behaves as a smooth macroscopic quantity because the number of individual dipoles in a macroscopically small volume is so vast that the addition or subtraction of a small number of dipoles (as $\mathbf{r}'$ is varied) appears to give rise to a continuous change in value of $\mathbf{P}(\mathbf{r}')$. Hence, (6.11) becomes

$$dV(\mathbf{r}) = dv' \frac{\mathbf{P}(\mathbf{r}') \cdot (\mathbf{r} - \mathbf{r}')}{4\pi\varepsilon_0 |\mathbf{r} - \mathbf{r}'|^3} \tag{6.13}$$

Remembering that this is incorrect only for the immediate environment of $\mathbf{r}$ (but in fact – as we have shown – the immediate environment of $\mathbf{r}$ does not contribute appreciably to $dV(\mathbf{r})$), we can integrate over the entire volume v of the dielectric material to obtain

$$V_\mathrm{d}(\mathbf{r}) = \int_v dV(\mathbf{r}) = \frac{1}{4\pi\varepsilon_0} \int_v dv' \frac{\mathbf{P}(\mathbf{r}') \cdot (\mathbf{r} - \mathbf{r}')}{|\mathbf{r} - \mathbf{r}'|^3} \tag{6.14a}$$

The expression $V_\mathrm{d}(\mathbf{r})$ is the contribution to the potential (and, via its gradient, to the electrostatic field) due to the polarization of the dielectric material. It has to be added to the potential $V(\mathbf{r})$ produced by the imposed electrostatic field. It is identically zero in free space, or in any material for which $\mathbf{P}$ is zero everywhere. If we use (4.4), we can rewrite (6.14a) as

$$V_\mathrm{d}(\mathbf{r}) = \frac{1}{4\pi\varepsilon_0} \int_v dv'\, \mathbf{P}(\mathbf{r}') \cdot \nabla' \frac{1}{|\mathbf{r} - \mathbf{r}'|}. \tag{6.14b}$$

Note that here the derivatives are with respect to x', y', z', i.e. with respect to the primed coordinates.

We will transform (6.14b) in such a way that the outcome resembles (4.8a), but with the charge density replaced by quantities derived from the polarization vector $\mathbf{P}(\mathbf{r}') \equiv \mathbf{P}$. To do so, we use the vector identity 2 of Appendix A.3:

$$\mathbf{A} \cdot \nabla B = \nabla \cdot (\mathbf{A}B) - B(\nabla \cdot \mathbf{A})$$

When applied to the integrand in (6.14b) this gives

$$\mathbf{P}' \cdot \nabla' \frac{1}{|\mathbf{r} - \mathbf{r}'|} = \nabla' \cdot \left(\frac{\mathbf{P}'}{|\mathbf{r} - \mathbf{r}'|} \right) + \frac{1}{|\mathbf{r} - \mathbf{r}'|} (\nabla' \cdot \mathbf{P}') \tag{6.15}$$

and so (6.14b) becomes

$$V_d(\mathbf{r}) = \frac{1}{4\pi\varepsilon_0}\int_v dv'\nabla' \cdot \left(\frac{\mathbf{P}'}{|\mathbf{r}-\mathbf{r}'|}\right) + \frac{1}{4\pi\varepsilon_0}\int_v dv' \frac{(-\nabla' \cdot \mathbf{P}')}{|\mathbf{r}-\mathbf{r}'|} \tag{6.16a}$$

We now apply Gauss's divergence theorem to the first integral, obtaining a closed-surface integral over the interface of the volume and free space:

$$\int dv'\nabla' \cdot \left(\frac{\mathbf{P}'}{\mathbf{r}-\mathbf{r}'}\right) = \oint d\mathbf{S} \cdot \frac{\mathbf{P}'}{\mathbf{r}-\mathbf{r}'}$$

Hence,

$$V_{\mathrm{d}}(\mathbf{r}) = \frac{1}{4\pi\varepsilon_0}\oint_S dS' \frac{(\hat{\mathbf{n}} \cdot \mathbf{P}')}{|\mathbf{r}-\mathbf{r}'|} + \frac{1}{4\pi\varepsilon_0}\int_V dv' \frac{(-\nabla' \cdot \mathbf{P}')}{|\mathbf{r}-\mathbf{r}'|} \tag{6.16b}$$

which, comparing with (4.8a), is formally equivalent to

$$V_{\mathrm{d}}(\mathbf{r}) = \frac{1}{4\pi\varepsilon_0}\oint_S dS' \frac{\varrho_{\mathrm{sp}}(\mathbf{r}')}{|\mathbf{r}-\mathbf{r}'|} + \frac{1}{4\pi\varepsilon_0}\int_V dv' \frac{\varrho_{\mathrm{vp}}(\mathbf{r}')}{|\mathbf{r}-\mathbf{r}'|} \tag{6.16c}$$

in which expressions we identify two polarization-charge densities:

$$\begin{aligned} &\textit{a surface-charge density } \varrho_{\mathrm{sp}}(\mathbf{r}') \equiv \hat{\mathbf{n}}' \cdot \mathbf{P}(\mathbf{r})' \quad \text{in C/m}^2 \\ &\textit{a volume-charge density } \varrho_{\mathrm{vp}}(\mathbf{r}') \equiv -\nabla' \cdot \mathbf{P}(\mathbf{r}') \quad \text{in C/m}^3 \end{aligned} \tag{6.16d}$$

It should be noted that $\hat{\mathbf{n}}'$ points out of the dielectric. It is easy to see that the separation of + and − charges in arrays of aligned dipoles will give rise to a surface-charge density; there is no cancellation of the opposite charge at the surface. It is probably somewhat harder to understand how there will be a local volume-charge density ϱ_{vp}, but if the density of aligned dipoles were to decrease, say, in a particular direction, then it becomes clear that there is not a complete overlap of interior + and − charges, and the existence of a volume-charge density $-\nabla \cdot \mathbf{P} \neq 0$ is consistent with that fact.

Remember now that we expressed some doubt in the discussion after (6.13) about the accuracy of (6.13)–(6.16c) for points $\mathbf{r}'$ close to $\mathbf{r}$. For such points $\mathbf{r}'$, the values of the two apparent charge densities $\varrho_{\mathrm{sp}}(\mathbf{r}')$ and $\varrho_{\mathrm{vp}}(\mathbf{r}')$ will be so close to the value at $\mathbf{r}$ that the two integrands in (6.16b) are functions only of $1/|\mathbf{r}-\mathbf{r}'|$ for, say, a macroscopically small but microscopically large sphere around $\mathbf{r}$. It can easily be seen that these two integrals around such a sphere are zero. The following optional text clarifies this issue mathematically, starting from (6.14a).

In (16.14a), let $\mathbf{R} \equiv \mathbf{r} - \mathbf{r}'$. Then, for a small sphere of radius ϵ around $\mathbf{r}$, we may set $P(\mathbf{r}') \approx P(\mathbf{r})$. The integral over the small sphere becomes $P(\mathbf{r}) \cdot \int_0^{2\pi} d\varphi \times \int_0^{\pi} d\theta \sin\theta \int_0^{\epsilon} dR\, R^2 \hat{\mathbf{R}}/R^2$, and we see that the $1/R^2$ factor in the integrand is canceled by the R^2 preceding it. The remaining volume has all $R > \epsilon$ so that there is no singularity problem.

The above development, starting from (6.9) and ending at (6.16), shows us that the presence of dipoles – polarization – in a dielectric material in general gives rise to additional charge densities ϱ_{vp} within the volume and ϱ_{sp} on the interface with the surrounding free space. Thus, the Maxwell equation (5.23b), which we read as $\nabla \cdot \mathbf{E} = \varrho_{\mathrm{v}}/\varepsilon_0$, now becomes, within the dielectric,

$$\nabla \cdot \mathbf{E} = (\varrho_{\mathrm{v}} + \varrho_{\mathrm{vp}})/\varepsilon_0 = (\varrho_{\mathrm{v}} - \nabla \cdot \mathbf{P})/\varepsilon_0 \tag{6.17a}$$

which can be rearranged to read as

$$\nabla \cdot (\varepsilon_0 \mathbf{E} + \mathbf{P}) = \varrho_{\mathrm{v}} \tag{6.17b}$$

where $\mathbf{E}$ is still the actual force on a unit test charge. This force is *reduced* in a dielectric owing to the polarization field which opposes the applied field by the 'negative feedback' effect illustrated in Fig. 6.6; note that the polarization field is in the *opposite direction* to the dipole moment $\mathbf{p}$. In other words, the number of $\mathbf{E}$ lines is less in a dielectric. But if we redefine the *electric flux density vector* as

$$\mathbf{D} \equiv \varepsilon_0 \mathbf{E} + \mathbf{P}$$

we can maintain Eq. (5.23b) as

$$\nabla \cdot \mathbf{D} = \varrho_{\mathrm{v}}$$

Here we emphasize that ϱ_{v} refers only to *free* ('real') charges, not to bound (polarization) charges. This equation thus tells us that $\mathbf{D}$ lines begin and end on free charges: the $\mathbf{D}$ lines are unaffected by the presence of a dielectric. The definition of $\mathbf{D}$ reduces to the free space one when the polarization vector $\mathbf{P}$ is zero. Its dimension must be the same as that of $\mathbf{P}$, which is $\mathrm{C/m^2}$, as is most easily seen from (6.16c). As a check, the dimension of ε_0 must then be

$$\frac{(\mathrm{C/m^2})}{(\mathrm{V/m})} = \frac{(\mathrm{C/V})}{\mathrm{m}} = \frac{\mathrm{F}}{\mathrm{m}}$$

(because capacitance = charge/voltage, and the unit of capacitance is the farad, F).

We shall see shortly that the polarization surface-charge density $\varrho_{\mathrm{sp}} = \hat{\mathbf{n}} \cdot \mathbf{P}$ gives rise to an extension of the boundary condition, (6.2b), that applies at the interface between free space (vacuum) and a perfect conductor; this extension occurs when we replace the vacuum by a dielectric. Let us first continue with what we have found so far, namely that we now have extended the vector **D** to include the vector **P** along with $\varepsilon_0\mathbf{E}$, in a dielectric.

The next question is, exactly how is **P** related to the imposed electrostatic field? This is a trickier question than it may seem to be, because, first, we need to know how *one* dipole moment $\mathbf{p}_m$ is related to the electric field **E**. The exterior field $\mathbf{E}_0$ is imposed from outside the medium. Together with the effect of all the polarized bound charges (the dipoles), it forms an *effective field* **E** in the dielectric, as mentioned above. This is the field that would be experienced by a test charge introduced into the dielectric. But the *local field* $\mathbf{E}_{\text{loc}}$ working on the dipole excludes the dipole $\mathbf{p}_m$ itself, and its calculation is a classical problem treated in more advanced textbooks. It is found that $\mathbf{E}_{\text{loc}} \propto \mathbf{P}$ (which is assumed to be constant in a macroscopically small but microscopically large volume).

Finally, it seems plausible that the effective field **E** must be proportional to the local field $\mathbf{E}_{\text{loc}}$, and calculations based on matters that we have not yet discussed bear that out. Thus, $\mathbf{p}_m \propto \mathbf{E}_{\text{loc}} \propto \mathbf{E}$. (We discuss this topic again, and in a little more detail, below Eq. (6.30).) So, in an isotropic dielectric for fields that are not strong, we may assume that **P** *is proportional to* **E**, i.e.

$$\mathbf{P} = \chi_e \varepsilon_0 \mathbf{E} \tag{6.18}$$

This assumption might also be considered independent of the discussion about local fields, as it seems an obvious one to make for weak fields: an increase in field strength turns more polar dipoles into the field or induces greater charge separation in each nonpolar molecule. Note that the constant of proportionality, the *dielectric susceptibility* χ_e, is dimensionless. In the weak field case it is usually a constant. If the field were to be increased in strength, then a point would be reached at which one or more electrons would be pulled off the molecular or atomic structure to which they are attached; the energy imposed by the field has exceeded the binding energy of (some or all of) the electrons. At this stage, the process becomes *nonlinear*; the susceptibility χ_e itself becomes a function of **E**. At still higher fields, negative ions can even be separated from positive ions, and a *plasma* medium consisting of separated positive and negative ions ensues. The properties of plasmas are fundamentally different from dielectrics, and plasma is considered to be a fourth state of matter (besides solid, liquid, and gas). If the field is sufficiently strong, then even a *flow of charge* (*a current*) can come into being, and this is called *dielectric breakdown*. In air at normal temperature $\sim 20\,^\circ\text{C}$ and 1 atmosphere pressure, *the breakdown field strength* causing a current (lightning bolts, sparks) is $\sim 3 \times 10^6$ V/m. Lightning bolts typically traverse a distance of several hundred meters, so that the *breakdown voltage* from clouds to ground can exceed 10^9 V. While that may seem very high, one must remember that the normal electric field at the Earth's surface is ~ 100 V/m, so that the normal cloud-to-ground voltage difference is around 5×10^4 V.

Richard Feynman gives a very readable account at an elementary level of lightning and the ambient voltage differences in the atmosphere ($\sim$ 400 000 V from the top of the atmosphere to the ground) in his *Lectures on Physics*.

If (6.18) is inserted into the definition (6.17c) of **D** we find

$$\mathbf{D} = \varepsilon_0(1 + \chi_e)\mathbf{E} \tag{6.19a}$$

Comparing this with our earlier free-space definition, $\mathbf{D} = \varepsilon_0\mathbf{E}$, we see that (6.19) represents an extension, namely

$$\mathbf{D} = \varepsilon\mathbf{E} \qquad \text{with} \qquad \varepsilon \equiv \varepsilon_0(1 + \chi_e)$$

The dielectric permittivity has changed from its vacuum value ε_0, to a new value ε, that differs from ε_0 by the factor $1 + \chi_e \equiv \varepsilon_r$, the *relative dielectric permittivity* of the dielectric medium. Thus, for practical purposes, the behavior of a dielectric medium under the influence of weak electrostatic fields is characterized by a relative permittivity factor ε_r that differs from unity (its value in a vacuum). Many handbooks will provide a table of values of ε_r for different dielectric materials. Water at 0 °C, for example, has $\varepsilon_r \approx 81$, whereas ice has $\varepsilon_r \approx 3$. Most common materials have $1 < \varepsilon_r < 100$.

A relative dielectric permittivity also can be defined for alternating-current (ac) fields, and many common materials exhibit a marked dependence of ε_r upon the frequency of the imposed field. This is known as *dispersion*, and it has important consequences for communications by means of electromagnetic signals because it distorts signals of given shape. That is, if a pulse-like signal has a certain shape in the time domain at $z = 0$, then it will have another shape at a distance $z = L$ inside a *dispersive* dielectric. However, we will discuss that phenomenon later, in the context of waves propagating through dielectric media.

In summary, we have found that in a dielectric medium, where the effective field is **E**,

$$\begin{aligned} &\mathbf{D} = \varepsilon_0\mathbf{E} + \mathbf{P} \qquad \mathbf{P} = \chi_e\varepsilon_0\mathbf{E} \\ &\mathbf{D} = \varepsilon\mathbf{E} \qquad \varepsilon = \varepsilon_0\varepsilon_r \qquad \varepsilon_r = 1 + \chi_e \end{aligned} \tag{6.20a}$$

The first equation is quite general. The others hold for *linear isotropic* media under the influence of weak fields. If the field is no longer weak, then the medium becomes nonlinear because ε_r changes as **E** is increased (whereas it is constant for weaker fields). We label the medium as isotropic because we assume that the relationship between **D** and **E** in the second and third equations of (6.20a) is the same at any fixed location in the dielectric, no matter in what *direction* the **E** field points. There are *anisotropic* media, in which ε_r is a function of direction; such media, to the class of which certain crystalline materials belong, have different dielectric properties in different directions. We hope the reader will understand the difference between *uniform* media and *isotropic* media: in *uniform* media, the value of ε_r is the same from one location to the next; in *isotropic* media, ε_r may well differ from point to point but at each location the ratio of **D** to **E** is independent of the direction of **E**.

Thus, Gauss's law in differential and in integral form has become

$$\nabla \cdot \mathbf{D} = \varrho_v \quad \text{and} \quad \oint d\mathbf{S} \cdot \mathbf{D} = q_{encl} \tag{6.20b}$$

Equations (6.20a) and (6.20b) summarize what is needed to characterize electrostatic fields in dielectric media. Later, we will discuss what happens at a sharp interface with another medium. Let us first discuss Example 6.1 again, but with a dielectric rather than a conducting shell between the two concentric spherical surfaces.

Example 6.3 A spherical shell of dielectric material surrounding a point charge

A point charge q at the origin (see Fig. 6.4 again) is surrounded by a concentric spherical shell of dielectric with inner radius r_i and outer radius r_o. The point charge produces a field $\mathbf{E}(\mathbf{r}) = (q/4\pi\varepsilon_0 r^2)\hat{\mathbf{r}}$ in the absence of the shell, as before.

We apply the integral form of Gauss's law (6.20b) to a spherical Gaussian surface around the origin with radius $r < r_i$. The enclosed free charge can only be q, the point charge at the center. As mentioned earlier, bound charges are not included in the right-hand side of (6.20b); they are included in ε_r. The symmetry of the situation tells us that $\mathbf{D}(r)$ has the same value everywhere on that surface and points outward (if $q > 0$). Hence $\oint d\mathbf{S} \cdot \mathbf{D}$ must be equal to $4\pi r^2 D_r(r)$. This, in turn, is equal to q, as (3.15a, b) confirm. Hence the field in the space *inside* the shell is the same that it would have been in the absence of that shell; we have free space here, hence $D_r = q/4\pi r^2$ and $E_r = D_r/\varepsilon_0 = q/4\pi\varepsilon_0 r^2$.

Now consider a spherical Gaussian surface with r such that $r_i < r < r_o$, i.e. within the dielectric shell itself. There will be a nonzero field which is calculated in the same way as for $r < r_i$, but here we have permittivity $\varepsilon_0\varepsilon_r$, so that $D_r = q/4\pi r^2$ and $E_r = D_r/\varepsilon_0\varepsilon_r = q/4\pi\varepsilon^0\varepsilon_r r^2$. Unlike the perfect-conductor case, the **E** field is not zero in this region, but it is reduced beneath the free-space value by a factor $1/\varepsilon_r$.

When the spherical Gaussian surface is placed outside the dielectric shell ($r > r_o$), then the calculation is identical to that for $r < r_i$, except that the value of r is larger than r_o in this region. The field jumps back to its free-space value when r increases infinitesimally above r_o from infinitesimally below that radial distance. Summarizing,

$$\mathbf{E}(\mathbf{r}) = \frac{q}{4\pi\varepsilon_0 r^2}\hat{\mathbf{r}} \qquad \mathbf{D}(\mathbf{r}) = \frac{q}{4\pi r^2}\hat{\mathbf{r}} \qquad \text{for } r < r_i \text{ and } r > r_o \tag{6.21a}$$

$$\mathbf{E}(\mathbf{r}) = \frac{q}{4\pi\varepsilon_r r^2}\hat{\mathbf{r}} \qquad \mathbf{D}(\mathbf{r}) = \frac{q}{4\pi r^2}\hat{\mathbf{r}} \qquad \text{for } r_i < r < r_o \tag{6.21b}$$

The potentials in each of the regions can be obtained as in Example 6.1. Again, in this spherically symmetric case, we invoke the property (4.5) of the gradient of $1/r$, namely $\nabla(1/r) = -\hat{\mathbf{r}}/r^2$. Starting with the outside region $r > r_o$, it follows that $V(r) = q/4\pi\varepsilon_0 r + V_1$ (V_1 is an integration constant). As before, we choose $V(\infty) = 0$, which results here in $V_1 = 0$. Thus, $V(r) = q/4\pi\varepsilon_0 r$ outside the shell. This must hold even at $r = r_o$, as the potential must be continuous everywhere.

In the dielectric region, $r_i < r < r_o$, indefinite integration as above of the field in (6.21b) yields $V(r) = q/4\pi\varepsilon_0\varepsilon_r r + V_2$ on and in the shell. Another way of seeing this is to note that it makes $-\nabla V(\mathbf{r})$ correct. We obtain the value of the constant V_2 by applying this formula to the surface $r = r_o$, for which we have already found $V(r_o) = q/4\pi\varepsilon_0 r_o$. Thus we find

$$\begin{aligned} V(r_o) &= \frac{q}{4\pi\varepsilon_0 r_o} = \frac{q}{4\pi\varepsilon_0\varepsilon_r r_o} + V_2 \\ V_2 &= \frac{q}{4\pi r_o}\left(\frac{1}{\varepsilon_0} - \frac{1}{\varepsilon}\right) \end{aligned} \tag{6.22a}$$

As a result, we obtain

$$V(r) = \frac{q}{4\pi\varepsilon_0}\left[\frac{1}{\varepsilon_r r} + \frac{1}{r_o}\left(1 - \frac{1}{\varepsilon_r}\right)\right] \qquad \text{for } r_i < r < r_o \tag{6.22b}$$

In the free-space region inside the shell, where $r < r_i$, we again have $V(r) = q/4\pi\varepsilon_0 r + V_3$, with a third integration constant V_3. The potential must be continuous at $r = r_i$, hence $q/4\pi\varepsilon_0 r_i + V_3$ must be equal to the value $V(\mathbf{r}_i)$ obtained from (6.22b). This gives

$$V_3 = -\frac{q}{4\pi\varepsilon_0}\left(\frac{1}{r_i} - \frac{1}{r_o}\right)\left(1 - \frac{1}{\varepsilon_r}\right) \tag{6.23a}$$

which yields

$$V(r) = \frac{q}{4\pi\varepsilon_0}\left[\frac{1}{r} - \left(\frac{1}{r_i} - \frac{1}{r_o}\right)\left(1 - \frac{1}{\varepsilon_r}\right)\right] \qquad \text{for } r < r_i \tag{6.23b}$$

Summarizing the results for the potential, we have found

$$\begin{aligned} V(r) &= \frac{q}{4\pi\varepsilon_0}\left[\frac{1}{r} - \left(\frac{1}{r_i} - \frac{1}{r_o}\right)\left(1 - \frac{1}{\varepsilon_r}\right)\right] && \text{for } r < r_i \\ V(r) &= \frac{q}{4\pi\varepsilon_0}\left[\frac{1}{\varepsilon_r r} + \frac{1}{r_o}\left(1 - \frac{1}{\varepsilon_r}\right)\right] && \text{for } r_i < r < r_o \\ V(r) &= \frac{q}{4\pi\varepsilon_0\varepsilon_r r} && \text{for } r > r_o \end{aligned} \tag{6.24}$$

There will be polarization surface charges at $r = r_i$ and at $r = r_o$, because charge separation in the dielectric produces a surface-charge density $\varrho_{sp} = \hat{\mathbf{n}} \cdot \mathbf{P}$, with

$\hat{\mathbf{n}}(\mathbf{r}_i) = -\hat{\mathbf{r}}$ and $\hat{\mathbf{n}}(\mathbf{r}_o) = +\hat{\mathbf{r}}$. We can obtain the polarization vector $\mathbf{P}$ from the difference $\mathbf{D} - \varepsilon_0\mathbf{E}$, and from (6.21b) it is easily seen that the difference yields

$$\mathbf{P}(r) = \frac{q}{4\pi r^2}\left(1 - \frac{1}{\varepsilon_r}\right)\hat{\mathbf{r}} \qquad \text{for } r_i < r < r_o \tag{6.25}$$

Consequently, from (6.16d) the two surface-charge densities are

$$\begin{aligned} \varrho_{sp}(r_i) &= -\frac{q}{4\pi r_i^2}\left(1 - \frac{1}{\varepsilon_r}\right) \\ \varrho_{sp}(r_o) &= +\frac{q}{4\pi r_0^2}\left(1 - \frac{1}{\varepsilon_r}\right) \end{aligned} \tag{6.26a}$$

The total polarization charge on each surface is found by multiplying each of the densities (6.26a) by the corresponding surface area. The inner surface then has charge $q_i = -q(1 - 1/\varepsilon_r)$ and the outer surface has $q_o = +q(1 - 1/\varepsilon_r)$. These two charges are equal and opposite to each other.

Is there any polarization charge within the dielectric region? To see if there is, we must look at $\varrho_{vp} = -\nabla \cdot \mathbf{P}$ (see again (6.16d)). Here, spherical coordinates are the obvious choice for the Laplacian operator: using (6.25),

$$\varrho_{vp} = -\nabla \cdot \mathbf{P} = -\frac{1}{r^2}\frac{\partial}{\partial r}(r^2 P_r) = 0 \tag{6.26b}$$

This is zero because (6.25) has shown that $P_r \propto 1/r^2$, so that $r^2 P_r$ does not depend upon the radial coordinate r. There is *no* polarization-charge density in the interior even though there is a $\mathbf{P}$ vector everywhere. Hence there are only surface charges, as given by (6.26a).

An *insulator* is characterized by high values of ε_r. In the limit $\varepsilon_r \to \infty$, we observe that the formulas (6.21), (6.24), and (6.26a) all become identical to what we found in Example 6.1 for a perfect conductor. Note however, from (6.25), that vector $\mathbf{P}$ is then maximal, and is equal to $\mathbf{D}$ in the insulator region. Even though the electrostatic fields are the same as for a conductor here, we shall see later that differences occur for alternating-current (ac) fields.

6.4.2 Boundary conditions

This subsection deals with boundary conditions for an interface between two dielectric materials with relative dielectric permittivities ε_1 and ε_2, in an external field. The interface may be curved, but we will look at a section of it that is so small that we may neglect the curvature, e.g. as for the contour in Fig. 6.8.

The first boundary condition corresponds to the Maxwell equation $\nabla \times \mathbf{E} = 0$, which in integral form is $\oint d\boldsymbol{l} \cdot \mathbf{E} = 0$ for any contour whatever, as discussed in connection with (4.9). If we apply this to the infinitesimal contour of Fig. 6.8, in

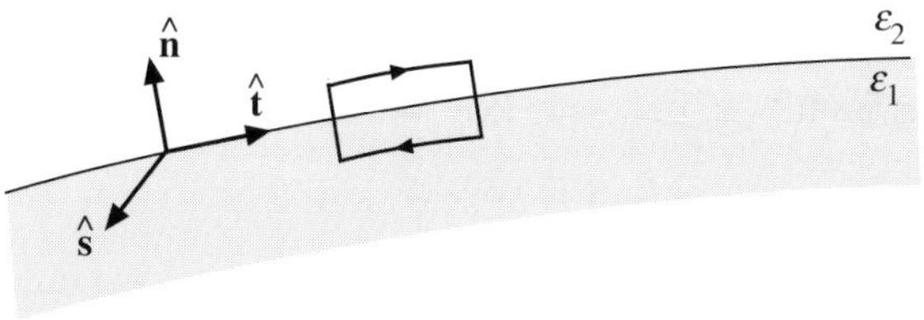

Fig. 6.8 A local orthogonal coordinate system at a dielectric–dielectric interface.

which it is assumed that the two sides $\perp$ the interface are negligibly short compared to the two sides $\|$ the interface, then, writing $d\boldsymbol{l} = dl\,\hat{\mathbf{t}}$, the contour integral has contributions from the latter sides only:

$$dl\,\hat{\mathbf{t}} \cdot (\mathbf{E}_2 - \mathbf{E}_1) = 0 \tag{6.27a}$$

Here, the length dl of each side $\|$ to the interface is small enough that the **E** fields at the interface do not change appreciably in that distance. That is why no integral sign is needed here; the integral becomes a dot product of field vector times the distance vector $\hat{\mathbf{t}}\,dl$. As $dl \neq 0$, we can divide by it, and thus omit it. We can also write $\hat{\mathbf{t}} = \hat{\mathbf{n}} \times \hat{\mathbf{s}}$ so that

$$(\hat{\mathbf{n}} \times \hat{\mathbf{s}}) \cdot (\mathbf{E}_2 - \mathbf{E}_1) = 0$$

From the vector-algebra rule (2.6), we see that this can be written as $[(\mathbf{E}_2 - \mathbf{E}_1) \times \hat{\mathbf{n}}] \cdot \hat{\mathbf{s}} = 0$, which is the same as $-[\hat{\mathbf{n}} \times (\mathbf{E}_2 - \mathbf{E}_1)] \cdot \hat{\mathbf{s}} = 0$, and because $\hat{\mathbf{s}} \neq 0$ it follows that

$$\hat{\mathbf{n}} \times (\mathbf{E}_2 - \mathbf{E}_1) = 0 \tag{6.27b}$$

Note that $\hat{\mathbf{n}}$ is the unit vector pointing from the interface *into* medium 2 (although *this* boundary condition is obviously unchanged if $\hat{\mathbf{n}}$ were to point the other way). Alternatively, from (6.27a) it follows directly that

$$E_{2\mathrm{t}} - E_{1\mathrm{t}} = 0 \tag{6.27c}$$

$E_{i\mathrm{t}}$ being the tangential component of $\mathbf{E}_i$ $(i = 1, 2)$. Either form of (6.27) lends itself as the surface form of the Maxwell equation curl $\mathbf{E} = 0$.

The second boundary condition corresponds to the Maxwell equation $\nabla \cdot \mathbf{D} = \varrho_\mathrm{v}$. To obtain it, we construct a pillbox of infinitesimal height h intersected midway by the interface in a surface area S, as shown in Fig. 6.9. We apply the integral form of this Maxwell equation, which, of course, is Gauss's law, $\oint d\mathbf{S} \cdot \mathbf{D} = q_\mathrm{encl}$, and note again that q_encl is nonzero only if there is free charge on the interface. We also note that the normal to the bottom surface is $-\hat{\mathbf{n}}$. Consequently,

$$S\hat{\mathbf{n}} \cdot (\mathbf{D}_2 - \mathbf{D}_1) = S\varrho_\mathrm{s}$$

where ϱ_s is the surface density of real (i.e. free) charge at the interface.

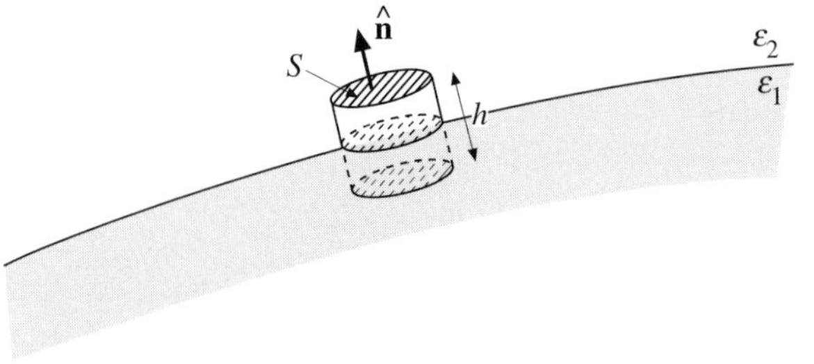

Fig. 6.9 A Gaussian 'pillbox' at a dielectric–dielectric interface.

As this holds for any infinitesimal dS, it is therefore found that

$$\hat{\mathbf{n}} \cdot (\mathbf{D}_2 - \mathbf{D}_1) = \varrho_s \tag{6.28a}$$

However, we know that $\hat{\mathbf{n}} \cdot \mathbf{D}$ is the length of the component of $\mathbf{D}$ that is perpendicular to the interface, i.e. the normal component, so that, equivalently,

$$D_{2n} - D_{1n} = \varrho_s \tag{6.28b}$$

We note that if $\varrho_s = 0$ then the normal components of $\mathbf{D}$ are equal. Unlike for the boundary condition (6.27), it is important to remember that the *direction* of $\hat{\mathbf{n}}$ in (6.28a) *must* point from the interface into the medium of the *first*-mentioned $\mathbf{D}$ on the left-hand side; here, it points into medium 2 because $\mathbf{D}_2$ appears first. It should be emphasized that the ϱ_s on the right-hand side of (6.28) is the *real* surface-charge density and does not include the polarization-charge density. In order to use (6.28a), one must first write down the real surface-charge density of the interface, with whatever sign it has, and then decide which $\mathbf{D}$ comes first on the left-hand side so that $\hat{\mathbf{n}}$ can be chosen as indicated. In terms of the $\mathbf{E}$ field, we can rewrite (6.28b) as

$$\varepsilon_2 E_{2n} - \varepsilon_1 E_{1n} = \varrho_s \tag{6.28c}$$

Example 6.4 The field discontinuity at a planar dielectric interface between two dielectric media

Consider the situation in Fig. 6.10, showing the discontinuity in length and direction of the field $\mathbf{E}_2$ compared to $\mathbf{E}_1$. Let us suppose that both fields are created by a single source, say in medium 1; we shall see that the discontinuity is due to the boundary conditions. Suppose also that there is no real charge on the interface (there will not be any, unless it has been placed there by someone). Then (6.27) and (6.28) predict for the tangential and normal components of $\mathbf{E}_2$

$$\begin{aligned} E_{2t} &= E_{1t} \\ E_{2n} &= \frac{\varepsilon_1}{\varepsilon_2} E_{1n} \end{aligned} \tag{6.29a}$$

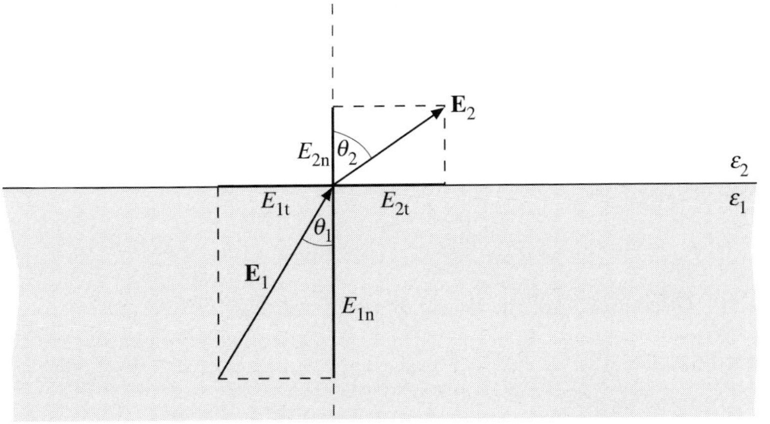

Fig. 6.10 Discontinuity of the **E** field at a dielectric–dielectric interface.

As shown in Fig. (6.10), there is a discontinuity in **E** at the interface if $\varepsilon_2 \neq \varepsilon_1$. If we wish to calculate the $\mathbf{E}_2$ field at the interface, given that we know the $\mathbf{E}_1$ field there, we might wish to use the given vector lengths and angles, rather than the tangential and normal components. In that case, (6.29a) is written as

$$E_2 \sin\theta_2 = E_1 \sin\theta_1$$
$$E_2 \cos\theta_2 = \frac{\varepsilon_1}{\varepsilon_2} E_1 \cos\theta_1 \tag{6.29b}$$

Divide the first equation by the second to obtain

$$\tan\theta_2 = \frac{\varepsilon_2}{\varepsilon_1} \tan\theta_1 \tag{6.29c}$$

This gives us the angle θ_2, and we see immediately from (6.29c) that $\theta_2 > \theta_1$ if $\varepsilon_2 > \varepsilon_1$, i.e. the field line deflects *away* from the normal when the second medium has a higher-valued permittivity. The length of vector $\mathbf{E}_2$ is easily obtained by inserting (6.29c) into either of (6.29b), e.g.

$$E_2 = \frac{E_1 \sin\theta_1}{\sin[\arctan(\varepsilon_2 \tan\theta_1/\varepsilon_1)]} \tag{6.29d}$$

Numerical calculations are now easily performed using a calculator.

Example 6.5 Field discontinuities at a planar dielectric slab

A uniform field $\mathbf{E}_0$ in free space is in the $\hat{\mathbf{z}}$ direction; a planar slab of uniform dielectric material with permittivity ε is inserted into the field, as shown in Fig. 6.11. The **E** field in the slab will be modified because the dielectric is polarized. As $\nabla \cdot \mathbf{P} = 0$ here, because both field and medium are uniform, there will be only surface polarization charges $\pm\hat{\mathbf{z}} \cdot \mathbf{P}$ (the minus sign is at the left interface because the surface normal must point *out* of the dielectric, see after (16.16d), and because the negative

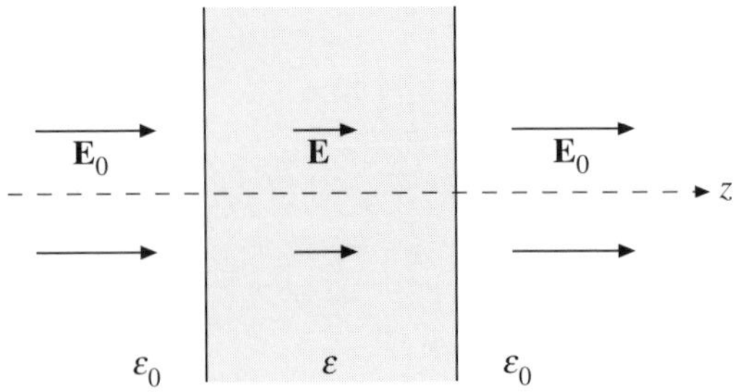

Fig. 6.11 Discontinuity of the **E** field in a dielectric slab.

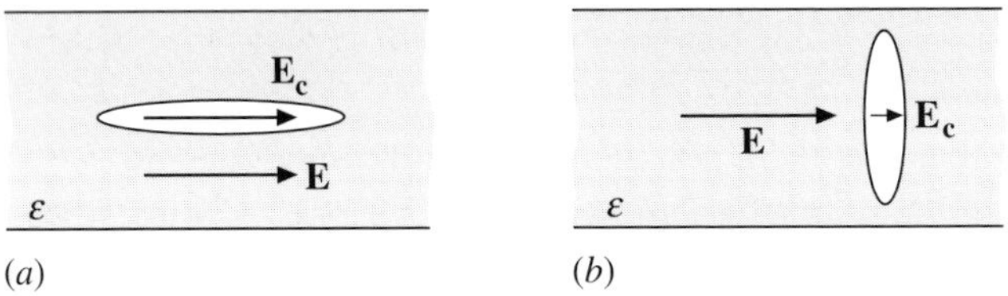

Fig. 6.12 (*a*) Needle and (*b*) disk cavities inside a dielectric medium.

charges move against the field direction. The surface polarization charges make the field smaller in the dielectric, as mentioned earlier. From the boundary conditions at each interface, we have $E = (\varepsilon_0/\varepsilon)E_0$ (we may leave off the subindex 'n' because there are only normal, no tangential, fields here). If $\varepsilon > \varepsilon_0$, as is usually the case,[1] it follows that indeed $E < E_0$. Furthermore, (6.26b) shows that here the **D** vector is continuous and constant everywhere. All fields point in the $\hat{\mathbf{z}}$ direction.

Example 6.6 Needle-like and disk-like cavities inside a dielectric medium

Consider a dielectric medium, say a slab, as in Example 6.5, in which there is a uniform effective field $\mathbf{E} = E\hat{\mathbf{z}}$. Suppose a long thin needle-like cavity (preferably in the shape of a prolate spheroid) is introduced in the $\hat{\mathbf{z}}$ direction, as shown in Fig. 6.12(*a*). We will now show, using the boundary conditions, that the field in it will be $\mathbf{E}_c \approx \mathbf{E}$.

The fact that the (slightly curved) interface of the needle cavity with the dielectric is practically parallel to the $\hat{\mathbf{z}}$ direction brings with it that there are practically no normal field components, and thus that (6.27) holds approximately here: the tangential components are equal. A precise calculation, which goes beyond the scope of this text, shows that $\mathbf{E}_c$ is exactly equal to **E**, even though the interface is not planar. This is a special property of an ellipsoid cavity thus oriented inside a dielectric, and a prolate spheroid (i.e. a needle) is a special case of an ellipsoid. Such a cavity could, in principle, be used to measure the **E** field inside a dielectric!

[1] As implied, there are exceptions: $\varepsilon_r < 0$ in a cold plasma.

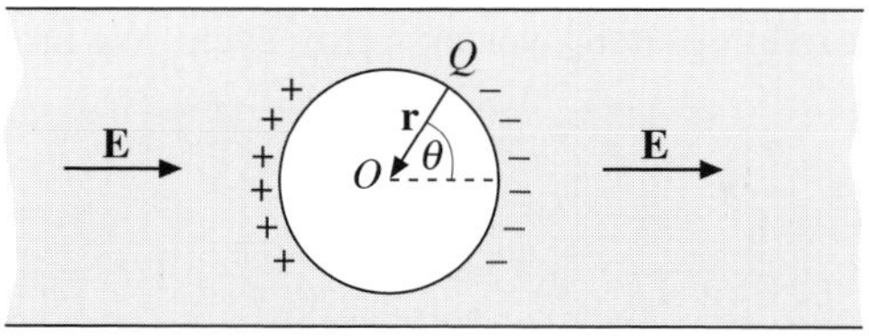

Fig. 6.13 Spherical cavity inside a dielectric medium.

In Fig. 6.12b we orient a disk-like cavity perpendicularly to the field direction. Now we conclude from (6.29a), and the almost complete lack of tangential components, that here $\mathbf{E}_c \approx (\varepsilon/\varepsilon_0)\mathbf{E}$. As mentioned above, a rigorous calculation shows that this is exact if the cavity is indeed a spheroid, as in this case. Overall, this gives us a method of measuring the permittivity $\varepsilon_r = \varepsilon/\varepsilon_0$ of the dielectric, because the first cavity shape allowed us to measure E and the second cavity shape gives E_c. The ratio E_c/E then gives ε_r.

Example 6.7 Spherical cavity inside a dielectric medium

Consider now a spherical cavity cut out of a dielectric, as shown in Fig. 6.13. We will try to calculate the field at its center point O (which we take as the coordinate origin). A uniform effective field $\mathbf{E}$ has been created in the dielectric (with permittivity ε). In the dielectric, there is a uniform polarization vector, $\mathbf{P} = \mathbf{D} - \varepsilon_0\mathbf{E} = \varepsilon_0(\varepsilon_r - 1)\mathbf{E}$. Clearly $\nabla \cdot \mathbf{P} = 0$ everywhere, owing to the uniformity of medium and field. But, at the spherical interface, there is a surface polarization-charge density $\varrho_{sp} = \hat{\mathbf{n}} \cdot \mathbf{P} = -\hat{\mathbf{r}} \cdot P$.

At a point Q on the rim, we have $-\hat{\mathbf{r}} \cdot \mathbf{P} = -P\cos\theta$. We will use Coulomb's law for a surface-charge distribution to obtain the $\mathbf{E}$ field at O due solely to the polarization charge on the interface:

$$\delta\mathbf{E}_O = -\oint dS \frac{\varrho_{sp}}{4\pi\varepsilon_0 r^2}\hat{\mathbf{r}}$$

The minus sign is due to the fact that the unit vector pointing from charge to observation point is $-\hat{\mathbf{r}}$. Because $dS = r^2 d\theta \sin\theta\, d\varphi$ for spherical coordinates, it follows that

$$\delta\mathbf{E}_O = -r^2 \int_0^{2\pi} d\varphi \int_0^{\pi} d\theta \sin\theta \frac{(-P\cos\theta)}{4\pi\varepsilon_0 r^2}(\hat{\mathbf{x}}\sin\theta\cos\varphi + \hat{\mathbf{y}}\sin\theta\sin\varphi + \hat{\mathbf{z}}\cos\theta)$$

where we have used the Cartesian expansion (2.35a) for the unit spherical-coordinate vector $\hat{\mathbf{r}}$. The $d\varphi$ integration of the $\hat{\mathbf{x}}$ and the $\hat{\mathbf{y}}$ components is zero because both $\sin\varphi$ and $\cos\varphi$ are integrated over an entire cycle. We are left with

$$\delta\mathbf{E}_O = \frac{P}{2\varepsilon_0}\int_0^{\pi} d\theta \sin\theta\cos^2\theta\,\hat{\mathbf{z}} = \frac{P}{3\varepsilon_0}\hat{\mathbf{z}} \tag{6.30}$$

The *local field* at O is in total $\mathbf{E} + \delta\mathbf{E}_O = (E + P/3\varepsilon_0)\hat{\mathbf{z}}$, which is an important result used in deriving the microscopic underpinnings of the dielectric permittivity.

The relative dielectric permittivity ε_r is determined from the microscopic or molecular quantities of the dielectric. First, (6.20a) shows that ε_r is determined by the polarization $\mathbf{P}$, which is $\varepsilon_0(\varepsilon_r - 1)\mathbf{E}$. In the second place, it is also true for a uniform medium that $\mathbf{P} = N\mathbf{p}$ (if there are N dipoles per unit volume). The *molecular* dipole moment $\mathbf{p}$ is determined by the *local field*, $\mathbf{E}_{\text{loc}} = \mathbf{E} + \mathbf{P}/3\varepsilon_0$, since $\mathbf{p} = \alpha\mathbf{E}_{\text{loc}}$, where α is the *molecular polarizability*. As mentioned earlier, in the discussion before (6.18), the local field must exclude the ith dipole itself, because that dipole is caused by the electrostatic field produced by all *other* sources. It follows from these simple equations that

$$\mathbf{E}_{\text{loc}} = (1 - N\alpha/3\varepsilon_0)^{-1}\mathbf{E}$$

Now we use $\mathbf{P} = \varepsilon_0(\varepsilon_r - 1)\mathbf{E}$ and $\mathbf{P} = N\alpha\mathbf{E}_{\text{loc}}$. By eliminating $\mathbf{P}$, $\mathbf{E}$, and $\mathbf{E}_{\text{loc}}$, we obtain the *Clausius–Mosotti* relationship

$$\varepsilon_r = \frac{1 + 2N\alpha/3\varepsilon_0}{1 - N\alpha/3\varepsilon_0} \tag{6.31}$$

which expresses ε_r in terms of the molecular polarizability α (under the assumption here that all molecules are identical in size, shape, etc.). It is often possible to establish a relationship between α and other more fundamental parameters of the molecule. As an example, ε_r might pertain not to a solid dielectric but to an aerosol consisting of a collection of N minuscule dielectric particles per unit volume, all of which have permittivity ε. The polarizability α then pertains not to a molecule, but to a small dielectric particle with permittivity ε. It is possible to find the dependence of α upon ε, and thus to describe ε_r in terms of ε_0 and ε, and ε_r would be the *effective* relative dielectric permittivity of the aerosol.

6.4.3 Anisotropic dielectric materials

There are materials for which the separation of bound charges is subject to spatial constraints in certain directions more than in others. Or, for the case of polar molecules, the shape of these molecules and their positioning with respect to each other in solid materials may restrict the possible orientation of the dipoles i when an exterior field $\mathbf{E}$ is imposed upon them. It may not be the case that $\mathbf{p}_i \parallel \mathbf{E}$, perhaps because a polar molecule is hampered in rotating under the influence of an $\mathbf{E}$ field by neighboring molecules. This is typically the case for certain crystalline materials. Such materials may still have $\mathbf{p}_i \propto \mathbf{E}$ for weak fields, and this implies that

$$\mathbf{P} = \varepsilon_0 \boldsymbol{\chi}_\mathrm{e} \mathbf{E} \tag{6.32a}$$

where the dielectric susceptibility $\boldsymbol{\chi}_\mathrm{e}$ is now a *tensor* not a scalar quantity, which implies that it is a 3×3 matrix if $\mathbf{E} = (E_1, E_2, E_3)$ is given in terms of the components of a coordinate system. In terms of Cartesian coordinates, (6.32a) is a short-hand notation for

$$\mathbf{P} = \varepsilon_0 \begin{bmatrix} \chi_{xx} & \chi_{xy} & \chi_{xz} \\ \chi_{yx} & \chi_{yy} & \chi_{yz} \\ \chi_{zx} & \chi_{zy} & \chi_{zz} \end{bmatrix} \begin{bmatrix} E_x \\ E_y \\ E_z \end{bmatrix} \tag{6.32b}$$

Unless all off-diagonal terms are zero, and all three diagonal terms are the same, this tensor form implies that **P** is not parallel to **E**. Equations (6.32) also express the polarization due to N ellipsoidal particles per unit volume, all of which are oriented in the same direction. The off-diagonal terms are zero if one of the three ellipsoid main axes is parallel to the electrostatic field, but the diagonal terms generally will differ in magnitude.

As $\varepsilon_\mathrm{r} = 1 + \chi_\mathrm{e}$ for isotropic materials, the obvious extension is to define $\boldsymbol{\varepsilon}_\mathrm{r} = \mathbf{1} + \boldsymbol{\chi}_\mathrm{e}$, where **1** is the unit 3×3 matrix in Cartesian coordinates. Then

$$\mathbf{D} = \varepsilon_0 \boldsymbol{\varepsilon}_\mathrm{r} \mathbf{E}$$

for anisotropic materials. Note again that, even though the material is *anisotropic* if $\boldsymbol{\varepsilon}_\mathrm{r}$ is nonscalar, the material may well be *uniform* ($\boldsymbol{\varepsilon}_\mathrm{r}$ is the same at all locations).

If the field is oriented along one of the main axes of a *biaxial* crystal, then

$$\mathbf{D} = \varepsilon_0 \boldsymbol{\varepsilon}_\mathrm{r} \mathbf{E}$$

implies a relative dielectric permittivity matrix as follows:

$$\boldsymbol{\varepsilon}_\mathrm{r} = \begin{bmatrix} \varepsilon_1 & 0 & 0 \\ 0 & \varepsilon_2 & 0 \\ 0 & 0 & \varepsilon_3 \end{bmatrix} \tag{6.33}$$

A *uniaxial* crystal will have $\varepsilon_2 = \varepsilon_1$ if the field is oriented along the main axis (the z axis). One consequence of this is that such media will support waves of different wavelengths in different directions. But this, again, is a topic that can be treated more fully only after electromagnetic waves have been discussed.

Problems

6.1. The electrostatic field at a point **P** on the surface of a perfect conductor is $\mathbf{E} = 3\hat{\mathbf{x}} - 2\sqrt{2}\,\hat{\mathbf{y}} + 2\sqrt{2}\,\hat{\mathbf{z}}$ V/m. What is the surface-charge density ϱ_s at P?

6.2. The electrostatic field at the surface of a perfect conductor in $z \leq 0$ has a magnitude $E = 30\,\text{V/m}$. There is free space in $z > 0$ and the surface charge is positive.
(a) Specify the vector $\mathbf{E}$ at the interface.
(b) Specify the surface-charge density at the interface.
(c) Suppose that there also is free space in $z < -5\,\text{m}$. Specify the electric field vector at $z = -5\,\text{m}$ and explain how you arrived at that result.

6.3. An electrostatic field line in free space ends on a perfect conductor, at which location $\mathbf{E} = 5\hat{\mathbf{x}} - 2\hat{\mathbf{y}} + 3\hat{\mathbf{z}}\,\text{V/m}$. There is a planar interface containing the point $(1, 2, 3)$. What is the equation describing the plane of the interface?

6.4. The interface between free space and a conductor containing the origin is $7.5x + 3y + 5z = 15$. It contains a surface charge with density $\varrho_s = 4.427 \times 10^{-11}\,\text{C/m}^2$. Find the electrostatic field vector $\mathbf{E}$ at $P(0, 5, 0)$.

6.5. A charge of $10^{-10}\,\text{C}$ is placed on a sphere of radius R in air. At what value of R does the surface charge cause a breakdown electrostatic field in the surrounding air ($E_{\text{breakdown}} = 3 \times 10^6\,\text{V/m}$)?

6.6. Use the integral form of $\nabla \times \mathbf{E} = 0$ to prove that the tangential component of the electrostatic field is zero at the interface between free space and a perfect conductor.

6.7. Use the concept of field lines to demonstrate that perfectly conducting walls entirely surrounding an interior cavity free of charge will 'shield' that interior from exterior electrostatic fields. Why, though, will a free-of-charge exterior not be shielded from an interior charge?

6.8. A perfectly conducting material exists in a cylindrical shell defined by $a < \rho < b$. There is a uniform line charge ϱ_ℓ on the z axis. Find the $\mathbf{E}$ fields in all three regions.

6.9. The dielectric susceptibility of a uniform and isotropic material is $\chi_e = 2.5$. It contains an electric field of magnitude $E = 17.71\,\text{V/m}$.
(a) What is the relative dielectric permittivity?
(b) What is the magnitude of the polarization P?
(c) What is the magnitude of the electric flux density D?

6.10. An imposed electric field $\mathbf{E} = 5\hat{\mathbf{x}}\,\text{V/m}$, uniform everywhere, produces a polarization vector $\mathbf{P} = \frac{1}{2}\varepsilon_0 \sin(3x)\hat{\mathbf{x}}$ in the uniform dielectric that fills $x > 0$. What is the relative permittivity in $x > 0$?

6.11. A uniform and isotropic dielectric material with relative permittivity $\varepsilon_r = 4.5$, subject to a uniform electrostatic field, is measured to have a polarization $P = 3 \times 10^{-8}\,\text{C/m}^2$. Find the magnitude of the effective field (see the text before Eq. (6.18)).

6.12. A uniform slab of dielectric material with $\varepsilon_r = 3.5$, sandwiched between $x = 0$ and $x = d$, is subjected to an exterior uniform field $\mathbf{E} = E_0\hat{\mathbf{x}}$. Describe the equivalent polarization-charge distribution in the slab that is induced by $\mathbf{E}$.

6.13. Calculate the *local field* E_{loc} at a point P inside a slab of dielectric. To do this, introduce a small hollow sphere around P and calculate the field produced at P by the polarization charge at the surface of this sphere. Add this field to the mean field in the dielectric to obtain E_{loc}.

6.14. Two uniform dielectric media are characterized by $\varepsilon_{r1} = 1.5$ in $z < 0$ and $\varepsilon_{r2} = 3.5$ in $z > 0$. The electrostatic flux density in $z < 0$ is

$$\mathbf{D}_1 = 4.427 \times 10^{-11}(2\hat{\mathbf{x}} - 5\hat{\mathbf{y}} + 3\hat{\mathbf{z}})\ \text{C/m}^2$$

(a) Find the $\mathbf{E}_2$ field in $z > 0$.
(b) Find the $\mathbf{D}_2$ field in $z > 0$.
(c) Find the polarization charge ϱ_{sp} on the interface in terms of ε_{r1}, ε_{r2}, and $\mathbf{E}_1$.

6.15. An interface $4x + 5y + 10z = 20$ separates an $\varepsilon_{r1} = 1.75$ dielectric medium (containing the origin) from an $\varepsilon_{r2} = 3.25$ dielectric medium. It is specified that $\mathbf{E}_1 = 4.2\hat{\mathbf{z}}$ uniformly in medium 1. Calculate the field $\mathbf{E}_2$ in medium 2.

6.16. Two uniform dielectric media are characterized by $\varepsilon_{r1} = 2.5$ in $z < 0$ and $\varepsilon_{r2} = 5$ in $z > 0$. The electrostatic field density in $z < 0$ is $\mathbf{E}_1 = 2\hat{\mathbf{x}} - 5\hat{\mathbf{y}} + 3\hat{\mathbf{z}}$ V/m. The interface contains a uniform real surface-charge density $\varrho_s = 4.427 \times 10^{-11}\ \text{C/m}^2$. Calculate $\mathbf{E}_2$ at the interface.

6.17. The half-space $y > 0$ contains a uniform dielectric with $\varepsilon_r = 1.25$, whereas $y < 0$ is free space. A half-cylinder with axis in the z direction and with radius $\rho_0 = \sqrt{x^2 + y^2} = 5$ m (and $y > 0$) has been removed from the dielectric half-space, leaving free space. There is no variation in the z direction. To a good approximation the free-space field can be set as $\mathbf{E}_0 = 4\hat{\mathbf{x}}$ V/m. Where, on the half-circle interface, is the $\mathbf{E}$ field in the dielectric purely in the direction $(\hat{\mathbf{x}} - 0.1\hat{\mathbf{y}})/\sqrt{2}$?

6.18. Given that $D_x = \varepsilon_0(E_x + 2E_y + 3E_z)$, $D_y = \varepsilon_0(3E_x + E_y + 2E_z)$, $D_z = \varepsilon_0(2E_x + 3E_y + E_z)$, state the dielectric susceptibility tensor.

7

Electrostatic energy, electromechanical force, and capacitance

7.1 N point charges

The engineering interest in electrostatic energy derives from the ability to extract from it mechanical work and/or other forms of 'useful' energy. To do so, we need to know just how much electrostatic energy can be extracted in principle from a distribution of charges, because these are the *source* of electrostatic fields. The most direct way to go about this is to calculate how much mechanical energy was converted into electrostatic energy in bringing the charge distribution into place. Let us consider a set of N charges $q_1, q_2, \ldots, q_N$ at locations $\mathbf{r}_1, \mathbf{r}_2, \ldots, \mathbf{r}_N$ respectively. These charges act electrostatically upon each other and also upon a test charge q put anywhere in space. The field force will cause that test charge to move if it is not constrained by external forces. Mechanics teaches us that the field force $\mathbf{F} = -\nabla \cdot U$, if U is the potential energy of the charge distribution. In the electrostatic case, $\mathbf{F} = q\mathbf{E} = -q\nabla \cdot V$, so that – as we saw in the chapter on potential – we may identity qV with the potential energy U.

Now consider the N charges at their locations to have been brought there from a state of zero energy corresponding to initial positions of those charges at infinity. That is to say, let us start with all charges infinitely distant from all other charges. Now move charge q_1 from infinity to its location at $\mathbf{r}$ in the absence of any neighboring charge. This does not cost any electrostatic energy whatever; it might cost some mechanical energy but let us assume a universe of free space so that mechanical friction, etc., can be ignored entirely. Hence the work required $w_1 = 0$.

With charge q_1 remaining fixed at $\mathbf{r}_1$, let us now move charge q_2 from infinity to $\mathbf{r}_2$, as illustrated in Fig. 7.1. The work required w_2, as expressed by the definition of the electrostatic potential (see Section 4.1), is equal to the gain in potential energy U, i.e.

$$w_2 = q_2 V_{12} \tag{7.1}$$

Here, $V_{ij} \equiv q_i/4\pi\varepsilon_0 R_{ij}$ represents the electrostatic potential at $\mathbf{r}_j$ due to a charge q_i at $\mathbf{r}_i$, and $R_{ij} \equiv |\mathbf{r}_j - \mathbf{r}_i|$. Thus w_2 represents the work done in moving a charge q_2 from

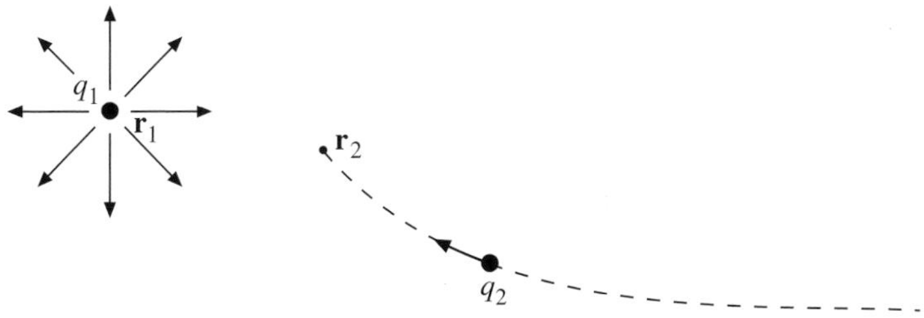

Fig. 7.1 Illustration depicting work done in placing q_2 at $\mathbf{r}_2$ when q_1 already is at $\mathbf{r}_1$.

infinity to $\mathbf{r}_2$ in the presence of an electrostatic field $\mathbf{E}_1$ produced by charge q_1 located at $\mathbf{r}_1$.

Now keep charges q_1 and q_2 fixed at their locations $\mathbf{r}_1$ and $\mathbf{r}_2$, and move charge q_3 from infinity to $\mathbf{r}_3$. In this case, the amount of work done in moving $\mathbf{r}_3$ from infinity to its final location is

$$w_3 = q_3(V_{13} + V_{23}) \tag{7.2a}$$

This represents the work done in moving charge $\mathbf{q}_3$ to $\mathbf{r}_3$ under the *combined* effect of fields $\mathbf{E}_1$ due to q_1 and $\mathbf{E}_2$ due to q_2. With charges q_1, q_2, and q_3 kept fixed at their locations, the move of q_4 from infinity to $\mathbf{r}_4$ requires work

$$w_4 = q_4(V_{14} + V_{24} + V_{34}) \tag{7.3}$$

Continuation of this process up to the Nth and last charge gives

$$w_N = q_N \sum_{n=1}^{N-1} V_{nN} \tag{7.4}$$

Hence the *total* stored energy in the charge distribution must be the sum of all the work included in (7.1)–(7.4), i.e. it is

$$W_N = \sum_{j=2}^{N} q_j \sum_{i=1}^{j-1} V_{ij} = \sum_{j=2}^{N} \sum_{i=1}^{j-1} \frac{q_i q_j}{4\pi\varepsilon_0 R_{ij}} = \sum_{\substack{i,j=1 \\ i<j}}^{N} \frac{q_i q_j}{4\pi\varepsilon_0 R_{ij}} \tag{7.5}$$

In the last term, nothing changes in the numerator and denominator if we interchange j and i, because $R_{ji} = R_{ij}$, but the condition $i < j$ changes into $i > j$. If we take half the sum of the interchanged term and the last term of (7.5), we obtain a more symmetric expression,

$$W_N = \frac{1}{2} \sum_{i=1}^{N} \sum_{\substack{j=1 \\ j \neq i}}^{N} \frac{q_i q_j}{4\pi\varepsilon_0 R_{ij}} \tag{7.6a}$$

A short way of writing this is to introduce the potential V_i at $\mathbf{r}_i$ due to *all* the other charges, $q_1, \ldots, q_{i-1}, q_{i+1}, \ldots, q_N$, because that translates (7.6a) into

$$W_N = \frac{1}{2}\sum_{i=1}^{N} q_i V_i \qquad \text{with} \quad V_i = \sum_{\substack{j=1 \\ j\neq i}}^{N} \frac{q_j}{4\pi\varepsilon_0 R_{ij}} \tag{7.6b}$$

Thus, W_N is the *total energy stored in these N charges*, and it represents the maximum electrostatic energy that can be extracted from them.

7.2 Continuous charge distribution

Equations (7.6) are useful for a collection of discrete (i.e. well-separated) point charges, but how do we calculate the potential energy when we have a continuous charge distribution? The answer derives from (7.6a) when N is a very large number, corresponding perhaps to a number density of 10^{20} particles/cm^3. Let us look at the infinitesimal contribution to the potential energy, $d^2 W_e$, formed from (7.6a) by including only the interactions between pairs of particles in infinitesimal volume elements dv and dv', as shown in Fig. 7.2:

$$d^2 W_e = \frac{1}{2}\sum_{i\in dv}\sum_{j\in dv'} \frac{q_i q_j}{4\pi\varepsilon_0 R_{ij}} \tag{7.7a}$$

As before, we may replace R_{ij} by $|\mathbf{r}-\mathbf{r}'|$ without serious error if we presume it to be much larger than a side of the volume elements dv, dv'. The summations are then independently over the charges only, and we may set

$$\sum_{i\in dv} q_i = \varrho_v(\mathbf{r})\, dv \qquad \sum_{j\in dv'} q_j = \varrho_v(\mathbf{r}')\, dv'$$

in terms of the volume-charge density and thus obtain

$$d^2 W_e = \frac{1}{2}\frac{\varrho_v(\mathbf{r})\varrho_v(\mathbf{r}')}{4\pi\varepsilon_0|\mathbf{r}-\mathbf{r}'|}\, dv\, dv' \tag{7.7b}$$

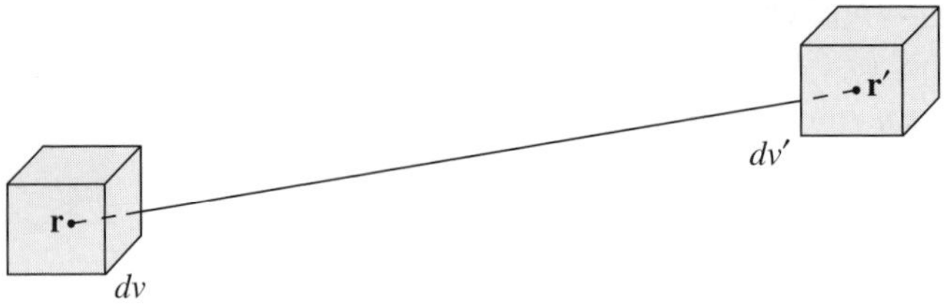

Fig. 7.2 Two volume elements dv and dv' for stored-energy calculation.

We find, upon integration over the entire charge distribution,

$$W_e = \frac{1}{2}\int dv\, \varrho_v(\mathbf{r}) \int dv' \frac{\varrho_v(\mathbf{r}')}{4\pi\varepsilon_0|\mathbf{r}-\mathbf{r}'|} \qquad \Rightarrow$$

$$W_e = \frac{1}{2}\int_v dv\, \varrho_v(\mathbf{r}) V(\mathbf{r}) \qquad \text{with} \quad V(\mathbf{r}) \equiv \int_v dv' \frac{\varrho_v(\mathbf{r}')}{4\pi\varepsilon_0|\mathbf{r}-\mathbf{r}'|} \tag{7.8}$$

This represents the total energy stored in a charge distribution in volume v. Note that the energy is measured in joules (J) with 1 J = 1 W s (watt second). Note also that the integration in (7.8) should be over *all* space; the fact that $\varrho_v(\mathbf{r})$ is zero when $\mathbf{r}$ is outside effectively restricts the integral to the volume v. It is important, however, for what follows to remember that we must integrate over all space to include all possible charges.

It appears as if the point $\mathbf{r}' = \mathbf{r}$ has a significant influence upon the potential $V(\mathbf{r})$, but in fact it does not. If as before we take an infinitesimal sphere of radius ϵ around $\mathbf{r}$, then $|\mathbf{r}-\mathbf{r}'| = R$ for $R < \epsilon$ and $dv' = 4\pi R^2\, dR$, whereas $\varrho_v(\mathbf{r}') = \varrho_v(\mathbf{r})$ everywhere in the infinitesimal sphere (unless there is a discontinuity in charge distribution inside the volume *and* $\mathbf{r}$ is placed right on it). We shall not concern ourselves here with discontinuities in charge distribution, so we find

$$\int dv' \frac{\varrho_v(\mathbf{r}')}{4\pi\varepsilon_0|\mathbf{r}-\mathbf{r}'|} = \frac{\varrho_v(\mathbf{r})}{\varepsilon_0}\int_0^\epsilon dR\, R = \frac{\varrho_v(\mathbf{r})}{2\varepsilon_0}\epsilon \quad \longrightarrow \quad 0 \tag{7.9}$$

Thus the contribution of the immediate environment of $\mathbf{r}$ to the potential $V(\mathbf{r})$ tends to zero as the radius $\epsilon \to 0$, which proves that the point $\mathbf{r}' = \mathbf{r}$ does not produce a singularity in (7.8) and has essentially zero contribution to the integral.

Example 7.1 Energy stored in a sphere of uniform charge density: method starting from the charge distribution

Consider a spherically symmetric charge distribution $\varrho_v(\mathbf{r}) = \varrho_v$ for $r < r_0$ and $\varrho_v = 0$ for $r > r_0$. Let us calculate the electrostatic potential energy W_e, using (7.8), which implies that we need to calculate $V(\mathbf{r})$ for $r < r_0$.

(1) For $r > r_0$ we find from Gauss's law and spherical symmetry that

$$4\pi\varepsilon_0 r^2 E_r(r) = q_{\text{encl}} = \frac{4\pi r_0^3}{3}\varrho_v$$

so that

$$E(r) = \frac{\varrho_v r_0^3}{3\varepsilon_0 r^2} \quad \Rightarrow \quad V(r) = \frac{\varrho_v r_0^3}{3\varepsilon_0 r} \qquad \text{for } r > r_0$$

(2) For $r < r_0$, we obtain

$$4\pi\varepsilon_0 r^2 E_r(r) = q_{\text{encl}} = \frac{4\pi r^3}{3}\varrho_v$$

so that

$$E(r) = \frac{\varrho_v r}{3\varepsilon_0} \quad \Rightarrow \quad V(r) = -\frac{\varrho_v r^2}{6\varepsilon_0} + C$$

where C is a constant. The potential must be continuous at $r = r_0$, hence the previous two expressions for $V(r)$ must be equal there:

$$\frac{\varrho_v r_0^3}{3\varepsilon_0 r_0} = -\frac{\varrho_v r_0^2}{6\varepsilon_0} + C \quad \Rightarrow \quad C = \frac{\varrho_v r_0^2}{2\varepsilon_0}$$

Thus, it follows that

$$E(r) = \frac{\varrho_v r}{3\varepsilon_0} \quad \Rightarrow \quad V(r) = \frac{\varrho_v}{2\varepsilon_0}\left(r_0^2 - \frac{1}{3}r^2\right) \qquad \text{for } r < r_0$$

Having found expressions for $V(\mathbf{r})$ in this fashion in the region where there is a charge distribution ($r < r_0$), we are now able to apply (7.8), using the spherically symmetric volume element $dv = 4\pi r^2 dr$:

$$W_e = 2\pi\varrho_v \int_0^{r_0} dr\, r^2 \frac{\varrho_v}{2\varepsilon_0}\left(r_0^2 - \frac{1}{3}r^2\right) = \frac{4\pi\varrho_v^2}{15\varepsilon_0} r_0^5 \tag{7.10}$$

The answer has required only the evaluation of $\int dr\, r^2$ and $\int dr\, r^4$.

7.3 Stored energy in terms of the fields

The expression (7.8) gives the stored energy in a continuous charge distribution. We know that charges are the sources of fields and that mechanical energy can be extracted from the fields because these exert a force $\mathbf{F} = q\mathbf{E}$ upon a charge q, which causes that charge to move (if there are insufficient restraints) and thus expend electrostatic potential energy.

It is possible to express the stored energy in terms of the fields produced by the charge distribution. In the integral for W_e in (7.8) we use Gauss's law and substitute $\varrho_v(\mathbf{r}) = \nabla \cdot \mathbf{D}(\mathbf{r})$; consequently

$$W_e = \frac{1}{2}\int_\infty dv\, V(\mathbf{r})\nabla \cdot \mathbf{D}(\mathbf{r}) = \frac{1}{2}\int_\infty dv \{\nabla \cdot [V(\mathbf{r})\mathbf{D}(\mathbf{r})] - [\mathbf{D}(\mathbf{r}) \cdot \nabla V(\mathbf{r})]\} \tag{7.11a}$$

The second equality uses the vector identity $V\nabla\cdot\mathbf{D} = \nabla\cdot(V\mathbf{D}) - \mathbf{D}\cdot(\nabla V)$ in the integrand. The integration bound is now explicitly over all space (it really was so already in (7.8), in which expression the charge density is zero outside V). The volume integral of the divergence of $V\mathbf{D}$ yields a surface integral over $V\mathbf{D}$ at the infinite surface $(r_s \to \infty, \theta, \varphi)$ by virtue of Gauss's divergence theorem; thus

$$W_e = \frac{1}{2}\oint_\infty d\mathbf{S}\cdot V(\mathbf{r}_s)\mathbf{D}(\mathbf{r}_s) - \frac{1}{2}\int_\infty dv\,\mathbf{D}(\mathbf{r})\nabla V(\mathbf{r}) \tag{7.11b}$$

In the surface-integral term, each element dS of which is so far away that the source charge distribution appears from that vantage point to be a point charge, we therefore have $V(\mathbf{r}_s) \propto 1/r_s$ and $D(\mathbf{r}_s) \propto 1/r_s^2$; $dS = r_s^2 \sin\theta\, d\theta\, d\varphi$. It follows that $d\mathbf{S}\cdot V\mathbf{D} \propto 1/r_s \to 0$ as $r_s \to \infty$. Thus the entire surface-integral term in (7.11b) vanishes in the limit $r_s \to \infty$, and so the result is

$$W_e = \frac{1}{2}\int_\infty dv\,\mathbf{D}(\mathbf{r})\cdot\mathbf{E}(\mathbf{r}) \tag{7.12a}$$

because $\nabla V = -\mathbf{E}$. An alternative form is obtained by using the dielectric permittivity $\varepsilon(\mathbf{r})$:

$$W_e = \frac{1}{2}\int_\infty dv\,\varepsilon(\mathbf{r})|\mathbf{E}(\mathbf{r})|^2 \tag{7.12b}$$

These two expressions give the stored energy in terms of the fields produced by the sources. This must really be borne in mind when using them. If, for example, we have *two* well-separated source distributions ϱ_{v1} and ϱ_{v2}, then the field energy represented by (7.12a, b) will be the same as that given by (7.8) provided we integrate the latter over all of *both* charge distributions. Equations (7.8) and (7.12) describe different situations if the integral in (7.8) were to be only over *part* of the charge distribution.

However, we can assign meaning to

$$W_e = \frac{1}{2}\int_v dv\,\varepsilon(\mathbf{r})|\mathbf{E}(\mathbf{r})|^2 \tag{7.13}$$

where v is a *finite* volume: the energy W_e then represents the theoretical maximum of kinetic energy that can be extracted from the $\mathbf{E}$ fields within v (assuming that these are not replenished).

Example 7.2 Energy stored in a sphere of uniform charge density: method starting from the field

Returning to Example 7.1, let us recalculate W_e using (7.12b). We found

$$E(r) = \frac{\varrho_v r}{3\varepsilon_0} \quad \text{for } r < r_0 \qquad \text{and} \qquad E(r) = \frac{\varrho_v r_0^3}{3\varepsilon_0 r^2} \quad \text{for } r > r_0 \tag{7.14}$$

for the given spherically symmetric charge distribution. Using $dv = 4\pi r^2 dr$ in (7.12b), we find for the fields in region 1, $r < r_0$, that

$$W_e^{(1)} = 2\pi\varepsilon_0 \int_0^{r_0} dr\, r^2 \left|\frac{\varrho_v r}{3\varepsilon_0}\right|^2 = \frac{2\pi\varrho_v^2 r_0^5}{45\varepsilon_0} \tag{7.15a}$$

and for the fields in region 2, $r > r_0$, that

$$W_e^{(2)} = 2\pi\varepsilon_0 \int_{r_0}^{\infty} dr\, r^2 \left|\frac{\varrho_v r_0^3}{3\varepsilon_0 r^2}\right|^2 = \frac{2\pi\varrho_v^2 r_0^5}{9\varepsilon_0} \tag{7.15b}$$

and we can easily see that the sum $W_e^{(1)} + W_e^{(2)}$ is equal to the W_e of (7.10). The meaning of each of the parts has little connection with the charge distribution, as each small element of charge in $r < r_0$ produces fields that contribute not only to (7.15a) but also to (7.15b), where the region lies entirely outside the charge distribution!

7.4 Extraction of energy by means of the field forces

This section deals with the mechanics of converting the stored energy to work, i.e. with the calculation of electromechanical forces produced by the available energy in the charge distribution. We saw in Section 7.1 that the gradual building-up of the charge distribution from an original situation of zero energy led to one in which potential energy W_e is available. How can that energy be used? Let us consider two special cases.

7.4.1 N isolated conductors, with charges $q_1, \ldots, q_N$

Consider a system of N isolated conductors with charges $q_1, \ldots, q_N$. In this way, the *charges* on each conductor are kept constant. The infinitesimal amount of work done by the field in moving charge q_j $(j = 1, \ldots, N)$ an infinitesimal vectorial distance $d\boldsymbol{l}$ is $dW_j = \mathbf{F}_j \cdot d\boldsymbol{l} = q_j \mathbf{E}_j \cdot d\boldsymbol{l}$. The field $\mathbf{E}_j$ is the field due to the charges $q_1, \ldots, q_{j-1}, q_{j+1}, \ldots, q_N$. Thus, the force acting on q_j is

$$\mathbf{F}_j = q_j \mathbf{E}_j = \sum_{\substack{i=1 \\ i \neq j}}^{N} \frac{q_j q_i \hat{\mathbf{R}}_{ij}}{4\pi\varepsilon_0 R_{ij}^2} \tag{7.16a}$$

We see from (7.6a) that

$$-\nabla_j W_N = -\nabla_j \left(\frac{1}{2} \sum_{i=1}^{N} \sum_{\substack{k=1 \\ k \neq i}}^{N} \frac{q_i q_k}{4\pi\varepsilon_0 R_{ik}} \right) = \sum_{\substack{i=1 \\ i \neq j}}^{N} \frac{q_i q_j \hat{\mathbf{R}}_{ij}}{4\pi\varepsilon_0 R_{ij}^2} \tag{7.16b}$$

Note that the double sum contains *twice* every term upon which ∇_j acts, either $i = j$ or $k = j$. Whether we have $1/R_{mj}$ or $1/R_{jm}$, the negative gradient operator $-\nabla_j$ acting on either results in $\hat{\mathbf{R}}_{mj}/R_{mj}^2$. From a comparison of (7.16b) and (7.16a) we see that

$$\mathbf{F}_j = -\nabla_j (W_N)_q, \tag{7.17}$$

where the subindex 'q' indicates that the gradient is to be taken upon a system in which the individual charges are kept constant on each conductor.

7.4.2 N conductors with potentials maintained at $V_1, V_2, \ldots, V_N$

Consider a second system of N conductors, where the potentials V_i on each are kept constant by means of batteries or other voltage sources. Let us allow the position of a conductor at potential V_j to change, under the action of the field force on it, $\mathbf{F}_j$, by a small displacement $d\boldsymbol{l}$; a charge δq_j is added to the plate in order to maintain the potential V_j. The work done by the batteries is $V_j \delta q_j$. Because the relative distances are now different, each of the other conductors i will experience a change in charge δq_i. Therefore the total energy supplied by the sources is

$$dW_s = \sum_i V_i \delta q_i \tag{7.18a}$$

where the sum includes all the conductors. Half of this energy is stored as an increase in electrostatic energy, see (7.6b):

$$dW_N = \frac{1}{2} \sum_i V_i \delta q_i \tag{7.18b}$$

But by the conservation of energy we may then deduce that work $dW_j = dW_s - dW_N = \frac{1}{2} \sum_i V_i \delta q_i$ has been done on the outside world by the electrostatic field force $\mathbf{F}_j$, where $dW_j = \mathbf{F}_j \cdot d\boldsymbol{l}$. Consequently,

$$\mathbf{F}_j = \nabla_j W_j \qquad \text{or} \qquad \mathbf{F}_j = +\nabla_j (W_N)_V \tag{7.19}$$

Equations (7.17) and (7.19) appear to be each other's opposite. However, charges are kept constant in one case, whereas the potentials are maintained in the other. Consider a simple example of two parallel plates with potentials V_i and V_j, where the separation distance l is allowed to decrease slightly under the action of the field force. If the plate is attached to a fixed spring this will become extended and so external work $dW = \mathbf{F} \cdot d\boldsymbol{l}$ is done by the field force $\mathbf{F}$. The stored energy

$W_2 = \frac{1}{2}\varepsilon S l E^2$, which is proportional to $1/l$ because $E = (V_j - V_i)/l$, increases if the potential difference is maintained. If, however, the charges are kept constant then E is unchanged and W_2 decreases because l does. The value of $\mathbf{F}_j$ provided by (7.17) and (7.19) is, of course, identical. This example will be repeated later from the point of view of *capacitance*.

7.5 Capacitance – definition

The concept of *capacitance*, well-known from elementary circuit network theory, is useful in expressing the stored energy in a system of conductors and dielectrics. It is best introduced by looking at two perfect conductors, as shown in Fig. 7.3, between which a voltage difference $V = V_+ - V_-$ is maintained by means of a battery. We shall suppose that the conductors are immersed in a dielectric of varying permittivity $\varepsilon(\mathbf{r})$. In some cases, one conductor can be considered to be at infinity, which effectively defines a single-conductor system. However, we will consider that to be a special case of the two-conductor system of Fig. 7.3.

The capacitance C of the system is defined as the ratio Q/V of the total positive charge Q to the voltage difference V, and its unit is the farad (F). Note that $1\ \mathrm{F} \equiv 1\ \mathrm{C/V}$. The real point of interest for linear capacitors, however, is that the capacitance is *not* a function of either Q or V, but only of the geometry and the dielectric properties of the medium between the conductors. To see this, consider an initial voltage difference V leading to a positive charge Q on one conductor and to $-Q$ on the other, if charge neutrality for the whole system is to be maintained throughout. Now increase the voltage difference to $V + \delta V \equiv \kappa V$ (κ is a constant). Along any field line between the conductors, such as the one shown in Fig. 4.3, we have from (4.2)

$$\kappa V = V + \delta V = -\int_A^B d\boldsymbol{l} \cdot (\mathbf{E} + \delta\mathbf{E}) \tag{7.20}$$

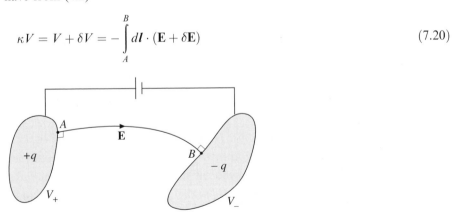

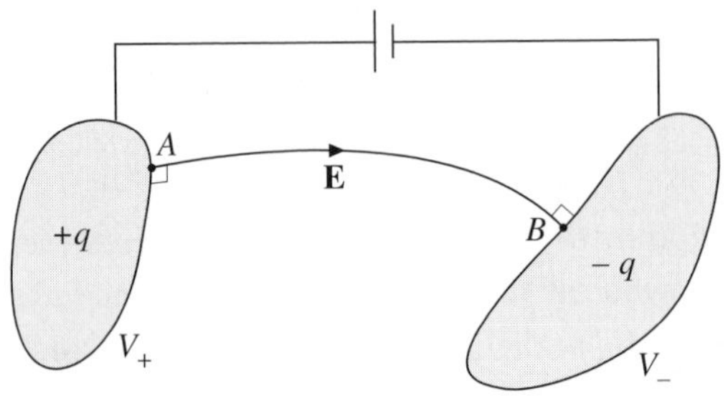

Fig. 7.3 An arbitrary two-body capacitor: the conductors are immersed in a medium with permittivity $\varepsilon(\mathbf{r})$. The potential difference is $V_+ - V_-$.

Because the geometry is totally unchanged, we expect the ratio between $\mathbf{E} + \delta\mathbf{E}$ and $\mathbf{E}$ to be constant at every point on the field line, and obviously it only can be κ. Hence $\mathbf{E} + \delta\mathbf{E} = \kappa\mathbf{E}$ everywhere between the conductors. Initially, before increasing the voltage, at the edge of the $+Q$ conductor we had the boundary condition $(\varrho_s)_i = \varepsilon(\mathbf{r}_s)\hat{\mathbf{n}} \cdot \mathbf{E}$. After the increase we have $(\varrho_s)_f = \varepsilon(\mathbf{r}_s)\hat{\mathbf{n}} \cdot (\mathbf{E} + \delta\mathbf{E}) = \kappa\varepsilon(\mathbf{r}_s)\hat{\mathbf{n}} \cdot \mathbf{E} = \kappa(\varrho_s)_i$. Thus, the surface charges everywhere on the conductors have increased by the factor κ. Hence the total charge is now $Q + \delta Q = \kappa Q$. So it is found that the increase in voltage from V to κV leads to an increase in total positive charge from Q to κQ, and as a result the new capacitance is $C = \kappa Q/\kappa V = Q/V$, i.e. *the capacitance is unchanged.* Thus, indeed, C cannot be a function of Q or V but only of their ratio. To obtain the functional dependence of C, we integrate over points $\mathbf{r}_s$ on the surface of A:

$$\begin{aligned} Q &= \oint dS\,\varrho_s = \oint dS\,\hat{\mathbf{n}} \cdot \varepsilon(\mathbf{r}_s)\mathbf{E}(\mathbf{r}_s) \\ V &= -\int_A^B d\mathbf{l} \cdot \mathbf{E} \end{aligned} \tag{7.21}$$

Then C is the ratio of the top surface integral (*only* over the positively charged conductor's surface) to any line integral V of the electrostatic field between the two conductors. For most practical purposes, the voltage difference V is a known quantity determined by the voltage source (battery), so that

$$C = \frac{1}{V}\oint dS\,\hat{\mathbf{n}} \cdot \varepsilon(\mathbf{r}_s)\mathbf{E}(\mathbf{r}_s) \tag{7.22}$$

As in (7.21), the surface integral is only over the positively charged conductor's surface.

Example 7.3 Capacitance of a parallel-plate conductor filled with uniform dielectric

Figure 7.4 shows the work-horse of elementary electrostatic theory: the parallel-plate capacitor, which consists of two parallel metallic plates separated by a

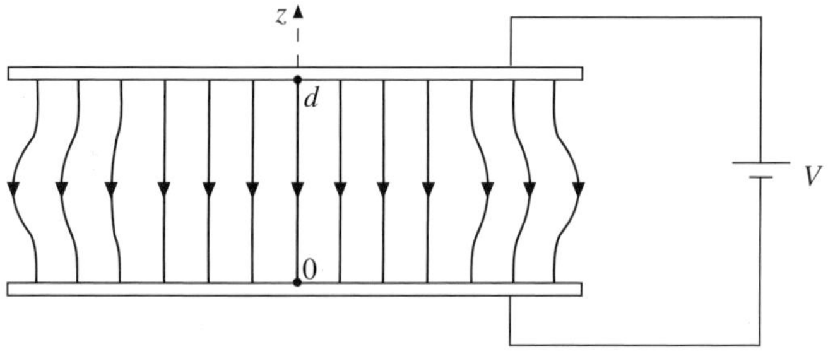

Fig. 7.4 A parallel-plate capacitor.

distance d, with a dielectric medium characterized by permittivity ε between and around the plates.

If the plates were infinitely extended in the x and y directions, we would use the symmetry arguments so often invoked before to convince ourselves that the **E** field between the plates can only point in the $\hat{\mathbf{z}}$ direction, and that E_z cannot be a function of x or of y. But Gauss's law, applied to a small cylinder placed in the $\hat{\mathbf{z}}$ direction between the plates, would convince us that E_z cannot even depend upon z, hence $\mathbf{E}(\mathbf{r}) = E_z\hat{\mathbf{z}}$ everywhere inside the capacitor, and E_z is a constant. Application of the voltage-difference integral in (7.21) would then give $E_z d = V$, because $d\boldsymbol{l} = -\hat{\mathbf{z}}\,dz$ here, so that we obtain

$$\mathbf{E} = -(V/d)\hat{\mathbf{z}} \tag{7.23}$$

If the plates do not extend to infinity, then (7.23) is approximately correct as long as it is not applied to regions near the rims of the plates, where significant 'fringing' takes place. An engineering rule of thumb is that fringing or curvature of the field lines is significant only within $1\frac{1}{2}$ separation distances d of the rim, so that capacitors with a separation distance much less than plate diameter have almost all the field region inside them governed by (7.23). Insertion of (7.23) into (7.22) for uniform ε then gives, for cross-sectional area S,

$$C = \varepsilon S/d \tag{7.24}$$

as an approximate expression for the capacitance of the parallel-plate capacitor. It is clear that the expression for C does not involve Q or V although it was derived from their ratio!

Example 7.4 Capacitance of a two-conductor coaxial cylinder structure filled with uniform dielectric

A similar calculation can be performed for a capacitor formed from two coaxial circular cylinders of length L, of which a cross section is shown in Fig. 7.5. Let us

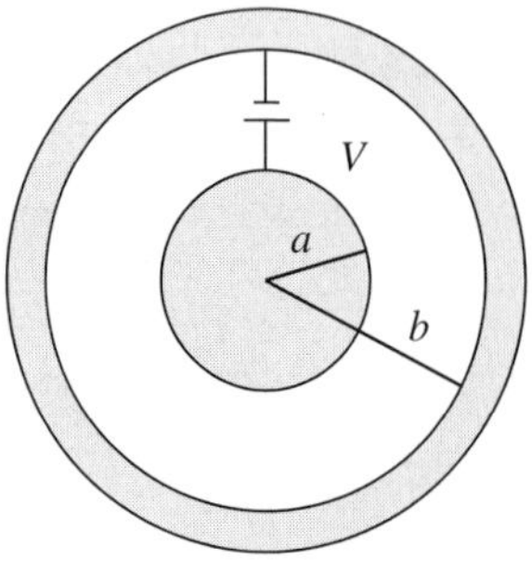

Fig. 7.5 A coaxial cylinder or a coaxial spherical-shell capacitor.

assume a charge Q on the surface of the inner cylinder at $\rho = a$: due to symmetry $Q = 2\pi a \varrho_s$, where ϱ_s is a uniform surface-charge density.

If we draw a circle around the z axis (which is the axis of the two cylinders) with radius ρ such that $a < \rho < b$ (b is the radius of the thin outer conductor), then the disk thereby defined is the cross section of a third circular cylinder of height L. This cylinder, with its top and bottom surfaces, entirely encloses the charge Q. Symmetry tells us that $\mathbf{E}(\rho, \varphi) = E_\rho(\rho)\hat{\boldsymbol{\rho}}(\varphi)$, if – as above in the previous example – we ignore fringing effects at the top and bottom. Consequently, Gauss's law predicts that

$$2\pi\rho L E_\rho(\rho) = \frac{Q}{\varepsilon} \qquad \Rightarrow \qquad E_\rho(\rho) = \frac{Q}{2\pi\varepsilon L\rho} \tag{7.25a}$$

Insert this into the line integral for potential difference, using $d\boldsymbol{l} = \hat{\boldsymbol{\rho}}(\varphi)\, d\rho$ for *fixed* angle φ:

$$V = -\int_b^a d\rho \frac{Q}{2\pi\varepsilon L\rho} = \frac{Q}{2\pi L\varepsilon}\ln\left(\frac{b}{a}\right) \tag{7.25b}$$

This yields the desired ratio $C = Q/V$:

$$C = \frac{2\pi L\varepsilon}{\ln(b/a)} \qquad \text{or} \qquad \frac{dC}{dL} = \frac{2\pi\varepsilon}{\ln(b/a)} \tag{7.26}$$

The latter expression gives the *capacitance per unit length* of the cylindrical structure (in F/m). This expression is also useful in understanding that the unit of dielectric permittivity ε must be F/m, as it has the same dimension as capacitance per unit length.

Example 7.5 Capacitance of two concentric conducting spheres immersed in uniform dielectric

Referring again to Fig. 7.5, we now interpret it as the cross section of two spherical metal shells, with radii $r = a$ and $r = b$. Symmetry arguments similar to prior cases teach us that $\mathbf{E}(r, \theta, \varphi) = E_r\hat{\mathbf{r}}(\theta, \varphi)$. Gauss's law now yields

$$4\pi r^2 E_r(r) = \frac{Q}{\varepsilon} \qquad \Rightarrow \qquad E_r(r) = \frac{Q}{4\pi\varepsilon r^2} \tag{7.27a}$$

and the line integral for potential difference yields

$$V = -\int_a^b dr \frac{Q}{4\pi\varepsilon r^2} = \frac{Q}{4\pi\varepsilon}\left(\frac{1}{a} - \frac{1}{b}\right) \tag{7.27b}$$

so that the spherical capacitor has

$$C = \frac{4\pi\varepsilon}{1/a - 1/b} = \frac{4\pi\varepsilon ab}{b - a} \tag{7.28}$$

Example 7.6 Capacitance of two parallel thin wires in a uniform dielectric

Consider two parallel infinite thin wires, each with radius a, separated by a distance $L \gg a$; see Fig. 7.6. We can make an *approximate* evaluation of the electrostatic field on the straight line joining the two centers, e.g. at point P, by using Gauss's law ($2\pi\rho h E_\rho = h\varrho_\ell$, where ϱ_ℓ is the charge per unit length on the surface of the wire, but here we let z and $L - z$, respectively, play the role of radial coordinate ρ) for each wire separately to obtain, by vector addition of the resulting fields,

$$\mathbf{E}_P \approx \frac{\varrho_\ell}{2\pi\varepsilon}\left(\frac{1}{z} - \frac{1}{L - z}\right)\hat{\mathbf{z}} \tag{7.29a}$$

The expression is approximate because the field due to one wire in fact induces a small surface charge on the other wire, and we have neglected that induced charge in (7.29a). Now perform the line integral to obtain the voltage difference between the wires:

$$V = -\frac{\varrho_\ell}{2\pi\varepsilon}\int_{L-a}^{a} dz\left(\frac{1}{z} - \frac{1}{L - z}\right) = -\frac{\varrho_\ell}{2\pi\varepsilon}\left[\ln\left(\frac{a}{L - a}\right) - \ln\left(\frac{L - a}{a}\right)\right] \quad \Rightarrow$$

$$V = \frac{\varrho_\ell}{\pi\varepsilon}\ln\left(\frac{L - a}{a}\right) \tag{7.29b}$$

and therefore, for a length z of the wire we find

$$C \approx \frac{\pi\varepsilon z L}{\ln(L/a - 1)} \quad \text{or} \quad \frac{dC}{dz} \approx \frac{\pi\varepsilon}{\ln(L/a - 1)} \tag{7.30}$$

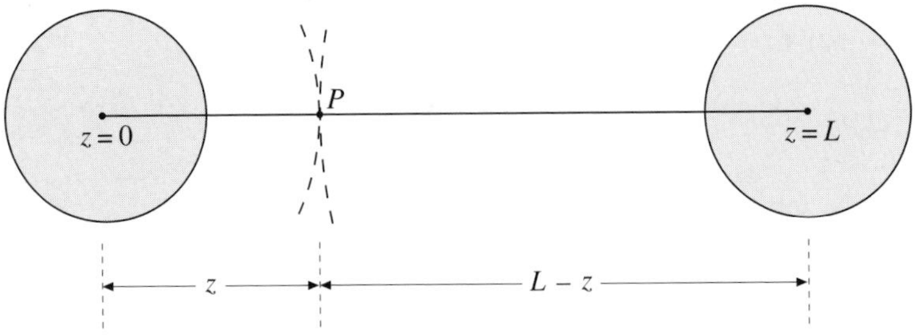

Fig. 7.6 A capacitor consisting of two identical parallel straight wires, each of radius a.

We shall return to find an exact expression at a later stage, when the *method of images* is discussed, but at this point let us state only that the value for C in (7.30) is in error by an amount $O(2a/L)$ compared to unity. This is of little practical consequence if indeed $L \gg a$.

The definition $C = Q/V$ leads, of course, to the well-known addition laws for capacitors in series or in parallel. These laws, for N capacitors, are:

$$\frac{1}{C} = \sum_{n=1}^{N} \frac{1}{C_n} \quad \text{(capacitors in series)}$$

$$C = \sum_{n=1}^{N} C_n \quad \text{(capacitors in parallel)} \tag{7.31}$$

and they follow from the fact that Q is the same for each capacitor in series and the voltages V_n add up, whereas in the case of parallel capacitors we find that V is the same and the charges Q_n need to be added.

Example 7.7 Parallel-plate capacitor with dielectric permittivity varying ⊥ to the plates

Consider a parallel-plate capacitor filled with a dielectric which has permittivity $\varepsilon(z)$ varying in the z direction, ⊥ the plates, as shown in Fig. 7.7(*a*). We can interpret a continuously varying $\varepsilon(z)$ as the limit of a number of infinitesimally thin parallel layers, each with its own uniform permittivity, as sketched in Fig. 7.7(*b*). Basically, this becomes a set of capacitors in series, because the layer between z and $z + \Delta z$ can be imagined to carry $+Q$ on its upper interface and $-Q$ on its lower interface. Thus, the total charge on each interface is zero, and only the plate charges remain ($+Q$ on the upper plate, $-Q$ on the lower plate). Hence, the first of (7.31) holds; if S is the

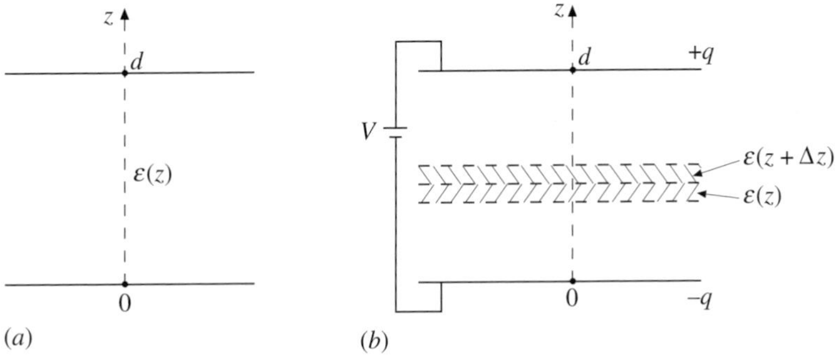

Fig. 7.7 A parallel-plate capacitor with vertically stratified permittivity profile.

surface area of the plates then

$$\frac{1}{C} = \lim_{\Delta z \to 0} \sum_{n=1}^{N=d/\Delta z} \frac{1}{C(z+(n-1)\Delta z)} = \lim_{\Delta z \to 0} \sum_{n=1}^{N=d/\Delta z} \frac{\Delta z}{S\varepsilon(z+(n-1)\Delta z)} \Rightarrow$$

$$\frac{1}{C} = \frac{1}{S}\int_0^d dz \frac{1}{\varepsilon(z)} \tag{7.32}$$

We recognize from the definition in ordinary calculus that the sum becomes an integral in the limit $\Delta z \to 0$. If there are only two layers with permittivities ε_1 of thickness d_1 and ε_2 of thickness d_2 and with $d_1 + d_2 = d$, then (7.32) becomes

$$\frac{1}{C} = \frac{1}{S}\left(\frac{d_1}{\varepsilon_1} + \frac{d_2}{\varepsilon_2}\right) \tag{7.33}$$

Example 7.8 Parallel-plate capacitor with dielectric permittivity varying $\parallel$ to the plates

Consider a parallel-plate capacitor filled with a dielectric which has permittivity $\varepsilon(x, y)$ varying in directions $\parallel$ to the plates, as shown in Fig. 7.8. Now the infinitesimal layers are $\perp$ to the plates and we have 'capacitors' in parallel. Instead of the situation of Fig. 7.7, we now have a set of rectangular boxes, of height d and of thicknesses Δx and Δy in the two other directions, so that we obtain a double sum over the integers m and n:

$$C = \lim_{\substack{\Delta x \to 0 \\ \Delta y \to 0}} \sum_m \sum_n C((m-1)\Delta x, (n-1)\Delta y)$$

$$= \lim_{\substack{\Delta x \to 0 \\ \Delta y \to 0}} \sum_m \sum_n \frac{\Delta x \Delta y}{d} \varepsilon((m-1)\Delta x, (n-1)\Delta y) \Rightarrow$$

$$C = \frac{1}{d}\int\int dx\, dy\, \varepsilon(x, y) \tag{7.34}$$

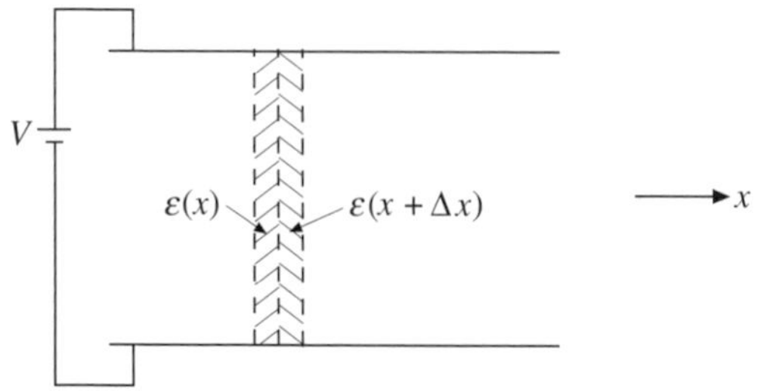

Fig. 7.8 A parallel-plate capacitor with horizontally stratified permittivity profile.

If $\varepsilon(x, y)$ is ε_1 for $x < 0$ and ε_2 for $x > 0$, the plane $x = 0$ being chosen inside the capacitor, then we have just two wide regions, and the capacitance is then

$$C = \frac{1}{d}(S_1\varepsilon_1 + S_2\varepsilon_2) \tag{7.35}$$

where S_1, S_2 and ε_1, ε_2 are the corresponding plate areas and permittivities.

7.6 Capacitance, stored energy, and forces

We see from (7.6b) in the case of a two-conductor capacitor that the stored energy $W = W_2 = \frac{1}{2}(q_1V_1 + q_2V_2)$, where $q_2 = -q_1 = -Q$. Therefore, if $V_1 - V_2 \equiv V$, the potential difference, we observe that

$$W = \tfrac{1}{2}Q(V_1 - V_2) = \tfrac{1}{2}QV = \tfrac{1}{2}CV^2 = \tfrac{1}{2}Q^2/C \tag{7.36}$$

The meaning of the word 'capacitance' is illuminated by the expression $W = \frac{1}{2}CV^2$, because the two-conductor capacitor (at given potential difference V) has a stored energy that is linearly proportional to the capacitance C.

The above formulas for stored energy in terms of capacitance allow us to illustrate the use of (7.17) and (7.19) for a parallel-plate capacitor, where it might seem, without sufficiently close inspection, that these formulas lead to different values for the force that one plate exerts upon another.

(1) Consider the situation when the plate charge Q is fixed. We should then use $\mathbf{F} = -\nabla(W)_Q$, and the most useful form to insert into this is $W = \frac{1}{2}QV$ so that $\mathbf{F} = -\frac{1}{2}Q\nabla V = \frac{1}{2}Q\mathbf{E}$.
(2) However, if V were fixed, we would then apply (7.19): $\mathbf{F} = +\nabla(W)_V$. In this case, we insert $W = \frac{1}{2}CV^2$ to obtain

$$\begin{aligned}\mathbf{F} &= \tfrac{1}{2}(\nabla C)V^2 = \tfrac{1}{2}[\nabla(\varepsilon S/z)]V^2 = -\tfrac{1}{2}\hat{\mathbf{z}}(\varepsilon S/z^2)V^2 \\ &= -\tfrac{1}{2}\hat{\mathbf{z}}(V/z)(\varepsilon S/z)V = \tfrac{1}{2}\mathbf{E}CV = \tfrac{1}{2}Q\mathbf{E}\end{aligned}$$

as before. So both formulas, used properly, lead to the same result. Was it correct to expect that the force in both these cases should be the same? The answer is of course yes, because the electric field $\mathbf{E}$ is the same whether we think of the capacitor as being isolated in space with charge $\pm Q$ on the plates or whether we consider it to be connected to a battery at voltage $V = Q/C$. In either case this field produces the same force on the positive plate, and the formulas (7.17) and (7.19) tell us how to calculate it.

We actually can prove quite generally (and quite easily!) that (7.17) and (7.19) should lead to the same result for a two-conductor capacitor. The proof goes as follows:

$$\begin{aligned}\mathbf{F}_Q &= -\nabla\left(\tfrac{1}{2}Q^2/C\right)_Q = \tfrac{1}{2}(Q^2/C^2)\nabla C = \tfrac{1}{2}V^2\nabla C \\ \mathbf{F}_V &= +\nabla\left(\tfrac{1}{2}CV^2\right)_V = \tfrac{1}{2}V^2\nabla C\end{aligned} \tag{7.37}$$

which proves the equality of the result. A proof for a larger number of capacitors follows similar lines of argument.

7.7 Capacitance of a system of many conductors

Consider a system of N conducting (metallic) bodies embedded in a uniform dielectric medium with permittivity ε (see Fig. 7.9). Let us see what would be analogous to the concept of capacitance as introduced in Section 7.5. Assume that a voltage difference $V_{ij} = V_j - V_i$ is maintained between bodies i and j (one or more of these can be at the ground or reference voltage $V_0 = 0$). We will show that the charge induced on the surface of the ith body is

$$q_i = \sum_{j \neq i} C_{ij} V_{ij} \qquad \text{with} \qquad C_{ij} = \frac{\partial q_i}{\partial V_{ij}} \tag{7.38}$$

This can be demonstrated by reasoning similar to that at the start of Section 7.5. Along any field line between a point P_j on body j and a point P_i on body i it must be true that $V_{ij} = -\int_{P_i}^{P_j} d\boldsymbol{l} \cdot \mathbf{E}$. It follows from this that an increase by a factor κ in $E(\mathbf{r})$ everywhere in the space between the conductors leads to an identical increase in all V_{ij}. Likewise, because the boundary condition at the surface of each conductor is $\varrho_s = \hat{\mathbf{n}} \cdot \varepsilon\mathbf{E}$, it follows that all charges q_i are modified by the same factor κ. Thus, there must be a linear relationship between any q_i and all the voltage differences V_{ij}

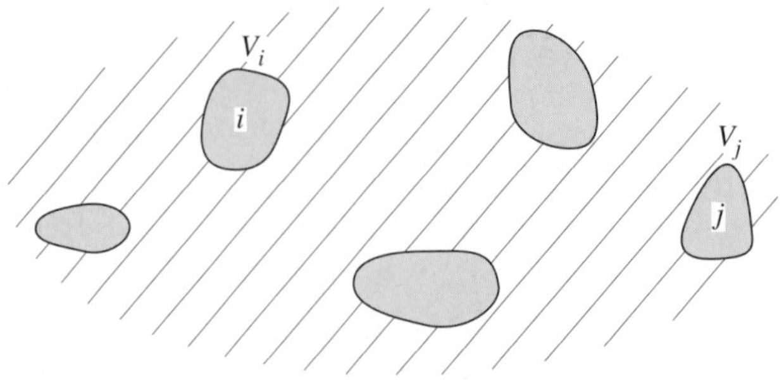

Fig. 7.9 Some of the conductors in a multicapacitor system. Each conductor is maintained at its own fixed potential; the potential difference between conductors i and j is denoted by $V_{ij} = V_j - V_i$.

between body i and any other body, as the charges on body i are determined by only those voltage differences. The C_{ij} defined as partial derivatives in (7.38) are *mutual capacitances*. If body i is grounded, then C_{ij} is known as a *self capacitance*. There are $N-1$ terms in the sum in (7.38) but N potentials V_i, so an alternative sum can be formed from (7.38):

$$q_i = \sum_{j=1}^{N} \tilde{C}_{ij} V_j \qquad \text{with} \qquad \tilde{C}_{ii} = \sum_{j \neq i} C_{ij} \quad \text{and} \quad \tilde{C}_{ij} = -C_{ij} \tag{7.39}$$

Similar reasoning leads to the reciprocity relationship $C_{ji} = C_{ij}$; it follows from the definitions $C_{ij} = \partial q_i / \partial V_{ij}$ and $C_{ji} = \partial q_j / \partial V_{ji}$. When the fields are increased by a factor κ, then V_{ij} and V_{ji} increase by a factor κ. The boundary condition $\varrho_s = \hat{\mathbf{n}} \cdot \varepsilon \mathbf{E}$ then ensures that q_i and q_j are modified by a factor κ; note however that $d_{qj} = -d_{qi}$ because field lines connecting the two bodies have factors $\hat{\mathbf{n}} \cdot \mathbf{E}$ with opposite signs. The same reciprocity holds for the coefficients $\tilde{C}_{ij}$, according to (7.39).

Equation (7.39) also shows directly that

$$\sum_{j=1}^{N} \tilde{C}_{ij} = 0 \qquad \text{and} \qquad \sum_{i=1}^{N} \tilde{C}_{ij} = 0 \tag{7.40}$$

The second of these follows from invoking reciprocity in the first. There are $\frac{1}{2}N(N+1)$ different coefficients C_{ij} (allowing $j = i$), and (7.40) gives N linear relationships among them.

How can the C_{ij} be calculated? The answer lies in the expression for the charge on body i:

$$q_i = \oint (dS \varrho_s)_i = \varepsilon \oint (d\mathbf{S} \cdot \mathbf{E})_i = \sum_{j \neq i} C_{ij} V_{ij} \tag{7.41}$$

We can use an appropriate Laplace-equation solver to calculate the $\mathbf{E}$ field in the dielectric medium between the conductors for given $V_1, V_2, \ldots, V_N$, and thus the surface integral in the third expression in (7.41). We then redo this calculation after changing V_i to $V_i + \delta V_i$ (an infinitesimal change). The surface integral then gives us

$$\varepsilon \oint [d\mathbf{S} \cdot (\mathbf{E}_i + \delta \mathbf{E})]_i = q_i + \delta q_i \qquad \Rightarrow \qquad \delta q_i = \varepsilon \oint (d\mathbf{S} \cdot \delta \mathbf{E})_i \tag{7.42}$$

It then follows that $C_{ij} = \delta q_i / \delta V_{ij}$. In fact the Laplace calculation is best done N times, each time with $V_i = 1$ and all the other $V_j = 0$. Superposition of these N solutions will yield all the information needed in (7.41) and (7.42). For example, the field $\mathbf{E}$ at an arbitrary location when we have potentials $V_1, V_2, \ldots, V_N$, measured in volts, is the superposition

$$\mathbf{E}(V_1, V_2, \ldots, V_N) = \left(\frac{V_1}{\text{volts}}\right)\mathbf{E}(1,0,0,\ldots,0) + \left(\frac{V_2}{\text{volts}}\right)\mathbf{E}(0,1,0,\ldots,0) + \cdots + \left(\frac{V_N}{\text{volts}}\right)\mathbf{E}(0,0,0,\ldots,1) \tag{7.43}$$

where each row of zeroes and a single 1 in the argument of **E** indicates the special field solution at that arbitrary location for bounding potentials all zero except for one, which equals 1 volt. Having obtained these N basic solutions for the purpose of calculating the third term of (7.41), we then form

$$\mathbf{E}(V_1 + \delta V_1, V_2, \ldots, V_N) = \left(\frac{V_1 + \delta V_1}{\text{volts}}\right)\mathbf{E}(1,0,0,\ldots,0) + \left(\frac{V_2}{\text{volts}}\right)\mathbf{E}(0,1,0,\ldots,0) + \cdots + \left(\frac{V_N}{\text{volts}}\right)\mathbf{E}(0,0,0,\ldots,1) \tag{7.44}$$

We arrive at the last integral in (7.42) by taking the difference of this field and that of (7.43) to obtain the C_{1j}. The C_{ij} for $i > 1$ are calculated in an obvious and similar fashion. The $\tilde{C}_{ij}$ then are easily found via (7.39).

A linear combination, as in (7.44), of N previously determined elementary solutions for **E** can represent a substantial saving in computer time, especially if many compound capacitor systems that are geometrically similar but have different voltages are considered simultaneously.

The energy stored in the system is obtained from (7.6b) and (7.39):

$$W_N = \frac{1}{2}\sum_{i=1}^{N} q_i V_i = \frac{1}{2}\sum_{i=1}^{N}\sum_{j\neq i}^{N} \tilde{C}_{ij} V_i V_j \tag{7.45}$$

so that the change in system energy upon changing potential V_i is

$$\frac{\partial W_N}{\partial V_i} = \sum_{j\neq i}^{N} \tilde{C}_{ij} V_j \tag{7.46}$$

This rounds off the discussion on capacitance and it demonstrates the essentials of calculating and using the capacitance coefficients for a (linear) system of conductors.

Problems

7.1. Reconsider the situation of Example 7.2: a sphere with charge density $\varrho_v =$ constant for $r \leq a$ and zero for $r > a$. Find an expression that gives the energy stored between $r = a$ and $r = 2a$.

7.2. A charge distribution in free space is given as $\varrho_{\mathrm{v}} = -2\varepsilon_0/r^3\,\mathrm{C/m^3}$ for $r > 1\,\mathrm{m}$ (there is no charge in the region $r < 1\,\mathrm{m}$). Calculate the stored electrostatic energy in the cylindrical shell defined by $1 < r < 3$, $-1 < z < 1$, where the distances are again in meters.

7.3. A spherically symmetric charge distribution is defined to be $\varrho_{\mathrm{v}}(r) = \rho_{\mathrm{v0}} r/a$ for $r < a$ and zero for $r > a$ (ρ_{v0} and a are constants). Calculate the energy stored in the charge distribution (a) using (7.8), (b) using (7.12b).

7.4. An infinite uniform line charge has a density of $5.35 \times 10^{-9}\,\mathrm{C/m}$. How much energy could one maximally extract from it by using the fields in $20 < \rho < 25$, $35^\circ < \varphi < 36^\circ$, $0 < z < 10$ (all distances in m)?

7.5. A point charge $q = 5\sqrt{\varepsilon_0}$ C is located at the origin. How much field energy is stored in $1\,\mathrm{m} < r < 2\,\mathrm{m}$?

7.6. Given that the classical radius of an electron is 2.81×10^{-15} m, assume that the charge 1.602×10^{19} C of the electron is distributed uniformly inside the electron, and calculate the electrostatic energy stored in the electron.

7.7. Find the electrostatic energy stored in a spherical region with radius $r > a$ around an infinitesimal electric dipole with moment $\mathbf{p}$ (a is a constant radius).

7.8. A parallel-plate capacitor is filled with a dielectric with $\varepsilon_{\mathrm{r}} = 1.5$. The plate separation distance $d = 10\,\mathrm{mm}$, and the plate area $S = 0.25\,\mathrm{m}^2$. The imposed voltage difference is 500 V. Calculate the amount of energy stored in the electric fields and compare it with the value of $\frac{1}{2}CV^2$.

7.9. A parallel-plate capacitor with plates at $z = 0$ and $z = d$ is filled with a dielectric having relative permittivity $\varepsilon_{\mathrm{r}}(x, y, z) = 1 + e^{-(x/a+y/b)}$. The horizontal extent of the plates is $-L_x < x < L_x$ and $-L_y < y < L_y$. Calculate the capacitance in terms of what is given, but ignore fringe effects.

7.10. A circularly coaxial cylindrical structure with its axis coinciding with the z axis, and with core radius a and sheath radius b, is filled in the sheath region $a < \rho < b$ with dielectric $\varepsilon_{\mathrm{r1}}$ in $0 < \varphi < \pi$ and with dielectric $\varepsilon_{\mathrm{r2}}$ in $\pi < \varphi < 2\pi$. Calculate the capacitance per unit length.

7.11. Repeat the calculation of problem 7.10 for the same coaxial cylindrical structure, but now with dielectric $\varepsilon_{\mathrm{r1}}$ in $a < \rho < c$ and with dielectric $\varepsilon_{\mathrm{r2}}$ in $c < \rho < b$, given that $a < c < b$.

7.12. A parallel-plate capacitor with plate area S and separation distance d is filled with a dielectric ε. A battery has charged the plates, which have a voltage V across them. After the battery is disconnected, the plates are moved to a separation distance $0.75d$.

(a) Express the capacitance C in terms of S, d, ε for the new situation.

(b) How have the charge on the plates and the voltage between them changed?

8

The Laplace and Poisson equations of electrostatics

So far, we have reduced the problem of obtaining electric fields to one of first obtaining the electrostatic potential V and then finding the field from $\mathbf{E} = -\nabla V$. Clearly, a general method for finding $V(\mathbf{r})$ is needed. While the integral equations (4.7) or (4.8a) are indeed quite general, they are usually not easy to solve, even by numerical methods, when the charge distributions are three-dimensional. Moreover, numerical solutions require great care when we need V_P in the immediate vicinity of charges at points Q, because the distance R_{QP} in the denominator of the integrand becomes small, so that inaccuracies in numerical sampling of the integrand are multiplied greatly compared to those arising from more distant charges. Experience shows that another approach is much more fruitful: direct solution of the differential equations that result from $\nabla \cdot \varepsilon\mathbf{E} = \rho$ after substitution of $\mathbf{E} = -\nabla V$. The ensuing equations have been studied extensively. This is not to say that integral-equation methods have been abandoned: as we shall show in Chapter 9, there are many situations in which the charge distribution is two dimensional, or can be considered as being so; in such situations the numerical solution of integral equations is even advantageous. However, in this chapter we shall study well-known aspects of the differential equations for $V(\mathbf{r})$ that result from Maxwell's equations. The best-known of these are Laplace's equation in charge-free regions and Poisson's equation in regions with charge distributions. It is often, if not always, possible to subdivide device regions in such a way that one of these equations holds for each region.

8.1 The general equation for electrostatic potential

Let us return to equations (5.23a, b), the Maxwell equations of electrostatics in a dielectric medium containing a charge distribution. From the Helmholtz theorem, discussed in connection with (5.18), we learned that the fact that $\nabla \times \mathbf{E} = 0$ everywhere in space implies that a scalar potential field $V(\mathbf{r})$ can be defined such that $\mathbf{E}(\mathbf{r}) = -\nabla V(\mathbf{r})$ everywhere. Of course we had already introduced the concept of potential separately in Chapter 4 and arrived at the relationship $E = -\nabla V$ via the concept of work upon a unit charge. Here we will invoke in a more abstract fashion the Helmholtz theorem and, for the purpose of this chapter, we need not know the physical meaning of the field $V(\mathbf{r})$. When $-\nabla V$ is inserted for $\mathbf{E}$ in the other electrostatic Maxwell equation, we obtain for an isotropic medium, in which $\mathbf{D} = \varepsilon(\mathbf{r})\mathbf{E}$,

$$\nabla \cdot [\varepsilon(\mathbf{r})\nabla V(\mathbf{r})] = -\varrho_{\mathrm{V}} \qquad \Rightarrow \qquad \varepsilon(\mathbf{r})\Delta V(\mathbf{r}) + \nabla\varepsilon(\mathbf{r}) \cdot \nabla V(\mathbf{r}) = -\varrho_{\mathrm{V}} \tag{8.1a}$$

where

$$\Delta \equiv \nabla^2 \equiv \nabla \cdot \nabla \equiv \operatorname{div} \operatorname{grad}$$

Equation (8.1a) is a second-order partial differential equation, which, in Cartesian coordinates, would be

$$\frac{\partial}{\partial x}\left[\varepsilon(\mathbf{r})\frac{\partial}{\partial x}V(\mathbf{r})\right] + \frac{\partial}{\partial y}\left[\varepsilon(\mathbf{r})\frac{\partial}{\partial y}V(\mathbf{r})\right] + \frac{\partial}{\partial z}\left[\varepsilon(\mathbf{r})\frac{\partial}{\partial z}V(\mathbf{r})\right] = -\varrho_{\mathrm{V}}$$

$$\Rightarrow \qquad \varepsilon(\mathbf{r})\sum_{i=1}^{3}\left(\frac{\partial^2 V}{\partial x_i^2}\right) + \sum_{i=1}^{3}\left(\frac{\partial \varepsilon}{\partial x_i}\frac{\partial V}{\partial x_i}\right) = -\varrho_{\mathrm{V}} \tag{8.1b}$$

It is not difficult to see that some extra conditions must be imposed to obtain a *unique* solution $V(\mathbf{r})$. Consider for example a collection of dielectrics, conductors, and a charge distribution in space, for which (8.1) holds. If we had exactly the same geometry and materials but *no spatial charge distribution* ϱ_{V}, then the equation for $V(\mathbf{r})$ would be

$$\nabla \cdot [\varepsilon(\mathbf{r})\nabla V(\mathbf{r})] = 0 \tag{8.2}$$

Let us assume that we have found a solution for (8.2); call it $V_0(\mathbf{r})$. If we now manage (with whatever difficulty is involved) to find a solution for (8.1) – call it $V_1(\mathbf{r})$ – then the combination $V_1(\mathbf{r}) + \text{constant} \times V_0(\mathbf{r})$ also satisfies (8.1). Any *real physical* situation, of course, would have to have a unique solution without an unspecified constant afloat in it.

In actual fact, the situation is more complicated. To illustrate this consider the one-dimensional analog of (8.1),

$$\frac{d}{dx}\left[\varepsilon(x)\frac{d}{dx}v(x)\right] = -\varrho_{\mathrm{V}} \tag{8.3}$$

This is a *second-order total differential equation*. The homogeneous equation (i.e. the same equation but with zero as the right-hand side), according to the theory of differential equations, has *two* independent solutions $v_1(x)$ and $v_2(x)$. One of the best-known examples of a second-order differential equation is the *harmonic equation*, $d^2v(x)/dx_2 + k^2v(x) = 0$, which has the two independent solutions $v_s(x) = \sin kx$ and $v_c(x) = \cos kx$. These two solutions are called *independent* because no linear combination of them, $c_1 v_s(x) + c_2 v_c(x)$ with two arbitrary nonzero constants c_1 and c_2, can be zero for *all* x. Thus the *general* solution of (8.3) must be $v_p(x) + c_1 v_1(x) + c_2 v_2(x)$, with two constants c_1 and c_2, if $v_p(x)$ is a particular solution of (8.3). This checks with the requirements prescribed by the theory for (8.3), namely that the solution for (8.3) in a region $a \leq x \leq b$ requires *initial conditions* such as knowledge of $v(a)$ and of the derivative $v'(a)$. In this case we have

$$\begin{aligned} v(a) &= v_p(a) + c_1 v_1(a) + c_2 v_2(a) \\ v'(a) &= c_1 v_1'(a) \end{aligned} \tag{8.4}$$

which are sufficient for solving uniquely for c_1 and c_2 and thus being assured of a unique solution to (8.3). This usually is also possible if we specify $v(a)$ and $v(b)$ as *boundary conditions.*

Whereas in the one-dimensional case initial or boundary conditions are needed, we will find that the three-dimensional case requires that we specify boundary conditions on an entire *surface*. Let us assume that we have two independent solutions of (8.1), namely $V_1(\mathbf{r})$ and $V_2(\mathbf{r})$. The difference $V_d \equiv V_1(\mathbf{r}) - V_2(\mathbf{r})$ must satisfy (8.2), i.e.

$$\nabla \cdot [\varepsilon(\mathbf{r}) \nabla V_d(\mathbf{r})] = 0 \tag{8.5}$$

We wish to discover the circumstances under which $V_d(\mathbf{r}) = 0$ everywhere in the volume v of the space in which (8.1) is presumed to hold, for only in this case where $V_1 = V_2$ is the solution unique. However, in some cases, we shall be satisfied with $V_d(\mathbf{r}) =$ constant because this means that the solution for $V(\mathbf{r})$ is unique except for a constant value, which is not a problem because potential *differences* and not potential values are of physical importance. Hence we will try to see under what circumstances ∇V_d is zero everywhere. The details are given in the following optional section.

Consider the following volume integral $\mathcal{I}$, which has a *positive definite* integrand, i.e. one that cannot be negative anywhere (if $\varepsilon/\varepsilon_0$ is positive everywhere); this is a condition that certainly holds for all loss-free dielectrics (a *lossy dielectric* is characterized by a complex permittivity; see Section 17.2).

$$\begin{aligned} \mathcal{I} &= \int_v dv\, \varepsilon |\nabla V_d|^2 = \int_v dv\, (\nabla V_d) \cdot (\varepsilon \nabla V_d) \\ &= \int_v dv\, \nabla \cdot (V_d \varepsilon \nabla V_d) - \int_v dv\, V_d [\nabla \cdot (\varepsilon \nabla V_d)] \\ &= \oint_S d\mathbf{S} \cdot (V_d \varepsilon \nabla V_d) \end{aligned} \tag{8.6}$$

The second line of (8.6) follows from the vector identity $(\nabla V)\cdot\mathbf{A} = \nabla\cdot(V\mathbf{A}) - V(\nabla\cdot\mathbf{A})$ for any scalar V and field $\mathbf{A}$. The third line recognizes that the divergence of $\varepsilon\nabla V_\mathrm{d}$ is zero in the second term by virtue of (8.5), and that Gauss's divergence theorem transforms the first term into a surface integral.

Now *if* $\mathcal{I}$ is to be zero, then the first equality in (8.6) tells us that this is only possible if the integrand is zero *everywhere* (because we already know that the integrand is positive definite). This requires $\nabla V_\mathrm{d}(\mathbf{r}) = 0$ everywhere in v, or, equivalently, that $V_\mathrm{d}(\mathbf{r})$ is constant in v. To state this again in another way: the last line of (8.6) tells us that $V_\mathrm{d}(\varepsilon\nabla V_\mathrm{d})$ must be zero everywhere on the bounding surface S around region v in order for $\mathcal{I}$ to be zero. This in turn tells us that V_d in the interior of v is a constant so that $V_2 = V_1 + \text{constant}$.

There are three alternative conditions, given below, that ensure a unique solution to (8.1). The derivation of those conditions is given above in the optional reading section, and they follow basically from (8.6).

(1) Let the value of $V(\mathbf{r})$ be specified at all points of the bounding surface S of the volume v in which (8.1) is to apply. This is known as the *Dirichlet* boundary condition for a unique solution. It would describe a situation such as that inside an infinite coaxial cylinder with one voltage on the interior and another on the exterior cylinder. We are assured of unique $V(\mathbf{r})$ in this case, because certainly $V_\mathrm{d} = 0$ at the specified boundary locations.

(2) Let the value of $\hat{\mathbf{n}}\cdot[\varepsilon(\mathbf{r})\nabla V(\mathbf{r})]$ be specified at all points of S, where $\hat{\mathbf{n}}$ is the unit surface-normal vector pointing *out* of volume v. This is the *Neumann* boundary condition, and it might describe the (somewhat artificial) situation inside an imaginary spherical surface at distance R from a point source q because we know that

$$\varepsilon(\mathbf{r})E_r = -\hat{\mathbf{r}}\cdot\varepsilon(\mathbf{r})\nabla V(\mathbf{r}) = q/4\pi R^2$$

there. Note that $\varepsilon(\mathbf{r})\nabla V(\mathbf{r}) = -\mathbf{D}(\mathbf{r})$, so that the Neumann condition for (8.1) implies that we specify $D_\mathrm{n}(\mathbf{r}) = \hat{\mathbf{n}}\cdot\mathbf{D}(\mathbf{r})$ – the normal component of the flux density – at all surface locations. The potential $V(\mathbf{r})$ is unique, except possibly up to a constant, in v. An example is a volume completely surrounded by perfectly conducting plates. To specify $V(\mathbf{r})$ everywhere, we need only specify the surface-charge density $\varrho_\mathrm{s} = \hat{\mathbf{n}}\cdot\mathbf{D}(\mathbf{r})$ over the bounding surface.

(3) Let us partition the surface S into two groups of parts, in such a way that each part belongs to either S_1 or S_2, with $S_1 + S_2 = S$. On S_1 we specify $V(\mathbf{r})$ and on S_2 we specify $D_\mathrm{n}(\mathbf{r})$. This is known as a *mixed* boundary condition. It is the one we encounter most often. Here, the reason that $V(\mathbf{r})$ is uniquely specified is that, as in the case of the Dirichlet condition, $V(\mathbf{r})$ is specified on S_1.

8.2 Laplace and Poisson equations for uniform-permittivity regions

If, in a region v of interest, the permittivity $\varepsilon(\mathbf{r})$ is a constant ε, then (8.1a) simplifies to

$$\Delta V = -\varrho_{\rm v}/\varepsilon \tag{8.7a}$$

This formula presents a form of the *Poisson equation*. The counterpart for zero charge density,

$$\Delta V = 0 \tag{8.7b}$$

is known as the *Laplace equation*. The uniqueness conditions for these equations are almost identical to the above ones for (8.1); the only difference is that we may specify $E_{\rm n} = -\hat{\mathbf{n}} \cdot \nabla V$ instead of specifying $D_{\rm n}$, at some or all locations of the bounding surface.

From (3.32b) inserted into (5.3), we see that the Laplacian $\Delta V \equiv \text{div grad } V$ has the general form

$$\Delta V(\mathbf{r}) = \frac{1}{h_1 h_2 h_3} \sum_{i=1}^{3} \frac{\partial}{\partial u_i} \left(\frac{h_j h_k}{h_i} \frac{\partial V}{\partial u_i} \right) \tag{8.8}$$

in the understanding that i, j, k are cyclical permutations of 1, 2, 3. In the three major coordinate systems, we obtain

$$\begin{aligned}
&\text{Cartesian} && \Delta V(\mathbf{r}) = \frac{\partial^2 V}{\partial x^2} + \frac{\partial^2 V}{\partial y^2} + \frac{\partial^2 V}{\partial z^2} \\
&\text{cylindrical} && \Delta V(\mathbf{r}) = \frac{1}{\rho} \frac{\partial}{\partial \rho} \left(\rho \frac{\partial V}{\partial \rho} \right) + \frac{1}{\rho^2} \frac{\partial^2 V}{\partial \varphi^2} + \frac{\partial^2 V}{\partial z^2} \\
&\text{spherical} && \Delta V(\mathbf{r}) = \frac{1}{r^2} \frac{\partial}{\partial r} \left(r^2 \frac{\partial V}{\partial r} \right) + \frac{1}{r^2 \sin\theta} \frac{\partial}{\partial \theta} \left(\sin\theta \frac{\partial V}{\partial \theta} \right) + \frac{1}{r^2 \sin^2\theta} \frac{\partial^2 V}{\partial \varphi^2}
\end{aligned} \tag{8.9}$$

An important case in which (8.1) *cannot* be reduced to the Poisson equation (8.7) is that of an *anisotropic* medium, for which the permittivity is a *tensor*; see subsection 6.4.3. That is to say, in Cartesian coordinates,

$$\begin{aligned}
D_x &= \varepsilon_{11} E_x + \varepsilon_{12} E_y + \varepsilon_{13} E_z \\
D_y &= \varepsilon_{21} E_x + \varepsilon_{22} E_y \varepsilon_{23} E_z \qquad \text{or} \qquad \mathbf{D} = \boldsymbol{\varepsilon}(\mathbf{r})\mathbf{E} \\
D_z &= \varepsilon_{31} E_x + \varepsilon_{32} E_y + \varepsilon_{33} E_z
\end{aligned} \tag{8.10}$$

With the second notation in (8.10) we are treating $\boldsymbol{\varepsilon}$ as a 3×3 matrix and $\mathbf{E}$, $\mathbf{D}$ as 3×1 (one-column) matrices. In such an anisotropic medium, the relationship between **D** and **E** at any point depends upon the *direction* of **E**. The relationship may be the same from point to point (i.e. the medium may be uniform), but at each point there is a dependence upon direction, so that **D** in general does not point in the same direction as **E** (for if $\mathbf{D} \parallel \mathbf{E}$, then $\varepsilon_{11} = \varepsilon_{22} = \varepsilon_{33}$ and all the other ε_{ij} are zero). Equation (8.1a) becomes

$$\nabla \cdot [\boldsymbol{\varepsilon}(\mathbf{r})\nabla V(\mathbf{r})] = -\varrho_{\mathrm{v}} \tag{8.11a}$$

If $\boldsymbol{\varepsilon}(\mathbf{r}) = \boldsymbol{\varepsilon}$ is a constant tensor, then the Cartesian form of (8.11a) would be

$$\sum_{i=1}^{3}\left[\varepsilon_{i1}\frac{\partial^2 V}{\partial x_i\,\partial x_1} + \varepsilon_{i2}\frac{\partial^2 V}{\partial x_i\,\partial x_2} + \varepsilon_{i3}\frac{\partial^2 V}{\partial x_i\,\partial x_3}\right] = -\varrho_{\mathrm{v}} \tag{8.11b}$$

if (x_1, x_2, x_3) corresponds to (x, y, z). This is more complicated in structure than (8.7), as it consists of nine terms rather than three.

Example 8.1 The parallel-plate capacitor

In Fig. 8.1, we show a parallel-plate capacitor of finite dimensions. In the Gaussian cylinder shown in the diagram, where all points are far from the edges, we can assume to good accuracy that the **E** field must be vertical (this certainly would be indisputable if the two plates extended to infinity in all directions $\perp \hat{\mathbf{z}}$).

Thus, from the symmetry inside the cylinder, we know that $V(z)$ and $\mathbf{E}(z) = -\hat{\mathbf{z}}\, dV/dz$ are functions only of z and, consequently, that the equation for the region *between* the plates (and inside the cylinder) would be the simplified Laplace equation

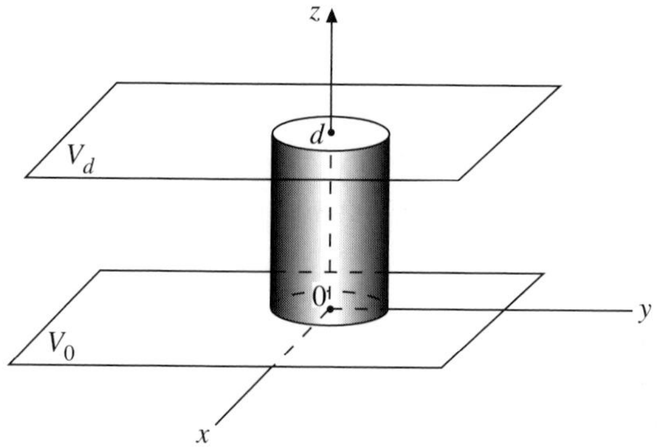

Fig. 8.1 Boundary conditions for a parallel-plate capacitor.

because there is no charge between the plates:

$$\frac{d^2V}{dz^2} = 0 \qquad \Rightarrow \qquad V(z) = az + b \tag{8.12a}$$

We now use the boundary conditions to determine the constants a and b. At $z = 0$ we have $V(0) = b$, so that as required $V(z) = az + b$. At $z = d$ we have $V(d) = ad + b$, which gives $a = [V(d) - V(0)]/d$. Consequently, the unique solution to the Laplace equation here is

$$V(z) = V(0) + \frac{z}{d}[V(d) - V(0)] \tag{8.12b}$$

If we were to extend the region v infinitesimally in the $\pm\hat{\mathbf{z}}$ directions to include the charges on the (insides of) the plates, we would have a Poisson equation, which we might solve in a similar fashion, but we will not introduce this needless complication here.

We need to discuss the boundary conditions we have used, because we have stressed above that boundary conditions are needed on a *closed* surface to obtain a unique solution. Here, we appear to have used only Dirichlet conditions on the upper and lower surfaces of the cylinder, so it may seem surprising that we have obtained a unique solution in (8.12b). The surprise is only if one does not realize that a boundary condition for the 'wrap-around' surface of the cylinder has been included implicitly: we used the symmetry to argue that $\mathbf{E}$ is vertical everywhere inside the cylinder. This implies an assumption that the dot product $\hat{\mathbf{n}} \cdot \mathbf{E}$ is zero, where $\hat{\mathbf{n}}$ is a unit vector $\perp$ the wrap-around surface (and therefore $\perp \hat{\mathbf{z}}$). This assumption is simply the Neumann condition for the wrap-around surface. Hence we have really applied *mixed boundary conditions* to the entire closed surface, well inside the edges of the capacitor, namely $V(\mathbf{r}) = V_d$ on the top surface, $V(\mathbf{r}) = V_0$ on the bottom surface, and $\hat{\mathbf{n}} \cdot \nabla V(\mathbf{r}) = 0$ on the wrap-around surface. That suffices to ensure that we obtain a unique solution (8.12b). At the edges of the plates we can retain the Dirichlet conditions, but we must relinquish the Neumann condition $\hat{\mathbf{n}} \cdot \mathbf{E} = 0$, because it certainly is not true: there is 'fringing', i.e. the electrostatic field lines are curved, as shown in Fig. 7.4. Thus, here varying techniques for incorporating the boundary conditions have to be used because we are not able to replace the Neumann condition by something leading to a direct analytical solution of (8.7b). We are only able to solve in the interior by approximating the boundary condition on a part of the bounding surface with unit surface vectors $\perp \hat{\mathbf{z}}$.

8.3 The method of images

The three alternative boundary conditions at the end of Section 8.1 for the uniqueness of solutions to Poisson's equation provide a powerful method for solving a set of seemingly intractable problems: the *method of images*. The method is perhaps best explained by first doing an example and then commenting on what seems to be the guiding principle.

Example 8.2 A point charge above a grounded planar surface

In Fig. 8.2, we show field lines from a point charge at location $(0, 0, h)$ above a grounded $z = 0$ plane. Symmetry tells us that the field line to the coordinate origin O is a vertical straight line. Other field lines will be curved, as shown, but they must be $\perp$ to the plane at the point of contact, as field lines are always perpendicular to equipotential planes. It is not at all clear how we can solve this problem using Poisson's or Laplace's equation. However, we do know the boundary values of $V(\mathbf{r})$ on the plane (it is zero there) and in the upper hemispherical surface S_∞, obtained by letting $r \to \infty$, where also $V(\mathbf{r}) = 0$. Hence we have well-defined Dirichlet conditions on the closed surface consisting of $z = 0$ and the hemispherical surface S_∞.

The method of images *replaces* the grounded surface at $z = 0$ by an image charge $-q$ at $(0, 0, -h)$; the two charges $+q$ and $-q$ have the same pattern of field lines in the half-space $z \geq 0$. This is shown in Fig. 8.3. The solution for the potential at any point P in the field of two such charges $+q$ and $-q$ is

$$V_P = \frac{q}{4\pi\varepsilon_0}\left(\frac{1}{R_+} - \frac{1}{R_-}\right) \tag{8.13}$$

given that R_+ and R_- are the distances of a point P to the charges $+q$ and $-q$. This, of course, is an easy solution, which we were able to find because we can just add up the potentials from point charges as calculated in Example 4.1. But is this also the solution to the potential in the upper hemisphere of Fig. 8.2? This would be the case if only we were certain that the boundary potentials for the upper half of Fig. 8.3 were identical to those for Fig. 8.2. Fortunately, it is easy to show that they are. For a point Q lying on the $z = 0$ plane, it is obvious from the symmetry that $R_- = R_+$,

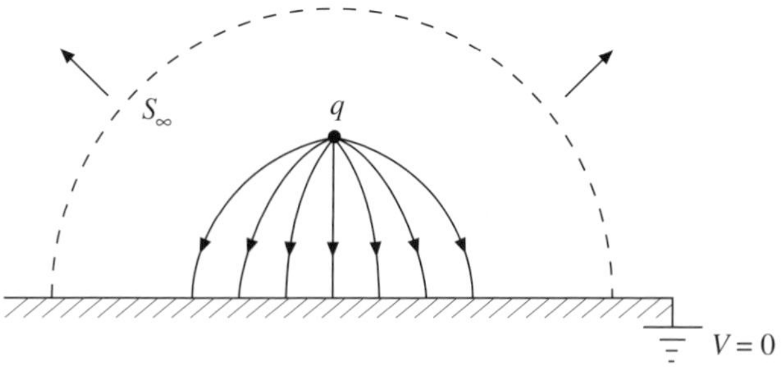

Fig. 8.2 A point charge above a grounded planar surface.

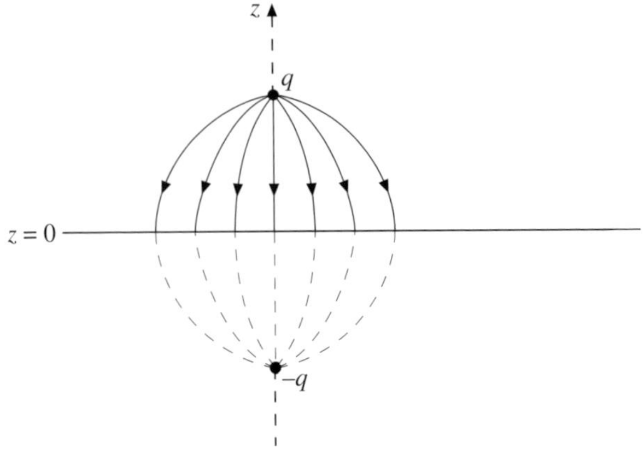

Fig. 8.3 The equivalent image charge for the situation of Fig. 8.2.

hence $V_Q = 0$. It is also obvious that $R_+ \to \infty$ and $R_- \to \infty$ if we allow the point P to recede to infinity. Hence the potential on the infinite hemispherical surface S_∞ of Fig. 8.3 is also zero. Thus the situations of Fig. 8.2 and Fig. 8.3 have an identical charge $+q$ at the same location, and they have identical bounding surfaces on which identical Dirichlet conditions are specified. It follows that (8.13) must be the solution also in the upper hemisphere of Fig. 8.2. Of course this tells us nothing about the lower hemisphere.

The power of the method of images lies in the ability to replace a complicated situation consisting of charges and equipotential surfaces by the same charges and image charges – which allows for a highly simplified analytical solution. That is, the real equipotential surfaces, usually consisting of metallic plates, are replaced by appropriately situated image charges. We will discuss several further examples.

Example 8.3 A point charge and two grounded planes intersecting at right angles

In Fig. 8.4, a point charge q is situated in a quadrant formed by two semi-infinite metallic planar plates that form a right angle with each other along the z axis – a right-angled wedge. The potential on the wedge is kept at zero volts ($V = 0$). We will show that we can replicate this situation if we replace the wedge by *three* image charges (two of $-q$ and one of $+q$). Then the potential at a point P within the quadrant is

$$V_P = \frac{q}{4\pi\varepsilon_0}\left(\frac{1}{R_1} - \frac{1}{R_2} + \frac{1}{R_3} - \frac{1}{R_4}\right) \tag{8.14}$$

The distances R_j ($j = 1, 2, 3, 4$) are the distances from P to each of the four charges, starting from the real charge q at R_1 and going round clockwise. If P is on the $+x$ axis

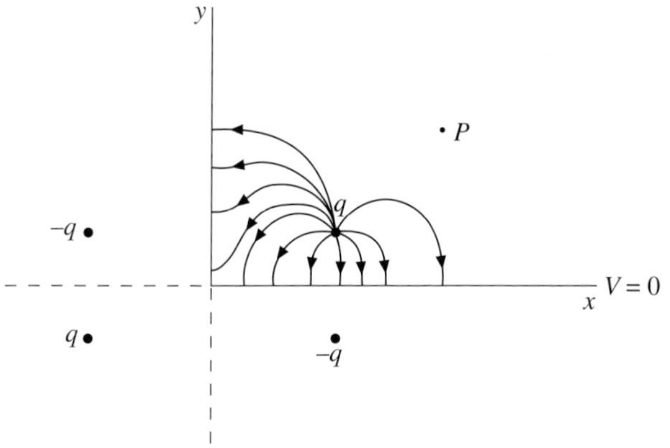

Fig. 8.4 A point charge in a quadrant formed by two perpendicular grounded planes, and its image charges. The 'virtual' field lines can be generated by reflection in the axes.

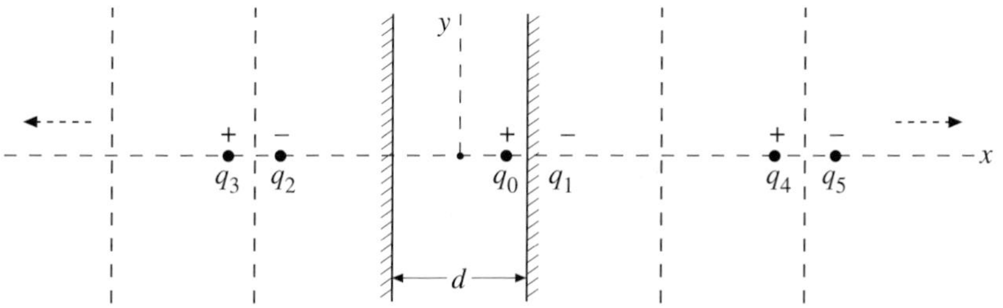

Fig. 8.5 A point charge between two parallel grounded planes, and image charges.

then $R_2 = R_1$ and $R_4 = R_3$, which as required makes $V = 0$ according to (8.14). If P is on the $+y$ axis then $R_4 = R_1$ and $R_3 = R_2$, which also makes $V = 0$. Clearly $V \to 0$ as $\mathbf{r}_P \to \infty$, so that the image situation has boundary Dirichlet conditions in the $x > 0, y > 0$ quadrant that are identical to those for the actual situation. Hence (8.14) must be the correct solution for the original situation.

Example 8.4 A point charge between two parallel infinite grounded planes

Figure 8.5 shows a point charge $q_0 = q$ at a location $x = x_0 (< d/2)$ between two infinite grounded equipotential plates placed at $x = \pm d/2$. The image charge $q_1 = -q$, at $x = d - x_0$, helps ensure that $V = 0$ on the right-hand plane $x = d/2$, but now we need two additional charges $q_2 = -q$ and $q_3 = +q$ (as shown in Fig. 8.5) to make the potential zero on the left-hand plane $x = -d/2$. But then things are wrong again on the right-hand plane, and we need $q_4 = +q$ and $q_5 = -q$ (as shown) to get things right there! It does not end there; in fact we need an infinite number of image charges to get perfect balance. We need $+q$ charges at $x = \pm 2md + x_{\mathrm{c}}$ and $-q$ charges at $x = \pm(2m+1)d - x_0$, where $m = 0, 1, 2, \ldots$. Thus we find, for a point

$P = (x, 0, 0)$ on the axis,

$$V_P = \frac{q}{4\pi\varepsilon_0} \sum_{m=-\infty}^{\infty} \left(\frac{1}{|x - 2md - x_0|} - \frac{1}{|x - (2m+1)d + x_0|} \right) \tag{8.15}$$

Although slowly converging, this infinite series is sufficiently simple that for a finite number of terms it is easily evaluated on a personal computer to sufficient accuracy.

Example 8.5 A uniform infinite line charge parallel to a grounded circular cylinder

This situation is illustrated in Fig. 8.6. We have $V = 0$ on a circular cylinder around the z axis, with radius $\rho = a$. There is a uniform infinite line charge on $(L_0, 0, z)$, i.e. on a line $\parallel$ the z axis at a distance L from the axis of the cylinder. We know that the potential at distance ρ from an infinite uniform line charge ϱ_ℓ is proportional to $(\varrho_\ell / 2\pi\varepsilon_0) \ln \rho$. Let us choose an image *line charge* $-\varrho'_\ell$ at $x = l$ and try to determine l and ϱ'_ℓ. The potential at point P (on the cylinder's surface) is

$$\begin{aligned} V_P &= \frac{1}{2\pi\varepsilon_0} (\varrho_\ell \ln R_1 - \varrho'_\ell \ln R_2) + V_0 \qquad \text{with} \\ R_1^2 &= (L_0 - a\cos\varphi)^2 + (a\sin\varphi)^2 \\ R_2^2 &= (a\cos\varphi - l)^2 + (a\sin\varphi)^2 \end{aligned} \tag{8.16a}$$

To see this, drop a perpendicular line from P to the x axis. Its length is $a\sin\varphi$ and it intersects at $x = a\cos\varphi$. A constant potential V_0 has been added; we shall see why shortly. Let us try $\varrho'_\ell = \varrho_\ell$, so that

$$V_P = \frac{\varrho_\ell}{2\pi\varepsilon_0} \ln \frac{R_1}{R_2} + V_0 \tag{8.16b}$$

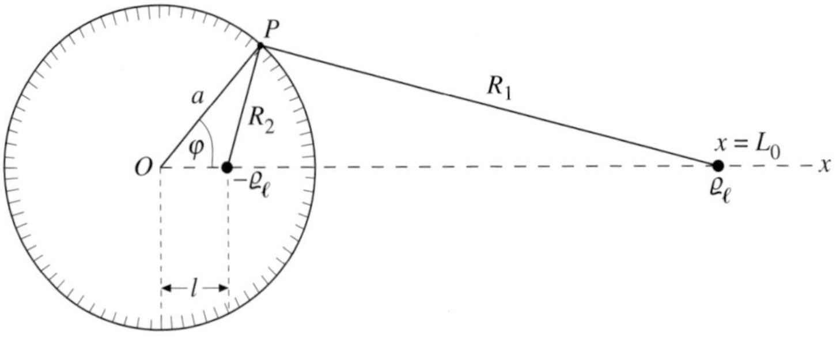

Fig. 8.6 A uniform infinite line charge ϱ_ℓ near a grounded parallel circular cylinder, and the image line charge.

We want the ratio R_1/R_2 to be independent of the angle φ. That is to say, we require that $R_1/R_2 = \kappa$, where κ is a constant, in which case we must choose $V_0 = -(\varrho_\ell/2\pi\varepsilon_0)\ln\kappa$ in order to obtain zero potential on the cylinder $\rho = a$. Equations (8.16a) and (8.16b) then tell us that we require

$$L_0^2 - 2aL_0\cos\varphi + a^2 = \kappa^2(l^2 - 2al\cos\varphi + a^2) \tag{8.16c}$$

for all φ; the choice $\kappa = \sqrt{L_0/l}$ achieves this because it makes $2aL_0\cos\varphi$ equal to $\kappa^2(2al\cos\varphi)$. If we insert this into (8.16c), we obtain $l = a^2/L_0$. Consequently, at an arbitrary point (ρ, φ, z), in cylindrical coordinates, we have found that

$$V_P = \frac{\varrho_\ell}{4\pi\varepsilon_0}\left\{\ln\left(\frac{\rho^2 - 2\rho L_0\cos\varphi + L_0^2}{\rho^2 - 2\rho\ell\cos\varphi + l^2}\right) - \ln\left(\frac{L_0}{a}\right)\right\} \tag{8.17}$$

and the image line charge is $-\varrho_\ell$ at $x = l = a^2/L_0$. We have already seen that this solution leads to $V = 0$ on $\rho = a$. As $\rho \to \infty$ we see from (8.17) that $V_P \to V_0$. It may seem surprising that the potential does not go to zero at infinitely distant locations (which is what would happen if the cylinder were replaced by a line charge $-\varrho_\ell$ at the origin, because then $R_1/R_2 \to 1$). The reason is that the image problem has replaced the cylinder at $x = 0$, with radius a, by a line charge $-\varrho_\ell$ at $x = l$ plus a constant V_0 everywhere, so that the potential at $\rho = a$ is zero. Two line charges $+\varrho_\ell$ and $-\varrho_\ell$ at a separation distance $L_0 - l$ will give zero potential half-way between them, but will *not* give zero potential on some sphere around one of them; it is clear that a sphere of radius $a \ll L_0 - l$ around one line charge will be influenced much more strongly by that line charge than by the more distant one. In any case, omission of the constant term from (8.17) does not affect the *electrostatic fields*.

Example 8.6 Two-wire transmission line

As an extension consider the situation of Fig. 8.7, which differs from Fig. 8.6 only in that the infinitesimally thin line charge at $x = L_0$ has been replaced by a cylinder

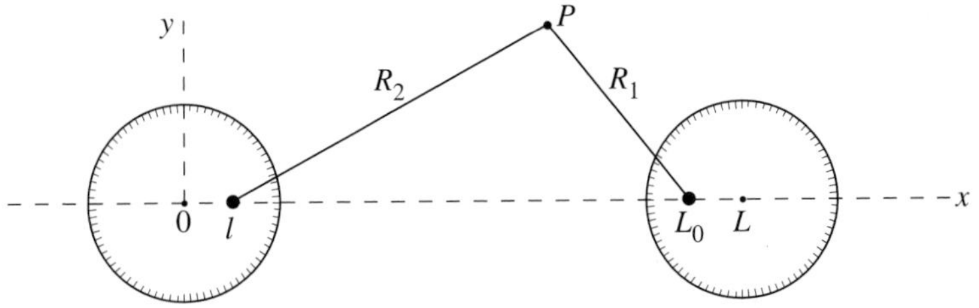

Fig. 8.7 Two-wire transmission line; the image line charges are at $x = l$ and $x = L_0 = L - l$.

around $x = L$ similar to that around $x = 0$. Note that we choose $L = L_0 + l$. The symmetry of the situation tells us that *both* cylinders can be replaced by image line charges, $-\varrho_\ell$ at $x = l$ and $+\varrho_\ell$ at $x = L_0$. Example 8.5 tells us that the appropriate constant potentials occur on spherical surfaces around $x = 0$ and $x = L$. Figure 8.7 shows that

$$\begin{aligned} R_1^2 &= (L_0 - l - \rho\cos\varphi)^2 + (\rho\sin\varphi)^2 \\ R_2^2 &= (\rho\cos\varphi - l)^2 + (\rho\sin\varphi)^2 \end{aligned} \tag{8.18}$$

and consequently, that

$$V_P = \frac{\varrho_\ell}{4\pi\varepsilon_0}\left\{\ln\left(\frac{\rho^2 - 2\rho(L_0 - l)\cos\varphi + (L_0 - l)^2}{\rho^2 - 2\rho\ell\cos\varphi + l^2}\right)\right\} + V_1, \tag{8.19}$$

where V_1 is constant. As the potential on one surface is the negative of that on the other, owing to the symmetry of the situation, we know from (8.16b) that the potential difference is

$$V_d = \frac{2\varrho_\ell}{2\pi\varepsilon_0}\ln\kappa = \frac{\varrho_\ell}{\pi\varepsilon_0}\ln\frac{L_0}{a} \tag{8.20a}$$

and, as $L_0 = L - l = L - a^2/L_0$, we can solve to find $L_0 = \frac{1}{2}L + \sqrt{\frac{1}{4}L^2 - a^2}$. Consequently, the capacitance per unit length is

$$\frac{dC}{dz} = \frac{\pi\varepsilon_0}{\ln\left[L/2a + \sqrt{(L/2a)^2 - 1}\right]} \tag{8.21a}$$

We observed that this result is the exact answer to which (7.30) was an approximation for $a \ll L$. An alternative form for (8.21a) is

$$\frac{dC}{dz} = \frac{\pi\varepsilon_0}{\cosh^{-1}(L/2a)} \tag{8.21b}$$

This is easily shown by using $L/2a = \cosh w = \frac{1}{2}(e^w + e^{-w})$ and then solving the ensuing quadratic equation in e^w.

8.4 Some simple Poisson-equation situations

We now wish to treat several simple physical situations which – when idealized a little – allow us to find analytical solutions for the potential and thus the electric fields by means of solving the Poisson equation explicitly (i.e. without devices such as image charges).

8.4.1 Symmetric p–n semiconductor junction diode

A symmetric p–n semiconductor junction diode is a structure that, initially, is uniformly doped with excess holes (positive charges) in $x < 0$ and excess electrons (negative charges) in $x > 0$, as shown in Fig. 8.8(*a*). These excess charge carriers have some mobility. Consider an 'initial' situation in which there is local charge neutrality in each region (the excess charge carriers in each region are balanced by immobile carriers of opposite polarity). This situation changes rapidly because both holes and electrons have nonzero kinetic energy at room temperature, and hence drift more or less randomly. Thus, afterwards some electrons will have drifted into $x < 0$ and some holes into $x > 0$, as shown in Fig. 8.8(*b*).

If the situation in Fig. 8.8(*a*) is one in which there is on average charge neutrality everywhere, then Fig. 8.8(*b*) represents a later situation in which there is an excess of negative charge to the left and of positive charge to the right. The electrons that have drifted into the region $x < 0$ leave behind positive ions, and a similar situation with opposite charges occurs for the right-drifting holes. An electric field starts to build up, pointing to the left, which, ultimately, counteracts the drift and leads to an equilibrium situation. Figure 8.9 shows qualitatively what the charge distribution in equilibrium looks like. Let us try to calculate the potential and the electric field in equilibrium.

Obviously, the exact dependence of the charge distribution $\varrho_V(x)$ upon x is different from one material to the next and it is not likely to be a simple function. But, just to illustrate the kind of fields that are found, let us assign a mathematical function to $\varrho_V(x)$ with a shape as in Fig. 8.9. There are infinitely many mathematical functions to choose from, but the one we choose has an advantage to be discussed in

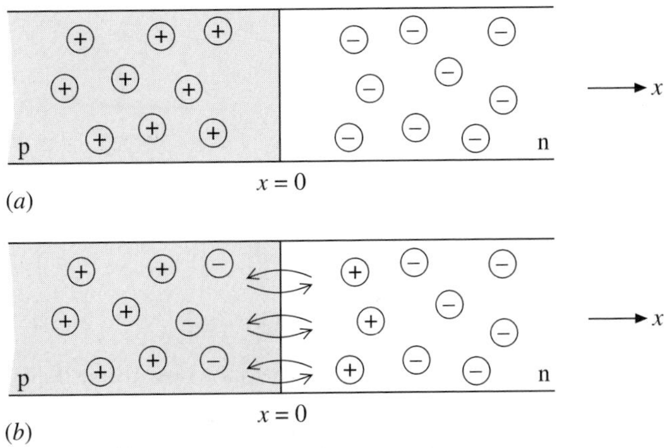

Fig. 8.8 Charge carrier diffusion in a symmetric p–n semiconductor junction diode.

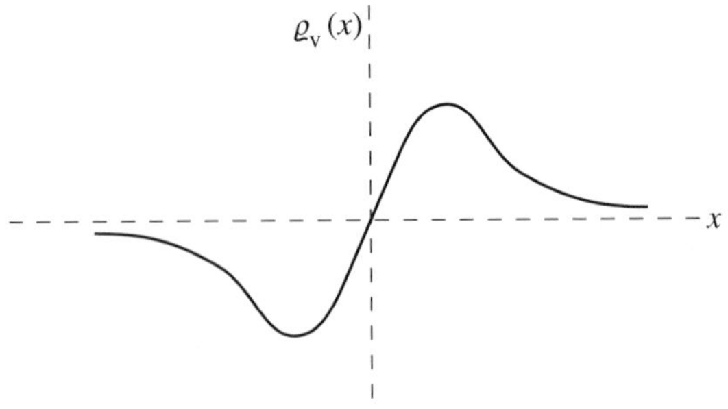

Fig. 8.9 Charge distribution in a symmetric p–n semiconductor junction diode.

connection with Poisson's equation in the form (8.23) below. It is

$$\varrho_v(x) = 2\varrho_{v0}\frac{\tanh(x/a)}{\cosh^2(x/a)} \tag{8.22}$$

Here ϱ_{v0} is a constant charge density (in C/m^3) and a is a constant distance, roughly of order of the width of the 'humps' in Fig. 8.9. A plot of (8.22), easily made on a personal computer, will convince the reader that it indeed simulates Fig. 8.9. Poisson's equation is one dimensional under the assumption that there is homogeneity in the y and z directions (so that there is a dependence of all variables only upon x), and thus it becomes

$$\frac{d^2V(x)}{dx^2} = -\frac{2\varrho_{v0}}{\varepsilon}\frac{\tanh(x/a)}{\cosh^2(x/a)} \tag{8.23}$$

We must integrate the right-hand side twice, and this particular function has analytical integrals that are easily sketched as functions of x/a. Integrate once to obtain

$$\frac{dV(x)}{dx} = \frac{a\varrho_{v0}}{\varepsilon}\frac{1}{\cosh^2(x/a)} - E_0 \tag{8.24a}$$

The constant E_0 can be taken as zero, so that the electric field $(-dV/dx)$ is zero as $|x| \to \infty$. Integrate once more to obtain

$$V(x) = \frac{a^2\varrho_{v0}}{\varepsilon}\tanh(x/a) + V_0 \tag{8.24b}$$

The second constant, V_0, also can be taken as zero so that we obtain $V(0) = 0$. This is desirable because we require an antisymmetric potential distribution: $V(-x) = -V(x)$. As the limit of $\tanh(x/a)$ is 1 as $x \to \infty$, we observe that $a^2\varrho_{v0}/\varepsilon = V_\infty$, the potential at $x \to \infty$. We have then found that the length $a = \sqrt{\varepsilon V_\infty/\varrho_{v0}}$.

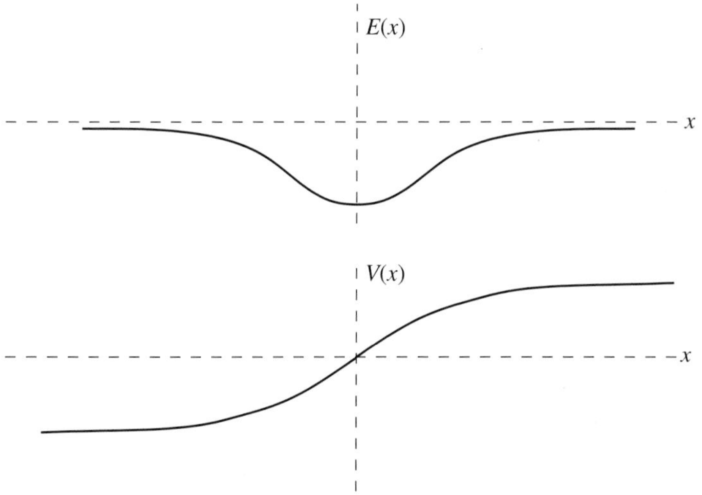

Fig. 8.10 Potential and field distribution in a symmetric p–n semiconductor junction diode.

Thus, we obtain from the above

$$V(x) = V_\infty \tanh(x/a) \qquad E(x) = -\frac{V_\infty}{a\cosh^2(x/a)} \qquad a = \sqrt{\varepsilon V_\infty/\varrho_{v0}} \quad (8.24c)$$

The potential and field dependence upon x are shown in Fig. 8.10.

The capacitance is easily calculated by calculating the total charge in $x > 0$:

$$Q = S\int_0^\infty dx\, \varrho_v(x) = 2S\varrho_{v0}\int_0^\infty dx \frac{\sinh(x/a)}{\cosh^3(x/a)} \quad (8.25a)$$

Let a dummy variable $t = \cosh(x/a)$, so that $adt = \sinh(x/a)\, dx$. The integral becomes

$$Q = 2Sa\varrho_{v0}\int_1^\infty dt \frac{1}{t^3} = Sa\varrho_{v0} \quad (8.25b)$$

Hence, from the third equality of (8.24c), we have a *nonlinear* connection between Q and the potential difference $V = V(\infty) - V(-\infty) = 2V_\infty$, namely

$$Q = S\sqrt{\varepsilon\varrho_{v0}V_\infty} = S\sqrt{\varepsilon\varrho_{v0}V/2} \quad (8.25c)$$

For such a nonlinear dependence, an expanded definition of capacitance is

$$C = dQ/dV$$

This would coincide with the old definition $C = Q/V$ for a linear capacitor because in that case we have

$$\frac{Q + dQ}{V + dV} = \frac{Q}{V} \quad \Rightarrow \quad V\,dQ = Q\,dV \quad \Rightarrow \quad \frac{dQ}{dV} = \frac{Q}{V} \tag{8.26}$$

For this nonlinear capacitor, we then obtain, again using (8.24c),

$$C = S\sqrt{\varepsilon \varrho_{v0}/8V_\infty} = \varepsilon S/(2a\sqrt{2}) \tag{8.27}$$

The similarity with the capacitance (7.24) of a parallel-plate capacitor is quite obvious when we equate $2a\sqrt{2}$ with an effective plate-separation distance, but here a is a function of the potential difference. From (8.22), we find that the maximum of $\varrho_v(x)$ occurs at $x \approx 0.88a$, which tells us that the length a is roughly 1.14 times the distance from $x = 0$ to the location where the charge is maximal.

8.4.2 Planar diode operating under space-charge-limited current

This, the simplest type of vacuum tube, is represented in idealized form by two infinite plates, separated at distance d, and maintained at a voltage difference V_d; see Fig. 8.11(*a*). The plate at the lower voltage is a *cathode*, i.e. it consists (at least partly) of a material, such as an oxide of cesium or barium, that is a good emitter of electrons under appropriate conditions. These conditions include a sufficient 'pull' by an external field or a high temperature in the cathode material, which increases the average velocity of the conduction electrons inside and allows them to overcome the potential *barrier*, known as the *work function*, posed by the interface with the

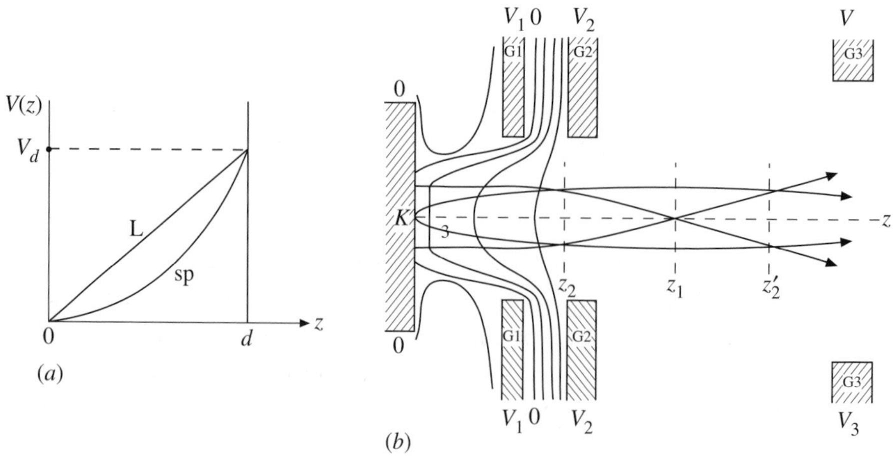

Fig. 8.11 (*a*) Laplace (straight line) and space-charge-limited (curved line) potentials in a planar diode. (*b*) Some equipotentials and electron trajectories in an electron gun.

interior of the diode. This barrier comes about because – unlike in the interior of the cathode – at the interface we do not have charge neutrality; the positive ions at the interface are only partly balanced by conduction electrons.

Thus, when electrons pass through the interface into the diode, they tend to be pulled back by the excess positive charge of the interface positive ions. The anode then attracts the electrons because it is at a potential $+V_d$ with respect to the cathode.

At very low temperatures, hardly any electrons are pulled into the diode, and the potential distribution in the diode is essentially that of a parallel-plate capacitor:

$$V(z) = V_d z/d \tag{8.28}$$

There is no loss of generality if we assume that the cathode voltage $V(0) = 0$ (otherwise, we simply add the cathode voltage as a constant to the potential $V(z)$ everywhere in the diode). This vacuum potential distribution is shown in Fig. 8.11(a) as a straight line.

As the temperature of the cathode is increased, more electrons are able to overcome the surface barrier and enter the region between the plates. Those electrons that have arrived there exert a repelling force on electrons still in the cathode, which tends to inhibit further extraction. In a *space-charge-limited* situation, the electric field at the cathode interface is driven back essentially to zero (actually the field is slightly negative at the interface, but the deviation from zero can be ignored), and there is a steady space-charge-limited current into the region between the plates. The potential distribution and the current can be determined in a self-consistent situation from the Poisson equation,

$$\frac{d^2V(z)}{dz^2} = -\frac{1}{\varepsilon_0}\varrho_v(z) \qquad \varrho_v(z) = -\frac{J(z)}{u(z)} \tag{8.29}$$

The *current density* $J(z)$ is related to the charge density by the velocity $u(z)$, as shown. Conservation of current I implies that $SJ(z)$, where S is the cross-sectional area of the current, is conserved and the straight-line motion of the electrons ensures that S is conserved. Therefore, $J(z) = J$ is a constant in this one-dimensional situation. That may seem to be contradicted by the fact that $J = -\varrho_v(z)u(z)$, (8.29), but as the electrons accelerate (their velocity $u(z)$ becomes larger) the density will decrease (ϱ_v becomes smaller), as a result of which the total current across any $z =$ constant plane will remain unchanged. Conservation of energy tells us that the sum of kinetic energy $\frac{1}{2}mu^2(z)$ and potential energy $-eV(z)$ must be a constant, i.e.

$$\tfrac{1}{2}mu^2(z) - eV(z) = \text{constant} \tag{8.30a}$$

The constant is zero, because both $V(0)$ and $u(0)$ equal zero at $z = 0$. Consequently

$$u(z) = \sqrt{2eV(z)/m} \tag{8.30b}$$

We obtain a total differential equation in $V(z)$ by inserting (8.30b) for $u(z)$ into (8.29):

$$\frac{d^2V(z)}{dz^2} = \frac{1}{2\pi\varepsilon_0}\sqrt{\frac{m}{2e}}\frac{2\pi J}{V^{1/2}(z)} \tag{8.31a}$$

It may seem odd to insert the extra factors 2π into the numerator and denominator, but the combination $(1/2\pi\varepsilon_0)\sqrt{m/2e} \equiv \beta$ is a handy constant, although with an unwieldy dimension; numerically $\beta \approx 30.307\ \mathrm{V}^{3/2}/\mathrm{mA}$. So, one must remember to give potentials in volts and currents in milliamperes (i.e. J in $\mathrm{mA/m^2}$) when using this value of β. The dimension of β is easily inferred from (8.31a), which now becomes

$$\frac{d^2V(z)}{dz^2} = (2\pi\beta J)V^{-1/2}(z) \tag{8.31b}$$

We multiply both sides by $dV(z)/dz$ and then obtain after one integration

$$\left(\frac{dV(z)}{dz}\right)^2 = (8\pi\beta J)V^{1/2}(z) + E_0^2 \tag{8.32a}$$

where E_0 is an integration constant (dimensionally equal to an electrostatic field), which we must choose as zero because at $z = 0$ we have both $V(0) = 0$ and $dV/dz = 0$ (zero field at the input in the space-charge-limited case). Verify this by taking the derivative of (8.32a) to obtain (8.31b). Hence

$$\frac{dV(z)}{dz} = \sqrt{8\pi\beta J}\,V^{1/4}(z) \tag{8.32b}$$

Multiply by $V^{-1/4}(z)$, and the resulting equation can be integrated once more to obtain

$$V(z) = V_0 + \left(\frac{9\pi\beta J}{2}\right)^{2/3} z^{4/3} \tag{8.33}$$

where V_0 is an integration constant, which must be the potential of the cathode. The current density J can be inferred by setting $z = d$, in which case $V(d) = V_0 + V_d$, to obtain

$$J = \left(\frac{2}{9\pi\beta}\right)\frac{V_d^{3/2}}{d^2} \tag{8.34}$$

This expression for the current density in a space-charge-limited diode is known as the *Child–Langmuir law*, and it predicts that the current density is proportional to the three-halves power of the anode–cathode voltage difference and is inversely proportional to the square of the separation distance.

A common application of a space-charge-limited diode occurs in the *television electron gun*, of which a cross section is shown in Fig. 8.11(*b*). The electron gun in a television set serves to extract electrons emitted by a heated cathode (coated by an electron-emitting material such as cadmium or bismuth oxide) and subsequently to guide those electrons into a well-formed *beam* that can be focussed into a small spot on the phosphor stripes at the screen location. The brightness of the produced spot is proportional to the spot current, and the electron gun is designed in such a way that the current can be modulated and controlled at very high frequencies. For this purpose the electron gun includes an *electronic aperture*. The G1 grid is designed to operate at variable potentials negative with respect to the cathode, whereas the G2 grid is strongly positive and fixed (both 'grids' are in fact circular apertures in metal plates). The more negative the G1 grid, the narrower is the region around the central axis in which electrons are pulled towards the G2 aperture, resulting in a lower current. The part of the gun immediately adjacent to the cathode behaves very much as a space-charge-limited diode (but of course there is no anode plate at fixed voltage). As the electrons accelerate towards the G2 grid, the current flow becomes less space-charge saturated. The computational difficulties close to the cathode arise because, as mentioned above, there is no anode plate and also because the geometry is cylindrical rather than planar. A television electron gun driven at $V_d = 600$ V and with $d = 2$ mm gives $J \approx 8.6\,\mathrm{mA/mm^2}$, from (8.34).

The space-charge-limited electric field is obtained by taking the derivative of (8.33), and replacing J by the Child–Langmuir expression. A similar replacement also gives us something simpler for the potential:

$$V_{\mathrm{sp}}(z) = V_d(z/d)^{4/3} \qquad E_{\mathrm{sp}}(z) = \tfrac{4}{3}E_{\mathrm{L}}(z)(z/d)^{1/3} \tag{8.35}$$

In the second expression $E_{\mathrm{L}}(z) = -V_d/d$ is the Laplace field, the slope of the straight line in Fig. 8.11(*a*). The space-charge-limited potential $V_{\mathrm{sp}}(z)$ is the curve shown in Fig. 8.11(*a*); it has zero slope at $z = 0$.

This example has illustrated a somewhat more complicated solution of Poisson's equation, but the space-charge-limited diode nevertheless is one of the simplest situations available. Most situations lead to a Poisson equation that needs to be solved numerically (numerical methods will be discussed shortly). One of the pleasing features of this simple treatment of the diode is that analytical solutions also can be found for cylindrical and spherical diodes, but discussion of these goes beyond the scope of this text.

8.5 Separation of variables

In certain geometrical configurations, a special technique known as *separation of variables* allows us to express the solution to Laplace's equation as a series of known terms. We may have to add these terms numerically, but at least the solution is otherwise analytical. The separation-of-variables technique exploits the geometry in a way that is reminiscent of 'putting a round peg in a round hole, and a square peg in a square hole'. We shall illustrate it with various examples.

8.5.1 Separation of variables in the Cartesian coordinate system

Suppose we have a structure consisting entirely of planar voltage plates at right angles to each other. It would be sensible to tackle the interior problem with Laplace's equation in Cartesian coordinates,

$$\frac{\partial^2 V}{\partial x^2} + \frac{\partial^2 V}{\partial y^2} + \frac{\partial^2 V}{\partial z^2} = 0 \tag{8.36}$$

since if the bounding plates are $\perp$ to the Cartesian axes, we expect that many of the bounding voltages – if not all – will be functions of only one of the three Cartesian coordinates. Let us try as a solution the product function

$$V(x, y, z) = X(x)Y(y)Z(z) \tag{8.37}$$

but we first need to work out the implications before we can justify this choice of trial solution, as follows. If (8.37) is substituted into (8.36) and we then divide by $V(x, y, z)$ we obtain

$$\frac{1}{X(x)}\frac{d^2X(x)}{dx^2} + \frac{1}{Y(y)}\frac{d^2Y(y)}{dy^2} + \frac{1}{Z(z)}\frac{d^2Z(z)}{dz^2} = 0 \tag{8.38}$$

Note that this has the form $f_1(x) + f_2(y) + f_3(z) = 0$. If, for example, we keep y and z fixed but vary x, then it must follow that $f_1(x)$ is a constant for the equation to be satisfied. Likewise the two other functions must be constants. Hence, each of the three terms on the left-hand side of (8.38) must be a constant! Thus we set

$$\begin{aligned} \frac{1}{X(x)}\frac{d^2X(x)}{dx^2} &= -k_x^2 \\ \frac{1}{Y(y)}\frac{d^2Y(y)}{dy^2} &= -k_y^2 \\ \frac{1}{Z(z)}\frac{d^2Z(z)}{dz^2} &= -k_z^2 \end{aligned} \tag{8.39a}$$

and we require for the three complex constants k_x, k_y, k_z

$$k_x^2 + k_y^2 + k_z^2 = 0 \tag{8.39b}$$

in order that the Laplace equation (8.38) is satisfied. Thus we can conclude that (8.37) is a solution of (8.36) provided $X(x)$, $Y(y)$, and $Z(z)$ satisfy equations (8.39a, b). But it is not the most general solution, as we shall see.

Each of the three equations (8.39a) has the form of the *harmonic equation*,

$$\frac{d^2\psi(w)}{dw^2} + \kappa^2\psi(w) = 0 \tag{8.40}$$

Here, κ^2 is one of the three constants k_i; ψ is one of the three functions X, Y, Z; and w is one of the three Cartesian variables x, y, z. The solution of the harmonic equation is well known:

$$\begin{aligned} \psi(w) &= A\sin\kappa w + B\cos\kappa w \qquad && \kappa \neq 0 \\ \psi(w) &= a + bw && \kappa = 0 \end{aligned} \tag{8.41}$$

Let $\kappa \neq 0$. Then, $\psi(w)$ is a sum of trigonometric functions if κ is real ($\kappa^2 > 0$). When $\kappa = jk$ is imaginary (with real κ), we use $\sin(jkw) = j\sinh(kw) = \frac{1}{2}j(e^{kw} - e^{-kw})$ and $\cos(jkw) = \cosh(kw) = \frac{1}{2}(e^{kw} + e^{-kw})$ to obtain linear combinations of hyperbolic functions or exponentials. It is more involved if κ is complex. The coefficients A, B are determined by the boundary conditions for potential. For clarity, the two-dimensional case of a rectangular box of infinite length in the z direction, with Dirichlet conditions, will be worked out to demonstrate the procedure in detail.

Consider the cross section of such a box in a $z =$ constant plane, as shown in Fig. 8.12. The boundary values are $V(x, 0) = V_1(x)$, $V(x, b) = V_2(x)$, $V(0, y) = V_3(y)$, and $V(a, y) = V_4(y)$, i.e. two functions of x and two functions of y form the four boundary potential distributions on the sides of the box. The principle of

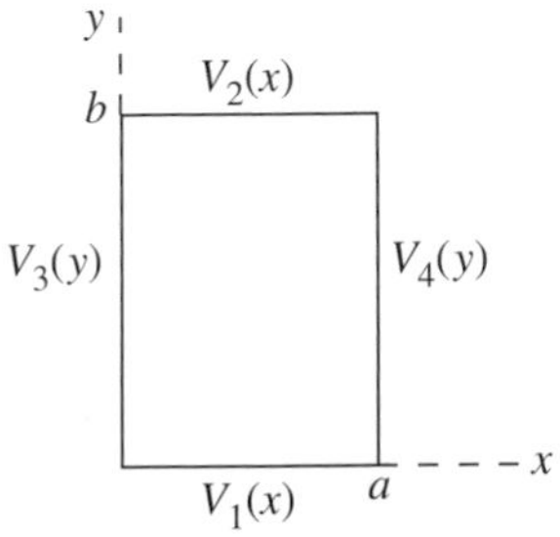

Fig. 8.12 A two-dimensional rectangular set of boundary conditions for the separation-of-variables method.

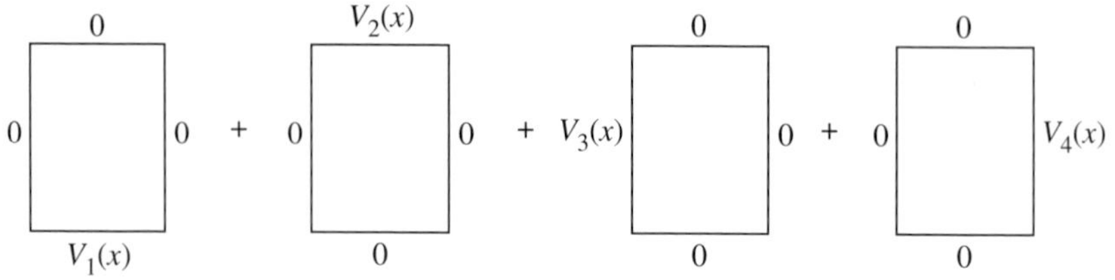

Fig. 8.13 Superposition of four elementary situations leading to that of Fig. 8.12.

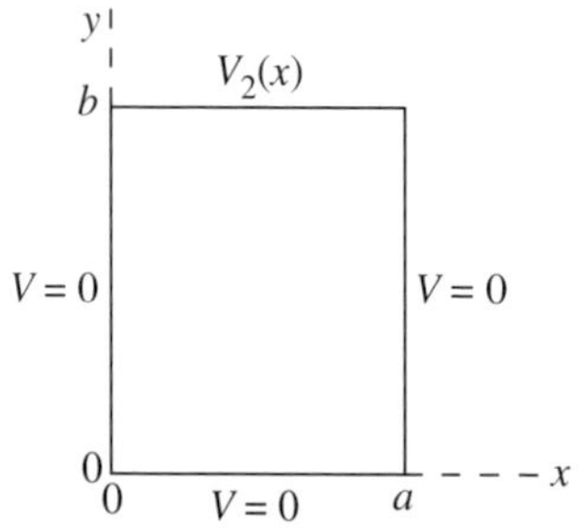

Fig. 8.14 One of the elementary situations of Fig. 8.13.

superposition tells us that the solution in the interior must be the same as the sum of the solutions in the four boxes shown in Fig. 8.13.

All four elementary solutions, as well as the superposition, are unique because Dirichlet boundary conditions are obeyed (see the end of Section 8.1).

We will look more closely at the situation of Fig. 8.14, which is the second diagram of Fig. 8.13. In words, the potential is a function of x at $y = b$ for $0 < x < a$, and it is zero on the other three sides. As $Z(z) \equiv 1$, we have by virtue of (8.39),

$$\begin{aligned} V(x,y) &= X(x)Y(y) \\ &= (A_1 \sin k_1 x + B_1 \cos k_1 x)(A_2 \sin k_2 y + B_2 \cos k_2 y) \end{aligned} \tag{8.42a}$$

with the condition that $k_1^2 + k_2^2 = 0$. Suppose we choose k_1 to be imaginary, i.e. $k_1 = jk$, with real k. Then $X(x) = jA_1 \sinh kx + B_1 \cosh kx$. But we should choose $jA_1 \equiv \tilde{A}_1$ to be a real constant, because we require real potential values. If we then try to plot $X(x) = \tilde{A}_1 \sinh kx + B_1 \cosh kx$, then there is no way that we can make $X(a) = 0$ *and* $X(0) = 0$. It just is not possible. As a consequence, we cannot satisfy the boundary conditions at $x = 0$ and at $x = a$ simultaneously by choosing k_1 to be imaginary. The same is true if k_1 is chosen complex with a nonzero imaginary part!

We therefore are forced to choose k_1 real, i.e. $k_1 = k$ and $k_2 = jk$ with k real, and therefore (8.42a) becomes

$$\begin{aligned} V(x, y) &= X(x)Y(y) \qquad \text{where} \\ X(x) &= A_1 \sin kx + B_1 \cos kx \\ Y(y) &= A_2 \sinh ky + B_2 \cosh ky \end{aligned} \tag{8.42b}$$

Of course, the meaning of the coefficients has changed, but that hardly matters as we have not yet determined them. The boundary condition $X(0) = 0$ can only be satisfied if we choose $B_1 = 0$. The boundary condition $Y(0) = 0$ is satisfied only if we set $B_2 = 0$. Hence we have left

$$V(x, y) = V \sin kx \sinh ky \tag{8.42c}$$

where $V \equiv A_1 A_2$ is a yet undetermined constant. But we still must satisfy a third boundary condition, $V(a, y) = 0$ for any y, and that means that we must make $\sin ka = 0$ because we certainly cannot choose $V = 0$, as that would produce a zero solution everywhere. Hence we are forced to choose ka to be an integer number m times π, i.e.

$$k = m\pi/a \qquad m = 1, 2, 3, \ldots \tag{8.43}$$

and consequently, the solution that satisfies *three* of the four boundary conditions is

$$V(x, y) = V \sin\left(\frac{m\pi x}{a}\right) \sinh\left(\frac{m\pi y}{a}\right) \tag{8.44}$$

We have excluded the $m = 0$ case because it is zero.

To satisfy the fourth boundary condition $V(x, b) = V_2(x)$, we need to take a linear combination of the solution (8.44) for various m-values. Therefore in (8.42c) we set $V = V_m$ and write[1]

$$V(x, y) = \sum_{m=1}^{\infty} V_m \sin\left(\frac{m\pi x}{a}\right) \sinh\left(\frac{m\pi y}{a}\right) \tag{8.45}$$

To fit (8.45) to the condition $V_2(x) = V(x, b)$, we now substitute $y = b$, obtaining

$$V(x, b) = \sum_{m=0}^{\infty} V_m \sinh\left(\frac{m\pi b}{a}\right) \sin\left(\frac{m\pi x}{a}\right) \tag{8.46}$$

[1] In the three-dimensional case, for which $V(x, y, z)$ is computed in a rectangular box, a double summation is needed.

Note that this equation is of the form

$$f(x) = \sum_{m=0}^{\infty} F_m \sin\left(\frac{m\pi x}{a}\right) \tag{8.47}$$

We can see that $f(-x) = -f(x)$, i.e. that $f(x)$ is an odd function of x, and also that $f(x+a) = f(x)$, i.e. that $f(x)$ is periodic in x modulo a; the theory of Fourier analysis then tells us that we can express the coefficients F_m as the following integrals, each of which can be evaluated, at least numerically:

$$F_m = \frac{2}{a}\int_0^a dx\, f(x) \sin\left(\frac{m\pi x}{a}\right) \tag{8.48}$$

From the theory of Fourier series it follows from (8.46) that

$$V_m \sinh\left(\frac{m\pi b}{a}\right) = \frac{2}{a}\int_0^a dx\, V(x, b) \sin\left(\frac{m\pi x}{a}\right) \tag{8.49}$$

and this expression gives us the coefficients V_m for all m. So we finally obtain from (8.45)–(8.49)

$$V(x, y) = \sum_{m=0}^{\infty} \tilde{V}_m \frac{\sin\left(\frac{m\pi x}{a}\right) \sinh\left(\frac{m\pi y}{a}\right)}{\sinh\left(\frac{m\pi b}{a}\right)} \quad \text{where}$$
$$\tilde{V}_m \equiv \frac{2}{a}\int_0^a dx\, V(x, b) \sin\left(\frac{m\pi x}{a}\right) \tag{8.50}$$

This is the analytical solution to the boundary-value Laplace problem depicted in Fig. 8.14. Unfortunately, it is in the form of a series, but the nature of Fourier series of physical functions ensures that the series decreases so that a series truncated at $m = N$ will give an answer with accuracy dependent upon how large N is chosen.

The solutions for the other three situations in Fig. 8.13 are found from (8.50) in similar fashion. We now evaluate (8.50) in two simple cases.

Example 8.7 Rectangular infinite box with one uniform nonzero boundary potential and three zero boundary potentials

Referring to Fig. 8.14, suppose $V(x, b) \equiv V_2(x) = V_0$, a constant. We then have from (8.50) that

$$\tilde{V}_m \equiv \frac{2V_0}{a}\int_0^a dx \sin\left(\frac{m\pi x}{a}\right) = \frac{2v_0}{m}(1 - \cos m\pi) \qquad \Rightarrow$$

$$\begin{aligned} \tilde{V}_m &= 0 \qquad &\text{for even } m \\ \tilde{V}_m &= \frac{4V_0}{m} \qquad &\text{for odd } m \end{aligned} \tag{8.51a}$$

Substitute into (8.50) to obtain

$$V(x, y) = 4V_0 \sum_{m=1,3,\ldots}^{\infty} \frac{\sin\left(\frac{m\pi x}{a}\right)\sinh\left(\frac{m\pi y}{a}\right)}{m \sinh\left(\frac{m\pi b}{a}\right)} \tag{8.51b}$$

Example 8.8 U-shaped infinite box with one uniform nonzero boundary potential and two zero boundary potentials

Figure 8.15 shows a shape similar to that of Fig. 8.14, with two differences: the side length $b \to \infty$, and the nonzero voltage V_0 is at $y = 0$. We obtain the solution from (8.51b) by making the replacement

$$\sinh\left(\frac{m\pi y}{a}\right) \to \sinh\left\{\frac{m\pi(b - y)}{a}\right\}$$

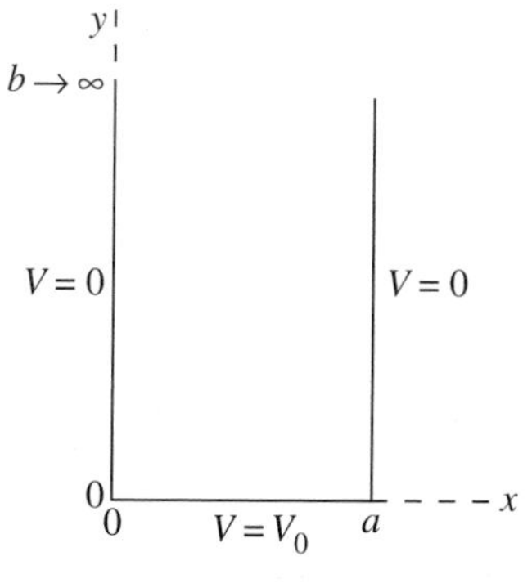

Fig. 8.15 A rectangular-trough set of boundary conditions.

and then allowing $b \to \infty$. Upon expanding the hyperbolic functions into their component exponentials we observe that

$$\lim_{b\to\infty} \frac{\sinh\left\{\dfrac{m\pi(b-y)}{a}\right\}}{\sinh\left(\dfrac{m\pi b}{a}\right)} = e^{-m\pi y/a}$$

Substitute into (8.51b) to obtain the result,

$$V(x,y) = 4V_0 \sum_{m=1,3,\ldots}^{\infty} \frac{e^{-m\pi y/a}}{m} \sin\left(\frac{m\pi x}{a}\right) \tag{8.52}$$

8.5.2. Separation of variables in the cylindrical coordinate system

In cylindrical coordinates we have

$$\frac{1}{\rho}\frac{\partial}{\partial\rho}\left(\rho\frac{\partial V}{\partial\rho}\right) + \frac{1}{\rho^2}\frac{\partial^2 V}{\partial\varphi^2} + \frac{\partial^2 V}{\partial z^2} = 0 \tag{8.53}$$

which suggests a trial solution that is of the form $R(\rho)\Phi(\varphi)Z(z)$. Rather than pursuing all possibilities, we will merely outline several two-dimensional cases without going into as much detail as above. We will consider circular cylindrical structures that have bounding potentials $V(\rho,\varphi,z)$ at *fixed* values of ρ, e.g. at $\rho=\rho_0$, and with variable z or φ. That is to say, the voltages on either one outer cylinder, or on one inner and one outer cylinder, are known and we wish to find $V(\rho,\varphi)$ in the fully bounded interior region.

1. No z-dependence

In this case we will consider circular cylindrical structures which have bounding potentials $V(\rho_0,\varphi)$ that are independent of z, but vary with φ. We set $V(\rho,\varphi) = R(\rho)\Phi(\varphi)$ and so split (8.53) into two equations,

$$\rho\frac{d}{d\rho}\left(\rho\frac{dR(\rho)}{d\rho}\right) - k^2R(\rho) = 0 \qquad \frac{d^2\Phi(\varphi)}{d\varphi^2} + k^2\Phi(\varphi) = 0 \tag{8.54}$$

The second equation yields solutions $\Phi(\varphi) = A\sin k\varphi + B\cos k\varphi$, and because we will require that $\Phi(2\pi) = \Phi(0)$, so that the potential is single valued on any cylinder surface, we observe that we should choose k to be real and integer. Thus we are forced to choose $k = m$ with $m = 0, 1, 2, 3, \ldots$. Hence, we find that

$$\Phi_m(\varphi) = A_m \sin m\varphi + B_m \cos m\varphi \qquad m = 0, 1, 2, \ldots \tag{8.55}$$

The remaining equation is

$$\rho^2 \frac{d^2 R(\rho)}{d\rho^2} + \rho \frac{dR(\rho)}{d\rho} - m^2 R(\rho) = 0 \tag{8.56}$$

This second-order total differential equation has two simple independent solutions, $R(\rho) = \rho^m$ and ρ^{-m}, and consequently its general solution is

$$R_m(\rho) = C_m \rho^m + D_m \rho^{-m}. \tag{8.57}$$

Structures inside a cylindrical bounding surface will most likely have a noninfinite potential at $\rho = 0$ (in the absence of a line charge on the z axis), so that we need to choose all $D_m = 0$ for $m > 0$ (and there is no loss of generality if we set $D_0 = 0$). Consequently, the most general solutions are of the form

$$V(\rho, \varphi) = \sum_{m=0}^{\infty} (A_m \sin m\varphi + B_m \cos m\varphi)\rho^m \tag{8.58}$$

and the boundary conditions at, say, $\rho = a$ will determine the coefficients. If the potentials $V(a, \varphi)$ are all known then the Fourier series

$$V(a, \varphi) = \sum_{m=0}^{\infty} (A_m \sin m\varphi + B_m \cos m\varphi)a^m \tag{8.59}$$

is just a generalization of (8.47) for functions (now of φ) that are neither even nor odd, and there is a straightforward generalization of (8.48) from the theory of Fourier series to obtain the coefficients:

$$\begin{aligned} A_m a^m &= \frac{2}{\pi}\int_0^{\pi} d\varphi V(a, \varphi) \sin m\varphi \\ B_m a^m &= \frac{2}{\pi}\int_0^{\pi} d\varphi V(a, \varphi) \cos m\varphi \end{aligned} \tag{8.60}$$

That determines A_m and B_m because a^m is a known quantity. If we have structures with an *inside* bounding surface $\rho = a$, then we replace a^m by a^{-m} in (8.60), and ρ^m by ρ^{-m} in (8.58). The reason is that now we must eliminate the C_m terms in (8.57) to retain finite solutions as $\rho \to \infty$. If we have *two* bounding cylindrical surfaces at $\rho = a$ and $\rho = b$, then we obtain two linear equations in A_m and B_m from the Fourier inverses of two equations such as (8.59), one for $\rho = a$ and one for $\rho = b$.

2. No φ-dependence (cylindrical symmetry)

In this case we set $V(\rho, z) = R(\rho)Z(z)$ and (8.53) becomes

$$\frac{1}{\rho}\frac{d}{d\rho}\left(\rho\frac{dR(\rho)}{d\rho}\right) - k^2 R(\rho) = 0 \qquad \frac{d^2 Z(z)}{dz^2} + k^2 Z(z) = 0 \tag{8.61}$$

The solutions of these equations depend critically upon whether k is real, imaginary, or complex. Cylindrical electrostatic structures of this type are known as *electrostatic lenses* because they possess the property that they can focus or defocus beams of charged particles, just as glass lenses do for beams of light. Exactly how they do so depends on the functional dependence of $Z(z)$ upon z. If, for example, for a long lens we have $Z(L) = Z(-L)$, then we have a so-called *unipotential lens* (in which case $Z(0)$ should be a different value). We can think of the lens as being infinitely long, but the potential is constant for $|z| > L$. As we saw before, in connection with (8.42a), this return from $Z(-L)$ at $z = -L$ to the same value at $z = L$ requires a trigonometric solution for $Z(z)$, so that $k^2 > 0$. In that case (8.61) yields

$$\begin{aligned} &\rho^2\frac{d^2R}{d\rho^2} + \rho\frac{dR}{d\rho} - \kappa^2\rho^2 R = 0 \\ &Z(z) = A\sin\kappa z + B\cos\kappa z \end{aligned} \tag{8.62a}$$

The first of equations (8.62a) is a special case of *Bessel's equation*, and its solutions are two functions known as *modified Bessel functions of order zero*, namely $I_0(\kappa\rho)$ and $K_0(\kappa\rho)$. (These are well-known tabulated functions with properties outlined below in the optional-reading section.) Thus the general solution of the first equation in (8.62a) is

$$R(\rho) = CI_0(\kappa\rho) + DK_0(\kappa\rho) \tag{8.62b}$$

The two functions $I_0(\kappa\rho)$ and $K_0(\kappa\rho)$ are shown in Fig. 8.16 as functions of $x \equiv \kappa\rho$. The solution $K_0(\kappa\rho) \to \infty$ as $\rho \to 0$, and we discard it because cylinder lenses do not have infinite potentials on their axis.

Bessel functions are solutions of the second-order total differential equation

$$x^2\frac{d^2f}{dx^2} + x\frac{df}{dx} \pm x^2 f - \nu^2 f = 0 \tag{8.63}$$

for arbitrary real ν. We will restrict ourselves to discussing these functions briefly for *integer* ν only, in which case ν is known as the *order* of the Bessel function. When the plus sign holds in (8.63), the two independent solutions are $J_\nu(x)$ and $Y_\nu(x)$. Apart from $\nu = 0$, all the $J_\nu(0) = 0$; $J_0(0) = 1$. As x grows, the $J_\nu(x)$ become oscillatory but the oscillations decay as $1/\sqrt{x}$ as $x \to \infty$. The functions $Y_\nu(x)$ also oscillate and decay in the same general way, as $x \to \infty$, but they

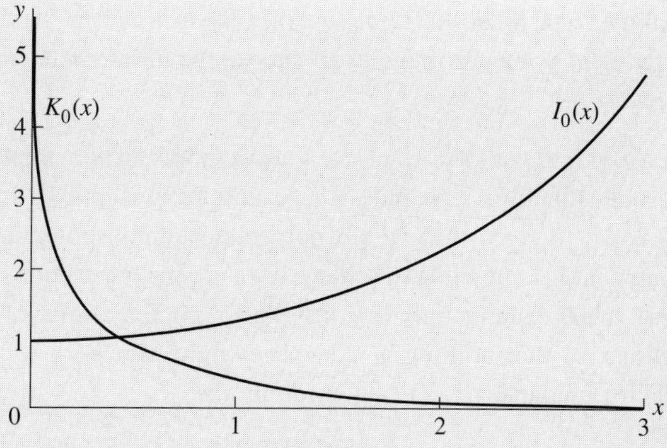

Fig. 8.16 Graphs of the two lowest-order modified Bessel functions, $I_0(x)$ and $K_0(x)$.

diverge logarithmically to $-\infty$ as $x \to 0$. These functions often show up in cylindrical structures in the same way that sines and cosines would show up in one-dimensional periodic structures.

When the minus sign holds in (8.63), we have the modified Bessel functions $I_\nu(x)$ and $K_\nu(x)$. Again, apart from $\nu = 0$, all $I_\nu(0) = 0$; $I_0(0) = 1$. All the $I_\nu(x)$ grow monotonically with x, and they grow as $e^x/\sqrt{x}$ as $x \to \infty$. The $K_\nu(x)$ all decay monotonically with x, and as $e^{-x}/\sqrt{x}$ as $x \to \infty$, but they diverge logarithmically to $\pm\infty$ as $x \to 0$. These functions are the analogs of the hyperbolic functions sinh and cosh.

All these functions have been studied and tabulated, and many function libraries in various computer languages include subroutines for them.

The most general solutions of (8.61) in the interior of interest are not sums but integrals over all possible κ,

$$V(\rho, z) = \int_0^\infty d\kappa \, I_0(\kappa\rho)[A(\kappa)\sin\kappa z + B(\kappa)\cos\kappa z] \tag{8.64}$$

The integrals are over only $\kappa \geq 0$ because negative κ does not add anything. The coefficients $A(\kappa)$ and $B(\kappa)$ can be determined by procedures that are analogous to (8.48). For example, we can multiply (8.64), taken at $\rho = a$, on both sides by either $\sin\kappa' z$ or $\cos\kappa' z$ and then integrate both sides over z from 0 to ∞. The right-hand side will yield essentially $A(\kappa')$ or $B(\kappa')$, aside from the (known) factor $I(\kappa' a)$ and a numerical factor, because $V(a, z)$ is the given boundary potential. It is not particularly useful to work out all the details here, although this is straightforward. However, we will explain a similar procedure for spherical symmetry in detail below, in subsection 8.5.3.

Another type of lens (e.g. the so-called *bipotential* lens) has $V(L) \neq V(-L)$. The different end voltages oblige us to choose the harmonic-equation solution

$$Z(z) = Ae^{\kappa z} + Be^{-\kappa z} \tag{8.65a}$$

with $\kappa > 0$, i.e. with $k = j\kappa$. Note that we prefer exponentials here instead of the hyperbolic functions sinh and cosh, because we must make sure (by setting either A or B equal to zero) that we do not obtain infinite potentials at $z \to \pm\infty$. This requirement is impossible to achieve if we use the hyperbolic functions directly. Of course these exponentials are just linear combinations of the sinh and cosh functions, so that nothing new has been done here with respect to the solutions (8.41) for imaginary k. The equation in ρ is

$$\rho^2 \frac{d^2 R}{d\rho^2} + \rho \frac{dR}{d\rho} + \kappa^2 \rho^2 R = 0 \tag{8.65b}$$

This is similar to the first equation in (8.62a), and the solutions are linear combinations of *Bessel functions of zero order* and argument $\kappa\rho$, i.e. in present-day notation $J_0(\kappa\rho)$ and $Y_0(\kappa\rho)$. These functions, too, are well-known and tabulated; they are discussed in the optional paragraph above, and shown in Fig. 8.17 as functions of $x \equiv \kappa\rho$.

As $Y_0(x) \to -\infty$ when $x \to 0$, we discard it and thus obtain

$$V(\rho, z) = \int_0^\infty d\kappa \, J_0(\kappa\rho) A(\kappa) e^{-\kappa z} \tag{8.66a}$$

for solutions inside a cylinder. Equation (8.66a) is a Laplace transform of $J_0(\kappa\rho)A(\kappa)$, which in turn is obtained from the inverse Laplace transform of $V(a, z)$ at the radius $\rho = a$, where this potential is a known function $V_0(z)$. The theory of Laplace transforms states that we can find $A(\kappa)$ from a contour integral in the complex z plane:

$$J_0(\kappa a) A(\kappa) = \frac{1}{2\pi j} \int_{c-j\infty}^{c+j\infty} dz \, V_0(z) e^{\kappa z} \tag{8.66b}$$

with the constant c chosen to lie to the right of any singularities generated by taking $V_0(z)$ as a function of complex values of z.

8.5.3 Separation of variables in the spherical coordinate system

In spherical coordinates, we have for Laplace's equation

$$\frac{1}{r^2} \frac{\partial}{\partial r} \left(r^2 \frac{\partial V}{\partial r} \right) + \frac{1}{r^2 \sin\theta} \frac{\partial}{\partial \theta} \left(\sin\theta \frac{\partial V}{\partial \theta} \right) + \frac{1}{r^2 \sin^2\theta} \frac{\partial^2 V}{\partial \varphi^2} = 0 \tag{8.67}$$

The trial solution here is $V(r,\theta,\varphi) = R(r)\Theta(\theta)\Phi(\varphi)$. A relevant physical situation would be a conical geometry with rotational symmetry, i.e. solutions $V(r,\theta)$ that are independent of φ. We then have

$$\begin{aligned} &\frac{d}{d\theta}\left(\sin\theta\,\frac{d\Theta(\theta)}{d\theta}\right) + k^2\sin\theta\,\Theta(\theta) = 0 \\ &\frac{d}{dr}\left(r^2\frac{dR(r)}{dr}\right) - k^2R(r) = 0 \end{aligned} \tag{8.68}$$

The solution to the $R(r)$ equation is $Ar^p + Br^{-(p+1)}$, provided that $p(p+1) = k^2$; this can be shown by substitution. We can generate a set of solutions that will allow us to match all kinds of boundary conditions if we set $k^2 = m(m+1)$, with $m = 0, 1, 2, 3, \ldots$. In this case, the solutions to the first equation are expressed better in terms of $\cos\theta \equiv \mu$ rather than θ, and we write $\Theta(\theta) \equiv P_m(\mu)$. If we note that $d/d\theta$ is equivalent to $-\sqrt{1-\mu^2}d/d\mu$, we observe that the first equation of (8.68) becomes

$$(1-\mu^2)\frac{d^2P_m(\mu)}{d\mu^2} - 2\mu\frac{dP_m(\mu)}{d\mu} + m(m+1)P_m(\mu) = 0 \tag{8.69}$$

The functions $P_m(\mu)$ are known as *Legendre polynomials*; like the Bessel functions they are well understood and tabulated. The first few are $P_0(\mu) = 1$, $P_1(\mu) = \mu$, $P_2(\mu) = \frac{1}{2}(3\mu^2 - 1)$; further Legendre polynomials are generated by the recursive relationship

$$(m+1)P_{m+1}(\mu) = (2m+1)\mu P_m(\mu) - \mu P_{m-1}(\mu) \tag{8.70}$$

The general solution of (8.67) thus generated is

$$V(r,\theta) = \sum_{m=0}^{\infty}[A_m r^m + B_m r^{-(m+1)}]P_m(\cos\theta) \tag{8.71a}$$

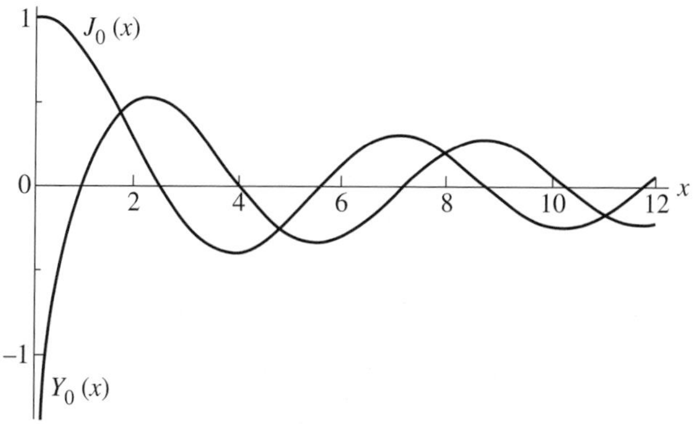

Fig. 8.17 Graphs of the two lowest-order Bessel functions $J_0(x)$ and $Y_0(x)$.

and we eliminate the B_m terms for situations including the origin because these terms become infinite at $r = 0$; thus

$$V(r,\theta) = \sum_{m=0}^{\infty} A_m r^m P_m(\cos\theta) \tag{8.71b}$$

The unknown coefficients are determined by matching to boundary conditions and making use of the *orthogonality relationship*

$$\left(m + \frac{1}{2}\right) \int_{-1}^{1} d_\mu P_m(\mu) P_n(\mu) = \delta_{mn}, \tag{8.72}$$

where the *Kronecker delta* δ_{mn} equals 1 when $m = n$ and 0 when $m \neq n$. To make good use of this orthogonality relationship, we multiply both sides of (8.71b) by $P_n(\cos\theta)$, then consider $\mu \equiv \cos\theta$ to be the variable of integration, and finally integrate with respect to μ over the range $-1 < \mu < 1$. Thus, as an example, we would find from (8.71b) and (8.72) that

$$A_n r^n = \frac{2}{n+2} \int_{-1}^{1} d(\cos\theta) V(r,\theta) P_n(\cos\theta) \tag{8.73}$$

which yields the coefficient A_n at a given radius $r = r_0$ for which we know $V(r_0, \theta)$.

If, however, we are analyzing a situation *outside* the sphere $r = r_0$, then we must eliminate all A_m terms in (8.71a), because we require a finite result when $r \to \infty$. In this case the left-hand side of (8.73) becomes $B_n r^{-(n+1)}$, and we apply it at $r = r_0$, where the voltage $V(r_0, \theta)$ is known.

Finally, in the general case in which we have bounding voltages at $r = r_\mathrm{i}$ and $r = r_\mathrm{f}$, with $r_\mathrm{i} < r < r_\mathrm{f}$, we obtain from the orthogonality relationship (8.72)

$$A_n r^n + B_n r^{-(n+1)} = \frac{2}{n+2} \int_{-1}^{1} d(\cos\theta) V(r,\theta) P_n(\cos\theta) \tag{8.74}$$

We apply this at $r = r_\mathrm{i}$ and at $r = r_\mathrm{f}$ to obtain two linear equations in two unknowns, which are easily solved to obtain A_n and B_n.

Problems

8.1. An important consequence of Laplace's equation is that the potential $V(x, y, z)$ cannot have an extreme value (a maximum or minimum) anywhere in the interior of a region v in which $\Delta V = 0$. Prove this. *Hint*: To determine extreme values for $f(x)$ requires one to look at df/dx and $d^2 f/dx^2$.

8.2. The Laplacian of a vector is defined by the relationship $\Delta \mathbf{A} = \nabla(\nabla \cdot \mathbf{A}) - \nabla \times (\nabla \times \mathbf{A})$. Prove that in Cartesian coordinates this is the same as the three equations $\Delta \mathbf{A} = (\partial^2 \mathbf{A}/\partial x^2) + (\partial^2 \mathbf{A}/\partial y^2) + (\partial^2 \mathbf{A}/\partial z^2)$.

8.3. A parallel-plate capacitor with plates at $z = 0$ and $z = 20\,\text{cm}$ is filled with a dielectric with permittivity $\varepsilon = 3.5\varepsilon_0$ in $0 < z < 8$ and $\varepsilon = 4.5\varepsilon_0$ in $8 < z < 20$ (all distances are in cm). The voltage across the plates is held at $V = 150$ volts.
(a) Calculate $V(z)$ in the interior, if $V(0) = 0$.
(b) Calculate the field $\mathbf{E}(z)$ in the interior.
(c) Check to see whether the boundary conditions for the fields hold at $z = 8\,\text{cm}$.

8.4. Consider the same parallel-plate capacitor as in problem 8.3, but now filled with a dielectric for which $\varepsilon(z) = e^{\beta z}$, with $\beta = 0.075\,\text{m}^{-1}$. Calculate $V(z)$, given that $V(0) = 0$. *Hint*: Use (8.1).

8.5. A coaxial circularly cylindrical capacitor has its metallic core at $0 < \rho < 2\,\text{cm}$ and its metallic sheath beyond $\rho = 16\,\text{cm}$. A potential difference of 85 volts is maintained between core and sheath. The region between core and sheath holds a dielectric with $\varepsilon(\rho) = 5\rho^2\varepsilon_0$, where all distances are in cm.
(a) Calculate the potential $V(\rho, z)$ in the interior, assuming that $V = 0$ in the core.
(b) Calculate $\mathbf{E}(\rho, z)$ in the interior.

8.6. Consider again the same coaxial capacitor of problem 8.5, but with two uniform dielectric media inside it (i.e. over the region $2\,\text{cm} < \rho < 16\,\text{cm}$). One medium, with $\varepsilon = 2.5\varepsilon_0$, occupies $0 < \varphi < \pi$ and the other, $\varepsilon = 4\varepsilon_0$, occupies $\pi < \varphi < 2\pi$.
(a) Calculate the potential $V(\rho, \varphi)$ in the interior, assuming that $V = 0$ in the core.
(b) Calculate $\mathbf{E}(\rho, \varphi)$ in the interior.

8.7. Return again to the coaxial capacitor of problem 8.5, this time with two uniform dielectric fillings: one medium, with $\varepsilon = 2.5\varepsilon_0$, occupies $2\,\text{cm} < \rho < 8\,\text{cm}$ and the other, with $\varepsilon = 4\varepsilon_0$, occupies $8\,\text{cm} < \rho < 16\,\text{cm}$.
(a) Calculate the potential $V(\rho)$ in the interior, assuming that $V = 0$ in the core.
(b) Calculate $\mathbf{E}(\rho)$ in the interior.

8.8. Two metallic cones, both with apex at the origin and axis coinciding with the z axis, are defined by the half-angles $\theta_1 = 30°$ and $\theta_2 = 60°$. The corresponding plate potentials are held at $V_1 = 20\,\text{V}$ and $V_2 = 90\,\text{V}$. Find $V(r, \theta, \varphi)$ for $\theta_1 < \theta < \theta_2$.

8.9. A capacitor consists of two metallic spherical surfaces at $r = 3\,\text{m}$ and $r = 9\,\text{m}$. The volume in between is filled with two dielectrics, $\varepsilon_{1r} = 2.5$ for $0 < \theta < \pi/2$ and $\varepsilon_{2r} = 4.5$ for $\pi/2 < \theta < \pi$. Find the capacitance.

8.10. A constant line charge $\varrho_\ell = 350\,\text{pC}$ exists at $P = (x, 0, 3)$, where all distances are in meters, above a grounded perfectly conducting plane at $z = 0$. Calculate the potential V at location $Q = (-4, 2, 1)$.

8.11. A grounded metallic right-angled wedge consists of the $y = 0$ plane and the $x = y$ plane, in both cases for $x \geq 0$. A point charge q is placed at $P = (3, 1, 0)$. Give the coordinates of the image charges needed for calculating the potential V within the wedge region.

8.12. A point charge q is separated by a distance d from the center of a grounded perfectly conducting sphere of radius a. What value does the image charge $\tilde{q}$ have, and where should it be located so that the potential and field outside the sphere are predicted accurately by just the two point charges q and $\tilde{q}$?

8.13. Two infinite metal planes are situated at $x = -0.5\,\text{m}$ and $x = +0.5\,\text{m}$. A charge q is placed at $x = 0.25\,\text{m}$. Calculate the potential V at $x = 0$ to three significant figures, in terms of a multiple of $q/4\pi\varepsilon_0$. Justify your accuracy.

8.14. A planar diode with a cathode–anode spacing of 1.85 mm is designed to produce a current of 3.5 mA over an area of $0.4\,\text{mm}^2$. What voltage difference will produce that current under space-charge-limited conditions?

8.15. Consider the symmetric p–n semiconductor junction diode discussed in subsection 8.4.1. Suppose that $\varrho_v(x) = \varrho_{v0}(x/a)\exp(-|x|/a)$, instead of what is specified there. Find

(a) the electric field $E(x)$ for all x,

(b) the potential $V(x)$ for all x,

(c) the total charge Q in $x > 0$,

(d) an expression for the capacitance C in terms of the area S, ε, and ϱ_{v0} or a.

8.16. In any source-free region bounded by a spherical surface S of radius R around $\mathbf{r}$, it is true, according to the *mean-value theorem*, that

$$V(\mathbf{r}) = \frac{1}{4\pi R^2}\oint_S dS\, V(\mathbf{r}_s)$$

where $\mathbf{r}_s$ is on the surface S. Use this theorem to prove that the potential $V(\mathbf{r})$ reaches extreme values (its maximum and its minimum) only at the bounding surfaces of a source-free region.

8.17. A rectangular box, infinite in the z direction, is bounded by sides at $x = 0, a$ and $y = 0, b$. The boundary potential values are: $V(x,0) = 0$, $V(0,y) = 0$, $V(a,y) = 0$, and $V(x,b) = V_0\sin(4\pi x/a)$.

(a) Use the method of separation of variables to find $V(x,y)$ in the interior.

(b) Use the result under part (a) to find $V(x,y)$ when $V(a,y) = V_1\sin(\pi y/b)$ instead of zero, the other three boundary conditions being unchanged. *Hint*: Use the principle of superposition.

8.18. The rectangular box of problem 8.17(a) now has $V(x,b) = V_0$, a constant potential. Find a series solution for the potential in the interior.

8.19. The potential distribution on an infinite circular cylinder of radius ρ_0 is $V(\varphi) = V_0\sin 2\varphi$. Use the method of separation of variables to find $V(\rho,\varphi)$ for $\rho < \rho_0$.

9

Numerical solutions of Laplace and Poisson equations

9.1 General potential equation in Cartesian coordinates

The general partial differential equation for the potential V, following from substitution of $\mathbf{E} = -\nabla V$ into the Maxwell equation $\nabla \cdot \mathbf{D} = \varrho_\mathrm{v}$, is

$$\nabla \cdot [\varepsilon(\mathbf{r})\nabla V(\mathbf{r})] = -\varrho_\mathrm{v}(\mathbf{r}) \tag{9.1}$$

It is obvious that a direct analytical solution of (9.1) is most unlikely in the general case for which both $\varepsilon(\mathbf{r})$ and $\varrho_\mathrm{v}(\mathbf{r})$ vary from point to point. We will outline a general method of solving (9.1) analytically. To do so, we take a volume v of interest, e.g. one with a bounding surface on which we know either the potential or the normal **E** field component at each point. The first step is to subdivide this volume into a three-dimensional grid of N small mesh cubes ($N \sim N_x N_y N_z$), as indicated in Fig. 9.1.

Each mesh cube should have a side of length h (the *grid parameter*) that is small compared to any distance l over which either $\varepsilon(\mathbf{r})$ or $\varrho_\mathrm{v}(\mathbf{r})$ changes appreciably, e.g. by 50% or so. Such a distance is often defined formally by the equations

$$l_1(\mathbf{r}) = \left|\frac{1}{\varepsilon(\mathbf{r})}\nabla\varepsilon(\mathbf{r})\right|^{-1} \qquad l_2(\mathbf{r}) = \left|\frac{1}{\varrho_\mathrm{v}(\mathbf{r})}\nabla\varrho_\mathrm{v}(\mathbf{r})\right|^{-1} \tag{9.2a}$$

For example, if $\varrho_\mathrm{v}(x) = \varrho_\mathrm{v}(0)e^{-x/d}$ then (9.2a) predicts that $l_2 = d$, the distance over which the charge density changes by a factor $1/e$. In general, these distances vary

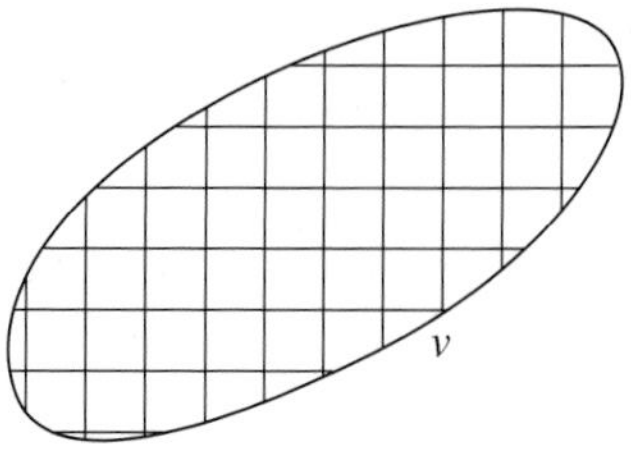

Fig. 9.1 A volume v subdivided into mesh cubes.

from point to point, and the idea is that we take h to be much smaller than the least of them. The ℓ_i are often called 'e-folding lengths', or 'relaxation lengths'. If, *locally*, the permittivity $\varepsilon(\mathbf{r})$ varies exponentially, as $e^{-(r-r_0)/d}$, in some direction around point r_0, then d turns out to be equal to $l_2(\mathbf{r})$. Therefore, just as in one-dimensional integration, we require $h \ll l$ everywhere in v.

However, when $\varepsilon(\mathbf{r})$ is a constant and there is no charge, it follows from (9.2a) that $l_1, l_2 = \infty$ everywhere. While that is true, that does not mean that we are free to take an arbitrarily large mesh size h. As we are going to require that potential $V(\mathbf{r})$ varies very slowly in a cube, we will also require

$$l_3(\mathbf{r}) = \left| \frac{1}{V(\mathbf{r})} \nabla V(\mathbf{r}) \right|^{-1} \tag{9.2b}$$

In this case, $l_3(\mathbf{r})$ is the distance with which to compare h, and, as described above, we must make sure that it is large compared to the chosen h. If, then, the grid parameter $h \ll l_3$ everywhere in v, then only small errors are made in approximating ε, ϱ_v, and the field $\mathbf{E}$ as constant inside each cube. Of course, a constant $\mathbf{E}$ field within a cube implies a linear variation of $V(\mathbf{r})$ within the cube, and the requirement $h \ll l_3$ ensures that this is a reasonable approximation.

Let us dwell somewhat longer on this last requirement and – for the sake of simplicity – consider only a one-dimensional case, so that

$$V(x+h) = V(x) + h\frac{dV}{dx} + \frac{h^2}{2}\frac{d^2V}{dx^2} + \frac{h^3}{6}\frac{d^3V}{dx^3} + \cdots \tag{9.3}$$

Actually, we only need all terms on the right-hand side beyond the second to be small. Hence we require

$$h^2\frac{d^2V}{dx^2} \ll V(x) \qquad \text{or} \qquad h^2 \ll \left|\frac{1}{V}\frac{d^2V}{dx^2}\right|^{-1} \tag{9.4}$$

Then, even though we have only required the third term on the right-hand side of (9.3) to be small compared to the first term, the chances are good that the higher-order terms are even smaller. Take as an example, $V(x) \propto e^{-x/l}$. We observe that $d^2V/dx^2 = V/l^2$, so that the second of Eqs. (9.4) becomes $h^2 \ll l^2$. The series (9.3) becomes

$$V(x+h) = V(x) - \frac{h}{l}V(x) + \frac{h^2}{2l^2}V(x) - \frac{h^3}{6l^3}V(x) + \cdots \tag{9.5}$$

Obviously, we have achieved our goal. Of course, not all $V(x)$ vary so simply, but the differences between exponential and other types of variation become quite small (locally!) when $V(x)$ varies slowly, which is what is required in the first place for accurate subdivision into mesh blocks.

Rather than working with (9.1), we will work with the original Maxwell equation $\nabla \cdot \mathbf{D} = \varrho_{\mathrm{v}}$ in its integral form

$$\oint d\mathbf{S} \cdot \varepsilon(\mathbf{r})\mathbf{E}(\mathbf{r}) = q_{\mathrm{encl}} \tag{9.6a}$$

We shall apply this equation to a mesh cube, the central one in Fig. 9.2; call it cube 0. The six cubes adjoining the sides of the mesh cube are the *nearest-neighbor cubes*. The 'digitized' version of (9.6a) is then

$$h^2 \sum_{i=1}^{6} \varepsilon_i E_{\mathrm{n}i} = \varrho_{\mathrm{v}0} h^3 \tag{9.6b}$$

given that $\varrho_{\mathrm{v}0}$ is the (constant) charge density in central cube 0, ε_i is the dielectric permittivity in nearest-neighbor cube i, and $E_{\mathrm{n}i}$ is the component of the $\mathbf{E}$ field in cube i that is normal to the interface with cube 0.

Using Fig. 9.3, we estimate that $E_{\mathrm{n}i} = -2(V_i - V_{0i})/h$, because this is what $-\hat{\mathbf{n}}_i \cdot (\nabla V)_i$ works out to be for a linear variation in potential. Likewise, the normal field inside cube 0 works out to be $E_{\mathrm{n}0} = -2(V_{0i} - V_0)/h$. The boundary condition at the interface between cubes 0 and i is

$$\varepsilon_i E_{\mathrm{n}i} = \varepsilon_{(0)} E_{\mathrm{n}0} \quad \Rightarrow \quad -2\varepsilon_i(V_i - V_{0i})/h = -2\varepsilon_{(0)}(V_{0i} - V_0)/h \tag{9.7}$$

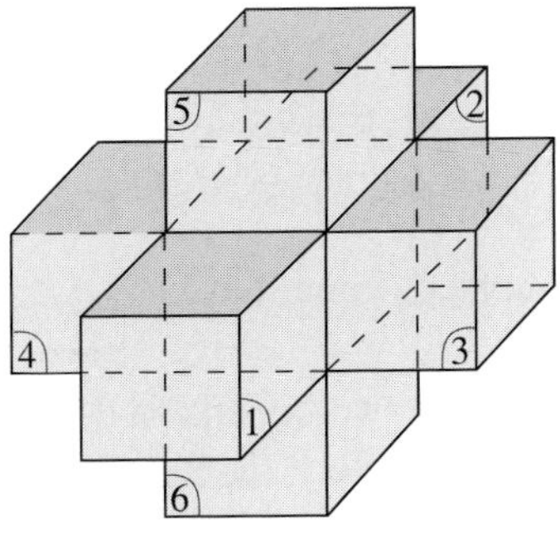

Fig. 9.2 A mesh cube (0) and its six nearest-neighbor cubes, labeled 1–6.

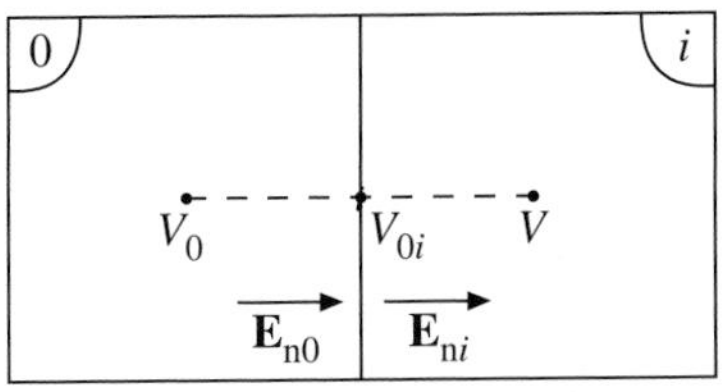

Fig. 9.3 The mesh cube (0) and nearest-neighbor cube i. The cube sides are of length h.

Note that $\varepsilon_{(0)}$ is the permittivity in cube 0. The parentheses in the subscript distinguish this permittivity from the free-space permittivity. From (9.7) we find the potential V_{0i} midway between the centers of the cubes:

$$V_{0i} = \frac{\varepsilon_i V_i - \varepsilon_{(0)} V_0}{\varepsilon_i - \varepsilon_{(0)}} \tag{9.8}$$

If we insert (9.8) into the estimate $E_{ni} = -2(V_i - V_{0i})/h$, we find from (9.6b) that

$$\sum_{i=1}^{6} \varepsilon_{(0)i}(V_i - V_0) = \frac{1}{2}\varrho_{v0}h^2 \qquad \text{where}$$
$$\frac{1}{\varepsilon_{(0)i}} \equiv \frac{1}{\varepsilon_{(0)}} + \frac{1}{\varepsilon_i} \tag{9.9a}$$

Equation (9.9a) involves the potential at the center of a mesh cube and the six potentials at the centers of the nearest neighbors. All other quantities in (9.9a) are known. We can write the equation in the equivalent form

$$\left(\sum_{i=1}^{6} \varepsilon_{(0)i}\right) V_0 = \frac{1}{2}\varrho_{v0}h^2 + \sum_{i=1}^{6} \varepsilon_{i0} V_i \tag{9.9b}$$

because this is useful in a method of solution known as *iteration* (see later).

If we consider a source-free region of uniform permittivity, then $\varrho_{v0} = 0$ everywhere in v and $\varepsilon_{(0)i} = \varepsilon/2$. Consequently, (9.9b) simplifies vastly in this special case to become

$$V_0 = \frac{1}{6}\sum_{i=1}^{6} V_i \qquad \text{or} \qquad \sum_{i=1}^{6} V_i - 6V_0 = 0 \tag{9.9c}$$

This, then, is an approximate solution to Laplace's equation $\Delta V(x, y, z) = 0$, at a particular location.

If we wish to apply the above reasoning to the two-dimensional case (i.e. to a three-dimensional case in which no changes occur in, say, the z direction), then the cubes become squares of area h^2, and there are only four nearest neighbors. The enclosed charge is $\varrho_{s0}h^2$, if ϱ_{s0} is the effective surface-charge density in the central square. Equation (9.9a) then reduces to

$$\left(\sum_{i=1}^{4} \varepsilon_{(0)i}\right) V_0 = \frac{1}{2}\varrho_{s0}h + \sum_{i=1}^{4} \varepsilon_{(0)i} V_i \tag{9.10a}$$

and, for the source-free uniform-ε case, to

$$V_0 = \frac{1}{4}\sum_{i=1}^{4} V_i \qquad \text{or} \qquad \sum_{i=1}^{4} V_i - 4V_0 = 0 \tag{9.10b}$$

9.2 Methods of solution for the potential

Equations (9.9), or (9.10) in two dimensions, usually constitute a formidable set. Recall that each mesh cube in Fig. 9.1 gives rise to an equation such as (9.9b) or (9.9c); such equations contain seven potential values, namely the potential V_0 at the center and six nearest-neighbor potentials V_i. Each nearest-neighbor V_i is itself the center of an adjacent cube (unless it lies on the boundary, where we assign the nearest-neighbor points beyond the boundary to have the same potential as the boundary point). Consequently, there must be N equations in N potentials – each equation having up to seven potentials. We can bring the charge-density terms and the boundary potentials to the right-hand side. If there are N_s boundary potentials, we arrive at $N - N_s \equiv N_i$ equations for N_i *interior* potentials and, because at least one nonzero boundary potential appears on the right-hand side, we can be sure – even in the charge-free region – that these are N_i inhomogeneous linear equations in the N_i interior potential values. This is important, because it ensures that we can always solve the set, and the fact that the solution is unique for given boundary conditions guarantees that the determinant of the left-hand side is not zero.

9.2.1 Matrix inversion

We consider the two-dimensional case corresponding to (9.10b). The form of the equations for N_i interior potential points, at the center of N_i mesh squares, is

$$\mathbf{MV} = \mathbf{R} \tag{9.11a}$$

Here, $\mathbf{V}$ is a single-column matrix that is the transpose of the row matrix

$$\mathbf{V} = (V_1 \; V_2 \cdots V_{N_i})$$

Likewise, $\mathbf{R}$ is the transpose of a single-row matrix with many zeroes and all the boundary-value potentials (which are known). Then $\mathbf{M}$ must be an $N_i \times N_i$ matrix in which each row has at most five nonzero elements, to conform to (9.10b). According to (9.10b) these nonzero elements are either 4 or 1. The formal solution to (9.11) is

$$\mathbf{V} = \mathbf{M}^{-1}\mathbf{R} \tag{9.11b}$$

which tells us that we need to find the inverse of matrix $\mathbf{M}$. This is always possible because the boundary conditions ensure that the determinant is nonzero. However, numerical mathematics informs us that it takes N_i^3 storage locations to invert the matrix. This easily can be an excessive demand upon a computer, because a 100×100 square mesh (certainly not excessive in two dimensions), yielding $N_i \sim 10^4$, would require $\sim 10^{12}$ storage locations. However, the matrix is *sparse* as most of its elements are zero, so that the actual number of storage locations is very much smaller. Nevertheless, direct inversion, or the method of *Cramer determinants* (essentially the same thing), is often unattractive owing to the high number of storage spaces still required. For that reason, other methods are often preferable.

9.2.2 Finite-mesh relaxation methods

One method to overcome storage limitations is to use the technique of *iteration*. Iteration techniques use a repetitive method to solve an equation. Consider as an example the equation

$$e^{-x} = x \tag{9.12a}$$

This equation cannot be solved analytically, but an alternative is to solve the equation set

$$e^{-x_n} = x_{n+1} \tag{9.12b}$$

for $n = 0, 1, 2, \ldots,$ as follows. If we choose some arbitrary value for x_0, say $x_0 = 0.3$, and calculate $\exp(-x_0) = x_1$, then calculate $\exp(-x_1) = x_2$, and continue on in that fashion, we are *iterating* the equation and we find that x_n moves toward the exact answer quite rapidly. In fact in 20 or so iterations we get three or more figures of accuracy. Moreover, the answer is reasonably *robust*; i.e. it will be obtained almost independently of the choice of x_0 (here provided it is positive).

In the case of the two-dimensional Laplace equation, we will work with a form of (9.10b) that shows explicitly how the iteration is to occur:

$$V_{ij}^{(n+1)} = \frac{1}{4}\left(V_{i+1,j}^{(n)} + V_{i-1,j}^{(n)} + V_{i,j+1}^{(n)} + V_{i,j-1}^{(n)}\right) \tag{9.13}$$

where the subindices i, j indicate the point $x = ih$, $y = jh$. Figure 9.4 shows how the four neighbor points at $x = (i \pm 1)h$, $y = (j \pm 1)h$ relate to the point where we wish

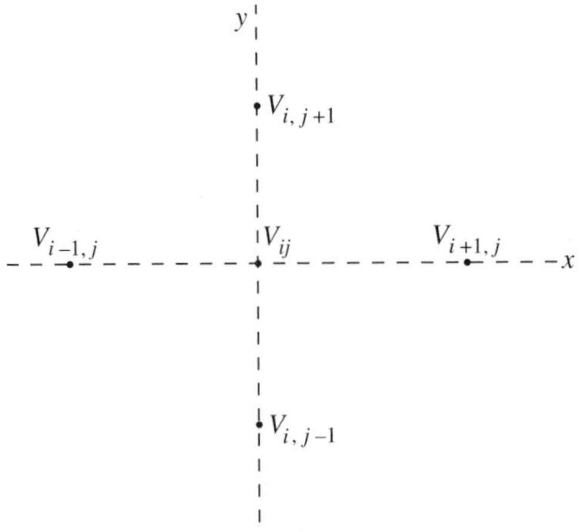

Fig. 9.4 The locations involved in the iteration of (9.13).

to calculate V_0. The method of finite relaxation iterates all the N_i equations of type (9.13) successively. It does this by first assigning arbitrary values $V_{ij}^{(0)}$ to the N_i centers of the mesh blocks. It then calculates the left-hand sides of (9.13) for $n = 0$ to obtain all $V_{ij}^{(1)}$. It continues in this fashion, continuously increasing the iteration parameter n by one, until the values of $V_{ij}^{(n+1)}$ no longer change appreciably. Here the word 'appreciably' requires some explanation. In almost all texts to date, examples are given of how to solve (9.13), and these examples uniformly tell one to stop iterating when all differences $|V_{ij}^{(n+1)} - V_{ij}^{(n)}|$, for fixed n and variable i, j, are less than a pre-chosen difference ϵ. Figure 9.5 illustrates why this can be quite wrong. It shows a case where a vertical ϵ has been chosen, but the iteration stops at the fifteenth step because the difference between that and the fourteenth step is already less than ϵ. Clearly, the asymptotic value $V_{ij}^{(\infty)}$ has not been reached at all. The actual error criterion is more complicated (see the optional paragraph below), and one usually must go on iterating considerably beyond this point for good accuracy.

In actual fact, if $V_{ij}^{(n)}$ is still changing rapidly with n, it is conceivable that the ultimate value $V_{ij}^{(\infty)}$ is many times ϵ removed in value from where we stopped, i.e. $|V_{ij}^{(\infty)} - V_{ij}^{(n)}|$ equals a multiple of ϵ even if $|V_{ij}^{(n+1)} - V_{ij}^{(n)}|$ equals a fraction of ϵ, as illustrated in Fig. 9.5. The reason for this follows from the fact that

$$V_{ij}^{(n)} = V_{ij}^{(\infty)} + c^{-n}\delta V_{ij}, \tag{9.14}$$

where both δV_{ij} and the constant c are functions of i, j. The equation holds provided n is reasonably large, say more than 50–100, in which case $V_{ij}^{(n)}$ may already be close to its ultimate $n \to \infty$ value. The equality (9.14) is not difficult

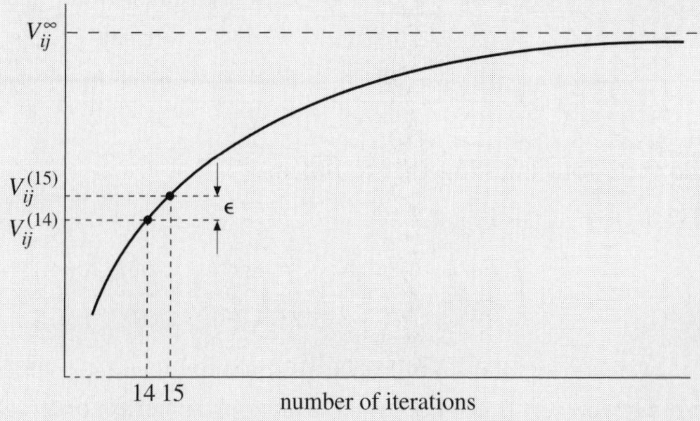

Fig. 9.5 A hypothetical plot of V_{ij} as a function of the number n of iterations of (9.13).

to prove, but we will not do so here. A simple calculation shows that

$$c = \frac{[V_{ij}^{(n-1)} - V_{ij}^{(n-2)}]}{[V_{ij}^{(n)} - V_{ij}^{(n-1)}]}$$
$$c^{-n}\delta V_{ij} = \frac{[V_{ij}^{(n)} - V_{ij}^{(n-1)}]}{[V_{ij}^{(n)} - V_{ij}^{(n-1)}] - [V_{ij}^{(n-1)} - V_{ij}^{(n-2)}]}\left[V_{ij}^{(n)} - V_{ij}^{(n-1)}\right] \tag{9.15}$$

The first of these equations indicates that gradual relaxation (i.e. a slow change in $V_{ij}^{(n)}$ with n) implies that c is close to unity; the second equation then implies that $c^{-n}\delta V_{ij} \equiv [V_{ij}^{(n)} - V_{ij}^{(\infty)}]$ greatly exceeds $[V_{ij}^{(n)} - V_{ij}^{(n-1)}]$, which means, according to (9.14), that we are not close enough to the infinitely iterated ideal answer, as we stated above, when $[V_{ij}^{(n)} - V_{ij}^{(n-1)}] \sim \epsilon$.

More importantly, the second of equations (9.15) gives us a sure way of computing the difference between the nth iterate $V_{ij}^{(n)}$ and the ideal answer $V_{ij}^{(\infty)}$. It requires only that we know three iterates at once, hence it requires storage of only $3N_i$ potentials – surely a large advantage over storing N_i^3 potentials!

Now consider the validity of the final value for the potential that we obtain from (9.13), namely

$$V_{ij} = \frac{1}{4}\left(V_{i+1,j} + V_{i-1,j} + V_{i,j+1} + V_{i,j-1}\right) \tag{9.16}$$

This answer is only an *approximate* solution of Laplace's equation. To see this, consider the following Taylor-series expansions of the four nearest-neighbor potentials:

$$\begin{aligned}
V_{i+1,j} &= V_{ij} + h\frac{\partial V_{ij}}{\partial x} + \frac{h^2}{2}\frac{\partial^2 V_{ij}}{\partial x^2} + \frac{h^3}{6}\frac{\partial^3 V_{ij}}{\partial x^3} + \frac{h^4}{24}\frac{\partial^4 V_{ij}}{\partial x^4} + \cdots \\
V_{i-1,j} &= V_{ij} - h\frac{\partial V_{ij}}{\partial x} + \frac{h^2}{2}\frac{\partial^2 V_{ij}}{\partial x^2} - \frac{h^3}{6}\frac{\partial^3 V_{ij}}{\partial x^3} + \frac{h^4}{24}\frac{\partial^4 V_{ij}}{\partial x^4} + \cdots \\
V_{i,j+1} &= V_{ij} + h\frac{\partial V_{ij}}{\partial y} + \frac{h^2}{2}\frac{\partial^2 V_{ij}}{\partial y^2} + \frac{h^3}{6}\frac{\partial^3 V_{ij}}{\partial y^3} + \frac{h^4}{24}\frac{\partial^4 V_{ij}}{\partial y^4} + \cdots \\
V_{i,j-1} &= V_{ij} - h\frac{\partial V_{ij}}{\partial y} + \frac{h^2}{2}\frac{\partial^2 V_{ij}}{\partial y^2} - \frac{h^3}{6}\frac{\partial^3 V_{ij}}{\partial y^3} + \frac{h^4}{24}\frac{\partial^4 V_{ij}}{\partial y^4} + \cdots
\end{aligned} \tag{9.17}$$

Add these four lines together to obtain

$$\begin{aligned}
&V_{i+1,j} + V_{i-1,j} + V_{i,j+1} + V_{i,j-1} \\
&\quad = 4V_{ij} + h^2\left(\frac{\partial^2 V_{ij}}{\partial x^2} + \frac{\partial^2 V_{ij}}{\partial y^2}\right) + \frac{h^4}{12}\left(\frac{\partial^4 V_{ij}}{\partial x^4} + \frac{\partial^4 V_{ij}}{\partial y^4}\right) + \cdots
\end{aligned} \tag{9.18}$$

If V_{ij} were a true solution of Laplace's equation, then the coefficient of h^2 would be zero. However, it is not because it is an approximation obtained from iteration of (9.13). The first error term is thus the h^2 term, which we identify as being equal to

$(h^2/4l_{ij}^2)V_{ij}$, with

$$l_{ij}^{-2} \equiv \frac{1}{V_{ij}}\left(\frac{\partial^2 V_{ij}}{\partial x^2} + \frac{\partial^2 V_{ij}}{\partial y^2}\right) \tag{9.19}$$

and thus we find from re-ordering (9.18) that

$$V_{ij} \approx \frac{1}{4}(V_{i+1,j} + V_{i-1,j} + V_{i,j+1} + V_{i,j-1}) - \frac{h^2}{l_{ij}^2}V_{ij} - \cdots \tag{9.20}$$

which expresses the need to make sure that the mesh parameter h is chosen sufficiently small compared to l_{ij}. Note that we do not really know how to calculate l_{ij} from (9.19), because we only have values for V_{ij} at centers of mesh squares, but of course we can estimate it, e.g. by fitting exponentials over a distance of a few mesh lengths h in various directions.

Example 9.1 A 10 × 14 rectangle with boundary voltages

To illustrate the above finite relaxation method in two dimensions, we consider a Dirichlet problem which we have already solved using the method of separation of variables. We take a rectangle with three zero-voltage sides and a sine function on the fourth side such that the sine goes through half a cycle from zero to zero along that fourth side. The solution for $\tilde{V}_1 = 50$ volts and all other $\tilde{V}_m = 0$ is given by (8.50). Thus, referring to the top boundary in Fig. 8.12, we know that $V(x, b) = 50\sin(0.1\pi x)$ and the other three boundary potentials are zero. Hence

$$V(x,y) = 50\frac{\sin(0.1\pi x)\sinh(0.1\pi y)}{\sinh(1.4\pi)} \tag{9.21}$$

Shown in Table 9.1 is a grid of $V(x, y)$ for integer values of x and y, $0 < x < 10$ and $0 < y < 14$. Compare these exact values with those generated by iteration via (9.13), shown in the grid of Table 9.2. You can easily check that (9.16) holds for this second grid at any interior point. The differences from the exact answers are obvious. Let us try a check on the accuracy. For example, the potential $V_{ij} \equiv V(5, 12)$ is exactly 26.66, but is found to be 26.80 by iteration. The second derivative in the y direction is approximated by

$$\frac{V_{i,j+2} - 2V_{ij} + V_{i,j-2}}{4h^2} \approx 2.69 \tag{9.22}$$

whereas that in the x direction is given by

$$\frac{V_{i+2,j} - 2V_{ij} + V_{i-2,j}}{4h^2} \approx -2.56 \tag{9.23}$$

The error term in (9.20) thus comes out to be $\sim\frac{1}{4}(2.69 - 2.56) \approx 0.033$. The difference actually inferred from the two grids is 0.14. The discrepancy between these differences comes from the higher-order terms neglected in (9.20) and it is

Table 9.1. Values of $V(x, y)$ from (9.21) at integer values of x and y, $0 < x < 10$ and $0 < y < 14$

0.00	15.45	29.39	40.45	47.55	50.00	47.55	40.45	29.39	15.45	0.00
0.00	11.28	21.46	29.54	34.73	36.52	34.73	29.54	21.46	11.28	0.00
0.00	8.24	15.67	21.57	25.36	26.66	25.36	21.57	15.67	8.24	0.00
0.00	6.02	11.44	15.75	18.51	19.47	18.51	15.75	11.44	6.02	0.00
0.00	4.39	8.35	11.49	13.51	14.21	13.51	11.49	8.35	4.39	0.00
0.00	3.20	6.09	8.38	9.85	10.36	9.85	8.38	6.09	3.20	0.00
0.00	2.33	4.43	6.10	7.17	7.54	7.17	6.10	4.43	2.33	0.00
0.00	1.69	3.22	4.43	5.21	5.48	5.21	4.43	3.22	1.69	0.00
0.00	1.22	2.33	3.20	3.76	3.96	3.76	3.20	2.33	1.22	0.00
0.00	0.87	1.66	2.29	2.69	2.83	2.69	2.29	1.66	0.87	0.00
0.00	0.61	1.17	1.61	1.89	1.99	1.89	1.61	1.17	0.61	0.00
0.00	0.41	0.79	1.08	1.27	1.34	1.27	1.08	0.79	0.41	0.00
0.00	0.25	0.48	0.67	0.78	0.82	0.78	0.67	0.48	0.25	0.00
0.00	0.12	0.23	0.32	0.37	0.39	0.37	0.32	0.23	0.12	0.00
0.00	0.00	0.00	0.00	0.00	0.00	0.00	0.00	0.00	0.00	0.00

Table 9.2. Values of $V(x, y)$ generated by iteration, using (9.13)

0.00	15.45	29.39	40.45	47.55	50.00	47.55	40.45	29.39	15.45	0.00
0.00	11.31	21.52	29.62	34.82	36.61	34.82	29.62	21.52	11.31	0.00
0.00	8.28	15.75	21.68	25.49	26.80	25.49	21.68	15.75	8.28	0.00
0.00	6.06	11.53	15.87	18.66	19.62	18.66	15.87	11.53	6.06	0.00
0.00	4.43	8.44	11.61	13.65	14.35	13.65	11.61	8.44	4.43	0.00
0.00	3.24	6.17	8.49	9.98	10.49	9.98	8.49	6.17	3.24	0.00
0.00	2.37	4.50	6.20	7.28	7.66	7.28	6.20	4.50	2.37	0.00
0.00	1.72	3.28	4.51	5.30	5.57	5.30	4.51	3.28	1.72	0.00
0.00	1.25	2.37	3.27	3.84	4.04	3.84	3.27	2.37	1.25	0.00
0.00	0.89	1.70	2.34	2.75	2.89	2.75	2.34	1.70	0.89	0.00
0.00	0.63	1.20	1.65	1.93	2.03	1.93	1.65	1.20	0.63	0.00
0.00	0.42	0.81	1.11	1.31	1.37	1.31	1.11	0.81	0.42	0.00
0.00	0.26	0.50	0.69	0.81	0.85	0.81	0.69	0.50	0.26	0.00
0.00	0.12	0.24	0.33	0.38	0.40	0.38	0.33	0.24	0.12	0.00
0.00	0.00	0.00	0.00	0.00	0.00	0.00	0.00	0.00	0.00	0.00

sizable in this example. The difference will not always be so large, but one must remember that our error estimate is based upon neglect only of the term that is quadratic in h in (9.20).

Finally, we present here two programs using which the grids in Example 9.1 could be produced. The reader should find these programs easy to use and very fast in execution.

The MATHEMATICA 3.0® program LAPL.NB produces a set of boundary-value potentials in a 10×14 grid. The input boundary values are as specified in

Example 9.1, but it is not difficult to modify the program for other boundary values of the potential.

LAPL.NB

```
(*Initial relaxation part of program*)
f = N[50*Sin[(n - 1)*Pi/10]];
R15 = Table[f, {n, 11}];
R14 = {0, 0, 0, 0, 0, 0, 0, 0, 0, 0, 0};
R12 = {0, 0, 0, 0, 0, 0, 0, 0, 0, 0, 0};
R11 = {0, 0, 0, 0, 0, 0, 0, 0, 0, 0, 0};
R10 = {0, 0, 0, 0, 0, 0, 0, 0, 0, 0, 0};
R9 = {0, 0, 0, 0, 0, 0, 0, 0, 0, 0, 0};
R8 = {0, 0, 0, 0, 0, 0, 0, 0, 0, 0, 0};
R7 = {0, 0, 0, 0, 0, 0, 0, 0, 0, 0, 0};
R6 = {0, 0, 0, 0, 0, 0, 0, 0, 0, 0, 0};
R5 = {0, 0, 0, 0, 0, 0, 0, 0, 0, 0, 0};
R4 = {0, 0, 0, 0, 0, 0, 0, 0, 0, 0, 0};
R3 = {0, 0, 0, 0, 0, 0, 0, 0, 0, 0, 0};
R2 = {0, 0, 0, 0, 0, 0, 0, 0, 0, 0, 0};
R1 = {0, 0, 0, 0, 0, 0, 0, 0, 0, 0, 0};
m = {R15, R14, R13, R12, R11, R10, R9, R8, R7, R6, R5, R4, R3, R2, R1};
mold = m;
Do[Do[m[[i, j]] = .25*(m[[(i - 1), j]] + m[[(i + 1), j]]+
m[[i, (j - 1)]] + m[[i, (j + 1)]]), {i, 2, 14}, {j, 2, 10}], {50}]
mnew = N[m, 3]
```

The execution of this part of the program is over 50 iterations, which in general is insufficient for accuracy. The next part of the program (below) allows one to increase the number of iterations by 50 at a time. One can repeat this part until the matrix of differences **mnew** – **mold** is sufficiently small.

```
(*50 iterations added to above part: this part can be repeated as needed,
or one can change {50} to {150}, etc. *)
m = mnew;
mold = m;
Do[Do[m[[i, j]] = .25*(m[[(i - 1), j]] + m[[(i + 1), j]]+
m[[i, (j - 1)]] + m[[i, (j + 1)]]), {i, 2, 14}, {j, 2, 10}], {50}]
mnew = N[m, 3]
N[mnew - mold, 3]
```

Finally, the third part of the program (below) presents a contour plot and some three-dimensional graphs to show the equipotential lines of interest.

```
(*Graphics part of program*)
n = 15; p = 11;
tbl = Table[mnew[[(n + 1 - i), j]], {i, n}, {j, p}];
tbl2 = -tbl;
plt2 = ListContourPlot[tbl2, ContourShading - >  False,
ContourSmoothing - > Automatic, Contours - >
{0, -1, -2, -3, -4, -5, -10, -15, -20, -25, -30, -35, -40, -45}, FrameLabel - >
{'V(x, 0)', 'V(0, y)', 'V(x, 14)', 'V(10, y)'}, PlotLabel - >
{'Contour for design project'}];
Show[SurfaceGraphics[plt2], ViewPoint - >  {3.3, 2.4, 2}]
Show[SurfaceGraphics[plt2], ViewPoint - >  {3.3, 2.5, 0}]
```

The user can modify the designated equipotentials easily.

Another high-level language option is given by the following MATLAB® program, which produces a contour plot and several three-dimensional plots of $V(x, y)$ for the same boundary conditions as above.

LAPLACE.M

```
clear;
for n = 1 : 15                          % Initialize array s to zero
  for m = 1 : 11
    s(n, m) = 0;
  end;
end

for m = 1 : 11
  s(1, m) = 50* sin(0.1*pi*(m - 1));    % Voltage on top plate
  s(15, m) = 0;                         % Voltage on bottom plate
end

for n = 2 : 14;                         % Voltages on side plate
  s(n, 1) = 0;
  s(n, 11) = s(n, 1);
end;

for i = 1 : 200;
  for n = 2 : 14;
    for m = 2 : 10;
    s(n, m) = 0.25*(s(n - 1, m) + s(n + 1, m) + s(n, m - 1) + s(n, m + 1));
    end;
  end;
end
```

```
for n = 1 : 15;
 for m = 1 : 11;
  ss(n, m) = -s(n, m);
  st(n, m) = s(16 - n, m);
 end;
end;

v1 = linspace(0, 10, 10);            % Generate range of voltages for contour plot
v2 = linspace(0, 50, 5);
V = [v1 v2];

figure(1)                            % Generate three-dimensional surface plot
surf(ss)
colormap(cool)
view(225, -45)
hold on
rotate3d                             % Put mouse on figure to rotate
zlabel ('volts, V')
xlabel ('x')
ylabel ('y')

figure(2)                            % Generate two-dimensional contour plot
contour(st,V);
C = contour(st, V);
clabel (C,V)
grid on
hold off
xlabel ('V(x, 0) = 0 V, V(x, 14) = 50 V')
```

This program does 200 iterations, and the three-dimensional plot can be rotated manually with a PC mouse. It is left up to the reader to modify the program for more iterations and for other designated equipotentials.

Let us summarize briefly the advantages of the finite-relaxation numerical method:

- small storage requirements
- easy to program
- robust (i.e. it will converge upon the correct answer from almost any starting values of $V_{ij}^{(0)}$)

Some disadvantages are:

- the mesh-block size is not easily variable
- a closed boundary, on which $V(\mathbf{r})$ is known, is required
- potentials are only given at fixed mesh points and not between mesh points

The second disadvantage makes itself felt severely in cases such as the finite parallel-plate capacitor, where we have no good way to close the boundary surface unless it is at a great distance, where $V(\mathbf{r}) \approx 0$, which would make the number of mesh cubes very large. The first disadvantage appears in regions where great accuracy is needed in a small subspace of the region but not elsewhere. We will now mention briefly several other numerical methods, which overcome some of these disadvantages, albeit at the cost of some of the advantages, of the finite-relaxation method.

9.2.3 Finite-element methods

The *finite-element* approach allows us to use a mesh of triangles in two dimensions (or tetrahedrons in three dimensions), which may be extremely variable in size. Thus, very small triangles can be used in a region where high accuracy is desirable, and larger ones elsewhere. The finite-element method consists of several steps, which we outline only very briefly for the two-dimensional case without details.

The potentials are to be calculated at the nodes (i.e. the vertices of the triangles) from the electrostatic energy W_e as given in (7.13) but adapted for the two-dimensional case:

$$W_e = \frac{1}{2}\int_S dS\, \varepsilon(\mathbf{r})|\nabla V(\mathbf{r})|^2 \tag{9.24}$$

This, of course, is an integral over the surface S. Instead of discretizing this integral in two dimensions, to turn it into a double sum over nodes, something else is done. In each triangle, the potential $V(x, y)$ is written as a linear combination $V(x, y) = v + e_x x + e_y y$, and the three coefficients v, e_x, and e_y are chosen so that values V_1, V_2, and V_3 are obtained at the three vertices. Of course these three values are not yet known, but inside each triangle we then have an *interpolation formula* $V(x, y) = \alpha_1(x, y)V_1 + \alpha_2(x, y)V_2 + \alpha_3(x, y)V_3$. We also allow these values V_i to hold *outside* the triangles. Thus at an arbitrary point in the region, $V(x, y)$ will be a weighted sum of the triple-sum contributions from *all* the triangles. The sum is weighted because there is always more than one triangle sharing one vertex, and the interpolation formula must assume the correct vertex potential values at the vertices.

Thus, $\nabla V(\mathbf{r})$ in (9.24) becomes a linear function of all the V_i and of the gradients of the α_i, and therefore W_e becomes a linear function of all the node potentials V_i.

A variational technique applied to the stored-energy integral (9.24) teaches that, if we allow arbitrary values for all the V_i, then W_e is at a minimum when these values correspond to the actual physical situation, i.e. when these V_i are proper solutions of Laplace's equation (in the absence of charges). Thus we must set $\partial W_e/\partial V_i = 0$ for all possible i. This yields N linear equations in the $V_i (1 \le i \le N)$, which can be solved by any of the usual matrix-inversion or other methods to finish the calculation.

The finite-element method can be extended similarly for three-dimensional applications of Poisson's equation, and sometimes more complicated interpolation formulas are used for greater accuracy. The linearity of the equations is then lost, but the more complicated N equations are still amenable to solution. The advantages of finite-element methods are

- variable mesh-element sizes
- better adaptation to nonrectangular shapes, as triangles (in two dimensions) are used
- nonclosed geometries are less of a problem because very large triangles can be used in regions where things change slowly (see end of previous subsection).

Two disadvantages are

- the need to set up a complicated interior mesh of triangles
- a more complicated numerical scheme than for finite relaxation methods.

9.2.4 Integral-equation methods

Sometimes the 'openness' of a situation makes it very hard to define a closed boundary on which reasonable boundary-value potentials or normal fields can be defined (unless an unreasonably large region is defined artificially). Integral-equation methods avoid this difficulty. Consider Fig. 9.6, in which two metallic plates of arbitrary shape, held at voltages V_1 and V_2, are placed in space. Certainly, a very large box would be needed around these to safely predict $V =$ constant to good accuracy on the surfaces of the box. Instead, we use the surface-charge-density form of the integral equation for $V(\mathbf{r})$ given in (4.8a); it is

$$V_P = \frac{1}{4\pi\varepsilon_0} \int dS_Q \, \frac{\varrho_s(\mathbf{r}_Q)}{R_{QP}} \tag{9.25}$$

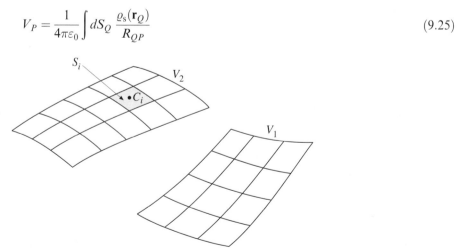

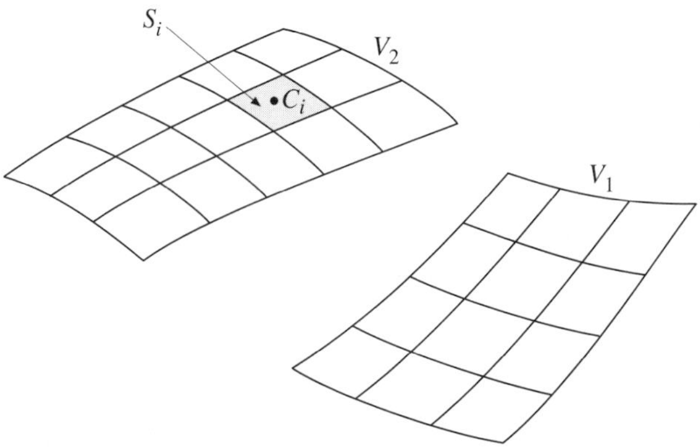

Fig. 9.6 Quantization of charge on two surfaces forming an 'open' geometry. Shown are surface elements on two curved metal plates. C_i is the center of element i of area S_i.

where $\mathbf{r}_Q$ is the position vector of a point Q on either of the plates. We now divide the two surfaces in Fig. 9.6 into N elementary areas S_i, sufficiently small that $\varrho_s(\mathbf{r}_Q)$ is essentially constant over surface element S_i (one of these is shaded in the upper surface); points within the surface element S_i are labeled Q. We can thus apply (9.25) to each surface element in turn, arriving at a sum over surface-element integrals:

$$V_P = \sum_{i=1}^{N} \alpha_i \varrho_{si} \qquad \alpha_i \equiv \frac{1}{4\pi\varepsilon_0} \int_{S_i} dS_Q \frac{1}{R_{QP}} \tag{9.26}$$

Here dS_Q is the infinitesimal surface area at Q within the element S_i and ϱ_{si} is the (constant) surface-charge density on that element. The coefficients α_i are merely integrals over a known geometrical configuration, and they can be evaluated numerically in principle. However, the N surface-charge densities ϱ_{si} are unknown. We determine them by taking point P to be at the center C_j of surface element S_j: we then obtain

$$V_{C_j} = \sum_{i=1}^{N} \alpha_{ij} \varrho_{si} \qquad \alpha_{ij} = \frac{1}{4\pi\varepsilon_0} \int_{S_i} dS_Q \frac{1}{R_{Q_j}} \tag{9.27}$$

where $\mathbf{R}_{Qj}$ is the displacement from the point Q on S_i to the center of S_j. The V_{C_j} are known because the points C_j lie on the two surfaces, on which we assume that we know the potentials. As the coefficients α_{ij} are just geometrical integrals which in principle can be evaluated, (9.27) again represents N equations in N unknown coefficients ϱ_{si}. We can therefore calculate the ϱ_{si} from (9.27) and so the α_i coefficients of (9.26) can be calculated for arbitrary P; thus V_P follows from the first equation of (9.26). It is numerically convenient to take the ratio $\varrho_{si}/\varepsilon_0 = \Omega_{si}$ as an effective surface-charge density, because ε_0 always occurs in this form. The integral-equation method has the following advantages:

- the geometry may be open, i.e. not enclosed by a surface on which $V(\mathbf{r})$ is known
- the mesh is two dimensional in three-dimensional situations, hence the number of equations needed to solve for the charge densities ϱ_{si} is only $N = N_x N_y$ instead of $N = N_x N_y N_z$.

The disadvantages are

- the calculation of the coefficients α_{ij} is not easy for mesh centers close to each other, especially if the surfaces are very curved
- each time P is placed differently, new coefficients α_i must be calculated.

Problems

9.1. Explain why even an infinite number of iterations for fixed mesh size h would yield only an approximation for the potential $V(x, y, z)$ when the method of finite iterations is applied in a finite-relaxation solution of Laplace's equation.

9.2. In the rectangle $0 \leq x \leq 10, 0 \leq y \leq 14$, the boundary conditions are $V(0, y) = 20$, $V(10, y) = 40$, $V(x, 0) = 10$, $V(x, 14) = 50$. Use a slightly modified version of the LAPLACE.NB or LAPLACE.M programs to calculate $V(4, 11)$. Alternatively, use a spreadsheet.

9.3. Work out an equivalent to Eq. (9.13) for a cylindrically symmetric case in which the Dirichlet boundary values are $V(\rho, \varphi, z) = f(z)$ on the cylindrical surface $\rho = a$. Here it suffices to work in a single plane containing the z axis.

10

Electric current

10.1 Concept of current density

10.1.1 Volume current density

We have seen that *electric charge* is the underlying source of electric fields. We shall see that *electric current* underlies magnetic fields, which we have yet to discuss and which form an integral part of modern electromagnetic field theory.

The concept of *current* is somewhat more complicated than is implied just by 'moving charge'. The formal definition of electric current is

$$I = \frac{dq}{dt} \tag{10.1}$$

This formula implies that the current I is the amount of charge q passing normally through some surface S per unit time. To understand it better, let us consider a surface element $d\mathbf{S} = \hat{\mathbf{n}}\, dS$, as shown in Fig. 10.1. Let us assume that each charge has the same velocity $\mathbf{u} = u\hat{\mathbf{z}}$. Hence, in a small time interval δt each charge is displaced by a distance vector $\delta \boldsymbol{l} = (u\,\delta t)\hat{\mathbf{z}}$. These displacement vectors are shown for three charges in Fig. 10.1(a), and we note that only one of them crosses surface element $d\mathbf{S}$ in δt. In fact, in the time interval δt only those charges in the volume δv, shown in Fig. 10.1(b), pass through the surface; the magnitude of this volume is

$$\delta v = \delta l\, dS \cos\theta = u\,\delta t\, dS \cos\theta$$

If we have N charges q per unit volume, then

$$\delta q_{\text{tot}} = Nq\,\delta v = Nqu\,\delta t\, dS \cos\theta$$

charges cross the surface $d\mathbf{S}$ in time interval δt. The current dI crossing $d\mathbf{S}$ thus must be the ratio of δq_{tot} to δt:

$$dI \equiv \frac{\delta q_{\text{tot}}}{\delta t} = (Nqu\cos\theta)\, dS \tag{10.2a}$$

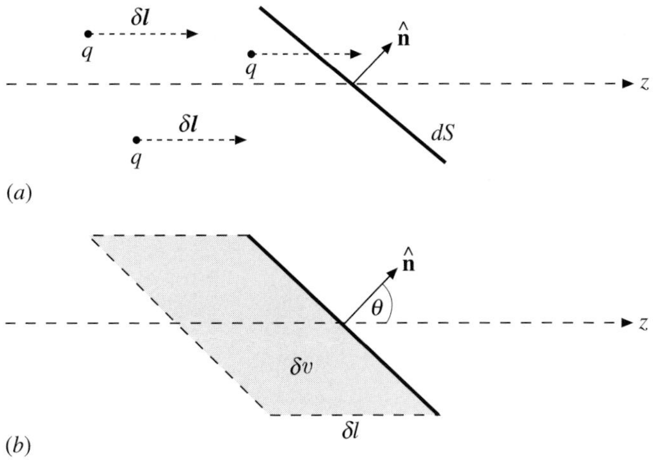

Fig. 10.1 Definition of electric current as the motion of charges through a surface element dS.

Because $\cos\theta = \hat{\mathbf{z}} \cdot \hat{\mathbf{n}}$, $\mathbf{u} = u\hat{\mathbf{z}}$, and $d\mathbf{S} \equiv \hat{\mathbf{n}}\, dS$, we can rewrite (10.2a) as

$$dI = Nq\mathbf{u} \cdot d\mathbf{S} \tag{10.2b}$$

From (10.1) we note that current is measured in coulombs per meter, or *amperes*, A. Hence $1\,\text{A} \equiv 1\,\text{C/m}$. The combination $Nq\mathbf{u}$ is a vector in the direction of charge flow, and its unit must be A/m^2. This vector is known as the *current density vector* and we denote it by the symbol $\mathbf{J}$. The current over a noninfinitesimal surface S is given by the integral

$$I = \int_S d\mathbf{S} \cdot \mathbf{J}(\mathbf{r}) \qquad \text{with} \quad \mathbf{J}(\mathbf{r}) \equiv N(\mathbf{r})q\mathbf{u}(\mathbf{r}) \tag{10.3}$$

We now see that $\mathbf{J} = Nq\mathbf{u}$ is a very special case; all the charges have the same magnitude and velocity vector. More generally we have N charges q_n per unit volume at $\mathbf{r}$, each with velocity vector $\mathbf{u}_n$. The current density vector at $\mathbf{r}$ then becomes

$$\mathbf{J}(\mathbf{r}) = \sum_{n=1}^{N} q_n \mathbf{u}_n \tag{10.4a}$$

If all the particles have the same charge q then this becomes

$$\mathbf{J}(\mathbf{r}) = q\sum_{n=1}^{N} \mathbf{u}_n = Nq\left[\frac{1}{N}\sum_{n=1}^{N} \mathbf{u}_n\right] = Nq\langle\mathbf{u}\rangle \tag{10.4b}$$

in which expression $\langle\mathbf{u}\rangle$ is an average velocity vector. This should clarify the meaning of $\mathbf{u}(\mathbf{r})$ in (10.3).

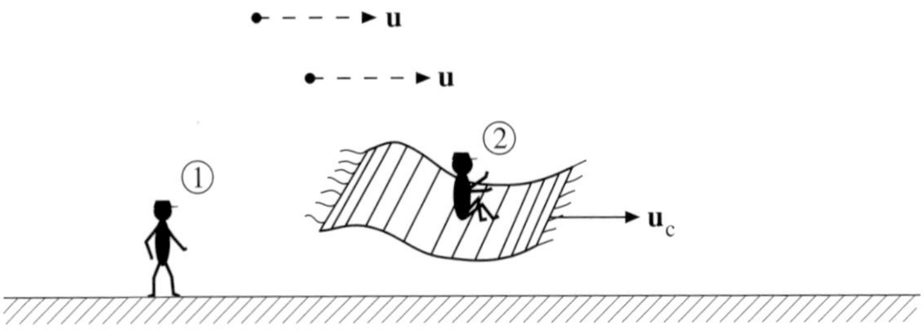

Fig. 10.2 Current as it appears to a motionless and a moving observer.

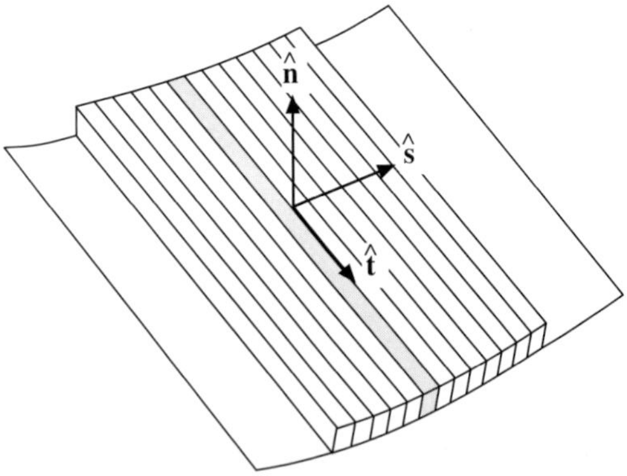

Fig. 10.3 Sheet-current (or surface-current) density on a surface.

Another interesting (but perhaps disturbing) fact is that the concept of *current* depends critically upon the situation of the observer. Consider Fig. 10.2: observer 1 stands on the ground and sees the charges q move with velocity $\mathbf{u}$. The current density, according to this observer, is $\mathbf{J} = Nq\mathbf{u}$. Observer 2 sits on a magic carpet moving with velocity vector $\mathbf{u}_c$, and in that observer's frame of reference the charges move with velocity $\mathbf{u} - \mathbf{u}_c$. To observer 2, the current density appears to be $\mathbf{J} = Nq(\mathbf{u} - \mathbf{u}_c)$. In fact, observer 2 would conclude there is no current at all if $\mathbf{u}_c = \mathbf{u}$. This important observation underlies an equally important truth: a static field is purely electric (only stationary charges), or purely magnetic (all charges are in motion such that the current densities are unchanged in time), or some mixture, depending upon the frame of reference of the observer! It will come back to haunt us in a future discussion of Faraday's law, the first law we will encounter that touches on the dynamic interplay between electric and magnetic fields.

10.1.2 Sheet-current and surface-current density

Consider a linear array of thin rectangular wires (cross-sectional area $h \times h$) lying on a surface (which may be curved), as shown in Fig. 10.3. Each wire carries a current I, and there are N wires per meter of surface. Consider the *curvilinear* orthogonal coordinate system x_s, x_t, x_n, with unit vectors $\hat{\mathbf{s}}$, $\hat{\mathbf{t}}$, $\hat{\mathbf{n}}$, centered at a point of the surface. At this location the current points in the direction $\hat{\mathbf{t}}$ and there are N wires per meter when one counts them in the $\hat{\mathbf{s}} = \hat{\mathbf{n}} \times \hat{\mathbf{t}}$ direction ($\hat{\mathbf{n}}$ is the surface-normal unit vector). Clearly the current density $\mathbf{J}(x_s, x_t, x_n)$ is $(I/h^2)\hat{\mathbf{t}}$ for $0 < x_n < h$ and zero elsewhere. As $h \to 0$ the current density becomes infinite, but the total current per meter in the $\hat{\mathbf{s}}$ direction is NI, regardless of how small h is. It would seem that the small but nonzero size of the region of current density along the $\hat{\mathbf{n}}$ direction is unimportant if all other physical sizes and distances are much larger.

Consequently, we often ignore the nonzero magnitude of the region of current density in the $\hat{\mathbf{n}}$ direction to say that the current is restricted to a surface, i.e. that we have a *sheet current* with density $\mathbf{K} = NI\hat{\mathbf{t}}$. The magnitude of this *surface-current density*, K, is measured in A/m. To find the total current along the sheet over a length $x_{s2} - x_{s1}$ we must integrate K along the x_s direction:

$$I = \int_{x_{s2}}^{x_{s1}} d\mathbf{x}_s \cdot [\hat{\mathbf{n}}(x_s) \times \mathbf{K}(x_s)] = \int_{x_{s1}}^{x_{s2}} dx_s \, K(x_s) \tag{10.5}$$

The second form is probably sufficient for most practical purposes, as it is usually obvious that we should count wires on the surface (but $\perp$ to the direction of the current). However, we may discard the wire picture entirely, and simply say that there is a sheet current on a surface characterized by one specific value of the curvilinear coordinate x_n, with surface-normal unit vector $\hat{\mathbf{n}}(x_s, x_t, x_n)$ and surface-current density $\mathbf{K}(x_s)$. In this case the first form of (10.5) describes more accurately how to obtain the total current by means of a line integral. The second form may at times be more difficult to use when there is curvature in the $\hat{\mathbf{s}}$ direction since in this case x_s is a function of the basic orthogonal coordinates x, y, z.

10.2 Various types of current

10.2.1 Convection currents

In subsection 8.4.2 we encountered the space-charge-limited planar diode. Electrons are pulled out of a cathode by a negative electrostatic field, produced by an anode (at distance $z = d$) at a potential that is V higher than that of the cathode at $z = 0$. At low fields, only relatively few electrons are extracted and N is therefore small. As a result, it is a good approximation to assume that the electrons ignore each other's Coulomb fields, and consequently each electron is accelerated only in the $\hat{\mathbf{z}}$ direction by the force of the Laplace potential $V(z) = Vz/d$. The potential energy at z is

$-eV(z) = -eVz/d$. The kinetic energy is $\frac{1}{2}mu^2(z)$, and thus conservation of total energy yields

$$\frac{1}{2}mu^2(z) - \frac{eVz}{d} = 0 \qquad u(z) = \sqrt{\frac{2eVz}{md}} \tag{10.6}$$

i.e. the velocity increases as $\sqrt{z}$. The total current at z over a surface $S \perp \hat{\mathbf{z}}$ is $I(z) = \int dS(z)J(z)$, and $dS(z)$ does not change with z because all electrons move only in the $\hat{\mathbf{z}}$ direction if we ignore their Coulomb interaction. We have not yet proven (but will do so shortly) that the total current is conserved in the static case; hence $I(z)$ is not a function of z, and therefore $J(z)$ also is not a function of z. That implies of course that the number density $N(z) \sim z^{-1/2}$; $N(z)$ decreases as the velocity increases (and the electrons spread out further from each other). This is an example of a *convection current*, which usually involves the motion of charges in a vacuum with acceleration.

10.2.2 Electrolytic currents

Another type of current was referred to implicitly in subsection 8.4.1 where we observed that in the semiconductor junction diode electrons and holes diffuse away from each other across the junction. This, too, involves the motion of charges, and here we have an *electrolytic current*. Another example is the current in a battery, where electrochemical forces induce positive and negative charges to separate from each other. In a liquid, positive ions would be induced to flow one way and negative ions or electrons the other way. Such currents are usually accompanied by electric fields due to the separated charges which inhibit further separation, and they come to a halt unless there is some other mechanism that removes the ions. In fact, this is how a battery works: when the poles of a battery are connected through a closed exterior circuit, with a resistive element in it, the excess electrons on one side no longer heap up but flow through the circuit to the other side, and a continuous current is then possible as long as the electrolyte of the battery still has charges that can separate. Thus work is done in the circuit: either heat is produced or energy is otherwise extracted, at the cost of exhausting the electrochemical energy available in the battery. In the semiconductor junction diode, the separation of holes and electrons continues until the opposing electric field caused by that separation further inhibits the diffusion.

10.2.3 Conduction currents

In Section 6.2, we explained how a conducting material can be considered as a lattice of positive ions with nearly free electrons (in the *conduction band*) that are not bound

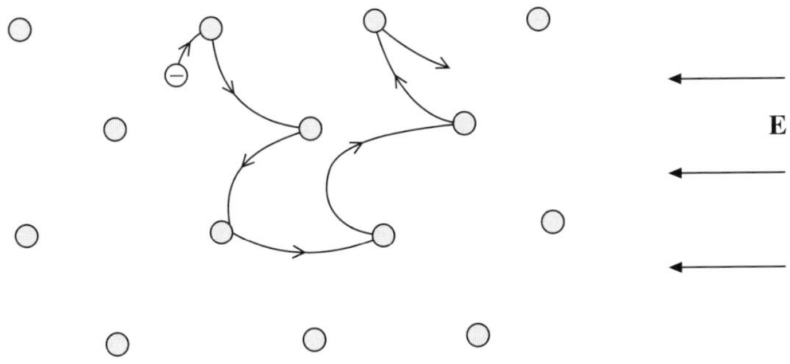

Fig. 10.4 Motion of a conduction electron in a metal.

to any particular lattice site. Figure 10.4 illustrates an interior region of such a lattice structure, with one conduction electron, initially moving as shown. An electrostatic field **E** tends to pull the electron into the direction $-\mathbf{E}$, but the electron is likely to collide with a much more massive and vibrating lattice ion. It will bounce off randomly, and so the new direction may bear no relationship to that before the collision. The electrostatic field again will tend to pull the electron into the direction $-\mathbf{E}$. Elastic collisions with lattice ions occur very many times per second, randomizing the velocity vector after each collision. The effect is of a 'drag' force on the electrons; their average motion is a *drift velocity* **u** against the field lines.

Let us calculate the average number of collisions of an electron per second ν_c, to show that it is very large indeed. The equation to do this is obtained from Newton's force equation, revised by adding a term for the 'drag' effect. It is

$$m\ddot{\mathbf{r}} = -e\mathbf{E} - m\nu_c\mathbf{u} \tag{10.7}$$

Here m, e are the electron mass and (absolute value of) charge and $\ddot{\mathbf{r}}$ is the acceleration. The last term on the right is the drag force, which counteracts $-e\mathbf{E}$. As can be seen, this term is proportional to the *drift velocity* **u** of the electron and in fact equals the total momentum lost per second by an electron. The above equation is one in which the terms represent *statistical averages* over a time that is macroscopically small, but in which millions of collisions take place.

In the situation explained above, the *average* acceleration $\ddot{\mathbf{r}}$ is zero. We then find that $m\nu_c\mathbf{u} = -e\mathbf{E}$, so that $\mathbf{u} = -e\mathbf{E}/m\nu_c$, and thus that

$$\mathbf{J} \equiv -Ne\mathbf{u} = \sigma\mathbf{E} \tag{10.8}$$

which is the point form of Ohm's law. Here σ is the *conductivity*, to be discussed below.

First, we can eliminate **E** from $\mathbf{u} = -e\mathbf{E}/m\nu_c$ and $-Ne\mathbf{u} = \sigma\mathbf{E}$ to obtain

$$\nu_c = \frac{Ne^2}{m\sigma} \tag{10.9}$$

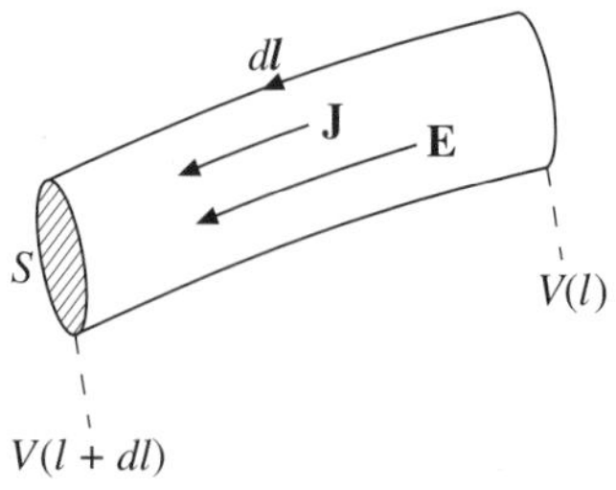

Fig. 10.5 Demonstration of $\mathbf{J}\cdot d\mathbf{S} = I\,dl$.

Rearranging (10.9) we obtain for the conductivity

$$\sigma = \frac{Ne^2}{m\nu_c} \tag{10.10}$$

as a consequence of (10.8). Thus, the collision frequency ν_c determines the dc conductivity σ, according to the relatively simple model given above. To connect σ to concepts more familiar in lumped-circuit analysis, let us consider Fig. 10.5, in which a short length $d\boldsymbol{l}$ of a conducting wire in a circuit has a cross-sectional area $S(l)$. The conductivity $\sigma(l)$ is essentially constant along dl. The total *voltage drop V* produces a constant current I in the wire, and a local current density $J(l) = I/S(l)$. In the short length shown, the voltage drop is $dV = V(l) - V(l + dl)$, where $dV = \mathbf{E}\cdot d\boldsymbol{l} = E\,dl$.

With Ohm's law (10.8) we obtain

$$dV = \frac{J}{\sigma}dl = \frac{I}{\sigma S}dl \quad \Rightarrow \quad V = I\int_0^l dl\frac{1}{\sigma(l)S(l)} \tag{10.11a}$$

This, of course, is the same as the macroscopic Ohm's law, $V = \mathcal{R}I$, in which $\mathcal{R}$ is the *resistance* in ohms (Ω). Thus, (10.11a) gives us a general expression for the resistance of a wire of uniform metal. If the cross-sectional area S is constant then we obtain from (10.11a)

$$\mathcal{R} = \frac{l}{\sigma S} \tag{10.11b}$$

and this connection between the macroscopic resistance $\mathcal{R}$ and the microscopic conductivity σ teaches us that the unit of conductivity is Ω^{-1}/m. The modern unit of conductivity σ is siemens per meter (S/m), and 1 S is the same as $1\,\Omega^{-1} \equiv 1$ mho. Given the typical density of lattice sites in a metal ($N \sim 10^{29}$ conduction electrons per cubic meter) and $\sigma \sim 5.8\times 10^7$ S/m for copper we can expect $\nu_c \sim 5\times 10^{13}$ collisions per second for copper at room temperature. Figures for other metals that are good conductors will not be drastically different.

The wire picture leads directly to the well-known formulas for addition of resistances. For two resistances in series we have $\mathcal{R} = \mathcal{R}_1 + \mathcal{R}_2$, the resistances add;

this follows immediately from (10.11a). For two resistances in parallel for which l and σ are the same, the total area is $S = S_1 + S_2$, and since from (10.11b) $\mathcal{R}^{-1} \propto S$, we conclude from this that $\mathcal{R}^{-1} = \mathcal{R}_1^{-1} + \mathcal{R}_2^{-1}$; more generally, any resistances add inversely when they are in parallel.

10.3 Power and conservation of current

10.3.1 Power

Let us consider a volume of moving charges q, e.g. as in the small piece of wire shown in Fig. 10.5. We interpret $\mathbf{J}$ as $Nq\mathbf{u}$. The force on one charge q is $q\mathbf{E}$, and that charge is on average displaced by $\mathbf{u}$ in one second. The work done in one second in displacing that one charge must be $p_1 = q\mathbf{u} \cdot \mathbf{E}$, and it is a power, which is measured in watts (W). The power involved in displacing N charges q per unit volume, i.e. the power per unit volume, must be $p_N = Nq\mathbf{u} \cdot \mathbf{E} = \mathbf{J} \cdot \mathbf{E}$, because there are N charges per unit volume. Thus we obtain for the total power in a volume v

$$P = \int_v dv\, \mathbf{J}(\mathbf{r}) \cdot \mathbf{E}(\mathbf{r}) \tag{10.12}$$

In the case of the wire we know that $dv = S\,dl$, so that $dv\,\mathbf{J} \cdot \mathbf{E} = dl\,S\mathbf{J} \cdot \mathbf{E} = d\mathbf{l} \cdot SJ\mathbf{E}$, because $\mathbf{J}$ and $d\mathbf{l}$ are in the same direction. Therefore $dv\,\mathbf{J} \cdot \mathbf{E} = I\,d\mathbf{l} \cdot \mathbf{E}$, and thus the total power in the wire is

$$P = I \int d\mathbf{l} \cdot \mathbf{E} = IV = \mathcal{R}^2 I \tag{10.13}$$

which is the customary form of Joule's law for the power dissipated in a resistive element by a current.

10.3.2 Conservation of current

A very general conservation law for current is formulated as follows. Consider a volume v in which there is a charge density $\varrho_v(\mathbf{r})$. The total charge $q = \int dv\, \varrho_v(\mathbf{r})$ changes in time for only one reason: there is current flowing in and/or out of the surface S surrounding v. Thus

$$\frac{dq}{dt} = -\oint d\mathbf{S} \cdot \mathbf{J} \tag{10.14a}$$

We rewrite this in terms of an integral over charge density:

$$\frac{d}{dt}\int dv\, \varrho_v(\mathbf{r}) = -\oint d\mathbf{S} \cdot \mathbf{J} = -\int dv\, \nabla \cdot \mathbf{J} \tag{10.14b}$$

Here, we have used Gauss's divergence theorem to transform the surface integral into a volume integral. If we consider the volume v to be unchanged in time then

$$\frac{d}{dt}\int dv\,\varrho_V(\mathbf{r},t) = \int dv\frac{\partial\varrho_V(\mathbf{r},t)}{\partial t} \tag{10.14c}$$

Comparing (10.14b) with (10.14c), and noting that the volume v is entirely arbitrary, we conclude from the equality of the right-hand sides that

$$\frac{\partial\varrho_V(\mathbf{r},t)}{\partial t} + \nabla\cdot\mathbf{J}(\mathbf{r},t) = 0 \tag{10.15}$$

This is the point form of (10.14a), and it is one of the basic equations of electromagnetics, together with Maxwell's equations in their (yet to be discussed) more general form. We can use this equation to return to the comment made in Section 6.2 that the time constant for charge deposited inside a conductor to vanish to the surface is very short. To see this, let us imagine that a small excess amount of charge $\varrho_V\,dv$ is deposited somewhere inside a conductor with local values of σ and ε that change with time slowly compared with the local charge density $\varrho_V(t)$. With the point form of Gauss's law, we then can replace $\nabla\cdot\mathbf{J}$ by $\sigma\nabla\cdot\mathbf{E} = (\sigma/\varepsilon)\varrho_V$ in (10.15) to obtain

$$\frac{\partial\varrho_V(\mathbf{r},t)}{\partial t} + \frac{\sigma}{\varepsilon}\varrho_V(\mathbf{r},t) = 0 \tag{10.16}$$

The quantity $\varepsilon/\sigma \equiv T$ is a time, as you can see from checking dimensions in (10.16), and this equation can be regarded as a total differential equation with respect to variable t. It is easy to solve: the solution can be written as

$$\varrho_V(\mathbf{r},t) = \varrho_V(\mathbf{r},0)\,e^{-t/T} \tag{10.17}$$

This equation states that the charge density in dv decays exponentially: it decreases by a factor $1/e$ in time $t = T \equiv \varepsilon/\sigma$ (which is known as a *relaxation time*). In copper, for example, $\varepsilon \approx \varepsilon_0$ and $\sigma \approx 5.8\times10^7$, and this gives $T \approx 1.5\times10^{-19}$ s. So we see that charge placed in a conductor does indeed flow very rapidly. The actual process by which charges flow to surface locations is quite complicated, and it is not described by (10.17). Furthermore, the whole process is nonstationary and the local parameters σ and ε are functions of time that change slowly during a time $\sim T$. Nevertheless, (10.17) gives at least an order-of-magnitude estimate of the relaxation time T.

Electrostatics deals with the electric fields produced by stationary charge distributions. In magnetostatics, we assume, rather, that the charge density at any location does not change in time. That is, there may well be current flowing, but there is as much charge flowing into, as out of, a volume element dv at any location. This implies that $dq/dt = 0$ in (10.14a), and therefore also that $\oint d\mathbf{S}\cdot\mathbf{J} = 0$

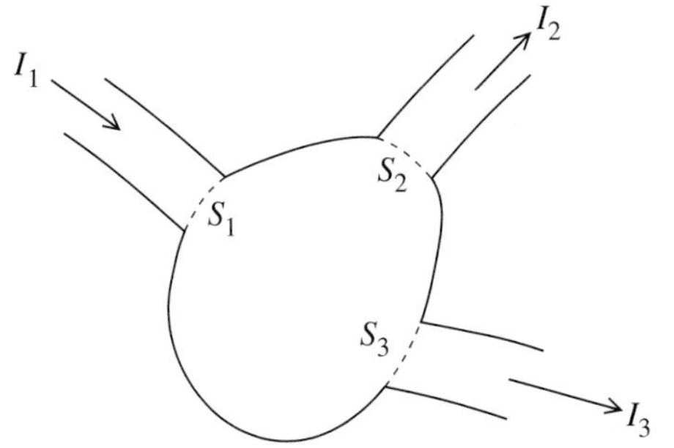

Fig. 10.6 Conservation of current at a node.

around dv. From (10.15) we obtain the static form of the equation of continuity:

$$\nabla \cdot \mathbf{J} = 0 \tag{10.18}$$

This is the point form of *Kirchhoff's current law* in circuits. If we apply its integral form, $\oint d\mathbf{S} \cdot \mathbf{J} = 0$, to the junction volume in Fig. 10.6, which is bounded by a surface partly indicated by the broken lines, we obtain

$$0 = \oint_S d\mathbf{S} \cdot \mathbf{J} = \int_{S_1} d\mathbf{S} \cdot \mathbf{J} + \int_{S_2} d\mathbf{S} \cdot \mathbf{J} + \int_{S_3} d\mathbf{S} \cdot \mathbf{J} = -I_1 + I_2 + I_3 \tag{10.19}$$

which is an example of the statement that all current entering a node of a circuit must equal all current leaving the node.

10.3.3 Open-circuit and closed-circuit voltage drop

Figure 10.7 shows a circuit between two poles P and Q of a current-producing device (e.g. a battery). The idea is that the two poles are relatively close to each other, so that the circuit between P and Q is almost closed. It consists of connecting wire with conductivity σ_c and a path through a resistive element or more complicated network.

We can certainly say that there is a variable conductivity $\sigma(l)$ in any almost-closed circuit path between P and Q, at least we can do so for the argument to follow. We can be sure that the field cannot be electrostatic everywhere in a *totally closed* circuit because $\oint d\boldsymbol{l} \cdot \mathbf{E} = 0$ if the field were electrostatic, which then would yield, as explained below,

$$\oint d\boldsymbol{l} \cdot \frac{\mathbf{J}(l)}{\sigma(l)} = 0 \qquad \Rightarrow \qquad \mathbf{J}(l) = 0 \tag{10.20}$$

for all points l in the circuit. The path integral is taken round the circuit external to the battery in the direction of the round. The conclusion that $\mathbf{J}(l) = 0$ everywhere in the circuit would follow from the facts that $\sigma(l)$ is positive and that $\mathbf{J}(l)$ is

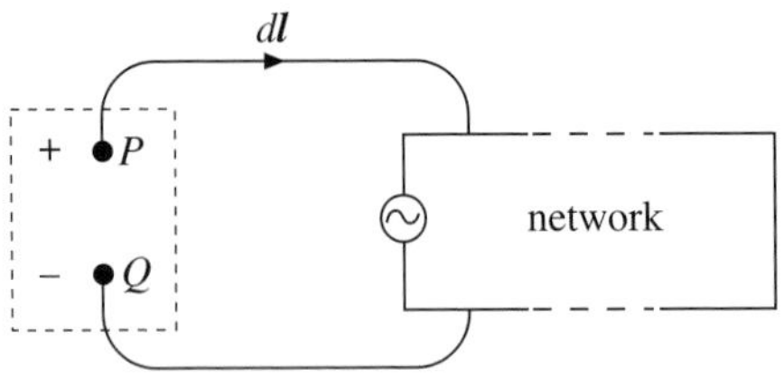

Fig. 10.7 A closed circuit with a power (current) source.

parallel to $\boldsymbol{dl}$ everywhere. The integrand is nonnegative at all points in the circuit, but the integral is zero. This requires the integrand to be zero everywhere. And that is a contradiction with the requirement that current be flowing! The resolution is to realize that the electric field in the current-producing device – the battery – between P and Q cannot be electrostatic. It is not; a battery separates charges inside it by means of electrochemical forces, which in turn produce an *electromotive force* V_{emf} (with the dimensions of voltage) between the poles given in general by

$$V_{\text{emf}} = \oint \boldsymbol{dl} \cdot \mathbf{E} \tag{10.21}$$

If the circuit is open, then the whole of the emf is presented across the terminals of the battery:

$$V_{\text{emf}} = \int_P^Q \boldsymbol{dl} \cdot \mathbf{E}$$

From the definition of potential V we have $\int_P^Q (-\boldsymbol{dl} \cdot \mathbf{E}) = V_Q - V_P$; hence it follows that

$$V_{\text{emf}} = V_P - V_Q$$

for an open circuit.

We now return to the general expression (10.21) and write it as

$$0 = V_{\text{emf}} - \oint \boldsymbol{dl} \cdot \mathbf{E} = V_{\text{emf}} - \oint \boldsymbol{dl} \cdot \frac{\mathbf{J}}{\sigma} = V_{\text{emf}} - \oint \boldsymbol{dl} \cdot \frac{\mathbf{I}}{\sigma S} \tag{10.22}$$

given that $\mathbf{I} \equiv S\mathbf{J}$. Because $\boldsymbol{dl}$ and $\mathbf{I}$ are in the same direction, we may write this as

$$0 = V_{\text{emf}} - \sum_m I_m \int dl_m \frac{1}{\sigma_m S_m} = V_{\text{emf}} - \sum_m \mathcal{R}_m I_m \tag{10.23}$$

where the summation is over all parts of a closed circuit. Internal resistances are also accounted for in this fashion. This last expression is *Kirchhoff's voltage law* for a single source, and it is easily extended for multiple sources. It may seem surprising that I can take on various values I_m in the closed circuit, but the requirement in

(10.23) is only that the circuit in question be *closed*; any part of this circuit may well be part of a larger circuit with many parallel paths branching on and off at various places.

10.4 Boundary conditions and consequences for lossy dielectrics

Under the condition of static currents, (10.18), we can find boundary conditions for $\mathbf{J}$ at the interface between two media with different conductivities σ and permittivities ε. In Fig. 10.8 such an interface is shown. The condition $\nabla \cdot \mathbf{J} = 0$ resembles the Maxwell equation $\nabla \cdot \mathbf{D} = 0$ for a region where there is no real charge. So we can carry through a line of reasoning parallel to that in Chapter 6.

The boundary condition for the $\mathbf{D}$ field in the absence of surface charge is given by

$$\hat{\mathbf{n}} \cdot (\mathbf{D}_2 - \mathbf{D}_1) = 0$$

Hence, by analogy

$$\mathbf{n} \cdot (\mathbf{J}_2 - \mathbf{J}_1) = 0 \qquad \text{or} \qquad J_{2\mathrm{n}} = J_{1\mathrm{n}} \tag{10.24}$$

This is the *only* boundary condition for $\mathbf{J}$ that is new, but of course we obtain a second one from (6.27a) by substituting $E = J/\sigma$:

$$\hat{\mathbf{n}} \times (\mathbf{J}_2/\sigma_2 - \mathbf{J}_1/\sigma_1) = 0 \qquad \text{or} \qquad J_{2\mathrm{t}} = \frac{\sigma_2}{\sigma_1} J_{1\mathrm{t}} \tag{10.25}$$

The algebra for calculating $\mathbf{J}_2$ from $\mathbf{J}_1$ is similar to (6.29):

$$\begin{aligned} J_2 \cos\theta_2 &= J_1 \cos\theta_1 \\ J_2 \sin\theta_2 &= \frac{\sigma_2}{\sigma_1} J_1 \sin\theta_1 \end{aligned} \tag{10.26}$$

and we obtain $\tan\theta_2 = (\sigma_2/\sigma_1)\tan\theta_1$ and thus θ_2 from division of the second of these two equations by the first. Then we easily obtain the magnitude J_2 from either

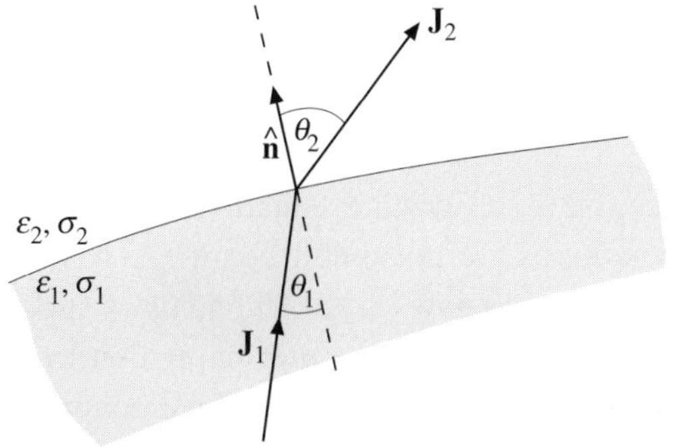

Fig. 10.8 Interface between two different ε, σ media: discontinuity of the current.

of these equations. There is a problem, however, that we did not encounter in the case of media with zero conductivity: the change in **J** at the interface requires the deposition of real surface charge on the interface! To see this, use $\mathbf{J} = \sigma\mathbf{E}$ (10.24) to obtain

$$\sigma_2 E_{2\mathrm{n}} = \sigma_1 E_{1\mathrm{n}} \tag{10.27}$$

and recall that the boundary condition for D_n predicts (6.28c),

$$\varrho_\mathrm{s} = \varepsilon_2 E_{2\mathrm{n}} - \varepsilon_1 E_{1\mathrm{n}} \tag{10.28}$$

Use (10.27) to eliminate $E_{2\mathrm{n}}$ from (10.28) and then substitute $E_{1\mathrm{n}} = J_{1\mathrm{n}}/\sigma_1$ to obtain

$$\varrho_\mathrm{s} = \left(\frac{\varepsilon_2}{\sigma_2} - \frac{\varepsilon_1}{\sigma_1}\right) J_{1\mathrm{n}} \tag{10.29}$$

Unless the unlikely coincidence $\varepsilon_2/\sigma_2 = \varepsilon_1/\sigma_1$ occurs, we observe that surface charge is deposited on the interface when $J_{1\mathrm{n}} \neq 0$. This is somewhat surprising at first sight, considering the fact that $J_{2\mathrm{n}} = J_{1\mathrm{n}}$, which means that the current flow *across* the interface is a constant. The following qualitative argument helps explain this somewhat: the different values of ε imply different fields E_n on either side of the interface, by virtue of (10.28), which means that the drift velocities u_n of the charge carriers also differ on either side. But $J_\mathrm{n} = Nqu_\mathrm{n}$ must be the same on either side so that the electron-charge densities Nq must differ on either side to compensate for the differing u_n. These charge densities are compensated for by oppositely charged lattice points in the interiors of both media, but not at the interface.

The condition (10.18) for static currents can be used to calculate the current density in a region in which the conductivity is constant. If we have such a region, say v, then in that region we conclude from $\nabla \times \mathbf{E} = 0$ and from Ohm's law $\mathbf{J} = \sigma\mathbf{E}$ that

$$0 = \nabla \times \frac{\mathbf{J}}{\sigma} = \frac{1}{\sigma}\nabla \times J \quad \Rightarrow \quad \nabla \times \mathbf{J} = 0 \quad \text{in the region } v \tag{10.30}$$

We then can introduce a scalar potential by setting $\mathbf{J} = -\nabla\psi$ in region v. This ensures that any ψ we obtain will lead to a **J** with zero curl. We now insert $\mathbf{J} = -\nabla\psi$ into (10.18) to obtain a Laplace equation,

$$\Delta\psi = 0$$

As we saw in Section 8.1, we are able to solve such an equation if favorable boundary conditions exist. Dirichlet conditions are not useful because ψ has no physical meaning, which implies that we cannot know $\psi(\mathbf{r})$ at all boundary-surface points. Neumann conditions, however, are useful because they imply that we know that $\hat{\mathbf{n}} \cdot \nabla\psi = -\hat{\mathbf{n}} \cdot \mathbf{J} \equiv \pm J_\mathrm{n}$ or 0 on the boundary surface; this corresponds to a realistic situation in which a known current enters and exits a metallic region. If, for example, the region v is the volume inside surface S (of which S_1, S_2, and S_3 are a part) in

Fig. 10.6, then clearly the normal component of the current density through S is zero except for S_1, S_2 and S_3, where it equals $\pm I_i/S_i$. $J_n(\mathbf{r})$ is thus known at all points $\mathbf{r}$ on the closed surface S. We then in principle could solve $\Delta\psi = 0$ inside v and so obtain the detailed density $\mathbf{J}(\mathbf{r})$ inside v.

10.5 Relationship between resistance and capacitance

The fundamental relationship $\mathcal{R}C = \varepsilon/\sigma$, which we will now derive, allows us to obtain the resistance between two surfaces of a uniform lossy dielectric, from the calculation of capacitance set forth in Sections 7.5 and 7.6. The relationship is easily shown as follows. If we cap both surfaces shown in Fig. 10.9 for such a dielectric with metal plates, and maintain a voltage difference V between them, then the capacitance $C = Q/V$ is given by

$$C = \frac{1}{V}\int dv\,\varrho_v = \frac{1}{V}\int dv\,\nabla\cdot\varepsilon\mathbf{E} = \frac{\varepsilon}{V}\int dv\,\nabla\cdot\mathbf{E} = \frac{\varepsilon}{V}\oint d\mathbf{S}\cdot\mathbf{E} \tag{10.31}$$

whereas the resistance of the lossy dielectric is defined by $\mathcal{R} = V/I$, which yields

$$\frac{1}{\mathcal{R}} = \frac{1}{V}\oint d\mathbf{S}\cdot\mathbf{J} = \frac{1}{V}\oint d\mathbf{S}\cdot\sigma\mathbf{E} = \frac{\sigma}{V}\oint d\mathbf{S}\cdot\mathbf{E} \tag{10.32}$$

Hence calculation of both $\mathcal{R}$ and C involves evaluation of the same surface integral $\oint d\mathbf{S}\cdot\mathbf{E}$ over the entire surrounding surface, and elimination of this integral from (10.31) and (10.32) yields

$$\mathcal{R}C = \varepsilon/\sigma$$

We saw in subsection 10.3.2 that $T = \varepsilon/\sigma$ is the relaxation time, during which the charge deposited inside a uniform lossy dielectric will decrease by a factor $1/e$. Here, we have just seen that also $T = \mathcal{R}C$, which is the relaxation time in which current or voltage difference in a macroscopic $\mathcal{R}C$ circuit decreases by the same amount. It is gratifying to learn that the microscopic picture corresponds accurately with the more-familiar macroscopic one!

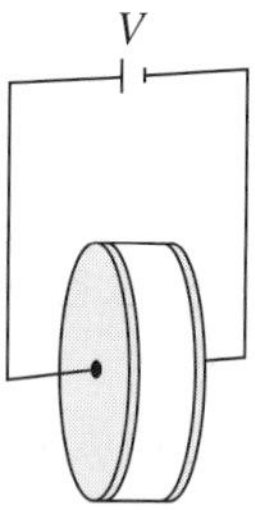

Fig. 10.9 Two conducting plates with a lossy dielectric sandwich in between.

The examples in Section 7.5 are easily extended to lossy dielectrics, and the resistance between the two conductors is then simply $\mathcal{R} = \varepsilon/\sigma C$, with capacitance C as calculated previously.

Problems

10.1. Current density is described by the vector $\mathbf{J} = a\hat{\mathbf{x}} + b\hat{\mathbf{y}} + c\hat{\mathbf{z}}$ in A/m^2. What is the current I crossing a circular area of radius ρ in the $z = 0$ plane?

10.2. Find the total current in a circular conductor of radius $\rho = 2$ mm if the current density is $J(\rho) = 5e^{-\rho^2/4}$ in A/mm^2.

10.3. Let $J(\rho) = 9e^{-81\rho}\hat{\boldsymbol{\varphi}}$ A/mm^2 be the current density in the region $1 < z < 2$ and $0.01 < \rho < 0.02$, where all distances are in mm. Find the total current through the intersection of this volume with any plane over which $\varphi =$ constant.

10.4. An electric field of strength 3.5 V/m creates a current density of 1.75 mA/m^2 in a certain material. What is the conductivity?

10.5. Find the resistance of an aluminum slab 6.5 cm square and 0.75 mm thick
(a) between the two square faces
(b) between opposite edges on a square face.

10.6. An electron gun is used to drive electrons from a heated cathode to the phosphor stripes in a TV tube. The input region of the gun is equivalent to a cylindrically shaped planar diode of length 1.8 mm operated at, say, 900 V. The cross-sectional radius of the maximum current is 0.25 mm. Find the maximum current strength when it is operated under space-charge-limited conditions. *Hint*: See subsection 8.4.2.

10.7. A sheet-current density $\mathbf{K} = 3\hat{\mathbf{x}} + 2\hat{\mathbf{y}}$ in A/m exists in the $z = 0$ plane. Find the direction of the line in the $z = 0$ plane across which a maximum current per meter can flow.

10.8. A planar distribution of current consists of 5000 closely packed parallel strands of insulated thin wires per meter in the plane $x - 2y = 0$; the wires are perpendicular to the line in that plane defined as $x = 2$, $y = 1$. Each wire carries 0.2 mA. Specify the equivalent sheet-current-density vector K.

10.9. Estimate the effective drift velocity in a metal wire of 50-mil radius (10^3 mil = 1 inch) at room temperature if it is given that there are $N = 8 \times 10^{28}$ electrons/m^3 and the current strength is 12 A.

10.10. If a small charge is placed inside a conductor, and it is found that the charge decays exponentially to $1/e$ of its value in 7×10^{-19} s, what then is the conductivity σ (use $\varepsilon = \varepsilon_0$)?

10.11. Given that in copper there are $N \approx 8.5 \times 10^{28}$ electrons per m^3 contributing to the current and that the conductivity is $\sigma \approx 5.8 \times 10^7$ S/m, estimate the number of collisions per second that the conduction electrons undergo.

10.12. A uniform circularly cylindrical wire of 75 cm length and of 5-mil radius (10^3 mil $=$ 1 inch) is made of aluminum. Find its resistance $\mathcal{R}$.

10.13. Consider a 50-mil aluminum wire wound in the form of a solenoid with radius $\rho = 5$ cm. How long would the solenoid need to be to obtain $\mathcal{R} = 100\,\Omega$? Ignore the 'pitch' of the windings, i.e. assume that each winding is $\perp$ the axis of the solenoid.

10.14. A composite circularly cylindrical wire 9 meters long consists of a copper core of 25-mil radius and an aluminum sheath of 25-mil thickness around it. What is the resistance $\mathcal{R}$ of the composite wire?

10.15. The Earth's atmospheric layer passes a current of approximately 1800 A from space to the surface. The total voltage difference is approximately 400 000 V.
(a) What is the effective resistance per m^2 at the surface?
(b) What is the power flux (power per m^2) through the surface?
(c) What current flows through 1 m^2 at the surface?

10.16. A square disk of a material has sides of length 2 cm and a thickness of 6 mm. Assume $\sigma = 6$ S/m. How should one assemble a set of identical such disks to produce a resistor of 1 Ω? Find at least two different assemblies.

10.17. Two lossy dielectrics with the same relative permittivity ε_0 have conductivities $\sigma_1 = 0.5$ S/m in $z < 0$ and $\sigma_2 = 2.5$ S/m in $z > 0$. The current density $\mathbf{J}_1$ in $z < 0$ makes an angle of 20° with the $+\hat{\mathbf{z}}$ direction.
(a) What is the angle of $\mathbf{J}_2$ with the $+\hat{\mathbf{z}}$ direction?
(b) Calculate the surface-charge density at $z = 0$, if $J_1 = 0.35\,\text{A/m}^2$.

10.18. At an interface $x = 0$ we have $\varepsilon_{1r} = 1.5$, $\sigma_1 = 3$ S/m at $x < 0$ and $\varepsilon_{2r} = 2.5$, $\sigma_2 = 6$ S/m at $x > 0$. The current density in $x < 0$ is $\mathbf{J}_1 = 4\hat{\mathbf{x}} + 2\hat{\mathbf{y}} + 5\hat{\mathbf{z}}$ A/m^2.
(a) Find the current density $\mathbf{J}_2$ in $x > 0$.
(b) Calculate the surface-charge density on $x = 0$.
(c) By what percentage should you increase σ_2 for there to be no surface charge at $x = 0$?

10.19. A metallic conductor with conductivity σ is shaped as a circularly cylindrical shell of thickness w around the z axis such that the inner radius is ρ and the outer radius is $\rho + w$. The height of the conductor is h (in the $\hat{\mathbf{z}}$ direction). An infinitesimal slit at $\varphi = 0$ prevents the conductor from presenting a closed circuit in the $\hat{\boldsymbol{\varphi}}$ direction. If a battery providing voltage V is connected to both sides of the slit, obtain formulas for
(a) the current in the $\hat{\boldsymbol{\varphi}}$ direction (ignore fringe effects)
(b) the resistance that the current undergoes (ignore the battery's internal resistance).

10.20. The formula $\mathcal{R}C = \varepsilon/\sigma$ was derived for uniform media. Can it hold in a non-uniform medium and, if so, under what conditions?

11

The magnetostatic field

11.1 Magnetostatic forces, fields, and the law of Biôt and Savart

To understand magnetostatic forces and fields, we will approach the subject in a way that differs from that sometimes presented elsewhere but which – we believe – appeals more to intuition. We start from a situation in which we have two charges in motion (see Fig. 11.1). It is found that the two charges, q_P and q_Q, with velocity vectors $\mathbf{u}_P$ and $\mathbf{u}_Q$, exert a magnetostatic force upon each other that differs from the Coulomb force of (3.1) and (3.2); the magnetostatic force on q_P due to q_Q is

$$\mathbf{F}_{QP} = \kappa_{\mathrm{m}} q_P \mathbf{u}_P \times \left(q_Q \mathbf{u}_Q \times \frac{\hat{\mathbf{R}}_{QP}}{R_{QP}^2} \right) \tag{11.1}$$

The resemblance to Coulomb's law is evident. However, there are two major differences. Here, *velocities* occur together with charges, and there are vector cross products, instead of simple products of scalars, but the same inverse-square dependence upon distance between the charges is there. If meters, seconds, coulombs, and newtons are used as units of length, time, charge, and force, then the constant $\kappa_{\mathrm{m}} = 10^{-7}\mathrm{H/m}$ precisely. Here, the unit H/m stands for henries per meter and it can be related to the more familiar units by (11.1). The reason for using H/m will be clarified later. The magnitude of κ_{m} is mainly a consequence of the choice of the coulomb as the unit of charge.

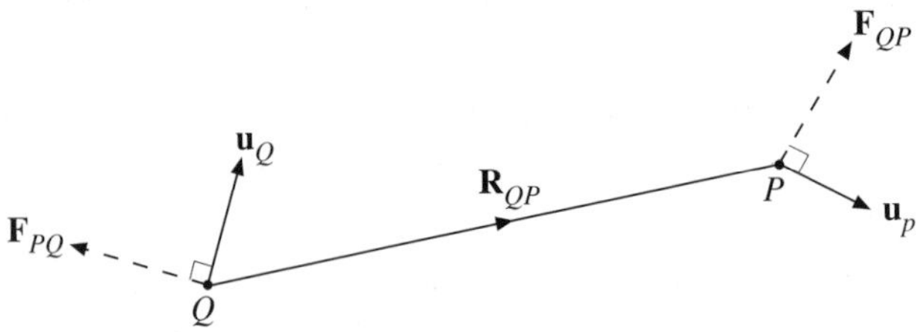

Fig. 11.1 Two moving charges at P and Q and the magnetostatic force between them.

Obviously this force is zero when either of the particle velocities is zero, and as we stated in Chapter 10, the frame of reference of the observer will determine whether a nonzero velocity is measured! However, let us assume that all velocities are with respect to the frame of reference in which the observer is, e.g. the Earth's surface locally. We then see that this magnetostatic force is determined by currents, not charges alone, because $q\mathbf{u}$ is the current associated with one moving point charge, the most elementary current we can think of. Its direction is relatively complicated: we must do two vector cross products to figure out the direction of $\mathbf{F}_{QP}$. The result is illustrated in Fig. 11.1 for the case in which both velocities lie in one plane. The force $\mathbf{F}_{PQ}$ is *not* the reverse of $\mathbf{F}_{QP}$, as you can see from Fig. 11.1.

An equivalent way of writing (11.1) is as follows:

$$\mathbf{F}_{QP} = q_P \mathbf{u}_P \times \mathbf{B}_P \qquad \text{where}$$
$$\mathbf{B}_P = \kappa_m q_Q \mathbf{u}_Q \times \frac{\hat{\mathbf{R}}_{QP}}{R^2_{QP}} \tag{11.2a}$$

We have thus introduced a new field $\mathbf{B}_P$ at the point P due to the point charge at Q having velocity $\mathbf{u}_Q$. This field is known as the *magnetic flux density* and it is measured in a derived unit known as the tesla (T) or weber per square meter (Wb/m^2). The relationship of the tesla or weber to known units follows from (11.2a).

It is useful, in connection with the ultimate Maxwell equations for magnetic fields, to replace the constant κ_m by $\mu_0/4\pi$, where μ_0 is the *permeability* of free space. Thus, $\mu_0 = 4\pi \times 10^{-7}$ H/m and we rewrite (11.2a) as

$$\mathbf{F}_{QP} = q_P \mathbf{u}_P \times \mathbf{B}_P \qquad \text{where}$$
$$\mathbf{B}_P = \mu_0 q_Q \mathbf{u}_Q \times \frac{\hat{\mathbf{R}}_{QP}}{4\pi R^2_{QP}} \tag{11.2b}$$

The first of these equations describes the *Lorentz force* (named after the Dutch physicist H.A. Lorentz, 1853–1928). Associated with the magnetic flux density in free space is the *magnetic field* $\mathbf{H} = \mu_0^{-1}\mathbf{B}$. The $\mathbf{H}$ field is the same as $\mathbf{B}$ in (11.2b) but without the factor μ_0. In magnetizable materials, the permeability μ has a different, much larger value. We will use either one or the other field, depending upon which is the most convenient. A distinction will be truly important only after magnetic materials are introduced.

To make closer correspondence with what was originally measured (namely the force of one current upon another current), we reformulate Eqs. (11.1) and (11.2) for two infinitesimal volume elements dv_P at P and du_Q at Q, with $N_P\, dv_P$ charges q_P in dv_P, each with velocity $\mathbf{u}_P$, and a similar set in dv_Q. This implies that N_P is the number of particles per unit volume at P. In place of (11.1), we obtain the force that dv_Q exerts upon dv_P simply by adding up all the two-particle forces of (11.1).

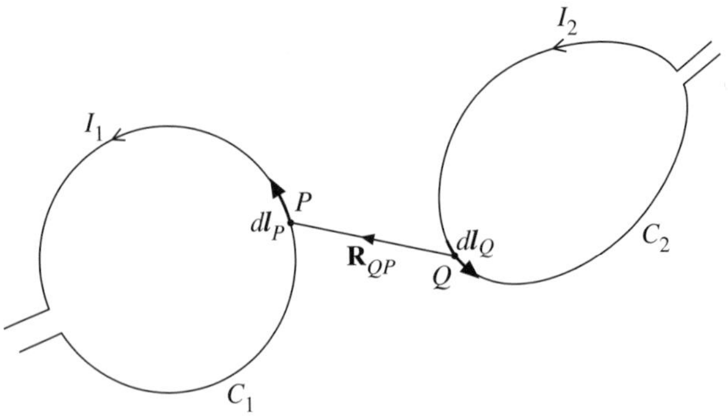

Fig. 11.2 Diagram for calculation of the magnetic interaction between two circuit elements.

The result is:

$$d^2\mathbf{F}_{QP} = \frac{\mu_0}{4\pi} N_P q_P \mathbf{u}_P \, dv_P \times \left(N_Q q_Q \mathbf{u}_Q \, dv_Q \times \frac{\hat{\mathbf{R}}_{QP}}{R_{QP}^2} \right) \tag{11.3a}$$

If we remember that the current density $\mathbf{J} = Nq\mathbf{u}$ then (11.3a) simplifies to

$$d^2\mathbf{F}_{QP} = \frac{\mu_0}{4\pi} \mathbf{J}_P \, dv_P \times \left(\mathbf{J}_Q \, dv_Q \times \frac{\hat{\mathbf{R}}_{QP}}{R_{QP}^2} \right) \tag{11.3b}$$

If both volume elements are each pieces of current-carrying wires (see Fig. 11.2), then $\mathbf{J}\, dv = I\, d\boldsymbol{l}$, as we observed in Chapter 10, and we obtain for the force on dl_P due to dl_Q

$$d^2\mathbf{F}_{QP} = \frac{\mu_0}{4\pi} I_1 \, d\boldsymbol{l}_P \times \left(I_2 \, d\boldsymbol{l}_Q \times \frac{\hat{\mathbf{R}}_{QP}}{R_{QP}^2} \right) \tag{11.3c}$$

where I_1, I_2 are the currents in the two loops. Let us now integrate $d\boldsymbol{l}_Q$ around the closed loop C_2 of which it forms a part. We then obtain the total force on $d\boldsymbol{l}_P$ due to the second loop:

$$d\mathbf{F}_P = \frac{\mu_0}{4\pi} I_1 \, d\boldsymbol{l}_P \times \left(I_2 \oint_{C_2} d\boldsymbol{l}_Q \times \frac{\hat{\mathbf{R}}_{QP}}{R_{QP}^2} \right) = I_P \, d\boldsymbol{l}_P \times \mathbf{B}_P \qquad \text{where}$$

$$\mathbf{B}_P \equiv \frac{\mu_0}{4\pi} I_2 \oint_{C_2} d\boldsymbol{l}_Q \times \frac{\hat{\mathbf{R}}_{QP}}{R_{QP}^2} \tag{11.4a}$$

The current I_2 in the loop C_2 is constant and thus it has been removed from the integrand. The expression for the magnetic flux density at P, $\mathbf{B}_P$, is known as the *law of Biôt and Savart* (named after the French researchers J.-P. Biôt, 1774–1862, and

F. Savart, 1791–1841). Equation 11.4 resembles most closely what was first measured. By working backwards from these macroscopic current-loop expressions, you can see how we arrived at the microscopic forms (11.1) and (11.2). The *Lorentz force* of (11.2b) holds for the integrated Biôt–Savart field $\mathbf{B}_P$ of (11.4a):

$$\mathbf{F}_P = q_P \mathbf{u}_P \times \mathbf{B}_P \tag{11.4b}$$

And this force has been inferred from measurements upon small charged particles.

Finally, the total force $\mathbf{F}_{21}$ of circuit 2 on circuit 1 is found by integrating $d\mathbf{F}_P$, Eq. (11.4), round circuit 1:

$$\begin{aligned} &\oint_{C_1} d\mathbf{F}_P = \mathbf{F}_{21} = I_1 \oint_{C_1} d\boldsymbol{l}_P \times \mathbf{B}_P \qquad \text{where} \\ &\mathbf{B}_P \equiv \frac{\mu_0}{4\pi} I_2 \oint_{C_2} d\boldsymbol{l}_Q \times \frac{\hat{\mathbf{R}}_{QP}}{R_{QP}^2} \end{aligned} \tag{11.5}$$

Example 11.1 Motion of a charged particle around a magnetic field

Let us consider a constant magnetic flux density field $\mathbf{B} = B\hat{\mathbf{z}}$, in which moves a point particle with charge q and initial velocity $\mathbf{u} = u_x\hat{\mathbf{x}} + u_y\hat{\mathbf{y}} + u_z\hat{\mathbf{z}}$. The Lorentz force (11.4b) on the particle is $\mathbf{F} = q\mathbf{u} \times B\hat{\mathbf{z}}$, and therefore there is no force due to the u_z component of the velocity; whatever else the charged particle does, the velocity in the $\hat{\mathbf{z}}$ direction remains unchanged. Let us look to see what happens in the other two directions. We will travel with the particle at velocity $u_z\hat{\mathbf{z}}$, and take $u_y = 0$; this is not a special case, because we can rotate our coordinate system around the z axis until indeed the x axis coincides with the direction of the initial velocity component $\perp \hat{\mathbf{z}}$. So we can look at the problem in which the initial velocity is just $\mathbf{u} = u\hat{\mathbf{x}}$. To find the trajectory of the particle, we must solve Newton's equation

$$m\ddot{\mathbf{r}} = \mathbf{F} = q\mathbf{u} \times \mathbf{B}$$

We obtain for the motion in a plane $\perp \hat{\mathbf{z}}$

$$\begin{aligned} m\ddot{x} &= F_x = qB\dot{y} \\ m\ddot{y} &= F_y = -qB\dot{x} \end{aligned} \tag{11.6}$$

We can write these two equations as the following functions of time t:

$$\begin{aligned} &\frac{d}{dt}(m\dot{x} - qBy) = 0 \\ &\frac{d}{dt}(m\dot{y} + qBx) = 0 \end{aligned} \tag{11.7}$$

We conclude that the contents of each pair of parentheses must be constant in time. Because $\dot{x}(0) = u$ and $\dot{y}(0) = 0$, and because we assume $x(0) = 0, y(0) = 0$, we have:

$$\begin{aligned} m(\dot{x} - u) - qBy &= 0 \\ m\dot{y} + qBx &= 0 \end{aligned} \tag{11.8}$$

By taking another derivative with respect to time in (11.8) and then eliminating y we obtain

$$\ddot{x} + \Omega^2 x = 0 \qquad \Omega \equiv \frac{qB}{m} \tag{11.9}$$

Note that Ω is a frequency; it is known as the *cyclotron frequency*, and we shall see why shortly. A solution of (11.9) is $x(t) = a \sin \Omega t + b \cos \Omega t$, so that $\dot{x}(t) = \Omega(a \cos \Omega t - b \sin \Omega t)$. One of our initial conditions was $x(0) = 0$, which implies that coefficient $b = 0$. Another was $\dot{x}(0) = u$, so that we know that $a = u/\Omega$. Thus we have found

$$x(t) = \frac{u}{\Omega} \sin \Omega t \tag{11.10}$$

Because $\dot{y}(t) = -\Omega x(t) = -u \sin \Omega t$, as is obtained from the second of (11.8) and (11.10), we find by simple integration that

$$y(t) = y_0 + \frac{u}{\Omega} \cos \Omega t \tag{11.11}$$

This satisfies the condition $\dot{y}(0) = 0$, and we need to set the integration constant $y_0 = -u/\Omega$ to ensure that $y(0) = 0$. So we have found

$$y(t) = -\frac{u}{\Omega}(1 - \cos \Omega t) \tag{11.12}$$

The trajectory given by (11.10) and (11.12) is plotted in Fig. 11.3; it is a circle with radius

$$R = \frac{u}{\Omega} = \frac{mu}{qB}$$

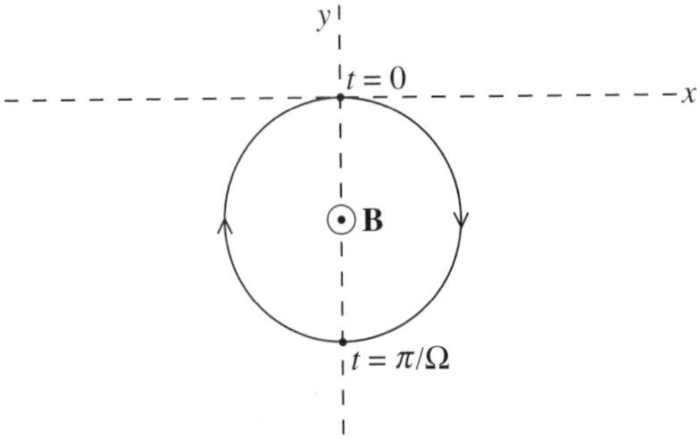

Fig. 11.3 Orbit of a point charge around the **B** lines, which are out of the page.

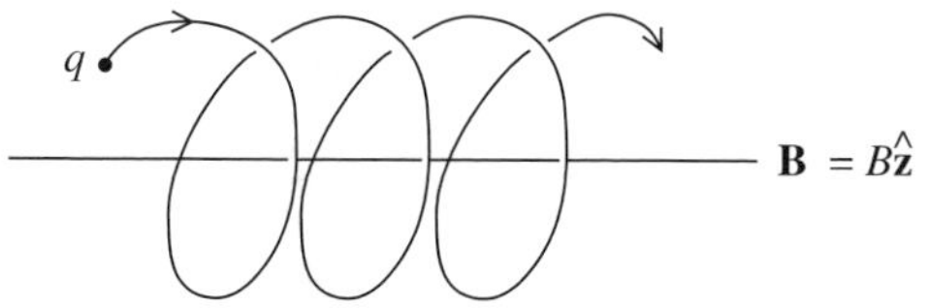

Fig. 11.4 Helical orbit of a point charge q around **B** lines when there is a velocity component parallel to **B**.

and the cycle time is $T = 2\pi/\Omega$. Thus, Ω is the angular frequency with which the charge q rotates in a circle around the **B** line (which in this case is at $x = 0, y = -R$). But, remember, if there is a velocity component in the direction of **B**, then that velocity component is unchanged in time. Hence the motion in general will resemble the *helix* shown in Fig. 11.4. The projection of the helix upon a z = constant plane is what we see in Fig. 11.3. The helical motion of Fig. 11.4 is that of a right-handed screw if charge q is positive. Negative charges rotate the other way while coasting in the $\hat{\mathbf{z}}$ direction.

This is the principle of the *cyclotron*, and other devices in which magnetic fields make charged particles rotate in circles – hence the name cyclotron frequency for Ω. In the following few paragraphs, we shall say something about the numbers and magnitudes that follow from comparisons between electric and magnetic phenomena. The connection between the two force constants, $\kappa_e = 1/4\pi\varepsilon_0$ and $\kappa_m = \mu_0/4\pi$, can be made in anticipation of something we have yet to learn. The combination $(\mu_0\varepsilon_0)^{-1/2}$ occurs in the wave equation that we will later derive from Maxwell's equations as a *velocity* c. The theory will show that c is the velocity of electromagnetic waves in free space. It has been measured to be 2.997925×10^8 m/s. This tells us (because $\mu_0 = 4\pi \times 10^{-7}$ H/m exactly) that $\varepsilon_0 = 1/\mu_0 c^2 \approx 8.854 \times 10^{-12}$ F/m. If, on the other hand, we approximate c as 3×10^8 m/s, then we will find $\varepsilon_0 \approx 10^{-9}/36\pi \approx 8.842 \times 10^{-12}$ F/m. The difference between the two values is $\sim 0.14\%$, which usually suffices in engineering practice. The approximation $1/4\pi\varepsilon_0 \approx 9.00$ $(8.988) \times 10^9$ m/F is often useful.

We also see that $\kappa_m/\kappa_e = \mu_0\varepsilon_0 = 1/c^2 \sim 10^{-17}\ \text{s}^2/\text{m}^2$. If we compare the magnetic force (11.1) for two particles, each moving with velocity of 1 m/s, with the Coulomb force (3.2a) for the same two particles at the same distance from each other, we then see that the magnetic force is less by this same enormous factor, $\sim 10^{-17}$. It takes near-relativistic velocities to make the forces comparable! So how is it that we can observe magnetic forces?

Consider two circuits again, as in Fig. 11.2. When current flows in the wires, what really happens is that conduction electrons drift, making a vast number of collisions against positive lattice sites, down the wire. In any small volume of the wire with cross-sectional area S and length element dl we have $\nabla \cdot \mathbf{J} = 0$, which means that $\mathbf{J}(l+dl) = \mathbf{J}(l)$. As the drift velocity $\mathbf{u}$ does not change in dl (for constant $E = V/l$), this implies that the electron density is constant over dl, $N(l+dl) = N(l)$. In any small volume Sdl, the net charge is zero before current starts flowing, because the charge of the conduction electrons balances the positive charge of the lattice sites missing one or more outer-shell electrons. It will remain so, when current flows under the condition $\nabla \cdot \mathbf{J} = 0$. The sum of local Coulomb forces will be small, as the two-by-two forces between conduction electrons and lattice ions will be in all possible directions, even with large variations in inverse-distance squared. The two-by-two magnetic forces will be close to zero because (a) the conduction electrons all drift at (nearly) the same velocity vector, and (b) the lattice-ion random velocity is essentially zero, so that $\mathbf{v}_1 \times \mathbf{v}_2 = 0$, in both cases. The only remaining magnetic forces are between conduction electrons in one wire and those in the other. Such forces are described by (11.5). A similar argument holds for almost any situation in which we produce a magnetic field; observable magnetic forces are experienced by currents acting upon other currents.

Example 11.2 Magnetic flux density around an infinite thin straight current wire

Consider in Fig. 11.5 the point $P = (\rho, \varphi, 0)$ in cylindrical coordinates, given that a current I flows in the $+\hat{\mathbf{z}}$ direction through an infinitesimally thin wire coinciding with the z axis. The point P at distance ρ from the wire is essentially an arbitrary point. Let us use the Biôt–Savart law (11.4a) to calculate the $\mathbf{B}$ field at P. It seems clear that $d\boldsymbol{l}_Q = \hat{\mathbf{z}}dz$ and that $\mathbf{R}_{QP} = \rho\hat{\boldsymbol{\rho}}(\varphi) - z\hat{\mathbf{z}}$. Thus we obtain

$$\mathbf{B}_P = \frac{\mu_0 I}{4\pi} \int_{-\infty}^{\infty} dz\, \hat{\mathbf{z}} \times \frac{\rho\hat{\boldsymbol{\rho}}(\varphi) - z\hat{\mathbf{z}}}{(\rho^2 + z^2)^{3/2}} \tag{11.13}$$

The $-z\hat{\mathbf{z}}$ term does not contribute because $\hat{\mathbf{z}} \times \hat{\mathbf{z}} = 0$. Because $\hat{\mathbf{z}} \times \hat{\boldsymbol{\rho}}(\varphi) = \hat{\boldsymbol{\varphi}}(\varphi)$, it follows that $\mathbf{B}_P$ points in a direction tangential to the circle around the z axis through P:

$$\mathbf{B}_P = \frac{\mu_0 I}{4\pi}\hat{\boldsymbol{\varphi}}(\varphi) \int_{-\infty}^{\infty} dz \frac{\rho}{(\rho^2 + z^2)^{3/2}} = \frac{\mu_0 I}{4\pi\rho}\hat{\boldsymbol{\varphi}}(\varphi) \int_{-\infty}^{\infty} dt \frac{1}{(1 + t^2)^{3/2}} \tag{11.14}$$

The second integral follows from the first by changing from integration variable z to the dimensionless integration variable $t = z/\rho$. It is not difficult to find (e.g. from integration tables) that the integral over t is just $[t/\sqrt{1+t^2}]_{t=\infty} - [t/\sqrt{1+t^2}]_{t=-\infty} = 2$. Hence we have found

$$\mathbf{B}_P = \frac{\mu_0 I}{2\pi\rho}\hat{\boldsymbol{\varphi}}(\varphi) \tag{11.15}$$

There is clearly no dependence upon z. The magnitude does not depend upon the polar angle φ, but the direction of $\mathbf{B}_P$ clearly does. We should bear in mind that real

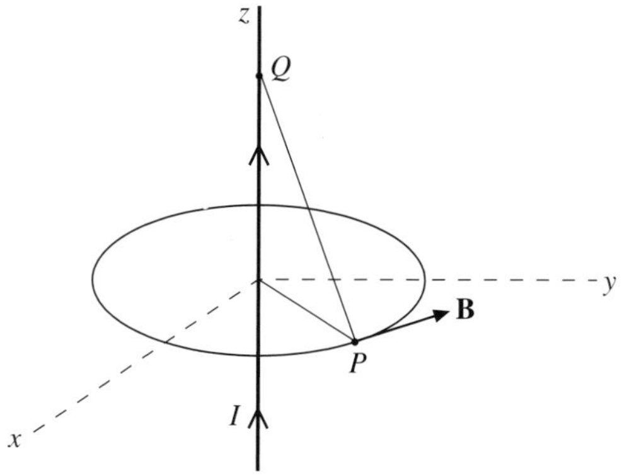

Fig. 11.5 Geometry for calculation of the magnetic flux density around an infinite straight current wire.

currents cannot flow infinitely far in one direction; they are always part of a circuit. This problem is an abstraction of a real situation in which the magnetic fields are calculated in locations that are very distant from the other 'arms' of a rectangular circuit but relatively close to the long arm that we may regard *locally* as part of an infinite straight wire.

Example 11.3 Axial magnetic flux density due to a circular loop of current

Figure 11.6 shows a circular loop of current I and a point P at height z on the axis through the center of the circle with radius ρ. It resembles Fig. 11.5, but now the roles of P and Q are reversed. Here, $\mathbf{R}_{QP} = z\hat{\mathbf{z}} - \rho\hat{\boldsymbol{\rho}}(\varphi)$ and $d\boldsymbol{l} = \rho\, d\varphi\, \hat{\boldsymbol{\varphi}}(\varphi)$, so that from (11.4a)

$$\mathbf{B}_P = \frac{\mu_0 I_\rho}{4\pi} \int_0^{2\pi} d\varphi\, \hat{\boldsymbol{\varphi}}(\varphi) \times \frac{z\hat{\mathbf{z}} - \rho\hat{\boldsymbol{\rho}}(\varphi)}{(\rho^2 + z^2)^{3/3}} \tag{11.16}$$

In this case, we can see that the $\hat{\boldsymbol{\varphi}}(\varphi) \times \hat{\mathbf{z}} = \hat{\boldsymbol{\rho}}(\varphi)$ contribution vanishes: as $\hat{\boldsymbol{\rho}}(\varphi)$ rotates through 360° as angle φ changes from 0 to 2π, whereas the rest of the integrand for this part is unchanged; thus, this part of $\mathbf{B}_P$ is a vector sum of equal-magnitude vectors pointing in all possible directions, which – of course – implies a zero sum. Equivalently, calculation of $\hat{\mathbf{x}} \cdot \mathbf{B}_P$ or $\hat{\mathbf{y}} \cdot \mathbf{B}_P$ produces a factor $\cos\varphi$ or $\sin\varphi$ as the only φ-dependent factor in the integrand of this part, and the 0 to 2π integral of $\cos\varphi$ or $\sin\varphi$ is zero. The remaining term in the integrand is not a function of φ because $-\hat{\boldsymbol{\varphi}}(\varphi) \times \hat{\boldsymbol{\rho}}(\varphi) = \hat{\mathbf{z}}$ for all angles φ. Thus

$$\mathbf{B}_P = \frac{\mu_0 I \rho^2}{2(\rho^2 + z^2)^{3/2}} \hat{\mathbf{z}} \tag{11.17}$$

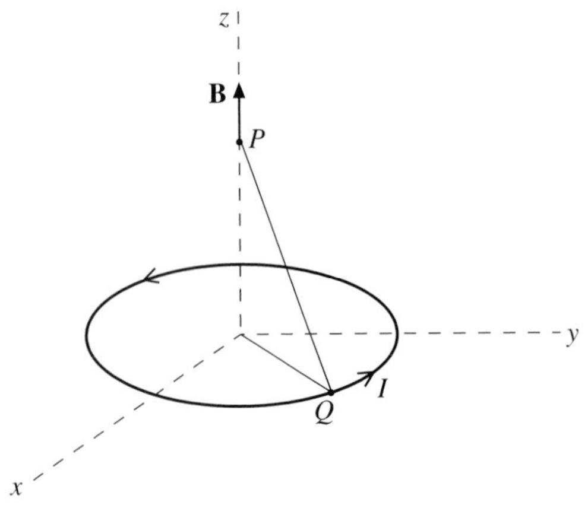

Fig. 11.6 Geometry for calculation of the axial magnetic flux density due to a circular current loop.

11.2 Basics of Maxwell's laws for magnetostatic fields

The Biôt–Savart law can be shown to lead to two magnetostatic Maxwell point-form equations. First consider the divergence of $\mathbf{B}_P$ as given by (11.4a):

$$\nabla_P \cdot \mathbf{B}_P \equiv \frac{\mu_0}{4\pi} I \nabla_P \cdot \oint_C d\boldsymbol{l}_Q \times \frac{\hat{\mathbf{R}}_{QP}}{R^2_{QP}} \tag{11.18}$$

In order to work towards a point form, we separate out the vector parts and use the vector identity 9 in Appendix A.4 (we are at liberty to move the divergence operator to the right of the integral sign since the variables of differentiation and integration are independent):

$$\nabla_P \cdot \left(d\boldsymbol{l}_Q \times \frac{\hat{\mathbf{R}}_{QP}}{R^2_{QP}} \right) = \frac{\hat{\mathbf{R}}_{QP}}{R^2_{QP}} \cdot (\nabla_P \times d\boldsymbol{l}_Q) - d\boldsymbol{l}_Q \cdot \left(\nabla_P \times \frac{\hat{\mathbf{R}}_{QP}}{R^2_{QP}} \right) \tag{11.19a}$$

The first term is zero because the derivatives are taken at point P whereas $d\boldsymbol{l}$ is a function only of $\mathbf{r}_Q$. We have already seen in (4.5) that $\hat{\mathbf{R}}_{QP}/R^2_{QP} = -\nabla_P(1/R_{QP})$ and therefore we observe that

$$\nabla_P \cdot \left(d\boldsymbol{l}_Q \times \frac{\hat{\mathbf{R}}_{QP}}{R^2_{QP}} \right) = -d\boldsymbol{l}_Q \cdot \left[\nabla_P \times \nabla_P \left(\frac{1}{R_{QP}} \right) \right] = 0 \tag{11.19b}$$

Here we have made use of the fact that the curl of a gradient is always zero; see (5.19). Thus, both parts of the right-hand side of (11.19a) are zero, and so, omitting the position subscript, we get

$$\nabla \cdot \mathbf{B} = 0 \tag{11.20}$$

This is the first of the Maxwell equations for magnetostatics, and we postpone discussion of it until we have found the second one. To do this, we take the curl of (11.4a),

$$\nabla_P \times \mathbf{B}_P \equiv \frac{\mu_0}{4\pi} I \nabla_P \times \left(\oint_C d\boldsymbol{l}_Q \times \frac{\hat{\mathbf{R}}_{QP}}{R_{QP}^2} \right) \tag{11.21}$$

If we replace $Id\boldsymbol{l}_Q$ by the $\mathbf{J}_Q dv_Q$, we obtain

$$\nabla_P \times \mathbf{B}_P \equiv \frac{\mu_0}{4\pi} \nabla_P \times \left(\int_v dv_Q\, \mathbf{J}_Q \times \frac{\hat{\mathbf{R}}_{QP}}{R_{QP}^2} \right)$$

$$= -\frac{\mu_0}{4\pi} \nabla_P \times \left[\int_v dv_Q\, \mathbf{J}_Q \times \nabla_P \left(\frac{1}{R_{QP}} \right) \right] \tag{11.22}$$

We use the vector identity 6 of Appendix A.4 in (11.22), again exchanging integration and differentiation:

$$\nabla_P \times \left[\mathbf{J}_Q \times \nabla_P \left(\frac{1}{R_{QP}} \right) \right] = \mathbf{J}_Q \nabla_P \cdot \left[\nabla_P \left(\frac{1}{R_{QP}} \right) \right] - (\mathbf{J}_Q \cdot \nabla_P) \nabla_P \left(\frac{1}{R_{QP}} \right) \tag{11.23}$$

Actually, the vector theorem predicts two more terms, but these are zero because they involve the derivatives at point P of $\mathbf{J}_Q$, which is a function only of $\mathbf{r}_Q$. We may replace ∇_P by $-\nabla_Q$ in both terms in (11.23) because the derivatives work on functions only of $\mathbf{r}_P - \mathbf{r}_Q$.

Let us look at the second term of (11.23). It appears in a dv_Q integral, which, as we explain below, we can transform into

$$\int_v dv_Q (\mathbf{J}_Q \cdot \nabla_Q) \nabla_Q \left(\frac{1}{R_{QP}} \right) = \int_v dv_Q\, \nabla_Q \cdot \left[\mathbf{J}_Q \nabla_Q \left(\frac{1}{R_{QP}} \right) \right]$$

$$- \int_v dv_Q (\nabla_Q \cdot \mathbf{J}_Q) \nabla_Q \left(\frac{1}{R_{QP}} \right) \tag{11.24}$$

The second term on the right-hand side is zero because $\nabla \cdot \mathbf{J} = 0$. Note that there is *no* dot product between $\mathbf{J}_Q$ and the gradient in the first term. Two vectors placed together like this are known as a *dyadic*, which we can consider as the product of a column matrix on the left and a row matrix on the right (yielding a 3×3 matrix as product). To clarify what has been done here, consider the following theorem about the divergence of such a dyadic:

$$\nabla \cdot (\mathbf{AB}) = \frac{\partial}{\partial x}(A_x \mathbf{B}) + \frac{\partial}{\partial y}(A_y \mathbf{B}) + \frac{\partial}{\partial z}(A_z \mathbf{B}) = (\nabla \cdot \mathbf{A})\mathbf{B} + (\mathbf{A} \cdot \nabla)\mathbf{B} \tag{11.25}$$

This statement is easily proved by checking it for the B_x, B_y, and B_z components separately. This is the theorem we have used in (11.24), albeit with one of the terms moved to the other side.

We now apply Gauss's divergence theorem to the first term on the right-hand side of (11.24):

$$\oint_S d\mathbf{S}_Q \cdot \left[\mathbf{J}_Q \nabla_Q\left(\frac{1}{R_{QP}}\right)\right] = \oint (d\mathbf{S}_Q \cdot J_Q)\nabla_Q\left(\frac{1}{R_{QP}}\right) = 0 \qquad (11.26)$$

We shall always assume that the surface S is drawn beyond all the sources; therefore it contains *no* current density (all $\mathbf{J}_Q = 0$) and that makes (11.26) equal to zero.

Now we have only the first term on the right-hand side of (11.23) left! When this is inserted into (11.22) we have

$$\nabla_P \times \mathbf{B}_P \equiv \frac{\mu_0}{4\pi}\int_v dv_Q\, \mathbf{J}_Q \nabla_Q \cdot \left[\nabla_Q\left(\frac{1}{R_{QP}}\right)\right] = \frac{\mu_0}{4\pi}\int_v dv_Q\, \mathbf{J}_Q\left[\Delta_Q\left(\frac{1}{R_{QP}}\right)\right] \qquad (11.27)$$

So we must now worry only about the Laplacian $\Delta(1/R_{QP})$, taken at all points Q in v, but for fixed P. If we evaluate the Laplacian, we will find that the outcome is zero *as long as* $Q \neq P$! However, zero is not the answer; there is a problem in the infinitesimal environment of P where $1/R_{QP} \to \infty$. A careful mathematical analysis, which we give in detail in the optional paragraph below, shows that $\Delta_Q(1/R_{QP})$ tends to infinity as $Q \to P$ in such a fashion that any volume integral of this quantity over even an *infinitesimal* volume completely around P yields the factor 4π.

Let us consider the integral

$$\mathbf{J}_P = \int_v dv_Q\, \mathbf{J}_Q\left[\Delta_Q\left(\frac{1}{R_{QP}}\right)\right] \qquad (11.28)$$

As long as $Q \neq P$, it is easy to show, e.g. by using Cartesian coordinates, that $\Delta_Q(1/R_{QP}) = 0$. The problem is in the immediate environment of point P. Let us define an infinitesimal sphere with radius ϵ around point P. Inside this sphere, we may make the approximation $\mathbf{J}_Q \approx \mathbf{J}_P$ to accuracy as great as we like in the limit $\epsilon \to 0$. So we have for that part of (11.28) *inside* the ϵ-sphere, volume δv,

$$\delta(\mathbf{J}_P) = \int_{\delta v} dv_Q\, \mathbf{J}_Q\left[\Delta_Q\left(\frac{1}{R_{QP}}\right)\right] \approx \mathbf{J}_P \int_{\delta v} dv_Q\, \nabla_Q \cdot \left[\nabla_Q\left(\frac{1}{R_{QP}}\right)\right] \qquad (11.29)$$

Using Gauss's divergence theorem and $R_{QP} = \mathbf{r}_P - \mathbf{r}_Q$, this becomes a surface integral over a sphere with radius $\epsilon = R_{QP}$ and volume δv:

$$\delta(\mathbf{J}_P) = \mathbf{J}_P \oint_{(\epsilon)} d\mathbf{S}_Q \cdot \left[\nabla_Q\left(\frac{1}{R_{QP}}\right)\right] \approx 4\pi\epsilon^2 \mathbf{J}_P\left(\frac{1}{\epsilon^2}\right) = 4\pi\mathbf{J}_P \qquad (11.30)$$

This proves that $\mathbb{J}_P = 4\pi \mathbf{J}_P$, and also that $\Delta_Q(1/R_{QP})$ acts as a *delta-function filter*, which eliminates all the integrand at points $Q \neq P$ and is infinite at $Q = P$ in such a way that (11.30) results.

When this is used in (11.27), we can set $\mathbf{J}_Q = \mathbf{J}_P$ (because the rest of the integrand is zero for $Q \neq P$) and then bring $\mathbf{J}_P$ outside the integral to obtain, again omitting the position subscript,

$$\nabla \times \mathbf{B} = \mu_0 \mathbf{J} \tag{11.31a}$$

This, then, is the second magnetostatic Maxwell equation.

At this point we introduce the *magnetic field* **H**, which is related to the magnetic flux density **B** in the same way that **D** and **E** are related: in free space we define

$$\mathbf{B} = \mu_0 \mathbf{H}$$

whereas in a simple magnetizable medium $\mathbf{B} = \mu \mathbf{H}$; for a ferromagnet at saturation $\mu \gg \mu_0$ and, in general, **B** and **H** are not parallel in ferromagnetic media.

We now use $\mathbf{B} = \mu_0 \mathbf{H}$ to reexpress (11.31a) in a form which holds in general, when magnetizable media are present:

$$\nabla \times \mathbf{H} = \mathbf{J} \tag{11.31b}$$

If the volume integral of this equation is taken, then using Stokes's law we obtain

$$\oint_C d\boldsymbol{l} \cdot \mathbf{H} = I_{\text{encirc}} \tag{11.31c}$$

where I_{encirc} is the total current encircled by the closed path C. Equation (11.31c) tells us that the path integral of **H** round any loop is determined only by the currents encircled; its value is indifferent to the presence of a magnetic medium.

We can now summarize what we have found:

$$\nabla \times \mathbf{H} = \mathbf{J} \tag{11.32a}$$

$$\oint_C d\boldsymbol{l} \cdot \mathbf{H} = I_{\text{encirc}} \equiv \int_S d\mathbf{S} \cdot \mathbf{J} \tag{11.32b}$$

$$\nabla \cdot \mathbf{B} = 0 \tag{11.32c}$$

$$\oint d\mathbf{S} \cdot \mathbf{B} = 0 \tag{11.32d}$$

$$\mathbf{B} = \mu_0 \mathbf{H} \qquad \text{in a vacuum} \tag{11.32e}$$

Let us discuss what these equations mean. We will use the *field line* concept for **B**, **H** in the same way as we did for **E**, **D**. Thus (11.32d) tells us that the *magnetic flux*, $d\mathbf{S} \cdot \mathbf{B}$, which we can interpret visually as the number of **B** lines crossing the area $d\mathbf{S}$, must be the same at one end of a thin tube of lines as at the other (cf. Fig. 5.7(a)). Field lines cannot cross (because the field direction is given by the tangent to a field line, and you cannot have two field directions at one location) so that $dS \neq 0$ at either end. Therefore $\mathbf{B}$ must be nonzero at the end if it is nonzero at the beginning of the tube, which implies that the field line for $\mathbf{B}$ continues on for ever. There is only one way in which this is possible:

Field lines for $\mathbf{B}$ *always close upon themselves.*

In other words, $\mathbf{B}$ lines form continuous loops.

Equation (11.32b) is known as *Ampère's law*. It pertains to a surface S, which may well be curved and which is bounded by a closed contour C. The surface S can be *any* simply curved surface that has C as its rim; by 'simply curved' we mean that we exclude surfaces with pieces that fold back and intersect other pieces. Upon some thought, one can see that *if* one such surface S with rim C is intersected by a $\mathbf{J}$ field line (current density vectors also have field lines), then *any other* such surface S with the *same* rim C will be intersected by that $\mathbf{J}$ field line. And furthermore, $\nabla \cdot \mathbf{J} = 0$ guarantees that $\int d\mathbf{S} \cdot \mathbf{J}$ is the same for all those surfaces because the ensuing difference integral for any two surfaces S_1 and S_2 with the same rim gives rise to

$$\oint_{S_1+S_2} d\mathbf{S} \cdot \mathbf{J} = 0 \tag{11.33}$$

In applying (11.32b) to a closed magnetic field line, it follows that $\oint d\boldsymbol{l} \cdot \mathbf{H} \geq 0$ because the $\mathbf{H}$ field is tangential to the field line everywhere. The integral is zero only if the right-hand side of (11.32b) is zero, i.e. if there are no encircled lines of current I_{encirc}. Ampère's law tells us that in a linear isotropic nonferromagnetic medium,[1]

Field lines for $\mathbf{B}$ *always encircle their source* $\mathbf{J}$ *lines.*

11.3 Ampère's law and applications

The general form of Ampère's law is given by (11.32b). In a vacuum or in a linear isotropic nonferromagnetic material, however, $\mathbf{H}$ and $\mathbf{B}$ differ only by a constant, so we are then free to put $\mathbf{H} = \mu^{-1}\mathbf{B}$ in Ampère's law. In fact, in what follows we assume that $\mu = \mu_0$ everywhere. Then, from (11.32c), $\mathbf{H}$ lines also will be continuous everywhere.

Application of equation (11.32b) requires special care when the magnetic field lines encircle only part of a current source, which can occur when the cross section of a wire is not infinitesimal. We will need to be very careful about which $\mathbf{J}$ lines lie inside and which do not lie inside a closed $\mathbf{H}$ line! Let us consider a simple example.

[1] In such a material the permeability μ is respectively independent of $\mathbf{H}$, independent of direction and single-valued.

Example 11.4 Magnetic field around an infinite thick straight current wire

This example resembles Example 11.2, but now we will allow the wire to have a finite radius a; again, it lies along the z axis. Figure 11.7 shows part of the wire.

The **H** field could not be a combination of $\hat{\mathbf{z}}$ and $\hat{\boldsymbol{\rho}}$ components because the result would then be closed **H** lines in a plane through the z axis, in which case they would not *encircle* any **J** lines at all. Equivalently, the symmetry of the situation excludes any field that has nonzero $\hat{\mathbf{z}}$ and/or $\hat{\boldsymbol{\rho}}$ components. Hence we can be sure that the **H** field is azimuthal, i.e. in the $\hat{\boldsymbol{\varphi}}$ direction, and furthermore that its magnitude is a function only of ρ (again owing to symmetry). For any closed circuit C that is a circle at a radius ρ around the z axis, Ampère's law (11.32b) can be applied with $d\boldsymbol{l} = \rho\, d\varphi\, \hat{\boldsymbol{\varphi}}$ to obtain for the left-hand side

$$\oint_C d\boldsymbol{l} \cdot \mathbf{H} = 2\pi\rho H_\varphi(\rho) \tag{11.34}$$

For the right-hand side of (11.32b) we need to calculate only the current due to the **J** lines enclosed by the circuit C. There are two cases, shown in cross section in Fig. 11.7: in the first, the broken curve C has radius less than a, in which case $I_{\text{encirc}} = (\pi\rho^2/\pi a^2)I$; in the second, C has radius larger than a, in which case $I_{\text{encirc}} = I$. The solution is now obtained, much more simply than in Example 11.2:

$$\begin{aligned} \mathbf{H} &= (\rho I/2\pi a^2)\hat{\boldsymbol{\varphi}}(\varphi) \quad \text{for } \rho < a \\ \mathbf{H} &= (I/2\pi\rho)\hat{\boldsymbol{\varphi}}(\varphi) \quad \text{for } \rho > a \end{aligned} \tag{11.35}$$

This example illustrates an important principle to which we shall return again and again: whenever we look closely enough to see a section of a **J** line as relatively straight, we find that the nearby **H** lines are circles $\perp$ **J**. The sense of rotation around

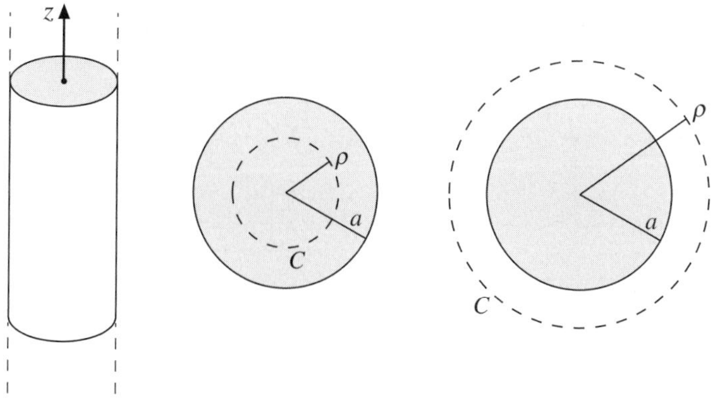

Fig. 11.7 Ampère's law for a uniform straight current wire of nonzero radius.

J is given by the right-handed screw convention; a right-handed screw moves away from you when you turn it clockwise with a screwdriver.

Example 11.5 Magnetic field due to a uniform sheet current in an infinite plane

Consider the surface-current density $\mathbf{K} = K\hat{\mathbf{y}}$, which is confined to the $z = 0$ plane and which is uniform in that plane (Fig. 11.8). To predict the **H** field, consider Fig. 11.8(b), in which we are looking in the $-\hat{\mathbf{y}}$ direction. We show some of the infinitesimally thin **J** lines, a continuum of which along the x axis forms the sheet current. As we have found above, each **J** line has circles of **H** lines around it. Upon some thought about the superposition of all such circles, each with the same radius but around different **J** lines, we can conclude that $\mathbf{H}(z) = H(z)\hat{\mathbf{x}}$ for $z > 0$ and $\mathrm{H}(z) = -H(z)\hat{\mathbf{x}}$ for $z < 0$, with positive $H(z)$. Now consider the rectangular contour C with two sides of length l at distance $\pm z$ from the $z = 0$ plane, and apply Ampère's law. Our argument shows that the sum of $d\boldsymbol{l} \cdot \mathbf{H} = 0$ for the two sides perpendicular to the sheet. Hence we conclude that

$$\oint_C d\boldsymbol{l} \cdot \mathbf{H} = l[H(z) - H(-z)] = 2lH(z) = I_{\text{encirc}} \tag{11.36}$$

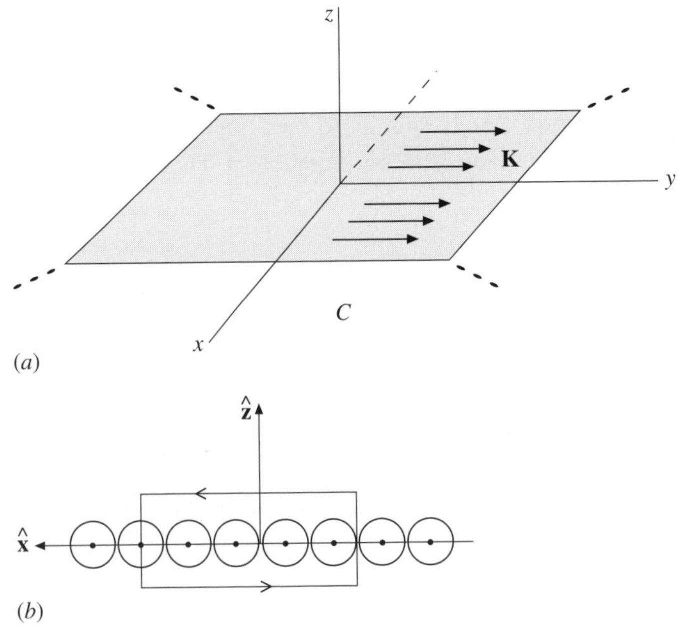

Fig. 11.8 Geometry for the calculation of the magnetic field due to a uniform sheet current on a plane.

It can be seen that $I_{\rm encirc} = lK$, the product of the current per unit length along the x axis and the length of sheet current encircled by C. Thus

$$\begin{aligned} \mathbf{H}(z) &= \tfrac{1}{2}K\hat{\mathbf{x}} \qquad \text{for } z > 0 \\ \mathbf{H}(z) &= -\tfrac{1}{2}K\hat{\mathbf{x}} \qquad \text{for } z < 0 \end{aligned} \tag{11.37a}$$

To generalize the problem and make it independent of the fact that we have restricted $\mathbf{K}$ to the $z = 0$ plane, we replace $\hat{\mathbf{x}}$ by $\hat{\mathbf{y}} \times \hat{\mathbf{z}}$ to obtain

$$\mathbf{H}(z) = \tfrac{1}{2}K\hat{\mathbf{y}} \times \hat{\mathbf{z}} = \tfrac{1}{2}\mathbf{K} \times \hat{\mathbf{n}} \qquad \Rightarrow \qquad \mathbf{H} = -\tfrac{1}{2}\hat{\mathbf{n}} \times \mathbf{K} \tag{11.37b}$$

This form summarizes *both* parts of (11.37a) in a *coordinate-free* fashion, because $\hat{\mathbf{n}}$ is the surface-normal unit vector, which always points *away* from the plane and therefore is either $+\hat{\mathbf{z}}$ or $-\hat{\mathbf{z}}$ in (11.37a), depending upon whether the point at which we require $\mathbf{H}$ is above or below the plane. Note especially that $\mathbf{H}$ is a constant on each side of the sheet current.

Example 11.6 Toroid with current windings

A rectangular toroid around the z axis is shown schematically in Fig. 11.9. It has inner radius a and outer radius b, and its thickness in the z direction is h. It is in fact tightly wound with $\mathcal{N}$ coils, each of which carries current I. We make a slight approximation by assuming that the coils all lie in planes through the z axis; this is

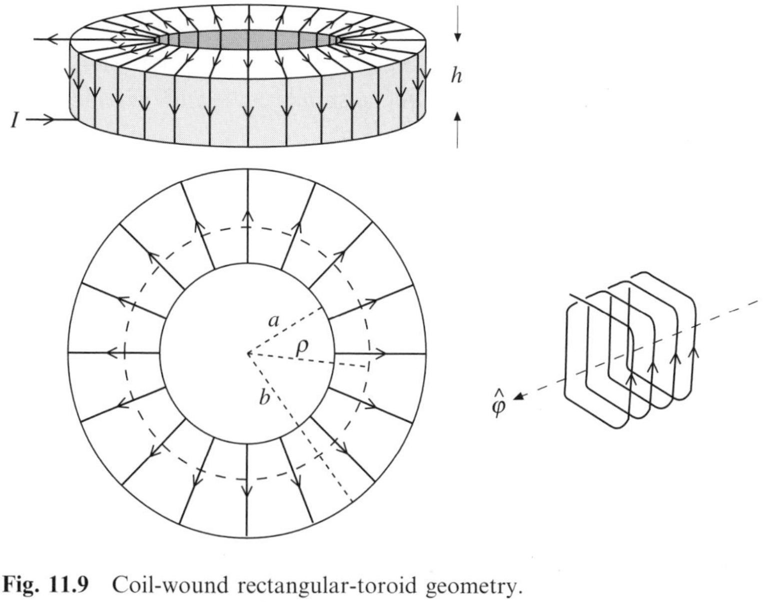

Fig. 11.9 Coil-wound rectangular-toroid geometry.

not a serious error provided that the number of coils in any small-angle sector is very large. With some thought, we can convince ourselves that the **H** lines can only lie in the azimuthal $\hat{\boldsymbol{\varphi}}$ direction, and that they must be in $-h/2 < z < h/2$ (if the origin lies at the center): the **H** lines of each coil are contours around that coil, and we can see, Fig. 11.9, that $\hat{\boldsymbol{\rho}}$ and $\hat{\mathbf{z}}$ components cancel out and only $\hat{\boldsymbol{\varphi}}$ components remain. By symmetry it would seem possible to have a perfectly circular **H** line around the z axis at $|z| > h/2$, but such a **H** line would not encircle current, hence cannot exist.

So we draw circles around the z axis for $-h/2 < z < h/2$, as shown in the lower left-hand figure. For each such circle, the contour integral of Ampère's law (11.32b) yields the factor $2\pi\rho H_\varphi$.

(1) For $\rho < a$ the contour does not enclose current. Hence $H = 0$ here.

(2) For $a < \rho < b$ the contour lies inside the toroid, and the enclosed current is found by counting the number of intersections with the plane of the contour. This number is $\mathcal{N} = n(2\pi a)$ if n is the number of windings per meter at the inner surface of the solenoid. Thus $I_{\text{encirc}} = 2\pi a n I$, which yields

$$\mathbf{H} = \frac{nIa}{\rho}\hat{\boldsymbol{\varphi}} \tag{11.38}$$

(3) For $\rho > b$, no current is enclosed, because for each intersection of current in the $+\hat{\mathbf{z}}$ direction at $\rho = a$, there is another intersection of current in the $-\hat{\mathbf{z}}$ direction at $\rho = b$. These two cancel each other when evaluating I_{encirc}. Hence $\mathbf{H} = 0$ here also.

It can be seen that the above argument is not affected by the cross-sectional shape of the toroid; this does not have to be rectangular. It is only necessary that the cross section should be the same for all angles φ around the z axis and that the toroid intersection with any $z =$ constant plane should have truly circular edges.

Example 11.7 The solenoid: a cylinder with current windings

The solenoid is a cylindrical structure with constant cross section S perpendicular to the axial z direction; see Fig. 11.10. If it is very long – essentially infinite – and we do not consider the field anywhere near the ends, then it can be thought of as a limiting case of a *sector* of the coil-wound toroid, of infinitesimal azimuthal-angle width $d\varphi$, obtained by allowing $a, b \to \infty$ while holding $b - a$ constant. In equation (11.38) let us rewrite $a < \rho < b$ as $1 < \rho/a < b/a$ or

$$1 < \frac{\rho}{a} < \left(1 + \frac{b-a}{a}\right) \tag{11.39}$$

Now as we make $a, b \to \infty$ while holding $b - a$ constant, we see that the ratio ρ/a is forced to go to unity, and consequently we find that *inside* the solenoid, using (11.38)

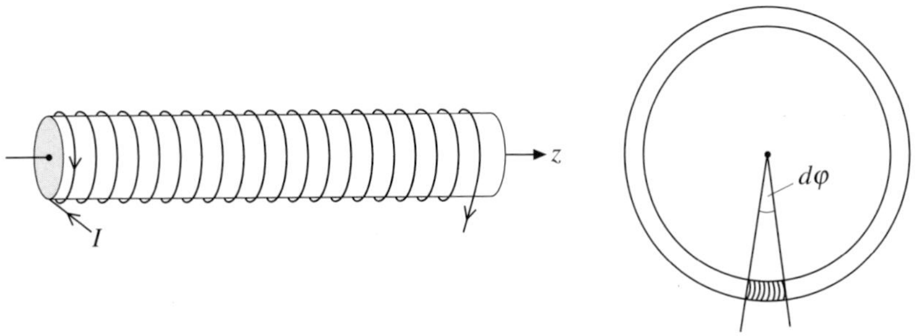

Fig. 11.10 The solenoid originating from the coil-wound rectangular toroid.

and replacing $\hat{\boldsymbol{\varphi}}$ in this situation by $\hat{\mathbf{z}}$,

$$\mathbf{H} = nI\hat{\mathbf{z}} \tag{11.38}$$

where n is the number of windings per meter. The field *outside* the solenoid is still zero. We must stress that (11.38) is approximately correct for a *finite* solenoid only for those regions inside that are at least several solenoid radii distant from the ends of the solenoid, and then only if the solenoid is thin compared to its length. This will be demonstrated explicitly in the following subsection.

11.4 The finite solenoid

We devote this entire subsection to a more exact calculation of the *axial field* inside a finite solenoid. Not only is the exercise in doing so of importance; what we can learn from the result is illustrative of much in magnetostatics. The calculation is best done in a number of steps.

Consider first a circular loop of current I and radius a, and an axial point P at height z above the plane of the loop; see Fig. 11.11. The Biôt–Savart integral for the **H** field at P is, from (11.4a),

$$\mathbf{H}_P \equiv \frac{I}{4\pi}\oint_C d\mathbf{l}_Q \times \frac{\mathbf{R}_{QP}}{R_{QP}^3} \tag{11.39}$$

with $\mathbf{R}_{QP} = z\hat{\mathbf{z}} - a\hat{\boldsymbol{\rho}}_Q$ and $d\mathbf{l}_Q = a\,d\varphi_Q\,\hat{\boldsymbol{\varphi}}$. The symmetry of the problem suggests that only the $\hat{\mathbf{z}}$ component of the field is going to be important. We can prove this rigorously by working out the cross product in (11.39):

$$\mathbf{H}_P \equiv \frac{Ia}{4\pi}\int_0^{2\pi} d\varphi_Q \frac{a\hat{\mathbf{z}} + z\hat{\boldsymbol{\varphi}}_Q}{R_{QP}^3} \tag{11.40}$$

The field components $\mathbf{x}\cdot\mathbf{H}$ and $\hat{\mathbf{y}}\cdot\mathbf{H}$ introduce the factor $\cos\varphi_Q$ or $\sin\varphi_Q$ in the second part of the numerator, which then integrates with $d\varphi_Q$ to zero. Thus only the

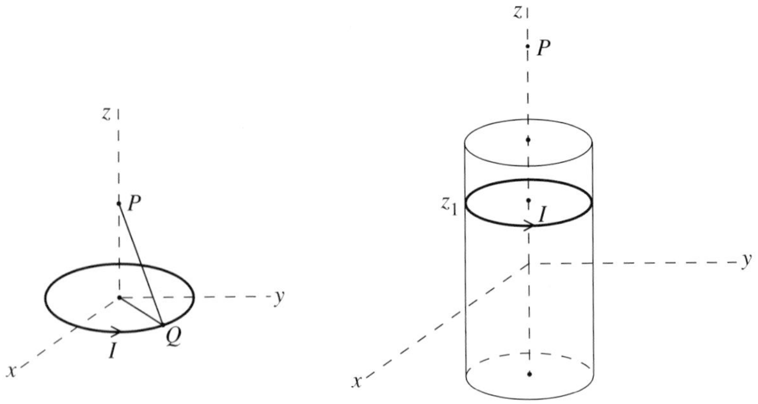

Fig. 11.11 The finite solenoid originating from a single turn.

z component remains and, because $\hat{\mathbf{z}} \cdot \hat{\boldsymbol{\varphi}}_Q = 0$ and $R_{QP} = \sqrt{z^2 + a^2}$, it follows that

$$\mathbf{H}_P = \frac{Ia^2}{2(z^2 + a^2)^{3/2}} \hat{\mathbf{z}} \tag{11.41}$$

The right-hand diagram in Fig. 11.11 shows this current loop now at height z_1 and point P still at height z; the current loop is part of a solenoid, with n windings per meter, that is situated between $-L/2$ and $+L/2$ on the z axis. Hence, in order to apply (11.41) to that loop as part of the solenoid, we replace I by Kdz_1 (with $K \equiv nI$ as the equivalent surface-current density), and z in the denominator by $z - z_1$; then

$$d\mathbf{H}_P = \frac{Ka^2}{2[a^2 + (z - z_1)^2]^{3/2}} dz_1 \, \hat{\mathbf{z}} \tag{11.42}$$

The replacement of I by $K\,dz_1$ is equivalent to using the Biôt–Savart law (11.4a), for a sheet current with element $\mathbf{K}\,dS$ such that $dS = dz_1\,dl$. We write $d\mathbf{H}_P$ because it is the contribution to the axial $\mathbf{H}$ field at P due to a band of sheet current at radius a around the z axis and located at z_1. The band has width dz_1. Now we integrate over z_1 in the interval $-L/2 < z_1 < L/2$, obtaining

$$\mathbf{H}_P = \int_{-L/2}^{L/2} dz_1 \frac{Ka^2}{2[a^2 + (z - z_1)^2]^{3/2}} \hat{\mathbf{z}} \tag{11.43a}$$

The most convenient way to carry out this integral is to introduce a new integration variable t to replace z_1, namely to let $z - z_1 = at$. Obviously t is dimensionless, and we obtain

$$\mathbf{H}_P = \frac{K}{2} \int_{z/a - L/2a}^{z/a + L/2a} dt \frac{1}{(1 + t^2)^{3/2}} \hat{\mathbf{z}} \tag{11.43b}$$

As mentioned earlier, the indefinite integral of $(1+t^2)^{-3/2}$ is $t/(1+t^2)^{1/2}$, hence we obtain as the final result

$$\mathbf{H}(z) = \frac{K}{2}\left[\frac{z+L/2}{\sqrt{a^2+(z+L/2)^2}} - \frac{z-L/2}{\sqrt{a^2+(z-L/2)^2}}\right]\hat{\mathbf{z}} \qquad \text{with } K \equiv nI \tag{11.44}$$

Obviously $\mathbf{H}(-z) = \mathbf{H}(z)$. Let us investigate various choices for point P.

(1) *The center of the solenoid.* Set $z = 0$ in (11.44). The result is

$$\mathbf{H}(z) = \frac{K}{2}\left[\frac{L}{\sqrt{a^2+(L/2)^2}}\right]\hat{\mathbf{z}} \tag{11.45}$$

If the solenoid is long and thin ($L \gg 2a$), then (11.45) reduces to $\mathbf{H} \approx K\hat{\mathbf{z}} = nI\hat{\mathbf{z}}$, the result found for the infinite solenoid in (11.38). The *same result* is also found for any z value inside the solenoid if in (11.44) we assume $a \ll L/2 - |z|$. It tells us what we surmised in the previous subsection; the infinite-solenoid result holds to within a few radii to reasonable accuracy.

(2) *The ends of the solenoid.* Now set $z = \pm L/2$. In either case we easily see that we obtain

$$\mathbf{H}(z) = \frac{K}{2}\left[\frac{L}{\sqrt{a^2+L^2}}\right]\hat{\mathbf{z}} \tag{11.46}$$

and therefore that a long thin solenoid ($a \ll L$) yields $\mathbf{H} \approx \frac{1}{2}K\hat{\mathbf{z}}$. The field at the ends is close to *one half* of the field at the center! That this must be the case can be seen if one imagines adding together *two* such solenoids; then two end points, each with field $\frac{1}{2}\mu_0 K\hat{\mathbf{z}}$, become the new middle point, with field $K\hat{\mathbf{z}}$.

(3) *Beyond the ends of the solenoid.* When $|z| \gg L/2$, (11.44) shows us that $\mathbf{H}(z)$ becomes very small. We can see this from numerical plots of the field as a function of z/L for various ratios a/L, shown in Fig. 11.12. These plots also show how the field decays at the solenoid ends to approximately one half of the value at the center.

(4) *Off-axis* **H** *field.* In the interior of the solenoid we have $\nabla \cdot \mathbf{B} = 0$ as in (11.32c), which is of course equivalent to (11.32d). Again we use **B** and **H** interchangeably here because $\mathbf{B} = \mu_0\mathbf{H}$. The integral form (11.32d) implies that $\mathbf{H} \cdot d\mathbf{S} = H\,dS$ is conserved along a tube of field lines if $d\mathbf{S}$ is a surface element vector $\parallel \mathbf{H}$. Very close to the axis, we may approximate $d\mathbf{S}$ as $\hat{\mathbf{z}}\,dS$ even though the surface elements may be curved. Therefore $\mathbf{H} \approx H\hat{\mathbf{z}}$ very close to the axis. The graphs in Fig. 11.12 show that the H field decreases

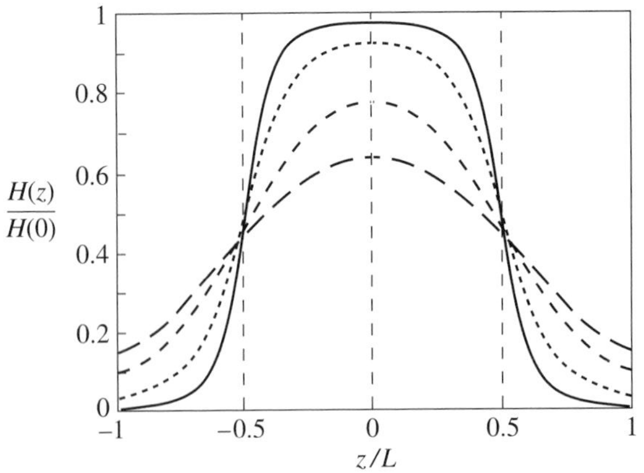

Fig. 11.12 Graphs of a solenoid's axial magnetic field as a function of axial location, for various ratios a/L of radius to length: going from the top curve to the bottom curve $a/L = 0.1, 0.2, 0.4, 0.6$.

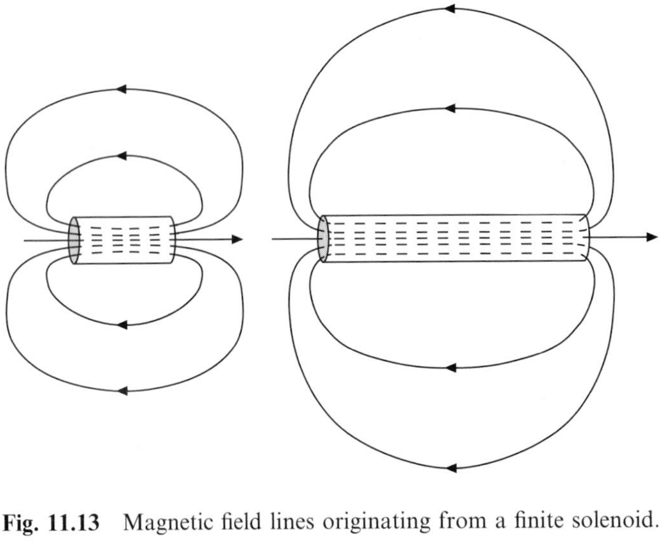

Fig. 11.13 Magnetic field lines originating from a finite solenoid.

as we move away from $z = 0$ in either the $+\hat{\mathbf{z}}$ or the $-\hat{\mathbf{z}}$ direction. The conservation of $H\,dS$ implies that dS becomes larger if H becomes smaller. Thus the field lines close to the axis must curve *away* from the axis if dS is increasing as we move away from the center of the solenoid. The field line picture therefore must be as shown in Fig. 11.13, because all field lines are closed. One diagram shows a short solenoid and the other a longer solenoid. The same field lines are shown in both. The diagrams show that a field line straightens out inside the solenoid, whereas its outside part moves away from the solenoid as we increase the length. In the limit of infinite length all the outside parts of the field lines have moved to $\rho = \infty$, and there is zero field outside!

11.5 The boundary conditions for B and H

Finally in this chapter we consider a situation where we have **B** and **H** lines passing from one medium to another. We shall work in analogy to subsection 6.4.2 on electrostatics. We assume an interface between two magnetic media with differing permeabilities μ_1 and μ_2. A real sheet current with density **K** may exist on the interface.

The Maxwell equation $\nabla \cdot \mathbf{B} = 0$ is analogous to $\nabla \cdot \mathbf{D} = \varrho_v$ in the special case where $\varrho_v = 0$ in the region of interest. The 'pillbox' argument, Fig. 6.9, then yields

$$\hat{\mathbf{n}} \cdot (\mathbf{B}_2 - \mathbf{B}_1) = 0 \qquad \text{or} \qquad B_{2n} = B_{1n} \tag{11.47}$$

The convention here is that the surface-normal unit vector $\hat{\mathbf{n}}$ points into medium 2, but it is not important because the right-hand side is zero. To obtain the condition on the tangential components, we consider the Maxwell equation $\nabla \times \mathbf{H} = \mathbf{J}$, (11.32a), which is the differential form of Ampère's law $\oint d\boldsymbol{l} \cdot \mathbf{H} = I_{\text{encirc}}$. If we apply this to the infinitesimal contour in Fig. 11.14, in which it is assumed that the plane of the contour $\perp$ **K** (the surface-current vector) and that the two sides $\perp$ the interface are negligibly short compared to the two parallel sides ($h \ll l$), then the contour integral becomes

$$l\hat{\mathbf{t}} \cdot (\mathbf{H}_2 - \mathbf{H}_1) = -l\hat{\mathbf{s}} \cdot \mathbf{K} \tag{11.48}$$

Here, the length l of each parallel side is small enough that the **H** fields do not change appreciably in that distance. That is why no integral sign is needed here; the integral becomes a dot product of the field vector times the distance vector $l\hat{\mathbf{t}}$.

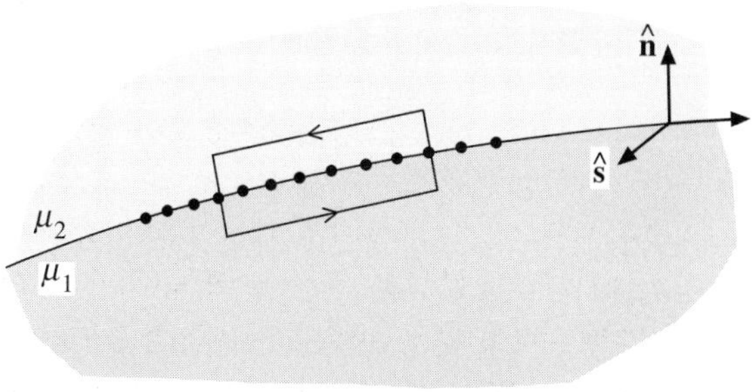

Fig. 11.14 Local orthogonal coordinate system for magnetic-field boundary conditions between magnetic media of differing permeabilities. The contour has height h and length l.

As $l \neq 0$, we can divide by it, and thus omit it. We can also write $\hat{\mathbf{t}} = \hat{\mathbf{n}} \times \hat{\mathbf{s}}$ so that

$$(\hat{\mathbf{n}} \times \hat{\mathbf{s}}) \cdot (\mathbf{H}_2 - \mathbf{H}_1) = -\hat{\mathbf{s}} \cdot \mathbf{K} \tag{11.49}$$

From the vector-algebra rule 3 in Appendix A.3, we see that this can be written as $[(\mathbf{H}_2 - \mathbf{H}_1) \times \hat{\mathbf{n}}] \cdot \hat{\mathbf{s}} = 0$, which is the same as $-[\hat{\mathbf{n}} \times (\mathbf{H}_2 - \mathbf{H}_1)] \cdot \hat{\mathbf{s}} = 0$, and because $\hat{\mathbf{s}} \neq 0$ it follows that

$$\hat{\mathbf{n}} \times (\mathbf{H}_2 - \mathbf{H}_1) = \mathbf{K} \qquad \text{or} \qquad H_{2\mathrm{t}} - H_{1\mathrm{t}} = K \tag{11.50}$$

Note again that $\hat{\mathbf{n}}$ is the unit vector pointing from the interface *into* medium 2.

Thus (11.47) and (11.50) represent the most general boundary conditions for magnetic fields, and the mechanics for solving for fields in medium 2, when fields in medium 1 are given, are very similar to those discussed in subsection 6.4.2.

Problems

11.1. A particle with charge $q = 2\,\mathrm{C}$ has velocity $\boldsymbol{u} = 3\hat{\mathbf{x}}\,\mathrm{m/s}$ and a mass 0.15 kg. It is subjected to a uniform magnetic flux density $\mathbf{B} = 7\hat{\mathbf{z}}\,\mathrm{T}$.
(a) In which direction is it deflected?
(b) What is the radius R of its orbit around the $\mathbf{B}$ field?
(c) What is the period T of its rotation?

11.2. A charge $q = -0.75\,\mathrm{C}$ has an initial velocity $\mathbf{u} = 2.5\hat{\mathbf{x}} - 4\hat{\mathbf{y}} + 3.5\hat{\mathbf{z}}\,\mathrm{m/s}$ in a magnetic flux density $\mathbf{B} = 5\hat{\mathbf{y}} - 2\hat{\mathbf{z}}\,\mathrm{T}$. What is the force vector that the $\mathbf{B}$ field exerts on it at this initial time?

11.3. An electron with energy $5 \times 10^3\,\mathrm{eV}$ (1 eV is the energy needed to take an electron to a point 1 V lower in potential) is incident perpendicularly upon the Earth's magnetic flux density, the magnitude of which we take to be 0.5 gauss (1 gauss = $10^{-4}\,\mathrm{T}$). Calculate the radius of the circle in which it moves around the $\mathbf{B}$ field. *Hint*: The sum of kinetic energy and potential energy is constant.

11.4. An infinitesimally thin wire carrying a current $I = 2\pi$ (in A) in the $\hat{\mathbf{z}}$ direction coincides with the z axis. Find the $\mathbf{B}$ field vector in the form $\mathbf{B} = \mu_0(a\hat{\mathbf{x}} + b\hat{\mathbf{y}} + c\hat{\mathbf{z}})$ at the points
(a) $P_1(x = 5, y = 0, z = 3)$
(b) $P_2(\rho = 1, \varphi = \pi/4, z = 0)$
(c) $P_3(r = 1, \theta = \pi/6, \varphi = \pi/2)$.
There is no need here to convert μ_0 to a number.

11.5. The lines $y = b$, $z = 0$ and $y = -b$, $z = 0$ both carry a current $\mathbf{I} = I\hat{\mathbf{x}}$. Find the $\mathbf{H}$ field at an arbitrary point on the z axis.

11.6. Reconsider Example 11.4 in Section 11.3; let the length of the wire now stretch from $z = -L/2$ to $z = +L/2$. What is the $\mathbf{B}$ field at point P in the $z = 0$ plane?

11.7. Use the result of problem 11.6 to calculate the **B** field at a point P at distance z above the center of a square loop of wire carrying a current I.

11.8. An infinite circularly cylindrical surface at $\rho = 10\,\text{cm}$ has a uniform sheet-current density $\mathbf{K} = 5\hat{\boldsymbol{\varphi}}\,\text{A/m}$ on its surface. Find the magnetic field **H** at an interior point.

11.9. Prove Eq. (11.25).

11.10. A sheet-current density is given as a vector in the plane $2x + y + 2z = 6$. The density is $\mathbf{K} = -3(\hat{\mathbf{x}} - \hat{\mathbf{z}})\,\text{A/m}$. Find the magnetic field vector **H** on either side.

11.11. A uniform current $\mathbf{I} = I\hat{\mathbf{z}}$ is confined to an infinitesimally thin wire. Calculate the magnetic flux through a rectangle $3 < \rho < 6$, $-1 < z < 1$ (all distances in cm), $\varphi = \pi/4$, given that $I = 1.91\,\text{A}$. What is the result for $\varphi = 3\pi/4$?

11.12. A classic problem is the following. Consider a uniform circularly cylindrical wire of radius b with uniform current density J. A nonconcentric but parallel smaller circular cylinder with radius $a < b$ is removed from the larger cylinder so that the metal is replaced by free space. Show that the **B** field inside the hollow area is uniform and that its magnitude depends upon the distance between the two centers of the cylinders. *Hint*: This problem is most easily handled by the superposition of two cylinders with $+J$ and $-J$ respectively as their current densities.

11.13. Another classic problem is that of a Helmholtz pair. Two circular loops in the planes $z = 0$ and $z = a$ share the same axis (the z axis) and have the same radius $\rho = a$; the currents I on each are the same and in the same rotational sense. Calculate the field on the z axis at the central point between the loops.

12

The magnetostatic potentials

This chapter deals with the magnetic analog of electrostatic potential. Unlike the electrostatic case, there are *two* potentials commonly used in magnetostatics: the *scalar* and the *vector* magnetic potentials. The advantages for calculation are similar to the electrostatic case: the equations for the potentials are well known and easier to solve in general than the Ampère's-law or Biôt–Savart integral expressions. A disadvantage lies in the fact that the magnetic potentials do not have as clearly defined a physical meaning, susceptible to direct measurement, as the electrostatic potential. Disadvantages have to be weighed against advantages in each particular case, but both magnetic potentials have been employed extensively in many basic calculations of magnetic fields, and some understanding of them is necessary. Unlike many other texts, we shall start with the less obvious one, the scalar potential.

12.1 The scalar magnetic potential

We found in Chapter 11 that $\nabla \cdot \mathbf{B} = 0$ and that $\nabla \times \mathbf{H} = \mathbf{J}$. In many practical applications, the current density $\mathbf{J}$ is restricted to wires or to other current carriers that occupy only a vanishingly small part of the domain in which one might wish to know the magnetic fields. If the media of the domain are linear, isotropic, nonferromagnetic materials, then we can use $\nabla \times \mathbf{B} = \mu\mathbf{J}$, with $\mu \neq \mu_0$, and we will restrict ourselves to a volume v, *outside* the current sources, in which μ is constant.

In such a restricted volume v, in which currents are excluded, we may set $\nabla \times \mathbf{H} = 0$, but we must be aware of the fact that *outside* v there are locations in which $\nabla \times \mathbf{H} = \mathbf{J}$, hence where $\nabla \times \mathbf{H} \neq 0$; these are current-source regions. We then may employ the Helmholtz theorem in the restricted region v and set

$$\mathbf{H} = -\nabla V_{\mathrm{m}} \tag{12.1}$$

which defines the scalar magnetic potential V_m, in analogy to what we did in the electrostatic case in Chapter 4. The fact that (12.1) holds implies that the curl of the magnetic flux density is zero. The remaining equation is $\nabla \cdot \mathbf{B} = 0$, or for constant permeability $\nabla \cdot \mathbf{H} = 0$; this then yields, when applied to (12.1),

$$\Delta V_m = 0 \tag{12.2}$$

In words: the scalar magnetic potential obeys Laplace's equation in a volume in which the permeability μ is constant and in which there are no current sources. Chapter 9 was devoted to numerical solutions of Laplace's and Poisson's equations, and we saw there that a unique solution is possible if we have enough information about boundary values. Specifically, we need to know either $V_m(\mathbf{r}_s)$ or $\hat{\mathbf{n}} \cdot \nabla V_m(\mathbf{r}_s)$ for all points $\mathbf{r}_s$ on the boundary surface S of the volume v. Unfortunately, $V_m(\mathbf{r}_s)$ has no direct physical meaning. But $\hat{\mathbf{n}} \cdot \nabla V_m(\mathbf{r}_s) = -\hat{\mathbf{n}} \cdot \mathbf{H} = -H_n$, and it is quite possible for normal components of $\mathbf{H}$ to be specified on a surface. We shall see in Example 12.1 below that we can solve the boundary-condition problem in yet another way by using Ampère's law!

There is a remaining problem: V_m is multivalued. How can we see this, and what does it mean? The answer to the first question is illustrated in Fig. 12.1. Figure 12.1(a) illustrates two paths, from a point Q to a point P, for obtaining, by analogy with (4.2),

$$V_{mP} - V_{mQ} = -\int_Q^P d\boldsymbol{l} \cdot \mathbf{H} \tag{12.3}$$

The issue here is that paths I and II go on opposite sides of the current I that flows along the z axis in the $\hat{\mathbf{z}}$ direction. Let us distort path II, as shown in (b). This distortion is allowed because $\int d\boldsymbol{l} \cdot \mathbf{H}$ is the same for any path from Q to P. The transition from (b) to (c) is completed when the two horizontal legs of path II

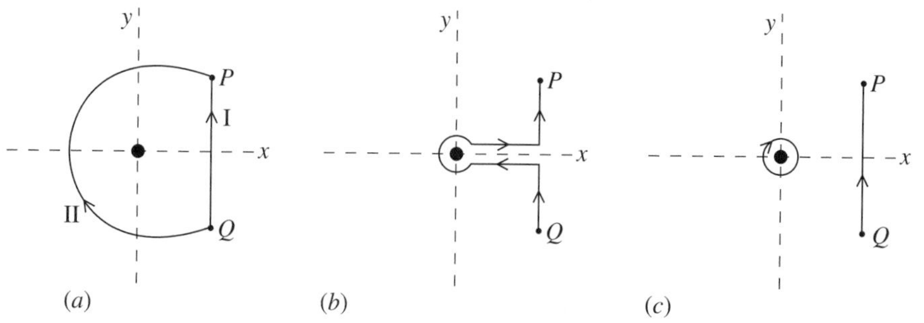

Fig. 12.1 Various paths for calculating the scalar magnetic potential difference $V_{mP} - V_{mQ}$.

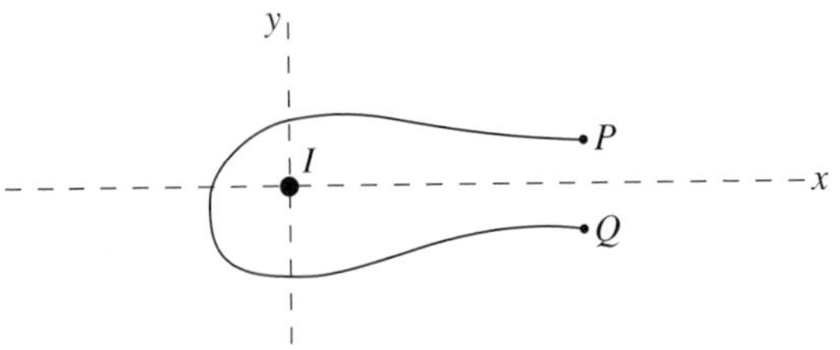

Fig. 12.2 Path-difference problem in solving $\Delta V_{\rm m} = 0$ in the presence of a current source I; $PQ = h$.

coincide and their contributions therefore cancel, as they are in opposite directions. It is clear that the right-hand part of the final two-part path is equal to path I. Consequently,

$$[V_{\mathrm{m}P} - V_{\mathrm{m}Q}]_{\mathrm{II}} = \left[-\int_Q^P dl \cdot \mathbf{H}\right]_{\mathrm{II}} = \left[-\int_Q^P dl \cdot \mathbf{H}\right]_{\mathrm{I}} - \oint dl \cdot \mathbf{H}$$
$$= [V_{\mathrm{m}P} - V_{\mathrm{m}Q}]_{\mathrm{I}} + I \tag{12.4}$$

The value of $V_{\rm m}$ at P differs from that at Q by a fixed amount plus or minus an integer number of times the current I, depending upon how many turns the path (clockwise or counterclockwise) from Q to P makes when distorted as in Fig. 12.1(*b*), (*c*). In the diagrams we have illustrated the situation for only a single turn clockwise around the current, but a generalization to more turns seems obvious. The above also reminds us that the unit of $V_{\rm m}$ must be ampères.

Now we return to the implications of multivaluedness. Consider two points P and Q in a μ_0 medium, separated by y-distance h, as shown in Fig. 12.2, with current I as shown. Any numerical Laplace-solving problem will progress, in some way, in finding $V_{\rm m}$, and the difference $V_{\mathrm{m}P} - V_{\mathrm{m}Q}$ ultimately will be given by a line integral of $\mathbf{H}$ that might partly encircle the current, as shown, or that might follow the shortest path of length h. The answer for $\mathbf{H} = -\nabla \cdot V_{\rm m}$ would depend strongly upon the path between P and Q followed by the numerical program in solving for $V_{\rm m}$. If the program indeed followed the shown longer route, then a numerical calculation of $\mathbf{H}$ from the difference of the $V_{\rm m}$ values at the points P and Q, divided by h, would be in serious error. One has to know what the solving routine is doing in order not to have the multivaluedness lead to incorrect answers for the fields!

Example 12.1 Magnetic field inside a coaxial cable

Figure 12.3 shows a cross section of a coaxial cable with core radius a and sheath radius b; we assume that the permeability is μ_0. A current I travels up the core, in the $\hat{\mathbf{z}}$ direction, and back down the sheath. We shall use the scalar magnetic potential to

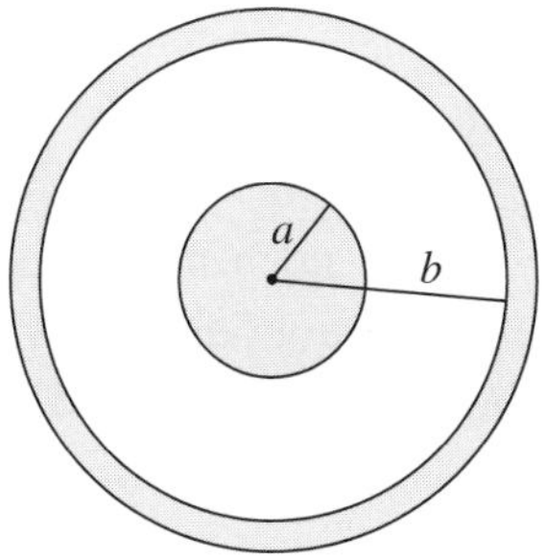

Fig. 12.3 Coaxial cable geometry for calculating V_{m}.

calculate the **H** field, and thus the **B** field, in $a < \rho < b$; the axis of the coaxial cable is also the z axis. Symmetry considerations would seem to suggest (incorrectly, as we shall see) that V_{m} is a function only of the radial distance from the axis, $V_{\mathrm{m}} = V_{\mathrm{m}}(\rho)$. Consequently it would appear that Laplace's equation for V_{m} becomes

$$\frac{1}{\rho}\frac{d}{d\rho}\left(\rho\frac{dV_{\mathrm{m}}}{d\rho}\right) = 0 \qquad \Rightarrow \qquad \frac{dV_{\mathrm{m}}}{d\rho} = \frac{\text{constant}}{\rho} \tag{12.5}$$

and this would lead to a solution $\mathbf{H} = H_\rho \hat{\boldsymbol{\rho}}$, with $H_\rho \propto 1/\rho$. This is an impossible solution because it yields field lines for **H** and **B** that are all in the radial direction. Such field lines cannot close, and thus they would violate $\nabla \cdot \mathbf{B} = 0$.

Let us instead consider the possibility that V_{m} is a function only of the polar angle φ, i.e.

$$\frac{1}{\rho^2}\left(\frac{d^2 V_{\mathrm{m}}}{d\varphi^2}\right) = 0 \qquad \Rightarrow \qquad V_{\mathrm{m}}(\varphi) = c_1\varphi + c_2 \tag{12.6}$$

The gradient of V_{m} then has only a φ component, $\nabla V_{\mathrm{m}} = dV_{\mathrm{m}}/\rho\, d\varphi$, so that

$$\mathbf{H}(\mathbf{r}) = -\frac{c_1}{\rho}\hat{\boldsymbol{\varphi}} \tag{12.7}$$

The constant c_1 is determined by applying Ampère's law:

$$\oint d\boldsymbol{l} \cdot \mathbf{H} = I_{\text{encirc}} \qquad \Rightarrow \qquad 2\pi\rho H_\varphi = I \quad \text{for } a < \rho < b \tag{12.8}$$

The combination of (12.7) and (12.8) yields $c_1 = -I/2\pi$. This completes the calculation, as we now find that if indeed $V_{\mathrm{m}} = V_{\mathrm{m}}(\varphi)$ then

$$\mathbf{B}(\mathbf{r}) = \mu_0\mathbf{H}(\mathbf{r}) = \frac{\mu_0 I}{2\pi\rho}\hat{\boldsymbol{\varphi}} \tag{12.9}$$

Two things are to be noted. First, we did *not* use the Neumann conditions for $V_{\mathrm{m}}(\mathbf{r}_{\mathrm{s}})$ at the boundary surfaces $\rho = a$ and $\rho = b$. As it turned out, we were able to eliminate one constant by using a condition on **H**, and we used Ampère's law to determine the second constant. In the second place, we were 'fooled' momentarily by the symmetry into thinking that V_m might be a function only of ρ. We should reflect upon this mistake! It was made because we thought of $V_m(\mathbf{r})$ in splendid isolation. The fact of importance is that it is the *gradient* of V_m that has physical meaning, not V_m itself. Thus, the gradient of V_m must be independent of φ, and V_m itself must be a linear function of φ. We have no *a priori* proof that V_m is not also a function of ρ, but we simply tried (12.5) as a possibility. In fact if V_m had also been a function of ρ, we would have failed in trying to solve for the constants.

12.2 The vector magnetic potential

Another potential, the vector magnetic potential, follows from the fact that $\nabla \cdot \mathbf{B} = 0$ everywhere. Then, Helmholtz's theorem (see Section 5.4) has shown that we may replace this Maxwell equation if we set $\mathbf{B} = \nabla \times \mathbf{A}$ and formulate the remaining Maxwell equation in terms of a new *vector potential* **A**. Let us now briefly review what the Helmholtz theorem tells us. It states that a vector field (**B** in this case) is determined by its divergence and curl fields. If we set $\mathbf{B} = \nabla \times \mathbf{A}$, we have at least made sure that the divergence of **B** is determined (it is zero everywhere; see identity 9 in Appendix A.4). Now we substitute $\mathbf{B} = \nabla \times \mathbf{A}$ into Ampère's law to obtain, for a medium in which $\mu = \mu_0$,

$$\nabla \times (\nabla \times \mathbf{A}) = \mu_0 \mathbf{J} \tag{12.10}$$

We may now use the vector identity $\nabla \times (\nabla \times \mathbf{A}) = \nabla(\nabla \cdot \mathbf{A}) - \Delta \mathbf{A}$; we then obtain after rearrangement of terms

$$\Delta \mathbf{A} - \nabla(\nabla \cdot \mathbf{A}) = -\mu_0 \mathbf{J} \tag{12.11}$$

The Helmholtz theorem tells us that the vector field **A** is determined by its divergence and curl fields. We know that its curl field is **B**, the magnetic flux density. But its divergence field is still totally open to choice, since whatever divergence we choose, we only need to know $\nabla \times \mathbf{A}$. The simplest choice is to try and solve (12.11) subject to the condition that $\nabla \cdot \mathbf{A} = 0$ everywhere. Hence we have

$$\Delta \mathbf{A} = -\mu_0 \mathbf{J} \qquad \text{subject to } \nabla \cdot \mathbf{A} = 0 \tag{12.12}$$

Here the Laplacian is actually *three* equations, one for each of the three components of **A**, and we must make sure that the solutions have zero divergence.

There is a formal solution to (12.12) if we notice that the equation $\Delta A_i = -\mu_0 J_i$ (for the ith component of vectors **A** and **J**) has the same form as that for the electrostatic potential in free space, $\Delta V = -\varrho_v/\varepsilon_0$. This is Eq. (4.7), and an explicit integral expression for it is also given in Eq. (4.7). Thus, by pure analogy we infer that

$$\mathbf{A}_P = \frac{\mu_0}{4\pi}\int dv_Q \, \frac{\mathbf{J}_Q}{R_{QP}} \tag{12.13}$$

where all sources are included in the range of integration. It is shown in the optional paragraph below that this expression satisfies $\nabla \cdot \mathbf{A} = 0$, hence any $\mathbf{A}(\mathbf{r})$ satisfying (12.13) obeys *both* parts of (12.12).

To see why (12.13) implies zero divergence, consider

$$\begin{aligned}\nabla_P \cdot \frac{\mathbf{J}_Q}{R_{QP}} &= \mathbf{J}_Q \cdot \nabla_P\left(\frac{1}{R_{QP}}\right) = -\mathbf{J}_Q \cdot \nabla_Q\left(\frac{1}{R_{QP}}\right) \\ &= -\nabla_Q \cdot \left(\frac{\mathbf{J}_Q}{R_{QP}}\right) + \frac{1}{R_{QP}}(\nabla \cdot \mathbf{J})_Q \end{aligned} \tag{12.14}$$

We have used the vector identity 2 of Appendix A.4 here, but there is one further important point: $\nabla \cdot \mathbf{J} = 0$ everywhere, in the case of static fields, and therefore from (12.13) and Gauss's divergence theorem

$$\nabla_P \cdot \mathbf{A}_P = -\frac{\mu_0}{4\pi}\int dv_Q \, \nabla_Q \cdot \left(\frac{\mathbf{J}_Q}{R_{QP}}\right) = -\frac{\mu_0}{4\pi}\oint_{S_\infty} d\mathbf{S}_Q \cdot \left(\frac{\mathbf{J}_Q}{R_{QP}}\right) = 0 \tag{12.15}$$

The zero at the end of (12.15) is there because S_∞ is chosen to be beyond all the sources $\mathbf{J}$. That completes the demonstration that the solution (12.13) has zero divergence.

From the definition of the vector magnetic potential we see that the unit of $\mathbf{A}$ is webers per meter (Wb/m). The calculation of $\mathbf{A}$ by means of (12.13) is usually not easy, but of course (12.12) represents three Poisson equations which can be solved numerically by techniques similar to those explained in Chapter 9. The vector magnetic potential is often useful for calculating magnetic fields due to relatively simple configurations of wires. If the thickness of the wires can be neglected, then (12.13) may be replaced for a circuit C of current I by

$$\mathbf{A}_P = \frac{\mu_0 I}{4\pi}\oint_C d\mathbf{l}_Q \, \frac{1}{R_{QP}} \tag{12.16}$$

since we can set $\mathbf{J}\,dv = \mathbf{J}S\,dl = JS\,d\mathbf{l} = I\,d\mathbf{l}$, when the current is restricted to a thin wire. The cross section does not have to be constant because current density J will decrease if area S increases in such a fashion that the product $JS = I$ stays constant. It should be noted $d\mathbf{l}_Q$ is the *only* vector in the right-hand side of (12.16).

Example 12.2 Vector magnetic potential for an infinite uniform straight-wire current

Figure 11.5 can be used again to help calculate the **A** field when we have a current $\mathbf{I} = I\hat{\mathbf{z}}$ restricted to the z axis along an infinitesimally thin wire. Application of (12.16) yields $dl_Q\,\hat{\mathbf{z}}\,dz$, and $R_{QP} = \sqrt{z^2+\rho^2}$. This yields

$$\mathbf{A}_P = \frac{\mu_0 I\hat{\mathbf{z}}}{4\pi}\int_{-\infty}^{\infty} dz\frac{1}{\sqrt{z^2+\rho^2}} = \lim_{Z\to\infty}\frac{\mu_0 I\hat{\mathbf{z}}}{2\pi}\int_0^Z dz\frac{1}{\sqrt{z^2+\rho^2}} \tag{12.17}$$

We have used the fact that the integrand is even in z and have replaced the top bound by Z, which we will allow to go to infinity after we have done the integral. It is easy to check that the integral gives

$$\mathbf{A}_P = \frac{\mu_0 I\hat{\mathbf{z}}}{2\pi}\lim_{Z\to\infty}\left[\ln\left(\sqrt{z^2+\rho^2}+z\right)\right]_{z=0}^{z=Z} = \frac{\mu_0 I\hat{\mathbf{z}}}{2\pi}\left[\lim_{Z\to\infty}\ln(2Z) - \ln\rho\right] \tag{12.18}$$

The first term in the final form here is essentially an additive constant, even though it becomes infinitely large as $Z\to\infty$. This constant is unimportant because $\mathbf{B} = \nabla\times\mathbf{A}$ is what we ultimately wish to know, and the ∇ operation removes the additive constant. Therefore, we have found

$$\mathbf{A}_P = \left(-\frac{\mu_0 I}{2\pi}\ln\rho + \text{constant}\right)\hat{\mathbf{z}} \tag{12.19}$$

Observe that the vector $\mathbf{A}_P$ is in the same direction as **I** (or in the opposite direction). As $\nabla\cdot\mathbf{A} = 0$, we know that the **A** field lines must close upon themselves. From (12.19) they may *seem* not to, here, but remember: the current along the z axis must also close in some way, so that (12.19) only describes the situation near to a long stretch of straight current-carrying wire, which is actually part of a closed circuit.

Example 12.3 Vector magnetic potential due to an infinite uniform planar sheet current

We assume a uniform surface-current density $\mathbf{K} = K\hat{\mathbf{y}}$ on the $z = 0$ plane, and now we have instead of (12.13) or (12.16)

$$\mathbf{A}_P = \frac{\mu_0}{4\pi}\int dS_Q\frac{\mathbf{K}_Q}{R_{QP}} \tag{12.20}$$

From Fig. 12.4, we see that the shaded area element $dS_Q = \rho\,d\rho\,d\varphi$ and $R_{QP} = \sqrt{z^2+\rho^2}$. We then obtain

$$\mathbf{A}_P = \frac{\mu_0 K}{4\pi}\hat{\mathbf{y}}\lim_{R\to\infty}\left[\int_0^{2\pi} d\varphi\int_0^R d\rho\frac{\rho}{\sqrt{z^2+\rho^2}}\right] = \frac{\mu_0 K|z|}{4}\hat{\mathbf{y}}\lim_{R\to\infty}\int_0^{R^2/z^2} dt(1+t)^{-1/2} \tag{12.21}$$

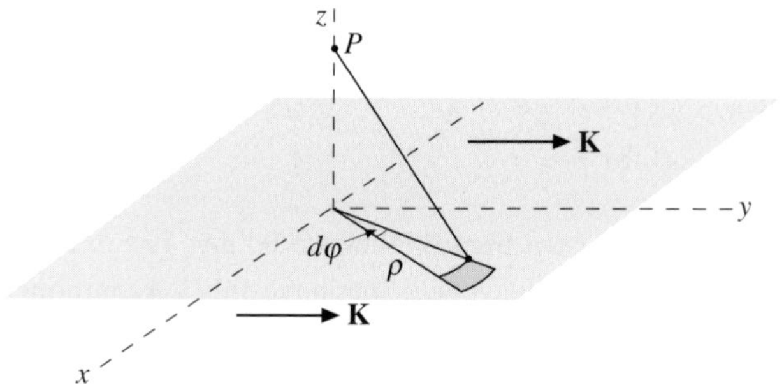

Fig. 12.4 Geometry for calculating V_{m} due to a uniform planar sheet current.

A change of variable $t = \rho^2/z^2$ has been made here. The integral is easily evaluated:

$$\mathbf{A}_P = \frac{\mu_0 K|z|}{2}\hat{\mathbf{y}} \lim_{R\to\infty}\left[(1+t)^{1/2}\right]_{t=0}^{t=R^2/z^2}$$

$$= \frac{\mu_0 K}{2}\hat{\mathbf{y}} \lim_{R\to\infty}(R-|z|) \quad \Rightarrow$$

$$\mathbf{A}_P(z) = -\frac{1}{2}\mu_0 K(|z| + \text{constant})\hat{\mathbf{y}} \tag{12.22}$$

The constant is an infinite length, but it *is* a constant, so that it plays no role in finding **B**. Again, **A** is in the same or opposite direction as the current (density K). If we ignore the constant, we see that **A** has opposite directions on either side of the current sheet, and that its magnitude actually *grows* with distance $|z|$ from the sheet.

Example 12.4 Vector magnetic potential due to an infinite solenoid

An infinite solenoid is an abstraction of the long but finite solenoid shown in Fig. 11.10. Let us consider it to be circular in cross section with radius a and to have current I in n windings per meter. In that case, the definition $\mathbf{B} = \nabla \times \mathbf{A}$, integrated over a cross-sectional surface S at radius ρ (and with application of Stokes's theorem) yields

$$\oint d\boldsymbol{l} \cdot \mathbf{A} = \int d\mathbf{S} \cdot \mathbf{B} \tag{12.23}$$

The symmetry suggests that the left-hand side is $2\pi\rho A_\varphi$, and the right-hand side follows from Example 11.8, where we deduced that well within a finite solenoid **B** is a constant vector, aligned along the axis, and that **B** drops to zero outside the solenoid. Thus $\int d\mathbf{S}\cdot\mathbf{B}$ equals $\mu_0\pi\rho^2 nI$ for $\rho < a$, and $\mu_0\pi a^2 nI$ for $\rho > a$.

Therefore we find

$$A(\mathbf{r}) = \tfrac{1}{2}\mu_0 nI\rho \qquad \text{for } \rho < a$$
$$A_\varphi(\mathbf{r}) = \tfrac{1}{2}\mu_0 nIa^2/\rho \qquad \text{for } \rho > a \tag{12.24}$$

Other components of **A** do not exist because they would give rise to **B** components that are not in the axial direction. Thus **A** field lines have only a $\hat{\boldsymbol{\varphi}}$ component and are therefore closed, as they should be according to $\nabla \cdot \mathbf{A} = 0$.

This raises an interesting philosophical argument: which is more 'real', the magnetic flux density **B** or the vector magnetic potential **A**? All we have said up to now seems to indicate that **B** is a real physical quantity, whereas **A** is a potential field without physical significance. An experiment done and discussed by Aharonov and Bohm may indicate the opposite; opinions are still somewhat divided. Their experiment concerned the quantum-mechanical effect on an optical interference pattern *outside* an infinite solenoid. There was a clear change between what one saw before and after a current was switched on in the windings around the solenoid. This is hard to understand in terms of **B**, as there is zero **B** outside the solenoid, but it seems easier to understand in terms of the nonzero **A** which merely falls off slowly to zero with distance ρ as $1/\rho$, and is distinct from zero at any finite distance.

12.3 Potential and field of an elementary magnetic dipole

The key to understanding the magnetism of materials lies in the infinitesimal magnetic dipole, just as our understanding of dielectrics was based upon the concept of an infinitesimal electric dipole. We shall see that a small current loop gives rise to the concept of a magnetic dipole, and therefore this section will deal with the potential and field that arises from current circulating around a small closed circuit.

A circular current loop lies around the origin in the $z = 0$ plane (Fig. 12.5). Its radius is a and it carries current I in the $\hat{\boldsymbol{\varphi}}$ direction. An arbitrary point P (which we take without loss of generality to lie in the $x = 0$ plane) is where we wish to observe the vector potential and the magnetic flux density **B**. The vector $\overrightarrow{OP}$ makes an angle θ with the z axis and an angle ψ with the line OQ. The point Q is an arbitrary location on the circuit defined by polar angle φ. We shall try to calculate

$$\mathbf{A}_P = \frac{\mu_0 I}{4\pi}\oint d\mathbf{l}_Q \frac{1}{R_{QP}} \tag{12.25}$$

where $d\mathbf{l}_Q$ lies on the current loop. The result of a calculation detailed below is (12.32).

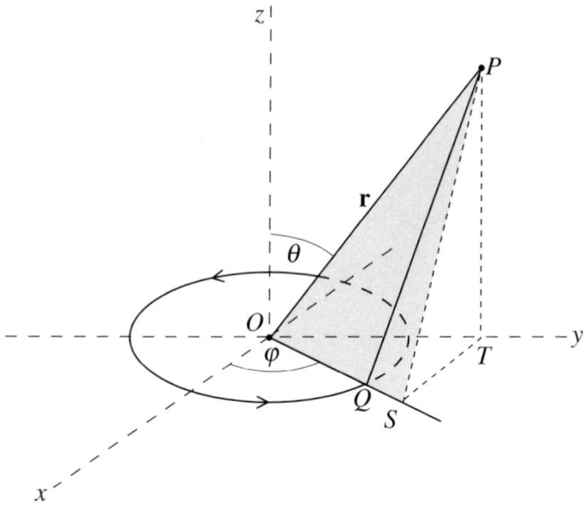

Fig. 12.5 A circular current loop as an elementary magnetic dipole.

It seems clear that $d\boldsymbol{l}_Q = a\,d\varphi\hat{\boldsymbol{\varphi}}$, but the asymmetry of point P (it lies off-axis) gives deviations from cylindrical symmetry so that we will do better to write $\hat{\boldsymbol{\varphi}} = -\hat{\mathbf{x}}\sin\varphi + \hat{\mathbf{y}}\cos\varphi$ (see (2.36a)). Hence

$$d\boldsymbol{l}_Q = a(-\hat{\mathbf{x}}\sin\varphi + \hat{\mathbf{y}}\cos\varphi)\,d\varphi \tag{12.26}$$

The real problem is that $R_{QP} = \sqrt{r^2 + a^2 - 2ra\cos\psi)}$, with ψ varying in a somewhat complicated fashion as point Q is rotated around the circular circuit. In fact, the calculation cannot be carried out any further without simplifying assumptions. Fortunately, the major interest in the behavior of the vector fields at P is for the case $r \gg a$. We may then apply (3.6) to $1/R_{QP}$ to obtain

$$\frac{1}{R_{QP}} = \frac{1}{r}\left(1 - \frac{2a}{r}\cos\psi + \frac{a^2}{r^2}\right)^{-1/2} = \frac{1}{r}\left[1 + \frac{a}{r}\cos\psi + O\left(\frac{a^2}{r^2}\right)\right] \tag{12.27}$$

In the rest of this calculation, we will drop terms of order a^2/r^2 compared to terms of order a/r. But first we need to make a connection between angle ψ and the angles θ, φ of the spherical coordinate system. We find, as explained below, that

$$\cos\psi = \sin\theta\sin\varphi \tag{12.28}$$

The geometrical calculation goes as follows. The line segment OS in Fig. 12.5 is the projection of line segment OP upon OQ. Therefore $OS = r\cos\psi$. But we can also obtain OS by first projecting OP on to the y axis (this gives $OT = r\sin\theta$ because OP lies entirely in the $x = 0$ plane), and then projecting the result, OT, onto OQ. This gives $OS = OT\sin\varphi$ and therefore also $OS = r\sin\psi\sin\varphi$.

Thus (12.28) is obtained. The result of (12.27) and (12.28) is

$$\frac{1}{R_{QP}} \approx \frac{1}{r}\left(1 + \frac{a}{r}\sin\theta\sin\varphi\right) \tag{12.29}$$

The next step is to return to the integral in (12.25) and to note that (12.26) and (12.29) combine to give

$$\begin{aligned}\oint d\mathbf{l}_Q \frac{1}{R_{QP}} &= a\int_0^{2\pi} d\varphi(-\hat{\mathbf{x}}\sin\varphi + \hat{\mathbf{y}}\cos\varphi)\frac{1}{r}\left(1 + \frac{a}{r}\sin\theta\sin\varphi\right)\\ &= \frac{a}{r}\int_0^{2\pi} d\varphi\left[-\hat{\mathbf{x}}\sin\varphi + \hat{\mathbf{y}}\cos\varphi + \frac{a}{r}\sin\theta(\hat{\mathbf{y}}\sin\varphi\cos\varphi - \hat{\mathbf{x}}\sin^2\varphi)\right]\end{aligned} \tag{12.30}$$

The first three terms integrate to zero, and the fourth term yields

$$\oint d\mathbf{l}_Q \frac{1}{R_{QP}} = -\left(\frac{a}{r}\right)^2 \hat{\mathbf{x}}\sin\theta \int_0^{2\pi} d\varphi \sin^2\varphi = -\pi\left(\frac{a}{r}\right)^2 \sin\theta\,\hat{\mathbf{x}} \tag{12.31}$$

Thus at a distant point P we find

$$\mathbf{A}_P = -\hat{\mathbf{x}}\frac{\mu_0 I}{4\pi r^2}\pi a^2 \sin\theta \tag{12.32}$$

At point P the vector $-\hat{\mathbf{x}}$ is just the unit vector $\hat{\boldsymbol{\varphi}} = \hat{\mathbf{z}} \times \hat{\mathbf{r}}$ for that location, given that $\mathbf{r}$ is the vector $\overrightarrow{OP}$. We therefore obtain

$$\mathbf{A}_P = \frac{\mu_0}{4\pi r^2}\mathbf{m} \times \hat{\mathbf{r}} \qquad \mathbf{m} = I\pi a^2 \hat{\mathbf{z}} \tag{12.33}$$

This expression is exact in the limits $a \to 0$, and $I \to \infty$ such that $I\pi a^2 = m$ is constant. The vector $\mathbf{m}$ is known as the *magnetic dipole moment*, and the expression for $\mathbf{A}_P$ reminds us of the parallel expression (4.15b) found for the electrostatic potential in terms of the electric dipole moment. In terms of spherical coordinates, (12.32) or (12.33) can be rewritten as

$$\mathbf{A}_P = \frac{\mu_0 m}{4\pi r^2}\sin\theta\,\hat{\boldsymbol{\varphi}} \tag{12.34}$$

so that with $\mathbf{B} = \nabla \times \mathbf{A}$, and since $\mathbf{A}$ has only a $\hat{\boldsymbol{\varphi}}$ component, we find

$$\begin{aligned} B_r &= \frac{\mu_0 m}{4\pi} \frac{1}{r \sin\theta} \frac{\partial}{\partial\theta}\left(\frac{\sin^2\theta}{r^2}\right) = \frac{\mu_0 m}{2\pi r^3}\cos\theta \\ B_\theta &= -\frac{\mu_0 m}{4\pi} \frac{1}{r} \frac{\partial}{\partial r}\left(r\frac{\sin\theta}{r^2}\right) = \frac{\mu_0 m}{4\pi r^3}\sin\theta \\ B_\varphi &= 0 \end{aligned} \tag{12.35}$$

This yields, when put together,

$$\mathbf{B} = \frac{\mu_0 m}{4\pi r^3}(2\hat{\mathbf{r}}\cos\theta + \hat{\boldsymbol{\theta}}\sin\theta) = \frac{\mu_0}{4\pi r^3}[3(\hat{\mathbf{r}}\cdot\mathbf{m})\hat{\mathbf{r}} - \mathbf{m}] \tag{12.36}$$

If we replace $\mu_0\mathbf{m}$ by $\mathbf{p}/\varepsilon_0$ we obtain (4.16b), the expressions for the electrostatic field due to an electric dipole. It thus appears that the infinitesimal loop with infinite current and magnetic moment $\mathbf{m}$ is the magnetic counterpart of $\mathbf{p}$: it is known as a *magnetic dipole*. The field lines for $\mathbf{B}$ due to a magnetic dipole are identical to those for $\mathbf{E}$ due to an electric dipole because the functional dependence of (12.36) upon the spherical coordinates is identical to that of (4.16).

We can expand the above argument to an infinitesimal loop of arbitrary shape without significant changes. In fact, we can even re-use Fig. 12.5, but we must then allow the *shape* of the circuit to be arbitrary, as shown in Fig. 12.6.

Referring again to Fig. 12.5, we set $\overrightarrow{OQ} = \boldsymbol{\rho}(\varphi)$ instead of $a\hat{\boldsymbol{\rho}}$, and the meaning of the angles ψ, θ, and φ is unchanged: whereas $\overrightarrow{OQ}$ was a vector with a constant magnitude for a circular loop, it now can vary in magnitude as well as direction

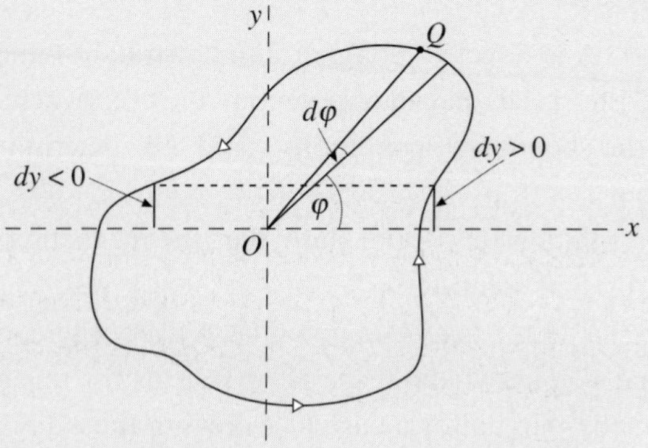

Fig. 12.6 Geometry for the general establishment of (12.32).

as Q moves around the loop. Instead of (12.27) we now have

$$\frac{1}{R_{QP}} = \frac{1}{r}\left[1 + \frac{\rho(\varphi)}{r}\cos\psi + O\left(\frac{\rho^2(\varphi)}{r^2}\right)\right] \tag{12.37}$$

and $\cos\psi = \sin\theta\sin\varphi$, as before. Because $\oint d\boldsymbol{l}_Q = 0$, it follows that the largest nonzero term in the integral replacing (12.30) is

$$\begin{aligned}\frac{1}{r^2}\oint d\boldsymbol{l}(\varphi)\,\rho(\varphi)\cos\psi &= \frac{1}{r^2}\oint[-\hat{\mathbf{x}}\,dx(\varphi) + \hat{\mathbf{y}}\,dy(\varphi)]\rho(\varphi)\sin\theta\sin\varphi \\ &= \frac{1}{r^2}\sin\theta\oint[-\hat{\mathbf{x}}y(\varphi)\,dx(\varphi) + \hat{\mathbf{y}}y(\varphi)\,dy(\varphi)]\end{aligned} \tag{12.38}$$

The $\hat{\mathbf{y}}$ contribution is zero here, because the loop always has an even number of values of $y(\varphi)$ in $0 < \varphi < \pi$ that are the same: half with $dy(\varphi) > 0$ and the other half with $dy(\varphi) < 0$ (and the same is true for $\pi < \varphi < 2\pi$). Note that

$$\oint dx(\varphi)\,y(\varphi) = \text{area inside loop} \equiv S \tag{12.39}$$

These facts reduce (12.38) to

$$\oint d\boldsymbol{l}_Q\frac{1}{R_{QP}} = -\hat{\mathbf{x}}\frac{S\sin\theta}{r^2} \tag{12.40}$$

which is the same as (12.31) because $S = \pi a^2$ for a circle with radius a. This reestablishes (12.33), but with $\mathbf{m} = IS\hat{\mathbf{z}}$.

Problems

12.1. A current $I = 3$ A is directed along an infinite straight infinitesimally thin wire.

(a) Calculate the scalar magnetic potential V_m everywhere, except on the wire. Specify the boundary conditions used to determine the constants of integration.

(b) Using the solution for $V_\mathrm{m}(\mathbf{r})$, show that the magnetic field $H(\mathbf{r})$ is predicted correctly by that solution.

(c) Calculate the vector magnetic potential $\mathbf{A}$ everywhere, except on the wire.

Hint: Use what you know about the $\mathbf{H}$ or $\mathbf{B}$ fields for this geometry.

12.1. An infinitesimally thin uniform current I flows in the $\hat{\mathbf{z}}$ direction along the z axis. It is given that $V_\mathrm{m}(Q) = 0$ at $Q = (1, 0, 0)$. Calculate the scalar magnetic potential $V_\mathrm{m}(P)$ at $P = (-1, 0, 0)$. Why is this problem incomplete?

12.3. A current of 15 A travels along an infinitesimally thin wire in the $+\hat{\mathbf{z}}$ direction. Figure 12.1 indicates paths I, II, by which we wish to calculate $V_m(P)$, given that $V_m(Q) = 1$ A; angle $QOP = 60°$ in all three diagrams. Calculate $V_m(P)$ for paths I and II.

12.4. A circularly cylindrical coaxial cable has inner radius a and outer radius b. The cable's axis coincides with the z axis. Calculate the scalar magnetic potential $V_m(\mathbf{r})$ in $a < \rho < b$, given that $V_m = 0$ at $y = 0$ and that a current I flows uniformly in the $\hat{\mathbf{z}}$ direction in the core.

12.5. The magnetic flux Φ through a surface with perimeter C is defined as the surface integral $\Phi \equiv \int d\mathbf{S} \cdot \mathbf{B}$ over any surface with C as its bounding closed contour. Show that Φ is a function only of the vector magnetic potential $\mathbf{A}$ on contour C.

12.6. Why cannot the scalar function $V(r) = a \ln r + \text{constant}$ (in spherical coordinates) be a scalar magnetic potential?

12.7. Use the differential relationships $\nabla \times \mathbf{A} = \mathbf{B}$ and $\nabla \cdot \mathbf{A} = 0$ to prove Eq. (12.19) for a current $\mathbf{I} = I\hat{\mathbf{z}}$ on the z axis.

12.8. Use the differential relationships $\nabla \times \mathbf{A} = \mathbf{B}$ and $\nabla \cdot \mathbf{A} = 0$ to prove Eq. (12.22) for a sheet current with density $\mathbf{K} = K\hat{\mathbf{y}}$ on the $z = 0$ plane.

12.9. Two sheet currents are specified, one with density $\mathbf{K}_1 = K\hat{\mathbf{y}}$ on $z = a$ and the other with density $\mathbf{K}_2 = -K\hat{\mathbf{y}}$ on $z = -a$. Find the vector potential $\mathbf{A}$ in $-a < z < a$.

12.10. An elliptical circuit in the $z = 0$ plane has half-major axis a and half-minor axis b. It carries a current $\mathbf{I} = -I\varphi$. Find the magnetic dipole moment.

12.11. Given an infinitesimal magnetic dipole $\mathbf{m} = m\hat{\mathbf{z}}$, at what locations in space is the $\mathbf{B}$ field perpendicular to $\hat{\mathbf{z}}$?

13

Inductance and magnetic stored energy

13.1 Mutual and self inductance

The concept of inductance will be coupled later to that of magnetic stored energy, because its role is very similar to that of capacitance in relation to electrostatic stored energy. To recapitulate, we found in electrostatics that the stored energy of a capacitor maintained at voltage V could be written as $W = \frac{1}{2}CV^2 = \frac{1}{2}Q^2/C$. This clarified that capacitance C is a measure of the energy that can be stored electrostatically. We will find, similarly, that *inductance* L is a measure of the energy that can be stored magnetostatically, so before discussing magnetostatic stored energy we will first introduce the *concept* of inductance. Since we need to use the magnetic flux (see the text after equations (11.32)) in this chapter, we work in terms of the magnetic flux density, **B**.

Consider the simple situation depicted in Fig. 13.1, where two coils or circuits of current, one with $\mathcal{N}_1$ windings of current I_1 and the other with $\mathcal{N}_2$ windings of current I_2, influence each other. The figure shows a number of the **B** field lines produced by coil C_1. Some of these field lines go through coil C_2, others do not. We are interested in those field lines that do. Such field lines are said to *link circuit* C_2 to *circuit* C_1. We speak of the *flux linkage* Λ_{12} of C_1 to C_2, and it is defined by

$$\Lambda_{12} \equiv \mathcal{N}_2\Phi_{12} = \mathcal{N}_2 \int_{S_2} d\mathbf{S} \cdot \mathbf{B}_1(\mathbf{r}) \tag{13.1}$$

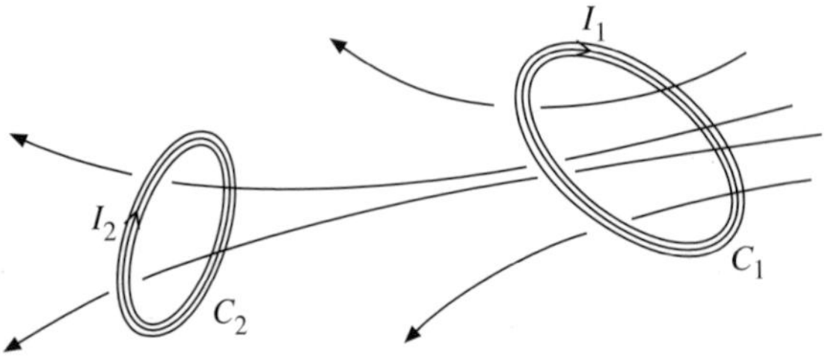

Fig. 13.1 One multiwound coil of current linking **B** fields produced by another coil. C_1 and C_2 carry $\mathcal{N}_1$ and $\mathcal{N}_2$ turns respectively.

where S_2 is any surface bounded by C_2 and Φ_{12} is the magnetic flux of the $\mathbf{B}_1$ field produced by coil C_1 that is captured by coil C_2. The flux linkage differs only by the number of windings from the magnetic flux. We could also look at

$$\Lambda_{11} \equiv \mathcal{N}_1 \Phi_{11} = \mathcal{N}_1 \int_{S_1} d\mathbf{S} \cdot \mathbf{B}_1(\mathbf{r}) \tag{13.2}$$

where S_1 is any surface bounded by C_1. We call Λ_{11} the *flux linkage of* C_1 *to itself.* The meaning of Λ_{22} is the same with respect to coil C_2.

The concept of *linkage* needs further explanation. Both current and flux lines are closed. Referring to Fig. 13.1, we see that all the flux lines circle around current loop C_1. Any change in the flux lines owing to a change in I_1 will cause an *induced electromotive force* (emf) around C_1, by Faraday's law; this will be in a direction such as to oppose the change producing it, by Lenz's law. The phenomenon is known as *self induction*; see Chapter 15 for a full discussion.

The two inner flux lines encircle both C_1 and C_2. The loops are therefore *linked* by these flux lines, and so a change in current in one will cause an induced emf in the other, again by Faraday's law. Here we have *mutual induction.*

To make quantitative the phenomenon of mutual induction we need to relate the flux linkage Λ_{12} to the current I_1 (or vice versa). In order to do this we substitute the vector magnetic potential (12.13) into (13.1), using $d\boldsymbol{l}_1$ for the first loop and $d\boldsymbol{l}_2$ for the second:

$$\Lambda_{12} = \mathcal{N}_2 \int_{S_2} d\mathbf{S} \cdot (\nabla \times \mathbf{A}) = \mathcal{N}_2 \oint_{C_2} d\boldsymbol{l}_2 \cdot \mathbf{A} = \frac{\mathcal{N}_2 \mathcal{N}_1 I_1}{4\pi} \oint_{C_2} d\boldsymbol{l}_2 \cdot \oint_{C_1} d\boldsymbol{l}_1 \frac{1}{R_{12}} \tag{13.3}$$

Here we have applied (12.16) to each of the $\mathcal{N}_1$ turns of current around C_1 to get to the last expression. Clearly the flux linkage Λ_{12} is proportional to the current I_1 in coil C_1. Thus, we have

$$\Lambda_{12} = \mathcal{L}_{12} I_1 \tag{13.4a}$$

where

$$\mathcal{L}_{12} = \frac{\mathcal{N}_2 \mathcal{N}_1}{4\pi} \oint_{C_2} d\boldsymbol{l} \cdot \oint_{C_1} d\boldsymbol{l} \frac{1}{R_{12}} \quad \text{(the Neumann formula)} \tag{13.4b}$$

The quantity $\mathcal{L}_{12}$ in (13.4a) is known as the *mutual inductance*; it is the coefficient giving the flux linkage Λ_{12} through coil C_2 due to a current I_1 in C_1. It is clear from (13.4b) that if we interchange indices we obtain

$$\mathcal{L}_{21} = \mathcal{L}_{12} \tag{13.5}$$

That is, the mutual inductance of coil C_1 with respect to fields produced by C_2 is the same as that of coil C_2 with respect to fields produced by C_1. Likewise, substituting for $\mathbf{B}_1(\mathbf{r})$ in (13.2) we obtain $\mathcal{L}_{11} \equiv \Lambda_{11}/I_1$ where $\mathcal{L}_{11}$ is the *self inductance* of coil C_1. Similarly, $\mathcal{L}_{22} = \Lambda_{22} I_2$.

It may seem that the above considerations are quite general, but in fact this is not so: there are circuits for which the flux linkage is a nonlinear function of the current I. For such nonlinear circuits, it is true only that the change in flux linkage $d\Lambda$ is linear in dI, for infinitesimal dI, and thus an expanded definition might be

$$\mathcal{L}(I) \equiv \frac{d\Lambda}{dI} \tag{13.6}$$

for both self and mutual inductance. The inductances will now depend upon the magnitude of the current because the derivative of Λ can vary with I. Clearly the previous definitions follow for linear circuits because then $d\Lambda/dI = \Lambda/I$.

Example 13.1 Self inductance of a long solenoid

We have seen in Section 11.3 that an infinite solenoid has a uniform interior field $\mathbf{B} = \mu_0\, nI\hat{\mathbf{z}}$ if it has n windings per meter with current I through the windings. From (13.2), $\Lambda_{11} = nzSB$, if z and S are the length and cross-sectional area of the solenoid. Thus, substituting for B in Λ_{11}, we obtain for the self inductance $\mathcal{L}_{11}$

$$\mathcal{L}_{11} = \frac{\Lambda_{11}}{I_1} = \mu_0 n^2 Sz$$

The self inductance per meter is then given by

$$\frac{d\mathcal{L}_{11}}{dz} = \mu_0 n^2 S \tag{13.7}$$

Note that one factor n is due to the $\mathbf{B}$ field, and the other occurs in the flux linkage. It should be borne in mind that the uniformity of the field, assumed here, is an idealization that does not hold near the ends, as we saw in Section 11.4.

Example 13.2 Mutual inductance of coaxial long and short solenoids with identical radius

In Fig. 13.2 we show two sets of windings around a long cylinder, of cross section S and length l_1, with n_1 windings per meter; a relatively short section in the middle also has a second set of windings of length l_2 with n_2 windings per meter.

Let us first calculate $\mathcal{L}_{12}$, the mutual inductance. We will consider the flux lines produced by the long set of windings, captured by the short set. As in the above example, we know that $B_1 = \mu_0\, n_1 I_1$ so, from (13.1), remembering that the total

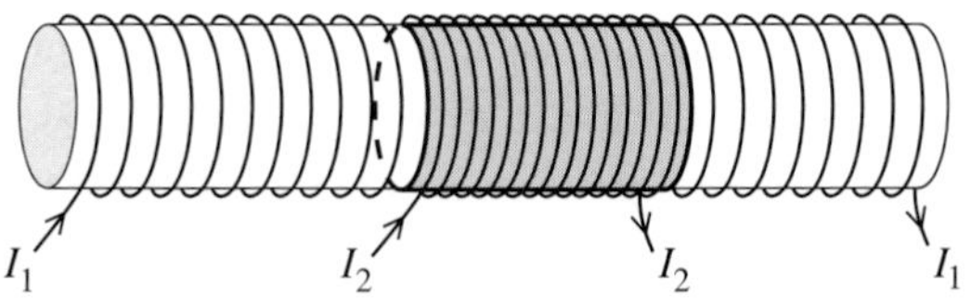

Fig. 13.2 Coaxial short and long solenoids. The longer coil has $\mathcal{N}_1$ turns, current I_1 and length l_1; the shorter coil has $\mathcal{N}_2$ turns, current I_2, and length l_2.

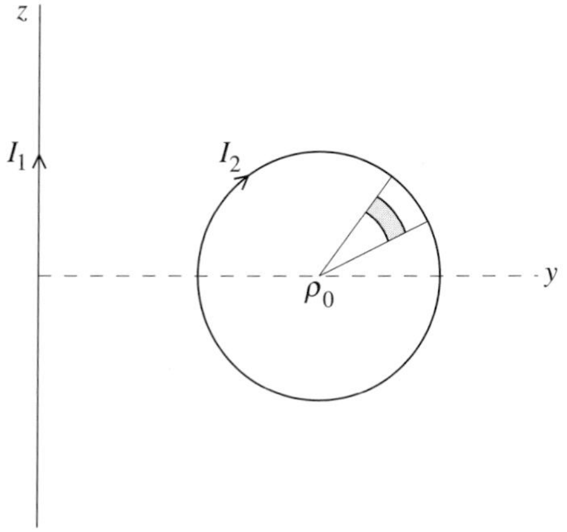

Fig. 13.3 Straight-line wire and circular loop of Example 13.3.

number of turns on the second coil, $\mathcal{N}_2$, is given by $n_2 l_2$, it follows that $\Lambda_{12} = n_2 l_2 B_1 S = \mu_0 n_2 n_1 l_2 S I_1$; therefore, from (13.4a),

$$\mathcal{L}_{12} = \mu_0 n_1 n_2 l_2 S = \mathcal{L}_{21} \tag{13.8}$$

The last equation follows by (13.5). If instead we considered the flux through the long coil due to the short coil then we would have to remember that away from the short coil the field of the latter quickly drops to zero, so that only along its length l_2 does its field $\mu_0 n_2 I_2$ link with the long coil. Thus only $l_2 n_1$ turns of the long coil are linked by this field and so $\Lambda_{21} = l_2 n_1 \mu_0 n_2 I_2 S$, giving again the result (13.8).

Example 13.3 Mutual inductance of a straight-line wire and a circular loop

A circular loop with radius a is situated in the $x = 0$ plane at distance ρ_0 from a wire along the z axis, which in turn represents one leg of a gigantic second circuit (with remaining sides essentially infinitely far from the loop). The situation is shown in Fig. 13.3. In the calculation of the mutual inductance we will consider the flux through the loop due to the straight-line wire, which is easily described as $\mathbf{B}_1(\rho, \varphi, z) = (\mu_0 I_1/2\pi\rho)\hat{\boldsymbol{\varphi}}$ at any point in space, and hence anywhere inside the loop.

In the plane of the loop we have $\hat{\boldsymbol{\varphi}} = -\hat{\mathbf{x}}$, as is easily inferred from the right-handed screw-movement rule in Ampère's law. Let us define polar coordinates, length l and angle α with respect to the center of the loop. The shaded area element in the loop between l and $l + dl$ and α and $\alpha + d\alpha$ receives the same field $\mathbf{B}$ at every point in it. Hence its contribution to the flux linkage is, from (13.1),

$$d\Lambda_{12}(l,\alpha) = \mathbf{B}(l,\alpha)\cdot d\mathbf{S}(l,\alpha) = \frac{\mu_0 I_1}{2\pi(\rho_0 + l\cos\alpha)}\hat{\mathbf{x}}\cdot(l\,d\alpha\,dl\,)\hat{\mathbf{x}}$$
$$= \frac{\mu_0 I_1 l}{2\pi(\rho_0 + l\cos\alpha)}dl\,d\alpha \tag{13.9}$$

Note that the $d\mathbf{S}$ direction has been chosen to be $-\hat{\mathbf{x}}$; the flux linkage is chosen to be a positive scalar quantity here so that the inductance also will emerge as a positive quantity. Thus we obtain

$$\Lambda_{12} = \frac{\mu_0 I_1}{2\pi}\int_0^a dl \int_0^{2\pi} d\alpha\, \frac{l}{\rho_0 + l\cos\alpha} = \frac{\mu_0 I_1}{2\pi}\int_0^a dl\, \frac{2\pi l}{\sqrt{\rho_0^2 - l^2}}$$
$$= \mu_0 I_1\left(\rho_0 - \sqrt{\rho_0^2 - a^2}\right) \tag{13.10}$$

Here, the $d\alpha$ integral was obtained from standard integral tables, and the remaining integral is solved by setting $l^2 = t$ as a dummy variable. The integral of $1/\sqrt{\rho_0^2 + t}$ is $-2\sqrt{\rho_0^2 - t}$, which gives rise to the final form in (13.10).

13.2 Inductance due to internal and external flux linkage

So far, we have dealt with cases in which the current windings have essentially zero radial extent with respect to the cross-sectional surface determining the flux linkage of those sources. This is not always the case. Consider again the situation of Fig. 13.1, and specifically the $\mathbf{B}$ line at the top right that encircles C_1. It contributes to Φ_{11} and Λ_{11} because it links entirely the currents on C_1. If one extra winding is added to C_1, then Φ_{11} will still be given by $\int d\mathbf{S}\cdot\mathbf{B}$, but $\mathbf{B}$ will have increased in magnitude by $(\mathcal{N}+1)/\mathcal{N}$, and Λ_{11} will be given now by $(\mathcal{N}+1)\int d\mathbf{S}\cdot\mathbf{B}$; this last expression also has the slightly larger $\mathbf{B}$ field in the integrand. So not only is the magnetic flux density larger; the *flux linkage* Λ_{11} has increased by a factor $[(\mathcal{N}+1)/\mathcal{N}]^2$. This brings us right to the situation of 'wide' current distributions: see Fig. 13.4, in which a flat coil C of current with ρ values between a and b is shown. The coil has zero thickness in the z direction but a nonnegligible width $b - a$ in any $\hat{\boldsymbol{\rho}}(\varphi)$ direction. One might consider this current source to consist of very many windings of very thin wire with current I wound in a spiral around the z axis, starting at $\rho = a$ and finishing at $\rho = b$.

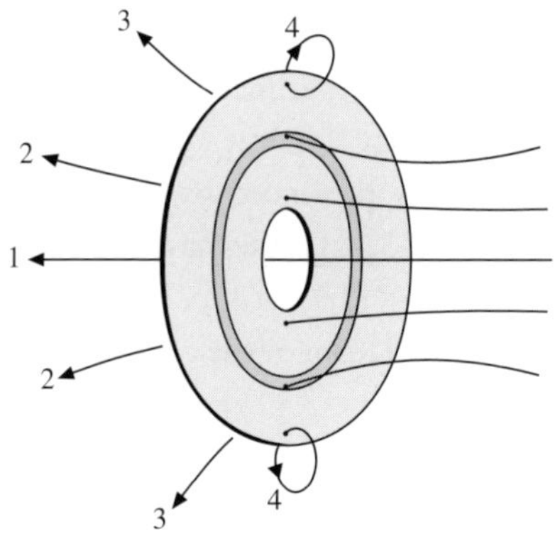

Fig. 13.4 Illustration of internal-inductance mechanism with spiral-wound circular coil.

Consider now the flux linkage of the current windings of the coil C with respect to its own **B** lines. Only one **B** line of those shown fully links C: line 1. The other **B** lines link only some of the windings. Note that an annulus of width $d\rho$ at ρ has been shown. It is fully linked by lines 2 and 3 (just), whereas the lines numbered 4 do not link it at all. The flux linkage $d\Lambda(\rho)$ *of only this annulus at* ρ is given by the following recipe, which is inspired by the paragraph above and which we will justify in subsection 13.4.1:

$$d\Lambda(\rho) = \frac{I(\rho)}{I(0)}\mathbf{B}(\rho)\cdot d\mathbf{S}(\rho) \qquad d\mathbf{S}(\rho) = 2\pi\rho\, d\rho \tag{13.11a}$$

Here, $I(\rho)$ is the current I times the number of windings linked by the **B** lines that go through the annulus: $I(\rho) = \mathcal{n}(b-\rho)I$ if there are $\mathcal{n}$ windings of current I per radial meter. The denominator, $I(0)$, represents the current times the total number of windings, so that $I(\rho)/I(0)$ is the *fraction of the total* number of windings of the coil linked by the field lines at the radius ρ. The *total* flux linkage must be

$$\Lambda = \int_{\rho=a}^{\rho=b} d\Lambda(\rho) \tag{13.11b}$$

In words, after we find the flux linkage of an infinitesimal annulus, we must add up all the annuli between a and b. As a result, (13.11a) becomes

$$\begin{aligned} d\Lambda(\rho) &= \frac{\mathcal{n}(b-\rho)I}{\mathcal{n}(b-a)I}\mathbf{B}(\rho)\cdot d\mathbf{S}(\rho) \\ &= \mathcal{n}(b-\rho)\left[\frac{\mathbf{B}(\rho)}{\mathcal{n}(b-a)}\cdot d\mathbf{S}(\rho)\right] \\ &= \mathcal{N}(\rho)\left[\frac{\mathbf{B}(\rho)}{\mathcal{n}}\cdot d\mathbf{S}(\rho)\right] \end{aligned} \tag{13.12}$$

where $\mathcal{N}(\rho)$ is the number of windings between ρ and b, and $\mathcal{N}$ is the total number of windings. To interpret (13.12), consider a thought experiment in which we shrink the thickness of the loop so that $b - a$ is infinitesimally thin but $\mathcal{N}$ is unchanged. In this case $\mathbf{B}(\rho)/\mathcal{N}$ will be the magnetic flux density produced by a single winding at $\rho = a = b$, and the square-bracketed factor in (13.12) is the magnetic flux increment $d\Phi(\rho)$, in agreement with the earlier definition (13.1) of flux linkage. It therefore appears that (13.11a, b) can be a useful generalization of the original definition in a number of situations. However a note of caution: (13.11a) is exact only for highly symmetric situations. One such situation is when the **B** field lines are perfect circles. A more precise justification of (13.11) is given in Section 13.5. It is not straightforward to extend (13.11) to less symmetric situations, or to current distributions in three dimensions, but there are other methods to obtain the inductance, e.g. from the stored magnetostatic energy, as we have yet to see.

The part of the inductance calculated in this way via (13.11) is due to those **B** lines that actually intersect the current distribution, and is known as the *internal inductance*. That part of the inductance due to **B** lines that totally enclose a current distribution is known as the *external inductance*. The following example will clarify this.

Example 13.4 Internal and external inductance of a coaxial cable

With the above introduction, we consider the internal and external inductance of a coaxial cable, shown in Fig. 13.5, that has an inner core of metal in $0 < \rho < a$ in which current I goes in the $+\hat{\mathbf{z}}$ direction and an outer sheath of metal in $b < \rho < c$ where current I returns in the $-\hat{\mathbf{z}}$ direction. We assume a uniform current density $\mathbf{J}_1 = (I/\pi a^2)\hat{\mathbf{z}}$ in the core and a corresponding uniform density $\mathbf{J}_3 = -[I/\pi(c^2 - b^2)]\hat{\mathbf{z}}$ in the sheath. These densities are found by division of $\pm I$ by the

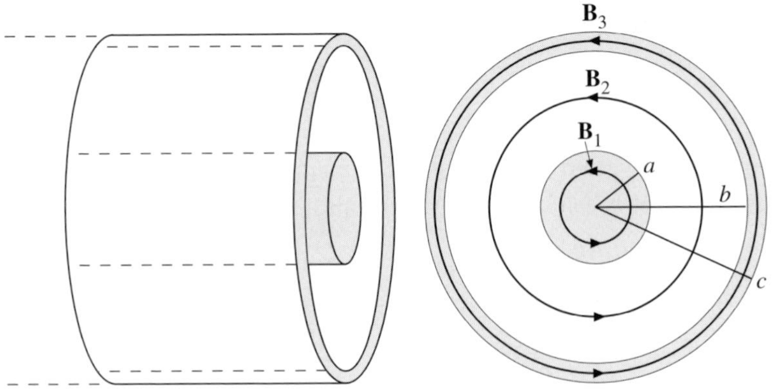

Fig. 13.5 Coaxial-cable situation for Example 13.4. The **B** lines 1, 2, 3 lie at three different radii.

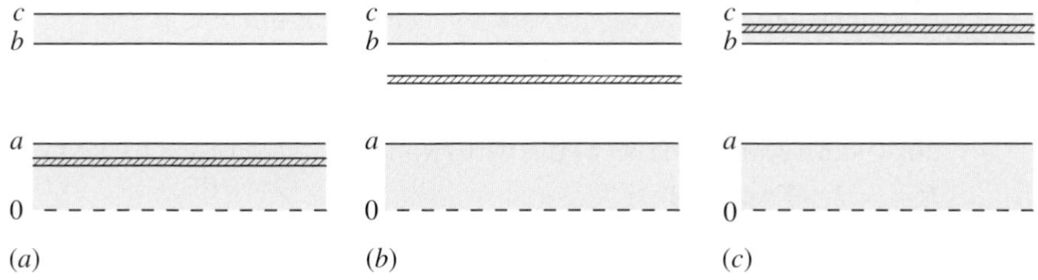

Fig. 13.6 Views of the coaxial cable in a plane through the axis. In three dimensions the hatched area dS corresponds to a cylindrical shell of thickness $d\rho$. (*a*) dS in core, (*b*) dS between core and sheath, (*c*) dS in sheath.

cross-sectional areas of core and sheath, respectively. The right-hand figure shows three **B** lines: the first, numbered 1, contributes to the internal inductance of the core because it lies within the current in the core. The second, numbered 2, is external to the core and completely links the current I of the core. The third, numbered 3, is external to the core but internal to the sheath, as it lies within the current $-I$ of the sheath. The only **B** lines that are completely external to the entire coaxial cable are those with radius $\rho > c$ (not shown in the sketch).

We now apply (13.11a, b) to the three regions of the coaxial cable, to find the total flux linkage. We show in Fig. 13.6 a cross section of the cable, of length l, in a plane containing the z axis. The figure is shown thrice, each time with the small hatched area $dS = l\,d\rho$ in a different region: our aim is to find the flux through dS and thereby its flux linkage, using (13.11a).

(a) Contribution of the core, $\rho < a$

The cylindrical symmetry allows us to use Ampère' law $\oint d\mathbf{l} \cdot \mathbf{B} = \mu_0 I_{\text{encirc}}$ for a circle around the z axis, and it tells us that in the core $2\pi\rho B_\varphi(\rho) = \mu_0 \pi \rho^2 J_1 = \mu_0 I \rho^2/a^2$, so that $B_\varphi(\rho) = \mu_0 I \rho / 2\pi a^2$. The flux lines are perpendicular to the hatched area $dS = l\,d\rho$ shown in Fig. 13.6, so that the magnetic flux through this area is just the scalar product $B_\varphi(\rho) dS$. The linked current must be $I(\rho) = \pi\rho^2 J_1 = I\rho^2/a^2$. Thus (13.11a) predicts

$$d\Lambda_1(\rho) = \left(\frac{\rho^2}{a^2}\right)\left(\frac{\mu_0 I \rho}{2\pi a^2}\right)(l\,d\rho) = \frac{\mu_0 I \rho^3}{2\pi a^4}\, l\,d\rho \tag{13.13a}$$

We now apply (13.11b) between $\rho = 0$ and $\rho = a$ to obtain for the flux linkage of the core region

$$\Lambda_1 = \int_0^a d\Lambda_1(\rho) = \frac{\mu_0 Il}{2\pi a^4}\int_0^a d\rho\,\rho^3 = \frac{\mu_0 Il}{8\pi} \tag{13.13b}$$

(b) Contribution of the region between core and sheath, $a < \rho < b$

Here, Ampère's law gives $2\pi\rho B_\varphi(\rho) = \mu_0 I$ because the entire core current is encircled by any **B** line in this region. Thus, $B_\varphi(\rho) = \mu_0 I/2\pi\rho$, and the magnetic flux is still a scalar product of this with $l\,d\rho$. The linked current in (13.11a) here is just $I(\rho) = I$, so we have

$$d\Lambda_2(\rho) = \frac{\mu_0 I}{2\pi\rho} l\,d\rho \tag{13.14a}$$

and thus obtain for the flux linkage of the region between the conductors

$$\Lambda_2 = \int_a^b d\Lambda(\rho) = \frac{\mu_0 Il}{2\pi} \int_a^b d\rho\,\frac{1}{\rho} = \frac{\mu_0 Il}{2\pi} \ln\left(\frac{b}{a}\right) \tag{13.14b}$$

(c) Contribution of the sheath, $b < \rho < c$

The only extra difficulty here is that the outermost circular **B** line, shown in Fig. 13.5, encloses the entire core current I but only a part of the sheath current (with opposite sign). Thus

$$I(\rho) = I - \pi(\rho^2 - b^2)J_3 = I\,[1 - (\rho^2 - b^2)/(c^2 - b^2)]$$

So we have

$$B_\varphi(\rho) = \frac{\mu_0 I}{2\pi\rho}\left[1 - \frac{(\rho^2 - b^2)}{(c^2 - b^2)}\right] = \frac{\mu_0 I}{2\pi\rho}\left(\frac{c^2 - \rho^2}{c^2 - b^2}\right)$$

Likewise, we can simplify the above expression for the linked current $I(\rho)$ to

$$I(\rho) = I\left(\frac{c^2 - \rho^2}{c^2 - b^2}\right)$$

and therefore (13.11a) predicts that

$$d\Lambda_3(\rho) = \frac{\mu_0 I}{2\pi\rho}\left(\frac{c^2 - \rho^2}{c^2 - b^2}\right)^2 l\,d\rho \qquad \Lambda_3 = \int_b^c d\Lambda_3(\rho) \tag{13.15a}$$

The calculation requires only the evaluation of the integral

$$\int_b^c d\rho\,\frac{(c^2 - \rho^2)^2}{\rho} = \int_b^c d\rho\left(\rho^3 - 2c^2\rho + \frac{c^4}{\rho}\right)$$

$$= \frac{(c^4 - b^4)}{4} - c^2(c^2 - b^2) + c^4 \ln\left(\frac{c}{b}\right)$$

so that the flux linkage of the sheath region is given by

$$\Lambda_3 = \frac{\mu_0 I l}{2\pi}\left[\frac{1}{4}\left(\frac{c^2+b^2}{c^2-b^2}\right) - \frac{c^2}{c^2-b^2} + \frac{c^4}{(c^2-b^2)^2}\ln\left(\frac{c}{b}\right)\right] \tag{13.15b}$$

The total flux linkage is given by the sum $\Lambda_1 + \Lambda_2 + \Lambda_3$, and the inductance is then found upon dividing by I. The result is

$$\mathcal{L} = \frac{\mu_0 l}{2\pi}\left[\frac{1}{4} + \ln\left(\frac{b}{a}\right) + \frac{1}{4}\left(\frac{c^2+b^2}{c^2-b^2}\right) - \frac{c^2}{c^2-b^2} + \frac{c^4}{(c^2-b^2)^2}\ln\left(\frac{c}{b}\right)\right] \tag{13.16}$$

The first term in (13.16) is often known as the inductance internal to the core. The second term is the inductance external to the core if the effect of the sheath (the remaining three terms) is ignored. Obviously, the inductance is proportional to the length of the cable, under the assumption that end effects are ignored. The **B** fields are quite different at the ends, so that (13.16) does require a reasonably long cable in order to be accurate (long being defined with respect to the diameter). If the sheath is very thin ($c - b \ll b$), then the sum of the last three terms is negligible, and the inductance per unit length is

$$\frac{d\mathcal{L}}{dl} \approx \frac{\mu_0}{2\pi}\left[\frac{1}{4} + \ln\left(\frac{b}{a}\right)\right] \tag{13.17}$$

The error in using (13.17) is of order $(c - b)/3b$ compared to 0.25, and therefore (13.17) is often a good approximation for a thinly sheathed coaxial cable.

Example 13.5 Internal and external inductance of a two-wire transmission line

The calculation of the inductance of a two-wire transmission line, of which a cross section is shown in Fig. 13.7(a), is based largely upon the calculation done in the previous example. Here, though, we have two wires, each with radius a separated by an axis-to-axis distance d, with equal currents I in opposite directions. We will consider a length l of this transmission line. The magnetic field at any point on the x axis between a and $d - a$ is given by a superposition of the field of each wire separately (calculated with Ampère's law):

$$\mathbf{B}(x) = \pm\left[\frac{\mu_0 I}{2\pi x} + \frac{\mu_0 I}{2\pi(d-x)}\right]\hat{\mathbf{x}} \tag{13.18}$$

The sign in (13.18) depends upon which wire has the current flowing in the $+\hat{\mathbf{z}}$ direction; we shall ignore the sign anyway for inductance calculations.

The magnetic field on the x axis outside this 0 to d interval vanishes as the distance to the wires increases, e.g. for $x > d + a$ and $x < -a$ it is

$$\mathbf{B}(x) = \pm\left[\frac{\mu_0 I}{2\pi x} - \frac{\mu_0 I}{2\pi(x-d)}\right]\hat{\mathbf{x}} \tag{13.19}$$

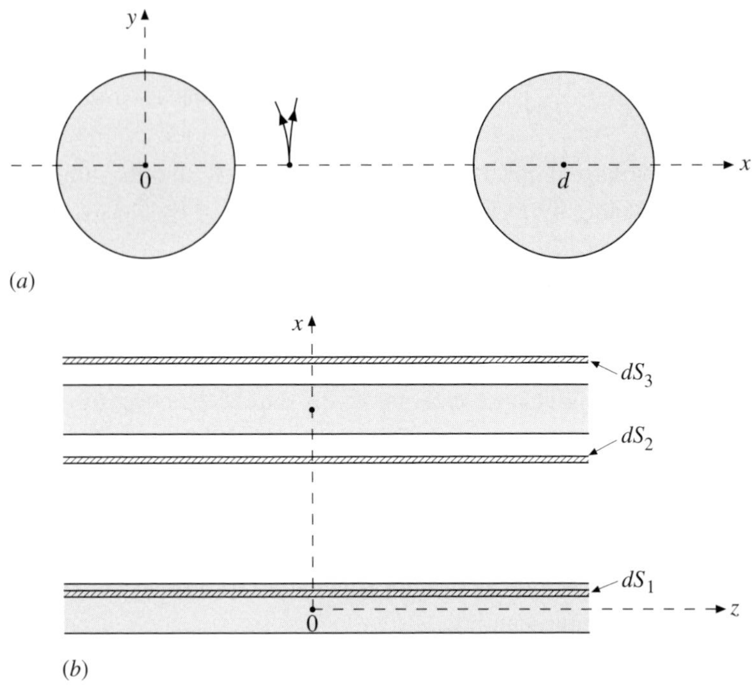

Fig. 13.7 Situation for a two-wire transmission line.

The flux linkage is again determined by considering the flux through a thin strip perpendicular to the **B** lines. Such a strip has area $l\,dx$ ($l \to \infty$) and lies in a plane containing the z and x axes. Three possible locations of such an area are shown in Fig. 13.7(*b*) as hatched strips dS_1, dS_2, dS_3.

The calculation involving areas of type dS_1 is complicated because the **B** field from the distant wire at $x = d$ has a complicated intersection with the current of the wire, and thus $I(x)$ is difficult to calculate. But if the distance $d \gg a$, then (13.18) shows that we may neglect the **B** field in dS_1 due to the distant wire, and then we have only the internal linkage for this wire, which is the same as that obtained in (13.13b). However, we get this contribution *twice* now because there is also a similar dS (not sketched) in $d - a < x < d$. Hence the internal inductance is approximately

$$\mathcal{L}_{\text{int}} \approx 2 \times \frac{\mu_0 l}{8\pi} = \frac{\mu_0 l}{4\pi} \tag{13.20a}$$

The area elements dS_2 between $x = a$ and $x = d - a$ give a contribution obtained by integrating (13.18):

$$\mathcal{L}_{\text{ext}} = \frac{\mu_0 l}{2\pi} \int_a^{d-a} dx \left(\frac{1}{x} + \frac{1}{d - x} \right) = \frac{\mu_0 l}{\pi} \ln\left(\frac{d - a}{a} \right) \approx \frac{\mu_0 l}{\pi} \ln\left(\frac{d}{a} \right) \tag{13.20b}$$

Finally, there are dS strips that lie beyond $x = d + a$ or $x = -a$. Such strips *either* have one $\mathbf{B}$ contribution that already has been considered as an external-inductance contribution, of dS_2 type, because the $\mathbf{B}$ line encircles one wire and intersects the x axis between a and $d - a$ and the other $\mathbf{B}$ contribution is negligibly small (the denominator is larger than $2\pi d$) *or* have negligible contributions from both $\mathbf{B}$ fields. Hence we obtain an *approximate* expression for the total inductance:

$$\mathcal{L} = \frac{\mu_0 l}{\pi}\left[\frac{1}{4} + \ln\left(\frac{d}{a}\right) + O\left(\frac{d}{a}\right)^2\right] \tag{13.21}$$

A more accurate solution is given below in problem 13.5. It follows from a transmission-line equivalence $LC = \mu\varepsilon$ (see Chapter 20) and the calculation in Example 8.6.

13.3 Magnetostatic stored energy

13.3.1 Expression in terms of currents

The calculation of the energy stored in the current circuits that produce the magnetostatic field is somewhat more involved than the calculations we did in Chapter 7 for the electrostatic energy stored in a charge distribution. We now need to look at time changes in currents. As we shall see in Chapter 15, these time changes give rise to electric fields, according to the full electromagnetic theory, and hence to potential differences, which in turn represent energies per unit charge as we have seen. The full electromagnetic theory teaches us that *wave phenomena* occur whenever there is an alternating current. Wave phenomena imply that this current is also a function of position $i(\mathbf{r}, t)$; it consist of superpositions of spatial sinusoids $i_n(\mathbf{r}, t)$, each of which has its own characteristic wavelength λ_n (the spatial length of one full cycle). For the nth sinusoid we have an 'electromotive force', with the dimensions of potential, that is given by the *inductive circuit-element law* $V_n = \mathcal{L}_n di_n/dt$. This electromotive force is essentially a potential difference. The addition of the V_n is complicated because the sinusoids add up differently at different locations. If, however, we assume that time changes of current $i(t)$ are very slow (i.e. are at a very low frequency f), so that the wavelengths λ associated with the frequency components are all very large compared to any dimension of the circuits, then we may use a *quasistatic approximation*. This approximation assumes that all frequency components $i_n(t)$ have the same sinusoidal factor, i.e. that we may set $V = \mathcal{L}\, di(t)/dt$, where $i(t)$ is the total current in the circuit and $\mathcal{L}$ is the inductance due to flux linkage of $\mathbf{B}$ lines that link the circuit. This latter relationship follows from Faraday's law for slow time changes in a circuit.

Consider a single isolated current loop C_1 fed by a current generator which builds up a current $i_1(t)$ from 0 at $t = 0$ to the final value I_1 at time $t = T$. At any time t between 0 and T, an inductive potential $v_1(t) = \mathcal{L}_{11}\, di_1(t)/dt$ is then created in the circuit, under the quasistatic assumption discussed above. The instantaneous power is $p_1(t) = v_1(t)i_1(t)$, and thus the magnetostatic energy that is stored up in the circuit by this self-inductive process must be the time integral of $p_1(t)$:

$$\delta W_\mathrm{s} = \int_0^T dt\, v_1(t)i_1(t) = \mathcal{L}_{11}\int_0^{I_1} di_1\, i_1 = \frac{1}{2}\mathcal{L}_{11}I_1^2 \tag{13.22}$$

This is relatively straightforward, but now consider the case of two loops C_1 and C_2, each of which is slowly fed by current generators from zero current to the final values I_1 in C_1 and I_2 in C_2. However, we do this in a particular order.

(a) Keep $i_2(t) = 0$ while turning on the $i_1(t)$ current from 0 to I_1. Following the reasoning leading to (13.22), this yields

$$\delta W_\mathrm{s}^{(\mathrm{a})} = \tfrac{1}{2}\mathcal{L}_{11}I_1^2 \tag{13.23a}$$

(b) Now turn on $i_2(t)$ from 0 to its final value I_2 while maintaining the current in C_1 at I_1. Not only is a voltage induced in C_2 through the self-inductance flux linkage of **B** lines; a voltage is also induced in C_1 due to the mutual-inductance flux linkage of **B** lines between the two circuits. We obtain $v_1(t) = \mathcal{L}_{21}di_2/dt$ in C_1 and $v_2(t) = \mathcal{L}_{22}\, di_2/dt$ in C_2. Upon integrating the instantaneous powers $v_1(t)I_1$ and $v_2(t)i_2(t)$ we obtain the stored energy due to mutual induction in C_1 and self induction in C_2:

$$\delta W_\mathrm{m}^{(\mathrm{b})} + \delta W_\mathrm{s}^{(\mathrm{b})} = \mathcal{L}_{21}I_1I_2 + \tfrac{1}{2}\mathcal{L}_{22}\, I_2^2 \tag{13.23b}$$

The integration steps are similar to those in (13.22).

Hence the total energy stored in these two circuits is the sum of (13.23a) and (13.23b):

$$\delta W = \frac{1}{2}\mathcal{L}_{11}I_1^2 + \mathcal{L}_{21}I_1I_2 + \frac{1}{2}\mathcal{L}_{22}I_2^2 = \frac{1}{2}\sum_{i=1}^{2}\sum_{j=1}^{2}\mathcal{L}_{ij}I_i\, I_j \tag{13.24}$$

We have used the reciprocal relationship $\mathcal{L}_{ji} = \mathcal{L}_{ij}$ here. If we generalize this argument to N loops we obtain

$$W_N = \frac{1}{2}\sum_{i=1}^{N}\sum_{j=1}^{N}\mathcal{L}_{ij}\, I_i\, I_j \tag{13.25a}$$

A similarity to (7.6b) for the electrostatic energy stored in N charges is found by using the definition of flux linkage (13.3b):

$$W_N = \frac{1}{2}\sum_{i=1}^{N} \Lambda_i I_i \qquad \text{with } \Lambda_i = \sum_{j=1}^{N} \mathcal{L}_{ij} I_j \tag{13.25b}$$

The flux linkage Λ_i represents the contributions from **B** lines linking circuit C_i with all N circuits (including itself), hence with the *total* **B** field lines linking C_i.

13.3.2 Expression in terms of fields

Consider the situation of N circuits as in (13.25) again. If current loop C_i consists of a single winding then from (13.1) the flux linkage Λ_i is given simply by the surface integral of the normal **B** field component, over any surface S_i bounded by C_i. Using $\mathbf{B} = \nabla \times \mathbf{A}$ and Stokes's theorem, we obtain for the contribution of the current I_i in C_i to the magnetostatic energy

$$\Lambda_i I_i = I_i \int_{S_i} d\mathbf{S} \cdot \mathbf{B} = I_i \oint_{C_i} d\boldsymbol{l} \cdot \mathbf{A} \tag{13.26}$$

Referring again to Fig. 10.5, we note that $I\, d\boldsymbol{l} \cdot \mathbf{A} = \mathbf{J}\, dv \cdot \mathbf{A}$, because $I = JS$ and $d\boldsymbol{l}$ is parallel to **J**. Here, dv is the volume of the infinitesimal current element $d\boldsymbol{l}$. It follows that

$$\Lambda_i I_i = \int_{v_i} dv\, \mathbf{J} \cdot \mathbf{A} \tag{13.27}$$

where v_i is the total volume of the circuit C_i. If circuit C_i has $\mathcal{N}$ windings, then the volume v_i must include all the windings. Upon insertion of this into (13.25b), we find

$$W_N = \frac{1}{2}\int_{v} dv\, \mathbf{J} \cdot \mathbf{A} \tag{13.28}$$

where v is the volume of all N circuits. Because **J** is nonzero only where there is current, we may replace v by the volume of all space. However, if one looks at some of the examples of **A** calculated in Chapter 12 it may seem that the constant that often appears in **A**, e.g. as in (12.19), leads to infinities in (13.28). This is resolved by noting that currents occur in closed circuits, and that the constant part of **A**, e.g. κz, leads to a closed-circuit integral $I \oint d\boldsymbol{l} \cdot \hat{\mathbf{z}}$, which is always zero, regardless of the shape of the circuit. We now eliminate in favor of the magnetostatic fields as follows, using (11.32a) and vector identity 4 of Appendix A.4:

$$\mathbf{J} \cdot \mathbf{A} = (\nabla \times \mathbf{H}) \cdot \mathbf{A} = -\nabla \cdot (\mathbf{A} \times \mathbf{H}) + (\nabla \times \mathbf{A}) \cdot \mathbf{H} = -\nabla \cdot (\mathbf{A} \times \mathbf{H}) + \mathbf{B} \cdot \mathbf{H}$$

Insert this into (13.28), and apply Gauss's divergence theorem to the divergence term:

$$W_N = -\frac{1}{2}\oint_\infty d\mathbf{S}\cdot(\mathbf{A}\times\mathbf{H}) + \frac{1}{2}\int_\infty dv\,\mathbf{B}\cdot\mathbf{H} \tag{13.29}$$

The surface element dS at distances extremely large compared to a typical linear dimension of the volume containing the N circuits is equal to $r^2\sin\theta\,d\theta\,d\varphi$. But $\mathbf{A}\propto 1/r$ and $\mathbf{H}\propto 1/r^2$ so that, in the limit $r\to\infty$, the surface integral becomes negligible. We thus find

$$W_N = \frac{1}{2}\int_\infty dv\,\mathbf{B}\cdot\mathbf{H} \tag{13.30a}$$

as the expression for the energy stored in the magnetostatic fields in all space. Note that we have derived this only for the infinite volume of all space. A corollary is the expression over a finite volume v:

$$W_B = \frac{1}{2}\int_v dv\,\mathbf{B}\cdot\mathbf{H} \tag{13.30b}$$

As in the electrostatic case, we must distinguish carefully between (13.25a, b) and (13.30b). Expressions (13.25a, b) give the energy stored in all or some of the current circuits that are counted in the sum. Expression (13.30b) is only the energy stored in the fields in volume v. These fields may very well have been produced by currents outside v. The two expressions are usually quite different numerically; they are equal only when (13.25) includes *all* currents, and when the volume of integration in (13.30) is over *all* space. This latter equality is another affirmation of the idea that the current sources create the fields! When (13.30b) is applied to a finite volume v, we often speak of the *magnetic energy density* $w_B \equiv \frac{1}{2}\mathbf{B}\cdot\mathbf{H}$, as this quantity is measured in $\mathrm{J/m^3}$ (joules per cubic meter). However, it is not directly measurable as such; we measure or infer only its integral (13.30b).

13.4 Inductance, stored energy, and forces

13.4.1 Inductance from stored energy

The availability of (13.25a) and (13.30a), and their equality, sometimes allow one to calculate inductance. If, for example, we have a single circuit, then from (13.25a) and (13.30a) we obtain the stored energy

$$\frac{1}{2}\mathcal{L}I^2 = \frac{1}{2}\int dv\,\mathbf{B}\cdot\mathbf{H} \tag{13.31}$$

which could provide a means of calculating $\mathcal{L}$, especially for linear isotropic non-magnetic media, for which $\mathbf{B} = \mu_0 \mathbf{H}$, so that

$$\mathcal{L} = \frac{\mu_0}{I^2} \int dv\, H^2 = \frac{1}{\mu_0 I^2} \int dv\, B^2 \tag{13.32}$$

Example 13.6 The inductance of a coaxial cable revisited

Referring again to Example 13.4, we have three expressions for the **B** fields:

$$\begin{aligned}
\mathbf{B} &= \frac{\mu_0 I \rho}{2\pi a^2} \hat{\boldsymbol{\varphi}} && \text{for } \rho < a \\
\mathbf{B} &= \frac{\mu_0 I}{2\pi \rho} \hat{\boldsymbol{\varphi}} && \text{for } a < \rho < b \\
\mathbf{B} &= \frac{\mu_0 I}{2\pi \rho} \left(\frac{c^2 - \rho^2}{c^2 - b^2} \right) \hat{\boldsymbol{\varphi}} && \text{for } b < \rho < c
\end{aligned} \tag{13.33}$$

To use (13.32) we need to find the volume element dv. In all three cases we have $dv = 2\pi \rho l\, d\rho$ (an infinitesimal cylindrical shell), and we may apply (13.32) three times to obtain three contributions to $\mathcal{L}$ that we will add together eventually. Thus for $\rho < a$ we find from (13.32):

$$\mathcal{L}^{(1)} = 2\pi l \frac{1}{\mu_0} \int_0^a d\rho\, \rho \left(\frac{\mu_0 \rho}{2\pi a^2} \right)^2 = \frac{\mu_0 l}{2\pi a^4} \int_0^a d\rho\, \rho^3 = \frac{\mu_0 l}{8\pi} \tag{13.34}$$

This is just the internal inductance found in Example 13.4. For $a < \rho < b$ we find

$$\mathcal{L}^{(2)} = 2\pi l \frac{1}{\mu_0} \int_a^b d\rho\, \rho \left(\frac{\mu_0}{2\pi \rho} \right)^2 = \frac{\mu_0 l}{2\pi} \int_a^b d\rho \frac{1}{\rho} = \frac{\mu_0 l}{2\pi} \ln\left(\frac{b}{a} \right) \tag{13.35}$$

and this is the external inductance if we ignore the sheath. Finally, we obtain for $b < \rho < c$

$$\begin{aligned}
\mathcal{L}^{(3)} &= 2\pi l \frac{1}{\mu_0} \int_b^c d\rho\, \rho \frac{1}{(2\pi \rho)^2} \left(\frac{c^2 - \rho^2}{c^2 - b^2} \right)^2 = \frac{\mu_0 l}{2\pi (c^2 - b^2)^2} \int_b^c d\rho \frac{(c^2 - \rho^2)^2}{\rho} \\
&= \frac{\mu_0 l}{2\pi} \left[\frac{1}{4} \frac{c^2 + b^2}{c^2 - b^2} - \frac{c^2}{c^2 - b^2} + \frac{c^4}{(c^2 - b^2)^2} \ln\left(\frac{c}{b} \right) \right]
\end{aligned} \tag{13.36}$$

for the sheath contribution, exactly as calculated in (13.15). We obtain the same result as in (13.16) when these three contributions to the inductance are added together. This calculation confirms in an independent way that our handling of internal and external inductance by means of (13.11) was correct. We return to this point in Section 13.5.

13.4.2 Magnetostatic forces

In (11.2a) we saw that the Lorentz force of a magnetic flux-density field **B** upon a moving charge is

$$\mathbf{F} = q\mathbf{u} \times \mathbf{B} \tag{13.37}$$

given that the point particle with charge q has a velocity vector **u**. Let there be $\mathcal{N}$ charges per unit volume, charge q_i having velocity vector $\mathbf{u}_i$, inside a conductor carrying current. Then the force per unit volume of conductor is

$$\frac{d\mathbf{F}}{dv} = \sum_{i=1}^{\mathcal{N}} q_i \mathbf{u}_i \times \mathbf{B} = \mathbf{J} \times \mathbf{B} \tag{13.38}$$

Here, **J** is the volume current density in the conductor at the location of interest. Thus the force upon a current-density distribution in a volume v of the conductor is

$$\mathbf{F} = \int_v dv \, \mathbf{J} \times \mathbf{B} \tag{13.39}$$

The volume element for a wire carrying current is $dv = S\,dl$, if S is the cross-sectional area and dl is an infinitesimal length element. The force on this element is

$$d\mathbf{F} = \mathbf{J} \times \mathbf{B}\, dv = S\mathbf{J} \times \mathbf{B}\, dl = I\, d\mathbf{l} \times \mathbf{B} \tag{13.40}$$

When we integrate over the entire current-carrying wire, we obtain the force on the circuit:

$$\mathbf{F} = I \oint d\mathbf{l} \times \mathbf{B} \tag{13.41}$$

This and (13.39) are expressions to which we will refer when calculating forces upon a circuit or any other current distribution.

From (11.4a) we can see that the force of a current loop C_1 upon another one, C_2, is

$$\mathbf{F}_{12} = \frac{\mu_o I_1 I_2}{4\pi} \oint_{C_2} d\mathbf{l}_2 \times \left(\oint_{C_1} d\mathbf{l}_1 \times \frac{\hat{\mathbf{R}}_{12}}{R_{12}^2} \right) \tag{13.42}$$

Indeed, this follows merely from the Biôt–Savart expression for the magnetic flux density produced by circuit C_2, after inserting it into (13.41) applied to circuit C_1. The use of vector identities allows us to transform (13.42) to a handier form:

$$\begin{aligned} d\mathbf{l}_2 \times \left(d\mathbf{l}_1 \times \frac{\hat{\mathbf{R}}_{12}}{R_{12}^2} \right) &= d\mathbf{l}_2 \times \left(d\mathbf{l}_1 \times \nabla_2 \frac{1}{R_{12}} \right) \\ &= \left(d\mathbf{l}_2 \cdot \nabla_2 \frac{1}{R_{12}} \right) d\mathbf{l}_1 - (d\mathbf{l}_1 \cdot d\mathbf{l}_2) \nabla_2 \frac{1}{R_{12}} \end{aligned}$$

The first term is zero because

$$\oint_{C_2} d\boldsymbol{l}_2 \cdot \nabla_2 \frac{1}{R_{12}} = 0$$

whenever $\mathbf{r}_1$ does not lie on or in C_2. We therefore obtain from the last term, after replacing $\nabla_2(1/R_{12})$ by $\hat{\mathbf{R}}_{12}/R_{12}^2$,

$$\mathbf{F}_{12} = -\frac{\mu_0 I_1 I_2}{4\pi} \oint_{C_1} d\boldsymbol{l} \cdot \oint_{C_2} d\boldsymbol{l}_2 \frac{\hat{\mathbf{R}}_{12}}{R_{12}^2} \tag{13.43}$$

Note that the dot product is between the two line elements. That makes this expression somewhat difficult to use because the unit vector $\hat{\mathbf{R}}_{12}$ changes direction as we integrate around the two circuits.

Example 13.7 Force of two parallel current-carrying wires upon each other

Consider the two parallel wires shown in Fig. 13.8. Each wire carries a current, and is itself, of course, part of an idealized rectangular circuit with two long sides parallel to each other, and the other two long sides very distant from each other (see Fig. 13.9). Let D be the distance between the wires. Let us consider the force $d\mathbf{F}_{12}/dz$ of the wire with current I_1 upon a *unit length* of the wire with current I_2.

The force upon a length dz on wire 2 is given by (13.40): it is $d\mathbf{F}_{12} = I_2(\hat{\mathbf{z}}\, dz) \times \mathbf{B}_1(z)$, given that $\mathbf{B}_1(z)$ is the magnetic flux density at $P = (D, 0, z)$. As the plan view in Fig. 13.8 suggests, we can find the $\mathbf{B}_1$ field from Ampère's law: it is $2\pi D B_\varphi = \mu_0 I_1$

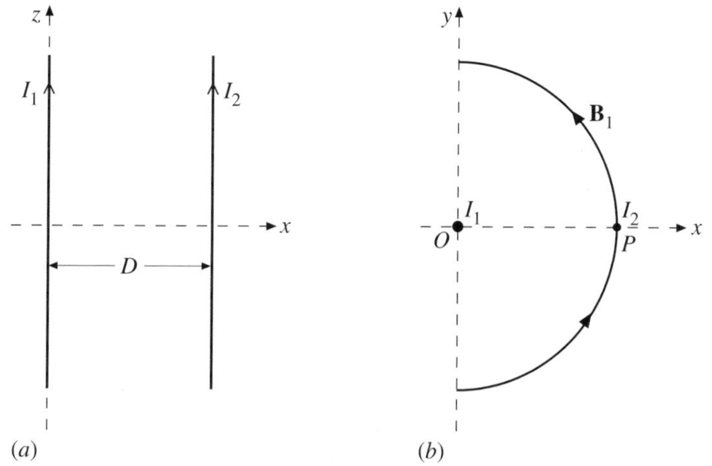

Fig. 13.8 Two parallel current-carrying wires: (*a*) view in $y = 0$ plane, (*b*) view in $z = 0$ plane.

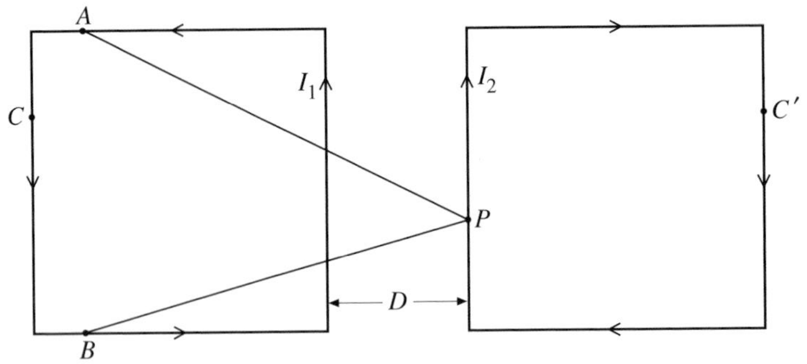

Fig. 13.9 Two parallel currents as parts of larger circuits.

and has unit vector $\hat{\boldsymbol{\varphi}} = \hat{\mathbf{y}}$ at the location P. Thus we find

$$d\,\mathbf{F}_{12} = I_2(\hat{\mathbf{z}}\,dz) \times \left(\frac{\mu_0 I_1}{2\pi D}\hat{\mathbf{y}}\right) = -\frac{\mu_0 I_1 I_2}{2\pi D}\hat{\mathbf{x}}\,dz \tag{13.44}$$

Thus the force of wire 1 upon a unit length of wire 2 is, in magnitude,

$$\frac{dF_{12}}{dl} = \frac{\mu_0 I_1 I_2}{2\pi D} \tag{13.45}$$

The force is *attractive* if both currents are in the same direction, and *repulsive* if the currents are in opposite directions. It now becomes clear why the other parts of the circuit play no role, if P lies sufficiently far away from them. Let us consider the force upon point P of the two 'legs' of the circuits that lie along the $\hat{\mathbf{x}}$ direction.

A line element dx at point A, see Fig. 13.9, yields a contribution to the total force upon dl at P given by (13.43):

$$d\,\mathbf{F}_{AP} = -\frac{\mu_0 I_1 I_2}{4\pi}(\hat{\mathbf{x}}\,dx)\cdot(\hat{\mathbf{z}}\,dl\,)\frac{\hat{\mathbf{R}}_{AP}}{R_{AP}^2} = 0 \tag{13.46}$$

The contribution is zero because $\hat{\mathbf{x}}\cdot\hat{\mathbf{z}} = 0$; likewise for the contribution due to point B. Thus (13.46) implies that the horizontal sides do not contribute to the total integral. The forces due to all points such as C and C' are also zero because $R_{CP} \to \infty$ and $R_{C'P} \to \infty$. So it seems clear that only the two vertical legs closest to each other exert a force upon each other, and this force is proportional to the length of the leg (if we ignore some fringe effects in regions where the legs join to the horizontal sections).

13.5 Internal inductance and equation (13.11)

Equation (13.11a) was presented without rigorous justification, and this section is intended to clarify the limits of its applicability. To do so, we note that a definition of inductance $\mathcal{L}$ is given by (13.32), which we now apply to the situation of Fig. 13.10. This figure shows a cross section of a 'thick' current with a series of magnetic field lines relevant to the internal inductance. The broken lines indicate surfaces perpendicular to the field lines. We introduce a local coordinate system such that any point P in the current is labeled by $(\mathbf{s}(l), l)$, where l is a parametric coordinate along a field line and $\mathbf{s}(l)$ is a two-dimensional vector indicating position on the surface $\perp \hat{\boldsymbol{l}}$ at parametric distance l.

We then obtain from (13.32)

$$\Lambda \equiv \mathcal{L}I = \frac{1}{\mu_0 I}\int dv\, B^2 = \frac{1}{\mu_0 I}\oint dl \int dS\,(l) B^2[\mathbf{s}(l), l] \tag{13.47}$$

where $dv = dl\, dS(l)$, dl is a line element along the field line, and dS is $\perp\, dl$. We now make the crucial assumption that, at any location $(\mathbf{s}(l), l)$ along a magnetic field line, $B(\mathbf{s}(l), l)dS(l)$ is not a function of l, i.e., that

$$B(\mathbf{s}(l), l)dS(l) = B(\mathbf{s})\, dS \tag{13.48}$$

This is a restrictive assumption requiring high symmetry, e.g. as in the circularly cylindrical coaxial cable, but if it holds then (13.47) reduces to

$$\Lambda = \frac{1}{\mu_0 I}\int dS\, B(\mathbf{s}) \oint dl\, B(\mathbf{s}) = \frac{1}{\mu_0 I}\int dS\, B(\mathbf{s})\mu_0\, I(\mathbf{s}) \tag{13.49}$$

in which expression the closed-contour integral over the magnetic flux density has been replaced, by virtue of Ampère's law, by the encircled or linked current $I(\mathbf{s})$ at the location $\mathbf{s}$ of the chosen surface $\perp \hat{\boldsymbol{l}}$. Consequently, (13.49) can be written as:

$$\Lambda = \int d\Lambda(\mathbf{s}) \qquad \text{with } d\Lambda(\mathbf{s}) \equiv \frac{I(\mathbf{s})}{I} B(\mathbf{s})\, dS \tag{13.50}$$

This formula is the basis for (13.11), and we see that it is rigorous under the assumption that $B(\mathbf{s})\, dS$ is independent in value of whatever perpendicular surface S

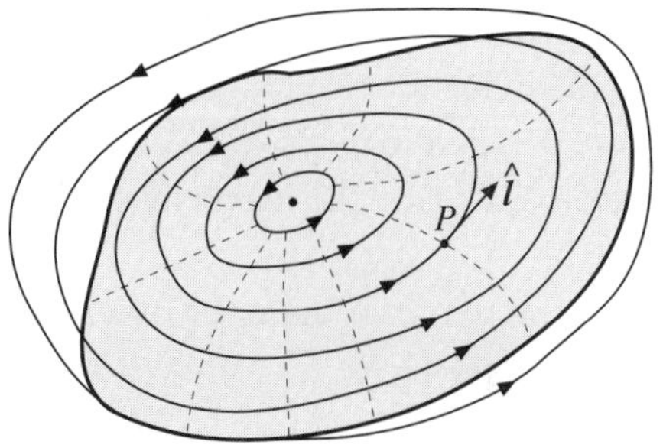

Fig. 13.10 Cross section of 'thick' current (shaded), with magnetic field lines (solid). The broken lines indicate surfaces perpendicular to the field lines.

is described when traveling around a field line. It will be approximately true for a uniform circular thin-wire circuit, provided the curvature of the length of the wire everywhere is gradual with respect to its thickness, and it is rigorously true for Example 13.4, the coaxial cable.

Problems

13.1. In Fig. 13.3 replace the circular loop by a metal rectangle with sides of length $b \parallel \hat{\mathbf{z}}$ and sides of length $a \perp \hat{\mathbf{z}}$ in the $x = 0$ plane. The nearest side is at distance d from the z axis. Find the mutual inductance.

13.2. Now replace, in problem 13.1, the rectangle by an equilateral metal triangle with sides b such that one side is parallel to $\hat{\mathbf{z}}$ but is further away from the z axis than the opposing corner (which is at distance d from the z axis). The triangle and the z axis lie in one plane. Find the mutual inductance.

13.3. A long solenoid of length L_1 and radius a is coaxial with one of length $L_2 \ll L_1$ and radius $b > 2a$. Calculate the mutual inductance, given that the numbers of windings are $\mathcal{N}_1$ and $\mathcal{N}_2$, respectively.

13.4. Consider the toroidal coil of Fig. 11.9, but let the toroid have a *circular*, not a rectangular, cross section with radius a. The larger radius of the toroid is b. There are $\mathcal{N}$ turns of current I in total. Find an expression for the self inductance. You may need to look up an integral or to do it with a computer program.

13.5. The mutual inductance of a pair of parallel infinite uniform circularly cylindrical wires of radius a separated by distance $d > 2a$ is given more exactly by

$$\mathcal{L} = \frac{\mu_0 l}{\pi} \cosh^{-1}\left(\frac{d}{2a}\right)$$

When does this lead to Eq. (13.20b)?

13.6. Find an expression for the magnetic energy stored in a long solenoid (assume n windings per unit length, length l, current I, and cross sectional area $S = \pi r^2$). Use what you have found to obtain the inductance. Ignore fringing effects.

13.7. A rectangular toroid is defined by $\rho_1 < \rho < \rho_2$ and $0 < z < h$. A current I flows in the positive $\hat{\mathbf{z}}$ direction along the z axis. Calculate the magnetic energy stored in the toroidal space.

13.8. Show directly from (13.42) that there is no net force upon the other circuit of the horizontal legs of the left-hand circuit in Fig. 13.9.

14

Magnetostatic fields in material media

14.1 Force and torque upon a current loop

The key to understanding magnetic materials is to consider the effect of a **B** field upon an infinitesimal current loop – a magnetic dipole – which has strong analogies to an electric dipole, although its physical manifestation as a current loop differs greatly from the pair of charges which constitute the electric dipole. Practical applications abound: motors, generators and transformers all are based on the principle that there are interactions between **B** fields and current loops. For those reasons, we embark upon a study of the forces upon a simple current loop, one which is circular with radius a as in Fig. 14.1.

Let a current I flow counterclockwise around the loop, which is in a uniform field $\mathbf{B} = (0,\ B_y = B\sin\theta,\ B_z = B\cos\theta)$. This is in fact quite general, because we can always rotate our coordinate system around the z axis until the x axis is perpendicular to the plane in which **B** lies. At point P, the magnetic force on a current element $(a\,d\varphi)\hat{\boldsymbol{\varphi}}$ is from (11.4a)

$$d\mathbf{F} = I(a\,d\varphi)\boldsymbol{\varphi} \times \mathbf{B} = (aI\,d\varphi)\hat{\boldsymbol{\varphi}} \times (B_y\hat{\mathbf{y}} + B_z\hat{\mathbf{z}}) \tag{14.1}$$

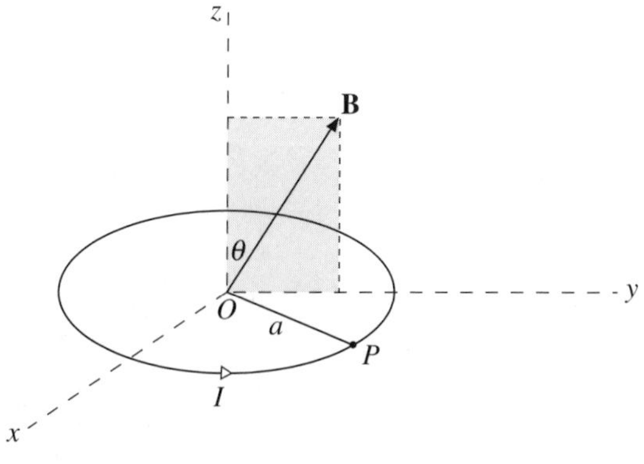

Fig. 14.1 A circular loop in a uniform **B** field.

There are *two* components to $d\mathbf{F}$: a component $d\mathbf{F}_t$ due to the first term in (14.1), and a component $d\mathbf{F}_s$ due to the second. We will use $\hat{\boldsymbol{\varphi}} = -\hat{\mathbf{x}}\sin\varphi + \hat{\mathbf{y}}\cos\varphi$, (2.36a), in evaluating these. The meaning of the indices will become apparent. Initially let us consider the second component of the force upon the line element at P:

$$\begin{aligned} d\mathbf{F}_s &= aI\,d\varphi(-\hat{\mathbf{x}}\sin\varphi + \hat{\mathbf{y}}\cos\varphi) \times B_z\hat{\mathbf{z}} \\ &= aIB_z\,d\varphi(\hat{\mathbf{y}}\sin\varphi + \hat{\mathbf{x}}\cos\varphi) = aIB_z\,d\varphi\,\hat{\boldsymbol{\rho}}(\varphi) \end{aligned} \tag{14.2}$$

This element of force is directed radially outward if B_z is positive (otherwise radially inward). It attempts to increase or decrease the radius of the circular loop. As such current loops are usually metallic, it would require a large magnetic flux density to *stretch* the loop to a larger radius (hence the index 's' on $d\mathbf{F}_s$). This component of the force simply increases the internal stress in the metal, and has no other electromagnetic effect. We shall not consider it further.

The other component is

$$d\mathbf{F}_t = aI\,d\varphi(-\hat{\mathbf{x}}\sin\varphi + \hat{\mathbf{y}}\cos\varphi) \times B_y\hat{\mathbf{y}} = -aIB_y\sin\varphi\,d\varphi\,\hat{\mathbf{z}} \tag{14.3}$$

This force is proportional to $\sin\varphi$, and is therefore zero along the x axis. The magnitude dF_t grows to its highest positive value at $\varphi = 270°$, i.e. at $y = -a$, and to its most negative value at $y = +a$, where $\varphi = 90°$. The total force upon the loop is $\oint d\mathbf{F}_t$, and it is zero here because the integral of $\sin\varphi$ from 0 to 2π over $d\varphi$ is zero. But the loop swivels around the x axis in such a way that the point $y = a$ goes down and the point $y = -a$ goes up (with respect to z). We say the loop experiences a *torque* around the x axis (hence the index t on $d\mathbf{F}_t$).

The swivel motion is depicted in Fig. 14.2 for the $y > 0$ half of the loop. An angular change $d\theta$ of the plane of the loop around the x axis (and 'downwards')

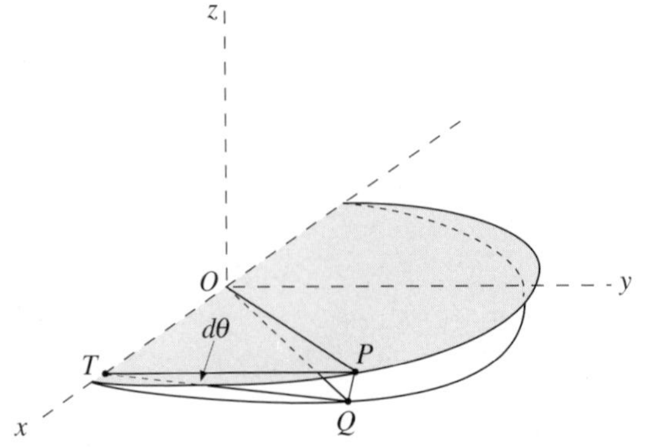

Fig. 14.2 A small angular rotation by an amount $d\theta$ of the loop of Fig. 14.1 around the x

is shown. The arc of motion at point P gives an infinitesimal displacement $d\boldsymbol{l} = -\hat{\mathbf{z}}[y(\varphi)\,d\theta]$, where $y(\varphi)$ is the length of the line TP, which is parallel to the y axis. The work done by the field in swiveling point P over this infinitesimal arc is

$$d^2W = d\boldsymbol{l}\cdot d\mathbf{F}_t = [-\hat{\mathbf{z}}y(\varphi)\,d\theta]\cdot d\mathbf{F}_t$$

If we apply (14.3), we get a positive quantity, the work d^2W done *at the cost* of the magnetostatic energy W_B stored in the field and circuit. This implies that the change in the stored magnetic energy is $d^2W_B = -d^2W$, which in turn is

$$d^2W_B = -aIB_y y(\varphi)\,d\theta \sin\varphi\,d\varphi = -(Ia^2B_y \sin^2\varphi\,d\varphi)\,d\theta \tag{14.4}$$

We have used $y(\varphi) = a\sin\varphi$ here; see Fig. 14.2. The change in stored energy in swiveling the *entire* loop over $d\theta$ is

$$dW_B = \left(-\int_0^{2\pi} d\varphi\, Ia^2B_y\sin^2\varphi\right)d\theta = -(I\pi a^2 B\sin\theta)\,d\theta \tag{14.5}$$

We observe that $dW_B < 0$ and note that the infinitesimal rotation has changed the orientation of the loop in such a way that the angle between the normal to the current loop and the **B** field has decreased. It can be seen that we can describe what is happening by imagining that the current loop stays fixed and the **B** field rotates towards the z axis so that θ decreases to $\theta - d\theta$, i.e. the angular coordinate of **B** decreases. The rotation *decreases* the stored energy, but this decrease becomes smaller because $\sin\theta$ is decreasing. This will continue until dW_B is zero, which occurs when θ is zero, that is to say when the loop is $\perp$ to the **B** field (and there is no longer any $d\mathbf{F}_t$ force). The system goes to a state of lowest energy and that in fact is what should happen!

We will now consider more closely the torque **T** on the circuit. If $d\mathbf{T}$ is the torque on $d\boldsymbol{l}$ at point P in the circuit, we have

$$\mathbf{T} = \int_0^{2\pi} d\mathbf{T}(\varphi) \qquad d\mathbf{T}(\varphi) \equiv \mathbf{r}_P \times d\mathbf{F} \tag{14.6}$$

where $d\mathbf{F}$ is given by (14.1). In this particular case, we observe that only the $d\mathbf{F}_t$ part of $d\mathbf{F}$ contributes to $d\mathbf{T}(\varphi)$. If the integral round the circuit $\oint d\mathbf{F} = 0$, i.e. if the total force is zero, then the total torque **T** does *not* depend upon where the origin is, for the following reason. If the origin is moved, then $\mathbf{r}_P$ becomes $\mathbf{r}_P + \delta\mathbf{r}$ (if $\delta\mathbf{r}$ is the vector between the old and new locations of the origin). Clearly,

$$\begin{aligned}\text{new}\,\mathbf{T} &= \oint(\mathbf{r}_P + \delta\mathbf{r})\times d\mathbf{F} = \oint(\mathbf{r}_P\times d\mathbf{F}) + \delta\mathbf{r}\times\oint d\mathbf{F} = \oint(\mathbf{r}_P\times d\mathbf{F})\\ &= \text{old}\,\mathbf{T}\end{aligned}$$

For our circular loop, for which, as discussed above, $\oint d\mathbf{F} = 0$, we find from (14.3) after substitution into (14.7) that $d\mathbf{T}(\varphi) = \hat{\mathbf{x}}\, dT_x(\varphi) + \hat{\mathbf{y}}\, dT_y(\varphi)$, and the components are

$$\begin{aligned} dT_x(\varphi) &= -Ia^2 B_y \sin^2\varphi\, d\varphi \\ dT_y(\varphi) &= +Ia^2 B_y \sin\varphi \cos\varphi\, d\varphi \end{aligned} \tag{14.7}$$

Note that the y component of the torque does not contribute to (14.7), because it would be

$$\begin{aligned} dT_y(\varphi) &= [\mathbf{r}_P \times d\mathbf{F}_t]_y \\ &= a\hat{\mathbf{y}} \cdot [\hat{\boldsymbol{\rho}} \times (-aIB_y \sin\varphi\, d\varphi\, \hat{\mathbf{z}})] = Ia^2 B_y \sin\varphi \cos\varphi\, d\varphi \end{aligned} \tag{14.8}$$

and the integral over $d\varphi$ of $\sin\varphi\cos\varphi$ from 0 to 2π is zero. Upon completing the integral in (14.7), we obtain

$$\mathbf{T} = \int_0^{2\pi} d\varphi[-(Ia^2 B_y \sin^2\varphi)\hat{\mathbf{x}} + (Ia^2 B_y \sin\varphi\cos\varphi)\hat{\mathbf{y}}] = -\hat{\mathbf{x}}(I\pi a^2)B_y \tag{14.9}$$

since the second term here does not contribute, and the integral of $\sin^2\varphi$ from 0 to 2π over $d\varphi$ is just π. We can easily relate the torque on the circuit to the stored energy, via (14.5): if we divide this equation by the rotation of the loop away from $\mathbf{B}$, $d\theta$, we get, using (14.9),

$$\frac{\partial W_B}{\partial \theta} = -I\pi a^2 B \sin\theta = T_x \tag{14.10}$$

The factor $I\pi a^2$ in (14.9) is the product of the current in the loop with its area $S = \pi a^2$. In Chapter 12, see (12.33), this was introduced as the *magnetic dipole moment* and it is given a vectorial direction perpendicular to the plane of the loop. We write

$$\mathbf{m} \equiv IS\hat{\mathbf{n}} \tag{14.11}$$

if $\hat{\mathbf{n}}$ is the unit vector normal to surface S in accord with the right-hand rule applied to the sense of the current around the loop; in Fig. 14.1, $\mathbf{m}$ is in the $+z$ direction.

Returning to (14.9), we observe that $-\hat{\mathbf{x}}$ can be replaced by $\hat{\mathbf{z}} \times \hat{\mathbf{y}}$, so that

$$\mathbf{T} = (I\pi a^2)\hat{\mathbf{z}} \times B_y\hat{\mathbf{y}} = (I\pi a^2)\hat{\mathbf{z}} \times (B_y\hat{\mathbf{y}} + B_z\hat{\mathbf{z}}) = \mathbf{m} \times \mathbf{B} \tag{14.12}$$

As you can see, we are able to replace $B_y\hat{\mathbf{y}}$ by the total $\mathbf{B}$ vector, because the other part of $\mathbf{B}$ does not contribute to the cross product. This is a specific example of a very general formula: whenever we have a loop of any shape whatever, we will find that $\mathbf{T} = \mathbf{m} \times \mathbf{B}$, where $\mathbf{m}$ is the magnetic dipole moment vector and $\mathbf{B}$ is the magnetic flux density vector. In the above case of a circular loop, the direction of $\mathbf{m} \times \mathbf{B}$ is $-\hat{\mathbf{x}}$.

The rotation of the loop is therefore *clockwise* when we look in the $-\hat{\mathbf{x}}$ direction (as shown in Fig. 14.2). And, using (14.10), we observe that

$$\hat{\mathbf{x}} \cdot \mathbf{T} = \frac{\partial W_B}{\partial \theta} \tag{14.13}$$

The left-hand side is the dot product of $\hat{\mathbf{x}}$ with the torque vector, which is in the opposite direction, and it therefore is negative (as it should be so that the stored energy decreases upon rotation).

We will now establish that all circuits have torques given by $\mathbf{m} \times \mathbf{B}$. To do so, first examine a rectangular loop.

Example 14.1 Torque of a rectangular loop

We still have $d\mathbf{F} = I d\boldsymbol{l} \times \mathbf{B}$, but the situation in Fig. 14.3 tells us that $d\boldsymbol{l} = \beta_x \hat{\mathbf{x}}\, dx + \beta_y \hat{\mathbf{y}}\, dy$, where either $\beta_x = \pm 1$ and $\beta_y = 0$ or $\beta_y = \pm 1$ and $\beta_x = 0$. There are four choices for the pair β_x, β_y and the choice depends upon the side of the rectangular circuit at which we are looking at. For example, for the side at $x = a$, we have $\beta_x = 0$ and $\beta_y = 1$.

As in the case of the circular loop, only the force due to B_y will be of interest (the other force just puts internal stresses upon the circuit). Thus

$$\begin{aligned} d\mathbf{F} &= I(\beta_x \hat{\mathbf{x}}\, dx + \beta_y \hat{\mathbf{y}}\, dy) \times B_y \hat{\mathbf{y}} = (\beta_y I B_y\, dx)\hat{\mathbf{z}} \\ &= -(I B_y\, dx)\hat{\mathbf{z}} \qquad \text{for } y = b \\ &= +(I B_y\, dx)\hat{\mathbf{z}} \qquad \text{for } y = -b \end{aligned} \tag{14.14}$$

The torque is nonzero only on the sides that are $\perp$ to the $B_y \hat{\mathbf{y}}$ vector. The torque on a point on the $y = b$ side is $dT_z = \mathbf{r} \times d\mathbf{F}$, where $d\mathbf{F}$ is given by the second line in

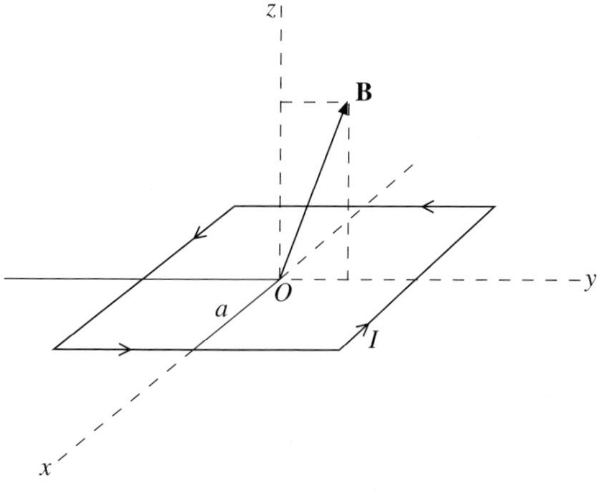

Fig. 14.3 A rectangular loop in a uniform **B** field.

(14.14) and $\mathbf{r} = x\hat{\mathbf{x}} + b\hat{\mathbf{y}}$, so that the torque on the entire $y = b$ side is

$$\mathbf{T}(y=b) = -\int_{-a}^{a} dx\,(x\hat{\mathbf{x}} + b\hat{\mathbf{y}}) \times (IB_y)\hat{\mathbf{z}}$$

$$= -b\int_{-a}^{a} dx\,(IB_y)\hat{\mathbf{y}} \times \hat{\mathbf{z}} = -2abIB_y\hat{\mathbf{x}} \tag{14.15a}$$

The $x\hat{\mathbf{x}}$ term in the first integral contributes zero to the dx integration because the $-a$ to 0 part cancels the 0 to $+a$ part. Likewise, for the $y = -b$ side,

$$\mathbf{T}(y=-b) = \int_{-a}^{a} dx\,(x\hat{\mathbf{x}} - b\hat{\mathbf{y}}) \times (IB_y)\hat{\mathbf{z}}$$

$$= -b\int_{-a}^{a} dx\,(IB_y)\hat{\mathbf{y}} \times \hat{\mathbf{z}} = -2abIB_y\hat{\mathbf{x}} \tag{14.15b}$$

Consequently, the total torque, being the sum of these two parts, must be

$$\mathbf{T} = -(4abI)B_y\hat{\mathbf{x}} = (IS)\hat{\mathbf{z}} \times (B_y\hat{\mathbf{y}} + B_z\hat{\mathbf{z}}) = \mathbf{m} \times \mathbf{B} \tag{14.16}$$

given that $S \equiv 4ab$ is the area of the rectangular loop and $\mathbf{m} \equiv (IS)\hat{\mathbf{z}}$ is its magnetic dipole moment. Here again, the torque yields a clockwise rotation with respect to the direction of $-\hat{\mathbf{x}}$, which is the direction of $\mathbf{T}$.

This raises the question of how to define $\mathbf{T}$ for an *arbitrary* shape of loop, and also when the loop does not lie in a plane, i.e. when the loop traces an asymmetric closed curve which may not lie upon a planar surface. Then, as shown in Fig. 14.4, we can subdivide the loop into a grid of infinitesimal rectangular loops, each of which is planar.

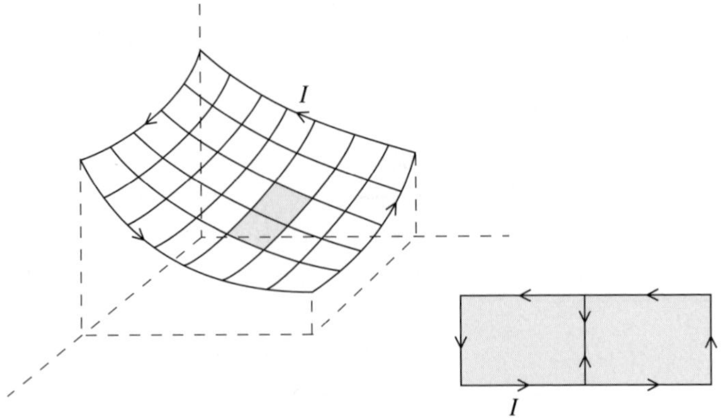

Fig. 14.4 An arbitrary loop in a uniform **B** field (not shown).

It is easily seen that we can consider the current I encircling the perimeter as if I goes around each of the infinitesimal loops in the same sense (clockwise or counterclockwise), because each piece of an infinitesimal internal loop is shared by one other internal loop with I going around in the opposite direction (and therefore canceling the current on that piece). Thus, as each infinitesimal loop is essentially planar with constant **B** field, we can define

$$\mathbf{T} = \sum_{i=1}^{N} (IS_i)\hat{\mathbf{n}}_i \times \mathbf{B}_i = \sum_{i=1}^{N} \mathbf{m}_i \times \mathbf{B}_i \tag{14.17}$$

This is a summation over all the infinitesimal loops. Thus for a circuit lying in a constant **B** field (14.17) predicts

$$\mathbf{T} = \mathbf{m} \times \mathbf{B} \tag{14.18}$$

14.2 Magnetization and magnetic permeability

14.2.1 Magnetic dipole moment and magnetic field

The concept of magnetic dipole moment **m** is the crucial starting point in the derivation of the characteristic behavior of magnetic materials under the influence of a magnetostatic field. Recall from (12.33) that the vector magnetic potential contribution at point P from an infinitesimal magnetic dipole (i.e. an infinitesimal current loop) at Q is

$$\mathbf{A}_P = \frac{\mu_0}{4\pi R_{QP}^2} \mathbf{m} \times \hat{\mathbf{R}}_{QP} \tag{14.19}$$

We shall apply this to an atom of a magnetic material. While the present-day quantum-mechanical model is somewhat more complicated, it is still useful to think of the atom as a nucleus with electrons revolving around it. In addition to this orbital motion the electrons have a property that corresponds to an intrinsic 'spin' motion. Thus each atomic electron not only has a magnetic moment due to its orbital motion but also has an *intrinsic* spin magnetic moment $\mathbf{m}_s$ with magnitude $eh/4\pi\mathfrak{m}_e \approx 9.3 \times 10^{-24}\ \mathrm{A\,m^2}$. The electronic nature of this constant is indicated by the quantity $\mathfrak{m}_e \approx 9.107 \times 10^{-31}$ kg (the mass of an electron), and the essential quantum-mechanical nature by the appearance of Planck's constant $h \approx 6.626 \times 10^{-34}$ J s.

The magnitude of the electron's spin angular momentum **L** is $\sqrt{3/4}(h/2\pi)$ but relevant here is the value of the component of **L** in any chosen direction of measurement, which equals $\pm h/4\pi$. Note that the spin component can *only* take these values. Moreover, classical mechanics predicts for a particle with charge

$-e$ rotating on its axis that the ratio of magnetic moment to angular momentum is $-e/2\mathfrak{m}_e$. But quantum mechanics leads to a ratio $-e/\mathfrak{m}_e$ for an electron, *twice* that which is predicted classically. The concept of a classical rotating charge therefore is one from classical mechanics and is not quantitatively supported by the more accurate quantum-mechanical description of an electron's motion.

The roles of spin and orbital magnetic moments will be clarified in Section 14.5, where some details of diamagnetism and paramagnetism are discussed. However, with respect to the spin, we cannot really say that there is an infinitesimal current loop inside each electron; we only can say that the electron possesses an intrinsic magnetic moment vector $\mathbf{m}_s$.

We write (14.19) for the vector magnetic potential at point P produced by an infinitesimal magnetic moment at point Q':

$$\mathbf{A}_P = \frac{\mu_0}{4\pi R^2_{Q'P}} \mathbf{m}_{Q'} \times \hat{\mathbf{R}}_{Q'P} \tag{14.20}$$

As in the case of dielectrics, we place a microscopically large but macroscopically small volume element around a center Q (such that there are very many electron magnetic moments $\mathbf{m}_s(\mathbf{r}_{Q'})$ in that box with volume dv_Q; see Fig. 14.5). See the discussion after (6.11) for more detail about this point. The contribution to $\mathbf{A}_P$ from this volume element is

$$d\mathbf{A}_P = \sum_{Q' \in dv_Q} \frac{\mu_0}{4\pi R^2_{Q'P}} \mathbf{m}_{Q'} \times \hat{\mathbf{R}}_{Q'P} \approx \frac{\mu_0}{4\pi R^2_{QP}} \left(\sum_{Q' \in dv_Q} \mathbf{m}_{Q'} \right) \times \hat{\mathbf{R}}_{QP} \tag{14.21}$$

The sketch clarifies our assumption that $\mathbf{R}_{Q'P} \approx \mathbf{R}_{QP}$ for all Q' in the volume element. This becomes untrue when point P is very close to Q, but we will see as before that nonnegligible contributions do not come from the immediate environment of P.

The major other assumption is that there are so many magnetic dipoles in dv_Q that we may write

$$\sum_{Q' \in dv_Q} \mathbf{m}_{Q'} = \mathbf{M}_Q \, dv_Q \tag{14.22}$$

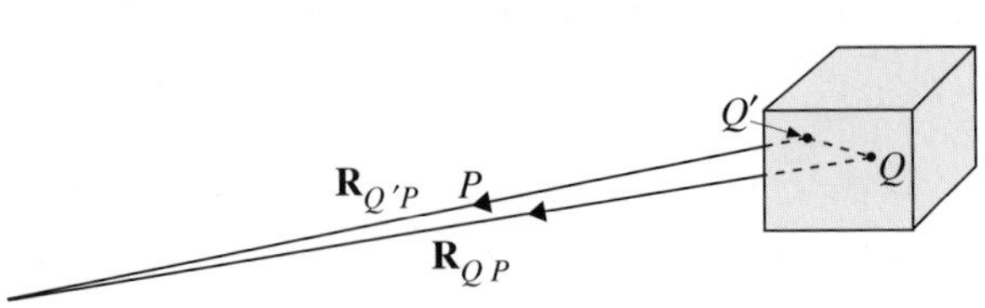

Fig. 14.5 Geometry for potential contributions at P from a collection of magnetostatic dipoles in a volume element dv_Q.

Here, $\mathbf{M}_Q$ is known as the *magnetization vector*; it is the magnetic dipole moment per unit volume, and it is a reasonably smooth function of position under the microscopic–macroscopic assumptions we have made. Hence after incorporating this into (14.21) and then integrating over the entire volume V of magnetic material, we obtain

$$\mathbf{A}_P = \frac{\mu_0}{4\pi}\int_V dv_Q\,\mathbf{M}_Q \times \frac{\hat{\mathbf{R}}_{QP}}{R_{QP}^2} = \frac{\mu_0}{4\pi}\int_V dv_Q\,\mathbf{M}_Q \times \nabla_Q\left(\frac{1}{R_{QP}}\right) \tag{14.23a}$$

The transition to the last step has been performed many times before, so that we will assume at this stage that the reader is familiar with it. We now use the vector identity 3 of Appendix A.4 to obtain

$$\mathbf{M}_Q \times \nabla_Q\left(\frac{1}{R_{QP}}\right) = -\nabla_Q \times \left(\frac{\mathbf{M}_Q}{R_{QP}}\right) + \frac{1}{R_{QP}}(\nabla_Q \times \mathbf{M}_Q) \tag{14.23b}$$

Insert this into (14.23a):

$$\mathbf{A}_P = -\frac{\mu_0}{4\pi}\int_V dv_Q\,\nabla_Q \times \left(\frac{\mathbf{M}_Q}{R_{QP}}\right) + \frac{\mu_0}{4\pi}\int_V dv_Q\,\frac{1}{R_{QP}}(\nabla_Q \times \mathbf{M}_Q) \tag{14.23c}$$

Apply a variant of Stokes's theorem to the first volume integral:

$$\mathbf{A}_P = \frac{\mu_0}{4\pi}\int_S dS_Q\left[\frac{-(\hat{\mathbf{n}} \times \mathbf{M})_Q}{R_{QP}}\right] + \frac{\mu_0}{4\pi}\int_V dv_Q\,\frac{(\nabla \times \mathbf{M})_Q}{R_{QP}} \tag{14.24a}$$

The first term in (14.24a) is a direct application of the following vector theorem: $\mathbf{C}\cdot\int dv(\nabla \times \mathbf{F}) = \int dv\,\nabla\cdot(\mathbf{F}\times\mathbf{C}) = \oint dS\,\hat{\mathbf{n}}\cdot(\mathbf{F}\times\mathbf{C}) = \mathbf{C}\cdot\oint dS(\hat{\mathbf{n}}\times\mathbf{F})$, given that $\mathbf{C}$ is a constant vector and $\hat{\mathbf{n}}$ is the normal at dS. Remove the constant $\mathbf{C}$ from the first and last integrals and we are there.

This expression is formally equal to

$$\mathbf{A}_P = \frac{\mu_0}{4\pi}\int_S dS_Q\,\frac{\mathbf{J}_{\text{ms}}(\mathbf{r}_Q)}{R_{QP}} + \frac{\mu_0}{4\pi}\int_V dv_Q\,\frac{\mathbf{J}_{\text{mv}}(\mathbf{r}_Q)}{R_{QP}} \tag{14.24b}$$

in which expressions we identify two apparent current densities:

$$\begin{aligned} &\textit{a surface-current density} \quad && \mathbf{J}_{\text{ms}}(\mathbf{r}') = -n(\hat{\mathbf{r}}') \times \mathbf{M}(\mathbf{r}') \quad \text{in A/m} \\ &\textit{a volume-current density} \quad && \mathbf{J}_{\text{mv}}(\mathbf{r}') = \nabla' \times \mathbf{M}(\mathbf{r}') \quad \text{in A/m}^2 \end{aligned} \tag{14.24c}$$

Thus, the effect of a magnetic field upon a magnetic material can be thought of as producing a surface-current density $\mathbf{J}_{\text{ms}}$ on the surface and, if the magnetization $\mathbf{M}(\mathbf{r}')$ is nonuniform in the interior, a volume-current density $\mathbf{J}_{\text{mv}}$ in the interior. The surface-current density due to magnetization, $\mathbf{J}_{\text{ms}}$, will appear later in a boundary condition that is the two-dimensional analog of (14.26b). We now consider the volume-current density $\mathbf{J}_{mv}$. If we restrict ourselves to points P in the interior we then see that Ampère's law (11.32a) must be modified, because now in addition to the external current density $\mathbf{J}$ we have $\mathbf{J}_{mv}$:

$$\nabla \times \mathbf{B} = \mu_0(\mathbf{J} + \mathbf{J}_{mv}) \tag{14.25a}$$

Replace $\mathbf{J}_{mv}$ by $\nabla \times \mathbf{M}$, (14.24c), and bring the term to the other side:

$$\nabla \times (\mu_0^{-1}\mathbf{B} - \mathbf{M}) = \mathbf{J} \tag{14.25b}$$

This then gives us the general definition of the magnetic field $\mathbf{H}$:

$$\mathbf{H} \equiv \mu_0^{-1}\mathbf{B} - \mathbf{M} \tag{14.26a}$$

so that (14.25b) becomes

$$\nabla \times \mathbf{H} = \mathbf{J} \tag{14.26b}$$

This restatement of Ampère's law holds everywhere even if magnetic media are present. We remind the reader that in integral form this is

$$\oint d\mathbf{l} \cdot \mathbf{H} = I_{encirc}$$

so it can be seen that $\mathbf{H}$ depends only on the macroscopic ('real') currents present.

Physicists sometimes consider $\mathbf{H}$ to be analogous to $\mathbf{D}$, which depends only on the real charge, and $\mathbf{B}$ to $\mathbf{E}$, since $\mathbf{E}$ is defined as the *electrostatic force* on unit charge and $\mathbf{B}$ as the *magnetic force* on unit length of unit current. From the conventional nomenclature ($\mathbf{E}$, $\mathbf{H}$ are *fields*, whereas $\mathbf{D}$, $\mathbf{B}$ are *flux densities*) we see that historically the analogies were considered differently: it is true that both $\mathbf{B}$ lines, and $\mathbf{D}$ lines in the absence of unbound charge, are continuous everywhere. In order to stress the latter view, we rewrite (14.26a) in analogy with (6.19a) as

$$\mathbf{B} = \mu_0(\mathbf{H} + \mathbf{M}) \tag{14.27}$$

It is clear that $\mathbf{M}$ and $\mathbf{H}$ have the same dimensional unit: A/m. It is also clear, as it was in the case of the polarization vector $\mathbf{P}$ in dielectric materials, that $\mathbf{M}$ sometimes is not a very useful quantity for characterizing magnetic materials, since it may vary with the impressed field.

14.2.2 Susceptibility and permeability

Now consider magnetic material that is inserted into a magnetic field $\mathbf{H}$ produced by a source. Whether $\mathbf{H}$ remains the same after insertion or is variable in space or time is immaterial. However, let us be specific to avoid some of the problems associated with this, and so let us assume that we insert a small specimen of magnetic material into a large region in which a uniform steady magnetic field $\mathbf{H}$ is produced.

Let us even assume that this field is essentially unchanged at large distances from the material (we shall see that it must change close to the material). We shall consider the situation inside the material, and write the magnetization as

$$\mathbf{M}(\mathbf{H}) = \chi_m(\mathbf{H})\mathbf{H} \tag{14.28}$$

That is, we act as though we expect $\mathbf{M}$ to be proportional to the outside field $\mathbf{H}$ (just as we expected the polarization $\mathbf{P}$ to be proportional to $\varepsilon_0\mathbf{E}$ in dielectrics), but we are more careful now and allow the constant of proportionality, χ_m, to be a function of $\mathbf{H}$ so as to allow for a general nonlinear dependence. The constant χ_m is known as the *magnetic susceptibility*, and it is dimensionless.

At least for weakly magnetic materials, and under the influence of relatively weak magnetic fields, we may treat the magnetic susceptibility as a constant. Under this assumption of linearity ($\mathbf{M} \propto \mathbf{H}$), (14.27) becomes

$$\mathbf{B} = \mu_0(1 + \chi_m)\mathbf{H} \tag{14.29}$$

and we see that $1 + \chi_m \equiv \mu_r$ is another dimensionless constant known as the *relative magnetic permeability*, which characterizes the magnetostatic aspect of the material adequately, just as the relative dielectric permittivity ε_r does for dielectric materials. The material's *magnetic permeability* is $\mu = \mu_r\mu_0$. Unfortunately χ_m, μ_r, and thus μ are functions of $\mathbf{H}$ for *ferromagnetic* materials, the most interesting case, and they then need to be given as functions of the impressed field to be useful. This will be discussed shortly.

14.2.3 Comparison of the magnetostatic and electrostatic cases

Table 14.1 summarizes some of the preceding ideas, and puts them in context with the dielectric-materials situation, stating the equations in the conventional way. However, the analogies between the magnetostatic and the electrostatic Maxwell equations are in closer harmony if we use $\mathbf{E} \leftrightarrow \mathbf{B}$ and $\mathbf{D} \leftrightarrow \mathbf{H}$, because the Maxwell equations are

$$\nabla \times \mathbf{E} = 0 \quad \leftrightarrow \quad \nabla \cdot \mathbf{B} = 0$$

$$\nabla \cdot \mathbf{D} = \varrho_v \quad \leftrightarrow \quad \nabla \times \mathbf{H} = \mathbf{J}$$

The analogies in the table are imperfect because the susceptibility χ_m was introduced via (14.27) rather than through (14.26). It would have been more in line with these analogies to define the susceptibility by e.g. $\mathbf{M} = \mu_0^{-1}\chi_m\mathbf{B}$, so that $\mathbf{H} = \mu_0^{-1}(1 - \chi_m)\mathbf{B}$, and then to have defined $\mu_r^{-1} \equiv 1 - \chi_m$. The physicists' analogy would then be somewhat clearer: $\mathbf{B}$ would then be known as the magnetic field and $\mathbf{H}$

Table 14.1. A comparison of the quantities in electrostatics and magnetostatics

Electrostatic	Magnetostatic
dipole moment $\mathbf{p} = qd\hat{\mathbf{z}}$	dipole moment $\mathbf{m} = IS\hat{\mathbf{z}}$
($\pm$charges along $\hat{\mathbf{z}}$ direction	(current loop $\perp$ $\hat{\mathbf{z}}$ direction)
polarization vector $\mathbf{P}$ (C/m^2)	magnetization vector $\mathbf{M}$ (A/m)
polarization charge density	magnetization current density
$\varrho_{\mathrm{vp}} = -\nabla \cdot \mathbf{P}$	$\mathbf{J}_{\mathrm{mv}} = \nabla \times \mathbf{M}$
$\mathbf{P} = \chi_{\mathrm{e}}\varepsilon_0\mathbf{E}$	$\mathbf{M} = \chi_{\mathrm{m}}\mathbf{H}$
$\mathbf{D} = \varepsilon_0\mathbf{E} + \mathbf{P} = \varepsilon_0(1 + \chi_{\mathrm{e}})\mathbf{E}$	$\mathbf{B} = \mu_0(\mathbf{H} + \mathbf{M}) = \mu_0(1 + \chi_{\mathrm{m}})\mathbf{H}$
$\mathbf{D} = \varepsilon_{\mathrm{r}}\varepsilon_0\mathbf{E} \equiv \varepsilon\mathbf{E}$	$\mathbf{B} = \mu_{\mathrm{r}}\mu_0\mathbf{H} \equiv \mu\mathbf{H}$

as the flux density. Perhaps there would be less confusion in that case, because the analogies

$$\varepsilon_0^{-1} \leftrightarrow \mu_0 \qquad \varrho_{\mathrm{v}} \leftrightarrow \mathbf{J} \qquad V \leftrightarrow \mathbf{A} \qquad \cdot \leftrightarrow \times$$

would also carry through to the potentials and their equations. But this would entail a needless struggle against the historical development.

14.3 Magnetic materials

We will restrict ourselves to three types of magnetic materials. First, *diamagnetic* materials are characterized by a magnetization $\mathbf{M}$ that is opposite in sign to the impressed magnetic field $\mathbf{H}$. The weak force of the $\mathbf{H}$ field upon a diamagnetic material is comparable to that of the $\mathbf{E}$ field upon a dielectric. However, because of the way the polarization vector is defined, the dielectric constant is always positive whereas the diamagnetic susceptibility is negative. Then there are *paramagnetic* materials in which the weak force is in the 'normal' direction, indicated by the fact that $\mathbf{M}$ is in the same direction as $\mathbf{H}$. And finally there are the *ferromagnetic* materials (e.g. iron) which have very pronounced – even strong – nonlinear $\mathbf{M}$ reactions to an impressed magnetic field $\mathbf{H}$. We all know how strong ferromagnetic forces are, as almost all 'permanent magnets' we see are ferromagnetics.

14.3.1 Diamagnetics

When a magnetic field is turned on, and a specimen of material, the atoms of which do not have a net magnetic moment (i.e. they have a complete outer shell of electrons around the nucleus), is present, then changes will occur in the *orbital* motions of the electrons around the nucleus. These changes are not described properly by classical mechanics; as stated at the start of Section 14.2 one needs quantum mechanics to supply an adequate description. However, some aid to understanding is given by the fact that a changing magnetic field acting upon charge carriers will induce a voltage difference (hence an electric field) that tends to *oppose* that change in the magnetic field (Lenz's law, which we will encounter in more detail shortly in connection with Faraday's law). Slight changes in the electron orbits are induced, which persist, in this resistanceless situation, and it follows that $\chi_m \leq -10^{-5}$, depending upon the material and the strength of the magnetic field. The time history of the magnetic field is responsible for this very weak diamagnetic effect, which manifests itself only in the absence of stronger effects yielding contributions to the susceptibility (with opposite sign in the case, for example, of paramagnetic materials). Even in bismuth, one of the stronger diamagnetic materials, the effect remains quite weak. In subsection 14.5.1 we show by means of a semiclassical argument how the slow (quasistatic) turning on of an external magnetic field brings about the diamagnetic effect.

14.3.2 Paramagnetics

Paramagnetic effects can occur in materials with net magnetic moments (for which an outer shell of electrons around an atomic nucleus must be incomplete). The effect is due entirely to the effect of a magnetic field **H** upon the atomic magnetic moments $m \sim 10^{-23}$ A m^2. From (14.13) and the discussion in the vicinity of that equation, we observe that[1]

$$dW/d\theta = T = mB \sin\theta$$

so that the stored magnetic energy $W \sim \int d\theta\, mB \sin\theta = -mB\cos\theta = -\mathbf{m}\cdot\mathbf{B}$. As the stored energy decreases when the field torque **T** works on the magnetic dipole, we interpret $\mathbf{m}_s \cdot \mathbf{B} = \mu_0 \mathbf{m}_s \cdot \mathbf{H}$ as the energy involved in turning the dipole so that it points along the field. This is the energy that the field expends upon doing the work. However, the turning is not done in splendid isolation; at room temperature vibrations and rotations in solids cause electron–atom collisions which tend to undo the orientation that would be produced under the influence of the **H** field alone. Thermodynamics and statistical mechanics teach that the energy per degree of freedom involved in the chaotic motion due to what we macroscopically describe as *temperature* is of the order of $\frac{1}{2}kT$, where T is the temperature in degrees kelvin and $k \approx 1.381 \times 10^{-23}$ J/K is Boltzmann's constant. We will not go into the details here of degrees of freedom, but basically $\frac{3}{2}kT$ is the average translational kinetic energy of a particle in a gas at temperature T. The collisions which 'randomize' the torques on the atoms or molecules come at the cost of that kinetic energy, and thus κT is a

[1] Here we take $d\theta$ as the increase in angle between **m** and **B**, in fact the negative of its meaning in Section 14.1 where it referred to the small angle through which the plane of the loop turns. As the latter small angle increases, the angle between the normal $\hat{\mathbf{n}}(\| \mathbf{m})$ and **B** *decreases* – see Figs. 14.1 and 14.2.

measure of the randomizing energy. The ratio $\alpha = mB/kT = \mu_0 mH/kT$ is a measure of the effectiveness of the **H** field in turning the dipoles into the field and thus creating a net vector **M**. Let us estimate it. At room temperature we find $kT \sim 4 \times 10^{-21}$ J. For the Earth's magnetic field ($B \sim 0.000\,05\ \mathrm{Wb/m^2}$) we find $mB \sim 5 \times 10^{-28}$ J, whereas for a medium-strength laboratory field ($B \sim 0.05\ \mathrm{Wb/m^2}$) we find $mB \sim 5 \times 10^{-25}$ J. Even at strong laboratory fields of $\sim 1\ \mathrm{Wb/m^2}$, we see that the ratio $\mu_0 mH/kT \ll 1$.

Statistical mechanics describes the probability of systems with a given energy having any particular distribution of magnetic dipole vectors. The most probable distribution at a given temperature then follows from a proper weighting of the sum of all possible **m**'s, and it yields for the magnetization of a gas

$$M(\alpha) = Nm\left(\operatorname{cotanh}\alpha - \frac{1}{\alpha}\right) \qquad \alpha \equiv \mu_0 mH/kT \tag{14.30}$$

Here, N is the number density of magnetic dipoles. The Langevin function, $\operatorname{cotanh}\alpha - 1/\alpha$, is shown in Fig. 14.6. A derivation is outlined in subsection 14.5.2.

We observe that $M(\alpha)$ tends to the maximum value Nm (all dipoles aligned along the field) as $\alpha \to \infty$. For $\alpha \ll 1$, it reduces to a linear relationship between M and H:

$$M(\alpha) = \frac{1}{3} Nm\alpha = \frac{\mu_0 Nm^2}{3\kappa T} H \tag{14.31}$$

which is known as *Curie's law*. In that regime, the magnetic susceptibility is a constant characteristic of paramagnetism: $\chi_m = \mu_0 Nm^2/3\kappa T$. Consider a metal, in which $N \sim 10^{28}\ \mathrm{m^{-3}}$, at a room temperature of 20–30 °C ($T \sim 300$ K); this yields $\chi_m \sim 0.9 \times 10^{-4}$, which is an upper bound on what one might expect, because the electron densities that contribute to net atomic dipoles are seldom larger than $10^{28}\ \mathrm{m^{-3}}$. Clearly, in metals the paramagnetic effect swamps the always-occurring diamagnetism. The theory for solid paramagnetics is more complicated, and will not be discussed further.

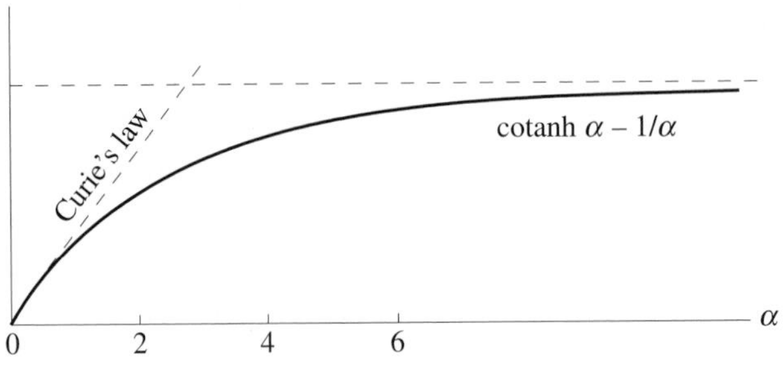

Fig. 14.6 The Langevin function $M(\alpha)$ vs. $\alpha = \mu_0 m_s H/kT$.

14.3.3 Ferromagnetics

The behavior of (solid) ferromagnetics is much more complicated below a critical temperature known as the *Curie point*, which is far above room temperature for the most common ferromagnetic materials such as iron. Very large values of the susceptibility ($\chi_m \sim 4000$) can occur at room temperature. Weiss (1906) found that *domains* of diameter ranging from 0.001 to 1 mm can occur in metals such as iron. With an atomic density of $8.5 \times 10^{28}\ \text{m}^{-3}$ in iron, this would yield an average number of $\sim 10^{16}$ atoms per domain. The peculiarity of each domain is that all the atomic dipole moments in it are aligned, owing to powerful internal molecular fields of a quantum-mechanical nature.

In the absence of an exterior **H** field the total dipole moments of the domains are oriented randomly, and a sum over any macroscopic region – even a quite small one – will give $\mathbf{M} \approx 0$. To understand, qualitatively at least, the behavior under the influence of an exterior magnetic field, let us slowly turn on an **H** field, in which a sample of a ferromagnetic material is immersed, from zero to a maximum value H_{max}. It will not be necessary to take the direction vector into account, but we will assume that the maximum field is in the $\hat{\mathbf{z}}$ direction. In fact, we will follow the behavior of the sample as H goes from zero to H_{max}, then through zero to $-H_{\text{max}}$, and finally to zero again. The flux density $B(H)$ will turn out to behave in a highly nonlinear way during this cycle.

Figure 14.7(*a*) is a graph of $B(H)$ versus $\mu_0 H$ during a time in which the field H increases slowly from zero to a small positive value H_1 in the $\hat{\mathbf{z}}$ direction. Note that the horizontal axis is $\mu_0 H$. The plot of B versus $\mu_0 H$ would be a straight line if the susceptibility were constant. Clearly it is not; as $B = \mu_0(H + M)$, it follows that $M(H) > 0$ in this graph (and that it grows in value) from the fact that the curve

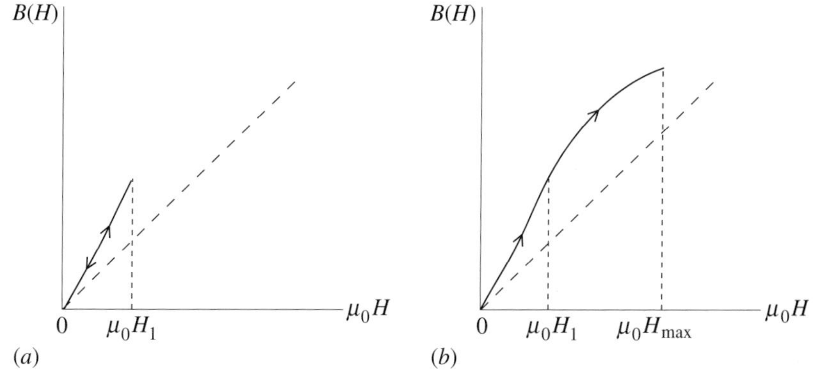

Fig. 14.7 The ferromagnetic $B(H)$ curve vs. $\mu_0 H$ for two stages of increasing H.

turns upwards. The physical explanation is that those domains that already have their net atomic dipole moments aligned with the field **H** grow in size at the expense of neighboring nonaligned domains. Hence a magnetization **M** parallel to **H** grows from zero to some magnitude giving rise to the magnitude H_1. This situation is *reversible*: if we now decrease the field H from H_1 to zero, $B(H)$ will decrease along the same curve.

If H then is increased from H_1 to H_{max}, the curve is continued as shown in Fig. 14.7(*b*). The value of H_1 is the highest value from which one can reverse the magnetization back to zero by turning the field **H** off again. Beyond H_1 the situation is *irreversible*. A different physical mechanism holds: the exterior field H will now exert a torque upon whole domains that are nonaligned and it will attempt to turn them. The domains have 'jagged' edge surfaces, and therefore the field must build up to some minimum value H_1 before it can overcome the friction forces preventing this turning. The curve now bends back towards the 45° direction if H_{max} is high enough because M will *saturate* ultimately at its maximum value Nm; after saturation, $B = \mu_0(H + M)$ would grow further only through an increase in H (i.e. at 45°).

If the field is now reversed, i.e. if H is now decreased from H_{max} to 0, the curve of Fig. 14.8(*a*) is followed. The irreversibility is obvious. We can regard the decrease in H as the result of adding an increasing negative component. This means that friction forces on other parts of the domain surfaces are being challenged. There is some analogy to the mechanical force we apply on a ratchet: we can only apply a torque in one rotational direction because the teeth 'slip' in the other. As the H component in the $-\hat{\mathbf{z}}$ direction grows from 0 to H_{max} in this part of the cycle, the initial torque in the opposite direction is small, and the curve stays *above* the earlier part. At $H = 0$ we retain a *remanant magnetism* M_r. That is how magnetic materials end up with magnetization after a field has been applied and turned off, and it therefore is the method by which permanent magnets are made.

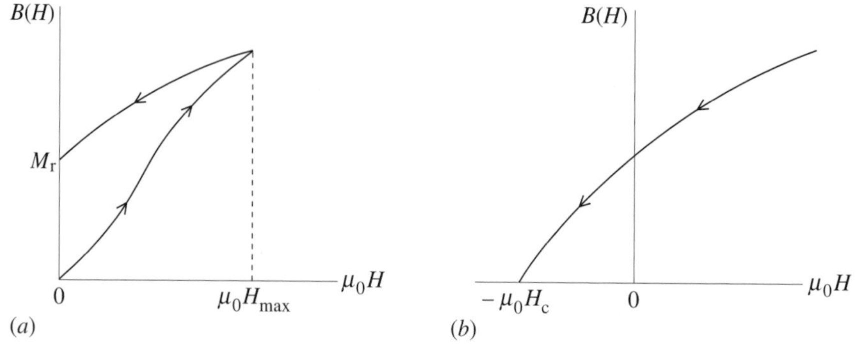

Fig. 14.8 The ferromagnetic $B(H)$ curve vs. $\mu_0 H$ for two stages of decreasing H.

Obviously a negative H is needed to bring the magnetic flux density B to zero. In Fig. 14.8(*b*), we show what happens when we try to do so. The field must be decreased from 0 to a value $-H_c$. The value H_c is known as the *coercive field*. Note that we still have $M > 0$ at $H = -H_c$; the reason is that $0 = B = \mu_0(H + M) = -\mu_0 H_c + \mu_0 M$ at this point of the curve. Let us now complete the entire cycle to obtain what is known as the *hysteresis curve* in Fig. 14.9.

At point P we have succeeded in making H sufficiently negative that $M = 0$. One might think happily that the sample has been demagnetized appropriately, but that is not so. The process of removing the sample from the field is the same as slowly bringing the H value from its negative value at P to zero, and the broken line shows what happens to $B(H)$. This broken line is similar to a portion of the curve in Fig. 14.7(*a*); the magnetization increases and at $H = 0$ there is an M remaining (shown here as a negative value). We will explain how in fact to *demagnetize* a magnet, but first note that the hysteresis curve is more or less symmetric so that the remaining portions can be explained by taking $-\hat{\mathbf{z}}$ as the reference direction.

We can demagnetize ('degauss' appears to be common jargon for this verb) a sample by immersing it in an exponentially (or otherwise) decaying alternating-current (ac) H field, as shown in Fig. 14.10(*a*). Such a field would have the time dependence

$$H(t) = H_{\max} e^{-\alpha t} \sin \omega t \tag{14.32}$$

The $\sin \omega t$ factor represents a sinusoid that in the absence of the exponential-decay factor corresponds in its first full cycle (plus an extra quarter cycle) to the hysteresis

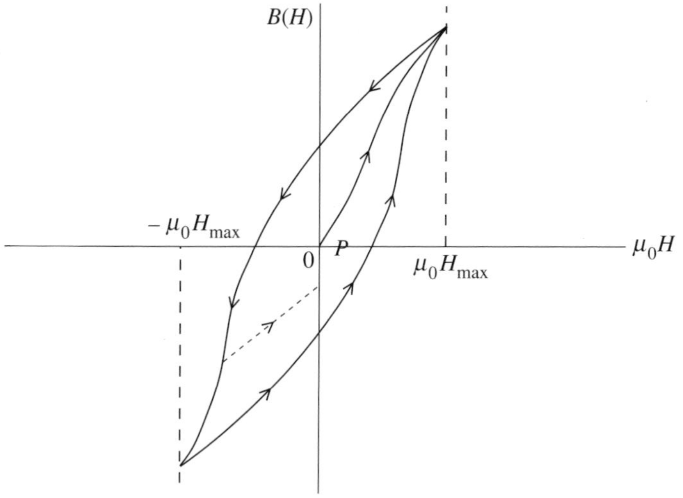

Fig. 14.9 The hysteresis curve of $B(H)$ vs. $\mu_0 H$ for one entire cycle of $H(t)$.

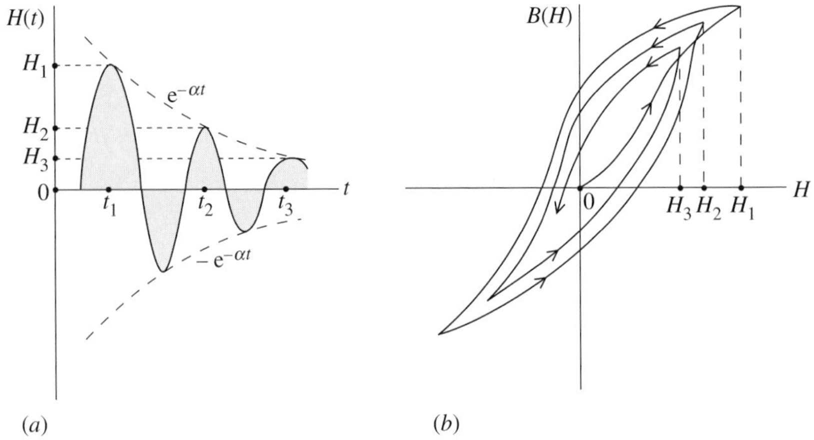

Fig. 14.10 The hysteresis curve of $B(H)$ vs. $\mu_0 H$ for several cycles of $H(t) = H_0 e^{-\alpha t} \cos \omega t$.

loop of Fig. 14.9. The corresponding hysteresis cycles, including the exponential factor, are shown in Fig. 14.10(*b*), and it should be clear that these shrink in area until they are negligibly close to the origin at every point (at which stage there effectively is no remanant magnetism or field left). Purely mechanical ways of removing remanant magnetism include subjecting the sample to severe shocks or to high temperatures, but neither of these alternatives guarantees preservation of the sample's integrity!

The shape of the hysteresis curve or loop is an aid in understanding why certain ferromagnetics ('hard' materials) are used for permanent magnets, whereas others ('soft' materials) are required for motors, transformers, generators, and the like. In subsection 13.3.2 we found that the energy stored in the totality of magnetic fields created by current sources is

$$W_B = \int_v dv \, \frac{1}{2} \mathbf{B} \cdot \mathbf{H} \tag{14.33}$$

The integrand is referred to as the *energy density* $w_B = \frac{1}{2}\mathbf{B} \cdot \mathbf{H}$, because it leads to the energy when integrated over volume. In a ferromagnetic sample (using scalar terminology) this would be $w_B(H) = \frac{1}{2} B(H) H$ because $B(H)$ varies nonlinearly with H, as Fig. 14.9 has demonstrated. Consider a small variation dH in the field H. The corresponding change in energy density is

$$dw_B = \frac{1}{2}\left(B + H\frac{dB}{dH}\right) dH = \frac{1}{2}(B\, dH + H\, dB) \tag{14.34}$$

where we have used the chain rule of differentiation. Referring to Fig. 14.11, it is clear that, except for a factor μ_0, $B\, dH$ is the area of rectangle $ABCG_1$ and $H\, dB$ is the

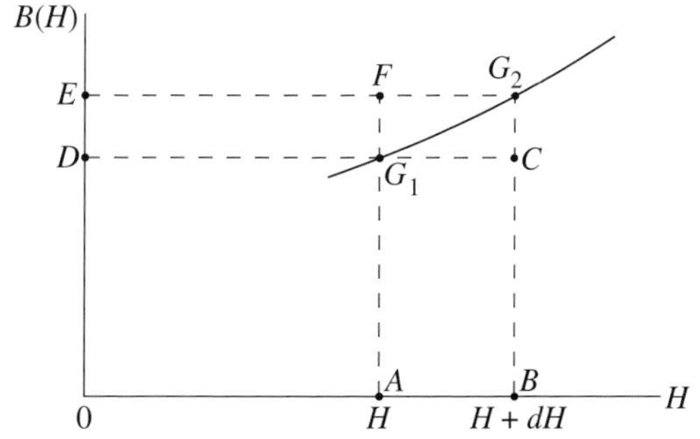

Fig. 14.11 $B(H)$ vs. H segment for evaluation of stored energy.

area of rectangle $DEFG_1$. The definition of an integral teaches that the triangular areas CG_1G_2 and FG_1G_2 are negligible compared to the rectangular areas if dH is an infinitesimal increment. Thus $B\,dH$ and $H\,dB$ represent, respectively, the areas *under* and to the *left of* the curve increment G_1G_2. Now, if we let H vary from $-H_{\mathrm{max}}$ to $+H_{\mathrm{max}}$ along the bottom part of the hysteresis curve in Fig. 14.9, and then back to $-H_{\mathrm{max}}$ along the top part, we will have traversed one entire loop of the hysteresis curve. If we apply what we have learned from Fig. 14.11 to this entire loop, we will see that $\oint dw_B$ corresponds to *twice* the area of the loop ($\oint B\,dH$ due to the sum of all rectangles $ABCG_1$ and $\oint HdB$ due to the sum of all rectangles $DEFG_2$). Thus,

$$\oint dw_B = \oint (B\,dH + H\,dB) = \frac{2}{\mu_0} \times \text{area of hysteresis loop} \tag{14.35}$$

Equation (14.35) represents energy lost due to the frictional forces during changes in the domains, resulting in turn in heat losses. Clearly, in soft materials, preferred for applications in which an ac field is applied for a lengthy period, we require those changes to be small so that a soft material would correspond optimally to as thin a hysteresis loop as possible, e.g. as in Fig. 14.12(*a*).

The design of a permanent magnet requires a large remanant magnetism, hence a hard material has a hysteresis loop as shown in Fig. 14.12(*b*). Large energy losses per cycle, due to the large area of the hysteresis loop, are not relevant because we only need to turn the field on to H_{max} and then back to zero to obtain the required magnetic sample. For this type of hysteresis loop we note that $\mu_{\mathrm{r}} = B/\mu_0 H$ is strikingly multivalued in addition to being nonlinear.

14.3.4 Some comments regarding permanent magnets and magnetism

What do the **B** field lines emanating from a permanent magnet look like? To answer this question, regard Fig. 14.13 as showing a short cylindrical magnet characterized by a uniform magnetization vector **M** that lies in the axial direction inside the

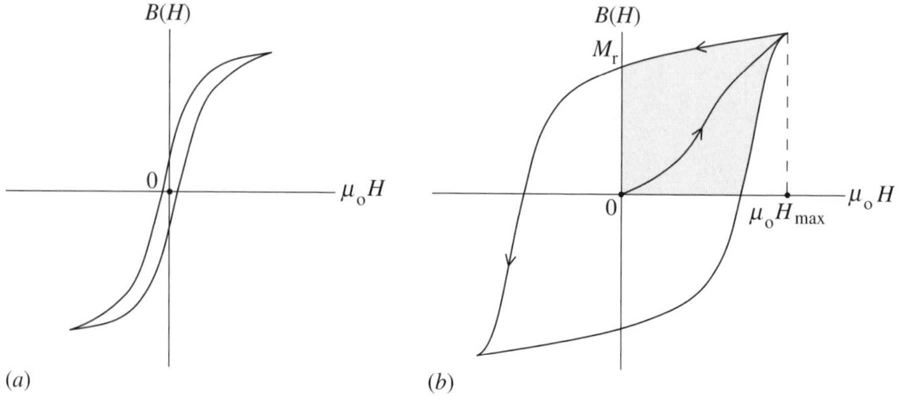

Fig. 14.12 Hysteresis curves (*a*) for generators and transformers, (*b*) for permanent

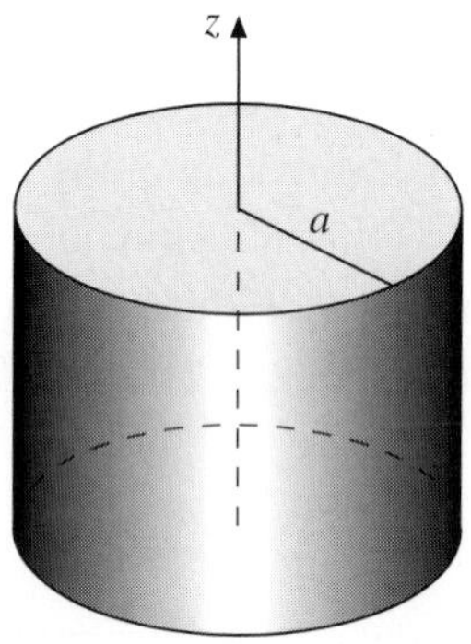

Fig. 14.13 A short cylindrical bar magnet; its field lines are like those of a solenoid having the same axis.

magnet and is zero outside. We have seen, (14.24c), that we can characterize magnetization by two apparent current sources, $-\nabla \times \mathbf{M}$ in the interior as a volume-current density and $\hat{\mathbf{n}} \times \mathbf{M}$ as a surface-current density at the surface with normal unit vector $\hat{\mathbf{n}}$. As $\mathbf{M}$ is uniform in an ideal cylindrical magnet, only the latter exists and then only on the wrap-around surface because $\hat{\mathbf{n}} \times \mathbf{M}$ is zero on the planar end surfaces. Hence, the magnet is entirely the same in its behavior with respect to magnetic properties, field lines, etc., as an identically shaped solenoid with surface current of density $\mathbf{K} = \hat{\mathbf{n}} \times \mathbf{M}$ (which in this case would be in the $\hat{\varphi}$ direction everywhere on the surface). For a solenoid, we had $\mathbf{K} = nI\hat{\varphi}$, where I is the current in one winding and n is the number of windings per unit length, as described in Section 11.3, Example 11.7. So we obtain all the results of Sections 11.3 and 11.4 for a solenoid which becomes a cylindrical magnet when we replace $K = nI$ on the surface by $M = K$ uniformly in the interior.

Historically, the fact that the **B** field lines seem to start at one end of a bar magnet and to circle around and terminate at the other end has led people to speak of two *poles* at the ends: a north and a south pole. We observe now that the **B** lines are continuous and that they close upon themselves; in fact they go through the inside of the bar magnet, as described for a finite solenoid in Section 11.4. The use of the word 'pole' is a little different here from how we have used it previously. In electrostatics, a dipole was introduced as a combination of two elementary charges of opposite sign; such single charges are known as *monopoles*. On the one hand, both electric monopoles and dipoles give rise to potentials and electric fields at a distance. On the other hand, the most elementary magnetic field appears to be due to a magnetic dipole (i.e. an infinitesimal current loop). There does not appear at this time to be a true magnetic *monopole*, but the possibility has been left open that one might exist. If so, the theory of electromagnetism would require revision, as Maxwell's equations do not allow for a magnetic monopole, only for an electric monopole.

14.4 Some vacuum–ferromagnet interface consequences

We discussed the boundary conditions for two magnetic media at their interface briefly in Section 11.5. To repeat the essentials, these conditions are, see (11.50) and (11.47),

$$\hat{\mathbf{n}} \times (\mathbf{H}_2 - \mathbf{H}_1) = \mathbf{K} \qquad \hat{\mathbf{n}} \cdot (\mathbf{B}_2 - \mathbf{B}_1) = 0 \tag{14.36}$$

This form of the expressions requires $\hat{\mathbf{n}}$ to be a surface-normal unit vector pointing *into* medium 2. There are some interesting engineering consequences when we deal with high-μ materials, e.g. ferrites, for which μ_r can exceed 2000. Consider for example a vacuum interface with a high-μ material, as shown in Fig. 14.14.

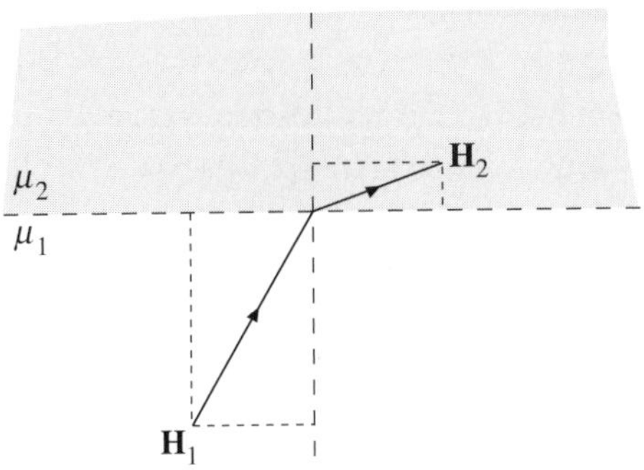

Fig. 14.14 Boundary situation for **H** at a vacuum–ferromagnet interface.

In the absence of a surface current at the interface ($K = 0$), these become

$$H_{2t} = H_{1t} \qquad H_{2n} = \frac{\mu_1}{\mu_2} H_{1n} \tag{14.37a}$$

or

$$B_{2t} = \frac{\mu_2}{\mu_1} B_{1t} \qquad B_{2n} = B_{1n} \tag{14.37b}$$

As we have $\mu_1 \ll \mu_2$ here, we observe that the normal component of $\mathbf{H}_2$ is a very small fraction of that of $\mathbf{H}_1$ and that B_{2t} greatly exceeds B_{1t}. Hence $\mathbf{H}_2$ is practically parallel to the interface, unless $\mathbf{H}_1$ is very close to being perpendicular to the interface. As a consequence, the magnetic field $\mathbf{H}$ in such materials tend to follow the shape of the interface (a sizable normal component in the material, at the interface, would require a huge one in free space on the other side of the interface).

Another consequence of (14.37) is the ability to produce large fields in an air gap between the poles of a horseshoe ferrite. Consider a circular solenoidal ferrite structure containing an air gap of length δl and with n windings of current I per meter; see Fig. 14.15. Ampère's law for a circular path at radius ρ states that

$$\oint d\boldsymbol{l} \cdot \mathbf{H} \equiv (2\pi\rho - \delta l) H_\varphi^{(m)} + \delta l H_\varphi^{(0)} = 2\pi\rho n I \tag{14.38}$$

As the interface in the air gap is essentially perpendicular to the azimuthally-oriented field lines (except for some fringe effects which we neglect), we replace the field in the ferrite, $H_\varphi^{(m)}$, by μ_0/μ_m times the field in the air gap, $H_\varphi^{(0)}$; this replacement requires use of the boundary condition for the normal components of $\mathbf{H}$.

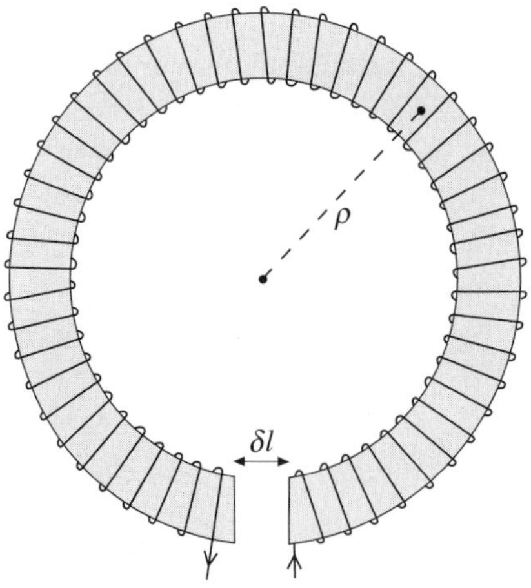

Fig. 14.15 Air gap in a current coil-wound toroidal geometry; $\mu_r \gg 1$.

Consequently,

$$H_\varphi^{(0)} = \frac{2\pi\rho n\mu_\mathrm{m} I}{(2\pi\rho - \delta l)\mu_0 + \delta l\mu_\mathrm{m}} \qquad H_\varphi^{(\mathrm{m})} = \frac{2\pi\rho n\mu_0 I}{(2\pi\rho - \delta l)\mu_0 + \delta l\mu_\mathrm{m}} \tag{14.39a}$$

Unless δl is extremely small, it is likely that the first term in the denominator is much smaller than the first, in which case

$$H_\varphi^{(0)} = \frac{2\pi\rho nI}{\delta l} \qquad H_\varphi^{(\mathrm{m})} = \frac{\mu_0}{\mu_\mathrm{m}}\frac{2\pi\rho nI}{\delta l} \tag{14.39b}$$

Comparing this field with the result for a diamagnetic toroidal coil, (11.38), with $a \approx \rho$ in that case, we see that the air-gap field has been increased by a factor $2\pi\rho/\delta l$. If, however, the opposite case holds, with δl so small that the first term in the denominator of (14.39a) is dominant, then we have

$$H_\varphi^{(0)} = \frac{\mu_\mathrm{m}}{\mu_0}\frac{2\pi\rho nI}{2\pi\rho - \delta l} \qquad H_\varphi^{(\mathrm{m})} = \frac{2\pi\rho nI}{2\pi\rho - \delta l} \tag{14.39c}$$

so that in both cases the introduction of an air gap greatly increases the **H** field above the value in the metal to either side.

14.5 Some details on diamagnetics and paramagnetics

We return in this section to give some more details on diamagnetic and paramagnetic substances. The calculation of diamagnetic and paramagnetic susceptibility actually requires material usually not taught at the undergraduate level, and therefore has been omitted in subsections 14.3.1 and 14.3.2. In fact, those calculations cannot be done classically for a system in thermal equilibrium, as pointed out by Feynman (see section 34.6 of Feynman *et al.*, 1964). They need to be done quantum mechanically. Nevertheless, a modified classical treatment leads to the correct formulas for the magnetization **M**. We think it might be useful to give some details here for those who are curious as to how one obtains a basis for the numerical values given in the cited sections. However, the next two sections can be omitted by those who wish to pass over these details.

14.5.1 Classical derivation of diamagnetic susceptibility

Although as mentioned above the calculation of the diamagnetic magnetization vector **M** cannot be done classically for a system in thermal equilibrium, we can obtain a correct expression for the magnetization vector from an appropriately modified classical treatment. We shall give one here which is similar to that given by Plonus[1] and by Elliott.[2] It starts from the quantum-mechanical fact that electrons are characterized by several quantum numbers. The *Pauli principle* states that no two electrons can have all quantum numbers identical.

In the older Bohr picture of the atom, electrons were visualized as orbiting around the nucleus in fixed orbits characterized by a quantum number. Electrons could change orbits only by emitting or absorbing one or more quanta of energy $h\omega/2\pi$, where ω is the angular frequency in rad/s that is involved. The quantity h is Planck's constant, and numerically $h \approx 6.63 \times 10^{-34}$ Js. Several orbits are arranged in a 'shell'. Each shell gets a principal quantum number, and each orbit in a shell gets a second and third quantum number. Finally, no more than two electrons can occupy the same orbit, but if they do, then a fourth quantum number distinguishes them. This last number corresponds to the z component of the spin angular momentum of the electron, known loosely as its 'spin', and quantum mechanics restricts any *measured* value of the 'spin' to $\pm h/4\pi$ along any direction. So if one electron in an orbit has 'spin' $h/4\pi$ in a given direction, then the other electron in that same orbit has 'spin' $-h/4\pi$ in the same direction.

Because the magnetic dipole moment $\mathbf{m}$ is proportional to angular momentum $\mathcal{L}$ (see below) it follows that the pair of electrons in one orbit have equal and opposite dipole moments $\mathbf{m}$ with magnitude $\sim 9.3 \times 10^{-24}$ A m^2.

A semiclassical picture of the electron magnetic dipole moment (see Fig. 14.16) is obtained by setting its magnitude equal to a current I time the orbital area:

$$m = I\pi a^2$$

where a is of the order of 5.29×10^{-11} m (the so-called Bohr radius of an atom: $\varepsilon_0 h^2/\pi \mathrm{m_e} e^2$ in terms of electron mass $\mathrm{m_e}$ and charge e). The 'current' $I \equiv dq/dt = -ef = -\omega e/2\pi$, if f is the rotation rate of the electron around its orbit (so that angular frequency $\omega = 2\pi f$). Thus

$$\mathbf{m} = -\frac{ea^2\omega}{2}\hat{\mathbf{n}} \tag{14.40}$$

where $\hat{\mathbf{n}}$ is a unit vector $\perp$ to the orbit, in a direction defined by the right-hand rule. This classical picture is somewhat fictitious, as we cannot really picture an electron as rotating around the nucleus in a fixed orbit. But it nevertheless serves a purpose: changes in $\mathbf{m}$ can be considered to occur because changes occur in I, which in turn induces changes in angular frequency ω.

[1] M.A. Plonus, *Applied Electromagnetics*, McGraw-Hill, New York, 1978.
[2] R.S. Ellior, *Electromagnetics – History, Theory and Applications*, IEEE Press series on Electromagnetic Waves, New York, 1993.

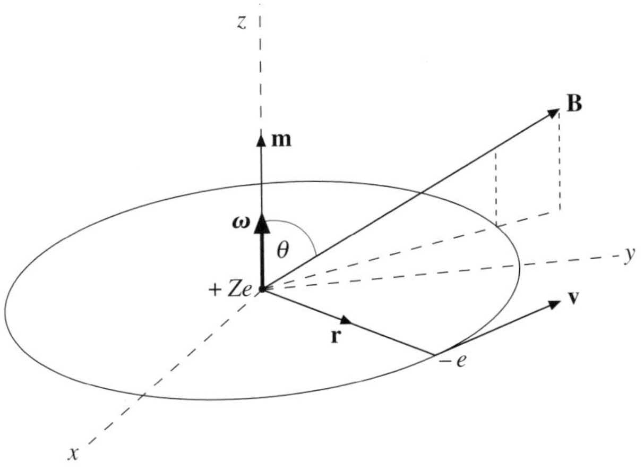

Fig. 14.16 An electron in a classical orbit around a nucleus.

The qualitative picture of diamagnetism is that it occurs when a **B** field is turned on quasistatically, i.e. gradually. According to Faraday's law (see Section 15.1) an identical V_{emf} is produced, in both electron-pair orbits, that opposes the change in magnetic flux **B**. In one of the orbits, that emf accelerates the electron, which increases ω, whereas in the other it decelerates the electron, which decreases ω. The two resulting changes in **m** reinforce each other and give rise to a small cumulative **M** opposed in direction to the **B** field.

The connection to the real quantum-mechanical picture follows from the fact that in that picture the orbital angular momentum is still $\boldsymbol{\mathcal{L}} = \mathfrak{m}\mathbf{r} \times \mathbf{v}$; in the semiclassical picture $\mathbf{v} = \mathbf{r} \times \boldsymbol{\omega} = a\hat{\mathbf{r}} \times \boldsymbol{\omega}$ for a circular orbit at constant angular frequency ω around an axis $\hat{\mathbf{n}}$ (i.e. $\boldsymbol{\omega} = \omega\hat{\mathbf{n}}$). Because $\mathbf{r} \perp \boldsymbol{\omega}$ it follows that $\boldsymbol{\mathcal{L}} = \mathfrak{m}a^2\omega\hat{\mathbf{n}}$, so that (14.40) yields for the semiclassical case

$$\mathbf{m} = -\frac{e}{2\mathfrak{m}}\boldsymbol{\mathcal{L}} \tag{14.41}$$

Quantum mechanics maintains the same relationship, but with a quantum-mechanical interpretation (see above) for the numerical values of $\mathcal{L}$. Changes in the dipole moment vector are due to (discrete) changes in the angular momentum vector.

To understand the changes in magnetic dipole moment, we should look at the balance of forces in an electron orbit. The 'centrifugal force' is given by

$$\begin{aligned}\mathbf{F}_c &= -\frac{\mathfrak{m}_e\,d\mathbf{v}}{dt} = -\mathfrak{m}_e\frac{d}{dt}(\mathbf{r} \times \boldsymbol{\omega}) = -\mathfrak{m}_e\frac{d\mathbf{r}}{dt} \times \boldsymbol{\omega} = -\mathfrak{m}_e\mathbf{v} \times \boldsymbol{\omega} \\ &= \mathfrak{m}_e(\mathbf{r} \times \boldsymbol{\omega}) \times \boldsymbol{\omega} = \mathfrak{m}_e\omega^2 a\hat{\mathbf{r}}\end{aligned} \tag{14.42}$$

in which expression we have assumed that $\boldsymbol{\omega}$ is constant. The rotating electron experiences an electrostatic force due to the nuclear charge Ze (Z is the atomic number of the atom):

$$\mathbf{F}_\mathrm{e} = -\frac{Ze^2}{4\pi\varepsilon_0 a^2}\hat{\mathbf{r}} \tag{14.43}$$

In the absence of other forces, these two balance, i.e. $\mathbf{F}_\mathrm{c} + \mathbf{F}_\mathrm{e} = 0$. This balance defines an angular frequency $\omega = \omega_0$, with

$$\omega_0^2 = \frac{Ze^2}{4\pi\varepsilon_0 a^3 \mathfrak{m}_\mathrm{e}} \tag{14.44}$$

Numerical insertion of the electronic charge and mass values yields $\omega_0 \approx 4.13 \times 10^{16}\, Z\,\mathrm{rad/s}$. However, a third force upon the rotating electron is felt when a $\mathbf{B}$ field has been switched on adiabatically, the Lorentz force $\mathbf{F}_\mathrm{L} = -e\mathbf{v} \times \mathbf{B}$. When an average over time is considered, the orbital motion of an electron could be clockwise or counterclockwise with equal probability. It follows that the balance of radial forces on each electron yields

$$\mathfrak{m}_\mathrm{e}\omega^2 a\hat{\mathbf{r}} - \frac{Ze^2}{4\pi\varepsilon_0 a^2}\hat{\mathbf{r}} \pm e(\mathbf{v} \times \mathbf{B})_r = 0 \tag{14.45}$$

with the plus sign for one possibility and the minus sign for the other. The Lorentz force component $e(\mathbf{v} \times \mathbf{B})_z$ contributes to the paramagnetic effect (see below). Now introduce the combination $e\mathbf{B}/\mathfrak{m}_\mathrm{e} = \boldsymbol{\Omega}$, which has the dimension of rad/s and therefore is an angular frequency in magnitude. Numerically $e/\mathfrak{m}_\mathrm{e} \approx 1.6 \times 10^{11}\,\mathrm{C/kg}$, so that even a magnetic flux density $B \leq 100\,\mathrm{T}$ yields $\Omega \leq 10^{13}\,\mathrm{rad/s}$. Equation (14.45) can be rewritten, after replacing $\mathbf{v} = a\hat{\mathbf{r}} \times \boldsymbol{\omega}$ and dividing by $\mathfrak{m}_\mathrm{e} a$, as

$$\omega^2 - \omega_0^2 \pm \boldsymbol{\omega} \cdot \boldsymbol{\Omega} = 0 \tag{14.46a}$$

where, numerically, $\Omega \ll \omega_0$. Therefore this relationship can only be satisfied if ω is of the same order as ω_0, i.e. $\omega = \omega_0 + \Delta\omega$, with $\Delta\omega \ll \omega_0$. In this case, (14.46a) reduces to

$$2\omega_0\Delta\omega \pm \omega_0\Omega\cos\theta = 0 \tag{14.46b}$$

if higher-order terms, of order Ω/ω_0, are ignored. Here, θ is the angle between the magnetic flux density $\mathbf{B}$ and the electron dipole moment $\mathbf{m} = m\hat{\mathbf{n}}$. So we obtain for the pair of angular frequency increments

$$\Delta\omega = \pm\frac{1}{2}\Omega\cos\theta = \pm\frac{e\hat{\mathbf{n}}\cdot\mathbf{B}}{2m_e} \tag{14.47}$$

From (14.40) we see that $\Delta\mathbf{m} = -ea^2\Delta\omega/2$, but each pair of clockwise and counterclockwise rotating electrons gives equal but opposite **m** vectors, which implies that $\cos\theta_2 = \cos(\pi - \theta_1) = -\cos\theta_1$. Thus $\Delta\omega$ as given by (14.47) is the same for *both* orbital electrons. Each such pair of orbital electrons thus produces a total change

$$\Delta\mathbf{m} = 2\Delta I\pi a^2\hat{\mathbf{n}} = -ea^2\Delta\omega\hat{\mathbf{n}} = -\frac{e^2a^2}{2m_e}(\hat{\mathbf{n}}\cdot\mathbf{B})\hat{\mathbf{n}} \tag{14.48}$$

Hence addition for all pairs in a unit volume yields

$$\mathbf{M} = -\frac{e^2a^2}{2m_e}\left(\sum\hat{\mathbf{n}}\hat{\mathbf{n}}\right)\cdot\mathbf{B} \tag{14.49}$$

The combination $\hat{\mathbf{n}}\hat{\mathbf{n}}$ is known as a *dyadic*. When multiplied into **B** it gives $\hat{\mathbf{n}}\hat{\mathbf{n}}\cdot\mathbf{B} = (\hat{\mathbf{n}}\cdot\mathbf{B})\hat{\mathbf{n}}$. Thus it is easy to see that $\hat{\mathbf{x}}\hat{\mathbf{x}} + \hat{\mathbf{y}}\hat{\mathbf{y}} + \hat{\mathbf{z}}\hat{\mathbf{z}} = \mathcal{I}$ is the unit dyadic, which does not modify **B** when you dot-multiply it from the left (or the right!) into **B**. A large sum over a more-or-less isotropic distribution of spin directions gives equal likelihood that $\hat{\mathbf{n}}\hat{\mathbf{n}}$ is in any direction: because the most general dyadic can be written as

$$\hat{\mathbf{n}}\hat{\mathbf{n}} = (a\hat{\mathbf{x}}\hat{\mathbf{x}} + b\hat{\mathbf{y}}\hat{\mathbf{y}} + c\hat{\mathbf{z}}\hat{\mathbf{z}})\Big/\sqrt{a^2+b^2+c^2})$$

it follows that it is equally likely that a, b, and c take on any particular value in an isotropic situation, and thus that $\langle a\rangle = \langle b\rangle = \langle c\rangle$, hence that $\hat{\mathbf{n}}\hat{\mathbf{n}} = \frac{1}{3}\mathcal{I}$. Therefore, we finally obtain

$$\mathbf{M} = -\frac{Ne^2a^2}{6m_e}\mathbf{B} \tag{14.50}$$

where N is the number of pairs of orbital electrons with equal but opposite spins; therefore

$$\chi_m = -\frac{\mu_0 Ne^2a^2}{6m_e} \sim -1.65\times10^{-35}\,(N \text{ in m}^{-3}) \tag{14.51}$$

From this we note that $\chi_m \le -10^{-5}$ whenever $N \le 6.5\times10^{29}\ \text{m}^{-3}$, as might occur in a metallic substance.

14.5.2 Derivation of the Langevin formula for paramagnetic magnetization

Paramagnetism, as is the case with diamagnetism, cannot really be treated correctly without quantum-mechanical reasoning, but again a semiclassical argument does lead to the correct formula. We saw above, in (14.45), that the component of the **B** field in the plane of the orbiting electron was not considered in the diamagnetic effect. But that component can be seen to produce a torque $\mathbf{T} = \mathbf{m} \times \mathbf{B}$, where now **m** is the total magnetic moment of the orbiting electron, which tilts **m** so that it orients itself in the direction of **B**. The work done by the field on the magnetic dipole is $W(\theta) = \mathbf{m} \cdot \mathbf{B} = mB\cos\theta$ and the energy of the collection of atoms and/or molecules in the field is lowered by $-W(\theta)$. The preferred situation arises when $\theta = 0$, but that occurs only in the absence of thermal motion, which contributes to a disorientation of the dipoles through induced collisions. These opposing forces give rise to paramagnetism.

Statistical mechanics predicts that the number of electrons with dipole moment **m** present in a system, with imposed **B** field, that is in thermal equilibrium at a temperature of T kelvins is given by

$$n(\theta) = n_0 e^{-W(\theta)/kT} = n_0 e^{\alpha\cos\theta} \qquad \text{with } \alpha \equiv mB/kT \tag{14.52}$$

Here $k \approx 1.38 \times 10^{-23}$ J/K is Boltzmann's constant. The quantity n_0 is obtained from the fact that the total number of particles per unit volume, N, is the integral of (14.52) over all solid angles:

$$N = \int_0^{2\pi} d\varphi \int_0^{\pi} d\theta \sin\theta\, n(\theta) = \frac{2\pi n_0}{\alpha}(e^{\alpha} - e^{-\alpha}) \tag{14.53}$$

Addition of all dipole moments per unit volume, weighted properly by their number-density $n(\theta)$, yields the magnetization:

$$\begin{aligned} M &= \int_0^{2\pi} d\varphi \int_0^{\pi} d\theta \sin\theta\, n(\theta) m\cos\theta \\ &= 2\pi n_0 m \int_0^{2\pi} d\varphi \int_0^{\pi} d\theta \sin\theta \cos\theta\, e^{\alpha\cos\theta} \end{aligned} \tag{14.54}$$

The integral is easily evaluated, and by using (14.53) to eliminate n_0 in favor of N we obtain

$$M = Nm\left(\operatorname{cotanh}\alpha - \frac{1}{\alpha}\right) \tag{14.55}$$

which is known as the *Langevin* formula for paramagnetism. At room temperature ($T = 27°$C) we have $kT \approx 4.14 \times 10^{-21}$ J, whereas $m \approx 10^{-23}$ A m^2 so that $\alpha \approx 2.5 \times 10^{-3}$ (B in T). Hence, unless the magnetic flux density is quite high, it follows that α is small, and then (14.55) is well approximated by

$$M = Nm^2 B/3kT \quad \longrightarrow \quad \chi_m = \frac{\mu_0 N m^2}{3kT} \tag{14.56}$$

This approximation is known as *Curie's law*, where the magnetization is a linear function of B. We refer the reader back to the discussion in subsection 14.3.2.

Problems

14.1. An element of a circuit carrying 4 A of current is 3 m long and is oriented in the $z = 0$ plane at $\varphi = 45^\circ$ to the x axis. It lies in a uniform magnetic field with $\mathbf{B} = 2\hat{\mathbf{z}} - 5\hat{\mathbf{x}}$ T. Calculate the force on the element, given that the current is moving away from the origin.

14.2. Another current element is 3 m long and carries a 2 A current in the $\hat{\mathbf{z}}$ direction. It is subject to a uniform magnetic flux-density field $\mathbf{B}$, which exerts a force of $0.5\hat{\mathbf{x}} - \hat{\mathbf{y}}$ N on it. Calculate $\mathbf{B}$.

14.3. A square loop in the $z = 0$ plane has its four corners on the two main axes instead of as shown in Fig. 14.3. The magnetic flux-density field is still $\mathbf{B} = B_y\hat{\mathbf{y}} + B_z\hat{\mathbf{z}}$.
(a) Explain how the loop rotates.
(b) Describe its end position.

14.4. Given a hypothetical value of 9.3×10^{-24} A m^2 for the orbital magnetic moment of an electron, use that value and the classical (Bohr) radius of an atom to calculate the rotation rate, i.e frequency, at which an electron would orbit around the atom in the (incorrect) classical picture.

14.5. Three parallel wires, each at a distance of 10 cm from each of the other two, carry a current of 15 A, all in the same direction. Calculate the force vector (magnitude and direction) per unit length on one of these wires due to the other two wires.

14.6. Consider a triangular loop in the $z = 0$ plane, with corners at $(x, y) = (-5, 0)$, $(5, 0)$, and $(0, 10)$, where all lengths are in meters. A constant current of 8 A flows counterclockwise (as seen from above) around the loop. There is a constant flux density $\mathbf{B} = 0.2\hat{\mathbf{y}}$ T. Find the forces (all three) and the torque on the loop. Explain carefully how the loop will rotate.

14.7. Consider a circuit defined on a cylindrical surface at distance ρ from the z axis. Four points on the circuit are $A(\rho, \varphi_1, 0)$, $B(\rho, \varphi_1 + \varphi, 0)$, $C(\rho, \varphi_1 + \varphi, h)$, $D(\rho, \varphi_1, h)$. Current I_2 runs in the direction $A \to D \to C \to B \to A$. The z axis carries a current I_1 in the $+\hat{\mathbf{z}}$ direction. Calculate the force on the four lengths AB, CD, AD, BC due to current I_1. How will the circuit rotate?

14.8. Consider a toroidal coil as shown in Fig. 14.15, with $n = 50$ windings per cm, $I = 4\,\text{mA}$, and an average radius $\rho = 20\,\text{cm}$. It contains a $\mu = 200\mu_0$ material and has a 5 mm air gap.
(a) Calculate the H field in the air gap.
(b) Which of the approximations (14.39b), (14.39c) can be used here, and why?

14.9. Considering that magnetic fields in high-μ materials ($\mu_r \geq 500$) are produced by current coils wrapped around the material, explain why the B, H fields produced in the material tend to be parallel to the interfaces with air.

14.10. A metallic circular cylinder of radius b around the z axis has relative permeability μ_r and current I_0 in the $+\hat{\mathbf{z}}$ direction.
(a) Give the magnetization vector $\mathbf{M}$ in terms of the relevant quantities.
(b) What equivalent current density $\mathbf{J}_{mv}$ can 'replace' the vector $\mathbf{M}$?

14.11. For a weak magnetic material it is given that $M = 0.00025\,\text{A/m}$ when $H = 10\,\text{A/m}$, and $M = 0.00040\,\text{A/m}$ when $H = 16\,\text{A/m}$. Calculate the relative permeability μ_r to six decimal places.

14.12. Assuming that the number of electrons contributing to the magnetization in a material is $N \approx 10^{30}$ per m^3, calculate the paramagnetic $\mathbf{M}$ induced in the material by the Earth's magnetic field with $B \approx 0.5$ gauss. You may assume a temperature of 27° C.

14.13. A B–H curve is given by $B(H) = 150\,\mu_0 H^2/(6400 + H^2)$, when H is turned on from $H = 0$ to $H = 150\,\text{A/m}$. Sketch the magnetization $M(H)$ and susceptibility $\chi_m(H)$ as functions of H in this regime.

15

Extension to electrodynamics

We have reached an important crossroads in the treatment of electromagnetic fields; we are at the stage at which time variations of the sources and/or the fields will play an important role. Let us summarize what we have found so far. This is done most succinctly with the following equations, which include the constitutive equations and the two equations of continuity as well as Maxwell's equations for static charges and currents:

$$\begin{array}{ll} \nabla \times \mathbf{E} = 0 & \nabla \times \mathbf{H} = \mathbf{J} \\ \nabla \cdot \mathbf{D} = \varrho_{\mathrm{V}} & \nabla \cdot \mathbf{B} = 0 \\ \partial \varrho_{\mathrm{V}} / \partial t = 0 & \nabla \cdot \mathbf{J} = 0 \\ \mathbf{D} = \varepsilon \mathbf{E} & \mathbf{B} = \mu \mathbf{H} \end{array} \tag{15.1}$$

Note that the analogies are somewhat different from those at the end of Section 14.2. Neither the curl **E** nor div **B** equations have a source term, which would be respectively a 'magnetic current density' and 'magnetic charge density'. Such sources do not appear to exist. More important, the magnetostatic equations to the right are completely separate from the electrostatic ones on the left: that is to say that no quantity on the right appears on the left, and vice versa. This is why we are able to handle magnetostatics separately from electrostatics, with no apparent connection between the two theories. Faraday's law and Maxwell's displacement current will interconnect these equations, and the full theory of electrodynamics results. This theory is one of the most perfected achievements in classical physics and its marriage to quantum mechanics, resulting in quantum electrodynamics, has left a similar mark upon modern physics. Even today – after more than 125 years of effort since Maxwell formulated the equations which now bear his name – the consequences of the full Maxwell equations seem to be inexhaustible, and a prodigious amount of analytical and numerical applications continue to provide work for numerous researchers.

15.1 Faraday's law – introduction

Faraday's law expresses the phenomenon Faraday found experimentally and reported in 1839: that a *change with time* in the flux linkage $\Lambda = \int_S d\mathbf{S} \cdot \mathbf{B}$ through a surface S gives rise to an *electric field* $\mathbf{E}$ around the boundary curve C. The quantity $\oint_C d\boldsymbol{l} \cdot \mathbf{E}$ is known as the *induced electromotive force*, V_{emf}. If a closed conductor lies around C then an *induced current* will flow. If the circuit is open then there is an induced electric potential difference V_{emf} at the terminals. This open circuit corresponds in the electrostatic case to a battery maintaining a potential difference V_{emf} between its poles. The mathematical form of Faraday's law is

$$V_{\text{emf}} = -\frac{d\Lambda(t)}{dt} = -\frac{d}{dt}\left(\int_S d\mathbf{S} \cdot \mathbf{B}\right) \tag{15.2}$$

Interpretation of this equation is not straightforward. In considering the change in time of the flux linkage, we note that either $\mathbf{B}$ or $d\mathbf{S}$ could be a function of time t – or both could be functions of t. Furthermore, the minus sign is the manifestation of a separate law, *Lenz's law*, which we can understand right away, without further ado, from a simple argument.

Lenz's law states that the *sense* (clockwise or counterclockwise) *of the induced emf created by a change in magnetic flux must be such as to oppose the change producing it.* Consider the following hypothetical case: a circular conductor with a steadily increasing uniform field $B(t)$ perpendicular to the plane of the circuit, as in Fig. 15.1. By (15.2) an emf is induced round the site of the conductor, so that an *induced current* will flow. If this current were to flow *counterclockwise* then it would give rise, through Ampère's law, to a further increase in the $\mathbf{B}$ field enclosed by the circuit. That further increase would in turn, through (15.2), give rise to yet a greater induced emf, and we would have a run-away effect, with V_{emf} becoming infinite. Needless to

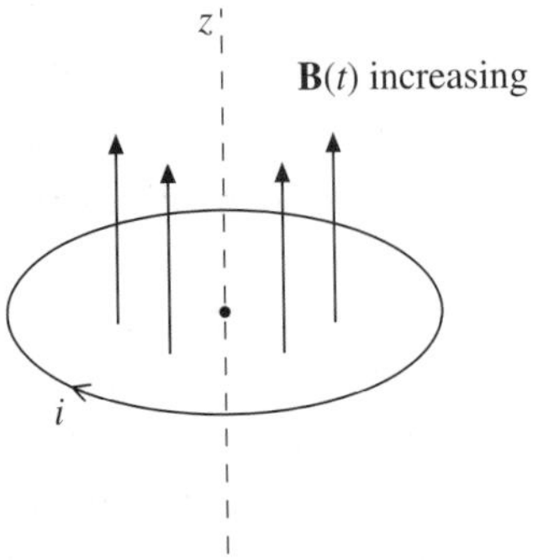

Fig. 15.1 A loop in a vertically upward and increasing **B** field.

say, by the conservation of energy this cannot happen, and so we have the minus sign in (15.2). Thus V_{emf} in this case must give rise to a *clockwise* current in the circuit, so that the **B** field this creates *opposes* the increase in $B(t)$. If, however, the field $B(t)$ were *decreasing*, then the induced emf and current would be in the *counterclockwise* sense, opposing this decrease.

Looking further at (15.2), it is obvious that the time change can work in two different ways. The first is through time changes in the magnetic flux density $\mathbf{B}(t)$, and the second is through changes in the shape and orientation of the circuit's surface area S. Combinations of both can occur. Nevertheless, it is useful to make a distinction between the two types of change.

15.1.1 Faraday's law: 'transformer' form

If the circuit is itself immobile in a frame of reference, but the field is not stationary, i.e. $\partial B(\mathbf{r}, t)/\partial t \neq 0$, then (15.2) becomes

$$V_{\text{emf}} = -\int_S d\mathbf{S} \cdot \frac{\partial \mathbf{B}(\mathbf{r}, t)}{\partial t} \tag{15.3a}$$

Note the partial derivative, to indicate that we differentiate **B** only with respect to the time variable. However, as mentioned above, we also have

$$V_{\text{emf}} = \oint d\boldsymbol{l} \cdot \mathbf{E} \tag{15.3b}$$

That is, V_{emf} equals the contour integral of the induced electric field.

Sign convention for $d\boldsymbol{l}$. Using what we have deduced from Faraday's law applied to Fig. 15.1, we can now interpret the directions of the quantities $d\mathbf{S}$ and $d\boldsymbol{l}$: $d\boldsymbol{l}$ is chosen in such a way that if $d\mathbf{S}$ points upwards, then the sense of $d\boldsymbol{l}$ is counterclockwise when the circuit is viewed from above. In other words, $d\mathbf{S}$ and $d\boldsymbol{l}$ are related by the right-hand screw rule.

As discussed above, in Fig. 15.1 the induced electric field **E**, and so the induced emf, are in the clockwise sense when **B** is increasing, and so if we choose $d\mathbf{S}$ in the direction of **B**, so that $d\boldsymbol{l}$ is counterclockwise, we will find that $d\boldsymbol{l} \cdot \mathbf{E}$ is always negative; therefore in this case the induced emf has a negative sign associated with it.

So this negative sign for induced emf, which follows from the sign convention we have chosen for $d\mathbf{S}$ and $d\boldsymbol{l}$, expresses the fact that the induced current has its own magnetic flux in a direction opposing the original flux. With the same sign convection, if **B** were instead *decreasing* the induced emf and current would be positive, expressing the fact that the corresponding induced flux is now in the same direction as the original field **B**.

We now develop (15.3b) by using Stoke's law:

$$V_{\text{emf}} = \oint d\boldsymbol{l} \cdot \mathbf{E} = \int_S d\mathbf{S} \cdot (\nabla \times \mathbf{E}) \tag{15.3c}$$

As S is an entirely arbitrary surface, the equalities (15.3a,c) yield

$$\nabla \times \mathbf{E} = -\frac{\partial \mathbf{B}}{\partial t} \tag{15.4}$$

and this differential form is the definitive modification of the curl **E** Maxwell equation, to account for changing magnetic flux density **B**.

Example 15.1 An exponentially time-varying B field in a rectangular circuit

Consider a field $\mathbf{B}(\mathbf{r}, t) = B_0 e^{-\alpha t}\hat{\mathbf{z}}$, with α, $B_0 > 0$, through the circuit of Fig. 15.2, which is a rectangle with sides a and b; it has a small break ***PQ*** as shown. The magnetic flux at time t is

$$\Lambda(t) = abB_0 e^{-\alpha t} \tag{15.5}$$

Its time derivative is simple (just multiply the right-hand side here by $-\alpha$); hence we find

$$V_{\text{emf}}(t) = +\alpha abB_0 e^{-\alpha t} \tag{15.6}$$

Thus, $V_P - V_Q = V_{\text{emf}}$ is positive, because the *change* in the magnetic flux in this example is *negative*. This negative change in the **B** field has to be opposed, according to Lenz's law, and, if the circuit were closed, the positive V_{emf} would produce a counterclockwise current, which in turn would tend to reinforce **B**.

The special theory of relativity teaches that formulations of the laws of physics cannot distinguish between frames of reference that differ only by a constant velocity from each other. Thus, a perfectly rigid circuit moving through a stationary but

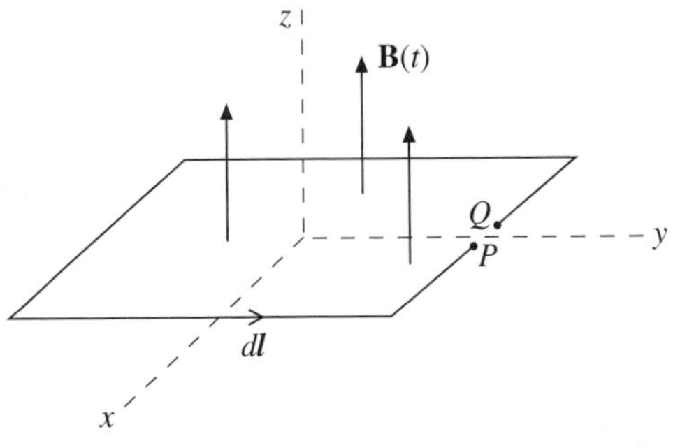

Fig. 15.2 A rectangular circuit in an exponentially time-varying **B** field.

spatially varying $\mathbf{B}(\mathbf{r})$ field would be regarded by an observer 'sitting' on the circuit as being in a time-varying $B(\mathbf{r}, t)$ field. Such a situation can be treated with the transformer form (15.3a) of Faraday's law. An example will suffice.

Example 15.2 A B field that varies exponentially in space in a rectangular circuit
Consider again the situation of Fig. 15.2, but now let $\mathbf{B} = B_0 e^{-\beta y}\hat{\mathbf{z}}$. Let the circuit move in the $+\hat{\mathbf{y}}$ direction with velocity u. Thus if a point on the circuit is at (x, y) at time $t = 0$, then it is at $(x, y + ut)$ at time $t > 0$. It experiences a field $B(t) = B_0 e^{-\beta(y+ut)} = B(0)e^{-\beta ut}$ at time t. The magnetic flux through this circuit, centered at the origin at time t, is

$$\Lambda(t) = -a \int_{-b/2}^{b/2} dy B_0 e^{-\beta(y+ut)} = -\frac{a}{\beta}(e^{\beta b/2} - e^{-\beta b/2})B_0 e^{-\beta ut} \tag{15.7}$$

Hence, we find for the induced emf

$$V_{\text{emf}}(t) = au(e^{\beta b/2} - e^{-\beta b/2})B_0 e^{-\beta ut} \tag{15.8}$$

An interesting comparison with Example 15.1 can be made if we allow $\beta \to 0$ in such a way that $\beta u = \alpha$ for a given nonzero value of α. This will still ensure that V_{emf} in (15.8) varies with time. That of course makes the velocity very large, even larger than 3×10^8 m/s, which is forbidden by the special theory of relativity, but let us simply assume that the velocity u is first chosen, and that we assume b is sufficiently small, and β sufficiently large, that $\beta b \ll 1$ and $\beta u = \alpha$. We then can approximate the difference in exponentials by βb, after which we set $\beta u = \alpha$ wherever we can in (15.8) to obtain

$$V_{\text{emf}}(t) = \alpha ab B_0 e^{-\alpha t} \tag{15.9}$$

This is identical to (15.6), which is precisely as it should be because we are now describing an essentially uniform field that appears to an observer, 'sitting' on the circuit, to be decaying exponentially in time with the factor $e^{-\beta ut} = e^{-\alpha t}$. This, of course, is exactly the case of Example 15.1. The special theory of relativity says that these two cases are indistinguishable, and it is borne out by this calculation.

The above example was sufficiently simple that it was possible to 'translate' from one frame of reference to another. The general case of an inhomogeneous time-varying $\mathbf{B}(\mathbf{r}, t)$ may not lend itself so easily to a similar treatment in an equivalent frame of reference in which the $\mathbf{B}$ field varies only spatially, and the circuit is in motion through it. For such cases, the *motional* form of Faraday's law is more suitable.

15.1.2 Faraday's law: 'motional' form

Now we consider a **B** field that is constant in time, linking a circuit part or all of which is in motion:

$$V_{\text{emf}} = -\int_S \frac{d}{dt}(d\mathbf{S}) \cdot \mathbf{B}(\mathbf{r}) \tag{15.10}$$

and try to decide what is meant by this expression. In this case, we assume that parts or all of the circuit are in motion, and let us for convenience allow the circuit to move only in its own plane (i.e. not to tilt its plane), although it could become distorted while doing so. Remember that each surface element $d\mathbf{S}_k$ (in direction $\hat{\mathbf{n}}_k$) can be described by a cross product $d\boldsymbol{l}_i \times d\boldsymbol{l}_j$, in a suitable orthogonal coordinate system with unit vectors $\hat{\mathbf{n}}_i, \hat{\mathbf{n}}_j, \hat{\mathbf{n}}_k$. Let us fill the surface of the circuit loop with small interior $d\mathbf{S}$ elements; we have seen earlier that a line element $d\boldsymbol{l}$ on the boundary of an interior surface element would be equal and opposite to one on a neighboring interior surface element if we defined all bounding contours to be counterclockwise. Thus the contributions from these interior surface elements cancel, and so we have for $d\mathbf{S} = d\mathbf{S}_k = \hat{\mathbf{n}}_k\, dS$ (we define $d\boldsymbol{l}_i = \hat{\mathbf{n}}_i\, dl_i$, and $d\boldsymbol{l}_j = \hat{\mathbf{n}}_j\, dl_j$)

$$\begin{aligned}\frac{d(d\mathbf{S}_k)}{dt} &= \frac{d}{dt}(d\boldsymbol{l}_i \times d\boldsymbol{l}_j) = \left(\frac{d}{dt}\, d\boldsymbol{l}_i\right) \times d\boldsymbol{l}_j + d\boldsymbol{l}_i \times \left(\frac{d}{dt}\, d\boldsymbol{l}_j\right) \\ &= \mathbf{u}_i \times d\boldsymbol{l}_j + d\boldsymbol{l}_i \times \mathbf{u}_j\end{aligned} \tag{15.11}$$

where $\mathbf{u} \equiv u_i, u_j, u_k$ is the velocity of the line element $d\boldsymbol{l}$.

Upon dot-multiplying the last expression on the right-hand side of (15.11) by **B** we see that

$$\frac{d}{dt}(d\mathbf{S}_k) \cdot \mathbf{B} = (\mathbf{u}_i \times d\boldsymbol{l}_j + d\boldsymbol{l}_i \times \mathbf{u}_j) \cdot \mathbf{B} = -d\boldsymbol{l}_j \cdot (\mathbf{u}_i \times \mathbf{B}) + d\boldsymbol{l}_i \cdot (\mathbf{u}_j \times \mathbf{B}) \tag{15.12}$$

Note that only the B_k component contributes to this because both velocities lie in the plane $\perp \hat{\mathbf{n}}_k$. To see this, note that only the last term remains for $\mathbf{B} = \mathbf{B}_i$; however, $\mathbf{u}_j \times \mathbf{B}_i$ points in the $-\hat{\mathbf{n}}_k$ direction and therefore is perpendicular to $d\boldsymbol{l}_i$, with which it then forms a zero dot product (and similarly for $\mathbf{B} = \mathbf{B}_j$). It then follows for this surface element that

$$\frac{d}{dt}[(d\mathbf{S}_k) \cdot \mathbf{B}] = dl_j(\mathbf{u} \times \mathbf{B})_j + dl_i(\mathbf{u} \times \mathbf{B})_i = d\boldsymbol{l} \cdot (\mathbf{u} \times \mathbf{B}) \tag{15.13}$$

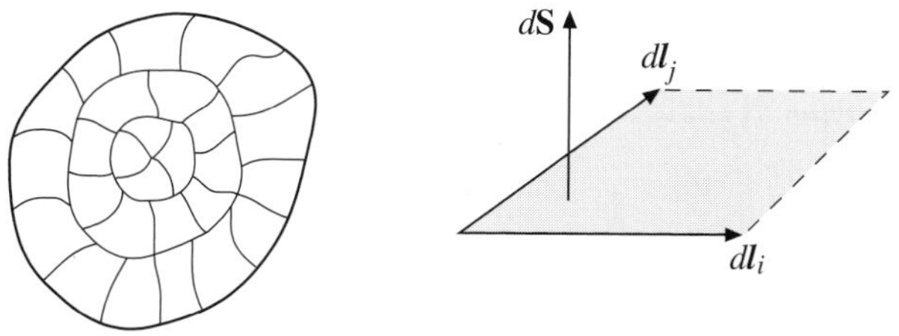

Fig. 15.3 A surface having as its boundary a current loop; to derive Faraday's motional law the surface is divided into elements $d\mathbf{S}$.

After addition of all the infinitesimal surface elements in S in similar fashion, (15.10) becomes

$$V_{\text{emf}} = \oint_C d\boldsymbol{l} \cdot (\mathbf{u} \times \mathbf{B}) \tag{15.14}$$

which is the motional form of Faraday's law. The same result holds for *arbitrary* motion of the line elements; to extend the above derivation to the more general case, we only need to reinterpret (15.12) somewhat.

To see what (15.14) implies, we note that qV_{emf} is the work done in moving a charge q around the circuit in Fig. 15.3, and this work is also $\oint d\boldsymbol{l} \cdot \mathbf{F}$ if $\mathbf{F}$ is the force on charge q at dl.

Consequently, we see using (15.14) that this yields for the force $\mathbf{F}$ on a charge q moving with velocity $\mathbf{u}$ relative to a field $\mathbf{B}$

$$\mathbf{F} = q\mathbf{u} \times \mathbf{B} \tag{15.15}$$

The motional form of Faraday's law thus corresponds to the presence of a *Lorentz* force upon charges in a circuit. That force is due to a motion of those charges *across* the magnetic field lines. We see that here such a motion causes an electromotive force, an emf, around the circuit. (It could also give rise to forces transverse to the circuit; cf. Section 14.1, where current gives rise to motion rather than vice versa as here.)

Example 15.3 A rectangular circuit with a bar that moves across a uniform B field

A classical example is shown in Fig. 15.4. A circuit has a bar on the far side that can slide freely in the $\pm\hat{\mathbf{x}}$ direction. We assume that it moves with uniform velocity $\mathbf{u} = u\hat{\mathbf{x}}$. We also assume a uniform and constant magnetic field $\mathbf{B} = B_0\hat{\mathbf{z}}$. We first use (15.2) to predict what we will get, because the magnetic flux is $\Lambda(t) = B_0bx(t)$, which follows from the fact that $\mathbf{S} = bx(t)\hat{\mathbf{z}}$. Therefore $d\Lambda/dt = B_0bu$, because the time derivative of $x(t)$ is just the velocity u. Hence we find $V_{\text{emf}} = -B_0bu$, and interpret $|V_{\text{emf}}|$ as the potential difference $V_{QP} = V_P - V_Q$. This latter is dictated by the fact that the flux is *increasing*; therefore, according to Lenz's law, if the circuit were

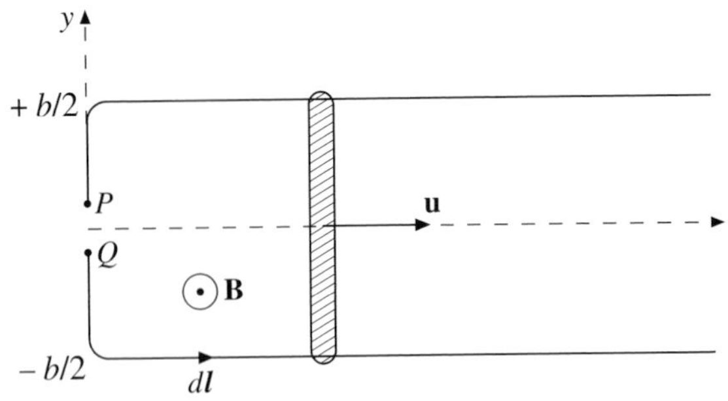

Fig. 15.4 A circuit with a moving bar of length b in a stationary **B** field that points out of the page.

closed the induced current would flow clockwise, producing a **B** contribution in the $-\hat{\mathbf{z}}$ direction and so opposing the increase in flux produced by the motion of the bar. From the motional form (15.14) of Faraday's law we have

$$V_{\text{emf}} = \oint dy\, \hat{\mathbf{y}} \cdot (u\hat{\mathbf{x}} \times B_0\hat{\mathbf{z}}) = -uB_0 \int_{-b/2}^{b/2} dy\, \hat{\mathbf{y}} \cdot \hat{\mathbf{y}} = -B_0 bu \tag{15.16}$$

which confirms what we found using the most general form (15.2) of Faraday's law.

To interpret the left-hand side of (15.14) in general, we remind the reader that according to the sign convention discussed earlier in the chapter, $d\boldsymbol{l}$ is chosen so that it is related to the field direction by the right-hand screw rule. In Fig. 15.4 the field is out of the paper so that $d\boldsymbol{l}$ must be counterclockwise.

Example 15.4 The Faraday-disk generator

A metal disk rotates at angular velocity ω (ω is the number of rotations per second multiplied by 2π) around its axis in a uniform **B** field perpendicular to the disk. A circuit between the edge of the disk and a point on the axis is formed by a pair of wire leads brushing against disk and axis, and part of the circuit goes through the metal disk. The motional form of Faraday's law tells us that an induced emf is generated if free charges in the circuit move *across* the **B** lines. Obviously, that is exactly what happens in the radial part of the circuit on the rotating disk.[1] Whereas the previous example could be analyzed using either (15.2) or (15.14), here we need the latter

[1] Application of (15.15) shows that the magnetic force on the charges is in the $\hat{\boldsymbol{\rho}}$ direction, giving rise to clockwise flow of current in the circuit shown. Moreover, *in this case* the circuit could be drawn differently, in fact in such a way that the current would then flow counterclockwise in it: the choice is arbitrary for the Faraday disk.

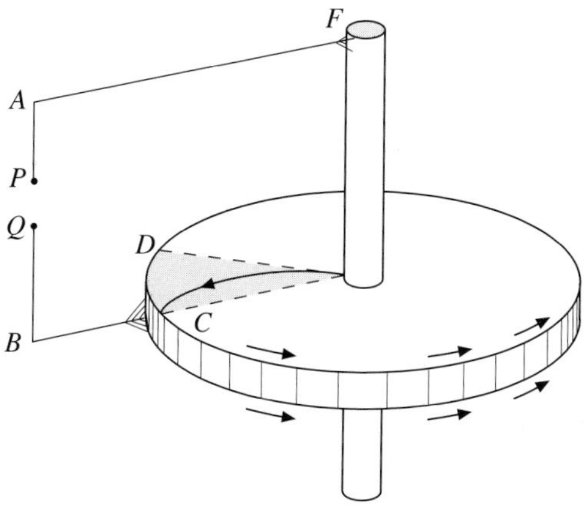

Fig. 15.5 A Faraday-disk generator.

expression; this gives us for the magnitude of the induced emf

$$V_{\text{emf}} = \left| \int_{\rho_0}^{0} d\rho\, \hat{\boldsymbol{\rho}}(\varphi) \cdot (\hat{\mathbf{u}} \times \mathbf{B}) \right| = \left| \int_{\rho_0}^{0} d\rho\, \hat{\boldsymbol{\rho}}(\varphi) \cdot (\rho\omega\hat{\boldsymbol{\varphi}} \times B\hat{\mathbf{z}}) \right|$$

$$= \left| \omega B \int_{\rho_0}^{0} d\rho\, \rho \right| = \frac{1}{2} B\rho_0^2\omega \tag{15.17}$$

We now relate (15.17) to the original statement (15.2) of Faraday's law. To do this we ask how the magnetic flux is changing. Consider the right-side of (15.17) as follows:

$$\frac{1}{2} B\rho_0^2\omega = \frac{1}{2} B\rho_0^2 \frac{d\varphi}{dt} = \frac{d}{dt}\left[B\left(\frac{1}{2}\rho_0^2\varphi \right) \right] \tag{15.18}$$

Circular geometry tells us that $\frac{1}{2}\rho_0^2\varphi$ is the area of a sector of the circle of angle φ. So, if φ is the angle of the sector CDE between $\rho_0(t_2)$ and $\rho_0(t_1)$ in Fig. 15.5, then $S = \frac{1}{2}\rho_0^2\varphi$ is the area that the radial part of the circuit has swept out between times t_1 and t_2. Hence (15.8) tells us that

$$|V_{\text{emf}}| = \frac{d}{dt}(BS) = \frac{d\Lambda}{dt} \quad \text{with} \quad \Lambda = BS \tag{15.19}$$

where Λ is the flux cut by DE between t_1 and t_2. Using the Faraday-disk method, it is possible to generate large dc voltages.

Example 15.5 A rotating loop in a uniform B field

A rectangular loop with sides a, b rotates around the x axis, as shown in Fig. 15.6, with constant angular velocity $\omega = d\theta/dt$ in rad/s in the presence of a uniform and stationary field $\mathbf{B} = B\hat{\mathbf{z}}$. Thus the angle $\theta(t) = \omega t$ if the circuit's $d\mathbf{S}$ vector points in

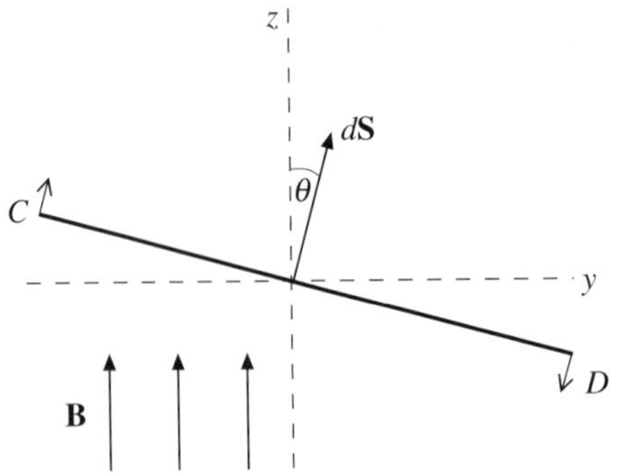

Fig. 15.6 A loop rotating about the x axis in a stationary **B** field. CD is of length b.

the $\hat{\mathbf{z}}$ direction at $t = 0$. The geometry tells us that

$$\Lambda(t) = \mathbf{B} \cdot \mathbf{S}(t) = Bab \cos \theta(t) = Bab \cos \omega t \tag{15.20}$$

Hence the general form (15.2) of Faraday's law predicts

$$V_{\text{emf}} = -\frac{d}{dt}(Bab \cos \omega t) = \omega Bab \sin \omega t \tag{15.21}$$

When $0 < \theta < \pi/2$, the figure shows that $\mathbf{B} \cdot \mathbf{S}$ is decreasing. Thus the induced current would flow counterclockwise, producing a **B** field in the direction of $d\mathbf{S}$ and thus reinforcing the decreasing flux as Lenz's law requires.

Now let us instead use the motional form of Faraday's law. The $q\mathbf{u} \times \mathbf{B}$ forces on the arms of the circuit of length b are directed perpendicularly to the coil and hence do not cause current flow. At angle θ, the velocity $\mathbf{u}(\theta) = (u \sin \theta)\hat{\mathbf{y}} + (u \cos \theta)\hat{\mathbf{z}}$ on the top arm C, and it is the opposite on the bottom arm D (this is easy to check for $\theta = 0$ and $\theta = \pi/2$). The line element $d\boldsymbol{l}$ must be chosen as counterclockwise, according to our sign convention; thus $d\boldsymbol{l} = (dx)\hat{\mathbf{x}}$ at the top and $-(dx)\hat{\mathbf{x}}$ at the bottom (at $\theta = 0$, and this does not change during rotation). Hence the top arm contributes

$$\int_0^a dx\ \hat{\mathbf{x}} \cdot [(u \sin \theta)\hat{\mathbf{y}} + (u \cos \theta)\hat{\mathbf{z}}] \times B\hat{\mathbf{z}} = \int_0^a dx\ B(u \sin \theta) = Bau \sin \theta(t) \tag{15.22}$$

to the emf. The bottom arm contributes exactly the same, because both $d\boldsymbol{l}$ and $\mathbf{u}$ change sign, so that $V_{\text{emf}} = 2Bau \sin \theta(t) = 2Bau \sin \omega t$. But velocity u is the *change of arc per unit time* $= \frac{1}{2}b(d\theta/dt) = \frac{1}{2}b\omega$. So we again find $V_{\text{emf}} = \omega Bab \sin \omega t$, consistent with (15.21).

Example 15.6 Feynman's rocking panels with uniform B field

Consider the situation sketched in Fig. 15.7 due to R.P. Feynman (together with R.B. Leighton and M. Sands, *The Feynman Lectures On Physics*, Addison-Wesley,

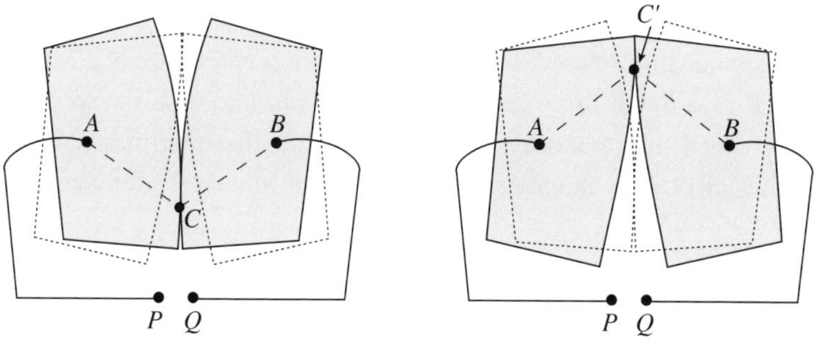

Fig. 15.7 The rocking panels of Example 15.6; the magnetic field is perpendicular to the plane of the page.

Reading MA, 1964). Two planar curved metal panels, which can swivel back and forth in the $z = 0$ plane around the fixed points A, B are shown in two positions. The point of contact C moves during swiveling from the position in the left sketch to C' in the right sketch and back. The magnetic field, perpendicular to the plane of the panels, is uniform. Let us suppose only very slight curvature of the curved edges, so that point C, and therefore the shape of the current path in the plates, moves quite rapidly for a very slow rocking movement of the panels. In the limit of zero curvature, point C moves infinitely fast, but we will not go to that limit.

What is the emf at the poles P, Q? The somewhat surprising answer is 'practically zero'. The reason is that, although part of the circuit changes rapidly from ACB to $AC'B$ etc., the actual charges on the plate move only very slowly across the field lines due to the rocking motion (which would be zero in the limit). Faraday's motional law says that *charges* must move across the magnetic flux density lines, whereas here the circuit path itself changes from one set of charges to another! The emf is not exactly zero when we do not go to the limit, because the charges do move, very slowly, with the rocking motion. However, they certainly do not move with the velocity suggested by the motion of C to C'.

15.1.3 Summary

Faraday's law is as follows: *the time rate of change in flux linkage across the surface of a circuit is equivalent to an induced emf around that circuit*. The sense of the emf is given by Lenz's law: *the induced current around the circuit (if any) must oppose the time change in the flux*. Faraday's law and Lenz's law are summed up by (15.2):

$$V_{\text{emf}} = -d\Lambda/dt$$

The transformer form (15.3a) leads to (15.4), which is an extension of the Maxwell $\nabla \times \mathbf{E} = 0$ law of electrostatics. The motional form (15.14) leads to (15.15), which tells us that the field $\mathbf{u} \times B$ may be interpreted as an *electric field* on a charge in the circuit. It is nonzero only when the charge has a component of motion **u** across the lines of magnetic flux density. Thus the apparent interpretation of a field as exerting either an **E** or a $\mathbf{u} \times \mathbf{B}$ force on a unit charge is in line with the special theory of relativity, which, like Faraday's law, implies that the interpretation of forces as being either electric or magnetic depends on the frame of reference.

15.2 Maxwell's displacement current

Faraday's seminal discovery of the influence of magnetism upon charges in conductors, which led to the new Maxwell equation $\nabla \times \mathbf{E} = -\partial \mathbf{B}/\partial t$, was followed by Maxwell's realization, around 1861, that variations in the position even of bound charges (as in a dielectric) could be regarded as being equivalent to a conduction current. Consider again a parallel-plate capacitor in an open circuit, as in Fig. 15.8. A voltage difference V is maintained across the plates, which carry charges q, $-q$. In the static situation, we know that $q = CV$, where $C = \varepsilon S/d$ is the capacitance for a uniform dielectric filling. If V becomes smaller, then q must become smaller so that C stays the same. Hence it follows for time changes in the voltage that

$$\frac{dq}{dt} = C\frac{dV}{dt} \tag{15.23}$$

This dq/dt is due to the fact that charges flow off one plate and (through the circuit connecting the plates) onto the other plate. An observer looking only at the plates, and not at the medium between them, might conclude from the change in charge that current flows between the plates! Maxwell came to the conclusion that the dq/dt in this equation, which amounts to a conduction current *in the circuit* between voltage source and capacitor, can be regarded as a *displacement current* inside the capacitor giving rise to a time change in the voltage difference. The current can be regarded as the integral of a current density J_D over the surface area S of each plate, so that $dq/dt = I_D(t) = -(\mathbf{J}_D \cdot \hat{\mathbf{z}})S$, and of course $V(t) = E(t)\,d = -(\mathbf{E} \cdot \hat{\mathbf{z}})d$. The minus signs serve to remind us that both $\mathbf{J}_D$ and $\mathbf{E}$ are in the $-\hat{\mathbf{z}}$ direction. Let us insert these

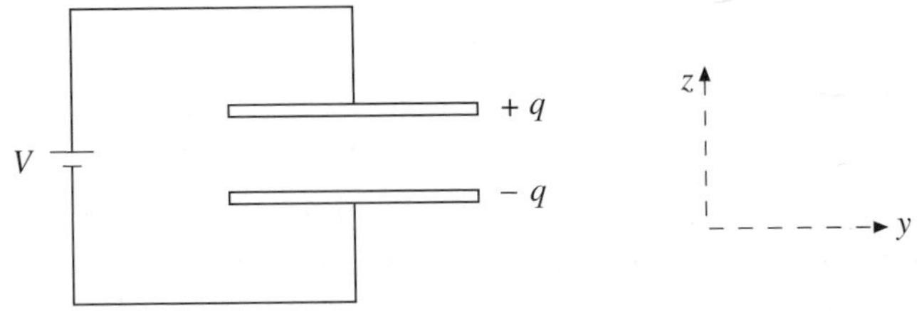

Fig. 15.8 A parallel-plate capacitor fed by a time-varying voltage difference V.

two changes into (15.23) to obtain

$$-S\mathbf{J}_{\mathrm{D}} \cdot \hat{\mathbf{z}} = -\frac{\varepsilon S}{d} d \frac{\partial \mathbf{E}}{\partial t} \cdot \hat{\mathbf{z}} \qquad \rightarrow \qquad \mathbf{J}_{\mathrm{D}} = \varepsilon \frac{\partial \mathbf{E}}{\partial t} = \frac{\partial \mathbf{D}}{\partial t} \tag{15.24}$$

The *displacement current density* $\mathbf{J}_{\mathrm{D}} = \partial \mathbf{D}/\partial t$ is thus defined by means of this example. Maxwell's idea amounts to inclusion of this new current density in the curl **H** equation:

$$\nabla \times \mathbf{H} = \mathbf{J} + \mathbf{J}_{\mathrm{D}} \qquad \rightarrow \qquad \nabla \times \mathbf{H} = \mathbf{J} + \frac{\partial \mathbf{D}}{\partial t} \tag{15.25}$$

It is possible that only one of the two terms on the right-hand side is nonzero, or that both are nonzero. In a capacitor with a lossy dielectric there may be a conduction current as well as a displacement current between the plates. At any rate, this is the final addition to the Maxwell equations.

The correctness of the displacement current term also can be seen in quite a different way. If we take the divergence of (15.25) we have

$$0 = \nabla \cdot \mathbf{J} + \nabla \cdot \frac{\partial \mathbf{D}}{\partial t} = \nabla \cdot \mathbf{J} + \frac{\partial}{\partial t}(\nabla \cdot \mathbf{D}) = \nabla \cdot \mathbf{J} + \frac{\partial \varrho_{\mathrm{v}}}{\partial t} \tag{15.26}$$

which is the equation of continuity (10.15) in the general dynamic case in which charge densities can vary in time. The displacement current term in (15.25) is clearly consistent with the fact that $\nabla \cdot \mathbf{J}$ is no longer zero but is now equal to $-\partial \varrho_{\mathrm{v}}/\partial t$.

15.3 The Maxwell equations

15.3.1 Fields

The Maxwell equations are now complete and we list them, together with the constitutive equations[1] and the equation of continuity:

$$\begin{array}{llll} \nabla \times \mathbf{E} = -\partial \mathbf{B}/\partial t & \text{(a)} & \nabla \times \mathbf{H} = \mathbf{J} + \partial \mathbf{D}/\partial t & \text{(d)} \\ \nabla \cdot \mathbf{D} = \varrho_{\mathrm{v}} & \text{(b)} & \nabla \cdot \mathbf{B} = 0 & \text{(e)} \\ \mathbf{D} = \varepsilon \mathbf{E} & \text{(c)} & \mathbf{B} = \mu \mathbf{H} & \text{(f)} \\ & \dfrac{\partial \varrho_{\mathrm{v}}}{\partial t} + \nabla \cdot \mathbf{J} = 0 \quad \text{(g)} & & \end{array} \tag{15.27}$$

Some comments on these equations now follow. The duality between **E** and **H** and between **D** and **B** is somewhat enhanced in the way these equations are written. There seems to be no counterpart to **J** in the curl **E** equation; that counterpart would be a *magnetic current density*. There seems not to be any

[1] for linear, isotropic, nonferromagnetic media.

physical existence of such a thing. Likewise, there is no counterpart to ϱ_v in the div **B** equation, and that indicates the non-existence of *magnetic charge*, as we have discussed before.

The reader will have noted that the equation of continuity has been placed between the two pairs of Maxwell equations in (15.27). The first four equations actually imply the equation of continuity, as noted above in finding (15.26). Moreover, the two divergence equations are implied by the two curl equations taken with the equation of continuity, as follows.

The div **B** *equation*: Take the divergence of the curl **E** equation to obtain

$$0 = -\nabla \cdot \frac{\partial \mathbf{B}}{\partial t} = -\frac{\partial}{\partial t}(\nabla \cdot \mathbf{B}) \tag{15.28}$$

This tells us that at *all* locations $\nabla \cdot \mathbf{B} = \kappa(\mathbf{r})$ is a constant in time. If, however, $\nabla \cdot \mathbf{B}$ actually equated 0 at some time in the distant past, then $\nabla \cdot \mathbf{B}$ would be zero at all subsequent times. Hence the div **B** equation is implied by the curl **E** equation provided we assume as an extra boundary condition only that all magnetic fields were zero in the distant past.

The div **D** *equation*: Take the divergence of the curl **H** equation to obtain

$$0 = \nabla \cdot \mathbf{J} + \frac{\partial}{\partial t}(\nabla \cdot \mathbf{D}) = -\frac{\partial \varrho_v}{\partial t} + \frac{\partial}{\partial t}(\nabla \cdot \mathbf{D}) = -\frac{\partial}{\partial t}(\nabla \cdot \mathbf{D} - \varrho_v) \tag{15.29}$$

By a similar argument to that used before, we see that $\nabla \cdot \mathbf{D} - \varrho_v = 0$ at all subsequent times, as long as this quantity is zero at all locations at some time in the (distant) past.

The integral forms of Maxwell's equations now become

$$\begin{aligned} &\oint dl \cdot \mathbf{E} = -\int_S d\mathbf{S} \cdot \frac{\partial \mathbf{B}}{\partial t} \qquad &&\oint dl \cdot \mathbf{H} = \int_S d\mathbf{S} \cdot \left(\mathbf{J} + \frac{\partial \mathbf{D}}{\partial t}\right) \\ &\int_S d\mathbf{S} \cdot \mathbf{D} = q_{\mathrm{encl}} \equiv \int_v dv \varrho_v \qquad &&\int_S d\mathbf{S} \cdot \mathbf{B} = 0 \end{aligned} \tag{15.30}$$

The contour integrals are around the rim C of the arbitrary surface S with respect to which the surface integrals are to be performed. Of course, the first term on the right-hand side of the curl **H** equation (in its integral form) is just I_{encirc}, as in Ampère's law, and doubtlessly one might replace the second term by dq_{encl}/dt, but that latter substitution would not clarify matters much as the idea is to separate conduction currents (in I_{encirc}) from displacement currents (with vector density $\mathbf{J}_D = \partial \mathbf{D}/\partial t$).

Equations (15.27) for the general time-varying case have as their counterpart the boundary conditions at the interface between media with differing permittivities and permeabilities. These conditions are precisely the same as found before, namely

$$\hat{\mathbf{n}} \times (\mathbf{E}_2 - \mathbf{E}_1) = 0 \qquad \hat{\mathbf{n}} \cdot (\mathbf{D}_2 - \mathbf{D}_1) = \varrho_s \tag{15.31a}$$

$$\hat{\mathbf{n}} \times (\mathbf{H}_2 - \mathbf{H}_1) = \mathbf{K} \qquad \hat{\mathbf{n}} \cdot (\mathbf{B}_2 - \mathbf{B}_1) = 0 \tag{15.31b}$$

The electric boundary conditions were discussed in subsection 6.4.2, and the magnetic ones in Section 11.5. Proofs of (15.31a,b) are substantially unaltered from these discussions.

15.3.2 Potentials

The electrostatic potential is given by $\mathbf{E}(\mathbf{r}) = -\nabla V(\mathbf{r})$, and the vector magnetostatic potential is given by $\mathbf{B}(\mathbf{r}) = \nabla \times \mathbf{A}(\mathbf{r})$. We will use (15.27) to see how these should change for electrodynamics. First, we can retain the vector magnetostatic potential without change because it was introduced to substitute for the div $\mathbf{B}$ equation and because that equation has not changed in the transition from magnetostatics to electrodynamics.

Second, let us set $\mathbf{E} = -\nabla V + \mathbf{e}$, and see if the Maxwell equations can predict what $\mathbf{e}$ must be. If we substitute this into the curl $\mathbf{E}$ equation we find

$$\nabla \times \mathbf{e} = -\frac{\partial \mathbf{B}}{\partial t} = -\frac{\partial}{\partial t}(\nabla \times \mathbf{A}) = \nabla \times \left(-\frac{\partial \mathbf{A}}{\partial t}\right) \tag{15.32}$$

which suggests that we set $\mathbf{e} = -\partial \mathbf{A}/\partial t$. This is not unique (because addition of the gradient of any scalar function would not change $\nabla \times \mathbf{e}$), but it is sufficient. Hence the new set of relationships between potentials and fields is

$$\mathbf{E}(\mathbf{r}, t) = -\nabla V(\mathbf{r}, t) - \frac{\partial \mathbf{A}(\mathbf{r}, t)}{\partial t} \qquad B(\mathbf{r}, t) = \nabla \times \mathbf{A}(\mathbf{r}, t) \tag{15.33}$$

Significantly, all fields now can be functions of the three spatial coordinates *and* of time t.

The *usefulness* of these new potentials depends upon whether we can find equations for them that are easier to solve than Maxwell's equations for the **E**, **D**, **B**, **H** fields. In electrostatics and in magnetostatics, we found that both $V(\mathbf{r})$ and $\mathbf{A}(\mathbf{r})$ could be made to obey Poisson equations under particular circumstances (regions with constant permittivity ε and permeability μ), provided we required the divergence of $\mathbf{A}$ to be zero. It will turn out that we must postulate *connections* between $V(\mathbf{r})$ and $\mathbf{A}(\mathbf{r})$ in electrodynamics in order to find sufficiently simple equations for these potentials. Such connections are called *gauge equations* and a variety of these are possible, with differing equations for each of the potentials. This is a matter for the next chapter.

Problems

15.1. A circuit with $\mathcal{N}$ turns captures a magnetic flux $\Phi(t) = \Phi(0)e^{-t/T}$. Calculate V_{emf}.

15.2. A single-turn circuit rotates in a constant $\mathbf{B}$ field in such a way that the projection of its area in a plane $\perp$ $\mathbf{B}$ is $S(t) = S_0 \cos \omega t$. Calculate V_{emf}.

15.3. Consider a rectangular circuit around the z axis, $-a < x < a$, $-b < y < b$, as in Fig. 15.2, with an infinitesimal opening at $x = a$, and therefore pole A at $(a, \epsilon, 0)$ and pole B at $(a, -\epsilon, 0)$ with $\epsilon \ll 1$. Let $\mathbf{B}(t) = -\mathbf{B} \sin(2\pi t/T)\hat{\mathbf{z}}$.
(a) Calculate $V_{\text{emf}}(t)$.
(b) Is $V_{\text{emf}} = +(V_A - V_B)$ or $-(V_A - V_B)$? Explain your answer.

15.4. A solenoid of length l has $\mathcal{N}$ turns of current $I = I_0 \sin \omega t$, and a cross-sectional area $S = \pi a^2$; $l = 50$ cm and $\mathcal{N} = 500$.
(a) Calculate the 'first-order' change in current induced by the magnetic field produced inside the solenoid. The entire length of coil has resistance $\mathcal{R}$.
(b) This change in current also produces a change in the magnetic field inside the solenoid, which in turn produces a further change in current. What is needed for this further change to be negligible?

15.5. Find the magnetic flux-density field $\mathbf{B}$ from the differential form of Faraday's law for the following $\mathbf{E}$ fields:
(a) $\mathbf{E} = E_0[\cos \omega t/\rho]\hat{\boldsymbol{\varrho}}$,
(b) $\mathbf{E} = (E_0 \theta t/r^2)\hat{\mathbf{r}}$.

15.6. The loop of problem 15.3 (at $t = 0$) moves with a steady velocity $\mathbf{u} = u\hat{\mathbf{y}}$ through a magnetic field $\mathbf{B} = \mathbf{B}_0\hat{\mathbf{z}}$ in $-b < y < b$ and $\mathbf{B} = 0$ for $|y| > b$. In a time interval $T = 2b/u$ the loop will have moved entirely out of the region where $\mathbf{B}$ is nonzero. What is the V_{emf} produced in the circuit during the exit time T?

15.7. The resistivity of iron is close to $10^{-7}\,\Omega\,$m. How large would a square loop (with cross-sectional area of 1 mm^2) have to be, if it is rotated optimally at 60 Hz in the Earth's magnetic field ($B \approx 0.5$ gauss), in order to produce a maximum current of 1 mA?

15.8. Explain why a rod of length L pulled at velocity $\mathbf{u}$ across lines of force of a $\mathbf{B}$ field appears to have an electric field $E = V/L$ along its length.

15.9. A parallel-plate capacitor with plate area S and separation distance d is connected by an ac source to a time-varying potential difference $V(t) = V_0 e^{-t/T}$. Calculate the $\mathbf{B}$ field in the midplane between the plates (this will be the zeroth-order field, because it is time-varying, and hence – in turn – will modify the charge distribution on the plates, etc.).

15.10. A magnetic field is characterized as $\mathbf{H} = (\mathbf{H}_0 e^{-t/T}/\rho)\hat{\boldsymbol{\varphi}}$ in cylindrical coordinates. Specify the accompanying electric field $\mathbf{E}(\mathbf{r})$. *Hint*: The 'source' of this field can be considered either as a current or as an $\mathbf{E}$ field on the z axis, but not as both.

16

How Maxwell's equations lead to waves and signals

16.1 Some forms for the constitutive equations

Equations (15.27), then, are the full-blown Maxwell equations for the general electrodynamic case. However, the word 'general' must be considered with some care because the constitutive relationships, $\mathbf{D} = \varepsilon\mathbf{E}$ and $\mathbf{B} = \mu\mathbf{H}$, are not really the most general possible. In fact for many dielectric materials it is proper to state that

$$\begin{aligned}\mathbf{D}(\mathbf{r},t) &= \int_0^\infty dt'\,\varepsilon(\mathbf{r},t-t')\mathbf{E}(\mathbf{r},t')\\ \mathbf{B}(\mathbf{r},t) &= \int_0^\infty dt'\,\mu(\mathbf{r},t-t')\mathbf{H}(\mathbf{r},t')\end{aligned} \tag{16.1}$$

These are known mathematically as *convolution integrals*. The lower bound in a convolution integral is in fact usually $-\infty$, but here has been set to zero. If we were to allow t' to be less than zero, then $t_2 = t - t'$ would be a time that is larger than t, and the field $\mathbf{D}(\mathbf{r}, t)$ would appear to be a function of $\varepsilon(\mathbf{r}, t_2)$ with some values of $t_2 > t$ (likewise for the magnetic flux density and the permeability). Thus, the fields would be caused by the value of the permittivity or permeability at some *future* time, and that seems to be unacceptable to our way of understanding events in our world! We say that such contributions are *acausal*. The same thing can be achieved in (16.1) with lower bounds equal to $-\infty$ if it is understood that $\varepsilon(\mathbf{r}, t - t')$ and $\mu(\mathbf{r}, t - t')$ are zero whenever $t' < 0$.

Equation (16.1) is typical for the dielectric media in which many communication signals propagate, such as the atmosphere, or the ocean, or layers of Earth. Also typical of many high-frequency communication signals is the restriction to a fairly narrow range (or *band*) of frequencies. The significance of that follows from the theory of Fourier synthesis, which states that an arbitrary signal $s(t)$ consists of a synthesis of ac (alternating current) frequency components as follows:

$$s(t) = \int_0^\infty dw[S_c(\omega)\cos\omega t + S_s(\omega)\sin\omega t] \tag{16.2}$$

The definitions of the frequency components are

$$S_c(\omega) \equiv \frac{2}{\pi}\int_0^\infty dt\, s_c(t)\cos\omega t \qquad S_s(\omega) \equiv \frac{2}{\pi}\int_0^\infty dt\, s_c(t)\sin\omega t \tag{16.3}$$

Here $s(t)$ could be either $\mathbf{E}(t)$ or $\mathbf{H}(t)$, i.e. one of the electromagnetic fields, in which case the frequency components $\mathcal{E}(\omega)$ or $\mathcal{H}(\omega)$ would be of special interest because (16.1) then translates into

$$\mathcal{D}(\mathbf{r},\omega) = \tilde{\varepsilon}(\mathbf{r},\omega)\,\mathcal{E}(\mathbf{r},\omega) \qquad \mathcal{B}(\mathbf{r},\omega) = \tilde{\mu}(\mathbf{r},\omega)\mathcal{H}(\mathbf{r},\omega) \tag{16.4}$$

This follows from a purely mathematical property of Fourier transforms. Thus, it is the *frequency components* $\tilde{\varepsilon}(\mathbf{r},\omega)$ and $\tilde{\mu}(\mathbf{r},\omega)$ for which the simplest constitutive relationships hold in most common dielectric materials. In common dielectrics, both these frequency components vary slowly as functions of ω, in the usual range of frequencies of interest. And when (16.4) holds, then the more complicated linear relationships of (16.1) hold for the *temporal* (time-dependent) counterparts! These more complicated forms, however, reduce to (16.6) – see below – if the frequency band of $\mathbf{E}(t)$ and $\mathbf{H}(t)$ is so narrow that $\tilde{\varepsilon}(\mathbf{r},\omega)$ and $\tilde{\mu}(\mathbf{r},\omega)$ are almost constant as functions of ω. We will return to this issue shortly, after we discuss (16.6).

Even more complicated constitutive relationships are needed for certain *anisotropic* materials, such as birefringent crystals, in which the *direction* of the fields influences the magnitude of what happens. In such cases, the permittivity and/or the permeability become *tensor quantities*, so that

$$\begin{aligned} D_x &= \varepsilon_{xx}E_x + \varepsilon_{xy}E_y + \varepsilon_{xz}E_z \\ D_y &= \varepsilon_{yx}E_x + \varepsilon_{yy}E_y + \varepsilon_{yz}E_z \\ D_z &= \varepsilon_{zx}E_x + \varepsilon_{zy}E_y + \varepsilon_{zz}E_z \end{aligned} \tag{16.5}$$

16.2 The gauge equations for the potentials

The nature of electromagnetic signals, consisting of propagating $\mathbf{E}$ and $\mathbf{B}$ fields, is best understood by first developing equations for the two potential fields V and $\mathbf{A}$. The new definitions (15.32) for the potentials $V(\mathbf{r},t)$ and $\mathbf{A}(\mathbf{r},t)$ can be substituted into the full Maxwell equations for a medium in which μ and ε are constants in space. In particular, if in (16.4) we also require that $\tilde{\mu}(\mathbf{r},\omega)$ and $\tilde{\varepsilon}(\mathbf{r},\omega)$

are independent of ω, then we also have

$$\mathbf{D}(\mathbf{r}, t) = \varepsilon \mathbf{E}(\mathbf{r}, t) \qquad \mathbf{B}(\mathbf{r}, t) = \mu \mathbf{H}(\mathbf{r}, t) \tag{16.6}$$

where $\varepsilon = \tilde{\varepsilon}(\mathbf{r}, \omega)$ and $\mu = \tilde{\mu}(\mathbf{r}, \omega)$ are true constants in space and time. Now the requirement that these constants are not functions of frequency ω is of course too stringent, but it is certainly plausible that they vary negligibly over a limited range of frequencies (as discussed above, before (16.5)). This is the case for many types of communication media, and is therefore a most useful approximation for communication signals, which, typically, have only a limited range of frequencies. In that case, (16.6) is a useful approximation, if not literally true, and the constants ε,μ take on the values that they have at the central (carrier) frequency.

In the general time-dependent case, it turns out that the equations of the two potentials $V(\mathbf{r}, t)$ and $\mathbf{A}(\mathbf{r}, t)$ are much simpler than the equations for $\mathbf{E}(\mathbf{r}, t)$ and $\mathbf{H}(\mathbf{r}, t)$. If we use (16.6) and (15.32) in the full Maxwell equations (15.27) a set of coupled equations is obtained for the potentials ('coupled' in the sense that both V and $\mathbf{A}$ occur in each equation):

$$\begin{aligned} &\left(\Delta \mathbf{A} - \mu\varepsilon \frac{\partial^2 \mathbf{A}}{\partial t^2}\right) - \nabla\left(\nabla \cdot \mathbf{A} + \mu\varepsilon \frac{\partial V}{\partial t}\right) = -\mu \mathbf{J} \\ &\left(\Delta V - \mu\varepsilon \frac{\partial^2 V}{\partial t^2}\right) + \frac{\partial}{\partial t}\left(\nabla \cdot \mathbf{A} + \mu\varepsilon \frac{\partial V}{\partial t}\right) = -\varrho_{\mathrm{v}}/\varepsilon \end{aligned} \tag{16.7}$$

Unfortunately, these equations are still too difficult to handle, especially as they are coupled. However, if we require the following relationship between the solutions for V and $\mathbf{A}$, which basically specifies the divergence of vector field $\mathbf{A}$ (remember, we only need the curl of $\mathbf{A}$, so that we are free to specify the divergence any way we desire),

$$\nabla \cdot \mathbf{A} + \mu\varepsilon \frac{\partial V}{\partial t} = 0 \qquad \text{(Lorentz condition)} \tag{16.8}$$

then (16.7) 'decouples' and simplifies to the equations

$$\begin{aligned} &\Delta \mathbf{A} - \mu\varepsilon \frac{\partial^2 \mathbf{A}}{\partial t^2} = -\mu \mathbf{J} \\ &\Delta V - \mu\varepsilon \frac{\partial^2 V}{\partial t^2} = -\varrho_{\mathrm{v}}/\varepsilon \end{aligned} \tag{16.9}$$

While (16.9) has indeed decoupled the four equations implied by (16.9), to the extent that three are only in $\mathbf{A}$ and the other is only in V, the Lorentz condition (16.8) still couples the solution of the first three equations to the solution of the fourth, and vice versa. The solutions are not independent.

16.3 Signals propagating at finite velocities

It is found that time changes in the sources ϱ_v and $\mathbf{J}$ cause electromagnetic fields to *propagate at a velocity* $v = 1/\sqrt{\mu\varepsilon}$. The meaning of this crucial statement will be explored in more detail below as we develop some of the consequences of Maxwell's equations. We refer those who wish to skip some of the details directly to Eqs. (16.17) for the potentials $V(\mathbf{r}, t)$ and $\mathbf{A}(\mathbf{r}, t)$, and we will explain there what this means.

The needed development starts with (16.9) for a 'signal' denoted here by the scalar potential $V(\mathbf{r}, t)$. We can proceed in the same way for each component of $\mathbf{A}$ provided we replace $\varrho_\mathrm{v}/\varepsilon$ by $\mu\mathbf{J}$. Hence we start with

$$\Delta V - \mu\varepsilon \frac{\partial^2 V}{\partial t^2} = -\varrho_\mathrm{v}/\varepsilon \tag{16.10}$$

This equation has a number of very special properties in the general time-varying case. First, let us consider the source distribution $\varrho_\mathrm{v}(\mathbf{r}, t)$ to be made up of a sum of small point-like charges, each of which takes up an infinitesimal part of space. Then, because equation (16.9) is *linear*, it follows that V is the sum of all fields calculated from (16.10), with the right-hand side restricted to that infinitesimal spherical volume in space. Figure 16.1 shows the source region and one such infinitesimal component as a dot. We have indicated a spherical surface S around the source region. Let us now imagine that *only one* infinitesimal volume dv has nonzero $\varrho_\mathrm{v}(t)$, at least for the moment, and let its location be the coordinate origin. Then we will have a *spherically symmetric* distribution of $V(r, t)$

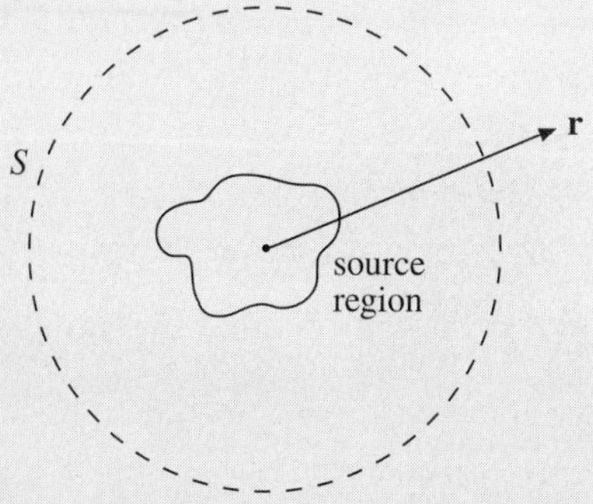

Fig. 16.1 A spherical surface S outside a source region in which signals are generated.

around the infinitesimal charge $q(0,t) = \varrho_v(0,t)dv$. If we exclude the infinitesimal source region dv from consideration in (16.10), that equation will become

$$\frac{1}{r^2}\frac{\partial}{\partial r}\left(r^2\frac{\partial V}{\partial t}\right) - \mu\varepsilon\frac{\partial^2 V}{\partial t^2} = 0 \tag{16.11}$$

It seems likely that the field V will decrease with distance r as $1/r$ (this is certainly the case for a dc potential field from a point charge), and therefore we *hypothesize* that $V = \psi/r$. When that is inserted into (16.10), which still excludes the source region dv, we find

$$\frac{\partial^2\psi}{\partial r^2} - \mu\varepsilon\frac{\partial^2\psi}{\partial t^2} = 0 \tag{16.12}$$

This equation contains the major result we are working towards because any function $\psi(r,t)$ that is a solution of (16.12) must be a direct function of only *one* variable, which is here a combination of r and t, namely $s = r \mp t/\sqrt{\mu\varepsilon}$. It could in fact be a linear combination of functions of the two variables $r - t/\sqrt{\mu\varepsilon}$ and $r + t/\sqrt{\mu\varepsilon}$. You can check this by using the chain rule of calculus when inserting such a function into (16.12). The fact that s has the dimension of length implies that $c \equiv 1/\sqrt{\mu\varepsilon}$ is a velocity, and the letter c is commonly used to indicate this velocity. We discussed this velocity in Section 11.1, after Eq. (11.12), and indicated that $c \approx 2.998 \times 10^8$ m/s in free space.

Now, we should add that to state that solutions of (16.12) are functions of $\tau = t \mp r/c$ is equivalent to saying that they are functions of $s = r \pm ct$ (because $\tau = \mp s/c$, and c is just a constant). We will discuss the difference between the + and the – sign. Starting with the minus sign, let us see if we can understand what it means when we say that $\psi(r,t) = f(t - r/c)$ is a solution of (16.12), hence of Maxwell's equations. Consider this $\psi(r,t)$ to be a significant part of some signal, for example a particular field strength (except for the factor $1/r$ which we split off above). Then

$$\psi(r+\delta r, t) = f(t - r/c - \delta r/c) = f(t - \delta r/c - r/c) = \psi(r, t - \delta r/c)$$

Thus

$$\psi(r+\delta r, t) = \psi(r, t - \delta r/c) \tag{16.13}$$

This important statement tells us that the signal strength of $V = \psi/r$, measured at $r + \delta r$ at time t, is the same as that measured at r at a *previous* time $t - \delta r/c$ (except that it has grown a little weaker because it is a little further away from the source). The difference in time is $\delta t \equiv \delta r/c$. We interpret this statement as saying that the signal *has traveled from r to $r + \delta r$ with the velocity $c \equiv 1/\sqrt{\mu\varepsilon}$*. It suggests that there is a point source at the origin, which is what we postulated to begin with, in (16.10), and that the signal emanates from there and travels outwards with velocity c, simultaneously weakening as $1/r$.

What about the possibility $\psi(r,t) = f(t + r/c)$? If the above reasoning is repeated, we will find that δr is replaced by $-\delta r$ (given that δr is a positive quantity). Then (16.13) tells us that the signal has traveled *inward* at a velocity c, and the source must be a spherically symmetric one at some distance *larger* than that part of space for which (16.11) holds. The key issue here is that a point source consisting of a time-varying current density emits signals that travel spherically symmetrically outward at a velocity $c = 1/\sqrt{\mu\varepsilon}$.

Any source that takes up a nonzero volume in space can be regarded as a superposition of point sources. We shall analyze what this implies, but first, returning to (16.13), let us place the point source not at the origin but at a location $\mathbf{r}'$. In that case the signal emanating from the source has $\psi(\mathbf{r},t) = f(t - |\mathbf{r} - \mathbf{r}'|/c)$, so that (16.13) is replaced by a somewhat more general statement that has the same meaning:

$$\psi(\mathbf{r},t) = \psi(\mathbf{r}', t - |\mathbf{r} - \mathbf{r}'|/c) \tag{16.14}$$

Thus, the solution $V(\mathbf{r},t)$ to Maxwell's equations for a point source at $\mathbf{r}'$ must be

$$V(\mathbf{r},t) = \frac{f(\mathbf{r},\tau)}{|\mathbf{r} - \mathbf{r}'|} \qquad \text{with} \quad \tau \equiv t - |\mathbf{r} - \mathbf{r}'|/c \tag{16.15}$$

The remaining question is, 'What is $f(\mathbf{r},\tau)$?' The answer is quite straightforward. As we slow down the time change of the source, namely of $\varrho_V(t)$, we approach a dc situation, for which we know from electrostatics that

$$V(\mathbf{r},t) = \frac{q}{4\pi\varepsilon|\mathbf{r} - \mathbf{r}'|} = \frac{\varrho_V}{4\pi\varepsilon|\mathbf{r} - \mathbf{r}'|}\, dv' \tag{16.16}$$

in which $q = \varrho_V dv'$, in the infinitesimal volume dv' around $\mathbf{r}'$, is a constant. However, as soon as ϱ_V at $\mathbf{r}'$ becomes time-varying, then we have learned that the right-hand side of (16.16) must be a function only of $\tau = t - |\mathbf{r} - \mathbf{r}'|/c$. The only way we can reconcile this with the dc situation of (16.16) is to require the time-varying situation to have the solutions shown below in Eqs. (16.17a, b).

As shown in some detail in the optional text immediately above, the first of equations (16.9) can be shown, for a source of infinitesimal volume dv', to lead to the solution

$$V(\mathbf{r},t) = \frac{\varrho_V(\mathbf{r}', t - |\mathbf{r} - \mathbf{r}'|/c)}{4\pi\varepsilon|\mathbf{r} - \mathbf{r}'|}\, dv' \tag{16.17a}$$

A solution for the vector potential is obtained in the very same way, by making the substitutions $V \leftrightarrow \mathbf{A}$, $\varrho_V \leftrightarrow \mathbf{J}$, and $\varepsilon^{-1} \leftrightarrow \mu$:

$$\mathbf{A}(\mathbf{r}, t) = \frac{\mu \mathbf{J}(\mathbf{r}', t - |\mathbf{r} - \mathbf{r}'|/c)}{4\pi|\mathbf{r} - \mathbf{r}'|} dv' \tag{16.17b}$$

To obtain the potentials for a distribution of charge and current, rather than for point sources, we only need to integrate these expressions:

$$V(\mathbf{r}, t) = \int dv' \frac{\varrho_V(\mathbf{r}', t - |\mathbf{r} - \mathbf{r}'|/c)}{4\pi\varepsilon|\mathbf{r} - \mathbf{r}'|} \tag{16.18a}$$

$$\mathbf{A}(\mathbf{r}, t) = \int dv' \frac{\mu \mathbf{J}(\mathbf{r}', t - |\mathbf{r} - \mathbf{r}'|/c)}{4\pi|\mathbf{r} - \mathbf{r}'|} \tag{16.18b}$$

So, to find fields at $\mathbf{r}$ at time t, we find the sources at $\mathbf{r}'$ at times $t - |\mathbf{r} - \mathbf{r}'|/c$. The formulas state that it takes the *travel time* $|\mathbf{r} - \mathbf{r}'|/c$ from the source point to the observation point before the observation point 'notices' the effect of the source point. Exactly the same holds for the fields **E** and **B**, but the source terms are then more complicated combinations of ϱ_V and **J**. The fields at **r** are thus found by locating all source points at times properly chosen *earlier* than t so that the time difference implies that some physical quantity has traveled at velocity $c = 1/\sqrt{\mu\varepsilon}$ from each source point to **r**. That physical quantity must be the fields! To see this in yet another way, suppose that we use (16.17a) for a source $\varrho_V(t)$ that is as before restricted to the origin in an infinitesimal volume dv' (so that $\varrho_V\, dv' = q$, a point charge at the origin), and let us assume that the point charge is turned on at $t = 0$ and off infinitesimally soon afterwards. We only get something nonzero in (16.17a) if $|\mathbf{r} - \mathbf{r}'| = ct$, because otherwise the charge was either not yet turned on or it was already turned off. But we chose $\mathbf{r}' = 0$ so we require $r = ct$. Then (16.17a) becomes

$$\begin{aligned} V(r, t) &= \frac{q}{4\pi\varepsilon r} \qquad \text{if} \quad r = ct \\ V(r, t) &= 0 \qquad\quad \text{if} \quad r \neq ct \end{aligned} \tag{16.18c}$$

So, a 'blip' of potential V travels out at velocity c from the 'instantaneous' point source equally strongly in all radial directions (i.e. as a spherically symmetric disturbance), and the same applies for all the other electromagnetic fields. This is the key fact, and all more complicated signals arising from more complicated sources, as in (16.18a), *propagate* in the same fashion! From (16.18a), we find for point sources

$$\mathbf{E}(\mathbf{r}, t) = \frac{1}{|\mathbf{r} - \mathbf{r}'|} \Phi(\mathbf{r}, t - |\mathbf{r} - \mathbf{r}'|/c) \tag{16.19}$$

where Φ is an appropriate vector function of the coordinates.

16.4 Frequency, wavelength, and all that

Let us suppose that we have a single-frequency source, a so-called *time-harmonic* source with time dependence given entirely by a factor $\cos \omega t$. Thus

$$\mathbf{J}(\mathbf{r}', t - |\mathbf{r} - \mathbf{r}'|/c) = \mathbf{J}(\mathbf{r}') \cos[\omega(t - |\mathbf{r} - \mathbf{r}'|/c)]$$

As before, the distance between source point and observation point is $|\mathbf{r} - \mathbf{r}'| \equiv R$. Hence the source-current density is $\mathbf{J}(\mathbf{r}') \cos[\omega(t - R/c)]$. If, then, the source-current density is nonzero only at the origin then all fields are a function of time only through the factor $\cos[\omega(t - R/c)]$.

Let us first look at the cosine factor at a *fixed* location $\mathbf{r}$ so that R is a constant. After one *cycle* of the cosine, as shown in Fig. 16.2, we have

$$\begin{aligned} &\omega(t_2 - R/c) = \omega(t_1 - R/c) + 2\pi \qquad \text{or} \\ &T \equiv t_2 - t_1 = 2\pi/\omega \end{aligned} \tag{16.20a}$$

The time T is known as the *period* of the time-harmonic signal, and its inverse $f = 1/T$ is the number of periods per second, which is called the *frequency* (expressed in $\mathrm{Hz} = \mathrm{s}^{-1}$). The second of (16.20a) also can be written as

$$\omega = 2\pi f \tag{16.20b}$$

Now consider the cosine factor at a given time t. The curve, a 'snapshot' of the wave, is shown in Fig. 16.3. One spatial cycle of the cosine stretches from $t - R_2/c$ to

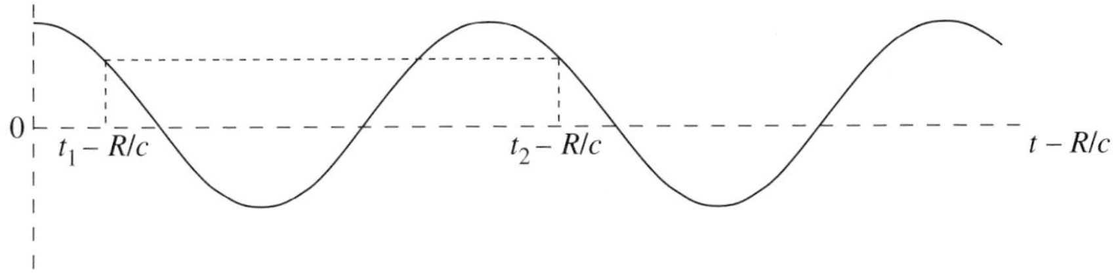

Fig. 16.2 Graph of $\cos[\omega(t - R/c)]$ as a function of t only.

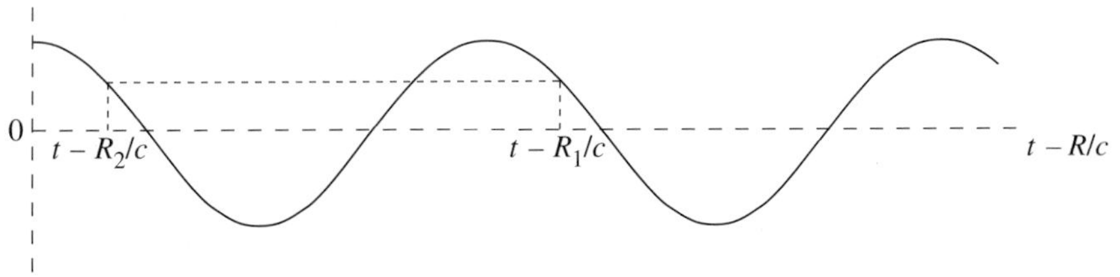

Fig. 16.3 Graph of $\cos[\omega(t - R/c)]$ as a function of R only.

$t - R_1/c$. Obviously, R_2 is bigger than R_1, so that $\mathbf{r}_2$ is further away from the point source than $\mathbf{r}_1$. Because this is one cycle of the cosine, we have

$$\omega(t - R_1/c) = \omega(t - R_2/c) + 2\pi \quad \text{or} \quad R_2 - R_1 = 2\pi c/\omega = c/f \tag{16.21}$$

The distance $R_2 - R_1$ is called the *wavelength* to which the Greek letter λ is assigned, and we then have found that

$$c = \lambda f \tag{16.22}$$

given that f is the frequency in Hz as expressed in (16.20). This is one of the most important relationships between wave quantities: *the product of wavelength and frequency is the velocity of the signal in the medium*!

In summary, point sources that vary in time as $\cos \omega t$ give rise to spherically symmetric sinusoidal fields (known as *waves*) that have a spatial period T and wavelength λ, propagate outwards at a velocity $c = 1/\sqrt{\mu\varepsilon}$, and at each location vary sinusoidally in time at a frequency $f = \omega/2\pi$. The cardinal relationship between these quantities is (16.22).

16.5 Plane waves – introduction

Another type of time-dependent signal helpful for our understanding is that with $\mathbf{E}$ or $\mathbf{H}$ fields that are a function only of one rectangular spatial coordinate. We might imagine a region $0 < z < Z$ which has a current density $\mathbf{J}(z, t)$ on it as a source. Note that $\mathbf{J}(z, t)$ does not depend upon x or y. From (15.27a, d) we then develop a new wave equation

$$\nabla \times (\nabla \times \mathbf{E}) = \nabla \times \left(-\mu \frac{\partial \mathbf{H}}{\partial t}\right) = -\mu \left(\frac{\partial \mathbf{J}}{\partial t} + \varepsilon \frac{\partial^2 \mathbf{E}}{\partial t^2}\right) \tag{16.23}$$

under the assumptions of spatially and temporally constant μ, ε. Using vector identity 10 in Appendix A.4 for the double curl, we find

$$\Delta \mathbf{E} - \nabla(\nabla \cdot \mathbf{E}) - \mu\varepsilon \frac{\partial^2 \mathbf{E}}{\partial t^2} = \mu \frac{\partial \mathbf{J}}{\partial t} \tag{16.24}$$

As the source depends spatially only upon z, we expect that the resulting fields depend only upon z. Furthermore, the absence of a charge distribution yields $\nabla \cdot \mathbf{D} = 0$ so that, for constant ε, $\nabla \cdot \mathbf{E}(z) \equiv dE_z(z)/dz = 0$, which eliminates the middle term of (16.24) (i.e. the z component of the $\mathbf{E}$ field is constant, in fact zero, in order that boundary conditions at infinity are met; the field is *transverse* to the direction of propagation). What remains is

$$\frac{\partial^2 \mathbf{E}(z,t)}{\partial z^2} - \mu\varepsilon \frac{\partial^2 \mathbf{E}(z,t)}{\partial t^2} = \mu \frac{\partial^2 \mathbf{J}(z,t)}{\partial t} \tag{16.25}$$

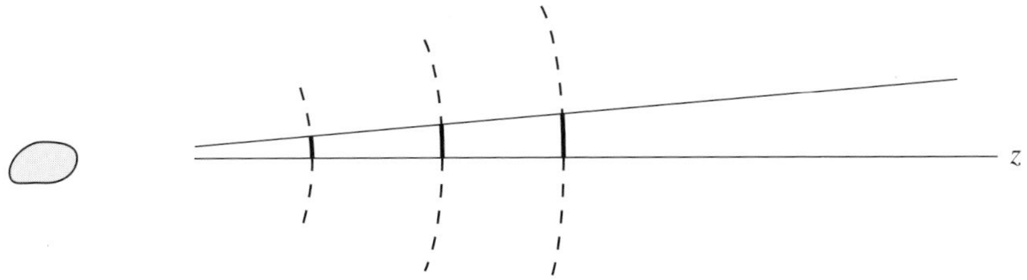

Fig. 16.4 A small-angle portion of a spherical wave distant from its source.

The vector field **E** is ⊥ to the z direction; let us take it in the x direction. Then, outside the source region ($z < 0$ or $z > Z$) we have

$$\frac{\partial^2 E_x(z,t)}{\partial z^2} - \mu\varepsilon \frac{\partial^2 E_x(z,t)}{\partial t^2} = 0 \tag{16.26}$$

This equation has the same form as (16.12), and consequently we find that $E_x(z, t)$ at location z is a function only of time through the combination $t \mp z/c$:

$$E_x(z,t) = f(z, t \mp z/c) \tag{16.27}$$

The minus sign indicates a *plane wave* propagating in the $+\hat{\mathbf{z}}$ direction, whereas the plus sign is one in the $-\hat{\mathbf{z}}$ direction. As before, a linear combination of both possibilities is also allowed. For waves that are *time-harmonic* (time dependence $\propto \cos \omega t$) and *linearly polarized* (**E** field always points in one linear dimension), we have

$$E_x(z,t) = A(z)\cos[\omega(t - z/c) + \varphi] \tag{16.28}$$

Here, again $c = 1/\sqrt{\mu\varepsilon}$, $A(z)$ is real and φ is a constant between 0 and 2π that is called the *phase* of the wave; $\cos\varphi$ is proportional to the field displacement when $t = z/c$. All conclusions drawn earlier for spherical waves formed from the potentials hold here directly for the **E** field (and can be shown to hold similarly for the **H** field), except that the fields do not fall off in strength with distance. Now there is only one direction of propagation (the $+\hat{\mathbf{z}}$ or $-\hat{\mathbf{z}}$ direction). Plane waves can be a good approximation to waves at a great distance from a source. The reason is that the source region then resembles a point charge, and if the lateral extent of the region where the wave is being measured or observed is small, then the curvature of any wavefront (i.e. locations where $\cos[\omega(t - R/c)]$ is constant for fixed z) will be negligible and $R \approx z$; see Fig. 16.4.

16.6 Use of phasors

The use of time-harmonic waves brings us to another important tool in physics and electrical engineering. Consider, for example, the curl **E** Maxwell equation,

$\nabla \times \mathbf{E} = -\partial \mathbf{B}/\partial t$. This equation yields $B \propto \sin \omega t$ if $\mathbf{E} \propto \cos \omega t$. Not only do we have to carry factors dependent upon ωt around; each equation has factors $\propto \cos \omega t$ as well as factors $\propto \sin \omega t$. While this is not a real problem, it is unwieldy because we need to formulate *two* equations from each one that has sines and cosines (all the coefficients of the cosines must add up to zero, and so must all the coefficients of the sines, in order that the resulting homogeneous equations hold for all t).

As an example, again considering the curl **E** equation, let

$$\begin{aligned}\mathbf{E} &= \mathbf{E}_1 \cos \omega t + \mathbf{E}_2 \sin \omega t \\ \mathbf{B} &= \mathbf{B}_1 \cos \omega t + \mathbf{B}_2 \sin \omega t\end{aligned} \tag{16.29}$$

We should be able to express the $\mathbf{B}_i$ vectors in terms of the $\mathbf{E}_i$ vectors (for $i = 1, 2$). We get

$$\begin{aligned}\nabla \times \mathbf{E}_1 \cos \omega t + \nabla \times \mathbf{E}_2 \sin \omega t &= -\frac{\partial}{\partial t}(\mathbf{B}_1 \cos \omega t + \mathbf{B}_2 \sin \omega t) \\ &= \omega [\mathbf{B}_1 \sin \omega t - \mathbf{B}_2 \cos \omega t]\end{aligned} \tag{16.30a}$$

Equating cosine and sine factors yields

$$\begin{aligned}\nabla \times \mathbf{E}_1 \cos \omega t &= -\mathbf{B}_2 \cos \omega t \quad \text{or} \quad \mathbf{B}_2 = -\nabla \times \mathbf{E}_1 \\ \nabla \times \mathbf{E}_2 \sin \omega t &= \mathbf{B}_1 \sin \omega t \quad \text{or} \quad \mathbf{B}_1 = \nabla \times \mathbf{E}_2\end{aligned} \tag{16.30b}$$

As intended, we have thus expressed the $\mathbf{B}_i$ in terms of the $\mathbf{E}_i$; the procedure works but it is unwieldy as we needed two equations. The use of *phasors* gets around this unwieldiness. Here follows a description of a step-by-step procedure for using phasors to express the **B** vector in terms of the **E** vector in the curl **E** equation (15.27a).

(1) Define $\mathbf{E}(\mathbf{r}, t) = \mathrm{Re}\{\mathbf{E}_\mathrm{p}(\mathbf{r})e^{j\omega t}\}$ and $\mathbf{B}(\mathbf{r}, t) = \mathrm{Re}\{\mathbf{B}_\mathrm{p}(\mathbf{r})e^{j\omega t}\}$. If, for example, $\mathbf{E}_\mathrm{p} = \mathbf{E}_1 - j\mathbf{E}_2$, with real vectors $\mathbf{E}_1$ and $\mathbf{E}_2$, then this amounts to the same as (16.29). $\mathbf{E}_\mathrm{p}(\mathbf{r})$ *is the phasor representation of* $\mathbf{E}(\mathbf{r}, t)$.
(2) Now replace $\mathbf{E}(\mathbf{r}, t)$ by $\mathbf{E}_\mathrm{p}(r)$, and $\mathbf{B}(\mathbf{r}, t)$ by $\mathbf{B}_\mathrm{p}(\mathbf{r})$, in the equation.
(3) Replace all $\partial/\partial t$ by $j\omega$ (i.e. replace all $\partial^n/\partial t^n$ by $(j\omega)^n$), and replace each $\int dt$ by $1/j\omega$. This turns the equation into a purely *algebraic* one!
(4) Solve this algebraic equation, giving $\mathbf{B}_\mathrm{p}(\mathbf{r})$ as a function of $\mathrm{E}_\mathrm{p}(\mathbf{r})$.
(5) 'Paste' the factor $e^{j\omega t}$ behind the expression for $\mathbf{B}_\mathrm{p}(\mathbf{r})$ and take the real part of the entire final expression.

We will carry this procedure out for the curl **E** equation

$$\nabla \times \mathbf{E}(\mathbf{r}, t) = -\frac{\partial \mathbf{B}(\mathbf{r}, t)}{\partial t}$$

First we obtain

$$\nabla \times \mathbf{E}_\mathrm{p} = -\frac{\partial}{\partial t}\mathbf{B}_\mathrm{p} = -jw\mathbf{B}_\mathrm{p} \tag{16.31}$$

We have used steps (1)–(3) here. We then do step (4):

$$\mathbf{B}_\mathrm{p}(\mathbf{r}) = \frac{j}{\omega}\nabla \times \mathbf{E}_\mathrm{p}(\mathbf{r}) \tag{16.32}$$

Finally we perform step (5) to obtain

$$\mathbf{B}(\mathbf{r}, t) = \mathrm{Re}\left\{\frac{j}{\omega}\nabla \times \mathbf{E}_\mathrm{p}(\mathbf{r})e^{j\omega t}\right\} \tag{16.33}$$

The answer cannot be written out more explicitly here because we have not carried out the curl operation yet, nor have specified whether $\mathbf{E}_\mathrm{p}(\mathbf{r})$ is real or complex. Nevertheless, (16.33) could be worked out further in a similar fashion if we did specify $\mathbf{E}_\mathrm{p}(\mathbf{r})$ completely.

The most important feature to note in using phasors in the way specified above is that *the equation must be linear in the phasor quantities*. We can use phasors in equations that are not linear (and we will, in talking about flow of energy) but then the above scheme will not work, and we must do something more complicated.

Example 16.1 A simple series *LRC* circuit

Consider, as a second example of the use of phasors, the circuit sketched in Fig. 16.5. A sinusoidal voltage source $v(t) = V_0 \cos(\omega t + \varphi)$ drives a current $i(t)$ through the circuit, which contains a resistor R, an inductance L, and capacitance C in series.

The extra term in the cosine, φ, is the phase, see Section 16.5, and it expresses the fact that $v(t)$ is not at its maximum value at $t = 0$. Alternatively, one could write $v(t) = V_1 \cos \omega t + V_2 \sin \omega t$, and it is not difficult to find the connection between V_1, V_2 and V_0, φ. Either way, Kirchhoff's voltage law predicts

$$v(t) = \frac{q(t)}{C} + L\frac{di(t)}{dt} + Ri(t) \tag{16.34}$$

In the unwieldy direct method, we would assume that $i(t) = I_1 \cos \omega t + I_2 \sin \omega t$, and then all four terms of (16.34) would produce sine and cosine terms, because

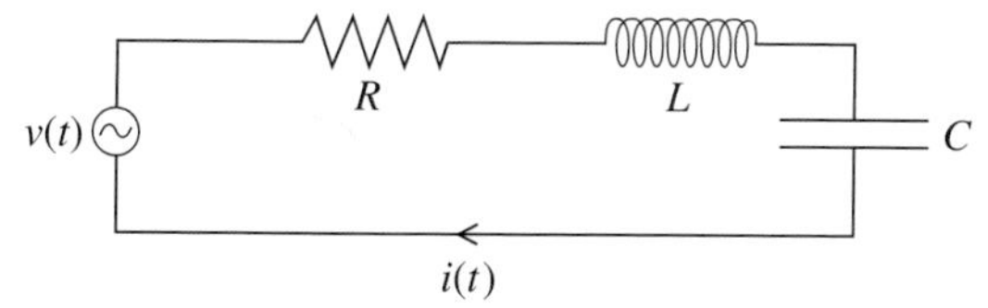

Fig. 16.5 A simple series *LRC* circuit.

$q(t) = \int dt\, i(t)$. The equation is linear in the quantities which will incorporate phasors. Two equations would be obtained if we do not use phasors, one from the cosine and the other from the sine terms.

With the phasor method, we rewrite the equation as

$$v(t) = \frac{1}{C}\int dt\, i(t) + L\frac{di(t)}{dt} + Ri(t) \tag{16.35}$$

then set $v(t) = V_0 e^{j\omega t}$, $i(t) = I_0 e^{j\omega t}$, and subsequently do the replacements of steps (2) and (3) above to obtain

$$v_0 = \left(\frac{1}{j\omega C} + j\omega L + R\right) i_0 \tag{16.36a}$$

The solution of this algebraic equation is straightforward:

$$i_0 = \frac{v_0}{(1/j\omega C) + j\omega L + R} \tag{16.36b}$$

The final step (5) yields

$$i(t) = \mathrm{Re}\left\{\frac{v_0 e^{j\omega t}}{(1/j\omega C) + j\omega L + R}\right\} \tag{16.37}$$

It is somewhat lengthy, but quite straightforward, to work out the rest and to find that indeed there is a cosine with coefficient I_1 and a sine term with coefficient I_2, but that does not add anything of relevance to this discussion. This procedure, up to (16.37), has advantages above the direct method as we do not have to differentiate or integrate sines and cosines.

16.7 Time-harmonic fields

16.7.1 Spherical time-harmonic fields

Let us write down the four Maxwell equations for time-harmonic fields, using the phasor recipe under (16.30). We obtain

$$\begin{aligned} \nabla \times \mathbf{E}_\mathrm{p} &= -j\omega \mathbf{B}_\mathrm{p} \qquad & \nabla \times \mathbf{H}_\mathrm{p} &= \mathbf{J}_\mathrm{p} + j\omega \mathbf{D}_\mathrm{p} \\ \nabla \cdot \mathbf{D}_\mathrm{p} &= \rho_{v\mathrm{p}} & \nabla \cdot \mathbf{B}_\mathrm{p} &= 0 \end{aligned} \tag{16.38}$$

All four of these equations are linear in the phasor quantities, so we can apply the phasor recipe to obtain the full space–time fields after solving a wave equation for either $\mathbf{E}_\mathrm{p}$ or $\mathbf{H}_\mathrm{p}$. The same holds true for the two potentials $\mathbf{A}$, V of Eqs. (16.9).

Let us see what (16.18a, b) lead to in the case of time-harmonic waves. In phasor notation, we apply step (1) of the recipe to obtain

$$V_{\mathrm{p}}(\mathbf{r}) = \int dv' \frac{\varrho_{\mathrm{vp}}(\mathbf{r}')e^{j\omega(t-|\mathbf{r}-\mathbf{r}'|/c)}}{4\pi\varepsilon|\mathbf{r}-\mathbf{r}'|}$$

$$\mathbf{A}_{\mathrm{p}}(\mathbf{r}) = \int dv' \frac{\mu\mathbf{J}_{\mathrm{p}}(\mathbf{r}')e^{j\omega(t-|\mathbf{r}-\mathbf{r}'|/c)}}{4\pi|\mathbf{r}-\mathbf{r}'|} \tag{16.39}$$

So, if we have a real point source of charge at $\mathbf{r}'$, then $\varrho_{\mathrm{vp}}(\mathbf{r}')\,dv' = q_{\mathrm{p}}(\mathbf{r}')$ only at location $\mathbf{r}'$ (is zero elsewhere), and the first of these equations gives a contribution

$$V_{\mathrm{p}}(\mathbf{r}) = \frac{q_{\mathrm{p}}(\mathbf{r}')e^{j\omega(t-R/c)}}{4\pi R} \qquad R \equiv |\mathbf{r}-\mathbf{r}'| \tag{16.40a}$$

and the space–time field for scalar potential then is obtained from this phasor by step (5) of the recipe below (16.30b),

$$V(\mathbf{r},t) = \frac{q(\mathbf{r}')}{4\pi R}\cos[\omega(t-R/c)] \tag{16.40b}$$

These are waves of the type described in Section 16.3, and the use of phasors has made it easier to arrive at the conclusions given there.

16.7.2 Plane time-harmonic fields

Invoking (16.26) for plane waves that are time-harmonic, the equation for the **E** field phasor becomes

$$\frac{d^2E_{x\mathrm{p}}(z)}{dz^2} + \omega^2\mu\varepsilon E_{x\mathrm{p}}(z) = 0 \tag{16.41}$$

This is the well-known harmonic equation, which appears again and again in physics and electrical engineering, and its solution for waves traveling in the *positive* z direction is $E_{x\mathrm{p}}(z) = E_{x0}e^{-jkz}$, with $k \equiv \omega\sqrt{\mu\varepsilon} = \omega/c$. Hence the space–time field is

$$\mathbf{E}(z,t) = \mathrm{Re}\left\{E_{x0}\,\hat{\mathbf{x}}e^{j(\omega t-kz)}\right\} \tag{16.42}$$

If E_{x0} is real, then it follows that $\mathbf{E}(z,t) = E_{x0}\hat{\mathbf{x}}\cos(\omega t - kz)$, which is what we found in (16.28) except for a phase term φ, which would arise here too if E_{x0} were complex instead of real! The quantity $k = \omega/c$ is known as the *wavenumber*, and it plays the same role with respect to the spatial coordinate z as ω does with respect to the temporal coordinate t: k depicts the number of spatial cycles (wavelengths λ) per unit length multiplied by 2π; just as the angular frequency ω gives the number of temporal cycles (periods T) per unit time, multiplied by 2π. It is easy to see that $k = 2\pi/\lambda$ either from these definitions or from $k = \omega/c$ and (16.22).

The magnetic field can be found easily from (16.42), using phasor representation. The curl **E** equation gives as in (16.31),

$$\mathbf{H}_\mathrm{p} = \frac{j}{\omega\mu}\nabla \times \mathbf{E}_\mathrm{p} \tag{16.43a}$$

and because $\mathbf{E}_\mathrm{p}$ is a function only of z, through the factor $\exp(-jkz)$, it follows by using the Cartesian representation of the curl that $\nabla \times \mathbf{E}_\mathrm{p} = -jk\hat{\mathbf{y}}E_{x0}e^{-jkz}$. Thus we find

$$\mathbf{H}_\mathrm{p}(z) = \frac{k}{\omega\mu}E_{x0}\hat{\mathbf{y}}e^{-jkz} = \sqrt{\frac{\varepsilon}{\mu}}E_{x0}\hat{\mathbf{y}}e^{-jkz} \tag{16.43b}$$

$$\mathbf{H}(z,t) = \mathrm{Re}\left\{\sqrt{\frac{\varepsilon}{\mu}}E_{x0}\hat{\mathbf{y}}e^{j(\omega t - kz)}\right\} \tag{16.43c}$$

The quantity $\sqrt{\mu/\varepsilon} \equiv \eta$ is known as the *intrinsic impedance* of the medium and its unit is Ω (ohms). For free space, and the approximate value $\varepsilon_0 \approx 10^{-9}/36\pi$ F/m, one obtains the value $\eta_0 \approx 120\pi \approx 377\,\Omega$, and for the more precise value $\varepsilon_0 \approx 8.854 \times 10^{-12}$ F/m one finds $\eta_0 \approx 376.7\,\Omega$. If E_{x0} is real, we then find from (16.42) and (16.43c) that

$$\begin{aligned}\mathbf{E}(z,t) &= E_{x0}\hat{\mathbf{x}}\cos(\omega t - kz)\\ \mathbf{H}(z,t) &= \frac{1}{\eta}E_{x0}\hat{\mathbf{y}}\cos(\omega t - kz)\end{aligned} \tag{16.44}$$

Figure 16.6 shows how these two fields vary for fixed x, y along the z direction, with an exaggeration in size of the **H** field (as if η were about unity instead of 377!). The two fields are *in phase*, i.e., both are zero at the same locations, and both reach their maxima at the same locations along the z axis. Furthermore, we see that the fields are perpendicular to each other and to the direction of propagation z. In fact, the

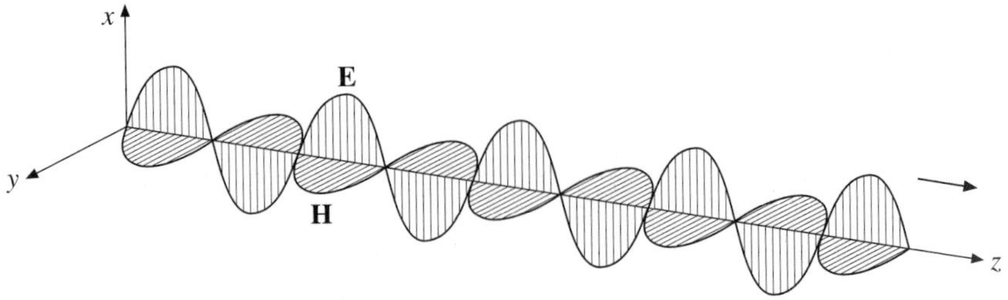

Fig. 16.6 The **E**, **H** fields along a 'ray' of an electromagnetic wave propagating in a lossless medium. The magnitudes of the fields are not to scale.

three vectors $\mathbf{E}$, $\mathbf{H}$, and $\mathbf{k} \equiv k\hat{\mathbf{z}}$ form a right-handed trio of mutually orthogonal vectors.

Note that kz is a special case of $\mathbf{k} \cdot \mathbf{r}$ for the vector $\mathbf{k} \equiv k\hat{\mathbf{z}}$. We use the name *wavefactor* for $\mathbf{k}$ and the magnitude of the wavevector is the wavenumber k. If (16.44) were to pertain to a plane wave moving in an arbitrary direction given by the unit vector $\hat{\mathbf{u}} = (\sin\theta\cos\varphi,\ \sin\theta\sin\varphi,\ \cos\theta)$ in spherical coordinates, then we would replace kz by $\mathbf{k} \cdot \mathbf{r}$ in (16.44), with $\mathbf{k} = k\hat{\mathbf{u}}$. But there is nothing special about the direction; z is simply the label of an axis that we allow to coincide with the direction of propagation, so that (16.44) corresponds anyway to an arbitrary direction. For that reason, we will often use kz instead of $\mathbf{k} \cdot \mathbf{r}$. Figure 16.6 can be somewhat misleading; it should be noted that the plane wave fields of (16.44) have the same displacement at any fixed z value for all possible x and y values.

Thus at regular spacings of λ in the z direction we have planes of maximum values of E and H and halfway in between two successive maxima is a plane where E and H are at minima. Such planes are called *wavefronts* and it is often advantageous to use not Fig. 16.6 but sketches of wavefronts of maximum and minimum E, H, as in Fig. 16.7. In the general case we think of *wavefronts of constant phase* ($\omega t - \mathbf{k} \cdot \mathbf{r} = \text{constant}$) moving at velocity c through space. These are plane for plane waves and spherical for waves emanating from 'point' sources.

As a last item for this section, note that

$$
\begin{aligned}
\nabla \times \mathbf{E}e^{-j\mathbf{k}\cdot\mathbf{r}} &= (-j\mathbf{k}) \times \mathbf{E}e^{-j\mathbf{k}\cdot\mathbf{r}} \\
\nabla \times \mathbf{E}e^{-j\mathbf{k}\cdot\mathbf{r}} &= (-j\mathbf{k}) \cdot \mathbf{E}e^{-j\mathbf{k}\cdot\mathbf{r}}
\end{aligned}
\tag{16.45}
$$

Clearly, whenever ∇ forms a dot or vector product with a phasor (with constant complex amplitude) proportional to $\exp(-j\mathbf{k} \cdot \mathbf{r})$ we simply replace the ∇ by $-j\mathbf{k}$.

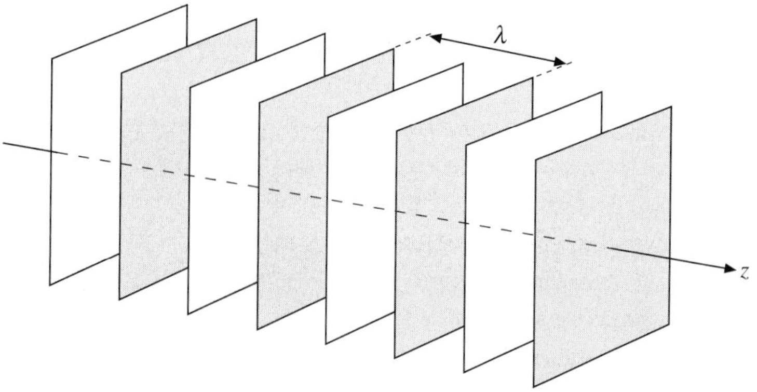

Fig. 16.7 Wavefronts of constant phase are perpendicular to the rays of a plane wave.

As a specific example, suppose that the phasor in (16.45) were $\mathbf{E}e^{-j\beta z}$, in which case $\mathbf{k} = \beta\hat{\mathbf{z}}$ because $\mathbf{k}\cdot\mathbf{r} = \beta z$. Then ∇ is replaced by $-j\beta\hat{\mathbf{z}}$ in the dot and vector products of (16.45).

Example 16.2 Use of phasors to calculate fields

Consider the electric time-harmonic field, $\mathbf{E}(\mathbf{r}, t) = 5\hat{\mathbf{y}}\cos(3\pi\times 10^9 t - \gamma x - 5\pi z)$ in V/m, propagating in free space. Suppose we are asked to calculate $\mathbf{H}(\mathbf{r}, t)$ and γ. It helps then to rewrite the $\mathbf{E}$ field as $\mathbf{E}(\mathbf{r}, t) = \mathbf{E}_0\cos(\omega t - \gamma x - \beta z) = \mathbf{E}_0\cos(\omega t - \mathbf{k}\cdot\mathbf{r})$. This indicates that $\mathbf{E}_0 = 5\hat{\mathbf{y}}\,(\mathrm{V/m})$, $\gamma = k_x$, and $\beta = k_z$ (both in m^{-1}), whereas $k_y = 0$ here. So the wavevector is $\mathbf{k} = (\gamma, 0, \beta)$, with $k = (\gamma^2 + \beta^2)^{1/2}$, and in free space we have $k = \omega\sqrt{\mu_0\varepsilon_0} = 3\pi\times 10^9/(3\times 10^8) = 10\pi\,\mathrm{m}^{-1}$. Combination of these two facts yields

$$\left(\frac{3\pi\times 10^9}{3\times 10^8}\right)^2 = \gamma^2 + (5\pi)^2 \quad\rightarrow\quad \gamma = 5\pi\sqrt{3}\,\mathrm{m}^{-1} \tag{16.46}$$

It will be useful to write $\mathbf{E}(\mathbf{r}, t) = \mathrm{Re}\{\mathbf{E}_0 e^{-j\mathbf{k}\cdot\mathbf{r}} e^{j\omega t}\}$ for the purpose of calculating $\mathbf{H}(\mathbf{r}, t)$. The phasor of the electric field is then $\mathbf{E}_\mathrm{p}(\mathbf{r}) = \mathbf{E}_0 e^{-j\mathbf{k}\cdot\mathbf{r}}$. At this point we use the time-harmonic form of Maxwell's curl $\mathbf{E}$ equation in (16.38) to find (with $\mathbf{B} = \mu_0\mathbf{H}$) that

$$\mathbf{H}_\mathrm{p} = \frac{j}{\mu_0\omega}\nabla\times\mathbf{E}_\mathrm{p}$$

and (16.45) then simplifies this to $\mathbf{H}_\mathrm{p} = (j/\mu_0\omega)(-j\mathbf{k})\times\mathbf{E}_\mathrm{p} = (k/\mu_0\omega\hat{\mathbf{k}})\times\mathbf{E}_\mathrm{p}$. Now we need to find out what $\hat{\mathbf{k}}$ is; we already know from (16.43) that $k/(\mu_0\omega) = \sqrt{\varepsilon_0/\mu_0} = 1/\eta_0$ with $k = 10\pi\,\mathrm{m}^{-1}$ (here $\eta_0 \approx 120\pi\,\Omega$ is the intrinsic impedance of free space). We obtain

$$\hat{\mathbf{k}} = \gamma\hat{\mathbf{x}} + \beta\hat{\mathbf{z}} = 5\pi\sqrt{3}\hat{\mathbf{x}} + 5\pi\hat{\mathbf{z}} = 10\pi\left(\tfrac{1}{2}\sqrt{3}\hat{\mathbf{x}} + \tfrac{1}{2}\hat{\mathbf{z}}\right) \equiv k\hat{\mathbf{k}} \tag{16.47}$$

so that we see that since $k = 10\pi\,\mathrm{m}^{-1}$, vector $\hat{\mathbf{k}} = 0.5(\hat{\mathbf{x}}\sqrt{3} + \hat{\mathbf{z}})$. The result of this is that

$$\hat{\mathbf{k}}\times\mathbf{E}_0 = 0.5(\hat{\mathbf{x}}\sqrt{3} + \hat{\mathbf{z}})\times E_0\hat{\mathbf{y}} = 0.5(-\hat{\mathbf{x}} + \hat{\mathbf{z}}\sqrt{3})E_0$$

So we now have

$$\mathbf{H}_\mathrm{p}(\mathbf{r}) = \frac{1}{\eta_0}E_0\left(\frac{1}{2}\hat{\mathbf{x}} + \frac{1}{2}\hat{\mathbf{z}}\sqrt{3}\right)e^{-j\mathbf{k}\cdot\mathbf{r}} \tag{16.48}$$

The calculation is finished by appending $\exp(-j\omega t)$ and then taking the real part, according to the procedure listed in Section 16.6, to obtain

$$\begin{aligned}\mathbf{H}(\mathbf{r},t) &= \frac{1}{\eta_0}E_0\left(\frac{1}{2}\hat{\mathbf{x}}+\frac{1}{2}\hat{\mathbf{z}}\sqrt{3}\right)\cos(\omega t-\mathbf{k}\cdot\mathbf{r})\\ &= \frac{1}{24\pi}\left(\frac{1}{2}\hat{\mathbf{x}}+\frac{1}{2}\hat{\mathbf{z}}\sqrt{3}\right)\cos(3\times 10^9\pi t-5\pi x\sqrt{3}-5\pi z)\quad \text{in A/m}\end{aligned} \tag{16.49}$$

where we have found $\mathbf{k}\cdot\mathbf{r}$ from (16.47).

Problems

16.1. Given that $D_x=\varepsilon_0(5E_x+0.3E_y+0.2E_z)$, $D_y=\varepsilon_0(0.1E_x+5E_y+0.5E_z)$, $D_z=\varepsilon_0(0.4E_x-0.25E_y+5E_z)$, specify the relative permittivity.

16.2. The relationship $\mathbf{D}(\mathbf{r},t)=\int_{-T}^{\infty}dt'\varepsilon(\mathbf{r},t-t')\mathbf{E}(\mathbf{r},t')$ for $T>0$ requires a condition on the electric field to satisfy causality. What condition is this?

16.3. Alternative potentials that lead to the same electromagnetic fields as do $\mathbf{A}$ and V are $\mathbf{A}_1=\mathbf{A}-\nabla\chi$ and $V_1=V+\partial\chi/\partial t$, where $\chi(\mathbf{r},t)$ is a function of location and time.

(a) Verify that indeed the new potentials lead to the same $\mathbf{E}$ and $\mathbf{B}$ fields as to do the old ones.

(b) Find the differential equation that the function χ must satisfy.

16.4. Prove that the Lorentz condition (16.8) contains the equation of continuity.

16.5. An electromagnetic signal is relayed to a neighbor via a geostationary satellite at 20 000 km above the Earth. What is the minimum time delay for reception by the neighbor?

16.6. An electromagnetic signal with

$$V(r,t)=\frac{V_0}{r}f(t+r/c)$$

would seem to violate causality in describing a signal emanating from a point source at the origin. Can this functional dependence on location and time describe another, more physical, situation?

16.7. An electromagnetic (em) wave in free space has a field $\propto\cos[\pi(0.6x+at)]$, with x in m, t in s.

(a) Specify the constant a, the frequency f and the wavelength λ.

(b) Which direction is the wave moving in, and what is its velocity in free space?

16.8. A time-harmonic plane wave has electric field $\mathbf{E}=\mathbf{E}_0\cos(\omega t+kz)$, with real $\mathbf{E}_0$.

(a) In which direction is the wave moving?

(b) If $f=3\,\text{GHz}$, what is the wavelength λ?

(c) Give the answer to part (b) again if $\varepsilon_r = 5$.

(d) Calculate the ratio E/H for cases (b) and (c), and specify its dimension.

(e) If $\mathbf{E}_0 = E_0\hat{\mathbf{x}}$, specify the direction of $\mathbf{H}$.

16.9. A time-harmonic wave has an electric field amplitude $A = 12\,\mathrm{V/m}$ (maximal at $t = 0$) and a wavelength of 1.2 m in free space. Write it in the form $E = A\sin(\omega t - kz + \varphi)$ by specifying k, ω and φ.

16.10. An em wave in free space has the form $\mathbf{E} = \mathbf{A}\cos(\omega t - 4x - 3y)$; all lengths are in m.

(a) Find its wavevector $\mathbf{k}$.

(b) Find its frequency f.

(c) Find two possible directions for $\mathbf{A}$.

16.11. A time-harmonic plane wave has $\mathbf{H} = H_0\hat{\mathbf{z}}\cos(\omega t - ky)$, where H_0 is real, in free space. Find the $\mathbf{E}$ field at $f = 0.8\,\mathrm{GHz}$ ($1\,\mathrm{GHz} = 10^9\,\mathrm{Hz}$).

16.12. A plane wave has the electric field $E = A\cos[4\pi(x - 1.50 \times 10^8 t)]$ where distances are in m. Specify the frequency f of the wave and the permittivity of the medium.

16.13. Prove Eqs. (16.45). What condition must $\mathbf{E}$ fulfill in order that these equations hold?

16.14. Given that in free space $\mathbf{E}(\mathbf{r}, t) = 6\hat{\mathbf{z}}\cos(5\pi \times 10^9 t + 3\pi x + \gamma z)$ in V/m (t in s, x in m), find γ and $\mathbf{H}(\mathbf{r}, t)$.

17

Important features of plane time-harmonic waves

Plane time-harmonic (i.e. monochromatic) waves are the simplest form of electromagnetic energy that can propagate through a medium, i.e. the simplest to describe mathematically. Nevertheless, their properties are surprisingly varied and worth looking into more closely, so that the main features can be recognized in more complicated forms of waves. In fact, the mathematical theorems of Fourier analysis teach that any solution of the wave equation (16.12) can be described as a linear superposition of plane monochromatic waves. The properties of these are therefore inherently present in more complicated waves. In this chapter, we shall discuss

(1) several kinds of velocities pertaining to the propagation of a plane time-harmonic wave
(2) the effect of dielectric permittivity upon the wave
(3) the orientation of the electric and magnetic fields in the plane $\perp$ propagation; this orientation is called the *polarization* of the wave
(4) the power and energy associated with the propagation of a wave.

17.1 Phase and group velocities

17.1.1 Phase velocity

In Section 16.3 the discussion centered upon the fact that electromagnetic fields produced by time-varying sources propagate through space at a finite velocity $c = 1/\sqrt{\mu\varepsilon}$. The concept of the *velocity of a signal* is more complicated than meets the eye at first sight. Consider first the infinite pure time-harmonic field, $E(t) = E_0 \cos(\omega t - kz)$, with real E_0. We omit the vector signs as it is understood we are dealing with a plane wave traveling in a direction that we label as the z direction, so that the field $\mathbf{E}_0$ is perpendicular to $\hat{\mathbf{z}}$. At a short time δt later (consider it to be infinitesimal), the field is

$$E(t + \delta t) = E_0 \cos[(\omega t + \delta t) - kz]$$

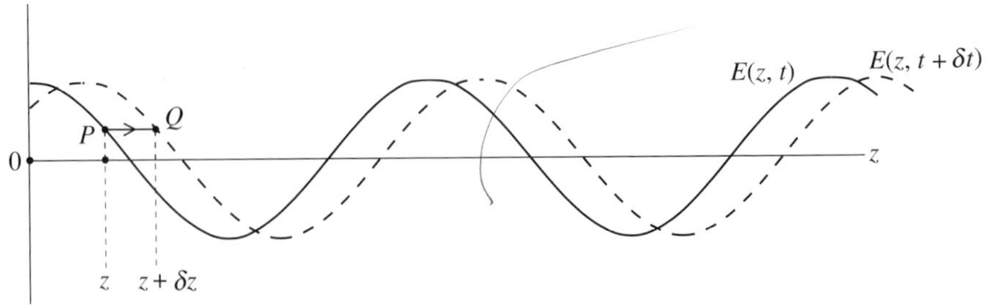

Fig. 17.1 Phase velocity deduced from considering $\cos(\omega t - kz)$ at (z, t) and at $(z + \delta z, t + \delta t)$.

The sinusoid is shown at both of these times in Fig. 17.1. Point P at location z on the sinusoid at time t becomes point Q at location $z + \delta z$ on the sinusoid at time $t + \delta t$ (δz is also infinitesimal). The value of the cosine has not changed. Therefore

$$\cos(\omega t - kz) = \cos[\omega(t + \delta t) - k(z + \delta z)] \tag{17.1}$$

and it follows that $\omega\,\delta t - k\,\delta z = 0$.

Hence it follows that

$$v_{\text{ph}} \equiv \lim_{\delta t \to 0} \frac{\delta z}{\delta t} = \frac{\omega}{k} = \frac{1}{\sqrt{\mu\varepsilon}} \tag{17.2a}$$

This is known as the *phase velocity*; it describes how far each *wavefront* progresses per unit time (see subsection 16.7.2 for a reminder of what a wavefront is). A problem with this is that phase velocities cannot be measured directly. Any signal that is a pure sinusoid must be infinitely long, and that contradicts the fact that we can measure finite-duration signals only. All finite-duration signals have more than one frequency component, as is expressed mathematically in (16.2). But, if $\varepsilon(\omega)$ is a function of frequency, as is usually the case in dielectric media, then each frequency component will have its own phase velocity $v_{\text{ph}}(\omega) = \omega/k(\omega) = 1/\sqrt{\mu\varepsilon(\omega)}$. The signal components will travel at different velocities, and the signal will become distorted at a distance because the integral (16.2) or sum of the various sinusoidal components changes with distance. Such dielectrics are dubbed *dispersive media*. If we define c as the velocity of a signal in free space ($c \equiv 1/\sqrt{\mu_0\varepsilon_0}$), then we have found, more precisely, for a plane time-harmonic wave in a dielectric medium

$$v_{\text{ph}} \equiv \lim_{\delta t \to 0} \frac{\delta z}{\delta t} = \frac{\omega}{k(\omega)} = \frac{1}{\sqrt{\mu\varepsilon(\omega)}} = \frac{c}{n(\omega)} \tag{17.2b}$$

In this notation, $n(\omega) \equiv \sqrt{\varepsilon_r(\omega)}$ is known as the *refractive index*; it is dimensionless. It is unity in free space (we remind the reader that $\varepsilon_r(\omega) \equiv \varepsilon(\omega)/\varepsilon_0$ is the *relative dielectric permittivity* of the medium). The phase velocity is the quantity previously labeled c in Section 16.3, but it has been defined more carefully here to distinguish it from other velocities. From now on, the letter c will indicate the velocity in free space only. In general, the refractive index is greater than unity in dielectrics so that waves travel more slowly than in free space. This appears to be quite satisfactory from the point of view of the theory of special relativity, which prohibits velocities greater than c. There are media, however, such as plasma (an ionized gas) in which $n < 1$, and then $v_{\mathrm{ph}} > c$. While this appears to violate special relativity, it is so only in appearance; the phase velocity in a medium is not a measurable quantity: it does not represent the motion of a parcel of energy. Examples of such 'illusory' velocities greater than c will be given below.

17.1.2 Group velocity

Having noted that signals become distorted in dispersive media (where $\varepsilon(\omega)$ is a function of frequency), and consequently that phase velocity is not usually a measurable quantity, the next question is, 'which velocity *is* measurable?' A partial answer to that question is given by considering two plane-wave sinusoids at frequencies very close to each other:

$$\begin{aligned} E_1(t) &= A_1 \cos(\omega_1 t - k_1 z) \qquad \omega_1 = \omega + \tfrac{1}{2}\delta\omega \quad k_1 = \omega\sqrt{\mu\varepsilon(\omega_1)} \\ E_2(t) &= A_2 \cos(\omega_2 t - k_2 z) \qquad \omega_2 = \omega - \tfrac{1}{2}\delta\omega \quad k_2 = \omega\sqrt{\mu\varepsilon(\omega_2)} \end{aligned} \tag{17.3}$$

The idea here is that the difference frequency $\delta\omega$ is very much smaller than the average frequency ω (also known as the *carrier frequency* in signal communications applications). For the sake of simplicity, we set $A_2 = A_1 \equiv A$ as a real quantity; both sinusoids then have the same amplitude A. We assume that a source generates the sum of these two signals. The mathematical sum of $E_1(t)$ and $E_2(t)$ is

$$\begin{aligned} E(t) &= 2A\cos\left[\tfrac{1}{2}(\omega_1 + \omega_2)t - \tfrac{1}{2}(k_1 + k_2)z\right]\cos\left[\tfrac{1}{2}(\omega_1 - \omega_2)t - \tfrac{1}{2}(k_1 - k_2)z\right] \\ &= 2A\cos(\omega t - kz)\cos(\delta\omega\, t - \delta k\, z) \end{aligned} \tag{17.4}$$

The meanings of k (the average of k_1 and k_2) and of δk (the difference in k_1 and k_2) should be clear from the first right-hand side of (17.4). This sum wave is shown in Fig. 17.2.

The first cosine in (17.4) varies much more rapidly as a function of z than the second because it is easy to see that $\delta k \ll k$ whenever $\delta\omega \ll \omega$. The sum signal is the rapidly oscillating curve in Fig. 17.2. The hypothetical curve labeled 'envelope' connects the peaks (both positive and negative) of the rapidly oscillating curve. The envelope is described mathematically by the second cosine factor in (17.4); its hump appears to be a feature that we might be able to follow physically (and therefore

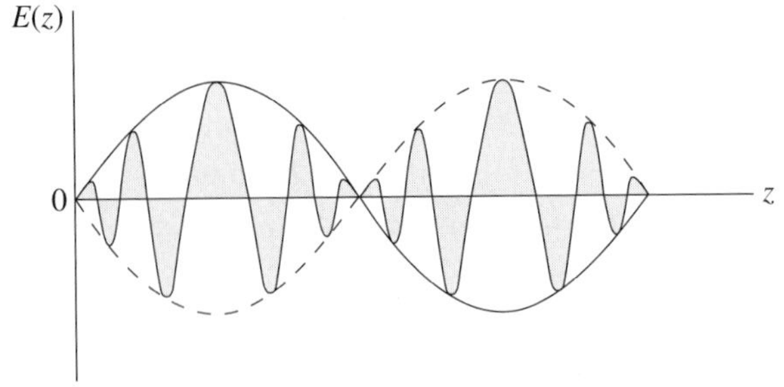

Fig. 17.2 Sum of two time-harmonic plane waves with slightly differing frequencies.

might correspond to a measurable velocity) as it corresponds to a very simple *amplitude-modulated* (AM) signal. If we followed the motion of its peak in time δt over distance δz, we would observe that

$$\cos(\delta\omega\, t - \delta k\, z) = \cos[\delta\omega(t + \delta t) - \delta k(z + \delta z)] \tag{17.5}$$

and thus we can define the *group velocity*

$$v_{\text{gr}} \equiv \lim_{\delta t \to 0} \frac{\delta z}{\delta t} = \lim_{\delta\omega \to 0} \frac{\delta\omega}{\delta k(\omega)} = \frac{d\omega}{dk} = \frac{1}{dk(\omega)/d\omega} \tag{17.6}$$

Only for *nondispersive media*, in which ε, μ are not functions of frequency, are the group and phase velocities equal. In this case, $k = \omega\sqrt{\mu\varepsilon}$ with $dk/d\omega = \sqrt{\mu\varepsilon}$, so that $v_{\text{gr}} = d\omega/dk = 1/\sqrt{\mu\varepsilon} = v_{\text{ph}}$. Stated in words, all frequency components travel at the same velocity in nondispersive media (so that there is no distortion as the wave propagates). It is often handy to use a simple relationship between these two velocities. We use $k = \omega/v_{\text{ph}}$, (17.2b), to find

$$\frac{dk}{d\omega} = \frac{d}{d\omega}\left(\frac{\omega}{v_{\text{ph}}}\right) = \frac{1}{v_{\text{ph}}} - \left(\frac{\omega}{v_{\text{ph}}^2}\right)\left(\frac{dv_{\text{ph}}}{d\omega}\right) = \frac{1}{v_{\text{ph}}}\left[1 - \left(\frac{\omega}{v_{\text{ph}}}\right)\left(\frac{dv_{\text{ph}}}{d\omega}\right)\right] \tag{17.7}$$

which yields

$$v_{\text{gr}} = \frac{v_{\text{ph}}}{1 - (\omega/v_{\text{ph}})(dv_{\text{ph}}/d\omega)} \tag{17.8a}$$

This immediately establishes the equality of the two velocities when v_{ph} does not depend upon frequency. It also shows that $v_{\text{gr}} < v_{\text{ph}}$ if the phase velocity *decreases* with increasing frequency. This is *normal dispersion*. The opposite case – *anomalous dispersion* – occurs when the phase velocity increases with frequency, for then we find $v_{\text{gr}} > v_{\text{ph}}$ from (17.8a). Another useful form of (17.8a) is found by noting that the phase velocity can be written as $v_{ph} = c/n$, where $n \equiv \sqrt{\varepsilon_r(\omega)}$ is the refractive index first introduced above. The derivative in (17.8a) is easily obtained, to yield

$$v_{gr} = \frac{v_{ph}}{1 + (\omega/n)(dn/d\omega)} \tag{17.8b}$$

So *normal dispersion* occurs when $n(\omega)$ increases with ω, otherwise we have *anomalous dispersion*. The very fact that the group velocity can exceed the phase velocity is cause for being careful with the idea of connecting it to a measurable quantity. The group velocity itself is only an approximation to what is really measurable: the velocity with which the *energy* represented by the wave travels through the medium. This cannot exceed c. We shall discuss energy velocities briefly below in connection with the Poynting vector (Section 17.4). Now we wish to discuss several illuminating cases in which phase velocities larger than c clearly are not real effects.

Example 17.1 Barber pole

Figure 17.3 shows a (now antiquated?) sign outside a barber shop. It consists of a rotatable cylinder on which a diagonal red stripe twists around in a helix. When the pole is rotated counterclockwise, the stripes appear to move upward. The velocity at which they move upward is dictated by the slant of the stripes and the speed of rotation, and it is equal to $\delta z/\delta t$ where $\delta z = v\,\delta t \tan\theta$ (see the diagram). If the slant is sufficiently far from horizontal and/or the speed of rotation sufficiently fast but still less than the speed of light, then $\delta z/\delta t$ could exceed c. However, we can be assured that this would then be a typical non-observable phase velocity: no material or energy is moving vertically! The *only* motion of material points is in horizontal circles around the axis of the pole, and the corresponding material velocity at the cylinder's surface is $v = a\omega$, given that a is the radius and ω is the angular velocity of rotation. It is not possible for v to exceed c.

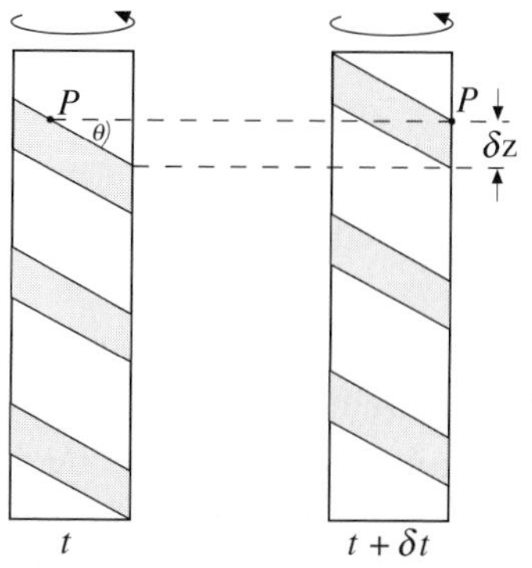

Fig. 17.3 The 'barber-pole' apparent violation of $v < c$.

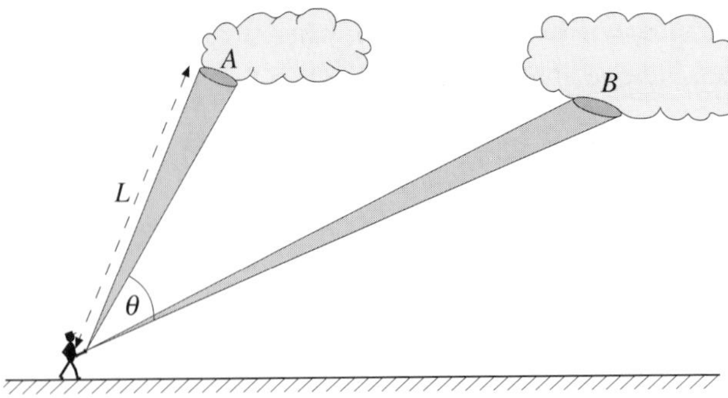

Fig. 17.4 An apparent violation of the maximum observable velocity of photons: a person moves a torch beam from one cloud to another.

Example 17.2 Flashlight aimed at the clouds

A person, as sketched in Fig. 17.4, aims a flashlight beam at a cloud overhead. The beam projects a spot at A. Upon swiveling the flashlight the spot moves rapidly to B. If the distance L and the speed of rotation $\omega = d\theta/dt$ are sufficiently large, then the speed at which the spot moves may appear to exceed c because that velocity is $\sim \omega L$ and it could exceed c. However, as before, that velocity would then be unobservable because no parcels of light energy (photons) are moving directly from A to B. A photon arriving at a point on spot B would have left the flashlight at a time $\delta t \sim L/c$ earlier. Its horizontal velocity is determined by the rotation of the hand, which imparts a horizontal velocity $\sim \omega l$ (given that l is of the order of the length of a flashlight). This obviously has to be much less than c, as it is due to the motion of material. Spot B is illuminated by photons that left the flashlight much later than those arriving at spot A.

In summary: the group velocity is a better approximation to the velocity with which wave energy propagates than the phase velocity. The latter is the same as the group velocity in nondispersive media.

17.2 Plane waves in lossy media

So far, we have considered waves in media with real dielectric permittivities. What happens when the permittivity is a complex number? To answer this, let $\varepsilon = \varepsilon_0(\varepsilon' - j\varepsilon'')$, so that the relative dielectric permittivity $\varepsilon_r = \varepsilon' - j\varepsilon''$, with real ε' and ε''. The minus sign will be explained shortly. We have seen that wavenumber k is defined by $k = \omega\sqrt{\mu\varepsilon} = \kappa_0\sqrt{\mu_r\varepsilon_r}$, given that the free-space wavenumber is $k_0 \equiv \omega\sqrt{\mu_0\varepsilon_0}$. We will assume that $\mu_r = 1$, because most propagation media are not magnetic. Thus we have $k = \beta - j\alpha$ in terms of real and imaginary parts, with

$$k = k_0\sqrt{\varepsilon' - j\varepsilon''} \tag{17.9}$$

If we square (17.9) and equate real and imaginary parts to obtain $\beta^2 - \alpha^2 = k_0^2\varepsilon'$ and $2\alpha\beta = k_0^2\varepsilon''$, we then find easily that

$$\beta = \frac{k_0}{\sqrt{2}}\sqrt{|\varepsilon_r| + \varepsilon'} \qquad \alpha = \frac{k_0}{\sqrt{2}}\sqrt{|\varepsilon_r| - \varepsilon'} \tag{17.10}$$

Here, $|\varepsilon_r|$ is the absolute value of the relative dielectric permittivity: $|\varepsilon_r| = \sqrt{\varepsilon'^2 + \varepsilon''^2}$. The phasor representation of a time-harmonic plane wave is given in the development between (16.41) and (16.42), and here it would be

$$\mathbf{E}_p(z) = E_{x0}\hat{\mathbf{x}}e^{-j(\beta - j\alpha)z} = E_{x0}\hat{\mathbf{x}}e^{-j\beta z - \alpha z} \tag{17.11}$$

so that the actual field (the real part of $\mathbf{E}_p(\mathbf{z})e^{j\omega t}$) would be

$$\mathbf{E}(z, t) = E_{x0}\hat{\mathbf{x}}e^{-\alpha z}\cos(\omega t - \beta z) \tag{17.12a}$$

There is an extra exponential *decay factor* in comparison with (16.44). If we had chosen a plus sign between the real and imaginary parts of ε and k, we would have found an exponential with a positive sign. That would indicate a *growth* in the strength of the wave as it propagates through the medium, which would be at the cost of energy of the medium. While this is not impossible, it is quite rare, and (17.12) is the normal situation, indicating loss of energy of the wave *to* the medium. The amplitude of the wave decays exponentially as the wave progresses in the z direction: the medium is *lossy*.

There is an important effect upon the magnetic field. First of all, note that (17.12a) contains purely real quantities. If the phasor amplitude E_{x0} were complex in (17.11), e.g. if $E_{x0} = |E_{x0}|e^{j\varphi}$, then (17.12) would be

$$\mathbf{E}(z, t) = |E_{x0}|\hat{\mathbf{x}}e^{-\alpha z}\cos(\omega t - \beta z + \varphi) \tag{17.12b}$$

so that the complex nature of E_{x0}, in this case, would only modify the *phase* of the wave (i.e. shift the entire cosine by φ radians). More complicated possibilities occur when we introduce the concept of *polarization*, for example, the possibility that $E_{x0}\hat{\mathbf{x}}$ is replaced by a vector with a complex $\hat{\mathbf{x}}$ and a complex $\hat{\mathbf{y}}$ component. This is postponed for now, but we will return to it in due time. The **H** field is obtained from the curl **E** equation for this time-harmonic plane wave having only an $E_x(z)$ component as follows:

$$\nabla \times \mathbf{E}_p = -j\omega\mu_0\mathbf{H}_p \tag{17.13}$$

From the form of (17.11) it is easy to see that '$\nabla\times$' in (17.13) is replaced by '$-j(\beta - j\alpha)\hat{\mathbf{z}}\times$' so that

$$\mathbf{H}_p = \left(\frac{\beta - j\alpha}{\omega\mu_0}\right)\hat{\mathbf{z}} \times \mathbf{E}_p = \frac{\beta - j\alpha}{\omega\mu_0}\hat{\mathbf{y}}E_{x0}e^{-j\beta z - \alpha z} \tag{17.14a}$$

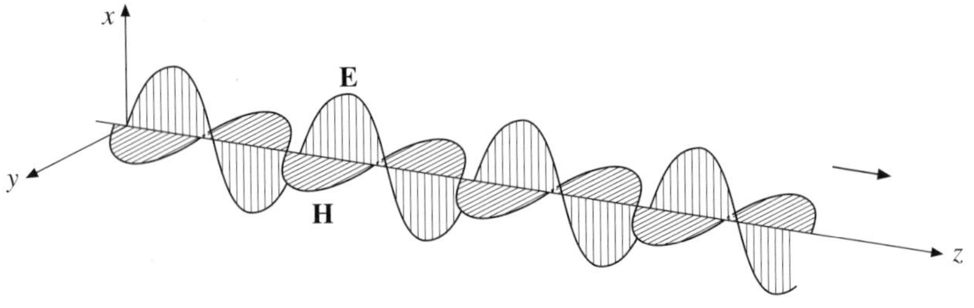

Fig. 17.5 The **E** and **H** fields along a 'ray' of an electromagnetic wave propagating in a lossy medium. The two fields are now out of phase. The magnitudes of the fields are not to scale.

The intrinsic impedance η is now complex, and we write this equation as

$$\mathbf{H}_\mathrm{p} = \frac{1}{\eta}\hat{\mathbf{z}} \times \mathbf{E}_\mathrm{p} \qquad \eta = \sqrt{\frac{\mu_0}{\varepsilon_0 \varepsilon_\mathrm{r}}} = \frac{\omega\mu_0}{\beta - j\alpha} \tag{17.14b}$$

Here, we have used $\beta - j\alpha \equiv \omega\sqrt{\mu_0\varepsilon_0\varepsilon_\mathrm{r}}$ in the last equality. We also can write the intrinsic impedance as

$$\eta = |\eta| e^{-j\theta_\eta} \qquad \text{with} \quad |\eta| \equiv \frac{\omega\mu_0}{\sqrt{\beta^2 + \alpha^2}} \qquad \text{and} \quad \tan\theta_\eta \equiv \frac{\alpha}{\beta} \tag{17.14c}$$

Summarizing the information in the three equations (17.14c) for the **H** phasor, and converting from phasor to real-field representation, we find

$$\mathbf{H}(z,t) = \mathrm{Re}\left\{\frac{1}{|\eta|}\hat{\mathbf{y}} E_{x0} e^{-\alpha z + j(\omega t - \beta z + \theta_\eta)}\right\} \qquad \text{or}$$

$$\mathbf{H}(z,t) = H_{y0}\hat{\mathbf{y}} e^{-\alpha z}\cos(\omega t - \beta z + \theta_\eta) \qquad H_{y0} \equiv \frac{1}{|\eta|}|E_{x0}| \tag{17.15}$$

and there would be an extra phase factor φ if E_{x0} were complex, as in (17.12b). In comparing (17.15) to (17.12a), we observe that not only do both the **E** and **H** fields decay in the z direction of propagation but also the two fields are *out of phase* by the angle θ_η. As shown in Fig. 17.5, the two fields are not zero at the same place (as they are, Fig. 16.6, for real intrinsic impedance η); one field lags in phase behind the other by θ_η. When $E(z,t)$ is maximal, then $H(z,t)$ differs from its maximal value by the factor $\cos\theta_\eta$.

17.2.1 Low-loss approximation

Media in which waves can propagate well are mostly characterized by the fact that $\varepsilon'' \ll \varepsilon'$, in which case one can make the binomial approximation (3.6) for $\sqrt{\varepsilon_\mathrm{r}}$,

$$\sqrt{\varepsilon' - j\varepsilon''} \approx \sqrt{\varepsilon'}\left(1 - \frac{1}{2}j\frac{\varepsilon''}{\varepsilon'}\right) \tag{17.16}$$

We use this in (17.10) to obtain the low-loss approximations

$$\beta \approx k_0\sqrt{\varepsilon'} \qquad \alpha \approx \frac{\varepsilon''}{2\varepsilon'}k_0\sqrt{\varepsilon'} \approx \frac{\varepsilon''}{2\varepsilon'}\beta \tag{17.17}$$

from which it is obvious that $\alpha \ll \beta$. The errors in (17.17) are of order $(\varepsilon''/\varepsilon')^2$ because that is the error in (17.16). The ratio $\varepsilon''/\varepsilon'$ is known as the *loss tangent* of the medium, and it needs to be very small for this approximation to be valid. The consequence for (17.15) is that the cosine will go through many cycles before its maximum amplitude decreases appreciably. The phase velocity retains its simple lossless form of $v_{\mathrm{ph}} = \omega/\beta \approx c/\sqrt{\varepsilon'}$. Matters are quite different, however, in the high-loss approximation.

17.2.2 High-loss approximation

In metals, we usually have very high values of the conductivity, namely $\sigma \sim 10^6 - 10^7$ S/m (or even higher, as is the case for copper). Consider the curl $\mathbf{H}$ equation for time-harmonic waves in such a metallic medium:

$$\nabla \times \mathbf{H} = \mathbf{J} + j\omega\varepsilon\mathbf{E} = (\sigma + j\omega\varepsilon_0\varepsilon')\mathbf{E} = j\omega\varepsilon_0\left(\varepsilon' - j\frac{\sigma}{\varepsilon_0\omega}\right)\mathbf{E} \tag{17.18}$$

This way of writing the equation shows that $\varepsilon' - j\sigma/\varepsilon_0\omega$ is the effective complex relative dielectric permittivity ε_{r} in such a medium, and therefore that $\varepsilon'' \equiv \sigma/\varepsilon_0\omega$ in a highly conducting medium. We have, of course, assumed that ε' in the metal is purely real, otherwise there would be an additional imaginary term in ε_{r}. As $\varepsilon_0 \sim 10^{-11}$ F/m and $\sigma \geq 10^6$ S/m, we see that $\sigma/\varepsilon_0\omega \geq 10^5$ for $\omega \leq 10^{12}\,\mathrm{s}^{-1}$ and hence is very much larger than ε' (which is of order unity). Thus we find $\varepsilon'' \gg \varepsilon'$ in metals. Instead of (17.16) we obtain

$$\sqrt{\varepsilon' - j\varepsilon''} \approx \sqrt{-j\varepsilon''}\left(1 + \frac{1}{2}j\frac{\varepsilon}{\varepsilon''}\right) \approx \sqrt{-j\varepsilon''} \tag{17.19}$$

and this then leads immediately via $k \approx k_0\sqrt{-j\varepsilon''}$ to

$$\alpha \approx \beta \approx \tfrac{1}{2}k_0\sqrt{2\varepsilon''} = \sqrt{\mu\omega\sigma/2} \tag{17.20a}$$

Thus, by the time $\beta z \approx 1$, which corresponds to only part of a cycle of the cosine, we see that $\alpha z \approx 1$; this means the maximum amplitude of the cosine would be down by a factor $1/e$. The wave is damped very rapidly in such a metal. In fact we often use the concept of *skin depth* in a metal to describe a typical length in which the wave is dampened by a factor $1/e$: the skin depth $\delta \equiv 1/\alpha \approx \sqrt{2/\mu\omega\sigma}$ in a high-loss medium by definition. After, say, 10 skin depths, the wave is attenuated by a factor $e^{-10} \approx 4.5 \times 10^{-5}$ in amplitude. For $\mu \approx \mu_0$ and in copper, where $\sigma \approx 5.8 \times 10^7$ S/m, we obtain

$$\delta \text{ in m} \approx \frac{0.0661}{\sqrt{f \text{ in Hz}}} \tag{17.20b}$$

Even at frequencies as low as $f = 10\,\text{kHz}$, the skin depth is already as small as 6.6 μm. This is the reason why electromagnetic fields impinging upon metals, and the currents they produce in those metals, are restricted to the surface region; thus the name *skin depth* for δ. A word of caution: both the approximations (17.20a, b) hold *only* for very high values of the loss tangent, $\varepsilon''/\varepsilon' \gg 1$. The skin depth is always $\delta = 1/\alpha$, but it is not always equal to $\sqrt{2/\mu\omega\sigma}$!

The phase velocity corresponding to (17.20a) is now $v_{\text{ph}} = \omega/\beta \approx \sqrt{2\omega/\mu\sigma}$. A comparison with its low-loss value, $c/\sqrt{\varepsilon'}$, shows us that

$$(v_{\text{ph}})_{\text{high-loss}} \approx (v_{\text{ph}})_{\text{low-loss}} \times \sqrt{\frac{2\varepsilon'}{\varepsilon''}} \tag{17.21}$$

Hence, the high-loss phase velocity is always much smaller than the low-loss phase velocity.

The high-loss approximation also has some consequences for the magnetic field expression. Going to (17.14), we note that $\eta^{-1} = k/\mu_0\omega = (\beta - j\alpha)/\mu_0\omega$ in a non-magnetic medium, and, with (17.20a), we obtain

$$\eta^{-1} \approx (1 - j)\frac{\beta}{\mu_0\omega} = \sqrt{2}e^{-\pi j/4}(\eta')^{-1} \tag{17.22}$$

given that $\eta' = \mu_0\omega/\beta$ is the intrinsic impedance of an equivalent lossless dielectric. The net result for $\mathbf{H}$ from (17.15) is then

$$\mathbf{H}(z, t) = \sqrt{2}H_{y0}\hat{\mathbf{y}}e^{-\alpha z}\cos(\omega t - \beta z - \pi/4) \qquad H_{y0} \equiv \frac{1}{\eta'}|E_{x0}| \tag{17.23}$$

We see that the $\mathbf{H}$ field differs in phase from the $\mathbf{E}$ field by $\sim 45^\circ$ in the high-loss approximation.

17.3 Polarization

In this section we explain how the electric field vectors of time-harmonic plane waves can trace out more complicated paths in space than sinusoids in a plane. This possibility is due to *polarization* of the wave. Let us return to the time-harmonic plane wave with electric and magnetic fields given by (16.44) in a dielectric medium that is not lossy. We know that the phasor $\mathbf{E}_{\text{p}}$ must be $\perp \hat{\mathbf{z}}$ (the direction of propagation), and whereas before we chose $\mathbf{E}_{\text{p}} = E_{x0}\hat{\mathbf{x}}e^{-jkz}$ (with real E_{x0}), we now will make the most general choice possible:

$$\mathbf{E}_{\text{p}} = (A_x\hat{\mathbf{x}}e^{-j\varphi_x} + A_y\hat{\mathbf{y}}e^{-j\varphi_y})e^{-jkz} \tag{17.24a}$$

Thus the vectorial factor in front of the phase exponential has four real constants in it: two amplitudes A_x, A_y and two phases φ_x, φ_y. One of these four constants, φ_x (or φ_y; you can choose either) plays a trivial role and therefore can be pulled out as a common factor to get

$$\mathbf{E}_\mathrm{p}(z) = (A_x\hat{\mathbf{x}} + A_y\hat{\mathbf{y}}e^{-j\delta\varphi})e^{-j(kz+\varphi_x)} \qquad \text{with} \qquad \delta\varphi \equiv \varphi_y - \varphi_x \tag{17.24b}$$

The real time-dependent field then becomes

$$\mathbf{E}(z,t) = \mathrm{Re}\Big\{(A_x\hat{\mathbf{x}} + A_y\hat{\mathbf{y}}e^{-j\delta\varphi})e^{j(\omega t-kz-\varphi_x)}\Big\} \tag{17.25}$$

You can see explicitly from this form that φ_x merely changes the phase of the sinusoidal argument $\omega t - kz$. Let us set $\omega t - kz - \varphi_x \equiv \varphi_\mathrm{ph}(z,t)$, which we will abbreviate as φ_ph, so that (17.25) finally becomes

$$\mathbf{E}(z,t) = A_x\hat{\mathbf{x}}\cos\varphi_\mathrm{ph} + A_y\hat{\mathbf{y}}\cos(\varphi_\mathrm{ph} - \delta\varphi) \qquad \text{with} \qquad \varphi_\mathrm{ph} \equiv \omega t - kz - \varphi_x \tag{17.26}$$

This is actually quite complicated. The **E** vector traces out an *ellipse* in a plane $\perp \hat{\mathbf{z}}$, either as a function of t for fixed z, or as a function of z for fixed t. It does not necessarily oscillate in a fixed plane through the z axis. In fact it traces an *elliptical helix* in space as the wave propagates (helical because the **E** vector's tip moves in the $\hat{\mathbf{z}}$ direction as it traces an ellipse in a $z =$ constant plane). Rather than tackle this, the most general, case at once, let us look at some simple cases first.

17.3.1 Linear polarization

If in (17.24b) and (17.26) we assume we are dealing with the case in which $\varphi_y = \varphi_x$, i.e. $\delta\varphi = 0$, then we find that

$$\mathbf{E}(z,t) = (A_x\hat{\mathbf{x}} + A_y\hat{\mathbf{y}})\cos\varphi_\mathrm{ph} \tag{17.27}$$

The direction of the **E** field does not change (except for a minus sign during one half of the cycle of the cosine). The **E** vector always lies on a line in a plane $\perp \hat{\mathbf{z}}$ that is inclined by an angle θ to the x axis, and $\tan\theta = A_y/A_x$ (see Fig. 17.6); it oscillates between the extremes $\overrightarrow{OP}$ (for $\varphi_\mathrm{ph} = 0$) and $\overrightarrow{OP}'$ (for $\varphi_\mathrm{ph} = \pi$). This is referred to as *linear polarization*, because the **E** vector always lies on the same inclined line.

17.3.2 Circular polarization

Suppose now that φ_x and φ_y differ by a quarter of a cycle, i.e. by $\pi/2$. Thus $\varphi_y = \varphi_x \pm \pi/2$ and $\delta\varphi = \pm\pi/2$. Then (17.26) becomes

$$\mathbf{E}(z,t) = A_x\hat{\mathbf{x}}\cos\varphi_\mathrm{ph} \pm A_y\hat{\mathbf{y}}\sin\varphi_\mathrm{ph} \tag{17.28}$$

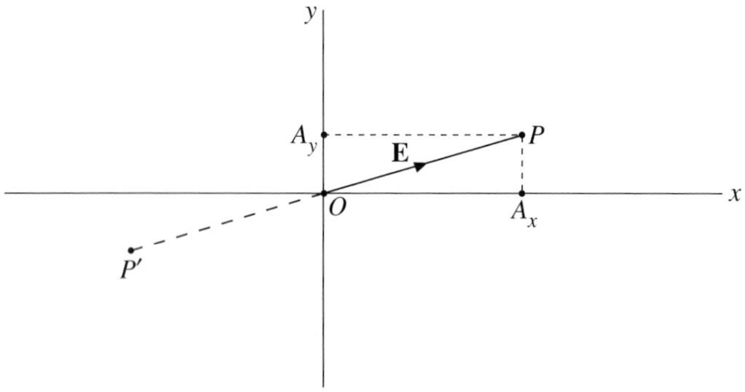

Fig. 17.6 An electric field linearly polarized in a plane through the propagation axis z: **E** oscillates between the extremes $\overrightarrow{OP}$ and $\overrightarrow{OP}'$.

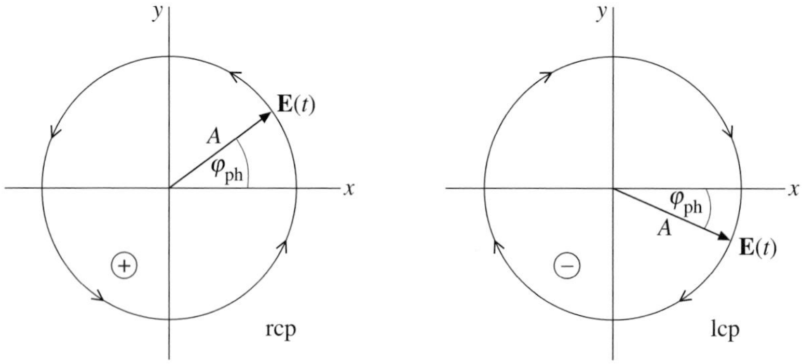

Fig. 17.7 Circularly polarized electric fields as seen in a plane perpendicular to the propagation axis z, which is upwards from the page.

If, in addition, the amplitudes are the same, i.e. $A_y = A_x \equiv A$, then we find

$$\mathbf{E}(z, t) = A(\hat{\mathbf{x}} \cos \varphi_{ph} \pm \hat{\mathbf{y}} \sin \varphi_{ph}) \tag{17.29}$$

As time progresses at fixed z we note that φ_{ph} increases. In that case (17.29), taking the $+$ sign, predicts that the $\mathbf{E}(t)$ field moves around the z axis in a *clockwise circle* (when looking along the $+z$ direction, as shown in the left-hand diagram of Fig. 17.7) with radius A. The convention in electrical engineering is to say that the wave is *right circularly polarized* (rcp). In the case of the $-$ sign, we obtain a *left circularly polarized* (lcp) wave; see the right-hand side of Fig. 17.7.

The rcp and lcp waves have a simple phasor representation, namely

$$\mathbf{E}_p(z) = A(\hat{\mathbf{x}} \pm j\hat{\mathbf{y}})e^{-j(kz+\varphi_x)} \tag{17.30}$$

with the upper sign for the rcp wave.

Table 17.1.

φ	E_x	E_y	point
0	A_x	$A_y \cos \delta\varphi$	P
π	$-A_x$	$-A_y \cos \delta\varphi$	P'
$\delta\varphi$	$A_x \cos \delta\varphi$	A_y	Q
$\delta\varphi + \pi$	$-A_x \cos \delta\varphi$	$-A_y$	Q'
$\delta\varphi - \pi/2$	$A_x \sin \delta\varphi$	0	R
$\pi/2$	0	$A_y \sin \delta\varphi$	S

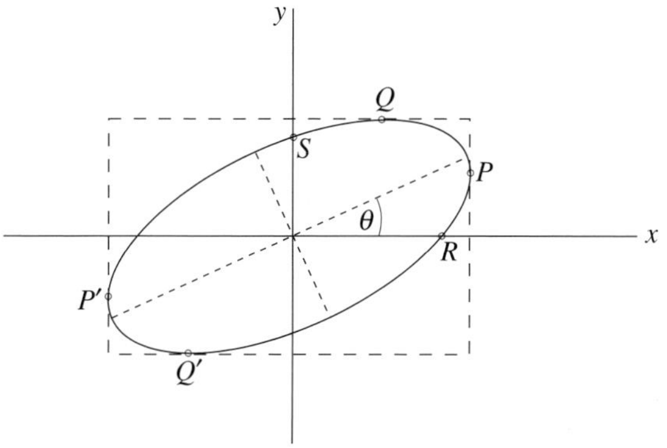

Fig. 17.8 An elliptically polarized wave field.

17.3.3 General elliptical polarization

In the most general case we have, according to (17.26),

$$\begin{aligned} E_x &= A_x \cos \varphi_{\text{ph}} \\ E_y &= A_y \cos(\varphi_{\text{ph}} - \delta\varphi) \end{aligned} \tag{17.31a}$$

By choosing fixed values for A_x, A_y, and $\delta\varphi$ and by allowing φ_{ph} to take on all values continuously between 0 and 2π one can see that (17.31a) generally defines an ellipse as the locus of the tip of $|E| = \sqrt{E_x^2 + E_y^2}$. To see this in the simplest way possible, let us look at (17.31a), and insert a number of values of φ_{ph}, referring also to Fig. 17.8; we then obtain Table 17.1.

Obviously, many more points can be drawn, and it is not difficult to use a simple spreadsheet on a personal computer to calculate E_x and E_y for fixed values of A_x, A_y, and $\delta\varphi$ and variable values of φ_{ph}. With special choices of the three parameters, the ellipse degenerates either to a line ($\delta\varphi = 0$) or a circle ($A_y = A_x$ and $\delta\varphi = \pm\pi/2$).

By expanding $\cos(\varphi_{ph} - \delta\varphi) = \cos\varphi_{ph}\cos\delta\varphi + \sin\varphi_{ph}\sin\delta\varphi$ in E_y, we can eliminate the phase φ_{ph} from (17.31a) to obtain an equation in E_x and E_y,

$$\frac{E_x^2}{(A_x \sin\delta\varphi)^2} + \frac{2E_xE_y\cos\delta\varphi}{A_xA_y\sin^2\delta\varphi} + \frac{E_y^2}{(A_y\sin\delta\varphi)^2} = 1 \qquad (17.31\text{b})$$

and this is one of the standard mathematical forms for an ellipse traced by the $\mathbf{E} = E_x\hat{\mathbf{x}} + E_y\hat{\mathbf{y}}$ vector in the $z = 0$ plane. Note in Fig. 17.8 that points P, P' represent extreme values of the ellipse in the $\pm x$ directions, and similarly so for points Q, Q' with respect to the $\pm y$ directions. The ellipse is tilted in general by an angle θ with respect to the x axis. From (17.31a) it follows that $|E|^2 = A_x^2\cos^2\varphi_{ph} + A_y^2\cos^2(\varphi_{ph} - \delta\varphi)$. Consider this as an equation for the square of the magnitude of vector $\mathbf{E}$ as a function of φ_{ph}. This magnitude will be an extreme (a maximum or minimum) at the phases $\varphi_{ph} = \psi$ and $\psi + \pi/2$ for which $d(|\mathbf{E}|^2)/d\varphi_{ph} = 0$. Reorganization of what we find by doing this yields

$$\tan[2(\psi \pm n\pi/2)] = \frac{A_y^2\sin(2\delta\varphi)}{A_x^2 + A_y^2\cos(2\delta\varphi)} \qquad n = 0, \pm1, \pm2, \ldots \qquad (17.32)$$

We can calculate the right-hand side of this unambiguously. The left-hand side is a little more difficult. As the tangent of an angle repeats itself every π radians, we observe that (17.32) cannot distinguish between ψ and $\psi \pm n\pi/2$ of the sketch in Fig. 17.8, and therefore have included that in the left-hand side of this formula. The easiest way to tell whether one half of the arctangent of the right-hand side of (17.32) is the phase angle corresponding to the maximum $|\mathbf{E}|$, namely ψ, or whether it represents $\psi + \pi/2$, is to set that half arctangent equal to φ_{ph} in $|\mathbf{E}|^2 = A_x^2\cos^2\varphi_{ph} + A_y^2\cos^2(\varphi_{ph} - \varphi)$, and then to do the same for an angle that differs by $\pi/2$. One of these evaluations gives a maximum, the other a minimum. The one giving a maximum then corresponds to ψ, the phase angle of an $\mathbf{E}$ field along the major axis of the ellipse.

However, ψ is *not* the angle with the x axis. It merely represents the value of $\varphi_{ph} \equiv \omega t - kz - \varphi_x$ for the maximum amplitude $|\mathbf{E}|$. At this location of the ellipse (the major axis) we have

$$\begin{aligned} E_x &= A_x\cos\psi \\ E_y &= A_y\cos\psi - \delta\varphi \end{aligned} \qquad (17.33)$$

and the angle with the x axis is then given by θ with

$$\tan\theta = \frac{A_y\cos(\psi - \delta\varphi)}{A_x\cos\psi} \qquad (17.34)$$

Hence we see that the **E** vector traces a helical curve in space; the helix is oriented in the $\hat{\mathbf{z}}$ direction, and its projection on a $z =$ constant plane is a tilted ellipse in general.

If, finally, the cosine of the difference angle in (17.26) is expanded, we obtain

$$\mathbf{E}(z,t) = (A_x\hat{\mathbf{x}} + A_y\hat{\mathbf{y}}\cos\delta\varphi)\cos(\omega t - kz - \varphi_x) + A_y\hat{\mathbf{y}}\sin\delta\varphi\sin(\omega t - kz - \varphi_x) \tag{17.35}$$

The first term is known as the *in-phase component*. The second term is the *quadrature component* and it lags behind the in-phase component by $\pi/2$ radians. The two components are generally in different directions. The quadrature component does not exist when either $A_y = 0$ or $\delta\varphi = 0$, in which case we have linear polarization.

17.4 Power flow; Poynting vector

17.4.1 Introduction

The concept of the flow of power (or *power flux*) is obtained somewhat indirectly from the Maxwell equations by invoking an analogy with the equation of continuity, discussed in subsection 10.3.2. That equation was obtained from the conservation of charge, $-dq/dt = \oint d\mathbf{S}\cdot\mathbf{J}$, and it led to

$$\frac{\partial\varrho_V}{\partial t} + \nabla\cdot\mathbf{J} = 0 \tag{17.36a}$$

We interpreted the first term as the quantity to be conserved (because it derived from dq/dt), and the second term as a 'flux' term due to the current density $\mathbf{J} = \varrho_V\mathbf{u}$ (with $\mathbf{u}$ as the velocity of the current carriers). The right-hand side is zero, but a 'source' term might have been included, in principle, in which case we would have

$$\frac{\partial\varrho_V}{\partial t} + \nabla\cdot\mathbf{J} = \left(\frac{\delta\varrho_V}{\delta t}\right)_{\text{source}} \tag{17.36b}$$

This would be the full equation of continuity in a universe in which charges are being produced at the location in question by some other means (e.g. by on-the-spot pair creation of charges which separate rapidly). To find the analogy, take Maxwell's curl **E** and curl **H** equations and dot multiply each with the other field:

$$\begin{aligned}\mathbf{H}\cdot(\nabla\times\mathbf{E}) &= -\mathbf{H}\cdot\frac{\partial\mathbf{B}}{\partial t}\\ \mathbf{E}\cdot(\nabla\times\mathbf{H}) &= \mathbf{E}\cdot\mathbf{J} + \mathbf{E}\cdot\frac{\partial\mathbf{D}}{\partial t}\end{aligned} \tag{17.37}$$

We then subtract these two equations and use vector identity 4 of Appendix A.4:

$$\mathbf{H}\cdot(\nabla\times\mathbf{E})-\mathbf{E}\cdot(\nabla\times\mathbf{H})=\nabla\cdot(\mathbf{E}\times\mathbf{H})$$

We then find that

$$\left(\mathbf{E}\cdot\frac{\partial\mathbf{D}}{\partial t}+\mathbf{H}\cdot\frac{\partial\mathbf{B}}{\partial t}\right)+\nabla\cdot(\mathbf{E}\times\mathbf{H})=-\mathbf{J}\cdot\mathbf{E} \tag{17.38a}$$

If we consider nonmagnetic media, with $\mathbf{B}=\mu_0\mathbf{H}$, and dielectric media with dispersive permittivities, but over restricted bandwidths so that essentially $\mathbf{D}\approx\varepsilon\mathbf{E}$ over a bandwidth as we discussed at the beginning of Section 16.2, then (17.38a) becomes

$$\frac{\partial}{\partial t}\left(\frac{1}{2}\mathbf{D}\cdot\mathbf{E}+\frac{1}{2}\mathbf{B}\cdot\mathbf{H}\right)+\nabla\cdot(\mathbf{E}\times\mathbf{H})=-\mathbf{J}\cdot\mathbf{E} \tag{17.38b}$$

This is the analogy of (17.36b). The meaning of the first term is well understood from what we discussed in Chapters 7 and 13; $\omega_e\equiv\frac{1}{2}\mathbf{D}\cdot\mathbf{E}$ is the energy density of the electric field and $\omega_m\equiv\frac{1}{2}\mathbf{B}\cdot\mathbf{H}$ is that of the magnetic field. Thus, 'energy density' replaces 'charge density' here.

In Section 10.3 it was shown that $\mathbf{J}\cdot\mathbf{E}$ is the ohmic power density (in $\mathrm{W/m^3}$) dissipated in a medium in which $\mathbf{J}$ is the current density and $\mathbf{E}$ is the electric field produced by a voltage difference. We therefore can interpret $-\mathbf{J}\cdot\mathbf{E}$ as the *source* $\delta\mathrm{W}/\delta\mathrm{t}$ of energy produced per unit time at the location for which (17.38b) is formulated. That makes this term analogous to the right-hand side of (17.36b).

It follows that, just as in (17.36b) $\mathbf{J}=\varrho_v\mathbf{u}$ is the 'charge flux' (which we refer to as the *current density vector*), we can identify $\mathbf{E}\times\mathbf{H}$ in (17.38b) with the power flux. The quantity $\mathbf{E}\times\mathbf{H}\equiv\mathbf{S}$ is known as the *Poynting vector* and it represents the power flux in $\mathrm{W/m^2}$.

The *total radiated power* leaving a volume v can be calculated from a surface integral:

$$P=\oint d\mathbf{S}\cdot\mathbf{S}=\oint d\mathbf{S}\cdot(\mathbf{E}\times\mathbf{H}) \tag{17.39}$$

in the understanding that the surface is one that totally encloses the volume v, with $d\mathbf{S}\equiv\hat{\mathbf{n}}\,dS$ such that the unit surface-normal vector $\hat{\mathbf{n}}$ points out of the surface. We interpret radiation as *entering* volume v if the result of calculating (17.39) is a negative value of the power P.

17.4.2 Power flow of time-harmonic waves

Consider a time-harmonic plane wave in which the phasor representations for **E** and **H** are given by (17.11) and (17.14b) respectively. We then would have

$$\mathbb{S} = \mathrm{Re}\{\mathbf{E}_\mathrm{p} e^{j\omega t}\} \times \mathrm{Re}\{\mathbf{H}_\mathrm{p} e^{j\omega t}\} \qquad \text{with} \qquad \mathbf{H}_\mathrm{p} = \frac{1}{\eta}\hat{\mathbf{z}} \times \mathbf{E}_\mathrm{p}$$
$$\eta = \sqrt{\frac{\mu_0}{\varepsilon_0 \varepsilon_\mathrm{r}}} = \frac{\omega\mu_o}{\beta - j\alpha} \equiv |\eta| e^{-j\theta_\eta} \qquad \mathbf{E}_\mathrm{p} = E_{x0}\hat{\mathbf{x}} e^{-\alpha z} e^{-j\beta z} \tag{17.40}$$

Working out the first right-hand side, we obtain two sets of terms:

$$\mathbb{S}(\mathbf{r}, t) = \tfrac{1}{4}\left(\mathbf{E}_\mathrm{p} \times \mathbf{H}_\mathrm{p}^* + \mathrm{cc}\right) + \tfrac{1}{4}\left(\mathbf{E}_\mathrm{p} \times \mathbf{H}_\mathrm{p} e^{2j\omega t} + \mathrm{cc}\right) \tag{17.41}$$

in which 'cc' is an abbreviation for 'complex conjugate' (which is also indicated by an asterisk as superscript). The Poynting vector $\mathbb{S}$ has a time-independent term and one that oscillates in time with frequency 2ω. This term vanishes if we average $\mathbb{S}$ over one or more full cycles; it represents energy that travels in one direction during half a cycle and then in the opposite direction during the other half cycle. It does not represent energy that flows steadily from one place to another. That is represented by the remaining term, which we indicate as $\mathbb{S}_\mathrm{av}$; it can be written as

$$\mathbb{S}_\mathrm{av}(\mathbf{r}) = \frac{1}{2}\mathrm{Re}\left(\mathbf{E}_\mathrm{p} \times \mathbf{H}_\mathrm{p}^*\right) \tag{17.42}$$

The decaying time-harmonic plane wave with the form specified in (17.40) then leads to

$$\mathbb{S}_\mathrm{av} = \frac{1}{2|\eta|}|E_{x0}|^2 e^{-2\alpha z}\cos\theta_\eta \hat{\mathbf{z}} \tag{17.43}$$

In free space we have no attenuation ($\alpha = 0$), no phase lag of **H** behind **E** ($\theta_\eta = 0$), and $\eta = 1/\sqrt{\mu_0\varepsilon_0}$, so that we then can write (17.43) as

$$\mathbb{S}_\mathrm{av} = \frac{1}{2\sqrt{\mu_0\varepsilon_0}}\varepsilon_0|E_{x0}|^2 = \left(\frac{1}{2}\varepsilon_0|E_{x0}|^2\right)c \tag{17.44}$$

and we see that the power flux (the Poynting vector) is indeed written as a product of the energy density times velocity, just as we interpreted the charge flux (current-density vector) as a product of charge density times velocity in (17.36a).

Let us consider (17.43) in more depth. If the medium represented by α and η is homogeneous, then that statement tells us that

$$\mathbb{S}_\mathrm{av}(z) = \mathbb{S}_\mathrm{av}(0)e^{-2\alpha z} \tag{17.45a}$$

Thus, the average power in the wave has decreased by a factor $\exp(-2\alpha z)$ when the wave has traversed the medium by a distance z. This decrease is often measured by looking at the ratio of the two power fluxes logarithmically. The quantity used specifically in engineering practice is

$$-10\log_{10}[\mathbb{S}_{av}(z)/\mathbb{S}_{av}(0)] = -10\log_{10}(e^{-2\alpha z}) \approx 8.686\alpha z \text{ in dB} \qquad (17.45b)$$

As can be seen, this quantity is expressed in *decibels* (dB). Here, we would say that *the power flux has decreased* (or *is down*) by $\approx 8.686\alpha z$ dB. The left-hand side of (17.45b) is a positive number when there is a decrease in power from denominator to numerator. The decibel is always a relative measure in that it expresses the ratio of two quantities (logarithmically). In engineering practice it is, aside from sign, either 10 times the logarithm-to-the-base-10 of the ratio of *powers* (as above) or 20 times the logarithm-to-the-base-10 of the ratio of *amplitudes*. The decibel is widely used, and therefore we recommend that some attention be paid to this convention.

The decay quantity in the exponential, the *attenuation coefficient* α, has the dimension m^{-1}, but that dimension is sometimes indicated as Np/m where 'Np' is an abbreviation of 'neper'. This extra designation, which adds nothing of use to the dimension, is perhaps somewhat outmoded, but it is still used and as such requires at least a passing mention. The decay in (17.45b) also can be referred to as a power drop of 8.686α dB/m over a distance z in meters.

Problems

17.1. A signal travels a distance of $l = 15$ cm in 1 ns without losses in a $\mu = \mu_0$ medium. What is the permittivity of the medium?

17.2. A dielectric medium is characterized by a real refractive index

$$n(\omega) = 1 + \frac{\omega}{\omega_0} e^{-\omega/\omega_0}$$

with $\omega_0 = 6\pi \times 10^9\ s^{-1}$.

(a) Calculate the phase velocity at the maximum value of $n(\omega)$.

(b) Calculate the group velocity at $\omega = 60\pi \times 10^9\ s^{-1}$. Does the medium have normal or anomalous dispersion here?

17.3. In a simple physical model of a *plasma*, i.e. a fully ionized but neutral gas, the dielectric permittivity is given by $\varepsilon_r(\omega) = 1 - \omega_p^2/\omega^2$, ω_p being an angular frequency determined solely by the number density of ions or electrons (assume it to be a constant and less than ω).

(a) Determine algebraic formulas for the phase and group velocities in this medium.

(b) Is the dispersion normal or anomalous? Explain.

17.4. The refractive index of water at 5 GHz is $n = 8.5 - 2.0j$. How far can a plane wave at this frequency travel through water before its (average) power is down by 60 dB?

17.5. A time-harmonic plane wave is down by 25 dB in power after propagating 1 km through a uniform dielectric.

(a) Calculate the attenuation coefficient α.

(b) If the power in the wave were 3.5 W initially, what would it be after 1 km?

17.6. If electron–neutral-atom collisions are included in the model for a plasma in problem 17.3, then ν_c represents the number of electron collisions with neutral atoms per second. The plasma is then not fully ionized, and the permittivity becomes complex:

$$\varepsilon_c(\omega) = \frac{1 - \omega_p^2}{\omega(\omega - j\nu_c)}$$

Assume $\omega_p \ll \omega$.

(a) Write down the condition for a low-loss approximation to $n(\omega)$ in this medium.

(b) If, however, the $\varepsilon' \ll \varepsilon''$ here (the high-loss approximation), write down an approximate expression for the skin depth.

17.7. In copper, the conductivity $\sigma \approx 5.8 \times 10^7$ S/m and the permeability $\mu \approx \mu_0$.

(a) Below what frequency f ($\omega = 2\pi f$) is the approximate expression for the skin depth, $\delta \approx \sqrt{2/\mu\omega\sigma}$, correct to within 1%?

(b) Express δ in mm as a function of f in GHz for copper, using the approximate expression.

17.8. For rain water at 0° C, the real part of the relative permittivity $\varepsilon' \approx 85$ and the conductivity $\sigma \approx 1.15$ S/m at $f \approx 1.5$ GHz (the values do vary with frequency, but we will not need to take this into account here).

(a) Specify the loss tangent at 1.5 GHz.

(b) How far has a wave penetrated at $f = 1.5$ GHz if at that point a decrease in power of 20 dB is registered?

(c) Given that

$$\mathbf{H}(z,t) = H_0 \hat{\mathbf{y}} e^{-\alpha z} \cos(\omega t - \beta z)$$

with $H_0 = 0.055$ A/m, find an expression for $\mathbf{E}(z,t)$ in the form

$$\mathbf{E}(z,t) = E_0 \hat{\mathbf{e}}\, e^{-\alpha z} \cos(\omega t - \beta z + \varphi)$$

with numerical values for α and E_0, and specify the unit vector $\hat{\mathbf{e}}$.

17.9. If the loss tangent $\varepsilon''/\varepsilon' = 2 \times 10^{-4}$, use the low-loss approximation to calculate the phase angle between $\mathbf{H}(z,t)$ and $\mathbf{E}(z,t)$ of a time-harmonic wave propagating in the z direction.

17.10. A time-harmonic plane wave is down by 5 dB of power in 1 km.
(a) Calculate the ratio $P(z)/P(0)$ for $z = 10$ km.
(b) Calculate the coefficient of attenuation α.
(c) If the loss tangent is 0.0001 and the phase velocity is $u_{\text{ph}} = 2.9 \times 10^8$ m/s, what is the frequency f of the wave? *Hint*: Use the result of part (b).

17.11. At room temperature and at a frequency of 90 GHz, rain water has permittivity $(9.47 - 16.7j)\varepsilon_0$. Calculate the propagation and attenuation constants at that frequency.

17.12. Show that the power attenuation in dB in a low-loss medium is essentially independent of frequency if the effective σ and ε' are not functions of frequency.

17.13. Show that any linearly polarized time-harmonic plane wave can be regarded as the sum of a right and a left circularly polarized wave.

17.14. A plane wave traveling in the z direction is characterized by $E_x = 5\cos\varphi$ and $E_y = 5\cos(\varphi - \pi/8)$.
(a) Use some special values of φ to sketch a locus of the tip of $E = \sqrt{E_x^2 + E_y^2}$ as a function of angle φ.
(b) Calculate the angles that the major and minor axes of the ellipse make with the x axis. *Hint*: Use (17.34).

17.15. Show that Eq. (17.34) leads explicitly to

$$\tan 2\theta = \frac{2A_x A_y \cos \delta\varphi}{A_x^2 - A_y^2}$$

Hint: Eliminate ψ from (17.34) using (17.32). This requires some work!

17.16. The phasor of a time-harmonic wave is specified as $\mathbf{E}_{\text{p}} = A_1\hat{\mathbf{z}} - jA_2\hat{\mathbf{x}}\, e^{j\beta y}$ where A_1 and A_2 are not necessarily real quantities, but β is real.
(a) What is the direction of propagation?
(b) What is the relationship between A_1 and A_2 for the wave to be linearly polarized?
(c) If we now choose A_1 and A_2 to be real, write down the real $E(\mathbf{r}, t)$ field and specify its polarization.

17.17. Consider a plane-wave with $\lambda = 2\pi$ m and with $\varepsilon' = 1.5$, $\varepsilon'' = 0.02$.
(a) Show that a low-loss approximation is reasonably accurate for the attenuation coefficient α. Write down the average Poynting vector, and specify $|\eta|$ and the impedance phase angle θ_η.
(b) Given that the medium is uniform, how far can the wave propagate before its power flux is down by 30 dB?

17.18. An infinitesimal dipole of strength $\mathbf{p} = p\hat{\mathbf{z}}$, oscillating at frequency ω, gives rise to the fields (in spherical coordinates),

$$\mathbf{E}(\mathbf{r}, t) = \hat{\boldsymbol{\theta}}\frac{\omega k\eta p}{4\pi r}\sin\theta\cos(\omega t - kr) \qquad \mathbf{H}(\mathbf{r}, t) = \hat{\boldsymbol{\varphi}}\frac{\omega k p}{4\pi r}\sin\theta\cos(\omega t - kr)$$

given that η is the intrinsic impedance of free space and $k = \omega\sqrt{\mu_0\varepsilon_0} = \omega/c$.

(a) Evaluate what the average Poynting vector is in terms of the given quantities.

(b) Then calculate the power emitted by the dipole in all directions. *Hint*: Integrate the power flux (Poynting vector) over the surface of any sphere of radius r.

(c) If such an infinitesimal dipole radiates 5 mW at 2.5 GHz, how much power does it radiate at 10 GHz?

(d) The above expressions for the fields are approximately valid for a dipole of nonzero size with separation distance d between the charges, provided $r \gg d^2/\lambda$ (this is the 'far zone' of radiation). By how much in dB does the radiated power change if d is halved?

17.19. A plane harmonic wave propagates in the direction $\hat{\mathbf{z}}$ in a uniform dielectric with attenuation coefficient α (in Np/m).

(a) Calculate how much power is absorbed in a slab of width d that is $\perp$ ($\hat{\mathbf{z}}$).

(b) Express briefly what has happened to the absorbed power.

18

Reflection and transmission of plane waves

This chapter deals with the situation that occurs when a time-harmonic plane wave is incident from one medium – we shall take it to be free space – onto another (either a conductor or a dielectric). The two media are separated from each other by an infinite planar interface, and each part is referred to as a *half-space*. When the incident wave travels perpendicularly to that interface we speak of *normal incidence*. When the incident wave travels in any other direction we speak of *oblique incidence*. At the interface part or all of the wave is *reflected* from the interface back into free space, and the remainder is *transmitted* through the interface into the other medium. We shall specify how much and at what angles these reflections and transmissions occur.

Of course, we will not really deal with actual reflections and transmissions. We only can speak literally of transmission and reflection when we have a short pulse of time duration τ, a signal that is nonzero at any location in free space only for some time around t_1 with $t_1 < t < t_1 + \tau$. Such a pulse will interact with the interface during a time interval of length $\sim \tau$. Before that time interval, it will not yet have reached the interface. After that time interval, there will be a transmitted pulse at some distance into the other medium (if it is a dielectric) and a reflected pulse that already has traveled some distance back in the original half-space.

Time-harmonic waves, however, are infinite in extent and have existed 'forever', and will continue to do so. The incident, reflected, and transmitted time-harmonic plane waves all exist at the same time as steady-state *modes* in the two-medium space. In fact, each point in free space may experience fields from the incident and reflected waves at the same time (the reflected field having been caused by the incident field at an earlier time). The transmitted wave exists at all points in the other half-space.

Hence, what we will do in this chapter is to survey the steady-state situation without concerning ourselves very much with how that situation was created. We can learn quite a bit about it by looking at the power fluxes (the time-averaged Poynting vectors). The fields and the Poynting vectors on each side of the interface – these are the main objects of our interest here.

18.1 Normal incidence upon a perfect conductor

The situation will be as shown in Fig. 18.1. The incident wave travels in the $+\hat{\mathbf{z}}$ direction and this direction is given by the wavevector $\mathbf{k}_\mathrm{i} = \beta\hat{\mathbf{z}}$. The reflected wave travels in the opposite direction and its wavenumber is $\mathbf{k}_\mathrm{r} = -\beta\hat{\mathbf{z}}$. For both these waves, $|\mathbf{k}| = \beta = \omega\sqrt{\mu_0\varepsilon_0} \equiv \omega/c$ is real, because we do not have attenuation in free space ($\alpha = 0$). The linear polarization of the incident $\mathbf{E}_\mathrm{i}$ field can be taken to be in the $-\hat{\mathbf{y}}$ direction without loss of generality because any direction $\perp$ $\hat{\mathbf{z}}$ will suffice. The $\mathbf{H}_\mathrm{i}$ field is then linearly polarized in the $+\hat{\mathbf{x}}$ direction because the Poynting vector $\mathbf{E} \times \mathbf{H}$ must be in the $+\hat{\mathbf{z}}$ direction. There is no $\mathbf{E}$ field in the conductor in the static-field case, as we explained in Section 6.2; this does not change for time-varying fields as long as the inverse of the time constant $T = \varepsilon/\sigma$, where σ is the conductivity, is very much greater than the frequency ω (because any interior net charge decays in a time that is a small fraction of one cycle). With $T \sim 10^{-19}$ s, it is almost always the case that $\omega\, T \ll 1$. The phasor representations of the incident fields must be

$$
\begin{aligned}
\mathbf{E}_\mathrm{i}(z) &= \mathbf{E}_{\mathrm{i}o}e^{-j\beta z} \qquad \text{with} \qquad \mathbf{E}_{\mathrm{i}o} \equiv -E_{\mathrm{i}o}\hat{\mathbf{y}} \\
\mathbf{H}_\mathrm{i}(z) &= \mathbf{H}_{\mathrm{i}o}e^{-j\beta z} \qquad \text{with} \qquad \mathbf{H}_{\mathrm{i}o} \equiv \frac{1}{\eta_0}E_{\mathrm{i}o}\hat{\mathbf{x}}
\end{aligned}
\tag{18.1}
$$

Here η_0 is the intrinsic impedance of free space. Note that we will no longer use a subscript 'p' to indicate a phasor; it will be obvious from the context whether we are dealing with phasors or actual fields, but it will be rare that we refer to fields in forms other than the phasor form. The expression for $\mathbf{H}_{\mathrm{i}o}$ follows from inserting the first expression in the second line of (18.1) into $\nabla \times \mathbf{H}_\mathrm{i} = -j\omega\varepsilon_0\mathbf{E}_\mathrm{i}$, and from using the mathematical device explained after (16.45). The reflected field is

$$
\begin{aligned}
\mathbf{E}_\mathrm{r}(z) &= \mathbf{E}_{\mathrm{r}o}e^{j\beta z} \\
\mathbf{H}_\mathrm{r}(z) &= \mathbf{H}_{\mathrm{r}o}e^{j\beta z} \qquad \text{with} \qquad \mathbf{H}_{\mathrm{r}o} \equiv -\frac{1}{\eta_0}\hat{\mathbf{z}} \times \mathbf{E}_{\mathrm{r}o}
\end{aligned}
\tag{18.2}
$$

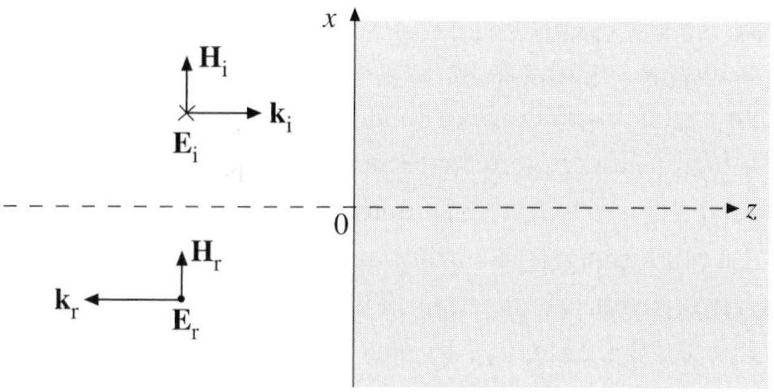

Fig. 18.1 Normal incidence of EM waves upon a vacuum–conductor interface.

Note the absence of a minus sign in the two exponents here; the reflected wave travels in the $-\hat{\mathbf{z}}$ direction. Observe also that we have used (16.45) again in the curl **E** equation to find the last relationship. We can find the values of the two amplitude vectors in (18.2) by using the boundary conditions at the interface, which we can place at $z = 0$ without any loss of generality. The continuity of tangential fields, (15.31), tells us that

$$\hat{\mathbf{z}} \times [\mathbf{E}_{\mathrm{i}}(0) + \mathbf{E}_{\mathrm{r}}(0)] = 0 \qquad \text{or} \qquad \mathbf{E}_{ro} = -\mathbf{E}_{io} \tag{18.3}$$

In other words, the reflected electric field vector *at the interface* is equal and opposite in sign to the incident electric field vector. As we know that $\mathbf{E}_{io}$ is in the $-\hat{\mathbf{y}}$ direction, it follows from (18.1)–(18.3) that

$$\mathbf{H}_{ro} = \mathbf{H}_{io} = \frac{1}{\eta_0} E_{io}\hat{\mathbf{x}} \tag{18.4}$$

The two **H** field vectors must be identical at the interface. In summary, we now have found that

$$\begin{aligned} \mathbf{E}_{\mathrm{i}}(z) &= -E_{io}\hat{\mathbf{y}}e^{-j\beta z} \qquad & \mathbf{H}_{\mathrm{i}}(z) &= \frac{1}{\eta_0} E_{io}\hat{\mathbf{x}}e^{-j\beta z} \\ \mathbf{E}_{\mathrm{r}}(z) &= +E_{io}\hat{\mathbf{y}}e^{j\beta z} \qquad & \mathbf{H}_{\mathrm{r}}(z) &= \frac{1}{\eta_0} E_{io}\hat{\mathbf{x}}e^{j\beta z} \end{aligned} \tag{18.5}$$

What about the other tangential boundary condition, $\hat{\mathbf{n}} \times (\mathbf{H}_{\mathrm{i}} + \mathbf{H}_{\mathrm{r}}) = \mathbf{K}$, applied at $z = 0$? The + sign may seem to be wrong, but, remember, the two fields on either side add up to the *total* field; this is zero in the conductor. So this boundary condition leads to

$$\mathbf{K} = \frac{2}{\eta_0} E_{io}\hat{\mathbf{x}} \tag{18.6}$$

according to (18.5). Although there are no fields within the conductor, together with the total reflection of the incident wave we have the creation of a *surface current* at the interface, with density **K**. One way of interpreting what happens is to say that the incident wave sets up a surface current, which in turn re-radiates a reflected wave. This statement is, in fact, a key one in understanding *antennas*. Antennas pick up radio waves in the air and convert them to surface currents. These surface currents will have the same information content that the radio waves carry, and some or all the energy in the currents is converted to audio signals (radio) or video signals (television). If this did not happen, the surface currents would re-radiate the energy as waves traveling away from the antenna. The shape of those waves would be dictated by the shape of the antenna. In the simple case described above, the re-radiated wave is a time-harmonic plane wave of the same strength traveling in the $-\hat{\mathbf{z}}$ direction.

Consider now the *total* field phasors in $z < 0$

$$\mathbf{E}(z) \equiv \mathbf{E}_{\rm i}(z) + \mathbf{E}_{\rm r}(z) = 2jE_{\rm io}\hat{\mathbf{y}}\sin\beta z$$
$$\mathbf{H}(z) \equiv \mathbf{H}_{\rm i}(z) + \mathbf{H}_{\rm r}(z) = \frac{2}{\eta_0}E_{\rm io}\hat{\mathbf{x}}\cos\beta z \tag{18.7a}$$

Therefore the total time-dependent fields, which we find by multiplying the factor $e^{j\omega t}$ into the phasors in (18.7a) and then taking the real parts, are

$$\mathbf{E}(z,t) = -2E_{\rm io}\hat{\mathbf{y}}\sin\beta z\sin\beta t$$
$$\mathbf{H}(z,t) = \frac{2}{\eta_0}E_{\rm io}\hat{\mathbf{x}}\cos\beta z\cos\beta t \tag{18.7b}$$

These can no longer be considered as propagating waves, because – for example – there are values of z, *nodes*, where the fields are zero at all times. We speak of *standing waves*; the reflected wave interferes with the incident wave to form a standing-wave pattern in $z < 0$. There are two complementary ways of looking at the situation: either as one standing wave or as an interference pattern produced by two waves propagating in opposite directions.

Yet another aspect of this elementary wave-interface situation is illustrated by the power-flux vectors (the Poynting vectors). Recalling that $\mathbb{S} = \frac{1}{2}{\rm Re}\{\mathbf{E}\times\mathbf{H}^*\}$ is the average power flux in a medium, we apply this formula to the incident wave and, separately, to the reflected wave. Unless a detector is angled so as to receive either the reflected or the incident wave, the two are not separable physically. But if they are, then we would find

$$\mathbb{S}_{\rm i} = \frac{1}{2\eta_0}|E_{\rm io}|^2\hat{\mathbf{z}} \qquad \mathbb{S}_{\rm r} = -\frac{1}{2\eta_0}|E_{\rm io}|^2\hat{\mathbf{z}} \tag{18.8}$$

The two power fluxes are equal and opposite in direction, hence cancel each other. If, however, we look at the power flux of the total fields, with (18.7a), we find

$$\mathbb{S} = {\rm Re}\left\{\frac{j}{2\eta_0}|E_{\rm io}|^2\sin\beta z\cos\beta z\right\} = 0 \tag{18.9}$$

The real part of a purely imaginary quantity is zero, so, consistent with the standing-wave pattern, we have found that there is no flow of power in $z < 0$.

18.2 Oblique incidence upon a perfect conductor

Matters become more complicated when we review the situation described above (free space separated from a perfect conductor by a planar interface at $z = 0$) for an obliquely incident wave. For a start, there now are two distinctly different situations, as shown in Fig. 18.2. In (*a*) the **E** fields lie in a plane $\perp$ to the *plane of incidence*.

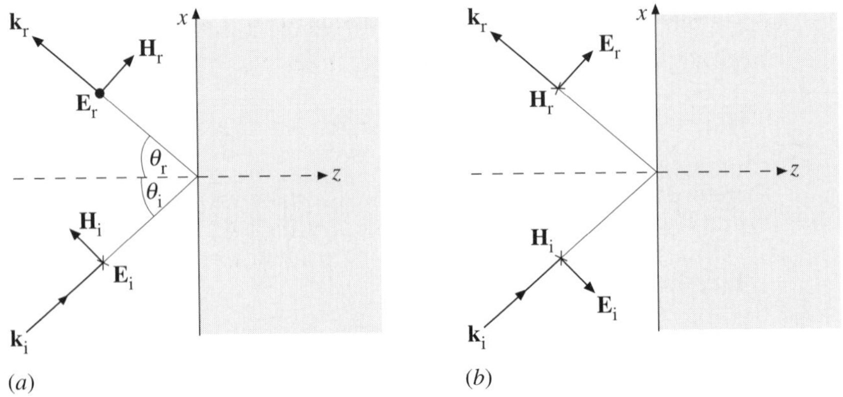

Fig. 18.2 Oblique incidence of EM waves upon a vacuum–conductor interface: (*a*) for the $\perp$ (TE) mode, (*b*) for the $\parallel$ (TM) mode.

This is indicated by a solid circle for the $\mathbf{E}_r$ vector, which points out of the page, and a cross for the $\mathbf{E}_i$ vector, which points into the page (the 'tail feathers'). The plane of incidence is defined by the vector $\mathbf{k}_i$ and its projection upon the interface. We speak of the $\perp$ or TE (transverse electric) mode. In (*b*) the **H** fields are $\perp$ to the plane of incidence, and one speaks here of the $\parallel$ or TM (transverse magnetic) mode. The $\parallel$ pertains to 'parallel', because the electric fields lie in, i.e. parallel to, the plane of incidence. Cases (*a*) and (*b*) need to be treated separately.

18.2.1 The $\perp$ (TE) modes obliquely incident upon a perfect conductor

Diagram (*a*) in Fig. 18.2 has been drawn so that the fields coincide with those of Fig. 18.1 when the *angle of incidence* $\theta_i = 0$. The direction of the incident wave is given by the wave vector $\mathbf{k}_i = \beta(\hat{\mathbf{x}}\sin\theta_i + \hat{\mathbf{z}}\cos\theta_i)$ and that of the reflected wave by $\mathbf{k}_r = \beta(\hat{\mathbf{x}}\sin\theta_r + \hat{\mathbf{z}}\cos\theta_r)$. The incident and reflected waves therefore must be

$$\begin{aligned} \mathbf{E}_i(\mathbf{r}) &= \mathbf{E}_{io}e^{-j\beta\hat{\mathbf{k}}_i\cdot\mathbf{r}} \qquad & \hat{\mathbf{k}}_i &\equiv \hat{\mathbf{x}}\sin\theta_i + \hat{\mathbf{z}}\cos\theta_i \\ \mathbf{E}_r(\mathbf{r}) &= \mathbf{E}_{ro}e^{-j\beta\hat{\mathbf{k}}_r\cdot\mathbf{r}} \qquad & \hat{\mathbf{k}}_r &\equiv \hat{\mathbf{x}}\sin\theta_r - \hat{\mathbf{z}}\cos\theta_r \end{aligned} \tag{18.10}$$

We have chosen $\mathbf{E}_{io} = -E_{io}\hat{\mathbf{y}}$ and $\mathbf{E}_{ro} = E_{ro}\hat{\mathbf{y}}$; the **H** fields that go with these must then be consistent with the rule that $\mathbf{E}\times\mathbf{H}$ points in the direction of the wave vector **k**.

At the interface $z = 0$ the boundary condition for the tangential **E** fields is

$$\mathbf{E}_{io}e^{-j\beta x\sin\theta_i} + \mathbf{E}_{ro}e^{-j\beta x\sin\theta_r} = 0 \tag{18.11}$$

This condition can be satisfied at *all* points x of the interface only if the exponents are the same for all x and the E coefficients add up to zero:

$$\theta_\mathrm{r} = \theta_\mathrm{i} \qquad \mathbf{E}_{\mathrm{r}o} = -\mathbf{E}_{\mathrm{i}o} \tag{18.12a}$$

Thus

The angle of reflection must be equal to the angle of incidence, and the reflected **E** *field vector is equal but opposite to the incident* **E** *field vector.* (18.12b)

Reminder: the second statement is true only at the interface. From here on we shall simply label $\theta_\mathrm{i} = \theta_\mathrm{r}$ by θ, the angle of incidence of the wave upon the interface. To find the directions of the $\mathbf{H}_\mathrm{i}$ fields, we need to invoke the curl **E** equation:

$$\begin{aligned} &\nabla \times (-E_{\mathrm{i}o}\hat{\mathbf{y}}e^{-j\beta\hat{\mathbf{k}}_\mathrm{i}\cdot\mathbf{r}}) = -j\omega\mu_0\mathbf{H}_\mathrm{i} \qquad \text{or} \\ &\mathbf{H}_\mathrm{i} = \frac{j}{\mu_0\omega}(-j\beta\hat{\mathbf{k}}_\mathrm{i}) \times (-E_{\mathrm{i}o}\hat{\mathbf{y}}e^{-j\beta\hat{\mathbf{k}}_\mathrm{i}\cdot\mathbf{r}}) \end{aligned} \tag{18.13}$$

To complete the calculation, we note that $\beta/\mu_0\omega = 1/\eta_0$, and

$$\hat{\mathbf{k}}_\mathrm{i} \times (-\hat{\mathbf{y}}) = \hat{\mathbf{x}}\cos\theta - \hat{\mathbf{z}}\sin\theta \equiv \hat{\mathbf{m}}_\mathrm{i}$$

We find

$$\begin{aligned} \mathbf{E}_\mathrm{i}(\mathbf{r}) &= -E_{\mathrm{i}o}\hat{\mathbf{y}}e^{-j\beta\hat{\mathbf{k}}_\mathrm{i}\cdot\mathbf{r}} & \hat{\mathbf{k}}_\mathrm{i} &\equiv \hat{\mathbf{x}}\sin\theta + \hat{\mathbf{z}}\cos\theta \\ \mathbf{H}_\mathrm{i}(\mathbf{r}) &= \frac{1}{\eta_0}E_{\mathrm{i}o}\hat{\mathbf{m}}_\mathrm{i}e^{-j\beta\hat{\mathbf{k}}_\mathrm{i}\cdot\mathbf{r}} & \hat{\mathbf{m}}_\mathrm{i} &\equiv \hat{\mathbf{x}}\cos\theta - \hat{\mathbf{z}}\sin\theta \\ \mathbf{E}_\mathrm{r}(\mathbf{r}) &= E_{\mathrm{i}o}\hat{\mathbf{y}}e^{-j\beta\hat{\mathbf{k}}_\mathrm{r}\cdot\mathbf{r}} & \hat{\mathbf{k}}_\mathrm{r} &\equiv \hat{\mathbf{x}}\sin\theta - \hat{\mathbf{z}}\cos\theta \\ \mathbf{H}_\mathrm{r}(\mathbf{r}) &= \frac{1}{\eta_0}E_{\mathrm{i}o}\hat{\mathbf{m}}_\mathrm{r}e^{-j\beta\hat{\mathbf{k}}_\mathrm{r}\cdot\mathbf{r}} & \hat{\mathbf{m}}_\mathrm{i} &\equiv \hat{\mathbf{x}}\cos\theta + \hat{\mathbf{z}}\sin\theta \end{aligned} \tag{18.14}$$

The calculation for $\mathbf{H}_\mathrm{r}$ is similar to that in (18.13). You can easily see that the directions found for $\hat{\mathbf{m}}_\mathrm{i}$ and $\hat{\mathbf{m}}_\mathrm{r}$ are consistent with Fig. 18.2(a); these unit vectors $\hat{\mathbf{m}}$ lie in the $y = 0$ plane and are perpendicular to the unit vectors $\hat{\mathbf{k}}$. The total fields in $z < 0$ are found by adding the incident and reflected fields to obtain, in phasor notation,

$$\begin{aligned} \mathbf{E}(\mathbf{r}) &= 2jE_{\mathrm{i}o}\hat{\mathbf{y}}e^{-j\beta x\sin\theta}\sin(\beta z\cos\theta) \\ \mathbf{H}(\mathbf{r}) &= \frac{2}{\eta_0}E_{\mathrm{i}o}e^{-j\beta x\sin\theta}[\hat{\mathbf{x}}\cos\theta\cos(\beta z\cos\theta) + j\hat{\mathbf{z}}\sin\theta\sin(\beta z\cos\theta)] \end{aligned} \tag{18.15}$$

Many interesting features show up in these equations. Let us as before move from phasor to actual wave notation by affixing $\exp(j\omega t)$ to (18.15) and then taking the real part (assuming for simplicity that E_{io} is real):

$$\begin{aligned}\mathbf{E}(\mathbf{r}, t) &= -2E_{io}\hat{\mathbf{y}}\sin(\beta z\cos\theta)\sin(\omega t - \beta x\sin\theta)\\ \mathbf{H}(\mathbf{r}, t) &= \frac{2}{\eta_0}E_{io}[\hat{\mathbf{x}}\cos\theta\cos(\beta z\cos\theta)\cos(\omega t - \beta x\sin\theta)\\ &\qquad - \hat{\mathbf{z}}\sin\theta\sin(\beta z\cos\theta)\sin(\omega t - \beta x\sin\theta)]\end{aligned} \tag{18.16}$$

It suffices to look at the field $\mathbf{E}(\mathbf{r}, t)$ in order to understand what this represents. We see that the *motion* of the wave is sinusoidal in the $\hat{\mathbf{x}}$ direction, with phase velocity ($v_{ph} = \omega/\beta\sin\theta$). The phase velocity exceeds the velocity $\omega/\beta = c$, but this velocity is another one of those nonobservable velocities discussed in subsection 17.1.1; it does not correspond to the motion of energy. The entire wave structure in $z < 0$ appears to move upward. The *spatial* structure of the wave in the $\hat{\mathbf{x}}$ direction (for $z < 0$) is a sinusoid with wavelength $\lambda_x = 2\pi/(\beta\sin\theta) = \lambda/\sin\theta$. If, however, we look at the spatial structure in the $\hat{\mathbf{z}}$ direction, we see that it is also sinusoidal but with wavelength $\lambda_z = 2\pi/(\beta\cos\theta) = \lambda/\cos\theta$.

In Fig. 18.3 we illustrate all this with an *interference pattern*. A sinusoidal **E** field with wavelength λ is incident at angle θ and is reflected at the same angle with the z axis. The planes of zero phase for the incident wave are labeled as $i_1, \ldots, i_8$, and the planes of zero phase for the reflected wave are labeled as $r_1, \ldots, r_8$. Because we have chosen $\theta < 45°$ it turns out that $\lambda_z < \lambda_x$, but this is not otherwise significant. In the middle of the diagram you will find two right-angled triangles. The smaller triangle has λ_z as its hypotenuse and the larger has λ_x as its hypotenuse. The spatial structure of the sum wave in the z direction is shown at $z = 0$ as a sine wave for one wavelength. Note that it repeats at distance λ_x in the $\pm\hat{\mathbf{x}}$ direction. The same sum wave is also shown at distance $\frac{1}{2}\lambda_x$ in the $\pm\hat{\mathbf{x}}$ direction; note the change in phase by π radians ($= 180°$). This entire sine curve thus moves in the $+\hat{\mathbf{x}}$ direction: $E(x, z)$, considered as a function of x for *fixed* z, is proportional to $\sin(\omega t - 2\pi x/\lambda_x)$.

The picture is further enhanced by considering the average power flux $\mathbb{S}$ from (18.15):

$$\mathbb{S} = \left\{\frac{2}{\eta_0}|E_{io}|^2\sin^2(\beta z\cos\theta)\sin\theta\right\}\hat{\mathbf{x}} \tag{18.17}$$

This confirms that any flow of power is only in the $\hat{\mathbf{x}}$ direction; we have a standing wave in the $\hat{\mathbf{z}}$ direction through the interference of incident with reflected waves. This flux becomes zero as $\theta \to 0$.

Now we refer in Fig. 18.3 to the broken vertical line between the labels i_5 and r_5. $\mathbf{E}_0 = 0$ everywhere on that line, which corresponds in (18.16) to $\beta z\cos\theta = -2\pi$, or $z = -\lambda_z$. The values of **H** and **E** are exactly the same at $z = -\lambda_z$ as they are at $z = 0$ (the interface). If we replace λ_z by $n\lambda_z$, with $n = 1, 2, 3, \ldots$, the same result would hold. Nothing would change in $-n\lambda_z < z < 0$ if we placed a metal plate (or perfect

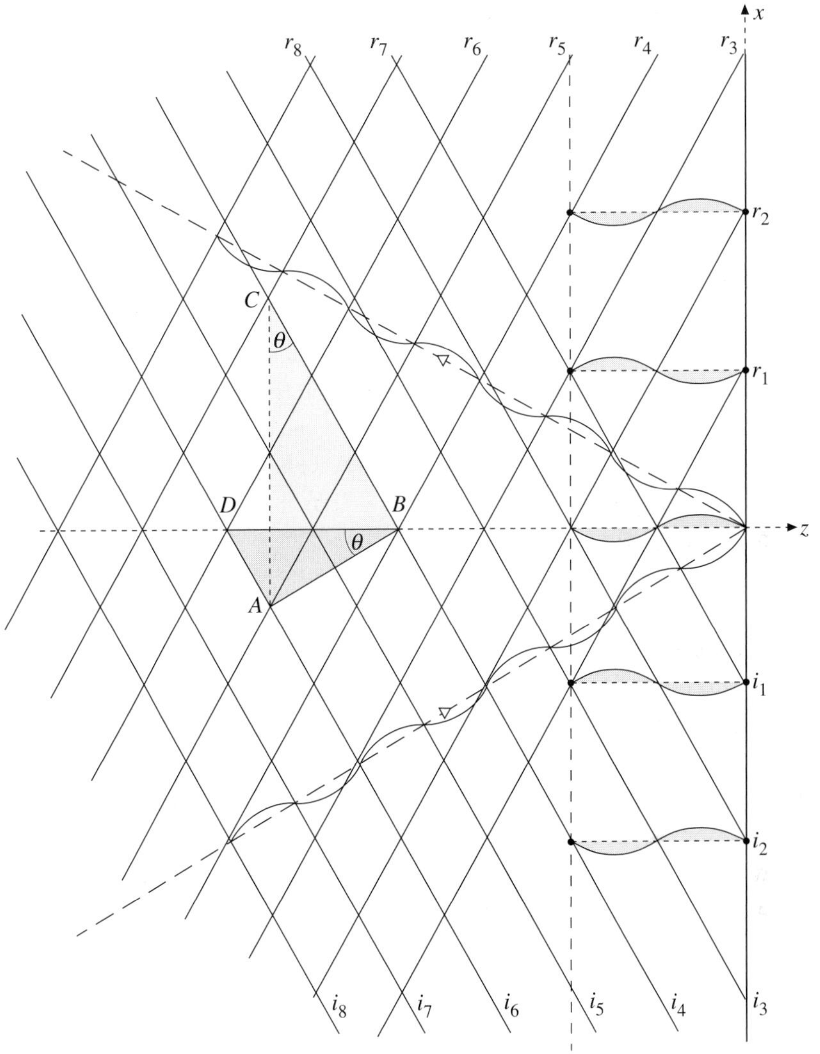

Fig. 18.3 A composite sketch of incident + reflected waves for oblique incidence upon a conducting medium, showing the various wavelengths involved in the sum wave: $AB = \lambda$, $AC = \lambda_x$, and $BD = \lambda_z$. The incident wavefronts are $i_1, i_2, \ldots$ and the reflected wavefronts are $r_1, r_2, \ldots$.

conductor, to be precise) so that its interface was at $z = -n\lambda_z$. We then would have a *planar waveguide* between the two conductors at $z = 0$ and $z = -n\lambda_z$. As long as $\beta z \cos\theta = -n\pi$, with $n = 1, 2, 3, \ldots$ and with *fixed* z but *variable* θ, we have a wave *mode* traveling between the two interfaces. A little thought will convince the reader that metal plates also can be placed at $z = -n\lambda_z/2$. Thus waveguide modes have only certain wavelengths that can exist, because $\beta x \sin\theta$ is then also discretized to a finite set of values. We shall return to this topic in the chapter on waveguides, but you already see the beginnings here in this simple picture. The modes also can be considered as plane waves bouncing back and forth between the two walls at discrete angles θ to the surface normal. These angles are determined solely by the width between the walls and the frequency, and are such that the reflected waves do not interfere destructively with the incident waves.

18.2.2 The $\parallel$ or TM modes obliquely incident upon a perfect conductor

The other possibility is shown in Fig. 18.2(*b*). The diagram shows that $\mathbf{E} \times \mathbf{H}$ is in the direction of the $\mathbf{k}$ vector for the incident and the reflected wave, and it reduces to Fig. 18.1 as $\theta \to 0$ (but with the $\mathbf{E}$ fields in the $\pm\hat{\mathbf{x}}$ direction). Comparison with (18.14) and with Fig. 18.2(*a*) tells us that the fields in Fig. 18.2(*b*) must be as follows:

$$\begin{aligned}
\mathbf{E}_\mathrm{i}(\mathbf{r}) &= -E_\mathrm{io}\hat{\mathbf{m}}_\mathrm{i}e^{-j\beta\hat{\mathbf{k}}_\mathrm{i}\cdot\mathbf{r}} & \hat{\mathbf{k}}_\mathrm{i} &\equiv \hat{\mathbf{x}}\sin\theta + \hat{\mathbf{z}}\cos\theta \\
\mathbf{H}_\mathrm{i}(\mathbf{r}) &= -\frac{1}{\eta_0}E_\mathrm{io}\hat{\mathbf{y}}e^{-j\beta\hat{\mathbf{k}}_\mathrm{i}\cdot\mathbf{r}} & \hat{\mathbf{m}}_\mathrm{i} &\equiv \hat{\mathbf{x}}\cos\theta - \hat{\mathbf{z}}\sin\theta \\
\mathbf{E}_\mathrm{r}(\mathbf{r}) &= E_\mathrm{io}\hat{\mathbf{m}}_\mathrm{r}e^{-j\beta\hat{\mathbf{k}}_\mathrm{r}\cdot\mathbf{r}} & \hat{\mathbf{k}}_\mathrm{r} &\equiv \hat{\mathbf{x}}\sin\theta + \hat{\mathbf{z}}\cos\theta \\
\mathbf{H}_\mathrm{r}(\mathbf{r}) &= -\frac{1}{\eta_0}E_\mathrm{io}\hat{\mathbf{y}}e^{-j\beta\hat{\mathbf{k}}_\mathrm{r}\cdot\mathbf{r}} & \hat{\mathbf{m}}_\mathrm{r} &\equiv \hat{\mathbf{x}}\cos\theta + \hat{\mathbf{z}}\sin\theta
\end{aligned} \tag{18.18}$$

This can be checked in many ways, of course. For example, $\hat{\mathbf{k}}_\mathrm{i} \times \hat{\mathbf{m}}_\mathrm{i}$ is easily seen to equal $\hat{\mathbf{y}}$ from the definitions of these unit vectors. We also see that the $\hat{\mathbf{x}}$ components of the $\mathbf{E}$ fields at the interface ($z = 0$) are equal but opposite because the angle of reflection is the same as the angle of incidence. The sum phasors in $z < 0$ are then seen to be

$$\begin{aligned}
\mathbf{E}(\mathbf{r}) &= 2E_\mathrm{io}e^{-j\beta x\sin\theta}[j\hat{\mathbf{x}}\cos\theta\sin(\beta z\cos\theta) + \hat{\mathbf{z}}\sin\theta\cos(\beta z\cos\theta)] \\
\mathbf{H}(\mathbf{r}) &= -\frac{2}{\eta_0}E_\mathrm{io}\hat{\mathbf{y}}e^{-j\beta x\sin\theta}\cos(\beta z\cos\theta)
\end{aligned} \tag{18.19}$$

and the space–time sum fields are

$$\begin{aligned}
\mathbf{E}(\mathbf{r},t) &= 2E_\mathrm{io}[-\hat{\mathbf{x}}\cos\theta\sin(\beta z\cos\theta)\sin(\omega t - \beta x\sin\theta) \\
&\qquad + \hat{\mathbf{z}}\sin\theta\cos(\beta z\cos\theta)\cos(\omega t - \beta x\sin\theta)] \\
\mathbf{H}(\mathbf{r},t) &= -\frac{2}{\eta_0}E_\mathrm{io}\hat{\mathbf{y}}\cos(\beta z\cos\theta)\cos(\omega t - \beta x\sin\theta)
\end{aligned} \tag{18.20}$$

The situation is essentially similar as for the TE modes: the sum wave moves sinusoidally in the $\hat{\mathbf{x}}$ direction, and the spatial structure of $\mathbf{H}$ in the $\hat{\mathbf{z}}$ direction is itself a sinusoid.

18.3 Normal incidence upon a dielectric half-space

Referring to Fig. 18.4, we now consider a time-harmonic plane wave incident perpendicularly on a planar interface at $z = 0$ with a uniform dielectric $(\varepsilon_1 \neq \varepsilon_0)$ in $z > 0$. It will involve little loss of generality to assume that $\varepsilon_1 > \varepsilon_0$ in the dielectric and we shall also at first assume that there are no losses, so that ε_1 is real. The major difference from the situation described in subsection 18.2.1 is that a wave ('mode') can exist in $z > 0$: we label this transmitted mode with index 't'. As the diagram shows, we choose as phasor representation

$$\begin{aligned} \mathbf{E}_\mathrm{i} &= -E_{\mathrm{i}o}\hat{\mathbf{y}}e^{-j\beta_0 z} \\ \mathbf{E}_\mathrm{r} &= E_{\mathrm{r}o}\hat{\mathbf{y}}e^{j\beta_0 z} \\ \mathbf{E}_\mathrm{t} &= -E_{\mathrm{t}o}\hat{\mathbf{y}}e^{-j\beta_1 z} \end{aligned} \tag{18.21}$$

As before, we are aided by the right-hand rule in going from $\mathbf{E}$ to $\mathbf{H}$ to $\mathbf{k} = \beta\hat{\mathbf{z}}$ in writing down these expressions for fields in phasor representation. The magnetic fields to go with these $\mathbf{E}$ fields are obtained as in (18.1)–(18.3), and they yield the following phasors:

$$\begin{aligned} \mathbf{H}_\mathrm{i} &= \frac{1}{\eta_0}E_{\mathrm{i}o}\hat{\mathbf{x}}e^{-j\beta_0 z} \\ \mathbf{H}_\mathrm{r} &= \frac{1}{\eta_0}E_{\mathrm{r}o}\hat{\mathbf{x}}e^{j\beta_0 z} \\ \mathbf{H}_\mathrm{t} &= \frac{1}{\eta_0}E_{\mathrm{t}o}\hat{\mathbf{x}}e^{-j\beta_1 z} \end{aligned} \tag{18.22}$$

Apply the two boundary conditions $\hat{\mathbf{n}} \times (\mathbf{E}_0 + \mathbf{E}_1) = 0$ and $\hat{\mathbf{n}} \times (\mathbf{H}_0 + \mathbf{H}_1) = \mathbf{K}$, where $\mathbf{E}_0, \mu_0, \mathbf{E}_1, \mu_1$ are the boundary fields in the first and second media and $\mathbf{K}$ is the

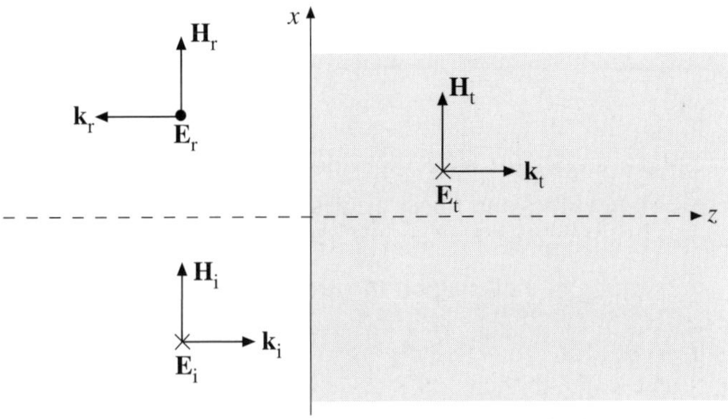

Fig. 18.4 Normal incidence of EM waves upon a dielectric–dielectric interface.

surface current; $\hat{\mathbf{n}} = -\hat{\mathbf{z}}$. We note that $\mathbf{E}_0 = \mathbf{E}_\mathrm{i} + \mathbf{E}_\mathrm{r}$ and $\mathbf{E}_1 = \mathbf{E}_\mathrm{t}$ (and likewise for the magnetic fields). We may set $\mathbf{K} = 0$ here because it will turn out that the incident and reflected $\mathbf{H}$ fields in the ε_0 medium at the interface can be matched entirely by the $\mathbf{H}$ field in the ε_1 medium, without postulating a surface current. This was not possible in the conductor–medium case, because $\mathbf{H}_1$ had to be zero yet the two $\mathbf{H}$ fields in the ε_0 medium added up constructively. The idea here is to do things with as few assumptions as possible, and it turns out we can manage without a surface current. The two boundary conditions then yield

$$\begin{aligned} E_{\mathrm{i}o} - E_{\mathrm{r}o} &= E_{\mathrm{t}o} && \text{or} \quad 1 - \mathcal{R} = \mathcal{T} \\ \frac{1}{\eta_0}(E_{\mathrm{i}o} + E_{\mathrm{r}o}) &= \frac{1}{\eta_1} E_{\mathrm{t}o} && \text{or} \quad 1 + \mathcal{R} = \frac{\eta_0}{\eta_1}\mathcal{T} \end{aligned} \tag{18.23}$$

These are two linear inhomogeneous equations in the two unknown ratios $\mathcal{R} \equiv E_{\mathrm{r}o}/E_{\mathrm{i}o}$ and $\mathcal{T} \equiv E_{\mathrm{t}o}/E_{\mathrm{i}o}$. They are easily solved to yield

$$\mathcal{R} = \frac{\eta_0 - \eta_1}{\eta_0 + \eta_1} \qquad \mathcal{T} = \frac{2\eta_1}{\eta_0 + \eta_1} \tag{18.24}$$

$\mathcal{R}$ is known as the *reflection coefficient* and $\mathcal{T}$ as the *transmission coefficient*. In many electrical engineering texts the symbol Γ is used for $\mathcal{R}$ and τ for $\mathcal{T}$ but $\mathcal{R}, \mathcal{T}$ are better mnemonics (aids to memory) so we will use them here. $\mathcal{R}$ and $\mathcal{T}$ are also known as the *Fresnel coefficients*. Please note the choice of polarization; we keep $E_{\mathrm{i}o}$ real in this entire section, i.e. we assume linear polarization. Matters become more complicated for other types of polarization, but not much new is learned by adding that complication to the treatment given here.

Suppose now that $\varepsilon_1/\varepsilon_0 = \varepsilon_1' - j\varepsilon_1''$ has an imaginary part, i.e. the $z > 0$ medium is lossy. Nothing changes in (18.21)–(18.24), but now $\eta_1 = \mu_0\omega/(\beta - j\alpha) = \sqrt{\mu_0/\varepsilon_1}$ is complex rather than real. In this case it follows that the two Fresnel coefficients both become complex. Specifically, we can write

$$\mathcal{R} = |\mathcal{R}|e^{j\theta_\mathcal{R}} \qquad \mathcal{T} = |\mathcal{T}|e^{j\theta_\mathcal{T}} \tag{18.25}$$

where $\theta_\mathcal{R}, \theta_\mathcal{T}$ are phase angles for reflection and transmission; when desirable, we express the two absolute values and phase angles in terms of the real and imaginary parts of η_0 and η_1, or even in terms of ε_1' and ε_1''. In the case of a perfect conductor, we note that $\varepsilon_1'' \to \infty$ so that $\eta_1 \to 0$, which in turn yields in (18.24)

$$\mathcal{R} \to -1 \qquad \mathcal{T} \to 0 \tag{18.26}$$

This result, of course, matches that of Section 18.1: there is no transmitted mode for a perfect conductor.

Let us again assume the lossless case and look at the situation in $z < 0$ first. The total fields in $z < 0$ are given by the sum of incident and reflected waves:

$$\mathbf{E}(\mathbf{r}) = -E_{io}\hat{\mathbf{y}}(e^{-j\beta_0 z} - \mathcal{R}e^{j\beta_0 z})$$
$$\mathbf{H}(\mathbf{r}) = \frac{1}{\eta_0}E_{io}\hat{\mathbf{x}}(e^{-j\beta_0 z} + \mathcal{R}e^{j\beta_0 z}) \qquad (18.27a)$$

With (18.23) it is straightforward to see that this pair is equivalent to

$$\mathbf{E}(\mathbf{r}) = -E_{io}\hat{\mathbf{y}}(\mathcal{T}e^{-j\beta_0 z} - 2j\mathcal{R}\sin\beta_0 z)$$
$$\mathbf{H}(\mathbf{r}) = \frac{1}{\eta_0}E_{io}\hat{\mathbf{x}}(\mathcal{T}e^{-j\beta_0 z} + 2\mathcal{R}\cos\beta_0 z) \qquad (18.27b)$$

and the full fields are

$$\mathbf{E}(\mathbf{r},t) = -E_{io}\hat{\mathbf{y}}(\mathcal{T}\cos(\omega t - \beta_0 z) + 2\mathcal{R}\sin\beta_0 z\sin(\omega t))$$
$$\mathbf{H}(\mathbf{r},t) = \frac{1}{\eta_0}E_{io}\hat{\mathbf{x}}(\mathcal{T}\cos(\omega t - \beta_0 z) + 2\mathcal{R}\cos(\beta_0 z)\cos(\omega t)) \qquad (18.27c)$$

A wave with fractional amplitude $\mathcal{T}E_{io}$ propagates in the $+\hat{\mathbf{z}}$ direction, and a *standing wave* with amplitude $2\mathcal{R}E_{io}$ exists in $z < 0$ alongside the propagating part. We will not be surprised to find, shortly, that the propagating part exactly matches in energy the wave transmitted beyond $z = 0$; where else can the energy in the propagating wave go? Let us examine the averaged Poynting vector for $z < 0$ obtained from (18.27a):

$$\mathbb{S}_0 = \frac{1}{2\eta_0}E_{io}^2(1 - \mathcal{R}^2)\hat{\mathbf{z}} \qquad (18.28)$$

Now consider the field in $z > 0$. It is given by

$$\mathbf{E}_1(\mathbf{r}) = -\mathcal{T}E_{io}\hat{\mathbf{y}}e^{-j\beta_1 z} \qquad \mathbf{H}_1(\mathbf{r}) = \frac{1}{\eta_1}\mathcal{T}E_{io}\hat{\mathbf{x}}e^{-j\beta_1 z} \qquad (18.29)$$

where the subscript 't' has been replaced by '1', because $\mathbf{E}_t$ is the only field in medium 1 in $z > 0$. The Poynting vector then works out to

$$\mathbb{S}_1 = \frac{1}{2\eta_1}\mathcal{T}^2E_{io}^2\hat{\mathbf{z}} \qquad (18.30)$$

It is straightforward to show from (18.23) that $1 - \mathcal{R}^2 = (\eta_0/\eta_1)\mathcal{T}^2$, and therefore that $\mathbb{S}_1 = \mathbb{S}_0$. The averaged Poynting vectors left and right are equal so that indeed the power flux in $z > 0$ results entirely from flux crossing the interface!

If medium 1 is lossy, then $k_1 = \beta_1 - j\alpha_1$ and the Fresnel coefficients are complex. The full fields in $z > 0$ are

$$\begin{aligned} \mathbf{E}(\mathbf{r}, t) &= -|\mathcal{T}|E_{io}\hat{\mathbf{y}}e^{-\alpha_1 z}\cos(\omega t - \beta_1 z + \theta_t) \\ \mathbf{H}(\mathbf{r}, t) &= -|\frac{1}{\eta_1}\mathcal{T}|E_{io}\hat{\mathbf{y}}e^{-\alpha_1 z}\cos(\omega t - \beta_1 z + \theta_t - \theta_\eta) \end{aligned} \tag{18.31}$$

where θ_η is the phase angle of the complex impedance η_1 and θ_t is defined in (18.25). The electric and magnetic fields are attenuated exponentially, and they are no longer in phase. The Poynting vector becomes

$$\mathbf{S}_1(z) = \frac{1}{2|\eta_1|}|\mathcal{T}|^2 E_{io}^2 \hat{\mathbf{z}} e^{-2\alpha_1 z}\cos(\theta_\eta) \tag{18.32}$$

We need to revisit the expressions on the $z < 0$ side because $\mathcal{R}$ is complex too. Expressions (18.27a) for the total fields on either side of the interface need to be corrected in this lossy case, but we will not write down the somewhat lengthy expressions that result.

The expressions (18.27a) predict that

$$\begin{aligned} |\mathbf{E}(r)| &= |E_{io}|\sqrt{1 + |\mathcal{R}|^2 - \mathcal{R}e^{-j\beta_0 z} - \mathcal{R}^* e^{j\beta_0 z}} \\ &= |E_{io}|\sqrt{1 + |\mathcal{R}|^2 - 2|\mathcal{R}|\cos(\theta_r - \beta_0 z)} \end{aligned} \tag{18.33}$$

The maximum and minimum values of the field amplitude occur when the cosine is ± 1, i.e. when $\theta_r - \beta_0 z = 0, \pm\pi, \ldots$. These extreme values are proportional to $1 \pm |\mathcal{R}|$. The *standing-wave ratio* (SWR) is defined as the ratio of the maximum to the minimum value of $|\mathbf{E}(\mathbf{r})|$, and it is given by

$$SWR = \frac{1 + |\mathcal{R}|}{1 - |\mathcal{R}|} \tag{18.34}$$

A measurement of SWR enables one to calculate the absolute value $|\mathcal{R}|$ of the reflection coefficient $\mathcal{R}$. The standing-wave ratio is usefully measured in transmission lines, where it is less easy to measure $\mathcal{R}$ directly.

Let us finally consider $\mathbf{H}_1 \equiv \mathbf{H}_t$ in $z > 0$ from (18.29), in the case where we allow the losses to become infinite. First we note that $\eta_1^{-1}\mathcal{T} = 1/(\eta_0 + \eta_1)$. Then we note that $\eta_1 \propto 1/\sqrt{\varepsilon_1/\varepsilon_0}$. The relative dielectric permittivity $\varepsilon_1/\varepsilon_0 \to \infty$ as the lossiness of the medium increases (the medium becomes a perfect conductor in the limit $\sigma \to \infty$). As a consequence, $\eta_1^{-1}\mathcal{T} \to 2\eta_0^{-1}$, so that an approximation for the transmitted magnetic field at very high losses becomes

$$\mathbf{H}_t(\mathbf{r}) \approx \frac{2}{\eta_0}E_{io}\hat{\mathbf{x}}e^{-(\alpha_1 + j\beta_1)z} \tag{18.35}$$

Note that $\alpha_1 z \to \infty$ for all $z > 0$ when $\alpha_1 \to \infty$ (as the medium tends towards becoming a perfect conductor). For extremely large α_1 we see that $\mathbf{H}_t$ is nonzero, with $H_t \approx 2E_{io}/\eta_0$, only for a very small region of extent $z \sim 1/\alpha_1$ beyond the interface. The continuity of tangential $\mathbf{H}$ components has thus turned into the boundary condition $\hat{\mathbf{n}} \times (\mathbf{H}_0 - \mathbf{H}_1) = \mathbf{K}$, in which we *interpret* $\mathbf{H}_1$ as being zero and $\mathbf{K}$ as $2\eta_0^{-1} E_{io}\hat{\mathbf{x}}$, in complete agreement with (18.6), so that $\mathbf{K} = \mathbf{n} \times \mathbf{H}_0$. This is the basis for a very useful *equivalence principle*: at the interface of a perfect conductor with free space, we may interpret the field $\hat{\mathbf{n}} \times \mathbf{H}$ as a surface-current density $\mathbf{K}$. It also amplifies the statement we made immediately below (18.6): radio or TV waves impinging upon an antenna 'translate' into a *surface current* with the same *modulation* characteristics, and the circuitry of the antenna then takes that rapidly modulating current and turns it into sound or video.

18.4 Oblique incidence upon a dielectric half-space

18.4.1 The ? or TE modes obliquely incident upon a dielectric

We are now ready to tackle the most complicated situation: a time-harmonic plane wave obliquely incident upon a uniform lossless dielectric medium separated from free space by the interface $z = 0$. While we will consider free space to the left, nothing in the argument would change if instead there were a uniform lossless dielectric. Figure 18.5 is an extension of Fig. 18.2, with a transmitted wave $\mathbf{E}_t$, $\mathbf{H}_t$ at angle θ_t with the surface normal. Only the $\perp$ or TE mode has been shown. We now look at

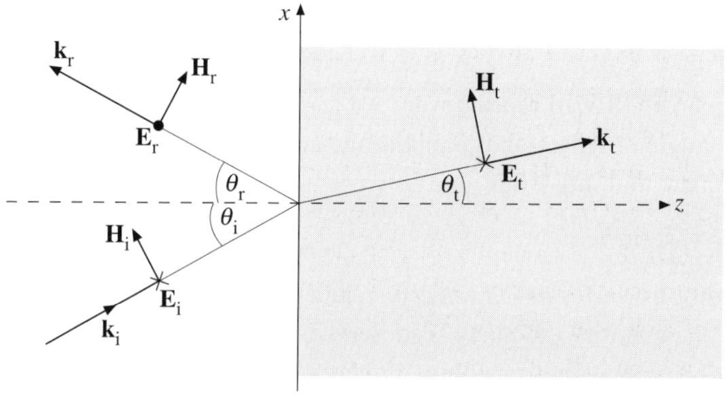

Fig. 18.5 Oblique incidence of EM waves upon a dielectric–dielectric interface for the $\perp$ (TE) mode.

Eqs. (18.10) and add to them similar expressions for the transmitted mode:

$$\begin{aligned}
\mathbf{E}_\mathrm{i}(\mathbf{r}) &= -E_\mathrm{io}\hat{\mathbf{y}}e^{-j\beta_0\hat{\mathbf{k}}_\mathrm{i}\cdot\mathbf{r}} & \hat{\mathbf{k}}_\mathrm{i} &\equiv \hat{\mathbf{x}}\sin\theta_\mathrm{i} + \hat{\mathbf{z}}\cos\theta_\mathrm{i} \\
\mathbf{E}_\mathrm{r}(\mathbf{r}) &= E_\mathrm{ro}\hat{\mathbf{y}}e^{-j\beta_0\hat{\mathbf{k}}_\mathrm{r}\cdot\mathbf{r}} & \hat{\mathbf{k}}_\mathrm{r} &\equiv \hat{\mathbf{x}}\sin\theta_\mathrm{r} - \hat{\mathbf{z}}\cos\theta_\mathrm{r} \\
\mathbf{E}_\mathrm{t}(\mathbf{r}) &= -E_\mathrm{to}\hat{\mathbf{y}}e^{-j\beta_1\hat{\mathbf{k}}_\mathrm{t}\cdot\mathbf{r}} & \hat{\mathbf{k}}_\mathrm{t} &\equiv \hat{\mathbf{x}}\sin\theta_\mathrm{t} + \hat{\mathbf{z}}\cos\theta_\mathrm{t}
\end{aligned} \tag{18.36a}$$

Note that there are *two* values of the wave vector magnitude β, one for each medium, and *three* unit wave vectors, one for each of the mode directions. Likewise, in analogy to (18.14) we can write down the $\mathbf{H}$ field phasors right away:

$$\begin{aligned}
\mathbf{H}_\mathrm{i}(\mathbf{r}) &= \frac{1}{\eta_0}E_\mathrm{io}\hat{\mathbf{m}}_\mathrm{i}e^{-j\beta_0\hat{\mathbf{k}}_\mathrm{i}\cdot\mathbf{r}} & \hat{\mathbf{m}}_\mathrm{i} &\equiv \hat{\mathbf{x}}\cos\theta_\mathrm{i} - \hat{\mathbf{z}}\sin\theta_\mathrm{i} \\
\mathbf{H}_\mathrm{r}(\mathbf{r}) &= \frac{1}{\eta_0}E_\mathrm{ro}\hat{\mathbf{m}}_\mathrm{r}e^{-j\beta_0\hat{\mathbf{k}}_\mathrm{r}\cdot\mathbf{r}} & \hat{\mathbf{m}}_\mathrm{r} &\equiv \hat{\mathbf{x}}\cos\theta_\mathrm{r} + \hat{\mathbf{z}}\sin\theta_\mathrm{r} \\
\mathbf{H}_\mathrm{t}(\mathbf{r}) &= \frac{1}{\eta_1}E_\mathrm{to}\hat{\mathbf{m}}_\mathrm{t}e^{-j\beta_1\hat{\mathbf{k}}_\mathrm{t}\cdot\mathbf{r}} & \hat{\mathbf{m}}_\mathrm{t} &\equiv \hat{\mathbf{x}}\cos\theta_\mathrm{t} - \hat{\mathbf{z}}\sin\theta_\mathrm{t}
\end{aligned} \tag{18.36b}$$

It is not difficult to ascertain that each triad $\hat{\mathbf{k}}$, $\mathbf{E}$, $\mathbf{H}$ is right-handed and that each member is mutually orthogonal to the other two, for the incident, reflected, and transmitted waves. The boundary condition for the tangential $\mathbf{E}$ components at $z = 0$ now becomes

$$E_\mathrm{io}e^{-j\beta_0 x\sin\theta_\mathrm{i}} - E_\mathrm{ro}e^{-j\beta_0 x\sin\theta_\mathrm{r}} - E_{t0}e^{-j\beta_1 x\sin\theta_\mathrm{t}} = 0 \tag{18.37a}$$

Think carefully about this equation! It has to be true for *all* x on the interface. Again, there is only one way that this is possible, namely if all three exponentials are the same and the coefficients add to zero:

$$\begin{aligned}
&\beta_0\sin\theta_\mathrm{i} = \beta_0\sin\theta_\mathrm{r} = \beta_1\sin\theta_\mathrm{t} \\
&E_\mathrm{io} - E_\mathrm{ro} - E_\mathrm{to} = 0
\end{aligned} \tag{18.37b}$$

The first line of (18.37b) repeats what we learned from (18.11) and (18.12): $\theta_\mathrm{r} = \theta_\mathrm{i} \equiv \theta$ (the angle of reflection is equal to the angle of incidence). However, it adds something new and important,

$$\beta_0\sin\theta_0 = \beta_1\sin\theta_1 \tag{18.38a}$$

This is known as *Snell's law of refraction*, and for nonmagnetic media ($\mu_1 = \mu_0$) it states that the quantity $\sqrt{\varepsilon_1}\sin\theta_1$ is conserved at any planar interface of medium 1 with any other nonmagnetic medium. If there is dielectric instead of free space to the left, then we have

$$\sqrt{\varepsilon_0}\sin\theta_0 = \sqrt{\varepsilon_1}\sin\theta_1 \tag{18.38b}$$

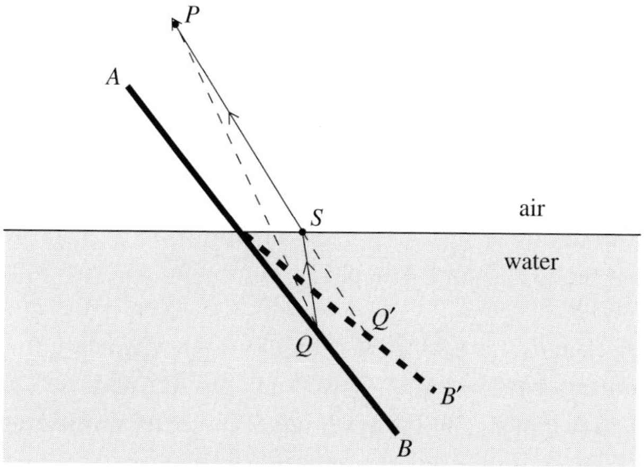

Fig. 18.6 Apparent bending of a pencil in water.

where ε_0 is the permittivity of the left-hand medium. We see that the $\hat{\mathbf{k}}$ vector refracts *towards* the surface-normal direction when it passes into a medium with higher permittivity. Likewise, there is refraction *away from* the surface normal when the second medium has a lower permittivity. The phenomenon is well known to us all from the apparent bending of a stick inserted obliquely into the surface of a placid pond of transparent water. (However, the insertion of a long thin stick into water leads to a somewhat surprising sight because the stick appears to bend *upwards*, even through water has a relative permittivity ~ 1.8 at room temperature, so that light rays entering the water would refract *downwards*. Figure 18.6 gives a partial explanation. The long thin stick AB is shown at its actual location. Consider a point Q on the underwater part of the stick. If water had relative permittivity $= 1$, then light from point Q would reach the eye at point P by means of the straight-line path QP. But the actual path must be QSP because the relative permittivity > 1 for water, and the light is refracted away from the normal. The eye at P then sees point Q in the direction **PS**, and hence at Q' (chosen so that $SQ = SQ'$). Thus, point Q' will lie closer to the surface than point Q. Likewise, every other point below the water level appears to lie higher, and so the stick seems to have been bent upwards!)

The boundary condition for tangential **H** components can be formulated using (18.36b), and it then leads to

$$\frac{1}{\eta_0}(E_{io} + E_{ro})\cos\theta_0 = \frac{1}{\eta_1}E_{to}\cos\theta_1 \tag{18.39}$$

Equations (18.37b) and (18.39), with definitions of reflection and transmission coefficients similar to before, namely $\mathcal{R}_\perp = -E_{ro}/E_{io}$, $\mathcal{T}_\perp = E_{to}/E_{io}$ (note the extra minus sign in the reflection coefficient to make $\mathcal{R}_\perp = -1$ when $\mathcal{T}_\perp = 0$, as in the case of oblique incidence upon a perfect conductor), then lead to

$$1+\mathcal{R}_\perp = \mathcal{T}_\perp \qquad \text{and} \qquad 1-\mathcal{R}_\perp = \frac{\eta_0 \cos\theta_1}{\eta_1 \cos\theta_0}\mathcal{T}_\perp \tag{18.40}$$

These are easily solved to yield

$$\mathcal{R}_\perp = \frac{\eta_1\cos\theta_0 - \eta_0\cos\theta_1}{\eta_1\cos\theta_0 + \eta_0\cos\theta_1} \qquad \mathcal{T}_\perp = \frac{2\eta_1\cos\theta_0}{\eta_1\cos\theta_0 + \eta_0\cos\theta_1} \tag{18.41}$$

and we can easily check at $\theta_0 = 0$ that we re-obtain (18.24), as we should. It helps to reiterate that we have chosen $\mathcal{R} = -E_{\mathrm{r}o}/E_{\mathrm{i}o}$ with a minus sign because we chose the sign of $\mathbf{E}_{\mathrm{r}o}$ opposite to that of $\mathbf{E}_{\mathrm{i}o}$, unlike our original choice for normal incidence in Section 18.3. This choice of sign makes the reflection coefficient consistent when we allow $\theta_0 \to 0$. Our three sets of fields are now given by

$$\begin{aligned}
\mathbf{E}_{\mathrm{i}}(\mathbf{r}) &= -E_{\mathrm{i}o}\hat{\mathbf{y}}e^{-j\beta_0\hat{\mathbf{k}}_{\mathrm{i}}\cdot\mathbf{r}} & \hat{\mathbf{k}}_{\mathrm{i}} &\equiv \hat{\mathbf{x}}\sin\theta_0 + \hat{\mathbf{z}}\cos\theta_0 \\
\mathbf{E}_{\mathrm{r}}(\mathbf{r}) &= -\mathcal{R}_\perp E_{\mathrm{i}o}\hat{\mathbf{y}}e^{-j\beta_0\hat{\mathbf{k}}_{\mathrm{r}}\cdot\mathbf{r}} & \hat{\mathbf{k}}_{\mathrm{r}} &\equiv \hat{\mathbf{x}}\sin\theta_0 - \hat{\mathbf{z}}\cos\theta_0 \\
\mathbf{E}_{\mathrm{t}}(\mathbf{r}) &= -\mathcal{T}_\perp E_{\mathrm{i}o}\hat{\mathbf{y}}e^{-j\beta_1\hat{\mathbf{k}}_{\mathrm{t}}\cdot\mathbf{r}} & \hat{\mathbf{k}}_{\mathrm{t}} &\equiv \hat{\mathbf{x}}\sin\theta_1 + \hat{\mathbf{z}}\cos\theta_1
\end{aligned} \tag{18.42a}$$

$$\begin{aligned}
\mathbf{H}_{\mathrm{i}}(\mathbf{r}) &= \frac{1}{\eta_0}E_{\mathrm{i}o}\hat{\mathbf{m}}_{\mathrm{i}}e^{-j\beta_0\hat{\mathbf{k}}_{\mathrm{i}}\cdot\mathbf{r}} & \hat{\mathbf{m}}_{\mathrm{i}} &\equiv \hat{\mathbf{x}}\cos\theta_0 - \hat{\mathbf{z}}\sin\theta_0 \\
\mathbf{H}_{\mathrm{r}}(\mathbf{r}) &= -\frac{1}{\eta_0}\mathcal{R}_\perp E_{\mathrm{i}o}\hat{\mathbf{m}}_{\mathrm{r}}e^{-j\beta_0\hat{\mathbf{k}}_{\mathrm{r}}\cdot\mathbf{r}} & \hat{\mathbf{m}}_{\mathrm{r}} &\equiv \hat{\mathbf{x}}\cos\theta_0 + \hat{\mathbf{z}}\sin\theta_0 \\
\mathbf{H}_{\mathrm{t}}(\mathbf{r}) &= \frac{1}{\eta_1}\mathcal{T}_\perp E_{\mathrm{i}o}\hat{\mathbf{m}}_{\mathrm{t}}e^{-j\beta_1\hat{\mathbf{k}}_{\mathrm{t}}\cdot\mathbf{r}} & \hat{\mathbf{m}}_{\mathrm{t}} &\equiv \hat{\mathbf{x}}\cos\theta_1 - \hat{\mathbf{z}}\sin\theta_1
\end{aligned} \tag{18.42b}$$

There are a number of things we can do easily with this set of equations. The most obvious is to compute the total field phasors in each of the two media. This is straightforward and leads to

$$\begin{aligned}
\mathbf{E}_0(\mathbf{r}) &= -E_{\mathrm{i}o}\hat{\mathbf{y}}e^{-j\beta_0 x\sin\theta_0}\left[e^{-j\beta_0 z\cos\theta_0} + \mathcal{R}_\perp e^{j\beta_0 z\cos\theta_0}\right] \\
\mathbf{H}_0(\mathbf{r}) &= \frac{1}{\eta_0}E_{\mathrm{i}o}e^{-j\beta_0 x\sin\theta_0}\left[\hat{\mathbf{x}}\cos\theta_0\left(e^{-j\beta_0 z\cos\theta_0} - \mathcal{R}_\perp e^{j\beta_0 z\cos\theta_0}\right)\right. \\
&\qquad \left. - \hat{\mathbf{z}}\sin\theta_0\left(e^{-j\beta_0 z\cos\theta_0} + \mathcal{R}_\perp e^{j\beta_0 z\cos\theta_0}\right)\right]
\end{aligned} \tag{18.43a}$$

$$\begin{aligned}
\mathbf{E}_1(r) &= -\mathcal{T}_\perp E_{\mathrm{i}o}\hat{\mathbf{y}}e^{-j\beta_1(x\sin\theta_1 + z\cos\theta_1)} \\
\mathbf{H}_1(\mathbf{r}) &= \frac{1}{\eta_1}\mathcal{T}_\perp E_{\mathrm{i}o}(\hat{\mathbf{x}}\cos\theta_1 - \hat{\mathbf{z}}\sin\theta_1)e^{-j\beta_1(x\sin\theta_1 + z\cos\theta_1)}
\end{aligned} \tag{18.43b}$$

Looking carefully at these expressions and remembering that the average Poynting vector $\mathbf{S} \propto \mathbf{E}\times\mathbf{H}^*$, we observe that the power flux has a $-\hat{\mathbf{y}}\times\hat{\mathbf{x}} = \hat{\mathbf{z}}$ component and a $\hat{\mathbf{y}} \times \hat{\mathbf{z}} = \hat{\mathbf{x}}$ component, so that

$$\begin{aligned} \mathbb{S}_{1z} &= \frac{1}{\eta_0}|E_{io}|^2 \cos\theta_0(1 - \mathcal{R}_\perp^2) \\ \mathbb{S}_{1x} &= \frac{1}{\eta_0}|E_{io}|^2 \sin\theta_0[1 + \mathcal{R}_\perp^2 + 2\mathcal{R}_\perp \cos(\beta_0 z \cos\theta_0)] \end{aligned} \tag{18.44a}$$

$$\begin{aligned} \mathbb{S}_{1z} &= \frac{1}{\eta_1}\mathcal{T}_\perp^2|E_{io}|^2 \cos\theta_1 \\ \mathbb{S}_{1x} &= \frac{1}{\eta_1}\mathcal{T}_\perp^2|E_{io}|^2 \sin\theta_1 \end{aligned} \tag{18.44b}$$

The continuity of the $\mathbb{S}_z$ components becomes obvious due to the equality implied by (18.40). That continuity implies that the *total* flux in the $\hat{\mathbf{z}}$ direction is constant, as it must be because the flux in medium 1 is caused entirely by the incidence of flux from medium 0.

This discussion becomes more difficult, and the results need modification, when the dielectric medium is lossy. For example, planes of constant amplitude no longer coincide with planes of constant phase. Also, the angle of refraction as it appears in Snell's law becomes complex, and the actual real angle of refraction is a more complicated function of the parameters.

18.4.2 The k or TM modes obliquely incident upon a dielectric

In this case the **E** fields are in the plane of incidence, and because everything is more or less a repeat of what we wrote down for the TE modes, with **E** and **H** fields exchanged (and some changes in sign), we will simply write down the fields:

$$\begin{aligned} \mathbf{H}_i(\mathbf{r}) &= -\frac{1}{\eta_0}E_{io}\hat{\mathbf{y}}e^{-j\beta_0\hat{\mathbf{k}}_i\cdot\mathbf{r}} & \hat{\mathbf{k}}_i &\equiv \hat{\mathbf{x}}\sin\theta_i + \hat{\mathbf{z}}\cos\theta_i \\ \mathbf{H}_r(\mathbf{r}) &= -\frac{1}{\eta_0}E_{ro}\hat{\mathbf{y}}e^{-j\beta_0\hat{\mathbf{k}}_r\cdot\mathbf{r}} & \hat{\mathbf{k}}_r &\equiv \hat{\mathbf{x}}\sin\theta_r - \hat{\mathbf{z}}\cos\theta_r \\ \mathbf{H}_t(\mathbf{r}) &= -\frac{1}{\eta_1}E_{to}\hat{\mathbf{y}}e^{-j\beta_1\hat{\mathbf{k}}_t\cdot\mathbf{r}} & \hat{\mathbf{k}}_t &\equiv \hat{\mathbf{x}}\sin\theta_t + \hat{\mathbf{z}}\cos\theta_t \end{aligned} \tag{18.45a}$$

$$\begin{aligned} \mathbf{E}_i(\mathbf{r}) &= -E_{io}\hat{\mathbf{m}}_i e^{-j\beta_0\hat{\mathbf{k}}_i\cdot\mathbf{r}} & \hat{\mathbf{m}}_i &\equiv \hat{\mathbf{x}}\cos\theta_i - \hat{\mathbf{z}}\sin\theta_i \\ \mathbf{E}_r(\mathbf{r}) &= E_{ro}\hat{\mathbf{m}}_r e^{-j\beta_0\hat{\mathbf{k}}_r\cdot\mathbf{r}} & \hat{\mathbf{m}}_r &\equiv \hat{\mathbf{x}}\cos\theta_r + \hat{\mathbf{z}}\sin\theta_r \\ \mathbf{E}_t(\mathbf{r}) &= -E_{to}\hat{\mathbf{m}}_t e^{-j\beta_1\hat{\mathbf{k}}_t\cdot\mathbf{r}} & \hat{\mathbf{m}}_t &\equiv \hat{\mathbf{x}}\cos\theta_t - \hat{\mathbf{z}}\sin\theta_t \end{aligned} \tag{18.45b}$$

The sign changes correspond to the differences in direction that we observe between the **H** (and **E**) fields of Fig. 18.7 and the **E** (and **H**) fields of Fig. 18.5. We shall not repeat all the details of subsection 18.4.1. Many of them are the same. Again we find

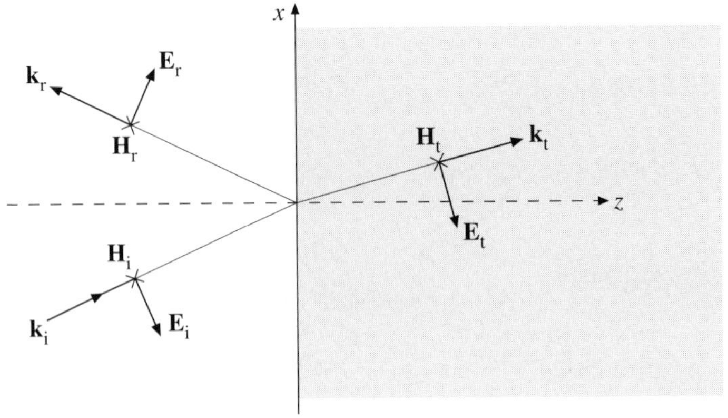

Fig. 18.7 Oblique incidence of EM waves upon a dielectric–dielectric interface for the $\|$ (TM) mode.

that the angle of reflection equals the angle of incidence and Snell's law, stating that $\beta_i \sin\theta_i$ is conserved as we go from medium 0 to medium 1.

However, the two boundary conditions for tangential fields lead now to reflection and transmission coefficients, $\mathcal{R}_\| = -E_{ro}/E_{io}$, $\mathcal{T}_\| = E_{to}/E_{io}$, with

$$1 - \mathcal{R}_\| = \frac{\eta_0}{\eta_1}\mathcal{T}_\| \qquad 1 + \mathcal{R}_\| = \frac{\cos\theta_1}{\cos\theta_0}\mathcal{T}_\| \tag{18.46}$$

which leads immediately to

$$\mathcal{R}_\| = \frac{\eta_1\cos\theta_1 - \eta_0\cos\theta_0}{\eta_1\cos\theta_1 + \eta_0\cos\theta_0} \qquad \mathcal{T}_\| = \frac{2\eta_1\cos\theta_0}{\eta_1\cos\theta_1 + \eta_0\cos\theta_0} \tag{18.47}$$

These are different from the coefficients (18.41) for the $\perp$ or TE modes.

18.5 Critical and Brewster angles, refractive phenomena

18.5.1 Critical angle

Snell's law indicates that for the transmission of waves from a low-ε to a high-ε medium the $\hat{\mathbf{k}}$ vector rays refract towards the normal. Even if we have a wave incident infinitesimally close to $\theta_i = 90°$, we will have a transmitted wave with $\theta_t < 90°$; there are transmitted waves at every incident angle between 0 and 90°. This is not at all the case in transmission the other way around, from a high-ε to a low-ε medium. Consider Fig. 18.8, in which a ray incident at angle θ_i is refracted away from the normal at angle $\theta_t > \theta_i$. The reason is that $\varepsilon_2 < \varepsilon_1$, and Snell's law for nonmagnetic media predicts that $\sqrt{\varepsilon_1}\sin\theta_1 = \sqrt{\varepsilon_2}\sin\theta_2$, which of course implies that $\sin\theta_1 < \sin\theta_2$. As we increase θ_1, we find that θ_t becomes 90° at some angle $\theta_{cr} < 90°$. This incident angle is known as the *critical angle*. Modes incident at angles

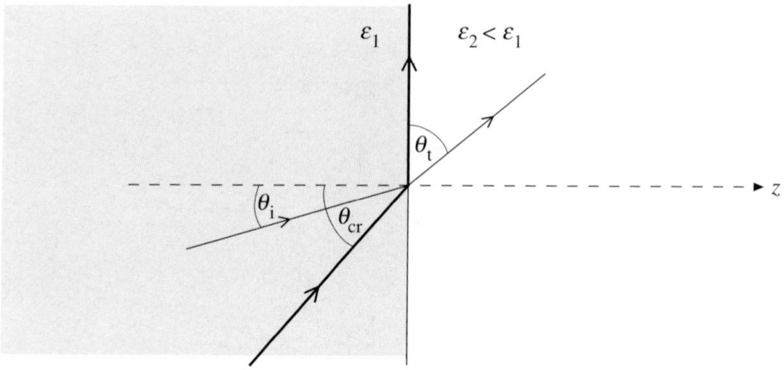

Fig. 18.8 Refraction away from the normal when $\varepsilon_2 < \varepsilon_1$.

$\theta_i > \theta_{cr}$ are totally reflected! The critical angle is given by the extreme value of Snell's law (in nonmagnetic media) in which $\sin\theta_2 = 1$:

$$\sin\theta_{cr} = \sqrt{\varepsilon_2/\varepsilon_1} \tag{18.48}$$

We conclude from (18.41) and (18.47) that total reflection can occur, because $\cos\theta_2 = \sqrt{1 - \sin^2\theta_2} = \sqrt{1 - \varepsilon_1 \sin^2\theta_1/\varepsilon_2}$, and this latter square root is purely imaginary when $\sin^2\theta_1 > \sin^2\theta_{cr}$, as follows from (18.48). Thus $\cos\theta_2$ is imaginary, and the two reflection coefficients are both of the form $\mathcal{R} = (a - jb)/(a + jb)$ with real a and b, which in turn implies that $|\mathcal{R}| = 1$. Equation (18.44a) for $\mathbb{S}_{0z}$ ($\equiv \mathbb{S}_{1z}$ here) will have the factor $1 - |\mathcal{R}|^2$ in it, hence indeed no power-flux z component in the less dense medium. Furthermore, the reflected-wave amplitude has the same magnitude as the incident-wave amplitude, and therefore fortifies the notion at the beginning of this paragraph that there cannot be a transmitted wave when the critical angle is exceeded (even though $\mathcal{T} \neq 0$).

Example 18.1 The underwater swimmer

An underwater swimmer is depicted in Fig. 18.9. His eye sees an object in the direction of point P in air (the upper medium); the $\hat{\mathbf{k}}$ vector is refracted toward the surface normal when entering the water as shown because $\varepsilon_w \sim 1.8\varepsilon_0$ at optical wavelengths, so that Snell's law predicts bending as shown. A point B lying at grazing angle with the surface is seen by the swimmer's eye via path BVS, and likewise point A on the other side via path AUS. A point P' above the water, as shown in the diagram, is also seen by the swimmer. If a light ray from P' enters the water outside a circle on the surface bounded by points U and V (broken line), then that light ray cannot enter the cone. But if the light ray from P' enters inside the circle then it can end up at the swimmer's eye!

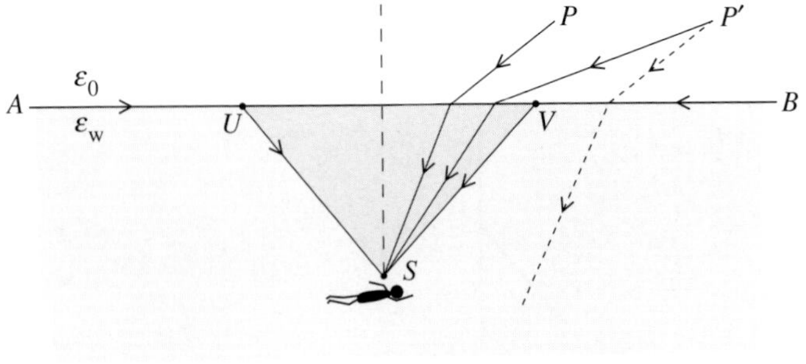

Fig. 18.9 The world view of the underwater swimmer.

Thus the underwater swimmer's view of the outside world comes via a cone of light that enters the circle depicted by the bounds U and V: the entire hemisphere in air can be seen to be apparently inside that cone by the swimmer's eye. The half-angle θ_{cr} of the cone obeys the approximate relationship $\sqrt{1.8}\sin\theta_{cr} = 1$, which yields $\theta_{cr} \approx 48°$. Clearly, this 'cone of light acceptance' is due to the higher value of water's permittivity compared to air.

Example 18.2 Surface (evanescent) waves

Consider a wave traveling at a small angle $\theta_1' = \pi/2 - \theta_0$ between two infinite planes containing a dielectric with $\varepsilon_1 > \varepsilon_0$, shown in Fig. 18.10. Such a sandwich is an example of a *dielectric waveguide*. It should be noted that the outer medium is not necessarily free space, so that in this case, as before, ε_0 does not necessarily refer to the vacuum dielectric constant.

Let us try time-harmonic plane-wave phasor solutions at the top interface ($z = 0$) for a $\perp$ mode, i.e. solutions such as (18.42). The angle of incidence is now θ_1 and the angle of refraction is θ_0 (not shown in the diagram). So, in applying the formulas (18.42) we have to exchange θ_0 and θ_1. Let us also assume that θ_1 *exceeds* the critical angle (so that θ_1' is small). Then for $z > 0$ equations (18.42) predict that

$$\mathbf{E}_t(\mathbf{r}) = -\mathcal{T}_\perp E_{io}\hat{\mathbf{y}}e^{-j\beta_0(x\sin\theta_0 + z\cos\theta_0)} \tag{18.49}$$

We can replace $\beta_0 \sin\theta_0$ by $\beta_1 \sin\theta_1$ (from Snell's law) and now repeat the calculation of $\cos\theta_0$, which shows that for large angles of incidence θ_1 it must be imaginary:

$$\cos\theta_0 = \sqrt{1 - \sin^2\theta_0} = \sqrt{1 - (\varepsilon_1/\varepsilon_0)\sin^2\theta_1} = \pm j\sqrt{(\varepsilon_1/\varepsilon_0)\sin^2\theta_1 - 1} \equiv \pm jp$$

The quantity inside the square-root sign is positive because $\sin^2\theta_1 > \sin^2\theta_{cr} = \varepsilon_0/\varepsilon_1$, and we will need to choose the sign in front of it in order to avoid the physically inconsistent situation of waves growing in amplitude as z increases.

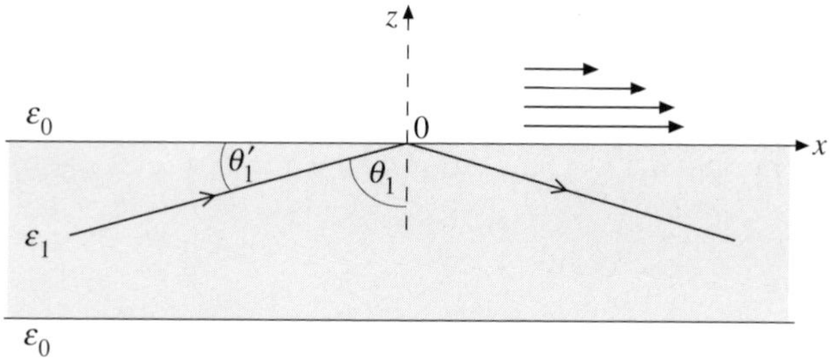

Fig. 18.10 Evanescent waves (the arrows above the interface $z = 0$) created at a dielectric–vacuum interface by modes at near-grazing angles θ_1'.

Hence (18.49) becomes

$$\mathbf{E}_t(\mathbf{r}) = -\mathcal{T}_\perp E_{io}\hat{\mathbf{y}}e^{-j(\beta_1 x \sin\theta_1 - j\beta_0 p z)} \qquad \text{or}$$
$$\mathbf{E}_t(\mathbf{r}, t) = -\mathcal{T}_\perp E_{io}\hat{\mathbf{y}}e^{-pz}\cos(\omega t - j\beta_1 x \sin\theta_1) \tag{18.50}$$
$$p \equiv \sqrt{(\varepsilon_1/\varepsilon_0)\sin^2\theta_1 - 1}$$

This is a mode that travels in the $\hat{\mathbf{x}}$ direction with wavelength $\lambda_x = \lambda/\sin\theta_1$, as we observe from the cosine factor, but it attenuates exponentially in the $+\hat{\mathbf{z}}$ direction with a factor $e^{-\beta_0 p z}$ for $z > 0$. This *evanescent wave* does not propagate significant energy from the inside ($z < 0$) to the outside ($z > 0$), but a little energy propagates along the surface. Thus in the steady-state condition the energy remains within or at the surface of the inner medium. However, if a second waveguide is brought above and very close to the $z = 0$ plane, then there is a possibility that some of the evanescent wave energy will leak (or 'tunnel') into that second waveguide. So now the evanescent wave travels parallel to the interface with the second waveguide but with reduced amplitude of the field: the evanescent wave at distance $z = d$ (d is the separation distance between the waveguides) above the first waveguide has had its power decreased by a factor $\propto e^{-2\beta_0 p d}$. A wave bouncing off the walls at angle θ_1', assuming the second waveguide also has dielectric constant ε_1, would be set up in the second waveguide with that reduced power (i.e. down by $8.686\beta_0 pd$ dB).

18.5.2 Brewster angles

Returning to our previous notation, we now consider how the reflection coefficient $\mathcal{R}$ varies with the angle of incidence θ_0. The angle of refraction is θ_1. When we look at the reflection coefficients given in (18.41) for the $\perp$ modes and in (18.47) for the $\|$ modes, we observe that $\mathcal{R}$ would become zero if the numerator vanished. In such a case, there would be complete transmission and zero reflection. Let us see if and when this can happen. For the $\perp$ mode we require $\eta_1 \cos\theta_0 - \eta_0 \cos\theta_1 = 0$. Snell's law yields $\beta_0 \sin\theta_0 = \beta_1 \sin\theta_1$. Together, these give

$$1 - \sin^2\theta_0 = \frac{\eta_0^2}{\eta_1^2}(1 - \sin^2\theta_1) \quad \longrightarrow \quad \sin^2\theta_0 = \left(1 - \frac{\eta_0^2}{\eta_1^2}\right) + \frac{\eta_0^2}{\eta_1^2}\sin^2\theta_1 \quad \text{and}$$

$$\sin^2\theta_0 = \frac{\beta_1^2}{\beta_0^2}\sin^2\theta_1$$

Solving these two equations yields

$$\sin^2\theta_{0B}^{\perp} = \frac{1 - \mu_0\varepsilon_1/\mu_1\varepsilon_0}{1 - \mu_0^2/\mu_1^2} \qquad \sin^2\theta_{1B}^{\perp} = \left(\frac{\mu_0\varepsilon_0}{\mu_1\varepsilon_1}\right)\frac{1 - \mu_0\varepsilon_1/\mu_1\varepsilon_0}{1 - \mu_0^2/\mu_1^2} \tag{18.50}$$

These are the two *Brewster angles for the* $\perp$ *mode*, and it is obvious that in general the angles do not exist for nonmagnetic media ($\mu_1 = \mu_0$). If the two dielectric permittivities are equal, then (18.50) becomes

$$\sin^2\theta_{0B}^{\perp} = \frac{1}{1 + \mu_0/\mu_1} \qquad \sin^2\theta_{1B}^{\perp} = \frac{\mu_0/\mu_1}{1 + \mu_0/\mu_1} \tag{18.51}$$

and the Brewster angles can exist. The first of expressions (18.51) is identical to the simpler form, $\tan\theta_{0B}^{\perp} = \sqrt{\mu_1/\mu_0}$. It appears as if μ_1 can be equal to μ_0 here, but we should remember that this implies that *both* permeability and permittivity are equal, in which case we have the trivial case of equal materials from an electromagnetic point of view and therefore no interface (hence zero $\mathcal{R}$ and no change in angle).

For the $\parallel$ mode we have $\eta_0 \cos\theta_0 = \eta_1 \cos\theta_1$. The roles of η_0 and η_1 are reversed, compared with the calculation for the $\perp$ mode, so we obtain

$$1 - \sin^2\theta_0 = \frac{\eta_1^2}{\eta_0^2}(1 - \sin^2\theta_1) \quad \longrightarrow \quad \sin^2\theta_0 = \left(1 - \frac{\eta_1^2}{\eta_0^2}\right) + \frac{\eta_1^2}{\eta_0^2}\sin^2\theta_1 \quad \text{and}$$

$$\sin^2\theta_0 = \frac{\beta_1^2}{\beta_0^2}\sin^2\theta_1$$

which results directly in

$$\sin^2\theta_{0B}^{\parallel} = \frac{1 - \mu_1\varepsilon_0/\mu_0\varepsilon_1}{1 - \varepsilon_0^2/\varepsilon_1^2} \qquad \sin^2\theta_{1B}^{\parallel} = \left(\frac{\mu_0\varepsilon_0}{\mu_1\varepsilon_1}\right)\frac{1 - \mu_1\varepsilon_0/\mu_0\varepsilon_1}{1 - \varepsilon_0^2/\varepsilon_1^2} \tag{18.52}$$

for the *Brewster angles for the* $\parallel$ *mode*. In nonmagnetic media, these become simplified to

$$\sin^2\theta_{0B}^{\parallel} = \frac{1}{1 + \varepsilon_0/\varepsilon_1} \qquad \sin^2\theta_{1B}^{\parallel} = \frac{\varepsilon_0/\varepsilon_1}{1 + \varepsilon_0/\varepsilon_1} \tag{18.53}$$

and a Brewster angle for the $\parallel$ mode clearly always exists. The first of expressions (18.53) is identical to the simpler form $\tan\theta_{0B}^{\parallel} = \sqrt{\varepsilon_1/\varepsilon_0}$. From this, or directly from

(18.53), we see that the Brewster angle is less than 45° when $\varepsilon_1 < \varepsilon_0$ and exceeds 45° in the opposite case.

Example 18.3 Brewster polarizer

The existence of a Brewster angle for the $\|$ mode suggests an idealized *polarizer* (a device for producing linearly polarized waves). Consider a block of dielectric material as shown in Fig. 18.11, with an incident elliptically polarized wave with angle of incidence equal to $\theta_{0B}^{\|}$; see (18.53).

The $\|$ (TM) component of the incident field will be completely transmitted, hence the reflected wave is purely $\perp$ (TE) mode. Thus the transmitted mode is still elliptical, but with a different mix of $\perp$ and $\|$ components, hence a different ellipticity. Furthermore, isolation of the reflected mode gives a linearly polarized mode that is TE with respect to the interface. In practice, one must remember that laboratory-produced waves are at best quasi-time-harmonic; as they must begin and end they will have some spread in frequency ω around a central or *carrier* value if the medium is *dispersive* ($d\varepsilon/d\omega$ is nonzero). This implies that the Brewster angle is a function of frequency if the medium is dispersive, so that transmission may not be total for all frequency components.

18.5.3 Phenomena associated with refractivity

A medium with a continuously varying permittivity can give rise to interesting refractive effects. The atmosphere is such a medium; the air at higher altitudes is thinner and its permittivity approaches unity (although not monotonically). The behavior of a wave propagating at an angle to the vertical is illustrated in Fig. 18.12.

The continuous medium can be seen as the limit $\Delta z \to 0$ of many contiguous layers of thickness Δz. At each interface we have $\sqrt{\varepsilon(z+\Delta z)}\,\sin\theta(z+\Delta z) = \sqrt{\varepsilon(z)}\,\sin\theta(z)$. The constant occurrence of $\sqrt{\varepsilon(z)}$ makes it useful to give this

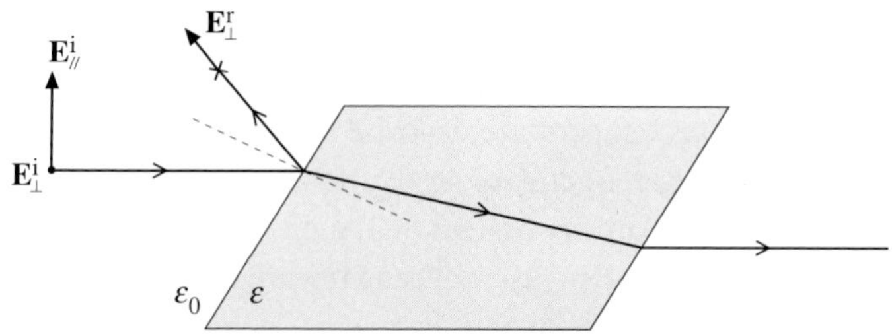

Fig. 18.11 A Brewster polarizer. The incident wave is a mixture of $\|$ and $\perp$ modes, but if the angle of incidence equals the Brewster angle $\theta_{0B}^{\|}$ then the reflected wave is a pure $\perp$ (TE) mode.

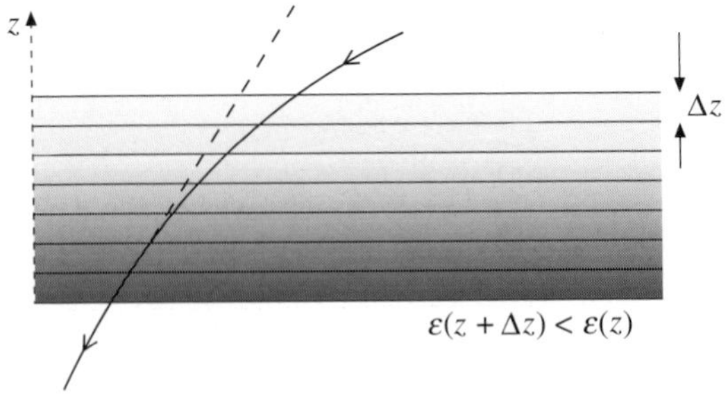

Fig. 18.12 Refraction of light through the atmosphere.

quantity its own name; we define $n(z) \equiv \sqrt{\varepsilon(z)/\varepsilon_0}$ as the *index of refraction*, or *refractive index*. In the limiting case, we obtain

$$\frac{d}{dz}[n(z)\sin\theta(z)] = 0 \tag{18.54}$$

The ray bends toward the normal at each interface. The result is a *curvature of the ray* towards the region of greater permittivity, as shown. In the atmosphere one finds the following dependence of the refractive index n, and therefore of its change with altitude δn, upon temperature T (K) and pressure p (mbar):

$$n(p, T) = 1 + \gamma\frac{p}{T} \quad \longrightarrow \quad \delta n = \gamma\left(\frac{1}{T}\delta p - \frac{p}{T^2}\delta T\right) \tag{18.55}$$

where γ is a constant of order 10^{-4}. The decrease in pressure usually dominates the decrease in temperature with increase in altitude close to the Earth's surface, and $n(z)$ decreases initially with altitude. Thus a star will appear to be somewhat higher than it really is (see the broken line in Fig. 18.12).

Example 18.4 The mirage (*fata morgana*)

Everyone has experienced the hot-summer's-day experience of driving on a long stretch of straight road and seeing what appears to be water or sky in the distance far ahead, along the road. The effect is explained by Fig. 18.13. The heated surface of the road heats the air, and a strong temperature decrease exists in the first 3–6 meters of air above the ground. This temperature decrease is strong enough to counter the pressure decrease, and so the refractive index $n(z)$ will initially *increase* with altitude for several meters or more. Any ray of light (and light is an electromagnetic wave) coming in at an angle to the gradient in n will bend towards larger n; in this case *away* from the normal when propagating with a downwards component. The driver sees, as if reflected from water, light that originated above the road from some point in the sky! This *mirage* or *fata morgana* is well-known to desert travelers who report seeing illusory lakes in the distance.

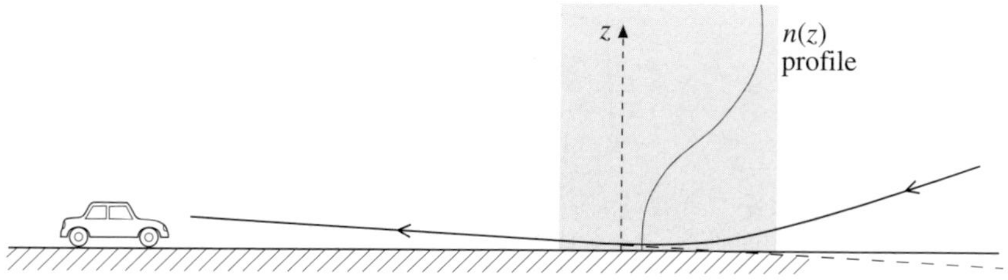

Fig. 18.13 A mirage (or *fata morgana* phenomenon) on a road.

Example 18.5 Super-refractivity

It has been found that the density of the atmosphere, and its gradient, on a planet such as Venus, is so great that the curvature of light rays can exceed the curvature of the planet's surface considerably. The phenomenon known as *super-refractivity* occurs. An observer on the surface of the planet (see Fig. 18.14(a)) sees more distant points on the planet's surface at slightly higher angles to the horizontal, owing to the strong curvature (with respect to that of the planet's surface) of the light reaching the observer's eye from those points. (In practice this effect would not really be observable on Venus due to the high absorption of light by its atmosphere.) The observer would have the sensation, depicted in Fig. 18.14(b), of standing at the bottom of a paraboloid-like depression, rather than on an apparently flat surface. Note that points B, C, and D are actually below the observer's horizon.

18.6 Optical filters; dielectric stacks

An *optical filter* can be constructed of a stack of thin layers of dielectric, each layer having its own uniform permittivity. A varying permittivity profile across the thickness of the stack can be constructed accordingly to have desirable properties, e.g. to filter out some of the frequencies in light with a broader frequency band. A stack of $N+1$ layers provides us with an example of a more general approach to solving for the electric (and magnetic) fields in a medium of varying permittivity. This approach consists of defining transmitted and reflected waves from layer n $(0 < n \leq N)$ to the next layer. A cascade of such definitions lets us move from the incident medium to the final medium and yields expressions for the waves reflected from and transmitted through the stack. Consider the stack shown in Fig. 18.15.

Medium 0 and medium $N+1$ are free space with permittivity ε_0, and medium n $(1 \leq n \leq N)$ is a lossless dielectric with permittivity ε_n. If we assume that a source at a distant $z < 0$ ($z = 0$ is the location of the first interface, between the ε_0 and ε_1 media) sends a time-harmonic plane wave in the $\hat{\mathbf{z}}$ direction, then we can define for

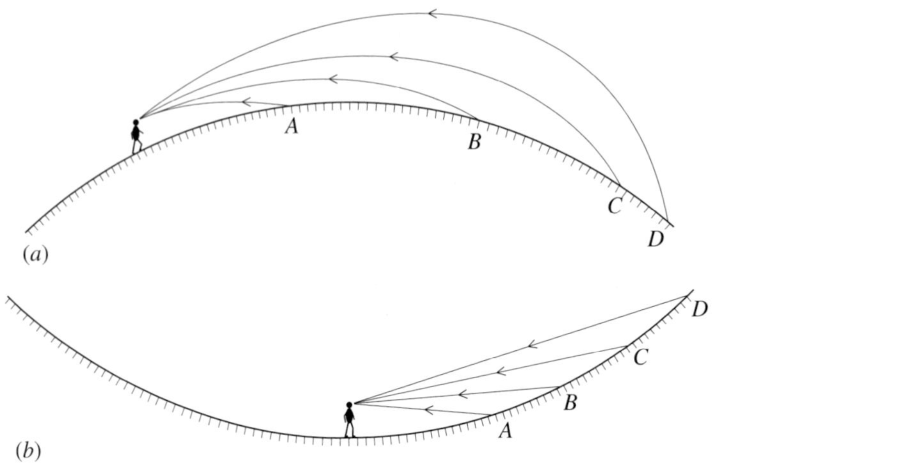

Fig. 18.14 (*a*) Super-refractivity on Venus; (*b*) the apparent shape of the surface to an observer.

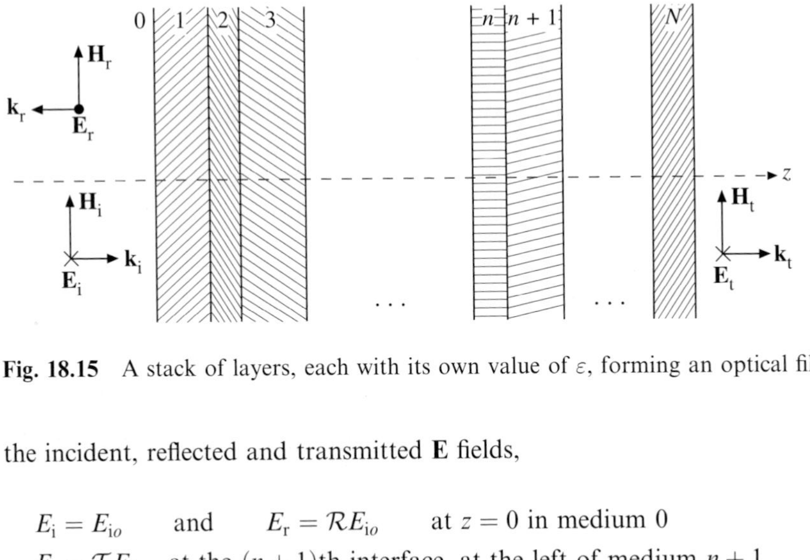

Fig. 18.15 A stack of layers, each with its own value of ε, forming an optical filter.

the incident, reflected and transmitted **E** fields,

$$\begin{aligned} &E_{\rm i} = E_{\rm io} \quad \text{and} \quad E_{\rm r} = \mathcal{R}E_{\rm io} \quad \text{at } z = 0 \text{ in medium } 0 \\ &E_{\rm t} = \mathcal{T}E_{\rm io} \quad \text{at the } (n+1)\text{th interface, at the left of medium } n+1 \end{aligned} \tag{18.56}$$

Fields $E_{\rm i}$ and $E_{\rm t}$ propagate in the $+\hat{\mathbf{z}}$ direction and $E_{\rm r}$ in the $-\hat{\mathbf{z}}$ direction. There cannot be a field in medium $N+1$ that propagates in the $-\hat{\mathbf{z}}$ direction as there is no further interface giving rise to the needed reflection. Vectors are not needed, as we will assume that all **E** fields are polarized in the $-\hat{\mathbf{y}}$ direction and the negative sign of a coefficient will then tell us when we have $+\hat{\mathbf{y}}$ instead. We *postulate* the situation shown in Fig. 18.16 for medium n for a solution and will show that the boundary conditions suffice to obtain a solution for $\mathcal{R}$ and $\mathcal{T}$.

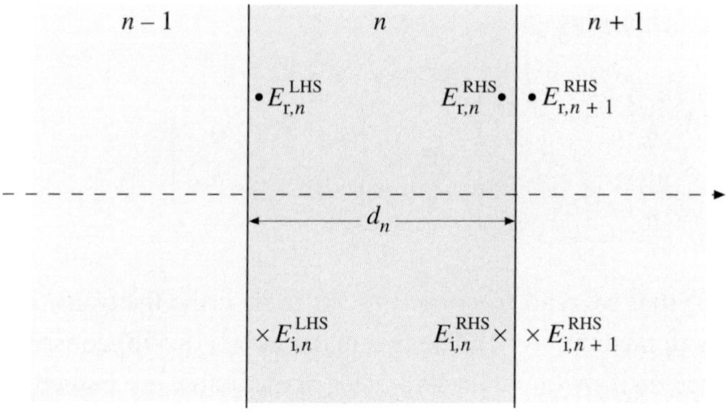

Fig. 18.16 The **E** field situation in one layer.

At the right-hand interface in medium n we have the tangential-field boundary conditions

$$
\begin{aligned}
&E_{\mathrm{i},n}^{\mathrm{RHS}} + E_{\mathrm{r},n}^{\mathrm{RHS}} = E_{\mathrm{i},n+1}^{\mathrm{LHS}} + E_{\mathrm{r},n+1}^{\mathrm{LHS}} \\
&\frac{1}{\eta_n}\left(E_{\mathrm{i},n}^{\mathrm{RHS}} - E_{\mathrm{r},n}^{\mathrm{RHS}}\right) = \frac{1}{\eta_{n+1}}\left(E_{\mathrm{i},n+1}^{\mathrm{LHS}} - E_{\mathrm{r},n+1}^{\mathrm{LHS}}\right)
\end{aligned}
\tag{18.57}
$$

where η_n is the intrinsic impedance of medium n; the second line describes the tangential H fields. Of course, all fields are and remain tangential at normal incidence. We can consider (18.57) as two linear equations in two unknown fields $E_{i,n+1}^{\mathrm{LHS}}$ and $E_{r,n+1}^{\mathrm{LHS}}$, which we relate to the fields just to the left of the interface as follows:

$$
\begin{bmatrix} E_{\mathrm{i},n+1}^{\mathrm{LHS}} \\ E_{\mathrm{r},n+1}^{\mathrm{LHS}} \end{bmatrix} = \frac{1}{2}\begin{bmatrix} 1+\dfrac{\eta_{n+1}}{\eta_n} & 1-\dfrac{\eta_{n+1}}{\eta_n} \\ 1-\dfrac{\eta_{n+1}}{\eta_n} & 1+\dfrac{\eta_{n+1}}{\eta_n} \end{bmatrix}\begin{bmatrix} E_{\mathrm{i},n}^{\mathrm{RHS}} \\ E_{\mathrm{r},n}^{\mathrm{RHS}} \end{bmatrix}
\tag{18.58a}
$$

The relationship between the fields at the right and left interfaces of medium n must be

$$
\begin{bmatrix} E_{\mathrm{i},n}^{\mathrm{RHS}} \\ E_{\mathrm{r},n}^{\mathrm{RHS}} \end{bmatrix} = \begin{bmatrix} e^{-j\beta_n d_n} & 0 \\ 0 & e^{j\beta_n d_n} \end{bmatrix}\begin{bmatrix} E_{\mathrm{i},n}^{\mathrm{LHS}} \\ E_{\mathrm{r},n}^{\mathrm{LHS}} \end{bmatrix}
\tag{18.58b}
$$

because these fields propagate over the length d_n of the uniform layer, with propagation constant $\beta_n = \omega\sqrt{\mu_n\varepsilon_n}$. Thus the fields immediately to the left of medium $n+1$ can be related to the field immediately to the left of medium n by the matrix-multiplication relationship

$$
\begin{bmatrix} E_{\mathrm{i},n+1}^{\mathrm{LHS}} \\ E_{\mathrm{r},n+1}^{\mathrm{LHS}} \end{bmatrix} = \mathbf{T}_n^{n+1}\mathbf{P}_n\begin{bmatrix} E_{\mathrm{i},n}^{\mathrm{LHS}} \\ E_{\mathrm{r},n}^{\mathrm{LHS}} \end{bmatrix}
\tag{18.59a}
$$

with the matrix definitions

$$\mathbf{T}_n^{n+1} \equiv \frac{1}{2}\begin{bmatrix} 1+\dfrac{\eta_{n+1}}{\eta_n} & 1-\dfrac{\eta_{n+1}}{\eta_n} \\ 1-\dfrac{\eta_{n+1}}{\eta_n} & 1+\dfrac{\eta_{n+1}}{\eta_n} \end{bmatrix} \qquad \mathbf{P}_n \equiv \begin{bmatrix} e^{-j\beta_n d_n} & 0 \\ 0 & e^{j\beta_n d_n} \end{bmatrix} \tag{18.59b}$$

Note in (18.59a) that we read *from right to left* to describe the transition of fields from medium n to medium $n+1$. The two matrices of (18.59b) consist entirely of known quantities, so if we know the two fields in medium n, we immediately know those in medium $n+1$. This will be exploited most fully in solving for $\mathcal{R}$ and $\mathcal{T}$. Thus, the solution for the **E** field transmitted through the entire stack of N layers would be

$$\begin{bmatrix} E_\mathrm{t} \\ 0 \end{bmatrix} = \mathbf{T}_N^{N+1}\mathbf{P}_N\mathbf{T}_{N-1}^{N}\mathbf{P}_{N-1}\cdots\mathbf{T}_n^{n+1}\mathbf{P}_n\cdots\mathbf{T}_1^2\mathbf{P}_1\mathbf{T}_0^1\begin{bmatrix} E_\mathrm{i} \\ E_\mathrm{r} \end{bmatrix} \tag{18.60a}$$

and with the substitutions of (18.56), this becomes,

$$\begin{bmatrix} \mathcal{T} \\ 0 \end{bmatrix} = \mathbf{T}_N^{N+1}\mathbf{P}_N\mathbf{T}_{N-1}^{N}\mathbf{P}_{N-1}\cdots\mathbf{T}_n^{n+1}\mathbf{P}_n\cdots\mathbf{T}_1^2\mathbf{P}_1\mathbf{T}_0^1\begin{bmatrix} 1 \\ \mathcal{R} \end{bmatrix} \tag{18.60b}$$

This is the final result, and as an example, we apply it to the following one-layer problem.

Example 18.6 A single dielectric layer between two dielectric half-spaces

We calculate the reflection from and transmission through a single layer of thickness d and permittivity ε_1, from (18.60b) with $N=1$, and $\beta_1 = \omega\sqrt{\mu_0\varepsilon_1} = k_0\sqrt{\varepsilon_{1\mathrm{r}}}$ with $k_0 \equiv \omega/c$. This layer we presume to be sandwiched between a dielectric with ε_0 in $z<0$ (not necessarily free space), and a dielectric with ε_2 in $z>d$. We assume no losses in any of the three media. Thus we have

$$\begin{bmatrix} \mathcal{T} \\ 0 \end{bmatrix} = \mathbf{T}_1^2\mathbf{P}_1\mathbf{T}_0^1\begin{bmatrix} 1 \\ \mathcal{R} \end{bmatrix} \equiv \mathbf{M}\begin{bmatrix} 1 \\ \mathcal{R} \end{bmatrix} \tag{18.61}$$

So we must compute the following matrix product:

$$\mathbf{M} = \frac{1}{4}\begin{bmatrix} 1+\dfrac{\eta_2}{\eta_1} & 1-\dfrac{\eta_2}{\eta_1} \\ 1-\dfrac{\eta_2}{\eta_1} & 1+\dfrac{\eta_2}{\eta_1} \end{bmatrix}\begin{bmatrix} e^{-j\beta_1 d_1} & 0 \\ 0 & e^{j\beta_1 d_1} \end{bmatrix}\begin{bmatrix} 1+\dfrac{\eta_1}{\eta_0} & 1-\dfrac{\eta_1}{\eta_0} \\ 1-\dfrac{\eta_1}{\eta_0} & 1+\dfrac{\eta_1}{\eta_0} \end{bmatrix}$$

in (18.61), with the following matrix elements:

$$\begin{aligned}
\mathbf{M}_{11} &= \frac{1}{4}\left\{\left(1+\frac{\eta_2}{\eta_1}\right)\left(1+\frac{\eta_1}{\eta_0}\right)e^{-j\beta_1 d_1} + \left(1-\frac{\eta_2}{\eta_1}\right)\left(1-\frac{\eta_1}{\eta_0}\right)e^{j\beta_1 d_1}\right\} \\
\mathbf{M}_{12} &= \frac{1}{4}\left\{\left(1+\frac{\eta_2}{\eta_1}\right)\left(1-\frac{\eta_1}{\eta_0}\right)e^{-j\beta_1 d_1} + \left(1-\frac{\eta_2}{\eta_1}\right)\left(1+\frac{\eta_1}{\eta_0}\right)e^{j\beta_1 d_1}\right\} \\
\mathbf{M}_{21} &= \frac{1}{4}\left\{\left(1-\frac{\eta_2}{\eta_1}\right)\left(1+\frac{\eta_1}{\eta_0}\right)e^{-j\beta_1 d_1} + \left(1+\frac{\eta_2}{\eta_1}\right)\left(1-\frac{\eta_1}{\eta_0}\right)e^{j\beta_1 d_1}\right\} \\
\mathbf{M}_{22} &= \frac{1}{4}\left\{\left(1-\frac{\eta_2}{\eta_1}\right)\left(1-\frac{\eta_1}{\eta_0}\right)e^{-j\beta_1 d_1} + \left(1+\frac{\eta_2}{\eta_1}\right)\left(1+\frac{\eta_1}{\eta_0}\right)e^{j\beta_1 d_1}\right\}
\end{aligned} \tag{18.62}$$

Then we find from (18.61) that

$$\mathcal{R} = -\mathbf{M}_{21}/\mathbf{M}_{22} \qquad \mathcal{T} = \mathbf{M}_{11} + \mathbf{M}_{12}\mathcal{R}$$

which leads to the following expressions:

$$\mathcal{R} = \frac{\mathcal{R}_{01} + \mathcal{R}_{12}e^{-2j\beta_1 d_1}}{1 + \mathcal{R}_{01}\mathcal{R}_{12}e^{-2j\beta_1 d_1}} \qquad \mathcal{R}_{ij} \equiv \frac{\eta_j - \eta_i}{\eta_j + \eta_i} \tag{18.63a}$$

$$\mathcal{T} = \frac{\mathcal{T}_{01}\mathcal{T}_{12}e^{-2j\beta_1 d_1}}{1 + \mathcal{R}_{01}\mathcal{R}_{12}e^{-2j\beta_1 d_1}} \qquad \mathcal{T}_{ij} \equiv \frac{2\eta_j}{\eta_j + \eta_i} \tag{18.63b}$$

$\mathcal{R}_{ij}$ and $\mathcal{T}_{ij}$ are the reflection and transmission coefficients, respectively, for propagation from a dielectric ε_i through its interface with a dielectric ε_j, as given in (18.24). In (18.63) the i, j pair is, respectively, 0, 1 and 1, 2.

The overall reflection coefficient $\mathcal{R}$ is shown in Fig. 18.17 as a function of d for the case of a single slab of dielectric in air.

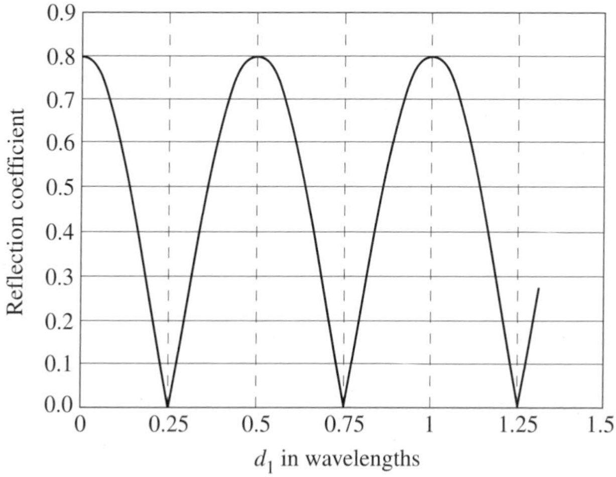

Fig. 18.17 The reflection coefficient $\mathcal{R}$ vs. width, for a single slab of dielectric in air.

If, for example, we wish to have perfect transmission through this intermediate layer for a certain frequency (no lossy media considered here), then we require $|\mathcal{R}|^2 = 0$. It follows from (18.63a) that

$$|\mathcal{R}|^2 = \frac{(\mathcal{R}_{01} + \mathcal{R}_{12})^2 - 4\mathcal{R}_{01}\mathcal{R}_{12}\sin^2 \beta_1 d_1}{(1 + \mathcal{R}_{01}\mathcal{R}_{12})^2 - 4\mathcal{R}_{01}\mathcal{R}_{12}\sin^2 \beta_1 d_1} \tag{18.64}$$

We write this in the following abbreviated fashion:

$$|\mathcal{R}|^2 = \frac{A - B\sin^2\varphi}{C - B\sin^2\varphi} = 1 + \frac{A - C}{C - B\sin^2\varphi}$$

and regard this expression as a function only of φ for fixed values of A, B, and C. Then

$$\frac{d}{d\varphi}\left(|\mathcal{R}|^2\right) = \frac{B(A - C)\sin 2\varphi}{(C - B\sin^2\varphi)^2}$$

This derivative is zero when $\sin 2\varphi = 0$, i.e. when $2\beta_1 d_1 = n\pi$ with $n = 1, 2, 3, \ldots$, and the power reflection coefficient $|\mathcal{R}|^2$ is then at an extreme value. The criterion amounts to the requirement that the thickness d_1 of the layer is a multiple of a *quarter wavelength*:

$$d_1 = n \times \tfrac{1}{4}\lambda_1$$

Then $\sin^2\varphi = 1$ for odd values of n and $\sin^2\varphi = 0$ for even values of n. We find after considerable algebraic manipulation

$$|\mathcal{R}|^2 = \frac{(\eta_2 - \eta_0)^2}{(\eta_2 + \eta_0)^2} \qquad \text{for even } n$$

$$|\mathcal{R}|^2 = \frac{(\eta_1^2 - \eta_2\eta_0)^2}{(\eta_1 + \eta_2\eta_0)^2} \qquad \text{for odd } n$$

For the odd n it can be seen $|\mathcal{R}|^2 = 0$ when the slab medium's intrinsic impedance is the *geometrical mean* of the impedances of the bounding media, $\eta_1 = \sqrt{\eta_0\eta_2}$. For even n, $|\mathcal{R}|^2 = 0$ when the bounding media have the *same* intrinsic impedance, $\eta_2 = \eta_0$.

So, while the power-reflection coefficient is at an extremum for $d_1 = n\lambda/4$, it follows that *perfect transmission* (zero reflected power) occurs whenever either of the following possibilities occurs.

The *first possibility* is to choose

$$\eta_1 = \sqrt{\eta_0\eta_2} \qquad \text{and} \qquad d_1 = \left(n + \frac{1}{2}\right)\lambda_1/2 \qquad \text{with } n = 0, 1, 2, \ldots$$

This possibility is included in the graph shown in Fig. 18.17, where we have chosen $\varepsilon_{0r} = 1$ and $\varepsilon_{2r} = 82$ (close to the permittivity of 'lossless' water at room temperature) and plotted $|\mathcal{R}|$ as a function of d_1/λ_1. Clearly $\mathcal{R} = 0$ at $d_1 = 0.25\lambda_1, 0.75\lambda_1, \ldots$. Note also that the minima are sharper than the maxima of the reflection coefficient; that is a feature we will observe again in transmission lines!

The second possibility requires $\eta_2 = \eta_0$ and $d_1 = n\lambda_1/2$, with $n = 0, 1, 2, \ldots$. Then $\sin(2\pi d_1/\lambda_1) = 0$, given that the wavelength in medium 1 is $\lambda_1 = 2\pi/\beta_1$. This is the case where a dielectric layer is sandwiched inside a large dielectric medium with permittivity ε_0 and its width d_1 is chosen as a multiple of a half-wavelength.

Optical filters can make use of such properties to change the balance of colors in *white light* away from a particular frequency. For the example chosen in Fig. 18.17 (specified above), we see that the reflection coefficient is maximal at half-wavelength widths of the filter, and the transmission is then smallest. Thus, one diminishes the role of a certain frequency by choosing the filter width to be an integer number of half wavelengths at that frequency. The maximal reflection coefficient approaches unity as we increase the value of ε_{2r}. A similar situation occurs when $\varepsilon_2 = \varepsilon_0$, and we allow ε_{1r} to become large. In both cases, a wider frequency band can be removed by using stacks with more than one layer and adjusting the parameters so that, successively, more frequencies are removed. In practice these considerations must be tempered somewhat by including lossiness in the layer, but the principle remains unchanged.

Problems

Unless stated explicitly, assume all permeabilities are μ_0 (the free-space value) in the following problems.

18.1. A TM plane wave with phasor representation $\mathbf{E} = \mathbf{E}_0 e^{-j\beta_0(x\sin\theta + z\cos\theta)}$ is incident at an angle of $\theta = 30°$ to the normal upon a perfectly conducting medium that is separated from free space by an interface $z = 0$. The frequency $f = 3.183$ GHz. The wave sets up a surface current at the interface. Give the expression for this surface current.

18.2. For the situation described in problem 18.1, it is also the case that a sum wave of incident plus reflected waves appears to propagate parallel to the interface.
(a) What is the wavelength associated with this parallel motion?
(b) What is the velocity associated with this motion?
(c) Can this velocity be observed, and if not, why not?
(d) Where can a metal plate be placed so that the wave structure between the two interfaces is unchanged?

18.3. A uniform time-harmonic plane wave with phasor $\mathbf{E}_{\text{ip}} = 10\hat{\mathbf{y}}e^{(6x+8z)}$ in V/m is incident from free space upon a perfect conductor beyond the interface $z = 0$.
(a) Find the frequency f and wavelength λ of the wave.
(b) Write down expressions for the actual fields $\mathbf{E}_{\text{i}}(x, z, t)$ and $\mathbf{H}_{\text{i}}(x, z, t)$.
(c) What is the angle of incidence θ?
(d) Find the phasors $\mathbf{E}_{\text{rp}}$ and $\mathbf{H}_{\text{rp}}$ of the reflected field.
(e) Find the $\mathbf{E}_{\text{p}}$ and $\mathbf{H}_{\text{p}}$ phasors of the *total* fields in $z < 0$.

18.4. A plane wave with phasor representation $\mathbf{E} = \mathbf{E}_0 e^{-\mathrm{i}\beta z}$ is normally incident from free space upon a medium with permittivity ε; the interface is $z = 0$. The transmission coefficient $T = 0.75$.
(a) Find the value of ε_{r}.
(b) Find the value of the reflection coefficient $\mathcal{R}$.

18.5. A TE plane wave with phasor representation $\mathbf{E} = \mathbf{E}_0 e^{-j\beta(x \sin\theta_1 + z\cos\theta_1)}$ is incident from a medium with relative permittivity 1.5 upon one with relative permittivity 3, separated by an interface at $z = 0$, at an angle of $\theta_1 = 30°$ to the normal. The frequency is 1.2 GHz.
(a) Find the angle of refraction θ_2.
(b) What percentage of the incident power is transmitted?

18.6. An underwater swimmer in a swimming pool views the surface of the pool from 2.5 m below it. With refractive index $n = 1.33$, what is the radius of the circle through which he can observe the world above the surface?

18.7. The Brewster angle for a TM wave is given by $\sin\theta_{1\text{B}} = \sqrt{\varepsilon_2/(\varepsilon_1 + \varepsilon_2)}$. Is there always a Brewster angle when $\varepsilon_2 < \varepsilon_1$, or is there a critical angle beyond which there is no Brewster angle?

18.8. An interface $z = 0$ separates two uniform dielectrics with $\varepsilon_1 = 3.5$ in $z < 0$, ε_2 in $z > 0$. There is no reflected TM wave when the angle of incidence is 40°. What is the angle of refraction?

18.9. A TE wave is incident upon a uniform dielectric medium beyond $z = 0$ at the Brewster angle $\theta_{1\text{B}}$ of a TM (!) wave. Assume permittivities ε_1 and ε_2 for the two uniform media. What is the angle between the reflected and the transmitted wave?

18.10. A plane wave is incident at an angle θ to the x axis upon a slab of glass ($\varepsilon_{\text{r}} = 3.2$) between $x = 0$ and $x = d$. Upon exit from the slab beyond $x = 0$ it is displaced parallel to itself by a perpendicular distance δ. Express δ in terms of d and θ.

18.11. The atmosphere normally has a refractive index dependent upon altitude z that can be approximated by $n(z) = 1 + Ne^{-z/h}$, where $h \approx 8$ km, $N \approx 2.6 \times 10^{-4}$. A star is seen at an angle of 40° with the horizon. What is the angular error due to refractive bending of light from the star?

18.12. An optical fiber consists of a circular cylinder of radius a filled with a uniform dielectric with refractive index n_1. Around this cylinder is a sheath, of thickness b, of a material with refractive index $n_2 < n_1$. The axis of the fiber coincides with the z axis, and the fiber starts at $z = 0$.

(a) What is the maximum angle of incidence θ_0 such that rays entering the core at $z = 0$ are trapped in the core without losing energy to the sheath?

(b) The *numerical aperture* $NA = \sin\theta_0$. Find NA and θ_0, given that $n_1 = 2.10$ and $n_2 = 1.89$.

18.13. Consider an interface $z = 0$ between free space in $z < 0$ and a lossy dielectric with $\varepsilon = \varepsilon_r - j\varepsilon_i$ in $z > 0$. A wave is incident upon the interface from $z < 0$ at an angle θ to the normal, in the $y = 0$ plane.

(a) Describe the planes of constant phase, i.e. find the *real* angle of refraction (see the end of subsection 18.4.1).

(b) Describe the planes of constant amplitude.

18.14. Let a uniform dielectric with $n = 3.2$ be in $z < 0$, and free space in $z > 0$. A time-harmonic plane wave with $\lambda = 2\pi$ m in this medium is incident from $z < 0$ at 45° to the normal, in the $y = 0$ plane. Find the ratio $E_t(x, z)/E_t(x, 0)$ for the transmitted wave.

18.15. A time-harmonic plane wave is normally incident upon a half-space of dielectric material, and the standing wave on the incident side has the power flux

$$S(z) = S_0[(1 - \mathcal{R})^2 + 4\mathcal{R}\cos^2\beta z]$$

where $\mathcal{R}$ is the reflection coefficient, β is the wavenumber, and S_0 is a constant power flux. A standing-wave amplitude $SWR = 3$ is measured.

(a) Use this information to deduce $|\mathcal{R}|$.

(b) How can one tell from $S(z)$ whether $\mathcal{R}$ is positive or negative?

(c) How can the wavelength λ be deduced from the $S(z)$ versus z graph?

18.16. A plane wave propagates through 174 layers of material. Slab layer m has $\mu_m = \mu_0$, $\varepsilon_m = (1 + 0.02m)\varepsilon_0$. The angle of incidence is $\theta_0 = 12°$. What is the angle of transmission through layer 174?

18.17. Pure water has refractive index $n = 1.33$ with negligible losses at visible wavelengths. Design a filter (i.e. find the thickness of the filter and ε' for the material) such that under water you can see light from above the water without reflection losses. Use $\lambda = 0.6\,\mu$m.

19

Waveguides

Chapters 17 and 18 have dealt with the propagation of waves – mostly time-harmonic waves – through unbounded media. It is true that we have also included the reflection of plane waves from and their transmission through layers of different materials; these layers may have been bounded in the $\hat{\mathbf{z}}$ direction, but such media nevertheless have been unbounded in the transverse directions. The underlying approach so far has been to deal with waves emanating from specific sources and to tailor our media so as to be able to deal with situations easy enough to analyze. Thus we have placed slab-like layers perpendicular to plane waves, and spherically symmetric media around point sources (producing spherical waves). We also have dealt with a few cases in which plane waves are obliquely incident upon infinite plane interfaces with a second medium. In all these cases, the media have been unbounded in the x and y directions.

We consider a different type of problem here. It concerns the guiding of waves in specific directions, rather than the following of waves produced by a source to wherever they may propagate. We speak of 'guiding' because the extent of the waves in directions perpendicular to the direction of propagation is now entirely or partially bounded by a *waveguide*. Waveguides are hollow devices relatively long compared to their transverse dimensions, and they are bounded by metals or dielectrics, inside which *wave modes* can be excited. Such wave modes can be made to follow the longitudinal or main direction of the waveguide. *Coaxial cables* and *optical fibers* are examples of cylindrical waveguides through which communication signals are guided from one place to another, and the *rectangular waveguide* is a common feature in antennas for guiding a signal from horn to feed (or vice versa). Some waveguides (e.g. certain transmission lines, microstrip lines) are not entirely closed; they may be partly open to the outside. The common feature of waveguides is the existence of a preferred direction. *Transmission lines* are special types of waveguide to which we will devote a separate chapter.

19.1 The waveguide equations

We will restrict ourselves to the special case where the waveguide is straight and uniform, i.e. its longitudinal axis is the z axis, and its transverse dimensions are the same at all z. The interior of the waveguide contains no sources of radiation and it is a medium with uniform ε and μ. Let us see how the Maxwell equations can be transformed to a useful form. Consider the Maxwell equations for time-harmonic waves at angular frequency ω in such a medium,

$$\begin{aligned} &\nabla \times \mathbf{E} = -j\mu\omega\mathbf{H} \qquad &&\nabla \times \mathbf{H} = j\varepsilon\omega\mathbf{E} \\ &\nabla \cdot \mathbf{E} = 0 &&\nabla \cdot \mathbf{H} = 0 \end{aligned} \tag{19.1}$$

The physical assumptions we have made, and our desire to examine waves that travel in the $+\hat{\mathbf{z}}$ direction, amount to the mathematical assumption that the field phasors $\mathbf{E}(\mathbf{r})$, $\mathbf{H}(\mathbf{r})$ are as follows (we now omit the subscript 'p' used earlier):

$$\mathbf{E}(\mathbf{r}) = E(x, y)e^{-jk_z z} \qquad \mathbf{H}(\mathbf{r}) = H(x, y)e^{-jk_z z} \tag{19.2a}$$

Note that the exponential has a wavenumber k_z, to be carefully distinguished from $k = \omega\sqrt{\mu\varepsilon}$. Thus, both fields are sinusoidal in the z direction with amplitudes that are constant along the z axis. The exponential has been chosen to correspond to waves moving in the $+\hat{\mathbf{z}}$ direction. The opposite sign in the exponent is needed for waves moving in the $-\hat{\mathbf{z}}$ direction, and two terms for each field are needed for a sum of waves in both directions. We assume a source at $z \to -\infty$, and thus will deal at this time only with waves propagating in the $+\hat{\mathbf{z}}$ direction.

Comment: Although we have not done so here, many texts follow the microwave-engineer custom of replacing (19.2a) by

$$\mathbf{E}(\mathbf{r}) = E(x, y)e^{-\gamma z} \qquad \mathbf{H}(\mathbf{r}) = H(x, y)e^{-\gamma z} \tag{19.2b}$$

so that $\gamma = jk_z = \alpha + j\beta$ is the complex propagation constant, of which the imaginary part describes propagation and the real part attenuation.

With this we find that

$$\begin{aligned} \nabla \times \mathbf{E} &= \left[\left(\frac{\partial \mathcal{E}_z}{\partial y} + jk_z\mathcal{E}_y\right)\hat{\mathbf{x}} - \left(jk_z\mathcal{E}_x + \frac{\partial \mathcal{E}_z}{\partial x}\right)\hat{\mathbf{y}} + \left(\frac{\partial \mathcal{E}_y}{\partial x} - \frac{\partial \mathcal{E}_x}{\partial y}\right)\hat{\mathbf{z}}\right]e^{-jk_z z} \\ \nabla \times \mathbf{H} &= \left[\left(\frac{\partial \mathcal{H}_z}{\partial y} + jk_z\mathcal{H}_y\right)\hat{\mathbf{x}} - \left(jk_z\mathcal{H}_x + \frac{\partial \mathcal{H}_z}{\partial x}\right)\hat{\mathbf{y}} + \left(\frac{\partial \mathcal{H}_y}{\partial x} - \frac{\partial \mathcal{H}_x}{\partial y}\right)\hat{\mathbf{z}}\right]e^{-jk_z z} \end{aligned} \tag{19.3}$$

Six equations follow from (19.2a), when (19.3) is used for the left-hand sides of the curl equations:

$$\begin{aligned}
&\frac{\partial \mathcal{E}_z}{\partial y} + jk_z\mathcal{E}_y = -j\mu\omega\mathcal{H}_x && \frac{\partial \mathcal{H}_z}{\partial y} + jk_z\mathcal{H}_y = j\varepsilon\omega\mathcal{E}_x \\
&\frac{\partial \mathcal{E}_z}{\partial x} + jk_z\mathcal{E}_x = j\mu\omega\mathcal{H}_y && \frac{\partial \mathcal{H}_z}{\partial x} + jk_z\mathcal{H}_x = -j\varepsilon\omega\mathcal{E}_y \\
&\frac{\partial \mathcal{E}_y}{\partial x} - \frac{\partial \mathcal{E}_x}{\partial y} = -j\mu\omega\mathcal{H}_z && \frac{\partial \mathcal{H}_y}{\partial x} - \frac{\partial \mathcal{H}_x}{\partial y} = j\varepsilon\omega\mathcal{E}_z
\end{aligned} \tag{19.4}$$

Using (19.1) and (19.2a), take an extra curl of one of the curl equations of (19.1) and apply $\nabla \times \nabla \times \mathbf{A} = \nabla(\nabla \cdot \mathbf{A}) - \Delta\mathbf{A} = -\Delta\mathbf{A}$, when $\mathbf{A}$ is either $\mathbf{E}$ or $\mathbf{H}$ in a region without source fields. Then, eliminating the other field with the remaining curl equation, it is straightforward to show that

$$\Delta_t E + (k^2 - k_z^2)E = 0 \qquad \Delta_t H + (k^2 - k_z^2)H = 0 \tag{19.5}$$

given that the transverse Laplacian operator Δ_t is $\partial^2/\partial x^2 + \partial^2/\partial y^2$ in Cartesian coordinates and $(1/\rho)(\partial/\partial\rho)\rho(\partial/\partial\rho) + (1/\rho^2)(\partial^2/\partial\varphi^2)$ in cylindrical coordinates. In (19.4) and (19.5), we have 12 equations in only six variables, clearly twice as many as we need. Eliminate $\mathcal{H}_x$ from the first and fourth of Eqs. (19.4) as follows:

$$\frac{\partial \mathcal{E}_z}{\partial y} + jk_z\mathcal{E}_y = -j\mu\omega\mathcal{H}_x$$

$$\frac{\partial \mathcal{H}_z}{\partial x} + j\varepsilon\omega\mathcal{E}_y = -jk_z\mathcal{H}_x$$

Subtract the second equation, after multiplication by $\mu\omega$, from the first, multiplied by k_z, to obtain

$$(k^2 - k_z^2)\mathcal{E}_y = -j\left(k_z\frac{\partial \mathcal{E}_z}{\partial y} + \frac{\partial \mathcal{H}_z}{\partial x}\right)$$

With three more such pair-wise combinations from the top four equations of (19.4), we obtain four equations expressing each of the four transverse field components in derivatives of the two longitudinal field components:

$$\begin{aligned}
(k^2 - k_z^2)\mathcal{E}_x &= -j\left(k_z\frac{\partial \mathcal{E}_z}{\partial x} + \omega\mu\frac{\partial \mathcal{H}_z}{\partial y}\right) \\
(k^2 - k_z^2)\mathcal{E}_y &= -j\left(k_z\frac{\partial \mathcal{E}_z}{\partial y} - \omega\mu\frac{\partial \mathcal{H}_z}{\partial x}\right) \\
(k^2 - k_z^2)\mathcal{H}_x &= j\left(\omega\varepsilon\frac{\partial \mathcal{E}_z}{\partial y} - k_z\frac{\partial \mathcal{H}_z}{\partial x}\right) \\
(k^2 - k_z^2)\mathcal{H}_y &= -j\left(\omega\varepsilon\frac{\partial \mathcal{E}_z}{\partial x} + k_z\frac{\partial \mathcal{H}_z}{\partial y}\right)
\end{aligned} \tag{19.6a}$$

So we need only two more equations for the fields on the right-hand side of these expressions, and these are obtained from (19.5):

$$\Delta_t \mathcal{E}_z + (k^2 - k_z^2)\mathcal{E}_z = 0 \qquad \Delta_t \mathcal{H}_z + (k^2 - k_z^2)\mathcal{H}_z = 0 \tag{19.6b}$$

If these two partial differential wave equations can be solved (with appropriate boundary conditions) to yield $\mathcal{E}_z$ and $\mathcal{H}_z$, then by substitution into the right-hand sides of (19.6a) we obtain purely algebraic equations that will yield the four fields $\mathcal{E}_x$, $\mathcal{E}_y$, $\mathcal{H}_x$, and $\mathcal{H}_y$. The six equations (19.6) are known as the *waveguide equations*, and here they are given in Cartesian coordinates. It is easy to give (19.6a) in cylindrical coordinates because we can interpret the denominators ∂x and ∂y as $h_1 \partial u_1$ and $h_2 \partial u_2$, in terms of the generalized orthogonal coordinates u_1, u_2 in two dimensions for the x and y component equations of (19.4) (this interpretation would be wrong for the z components but these are not needed here). Strictly speaking, we should reformulate (19.3) for cylindrical coordinates and then reformulate (19.4) from that, but the above shortcut suffices. Whichever procedure is followed, the transverse fields are then expressed as functions of the longitudinal fields in cylindrical coordinates as follows:

$$\begin{aligned}
(k^2 - k_z^2)\mathcal{E}_\rho &= -j\left(k_z \frac{\partial \mathcal{E}_z}{\partial \rho} + \omega\mu \frac{\partial \mathcal{H}_z}{\rho \partial \varphi}\right) \\
(k^2 - k_z^2)\mathcal{E}_\varphi &= -j\left(k_z \frac{\partial \mathcal{E}_z}{\rho \partial \varphi} - \omega\mu \frac{\partial \mathcal{H}_z}{\partial \rho}\right) \\
(k^2 - k_z^2)\mathcal{H}_\rho &= j\left(\omega\varepsilon \frac{\partial \mathcal{E}_z}{\rho \partial \varphi} - k_z \frac{\partial \mathcal{H}_z}{\partial \rho}\right) \\
(k^2 - k_z^2)\mathcal{H}_\varphi &= -j\left(\omega\varepsilon \frac{\partial \mathcal{E}_z}{\partial \rho} + k_z \frac{\partial \mathcal{H}_z}{\rho \partial \varphi}\right)
\end{aligned} \tag{19.7}$$

The longitudinal fields $\mathcal{E}_z$ and $\mathcal{H}_z$ are still determined by (19.5) and appropriate boundary conditions, and (19.7) then yields the four transverse fields $\mathcal{E}_\rho$, $\mathcal{H}_\rho$, $\mathcal{E}_\varphi$, $\mathcal{H}_\varphi$ as functions of the longitudinal fields. Equations (19.4) can be cast into the more general form

$$\begin{aligned}
\nabla_t \mathcal{E}_z + jk_z E_t &= -j\mu\omega \hat{\mathbf{z}} \times H_t \qquad & \hat{\mathbf{z}} \cdot (\nabla_t \times E_t) &= -j\mu\omega \mathcal{H}_z \\
\nabla_t \mathcal{H}_z + jk_z H_t &= j\varepsilon\omega \hat{\mathbf{z}} \times E_t \qquad & \hat{\mathbf{z}} \cdot (\nabla_t \times H_t) &= j\varepsilon\omega \mathcal{E}_z
\end{aligned} \tag{19.8a}$$

where $\nabla_t \equiv \hat{\mathbf{x}}\partial/\partial x + \hat{\mathbf{y}}\partial/\partial y$ and the equations following from (19.6b) remain:

$$\Delta_t \mathcal{E}_z + (k^2 - k_z^2)\mathcal{E}_z = 0 \qquad \Delta_t \mathcal{H}_z + (k^2 - k_z^2)\mathcal{H}_z = 0 \tag{19.8b}$$

These most general equations contain the Cartesian, cylindrical, and some other orthogonal-coordinate system forms. However, one will find that there may be still some degrees of freedom left in solving (19.8), and these are removed by using the two divergence equations, $\nabla \cdot \mathbf{E} = 0$ and $\nabla \cdot \mathbf{H} = 0$ (in uniform source-free regions).

19.2 Classification of modes for two-dimensional waveguides

Equations (19.6) or (19.7) allow us to classify a number of important categories of waveguide mode. These follow from the possibilities that either $\mathcal{E}_z$ or $\mathcal{H}_z$ is zero or both fields are zero. *Transverse electromagnetic* (TEM) modes follow when both longitudinal fields are zero (the name is obvious). *Transverse magnetic* (TM) modes are the result when $\mathcal{H}_z = 0$, and *transverse electric* (TE) *modes* result when $\mathcal{E}_z = 0$.

19.2.1 TEM modes

If both $\mathcal{E}_z = 0$ and $\mathcal{H}_z = 0$, then (19.6a) can be satisfied only if $k_z = k$. The wave-number of TEM modes is identical to that of free space: $k_z = \omega\sqrt{\mu\varepsilon}$. We must go to (19.8a) for the transverse fields; the equations on the left then yield

$$\mathcal{H}_\mathrm{t} = \frac{1}{Z}\hat{\mathbf{z}} \times \mathcal{E}_\mathrm{t} \tag{19.9}$$

Here, the *wave impedance* $Z = \sqrt{\mu/\varepsilon} = \mu$ is the same as the intrinsic impedance in the unbounded medium. Of course, E_t is the transverse electric field and H_t is the transverse magnetic field, regardless of which variables (Cartesian, cylindrical, etc.) are used in the two-dimensional transverse space.

The TEM modes cannot exist inside a single-conductor waveguide, e.g. a hollow rectangular or cylindrical waveguide. The reason has to do with the integral form of the curl **H** Maxwell equation,

$$\oint_C d\mathbf{l} \cdot \mathbf{H} = \int_S d\mathbf{S} \cdot \left(\mathbf{J} + \frac{\partial \mathbf{D}}{\partial t}\right) = 0 \tag{19.10}$$

When applied to any closed contour C lying $\perp$ to the $\hat{\mathbf{z}}$ direction totally inside the waveguide, inside which we have $\mathbf{J} = 0$ (no internal conductor), and also $d\mathbf{S} \cdot \partial\mathbf{D}/\partial z = dS\partial D_z/\partial z = 0$ (because there are supposed to be no longitudinal fields in a TEM model!). Thus the right-hand side of (19.10) is zero. Let the chosen contour correspond entirely to a magnetic field line. Then **H** is parallel to $d\mathbf{l}$ at all points of the closed field line and therefore the integral cannot be zero. This contradiction rules out the existence of a TEM mode.

TEM modes, which can exist in waveguides with cores or with more than one conducting guide wall, behave in all other respects as unbounded modes with the same phase velocity $v_\mathrm{ph} = 1/\sqrt{\mu\varepsilon}$ as in a uniform waveguide.

Those readers who only wish to learn as much as they need to understand transmission lines can skip ahead to Section 19.3, read that, and then skip the remaining part of this chapter and go directly to Chapter 20.

19.2.2 TE modes

The transverse electric modes follow by assuming that of the longitudinal fields only $\mathcal{E}_z$ is zero. From (19.8) we now obtain

$$\begin{aligned} jk_z E_t &= -j\mu\omega\hat{\mathbf{z}} \times H_t \\ \nabla_t \mathcal{H}_z + jk_z H_t &= j\varepsilon\omega\hat{\mathbf{z}} \times E_t \end{aligned} \tag{19.11}$$

and it is not difficult upon eliminating E_t from the second of these to show that

$$E_t = -Z_{TE}\mathbf{z} \times H_t \qquad H_t = -\frac{jk_z}{k^2 - k_z^2}\nabla_t \mathcal{H}_z \tag{19.12}$$

where $Z_{TE} \equiv \mu\omega/k_z$. It is necessary to solve the second of (19.8b) under the appropriate boundary conditions first to obtain $\mathcal{H}_z$; (19.12) will then provide the remaining fields.

19.2.3 TM modes

The transverse magnetic modes follow by assuming that of the longitudinal fields only $\mathcal{H}_z$ is zero. From (19.8) we now obtain

$$\begin{aligned} jk_z H_t &= j\varepsilon\omega\hat{\mathbf{z}} \times E_t \\ \nabla_t \mathcal{E}_z + jk_z E_t &= -j\mu\omega\hat{\mathbf{z}} \times H_t \end{aligned} \tag{19.13}$$

Similarly to the TE case, we now eliminate $\mathcal{H}_t$ from the second of these to obtain

$$H_t = \frac{1}{Z_{TM}}\mathbf{z} \times E_t \qquad E_t = -\frac{jk_z}{k^2 - k_z^2}\nabla_t \mathcal{E}_z \tag{19.14}$$

where $Z_{TM} \equiv k_z/\varepsilon\omega$. We start by solving the second of (19.8b) under the appropriate boundary conditions to obtain $\mathcal{E}_z$; (19.14) will then provide the remaining fields.

An interesting result for the product of the wave impedances is

$$Z_{TE}Z_{TM} = \eta^2 \tag{19.15}$$

Thus, the product of the TE times the TM impedance is just the square of the intrinsic impedance of the medium.

19.2.4 Cutoff frequencies, modes as bouncing waves

Both for the TE and the TM modes, we have found that the longitudinal component is a solution of the equation

$$\Delta_t \psi + (k^2 - k_z^2)\psi = 0 \tag{19.16}$$

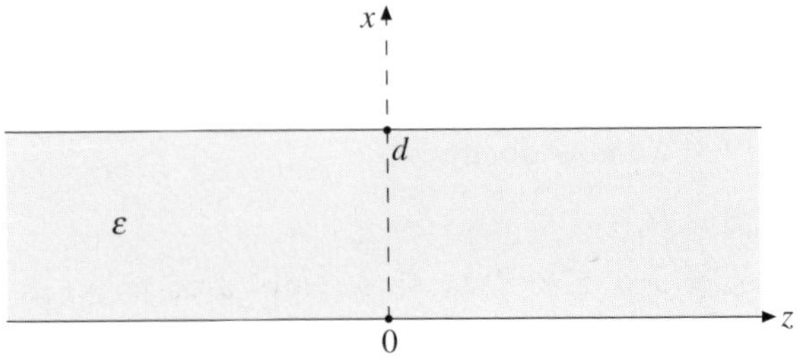

Fig. 19.1 A waveguide consisting of two long parallel plates.

where ψ stands for one of the two longitudinal fields: $\psi = \mathcal{E}_z$ for the TM mode and $\psi = \mathcal{H}_z$ for the TE mode. In this section we wish only to illustrate an important point about the TE and TM modes and we do not intend to give a complete coverage of the many types of mode that can exist, for the variety is a function of the various possible shapes of the waveguide.

Hence, consider a two-dimensional waveguide consisting of an infinitely wide conducting plate at $x = 0$ and another one at $x = d$, as in Fig. 19.1. Then (19.16) will become

$$\frac{d^2\psi}{dx^2} + K^2\psi = 0 \qquad \text{with} \qquad K^2 \equiv k^2 - k_z^2 \tag{19.17}$$

because there is no dependence upon y for so simple a waveguide. The solution to (19.17) must be

$$\psi(x) = A \sin Kx + B \cos Kx \tag{19.18}$$

We should remind ourselves here that $\psi(x)$ stands for either $\mathcal{E}_z(x)$ or $\mathcal{H}_z(x)$. Equations (19.6a) determine that the tangential E_t fields are proportional to $\partial\mathcal{E}_z/\partial x$ or $\partial\mathcal{H}_z/\partial x$ and the boundary conditions tell us that $\mathcal{E}_z = 0$ and $E_t = 0$ at $x = 0$ or $x = d$. Hence the boundary conditions require that either $\psi = 0$ or $\partial\psi/\partial x = 0$ at $x = 0$ and at $x = d$. The condition at $x = 0$ requires either $A = 0$ or $B = 0$, and the condition at $x = d$ requires that $Kd = n\pi$, where $n = 1, 2, 3, \ldots$. We find that $K = n\pi/d$ and therefore also that

$$k_z = \sqrt{k^2 - (n\pi/d)^2} \tag{19.19a}$$

Because $E_z(x, z) = \mathcal{E}_z(x)e^{-jk_z z}$, we observe that (19.19a) predicts a non-attenuating propagating wave when k_z is real, i.e. when $k > n\pi/d$, or when the frequency $f > nc/2d$. The wave is exponentially attenuated when k_z is imaginary, i.e. when $k < n\pi/d$, or when $f < nc/2d$. The reader is reminded that $c = 1/\sqrt{\mu\varepsilon}$ and we assume that the medium is lossless, i.e. that ε is real. The value $f_n = nc/2d$ is known as the *cutoff frequency*, because it describes the bounding frequency above which a TE or TM mode can propagate in the waveguide and below which it is attenuated, even though the medium itself would be lossless if unbounded. We also can write (19.19a) as

$$k_z = \frac{n\pi}{d}\sqrt{\left(\frac{f}{f_n}\right)^2 - 1} \tag{19.19b}$$

TE and TM waves can exist only above the lowest cutoff frequency $f_1 = c/2d$. If the operating frequency f lies between $f_n = nf_1$ and $f_{n+1} = (n+1)f_1$ then this particular waveguide, with parallel top and bottom plates of infinite extent, can support n TE and n TM modes, with indices m between 1 and n. Such modes are characterized as TE_m and TM_m modes for index m, with $1 \le m \le n$. Of course, the waveguide can support a TEM mode at any frequency (with the exception noted above for hollow waveguides).

Example 19.1 Possible TE and TM modes in a parallel-plate waveguide

Given that f_1 is 0.7 GHz and that we are operating a parallel-plate waveguide at 4.5 GHz, the question arises as to which TE and TM modes can exist. The answer is given by the critical value of n such that nf_1 is just below 4.5 GHz, whereas $(n+1)f_1$ is just above 4.5 GHz. Clearly $n = 6$ since $6f_1 = 4.2$ GHz and $7f_1 = 4.9$ GHz. Hence the possible modes are TE_1, TE_2, TE_3, TE_4, TE_5, TE_6, TM_1, TM_2, TM_3, TM_4, TM_5, TM_6.

As the wavelength in the waveguide would be $\lambda_z = 2\pi/k_z$ and the phase velocity is then $v_{ph} = 2\pi f/k_z$, we find a dependence upon frequency of both of these quantities for the index-n modes:

$$\lambda_z = \frac{2d/n}{\sqrt{(f/f_n)^2 - 1}} \qquad v_{ph} = \frac{2fd/n}{\sqrt{(f/f_n)^2 - 1}} \tag{19.20}$$

Figure 19.2 illustrates (right to left) the increase of λ_z as f decreases to the limiting value f_n at which the wavelength becomes infinite.

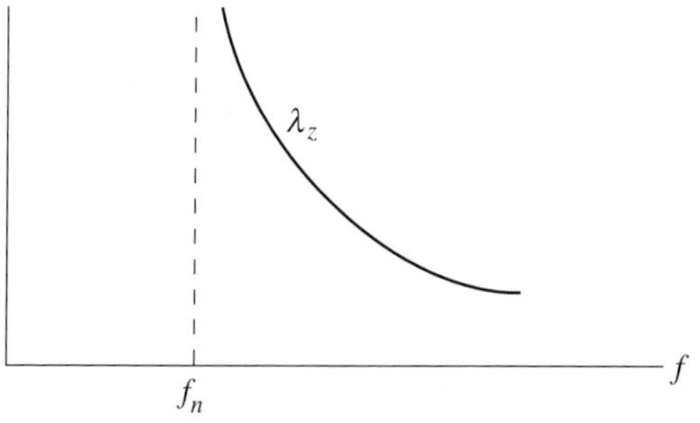

Fig. 19.2 The frequency dependence of the wavelength λ_z given in Eq. (19.20); f_n is the cutoff frequency.

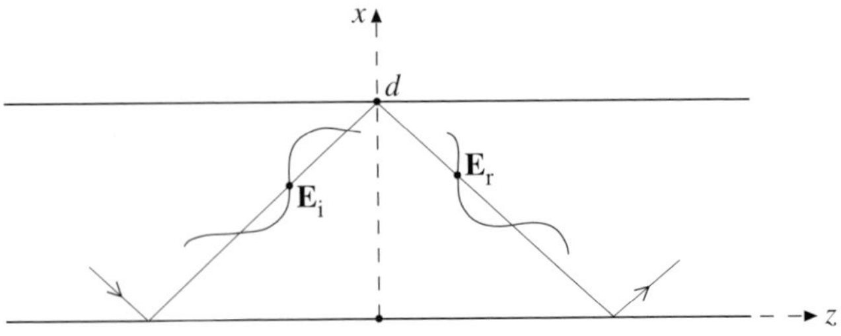

Fig. 19.3 Waveguide modes interpreted as plane waves propagating at an angle to and reflecting off the walls.

Figure 19.3 shows how one can interpret a TE_n or TM_n mode as a plane wave bouncing back and forth between two metallic planes, so that the reflection coefficient is $\mathcal{R} = -1$. The sum of the tangential electric fields at $x = d$ is proportional to

$$e^{-jk(d\cos\theta + z\sin\theta)} - e^{jk(d\cos\theta - z\sin\theta)} = -2je^{-jkz\sin\theta}\sin(kd\cos\theta) \tag{19.21}$$

The minus sign is due to the fact that $\mathcal{R} = -1$, as shown in (18.26), and the *tangential* field amplitude multiplying this sum is E_{i0} for TE and $E_{i0}\cos\theta$ for TM 'incident' and 'reflected' waves at $z = d$. The boundary condition that tangential fields must be zero implies that (19.21) must be zero at $z = 0$ (which is obviously satisfied if we replace $x = d$ by $x = 0$) and at $z = d$, hence that $kd\cos\theta = n\pi$, where $n = 1, 2, 3, \ldots$. Thus $\cos\theta = n\pi/kd = f_n/f$. The exponential on the right-hand side of (19.21) can be written as $e^{-jk_z z}$ with $k_z \equiv k\sin\theta$. We then find that

$$k_z = k\sqrt{1-\cos^2\theta} = \frac{kf_n}{f}\sqrt{\left(\frac{f}{f_n}\right)^2 - 1} = \frac{n\pi}{d}\sqrt{\left(\frac{f}{f_n}\right)^2 - 1} \tag{19.22}$$

which is exactly what (19.19b) predicts from the mode equations. At locations other than $x = d$, the sum field will still have a common factor $\exp(-jk_z z)$, as we have seen in Section 18.1, so that k_z is the wavenumber for propagation in the z direction along the waveguide.

Attenuating medium

If the medium itself is lossy, then the only item in (19.19)–(19.22) that changes is the cutoff frequency f_n, which now becomes

$$f_n = \frac{n}{2d\sqrt{\mu\varepsilon_0(\varepsilon' - j\varepsilon'')}} = \frac{n}{2d\sqrt{\mu\varepsilon_0\varepsilon'}\sqrt{(1 - j\varepsilon''/\varepsilon')}} = f_n'\frac{1}{\sqrt{1-j\delta}} \tag{19.23}$$

where $\delta = \varepsilon''/\varepsilon'$ is the loss tangent defined in Section 17.2 and $f_n' = nc'/2d$, $c' = 1/\sqrt{\mu\varepsilon_0\varepsilon'}$ being the velocity associated with the real part of the complex permittivity. For small loss tangents $\delta \ll 1$, we observe in (19.22) that $(f/f_n)^2$

is replaced by $(f/f_n')^2(1 - j\delta)$, and we may replace $(1 - j\delta)^{1/2} \approx 1 - j\delta/2$, so that $k_z = \beta - j\alpha$ and

$$k_z \approx \frac{n\pi}{d}\sqrt{\left(\frac{f}{f_n'}\right)^2 - 1 - j\left[\frac{\delta}{2}\left(\frac{f}{f_n'}\right)^2\right]} \tag{19.24}$$

There is no longer a sharp cutoff frequency, but to a good approximation (19.24) predicts an almost-cutoff at $f = f_n'$. At $f < f_n'$ the attenuation constant (the imaginary part of k_z) is much larger than the propagation constant (the real part of k_z).

19.3 TEM modes in two-dimensional planar geometry

Equation (19.9) yields possible forms for the TEM modes. Here, the transverse electric field component is $\mathcal{E}_\text{t} = \mathcal{E}_x\hat{\mathbf{x}} + \mathcal{E}_y\hat{\mathbf{y}}$; there is no $\mathcal{E}_z$ or $\mathcal{H}_z$ in a TEM mode. If the two-dimensional uniform-medium planar model, used above, is continued, then we require E_t to be zero at $x = 0$ and $x = d$. One of Eqs. (19.4) predicts that $\partial\mathcal{E}_y/\partial x = \partial\mathcal{E}_x/\partial y$. But the symmetry in the y direction dictates that $\mathcal{E}_x$ does not vary in that direction, so that $\partial\mathcal{E}_y/\partial x = 0$. The conclusion is that $\mathcal{E}_y(x)$ is constant. But the fact that it must be zero at the two interfaces forces one to conclude that $\mathcal{E}_y(x) = 0$ everywhere. Thus, $E_\text{t} = \mathcal{E}_x(x)\hat{\mathbf{x}}$ in this simple case, and (19.9) then predicts $H_\text{t} = Z^{-1}\mathcal{H}_y(x)\hat{\mathbf{y}}$ with $\mathcal{H}_y(x) = Z^{-1}\mathcal{E}_x(x)$, in which expressions $Z = \eta$, the intrinsic impedance of the medium. Finally the $\nabla\cdot\mathbf{D} = 0$ equation predicts, in a uniform-ε medium, that $\nabla\cdot\mathbf{E}$ and, in this case, that $d\mathcal{E}_x/dx = 0$ so that the only possible solutions are

$$E_\text{t}(x) = E_0\hat{\mathbf{x}} \qquad H_\text{t}(x) = \frac{1}{Z}E_0\hat{\mathbf{y}} \tag{19.25}$$

The full solutions for TEM waves moving in the $+\hat{\mathbf{z}}$ direction are

$$\mathbf{E}(\mathbf{r}) = E_0\hat{\mathbf{x}}e^{-jk_z z} \qquad \mathbf{H}(\mathbf{r}) = \frac{1}{Z}E_0\hat{\mathbf{y}}e^{-jk_z z} \tag{19.26}$$

These solutions will be important in Chapter 20 in discussing transmission lines.

19.4 TM modes in two-dimensional planar geometry

Here, we have no $\mathcal{H}_z$ component, and hence the ψ of subsection 19.2.4 is interpreted here as being $\mathcal{E}_z$. From (19.18) we easily conclude that

$$\mathcal{E}_z(x, y) = A\sin\left(\frac{n\pi x}{d}\right) \tag{19.27a}$$

Equations (19.6a), with $K^2 - k_z^2 = n\pi/d$, are handiest now in obtaining the transverse fields. We see from the second and third one that $\mathcal{E}_y = 0$ and $\mathcal{H}_x = 0$. The first predicts

$$\mathcal{E}_x(x, y) = -\frac{jk_z d}{n\pi} A \cos\left(\frac{n\pi x}{d}\right) \tag{19.27b}$$

The fourth predicts

$$\mathcal{H}_y(x, y) = -\frac{j\omega\varepsilon d}{n\pi} A \cos\left(\frac{n\pi x}{d}\right) \tag{19.27c}$$

Thus the only nonzero values for the field phasors of the TM_n modes are

$$\begin{aligned}
\mathcal{E}_z(x, y) &= A \sin\left(\frac{n\pi x}{d}\right) \\
\mathcal{E}_x(x, y) &= -\frac{jk_z d}{n\pi} A \cos\left(\frac{n\pi x}{d}\right) \\
\mathcal{H}_y(x, y) &= -\frac{j\omega\varepsilon d}{n\pi} A \cos\left(\frac{n\pi x}{d}\right)
\end{aligned} \tag{19.28a}$$

The actual fields are found, as before, by multiplying the exponential $e^{j(\omega t - kz)}$ factor into the expressions of (19.26) and then taking the real part of what ensues:

$$\begin{aligned}
\mathbf{E}_z(x, y, z, t) &= A \sin\left(\frac{n\pi x}{d}\right) \cos(\omega t - kz) \\
\mathbf{E}_x(x, y, z, t) &= \frac{k_z d}{n\pi} A \cos\left(\frac{n\pi x}{d}\right) \sin(\omega t - kz) \\
\mathbf{H}_y(x, y, z, t) &= \frac{\omega\varepsilon d}{n\pi} A \cos\left(\frac{n\pi x}{d}\right) \sin(\omega t - kz)
\end{aligned} \tag{19.28b}$$

The TM_1 mode ($n = 1$) is illustrated in Fig. 19.4, in which the arrows depict lines of electric field, the variously sized dots show H_y fields in the $+\hat{\mathbf{y}}$ direction, and the variously sized crosses show H_y fields in the $-\hat{\mathbf{y}}$ direction. Modes with $n > 1$ have more complicated structures. For example, whereas the **E** field structure in Fig. 19.4 for $n = 1$ fills the waveguide, modes with $n > 1$ show n such structures within the

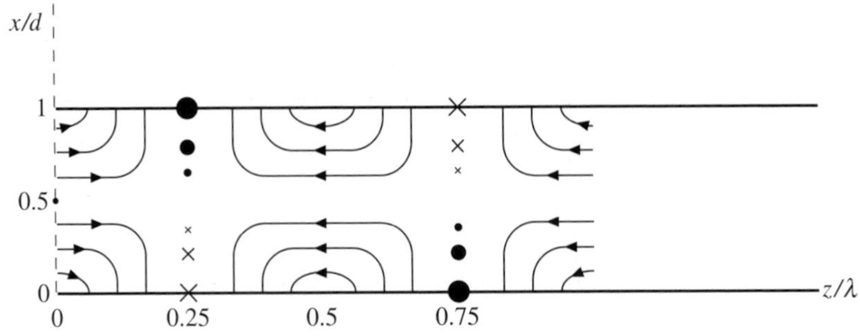

Fig. 19.4 The fields inside a TM_1 mode in a plane containing the propagation direction z.

width of the waveguide, corresponding to the fact that $\sin(n\pi x/d)$ has $n-1$ zeroes between $x=0$ and $x=d$.

19.5 TE modes in two-dimensional planar geometry

As there is no $\mathcal{E}_z$ component, we use one-dimensional solutions of the second of Eqs. (19.5) to find $\mathcal{H}(x) = A\cos(n\pi/d)$. The second of Eqs. (19.6a) shows that the tangential E_t field is proportional to $d\mathcal{H}_z/dx$, and (19.18) tells us that only the cosine term will satisfy $d\mathcal{H}_z/dx = 0$ at $x=0$ and $x=d$. We now label the coefficient of this term A rather than B. Equations (19.6a) also show immediately that $\mathcal{E}_x = \mathcal{H}_y = 0$ and that the remaining field phasors of the TE_n mode are

$$\begin{aligned} \mathcal{H}_z(x,y) &= A\cos\left(\frac{n\pi x}{d}\right) \\ \mathcal{H}_x(x,y) &= \frac{jk_z d}{n\pi} A\sin\left(\frac{n\pi x}{d}\right) \\ \mathcal{E}_y(x,y) &= -\frac{j\omega\mu d}{n\pi} A\sin\left(\frac{n\pi x}{d}\right) \end{aligned} \tag{19.29a}$$

and the concomitant actual fields are

$$\begin{aligned} \mathbf{H}_z(x,y,z,t) &= A\cos\left(\frac{n\pi x}{d}\right)\cos(\omega t - kz) \\ \mathbf{H}_x(x,y,z,t) &= -\frac{k_z d}{n\pi} A\sin\left(\frac{n\pi x}{d}\right)\sin(\omega t - kz) \\ \mathbf{E}_y(x,y,z,t) &= \frac{\omega\mu d}{n\pi} A\sin\left(\frac{n\pi x}{d}\right)\sin(\omega t - kz) \end{aligned} \tag{19.29b}$$

Some field lines for the TE_1 mode are shown in Fig. 19.5. In contrast with Fig. 19.4, the magnetic field lines with arrows lie in the plane of the paper, whereas the **E** field lines are perpendicular to the paper.

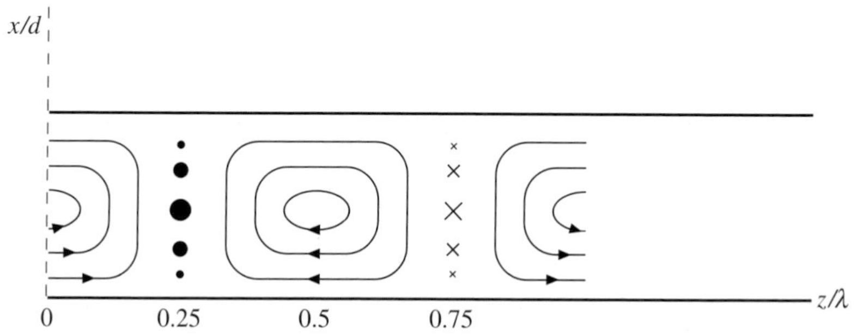

Fig. 19.5 The fields inside a TE_1 mode in a plane containing the propagation direction z.

19.6 Rectangular waveguides

We need to deal only with TM and TE modes because, as discussed above in subsection 19.2.1, single-conductor closed waveguides cannot support TEM modes.

19.6.1 TM modes in three-dimensional planar geometry

If we start with (19.6b), then the longitudinal $\mathcal{E}_z$ field is given by the following equation, for a waveguide in which there are four metal walls at $x = 0, x = a, y = 0$, and $y = b$:

$$\frac{\partial^2 \mathcal{E}_z}{\partial x^2} + \frac{\partial^2 \mathcal{E}_z}{\partial y^2} + K^2 \mathcal{E}_z = 0 \qquad \text{with} \qquad K^2 \equiv k^2 - k_z^2 \tag{19.30}$$

We try to solve this equation with a separation-of-variables method (see Section 8.5 for the electrostatic version of this method). In short, we set $\mathcal{E}_z(x, y) \equiv X(x)Y(y)$ and divide the ensuing equation by this product to obtain

$$\frac{1}{X}\frac{d^2 X}{dx^2} + \frac{1}{Y}\frac{d^2 Y}{dy^2} + K^2 = 0 \tag{19.31}$$

As in Section 8.5, we realize that the first term can only be a function of x, the second term only of y, whereas the third term is a constant. Accordingly, the only possible solutions for independent x and y must be

$$\frac{1}{X}\frac{d^2 X}{dx^2} = -k_x^2 \qquad \frac{1}{Y}\frac{d^2 Y}{dy^2} = -k_y^2 \qquad \text{with} \qquad K^2 - k_x^2 - k_y^2 = 0 \tag{19.32a}$$

Here, both k_x^2 and k_y^2 are constants, and thus we observe from the meaning of K^2 in (19.30) that the condition on the constants becomes

$$k_x^2 + k_y^2 + k_z^2 = k^2 \tag{19.32b}$$

While the most general solutions for $X(x)$ and $Y(y)$ are

$$\begin{aligned} X(x) &= A_x \sin k_x x + B_x \cos k_x x \\ Y(y) &= A_y \sin k_y y + B_y \cos k_y y \end{aligned} \tag{19.33a}$$

with complex values of k_x and k_y, the need for $\mathcal{E}_z$ to be zero at the locations $x = 0$ and $y = 0$ requires first that $B_x = 0$ and $B_y = 0$, yielding

$$\begin{aligned} X(x) &= A_x \sin k_x x \\ Y(y) &= A_y \sin k_y y \end{aligned} \tag{19.33b}$$

The need for zero $\mathcal{E}_z$ at $x = a$ and at $y = b$ now requires $k_x a = m\pi$ and $k_y b = n\pi$, where m, n are positive integers. A zero value of either m or n would make $\mathcal{E}_z = 0$, which is not allowed for the TM mode. The condition on k_z thus becomes

$$k_z = \sqrt{k^2 - \pi^2\left(\frac{m^2}{a^2} + \frac{n^2}{b^2}\right)} \tag{19.34a}$$

Compare this to (19.19a); it is essentially the same condition but in this case the condition takes into account *four* bounding walls of the waveguide instead of only two infinitely broad walls. If we again define a frequency for the mode characterized now by two integers m, n (we dub this a TM_{mn} mode), we can rewrite (19.34a) more handily as

$$k_z = \frac{2\pi f_{mn}}{c}\sqrt{\frac{f^2}{f_{mn}^2} - 1} \qquad f_{mn} \equiv \frac{c}{2}\sqrt{\left(\frac{m^2}{a^2} + \frac{n^2}{b^2}\right)} \tag{19.34b}$$

Clearly, k_z is real only for those frequencies f that exceed f_{mn} in value, but matters can be slightly more complicated than for the parallel-plate waveguide. Consider the following example.

Example 19.2 Possible TE, TM modes in a rectangular waveguide

Consider the case $b = 0.5a$, and let us enumerate the TM_{mn} modes for values of m, n less than or equal to 3. We use (19.34b) to evaluate the f_{mn} (to two decimal places only); these are given in Table 19.1.

If, for example, the operating frequency is $4.00 \times (c/2a)$, then the above table indicates that only the TM_{11}, TM_{21}, and TM_{31} modes will propagate. The TM_{12} and TM_{22} modes have f_{mn} values that exceed the operating frequency and hence cannot propagate according to (19.34b), which indicates imaginary k_z for those modes. This is what complicates matters in rectangular waveguides: when $a \neq b$ the ascending values of f_{mn} do not necessarily coincide with an ascending order of the indices m, n.

To find the TM_{mn}-mode fields, we revert to (19.6). From (19.30) we already have for the phasor $\mathcal{E}_z(x, y)$ that

$$\mathcal{E}_z(x, y) = E_0 \sin\left(\frac{m\pi x}{a}\right)\sin\left(\frac{n\pi y}{b}\right) \tag{19.35a}$$

The four equations of (19.6a) then yield directly (because $\mathcal{H}_z = 0$)

$$\begin{aligned}
\mathcal{E}_x(x, y) &= -\frac{jm\pi k_z/a}{k^2 - k_z^2} E_0 \cos\left(\frac{m\pi x}{a}\right)\sin\left(\frac{n\pi y}{b}\right)\\
\mathcal{E}_y(x, y) &= -\frac{jn\pi k_z/b}{k^2 - k_z^2} E_0 \sin\left(\frac{m\pi x}{a}\right)\cos\left(\frac{n\pi y}{b}\right)\\
\mathcal{H}_x(x, y) &= \frac{jn\pi\omega\varepsilon/b}{k^2 - k_z^2} E_0 \sin\left(\frac{m\pi x}{a}\right)\cos\left(\frac{n\pi y}{b}\right)\\
\mathcal{H}_y(x, y) &= -\frac{jm\pi\omega\varepsilon/a}{k^2 - k_z^2} E_0 \cos\left(\frac{m\pi x}{a}\right)\sin\left(\frac{n\pi y}{b}\right)
\end{aligned} \tag{19.35b}$$

Table 19.1. The minimum frequencies f_{mn} for the TM_{mn} modes, calculated from Eq. (19.34b), in units $c/2a$

m	n	f_{mn}
1	1	2.24
2	1	2.83
1	2	4.12
2	2	4.47
3	1	3.61
1	3	6.08
3	2	5.00
2	3	6.32
3	3	6.71

Using (19.2a), we now incorporate the $\exp(-jk_z z)$ factors, and taking the real parts of these full phasors after multiplication by $e^{j\omega t}$ we finally obtain the actual fields:

$$\begin{aligned} \mathbf{E}_z(x,y,z,t) &= E_0 \sin\left(\frac{m\pi x}{a}\right)\sin\left(\frac{n\pi y}{b}\right)\cos(\omega t - k_z z) \\ \mathbf{E}_x(x,y,z,t) &= \frac{m\pi k_z/a}{k^2-k_z^2} E_0 \cos\left(\frac{m\pi x}{a}\right)\sin\left(\frac{n\pi y}{b}\right)\sin(\omega t - k_z z) \\ \mathbf{E}_x(x,y,z,t) &= \frac{n\pi k_z/b}{k^2-k_z^2} E_0 \sin\left(\frac{m\pi x}{a}\right)\cos\left(\frac{n\pi y}{b}\right)\sin(\omega t - k_z z) \end{aligned} \tag{19.36a}$$

$$\begin{aligned} \mathbf{H}_x(x,y,z,t) &= -\frac{n\pi\omega\varepsilon/b}{k^2-k_z^2} E_0 \sin\left(\frac{m\pi x}{a}\right)\cos\left(\frac{n\pi y}{b}\right)\sin(\omega t - k_z z) \\ \mathbf{H}_y(x,y,z,t) &= \frac{m\pi\omega\varepsilon/a}{k^2-k_z^2} E_0 \cos\left(\frac{m\pi x}{a}\right)\sin\left(\frac{n\pi y}{b}\right)\sin(\omega t - k_z z) \end{aligned} \tag{19.36b}$$

The reader is reminded that k_z is specified by (19.34b) for the TM_{mn} mode. Consequently,

$$k^2 - k_z^2 = \pi^2\left(\frac{m^2}{a^2} + \frac{n^2}{b^2}\right)$$

19.6.2 TE modes in three-dimensional planar geometry

If we start again with (19.6b), then the longitudinal $\mathcal{H}_z$ field is given, similarly to (19.30), by

$$\frac{\partial^2 \mathcal{H}_z}{\partial x^2} + \frac{\partial^2 \mathcal{H}_z}{\partial y^2} + K^2 \mathcal{H}_z = 0 \qquad \text{with} \qquad K^2 \equiv k^2 - k_z^2 \tag{19.37}$$

and now we set $\mathcal{H}_z(x, y) \equiv X(x)Y(y)$ and divide the ensuing equation by this product again to obtain

$$\frac{1}{X}\frac{d^2X}{dx^2} + \frac{1}{Y}\frac{d^2Y}{dy^2} + K^2 = 0 \tag{19.38}$$

which is identical to (19.31). As in the TM case, we realize that the first term can only be a function of x, the second term only of y, whereas the third term is a constant. Accordingly, the only possible solutions again can be written as

$$\frac{1}{X}\frac{d^2X}{dx^2} = -k_x^2 \qquad \frac{1}{Y}\frac{d^2Y}{dy^2} = -k_y^2 \qquad \text{with} \qquad K^2 - k_x^2 - k_y^2 = 0 \tag{19.39a}$$

Here, both k_x^2 and k_y^2 are constants, and the condition on the constants remains

$$k_x^2 + k_y^2 + k_z^2 = k^2 \tag{19.39b}$$

The most general solutions for $X(x)$ and $Y(y)$ are still

$$\begin{aligned} X(x) &= A_x \sin k_x x + B_x \cos k_x x \\ Y(y) &= A_y \sin k_y y + B_y \cos k_y y \end{aligned} \tag{19.40a}$$

with complex values of k_x and k_y in general. While the need remains for $\mathcal{E}_z$ to be zero at the locations $x = 0$, $y = 0$ and at $x = a$, $y = b$ (the four walls of the waveguide), this information is best used in the first two of Eqs. (19.6a), in which we note that $\mathcal{E}_z = 0$ and in which we require $\mathcal{E}_x = \mathcal{E}_y = 0$ (no tangential E component at the interface with a perfect conductor). Thus we are left with the conditions $\partial\mathcal{H}_z/\partial x = 0$ and $\partial\mathcal{H}_z/\partial y = 0$ at the four interfaces. This makes the A's (not the B's!) zero in (19.40a) yielding

$$\begin{aligned} X(x) &= B_x \cos k_x x \\ Y(y) &= B_y \cos k_y y \end{aligned} \tag{19.40b}$$

for the TE modes, and the values of k_x, k_y and validity of condition (19.34) remain the same. Thus we have found

$$\mathcal{H}_z(x, y) = H_0 \cos\left(\frac{m\pi x}{a}\right) \cos\left(\frac{n\pi y}{b}\right) \tag{19.41a}$$

for the longitudinal magnetic field component of a TE mode. Straightforward application of (19.6a) requires only derivatives of (19.41a) on the right-hand side, and we find

$$\begin{aligned}
\mathcal{H}_x(x,y) &= \frac{jm\pi k_z/a}{k^2-k_z^2} H_0 \sin\left(\frac{m\pi x}{a}\right)\cos\left(\frac{n\pi y}{b}\right) \\
\mathcal{H}_y(x,y) &= \frac{jn\pi k_z/b}{k^2-k_z^2} H_0 \cos\left(\frac{m\pi x}{a}\right)\sin\left(\frac{n\pi y}{b}\right) \\
\mathcal{E}_x(x,y) &= \frac{jn\pi\omega\mu/b}{k^2-k_z^2} H_0 \cos\left(\frac{m\pi x}{a}\right)\sin\left(\frac{n\pi y}{b}\right) \\
\mathcal{E}_y(x,y) &= -\frac{jm\pi\omega\mu/a}{k^2-k_z^2} H_0 \sin\left(\frac{m\pi x}{a}\right)\cos\left(\frac{n\pi y}{b}\right)
\end{aligned} \tag{19.41b}$$

With (19.2a), we again incorporate the $\exp(-jk_z z)$ factors and with the real parts of these full phasors after multiplication by $e^{j\omega t}$ we finally obtain

$$\begin{aligned}
H_z(x,y,z,t) &= H_0 \cos\left(\frac{m\pi x}{a}\right)\cos\left(\frac{n\pi y}{b}\right)\cos(\omega t - k_z z) \\
H_x(x,y,z,t) &= -\frac{n\pi k_z/a}{k^2-k_z^2} H_0 \sin\left(\frac{m\pi x}{a}\right)\cos\left(\frac{n\pi y}{b}\right)\sin(\omega t - k_z z) \\
H_y(x,y,z,t) &= -\frac{n\pi k_z/b}{k^2-k_z^2} H_0 \cos\left(\frac{m\pi x}{a}\right)\sin\left(\frac{n\pi y}{b}\right)\sin(\omega t - k_z z)
\end{aligned} \tag{19.42a}$$

$$\begin{aligned}
E_x(x,y,z,t) &= -\frac{n\pi\omega\mu/b}{k^2-k_z^2} H_0 \cos\left(\frac{m\pi x}{a}\right)\sin\left(\frac{n\pi y}{b}\right)\sin(\omega t - k_z z) \\
E_y(x,y,z,t) &= \frac{m\pi\omega\mu/a}{k^2-k_z^2} H_0 \sin\left(\frac{m\pi x}{a}\right)\cos\left(\frac{n\pi y}{b}\right)\sin(\omega t - k_z z)
\end{aligned} \tag{19.42b}$$

Note that we can have here either $m = 0$ or $n = 0$; m and n cannot both be zero.

19.7 Cylindrical waveguides

When the waveguide is a circular cylinder with radius R and z along its axis, we can utilize (19.7), which we rewrite below:

$$\begin{aligned}
(k^2-k_z^2)\mathcal{E}_\rho &= -j\left(k_z\frac{\partial\mathcal{E}_z}{\partial\rho} + \frac{\omega\mu}{\rho}\frac{\partial\mathcal{H}_z}{\partial\varphi}\right) \\
(k^2-k_z^2)\mathcal{E}_\varphi &= -j\left(\frac{k_z}{\rho}\frac{\partial\mathcal{E}_z}{\partial\varphi} - \omega\mu\frac{\partial\mathcal{H}_z}{\partial\rho}\right) \\
(k^2-k_z^2)\mathcal{H}_\rho &= j\left(\frac{\omega\varepsilon}{\rho}\frac{\partial\mathcal{E}_z}{\partial\varphi} - k_z\frac{\partial\mathcal{H}_z}{\partial\rho}\right) \\
(k^2-k_z^2)\mathcal{H}_\varphi &= -j\left(\omega\varepsilon\frac{\partial\mathcal{E}_z}{\partial\rho} + \frac{k_z}{\rho}\frac{\partial\mathcal{H}_z}{\partial\varphi}\right)
\end{aligned} \tag{19.43a}$$

together with the wave equations for the longitudinal field components in cylindrical coordinates

$$\begin{aligned}&\frac{1}{\rho}\frac{\partial}{\partial\rho}\left(\rho\frac{\partial\mathcal{E}_z}{\partial\rho}\right)+\frac{1}{\rho^2}\frac{\partial^2\mathcal{E}_z}{\partial\varphi^2}+(k^2-k_z^2)\mathcal{E}_z=0\\&\frac{1}{\rho}\frac{\partial}{\partial\rho}\left(\rho\frac{\partial\mathcal{H}_z}{\partial\rho}\right)+\frac{1}{\rho^2}\frac{\partial^2\mathcal{H}_z}{\partial\varphi^2}+(k^2-k_z^2)\mathcal{H}_z=0\end{aligned}\tag{19.43b}$$

19.7.1 TM modes for cylindrical waveguides

The technique for solving, e.g. for $\mathcal{E}_z$, is similar to that discussed under the separation-of-variables method in electrostatics. We set $\mathcal{E}_z(\rho,\varphi)=\mathcal{R}(\rho)\Phi(\varphi)$, insert it into the wave equation, multiply all terms by ρ^2, and divide all terms by the product $\mathcal{R}(\rho)\Phi(\varphi)$:

$$\frac{1}{\mathcal{R}}\left(\rho^2\frac{d^2\mathcal{R}}{d\rho^2}+\rho\frac{d\mathcal{R}}{d\rho}\right)+(k^2-k_z^2)\rho^2+\frac{1}{\Phi}\frac{d^2\Phi}{d\varphi^2}=0\tag{19.44}$$

As before, the only possible solution to (19.44) for independent ρ and φ values is obtained by setting

$$\begin{aligned}&\frac{1}{\mathcal{R}}\left(\rho^2\frac{d^2\mathcal{R}}{d\rho^2}+\rho\frac{d\mathcal{R}}{d\rho}\right)+(k^2-k_z^2)\rho^2=\nu^2\\&\frac{1}{\Phi}\frac{d^2\Phi}{d\varphi^2}=-\nu^2\end{aligned}\tag{19.45}$$

where ν^2 is a constant. Let us name $k^2-k_z^2\equiv K^2$ as we did in (19.17). The above two equations then become

$$\begin{aligned}&\frac{d^2\mathcal{R}}{d\rho^2}+\rho\frac{d\mathcal{R}}{d\rho}+\left(K^2-\frac{\nu^2}{\rho^2}\right)\mathcal{R}=0\\&\frac{d^2\Phi}{d\varphi^2}+\nu^2\Phi=0\end{aligned}\tag{19.46}$$

The second of these equations is, as before, easily solved in terms of trigonometric functions of $\nu\varphi$:

$$\Phi(\varphi)=A_{\nu s}\sin\nu\varphi+A_{\nu c}\cos\nu\varphi=A_{\nu\varphi}\cos(\nu\varphi+\theta)\tag{19.47a}$$

where $A_{\nu s}$, $A_{\nu c}$, and θ_φ are constants. The first equation of (19.46) is known as *Bessel's equation*. We encountered this equation first in subsection 8.5.2, and it is shown there as (8.63) in a slightly different, but equivalent, form. The most general solution is

$$\mathcal{R}(\rho) = A_J J_\nu(K\rho) + A_Y Y_\nu(K\rho) \tag{19.47b}$$

with constants A_J and A_Y. When ν and K are real, the $J_\nu(K\rho)$ are *Bessel functions of the first kind*, which are 1 (for $\nu = 0$) or 0 (for all other values of ν) at $\rho = 0$, and then are oscillatory and decaying as the argument $K\rho$ grows. The $Y_\nu(K\rho)$ are *Bessel functions of the second kind*, and also are oscillatory and decaying as $K\rho \to \infty$, but – unlike the J_ν functions – go to $-\infty$ at the origin. The oscillatory nature of these functions implies that there are many values of $K\rho$ at which they are zero, and in that sense they are somewhat like the more familiar trigonometric functions. Figure 19.6 shows four of the lowest-order Bessel functions with real argument and real index ν. Unlike the trigonometric functions, the zero crossings do not lie at constant intervals.

The product of (19.47a) and (19.47b) that concerns us here would be

$$\mathcal{E}_z(\rho, \varphi) = A_z J_\nu(K\rho)\cos(\nu\varphi + \theta_\varphi) \tag{19.48a}$$

where A_z is a new constant. We cannot have a contribution from the $Y_\nu(K\rho)$ because it becomes infinite at $\rho = 0$, but the electric field inside the waveguide cannot become infinite anywhere. The next item to consider is that $\mathcal{E}_z(\rho, \varphi + 2\pi)$ must be the same value as $\mathcal{E}_z(\rho, \varphi)$. This requires the index ν to be an *integer*, $\nu = m$, with $m = 1, 2, 3, \ldots$. The constant θ_φ can be considered arbitrary because we can place the x axis in whatever direction we like in a plane $\perp z$ (thus defining the zero value of φ, the angle of $\boldsymbol{\rho}$ with $\hat{\mathbf{x}}$), hence we might just as well choose the x axis so that $\theta_\varphi = 0$. Then, the TM mode under consideration yields

$$\mathcal{E}_z(\rho, \varphi) = E_0 J_m(K\rho)\cos m\varphi \tag{19.48b}$$

with E_0 being a constant electric field value. Finally, we must ensure that $\mathcal{E}_z(R, \varphi) = 0$ because the tangential $\mathbf{E}$ field must vanish at the edge of the metallic conductor at $\rho = R$. Thus, $J_m(KR)$ must be zero, which implies that $KR = j_{mn}$, where j_{mn} is the nth zero of the mth-order Bessel function of the first kind (every mth-order Bessel function of the first kind has an infinite number of zeroes).

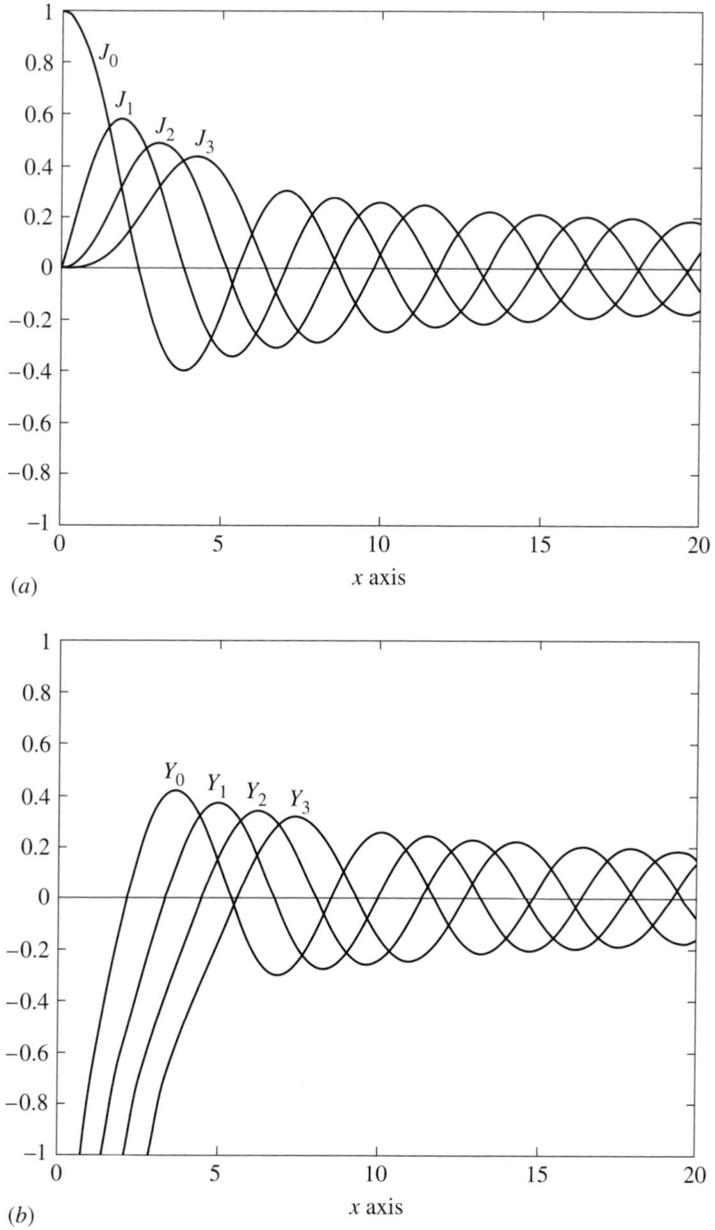

Fig. 19.6 The lowest four orders of the Bessel functions: (*a*) $J_n(x)$; (*b*) $Y_n(x)$.

So we finally have

$$\mathcal{E}_z(\rho, \varphi) = E_0 J_m(K_{mn}\rho) \cos m\varphi \qquad K_{mn} \equiv j_{mn}/R \qquad k_z \equiv \sqrt{k^2 - K_{mn}^2} \tag{19.49a}$$

We only need to apply (19.43a) to find the remaining components, and the straightforward result is

$$\begin{aligned}
\mathcal{E}_\rho(\rho,\varphi) &= -\frac{jk_z}{K_{mn}} E_0 J'_m(K_{mn}\rho)\cos m\varphi \\
\mathcal{E}_\varphi(\rho,\varphi) &= \frac{jmk_z}{K_{mn}^2\rho} E_0 J_m(K_{mn}\rho)\sin m\varphi \\
\mathcal{H}_\rho(\rho,\varphi) &= -\frac{jm\omega\varepsilon}{K_{mn}^2\rho} E_0 J_m(K_{mn}\rho)\sin m\varphi \\
\mathcal{H}_\varphi(\rho,\varphi) &= -\frac{j\omega\varepsilon}{K_{mn}} E_0 J'_m(K_{mn}\rho)\cos m\varphi
\end{aligned} \tag{19.49b}$$

In this expression, $J'_{mn}(K_{mn}\rho)$ is an abbreviation for the derivative $dJ_m(\xi)/d\xi$ evaluated at $\xi = K_{mn}\rho$. (These derivatives can be expressed in terms of a linear combination of J_m and either $J_{m+1,n}$ or $J_{m-1,n}$; see Abramowitz & Stegun, *Handbook of Mathematical Functions*, formulas (9.1.27).) The $\mathbf{E}(\rho,\varphi,z,t)$ field is obtained from (19.49) by once again 'pasting' the factor $\exp(\omega t - k_z z)$ behind each phasor of (19.49b) and then taking the real part. We leave this to the reader, as it is exactly the same procedure by which (19.36) follows from (19.35). Equations (19.49) give the phasors of the cylindrical TM_{mn} modes.

19.7.2 TE modes for cylindrical waveguides

We again go through the development from (19.43) to (19.49a), but allowing now for the existence of an $\mathcal{H}_z$ component instead of an $\mathcal{E}_z$ component; we then obtain

$$\mathcal{H}_z(\rho,\varphi) = H_0 J_m(K_{mn}\rho)\cos m\varphi \tag{19.50}$$

As we can see from the second of Eqs. (19.43a), the requirements that all tangential E fields vanish at $\rho = R$ and that $\mathcal{E}_z = 0$ everywhere tell us that $\partial\mathcal{H}_z/\partial\rho$ must be zero at $\rho = R$. From (19.50) we observe that $J'_{mn}(K_{mn}R) = 0$, i.e. the derivative of $J_{mn}(\xi)$ must be zero at $\xi = K_{mn}R$. The zeroes of $J'_{mn}(\xi)$ are at $\xi = j'_{m1}, j'_{m2}, j'_{m3}, \ldots$. Consequently, we find

$$\mathcal{H}_z(\rho,\varphi) = H_0 J_m(K_{mn}\rho)\cos m\varphi \qquad K_{mn} \equiv j'_{mn}/R \qquad k_z \equiv \sqrt{k^2 - K_{mn}^2} \tag{19.51a}$$

The rest of the TE$_{mn}$ modes are easily found from (19.43a):

$$\begin{aligned}
\mathcal{H}_\rho(\rho,\varphi) &= -\frac{jk_z}{K_{mn}} H_0 J'_m(K_{mn}\rho)\cos m\varphi \\
\mathcal{H}_\varphi(\rho,\varphi) &= \frac{jmk_z}{K_{mn}^2\rho} H_0 J_m(K_{mn}\rho)\sin m\varphi \\
\mathcal{E}_\rho(\rho,\varphi) &= \frac{jm\omega\mu}{K_{mn}^2\rho} H_0 J_m(K_{mn}\rho)\sin m\varphi \\
\mathcal{E}_\varphi(\rho,\varphi) &= \frac{j\omega\mu}{K_{mn}} H_0 J'_m(K_{mn}\rho)\cos m\varphi
\end{aligned} \tag{19.51b}$$

Problems

19.1. Show how Eqs. (19.8a) follow from Eqs. (19.4).

19.2. Find the cylindrical-coordinate equivalents of Eqs. (19.4).

19.3. Show that **E** and **H** are perpendicular to each other for *any* TE or TM wave.

19.4. Describe the TEM modes in a coaxial cable with inner radius a and outer radius b.

19.5. A parallel-plate waveguide with plate separation $d = 20\,\text{cm}$ supports wave modes with phasor fields $\mathbf{E}, \mathbf{H} \propto e^{-\gamma z}$ with $\gamma = \sqrt{(n\pi/d)^2 - \omega^2\mu_0\varepsilon_0}$. Express the cutoff frequency in terms of d, and give the labels of all TE, TM, TEM modes that can exist at $f = 0.78\,\text{GHz}$.

19.6. A parallel-plate waveguide with plate separation $d = 3\,\text{cm}$ carries modes at 19 GHz. Which TE and TM modes can exist in this waveguide at that frequency?

19.7. A rectangular waveguide has widths a and b with $a = 1.8b$. What values of m, n characterize the lowest six TM$_{mn}$ modes? Note that either $m = 0$, or $n = 0$, but not both. Index m refers to the side of length a; n to the side of length b.

19.8. For a rectangular waveguide, it is given that $E_z(x, y, z, t) = E_0 \sin\alpha x \sin\beta y \times \sin(\omega t - \gamma z)$, with $\alpha = 20$, $\beta = 5$, $\gamma = 102$ for the TM$_{21}$ mode in the free-space interior.

(a) What is the propagating frequency f?

(b) What are the widths a, b of the waveguide?

(c) What is the critical wavelength for this mode?

19.9. An air-filled rectangular waveguide has widths $a = 4.5\,\text{cm}$ and $b = 20\,\text{cm}$. The operating frequency is 2.26 GHz. List all possible TE modes.

19.10. A parallel-plate waveguide is filled with a lossy dielectric with $\varepsilon' = 2.5$, $\varepsilon'' = 0.000125$. The operating frequency $f = 2.5f_2$ and the plate separation is 3.5 cm. At what distance would the power be down by a factor $1/e$ for the TM$_2$ mode?

19.11. Obtain expressions for the phase and group velocities of a TM_{mn} or TE_{mn} mode in an air-filled rectangular waveguide.

19.12. Work out the expressions for the average power propagated by the TM_n mode of a parallel-plate waveguide at $f > f_n$. Use Eqs. (19.27), (19.28).

19.13. For a cylindrical waveguide with radius $R = 2\,\text{cm}$, find the lowest three cutoff frequencies and enumerate the modes to which they belong.

20

Transmission lines

This chapter deals with a class of waveguides known as *transmission lines* (TL), which guide waves of any desired frequency from one point to another but which restrict the waves more or less to a finite extent in directions perpendicular to that in which the propagation is guided. The main reason that not all waveguides are transmission lines is that transmission lines need to carry *transverse electromagnetic modes* (TEM *modes*, see Chapter 19) which can exist at all frequencies and which have no field components in the direction of propagation (in contrast to the TE and TM modes that can exist in other waveguides). In Chapter 19, it was clarified that waveguides that can accommodate TEM modes must have guiding walls that consist at least of two isolated pieces, e.g. a 'core' and an outer 'cladding' section, as shown in Fig. 20.1(*a*), or two long parallel flat plates, as shown in Fig. 20.1(*b*). A rectangular hollow box structure cannot serve as a transmission line because it

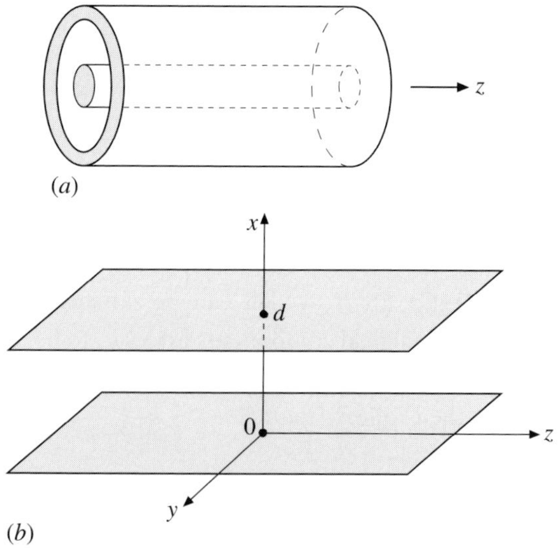

Fig. 20.1 Two typical transmission lines: (*a*) coaxial-cylinder; (*b*) parallel-plate.

cannot transmit TEM modes. Three common forms of transmission lines are parallel-plate waveguides, as in Fig 20.1(*b*), coaxial cables as in Fig. 20.1(*a*), and parallel-wire waveguides (a cross section of which is shown in Fig. 8.7). The first and the last of these have as a disadvantage that the electromagnetic fields are not shielded from the outside so that situations can occur in which they interfere with, or are interfered with by, other electromagnetic devices. Their relative simplicity, however, is an advantage.

A useful introduction to what is needed here from waveguide theory can be obtained from Chapter 19, Sections 19.1–19.3 (omitting subsections 19.2.2–19.2.4), but the present chapter stands by itself and can be read without referring to Chapter 19.

20.1 The TEM mode of a two-dimensional waveguide as transmission line

20.1.1 Transmission-line equations in E and H

We will recapitulate the essential items from Chapter 19 that are needed to understand the main points about transmission lines. First, we recognize the special role of the propagation-guidance direction and we shall assume it is a straight line along the z axis. Curvilinear waveguides could be considered too, but we shall not do so here. So we assume that the fields can be written as

$$\mathbf{E}(\mathbf{r}) = \boldsymbol{\mathcal{E}}(x, y)e^{-jk_z z} \qquad \mathbf{H}(\mathbf{r}) = \boldsymbol{\mathcal{H}}(x, y)e^{-jk_z z} \tag{20.1}$$

Under the assumption that there are *no longitudinal fields* ($\mathcal{E}_z = 0$, $\mathcal{H}_z = 0$), the curl E equations lead to

$$\frac{\partial E_y}{\partial z} = j\omega\mu H_x \qquad \frac{\partial H_x}{\partial z} = j\omega\varepsilon E_y \tag{20.2}$$

With substitution of (20.1) into (20.2), we find

$$-jk_z E_y = j\omega\mu H_x \quad -jk_z H_x = j\omega\varepsilon E_y \tag{20.3}$$

These two equations define a ratio E_y/H_x which can be satisfied only if $k_z = \omega\sqrt{\mu\varepsilon} \equiv k$; the absence of longitudinal field components forces k_z to be equal to the wavenumber valid in the μ, ε medium. The same result can be found also from the waveguide equations (19.6). It then follows that

$$\mathcal{H}_\mathrm{t} = \frac{1}{Z}\hat{\mathbf{z}} \times \mathcal{E}_\mathrm{t} \tag{20.4}$$

Here $\mathcal{E}_\mathrm{t} = E_0\hat{\mathbf{x}}$, $\mathcal{H}_\mathrm{t} = H_0\hat{\mathbf{y}}$, with $H_0 = E_0/Z$ for a *parallel-plate transmission line*, and there is no $\mathcal{E}_z$ or $\mathcal{H}_z$ in a TEM mode. The fields are constant for this type of transmission line because there is symmetry with respect to y, if we ignore 'fringing' effects close to the edges of the plates in the y direction, and also because $\nabla \cdot \mathcal{E}_t = \partial \mathcal{E}_x / \partial x = 0$. Furthermore, $Z = \sqrt{\mu/\varepsilon}$ is equal to the intrinsic impedance η of an ε, μ medium. The relationship (20.4) also follows more generally from the material discussed in Chapter 19, where (19.9) yields the possible forms for the TEM modes. For a coaxial cable, one finds similarly that $\mathcal{E}_t = \mathcal{E}_\rho \hat{\boldsymbol{\rho}}$ and $\mathcal{H}_t = \mathcal{H}_\varphi \hat{\boldsymbol{\varphi}}$. In fact the **E** field generally will be in the transverse direction, which is *perpendicular* to the transmission-line guide walls at the interfaces, and the **H** field will be *parallel* to that direction. This follows from the fact that there is no tangential **E** field at the interfaces, so that the **E** field lines must start and stop on the plates in a perpendicular direction. The divergence equation $\nabla_t \cdot \mathcal{E}_t = 0$, applied to a narrow tube of field lines, then determines the continuity of the field line between the plates. If there is enough symmetry, as in a coaxial cable, the field line will have a direction as simple as the $\hat{\rho}$ direction. The $\mathcal{H}_t$ field must obey (20.4), hence has the other desired direction.

In summary, the parallel-plate transmission line has

$$\mathcal{E}(x) = E_0 \hat{\mathbf{x}} \qquad \mathcal{H}(x) = \frac{1}{Z} E_0 \hat{\mathbf{y}} \tag{20.5}$$

The full solutions for TEM waves moving in the $+\hat{\mathbf{z}}$ direction are

$$\mathbf{E}(\mathbf{r}) = \mathcal{E}_t(x) e^{-jk_z z} = \mathbf{E}_0 \hat{\mathbf{x}} e^{-jk_z z} \qquad \mathbf{H}(\mathbf{r}) = \mathcal{H}_t(x) e^{-jk_z z} = \frac{1}{Z} E_0 \hat{\mathbf{y}} e^{-jk_z z} \tag{20.6}$$

The directions of the $\mathcal{E}(x)$ and $\mathcal{H}(x)$ fields are shown in Fig. 20.2.

Consider the remaining boundary conditions. The condition for the normal component of **D** becomes $\hat{\mathbf{n}} \cdot \varepsilon \mathbf{E} = \varrho_s$. With reference to Fig. 20.2 for the directions of surface normal unit vector $\hat{\mathbf{n}}$ at $x = 0$ and $x = d$, we find

$$\begin{aligned} \varrho_s(0, y, z) &= -\varepsilon^{-1} E_0 e^{-jk_z z} = -\varepsilon^{-1} E_x(z) \\ \varrho_s(d, y, z) &= \varepsilon^{-1} E_0 e^{-jk_z z} = \varepsilon^{-1} E_x(z) \end{aligned} \tag{20.7}$$

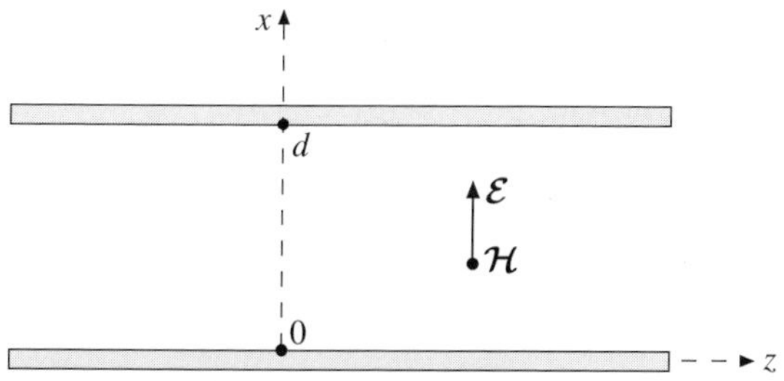

Fig. 20.2 $\mathcal{E}$ and $\mathcal{H}$ in a parallel-plate transmission line. $\mathcal{H}$ is $\perp$ to the page.

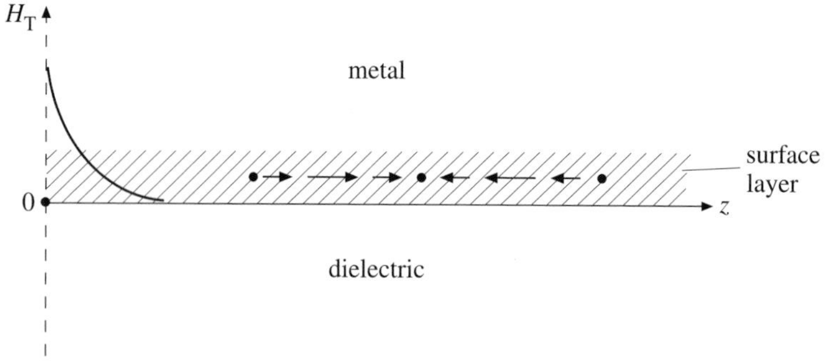

Fig. 20.3 Decay of fields and equivalent sheet current in the metal walls of a transmission line. The hatching shows the skin depth layer and the arrows parallel to the surface show the current $\mathbf{K}_m$.

The boundary condition for the tangential components of $\mathbf{H}$ is $-\hat{\mathbf{n}} \times (\mathbf{H}_m - \mathbf{H}) = \mathbf{K}_m$; the subscript m refers to quantities in the metal plates of the transmission line. In metals, impinging fields are rapidly attenuated (see subsection 17.2.2), so that in the limit of a perfect conductor we may set $\mathbf{H}_m = 0$. For real conductors, there is a skin depth layer in which $\mathbf{H}_m$ attenuates rapidly to unmeasurably low values. Beyond that layer there are no longer any measurable fields, and we can consider that very thin layer to contain only a sheet current $\mathbf{K}_m$ that obeys $-\hat{\mathbf{n}} \times \mathbf{H} = \mathbf{K}_m$ (with $\hat{\mathbf{n}} = +\hat{\mathbf{x}}$ at the top plate and $\hat{\mathbf{n}} = -\hat{\mathbf{x}}$ at the bottom plate); see Fig. 20.3.

$$\begin{aligned} \mathbf{K}_m(0, y, z) &= \hat{\mathbf{x}} \times \mathcal{H}_t e^{-jk_z z} = \frac{1}{Z} E_0(z) e^{-jk_z z} \hat{\mathbf{z}} \equiv H_0(z) \hat{\mathbf{z}} \\ \mathbf{K}_m(d, y, z) &= -\hat{\mathbf{x}} \times \mathcal{H}_t e^{-jk_z z} = -\frac{1}{Z} E_0(z) e^{-jk_z z} \hat{\mathbf{z}} \equiv -H_0(z) \hat{\mathbf{z}} \end{aligned} \tag{20.8}$$

The sheet currents on the upper and lower plate are equal in magnitude to the magnetic field in the medium inside the waveguide, but they are directed in the $\pm\hat{\mathbf{z}}$ direction.

20.1.2 Transmission-line equations in V and I

As the $\mathbf{E}$ field lines are perpendicular to the guide plates of the transmission line at the interfaces, it follows that there is a voltage difference $V(z)$ between the plates that is proportional to $\exp(-jkz)$. Thus

$$V(z) = \int_A^B d\mathbf{l}_1 \cdot \mathbf{E}(\mathbf{r}) = \int_A^B d\mathbf{l}_1 \cdot \mathcal{E}_t(x, y) e^{-jkz} \equiv V_0 e^{-jkz} \tag{20.9}$$

The bounds A and B are indicated in Fig. 20.4, but may lie elsewhere on the equipotential surfaces. Likewise, the total (sheet) current on each of the plates

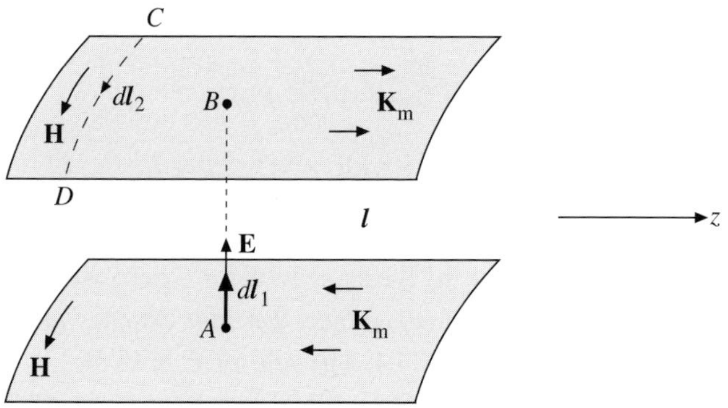

Fig. 20.4 V, I parameters derived from **E**, **H** fields.

is $\pm I(z)$, which we describe as a scalar line integral in a direction $\perp \mathbf{K}_m$ with

$$I(z) = \int_C^D dl_2 K_m(\mathbf{r}) = \int_C^D dl_2 \mathcal{H}_t(x, y)e^{-jkz} \equiv I_0 e^{-jkz} \tag{20.10}$$

See Fig. 20.4 for the directions of $d\boldsymbol{l}_1$ and $d\boldsymbol{l}_2$.

It is customary in transmission-line analysis to work with the voltage difference $V(z)$ and current $I(z)$ rather than with the interior fields. From these equations, it can be seen, by taking the derivative twice, that

$$\frac{d^2 V(z)}{dz^2} + k^2 V(z) = 0 \qquad \frac{d^2 I(z)}{dx^2} + k^2 I(z) = 0 \tag{20.11}$$

However, we do not deal directly with these equations, because the coupling between $V(z)$ and $I(z)$ has been lost. Let us restrict our attention, at least for the time being, to the parallel-plate transmission line. If (20.6) is used in (20.9) and (20.10), it is straightforward to find

$$V_0 = E_0 d \qquad I_0 = H_0 w = \frac{E_0 w}{Z} = \sqrt{\frac{\varepsilon}{\mu}} E_0 w \tag{20.12}$$

given that w is the lateral extent of either of the two parallel plates; the reader should be careful to distinguish this from the frequency ω. With single derivatives of Eqs. (20.9) and (20.10) we find

$$\begin{aligned} \frac{dV(z)}{dz} &= -jkV_0 e^{-jkz} = -j\omega\sqrt{\mu\varepsilon}\, E_0\, d e^{-jkz} = -j\omega\frac{\mu d}{w}\sqrt{\frac{\varepsilon}{\mu}} E_0 w e^{-jkz} \\ &= -j\omega L I_0 e^{-jkz} = -j\omega L I(z) \qquad \text{with} \qquad L \equiv \mu d/w \end{aligned} \tag{20.13}$$

and likewise

$$\frac{dI(z)}{dz} = -jkI_0e^{-jkz} = -j\omega(\varepsilon E_0 w)e^{-jkz} = -j\omega\left(\frac{w\varepsilon}{d}E_0 d\right)e^{-jkz}$$
$$= -j\omega CV_0e^{-jkz} = -j\omega CV(z) \qquad \text{with} \qquad C \equiv \varepsilon w/d \tag{20.14}$$

Substitutions from (20.12) have been made several times in the above equations for derivatives of $V(z)$ and $I(z)$. Comparing the expression for C here with (7.24), we see that $C = \varepsilon w/d$ in F/m expresses the *capacitance per unit distance along the axis* (as surface $S = wz$ for length z of the TL). The unit of C is F/m.

The magnetic flux linkage $\Lambda(z)$ for an infinitesimal length δz of transmission line is $\delta z\, d\mu H(z) = \delta z(\mu d/w)wH(z) = \delta z(\mu d/w)I(z)$, if we apply (20.12) for the purpose of replacing $H(z)$ by $I(z)$. But $\Lambda(z) = \mathcal{L}I(z)$, as we saw in the definition of inductance $\mathcal{L}$ in Section 13.1, so that $\mathcal{L} = L\delta z$ and consequently $L = \mu d/w$ expresses the *inductance per unit length along the axis*. The unit of L is H/m. The lossless transmission-line equations are therefore

$$\frac{dV(z)}{dz} = -j\omega LI(z) \qquad \frac{dI(z)}{dz} = -j\omega CV(z)$$
$$L = \frac{\mu d}{w} \quad (\text{H/m}) \qquad C = \frac{\varepsilon w}{d} \quad (\text{F/m}) \tag{20.15}$$

but while the equations are quite general when expressed in terms of C and L, the expressions for C and L are specific for the parallel-plate TL. Instead of (20.11) we obtain

$$\frac{d^2V(z)}{dz^2} + (\omega^2 LC)V(z) = 0 \qquad \frac{d^2I(z)}{dz^2} + (\omega^2 LC)I(z) = 0 \tag{20.16a}$$

So at least for the lossless TL it follows from a comparison of (20.16a) with (20.11) and from $k^2 = \omega^2\mu\varepsilon$ that $\mu\varepsilon = LC$ which is also seen directly from (20.15) for the parallel-plate TL. Thus, for lossless transmission lines we have

$$LC = \mu\varepsilon \tag{20.16b}$$

The ratio $V(z)/I(z)$ is known as the *characteristic impedance* $Z_0(z)$ of the TL, and (20.15) predicts

$$kV(z) = \omega LI(z) \qquad \text{and} \qquad kI(z) = \omega CV(z) \tag{20.17}$$

because voltage difference and current depend on z only through the exponential factor $\exp(-jkz)$. Division of the first of these two equations by the second yields

$$\frac{V(z)}{I(z)} \equiv Z_0 = \sqrt{\frac{L}{C}} \tag{20.18}$$

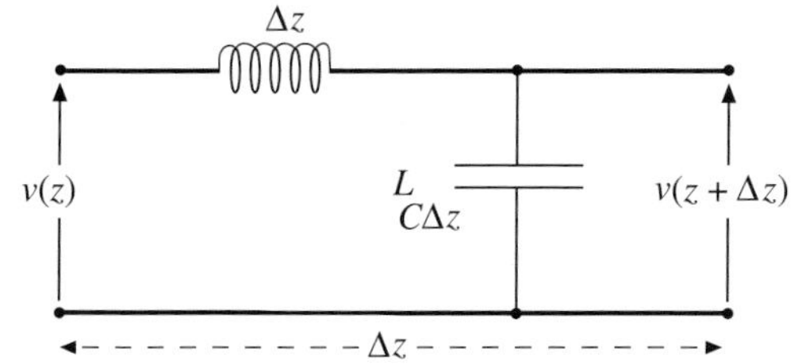

Fig. 20.5 The equivalent distributed circuit element for lossless transmission lines.

Hence the characteristic impedance Z_0 is a constant of the TL when waves are launched that travel only in the $+\hat{\mathbf{z}}$ direction. This is no longer the case when waves travel simultaneously in the $-\hat{\mathbf{z}}$ direction (a case yet to be discussed). Note for the parallel-plate TL that $Z_0 = (d/w)\eta$, i.e. the characteristic impedance is proportional to the intrinsic impedance η of the medium, and the constant of proportionality is the ratio of plate separation distance d to width w. That offers some degree of control over Z_0 for design purposes.

A circuit equivalent of (20.15) is shown in Fig. 20.5 for a distributed-circuit element of infinitesimal length Δz. The idea here is to consider the transmission line as an infinite sum (over finite distance) of infinitesimal pieces of length Δz, each of which is considered as an infinitesimal circuit element feeding input voltage $v(z, t)$ and current $i(z, t)$ into output $v(z + \Delta z, t)$ and $i(z + \Delta z, t)$. The two Kirchhoff equations for each infinitesimal element are

$$\begin{aligned} v(z,t) - L\Delta z \frac{\partial i(z,t)}{\partial t} &= v(z+\Delta z, t) \\ i(z,t) - C\Delta z \frac{\partial v(z+\Delta z,t)}{\partial t} &= i(z+\Delta z, t) \end{aligned} \tag{20.19}$$

In the first of these equations, put $v(z + \Delta z, t) \approx v(z, t) + \Delta z \partial v(z, t)/\partial z$, as the $(\Delta z)^2$ and higher-order terms are quite negligible, and likewise for $i(z + \Delta z, t)$ in the second equation. In the second of Eqs. (20.19) we may set $\partial v(z + \Delta z, t)/\partial t \approx \partial v(z, t)/\partial t$, because additional terms would merely produce terms of higher order in Δz, and thus obtain

$$L\frac{\partial i(z,t)}{\partial t} = -\frac{\partial v(z,t)}{\partial z} \quad \text{and} \quad C\frac{\partial v(z,t)}{\partial t} = -\frac{\partial i(z,t)}{\partial z} \tag{20.20}$$

Now solve by setting $v(z, t) = V(z)e^{j\omega t}$ and $i(z, t) = I(z)e^{j\omega t}$. This leads directly to (20.15), as is easily seen from the ensuing phasor equations after replacing $\partial/\partial t$ by $j\omega$ in (20.20). Equations (20.15) therefore are indeed represented by the equivalent circuit equations (20.19) of an infinitesimal distributed-circuit element.

20.2 Extension to lossy infinite transmission lines

In practice, transmission lines are not lossless, as assumed above. There are two major types of loss of electromagnetic energy to the materials of the transmission line: *bulk loss* and *surface loss*. Bulk loss is due to the fact that the dielectric medium inside the TL is lossy, i.e. $\varepsilon = \varepsilon_0(\varepsilon' - j\varepsilon'')$ with nonzero imaginary part ε''. This loss mechanism causes current with density **J** to leak from one plate through the medium to the other and is thus responsible for a *bulk conductance* G. The bulk conductance may be considered as due to the fact that electromagnetic energy is absorbed per m^3 in the lossy dielectric from the wave mode propagating inside the TL. The parameter G can be expressed as

$$G \equiv \frac{I}{V} = V^{-1}\oint d\mathbf{S}\cdot\mathbf{J} = V^{-1}\oint d\mathbf{S}\cdot\sigma\mathbf{E} \tag{20.21}$$

given that we write $\varepsilon'' = \sigma/\varepsilon_0\omega$, as we did in Section 17.2, so that the effective current density $\mathbf{J} = \sigma\mathbf{E}$. The surface integral extends over the surface of one interface with a bounding plate, i.e. over one of the surfaces $x = 0$ or $x = d$ of a parallel-plate TL, or over either the core or the sheath surface of a coaxial-cable TL (but not over both!). The parameter σ is the effective conductivity of the lossy dielectric. For dielectrics, we have already found the capacitance C to be

$$C \equiv \frac{Q}{V} = V^{-1}\oint dS\,\varrho_s = V^{-1}\oint d\mathbf{S}\cdot\varepsilon_0\varepsilon'\mathbf{E} \tag{20.22}$$

For uniform media both σ and ε' are constants, and thus we find when comparing (20.21) and (20.22) that

$$G = \frac{\sigma}{\varepsilon_0\varepsilon'}C \tag{20.23}$$

The requirement that the medium be uniform is overly restrictive. We really require (20.23) to apply to the capacitance and conductance *per unit length* in (20.21) and (20.22). Therefore, is suffices to require uniformity only in the direction $\perp \hat{\mathbf{z}}$, i.e. that $\sigma(z)$ and $\varepsilon'(z)$ vary only along the transmission-line axial direction.

The surface-loss mechanism is more complicated. While our idealization has been waveguides with perfectly conducting walls, the realistic situation is that such walls are made of materials with finite (not infinite) conductivities. Thus, the boundary condition that the tangential component of **E** must be zero at the interface is too stringent. A careful calculation that goes beyond the constraints of this text shows that a nonzero E_z component must exist. This nonzero component is directed in such a way that it combines with **H** to give an average Poynting vector that radiates *up* into the top metal plate at $x = d$ and *down* into the bottom metal plate at $x = 0$. Those two equal but opposite Poynting vectors represent the power flux lost from the TL at distance z (and that power flux is independent of z). We find (see the optional section below) for the power in W/m radiated per unit length of TL into each metal plate of width w

$$P = \frac{1}{2}\left(\frac{1}{w\delta_{\mathrm{m}}\sigma_{\mathrm{m}}}\right)I^2 \tag{20.24}$$

where σ_{m} is the conductivity and $\delta_{\mathrm{m}} \approx \sqrt{2/(\mu_0\omega\sigma_{\mathrm{m}})}$ is the skin depth of the nonmagnetic metal. The total power P_{tot} radiated per unit length of transmission line is twice as much

$$P_{\mathrm{tot}} = \left(\frac{1}{w\delta_{\mathrm{m}}\sigma_{\mathrm{m}}}\right)I^2 \equiv \frac{1}{2}RI^2 \quad \text{for a sinusoidal current} \tag{20.25}$$

and thus we define a *resistance parameter* $R = 2/(\omega\delta_{\mathrm{m}}\sigma_{\mathrm{m}})$ which works out to be

$$R = \frac{2}{w}\sqrt{\frac{\mu_0\omega}{2\sigma_{\mathrm{m}}}} \tag{20.26}$$

The power per unit length, P_{tot}, represents the loss due to power radiated into the metal plates, which in turn is converted into heat due to the high (but finite) conductivity of the metal. It is the *surface loss* to which we alluded at the beginning of this explanation.

We can make (20.24) plausible as follows. In subsection 20.1.1 we found that the idealized loss-free parallel-plate TL predicted the existence of a sheet-current density $\mathbf{K}_{\mathrm{m}} = \pm H_y\hat{\mathbf{z}}$ at the interface of the metallic walls with the dielectric (the minus sign holds at $x = d$ and the plus sign at $x = 0$). Here, $H_y = H_0(z)$, as specified in (20.8). Now that we have a situation with finite rather than infinite conductivity, this sheet-current density $\mathbf{K}_{\mathrm{m}}(z)$ in the *upper* plate is replaced by a rapidly decaying current density $\mathbf{J}_{\mathrm{m}}(x,z) = \mathbf{J}_{\mathrm{m}}(d,z)e^{-(x-d)/\delta_{\mathrm{m}}}$ that obeys

$$\mathbf{K}_{\mathrm{m}}(z) = \int_d^\infty dx\,\mathbf{J}_{\mathrm{m}}(x,z) = \int_d^\infty dx\,\mathbf{J}_{\mathrm{m}}(d,z)e^{-(x-d)/\delta_{\mathrm{m}}} \tag{20.27a}$$

with $\delta_{\mathrm{m}} \approx \sqrt{2/(\mu_0\omega\sigma_{\mathrm{m}})}$. Here δ_{m} is the skin depth in the nonmagnetic metal (in which we have replaced μ by μ_0), which is characterized by a large but finite conductivity σ_{m}; see Fig. 20.3 again. The integral is easily carried out to give

$$\mathbf{K}_{\mathrm{m}}(z) = \delta_{\mathrm{m}}\mathbf{J}_{\mathrm{m}}(d,z) = \delta_{\mathrm{m}}\sigma_{\mathrm{m}}\mathbf{E}_z(d,z) \tag{20.27b}$$

We have applied Ohm's law to the current density at the interface $x = d$ to obtain the last form of (20.27b). Because we already know that $\mathbf{K}_{\mathrm{m}}(z) = -H_0\hat{\mathbf{z}}$

at the top plate, we find additionally:

$$\mathbf{E}_z(d,z) = -\frac{H_0}{\delta_m \sigma_m}\hat{\mathbf{z}} \tag{20.28a}$$

So we have a small **E** component in the $-\hat{\mathbf{z}}$ direction at $z = 0$, and a large **H** component in the $+\hat{\mathbf{y}}$ direction at $x = d$. There is a component of the Poynting vector $S_{av} = \frac{1}{2}\mathrm{Re}\{\mathbf{E} \times \mathbf{H}^*\}$ that points in the $+\hat{\mathbf{x}}$ direction, and that component represents energy *radiated from the medium into the metal.* By means of a very similar calculation we find

$$\mathbf{E}_z(0,z) = \frac{H_0}{\delta_m \sigma_m}\hat{\mathbf{z}} \tag{20.28b}$$

which indicates a component of the average Poynting vector pointing *downwards* at $x = 0$, thus indicating equal and opposite radiation down into the plate at $x = 0$. These longitudinal electric fields vanish as $\sigma_m \to \infty$ and we return to the idealized loss-free case. But in reality we observe that such fields must exist in finite-conductivity media, and therefore there are *radiation losses* into the walls of the transmission line (TL). Since $\mathbf{H} = H_0\hat{\mathbf{y}} = |K_m|\hat{\mathbf{z}}$ then according to (20.8), which we have rewritten in an abbreviated form instead of using $\mathbf{K}_m(0,z) = -\mathbf{K}_m(d,z)$ with the $\exp(\pm jkz)$ factors, we find that the Poynting vectors at $x = 0$ and $x = d$ are given by

$$S(d)_{av} = \frac{1}{2}\hat{\mathbf{x}}\frac{|K_m|^2}{\delta_m \sigma_m} \qquad S(0)_{av} = -\frac{1}{2}\hat{\mathbf{x}}\frac{|K_m|^2}{\delta_m \sigma_m} \tag{20.29}$$

Thus, the power P radiated *per unit length* of parallel-plate TL is found by multiplying both power fluxes given in (20.29) by the width w and noting that $w|K_m|^2 = w^2|K_m|^2/w = I^2/w$, to obtain (20.24).

So, we have found two loss parameters. One, the *bulk-loss* parameter, can be written quite generally as $G = (\sigma/\varepsilon_0\varepsilon')C$. The other, the *surface-loss* parameter, is more specifically given for a parallel-plate TL in (20.26). Because $C = \varepsilon'\varepsilon_0 w/d$ is the capacitance per unit length of a parallel-plate TL, we find specifically for the parallel-plate transmission line:

$$G = \frac{\sigma w}{d} \ \ (\mathrm{S/m}) \qquad R = \frac{2}{w}\sqrt{\frac{\mu_0 \omega}{2\sigma_m}} \ \ (\Omega/\mathrm{m}) \tag{20.30}$$

The lossy transmission line is therefore governed by four parameters, L, C, G, and R. These are given specifically for the parallel-plate TL by (20.15) and (20.30), but G and C are connected as described above. In (20.16b) we also saw that

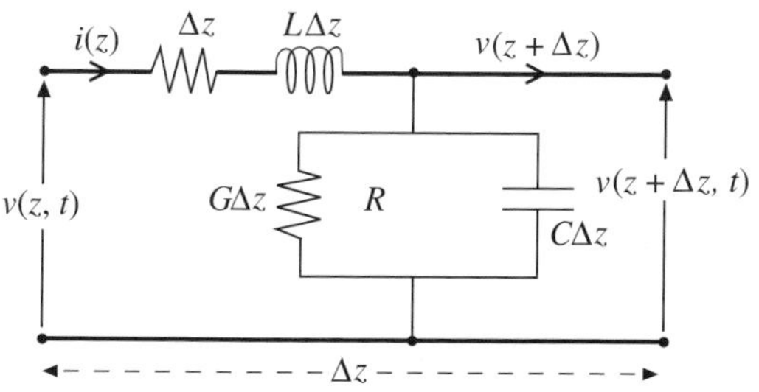

Fig. 20.6 The equivalent distributed circuit element for lossy transmission lines.

$LC = \varepsilon_0\varepsilon'\mu$, but this connection holds only for *lossless* transmission lines. Shortly, we shall see that it is approximately true for TLs with low-loss dielectric media; we will then have established two relationships between the four parameters.

The equivalent distributed-circuit diagram for an infinitesimal length Δz of TL, shown in Fig. 20.5, now can be extended to one for a lossy TL. It is shown in Fig. 20.6. The Kirchhoff voltage and current laws for the circuit are

$$\begin{aligned} v(z,t) - R\Delta z\, i(z,t) - L\Delta z\frac{\partial i(z,t)}{\partial t} &= v(z+\Delta z,t) \\ i(z,t) - G\Delta z\, v(z+\Delta z,t) - C\Delta z\frac{\partial v(z+\Delta z,t)}{\partial t} &= i(z+\Delta z,t) \end{aligned} \tag{20.31a}$$

In the first of these equations, set $v(z+\Delta z,t) \approx v(z,t) + \Delta z\partial v(z,t)/\partial z$, as the $(\Delta z)^2$ and higher-order terms are negligible, and do likewise for $i(z+\Delta z,t)$ in the second equation. We may replace $z+\Delta z$ by z in $\partial v(z+\Delta z,t)/\partial t$ and $\partial i(z+\Delta z,t)/\partial t$ (the other Taylor-series terms only produce higher-order terms in Δz). After canceling the infinitesimal factors Δz, which now are common to all terms, we obtain

$$\frac{\partial v}{\partial z} = -Ri - L\frac{\partial i}{\partial t} \qquad \frac{\partial i}{\partial z} = -Gv - C\frac{\partial v}{\partial t} \tag{20.31b}$$

and v, i both are functions of z and t. Now again solve by setting $v(z,t) = V(z)e^{j\omega t}$ and $i(z,t) = I(z)e^{j\omega t}$ to obtain the phasor equations

$$\frac{dV(z)}{dz} = -(R+j\omega L)I(z) \qquad \frac{dI(z)}{dz} = -(G+j\omega C)V(z) \tag{20.32}$$

These equations revert to the lossless TL equations (20.15), as they should, if we set $G = 0$ and $R = 0$.

In restricting ourselves temporarily to *infinite* transmission lines, we are merely restricting ourselves to those cases in which guided waves produced by a source inside a TL continue to travel in only one direction. Such waves will reach the end of a finite TL and be reflected from the end to produce waves propagating in the opposite direction, which interfere with the originally produced waves. Thus, there will be a combination of $\exp(jkz)$ with $\exp(-jkz)$ waves in general. For the time being this possibility will not be considered; we will assume that there are only $\exp(-jkz)$ waves. We also will adopt an engineering convention common to transmission-line analysis; *we set* $\gamma = jk$ so that in general $V(z)$ and $I(z)$ are proportional to $\exp(-\gamma z)$ in an infinite TL. As a result, (20.32) yields

$$\gamma \equiv \alpha + j\beta = \sqrt{(R + j\omega L)(G + j\omega C)} \tag{20.33}$$

In the lossless case we then have $\gamma \equiv j\beta = j\omega\sqrt{LC}$.

The *characteristic impedance* defined by the ratio $V(z)/I(z)$ yields a constant $V_0/I_0 \equiv Z_0$ given that $V(z) = V_0 \exp(-\gamma z)$ and $I(z) = I_0 \exp(-\gamma z)$. This constant is also found by inserting these two exponential relationships into (20.32), resulting in

$$\gamma V_0 = (R + j\omega L)I_0 \qquad \gamma I_0 = (G + j\omega C)V_0 \tag{20.34a}$$

Division of the first by the second of these equations yields

$$Z_0 = \sqrt{(R + j\omega L)/(G + j\omega C)} \tag{20.34b}$$

A comparison of equations (20.21) and (20.22) showed that $G/\omega C = \varepsilon''/\varepsilon'$, the loss tangent of the guiding medium inside or between the conducting plates of the TL. Such media are always chosen to have $\varepsilon'' \ll \varepsilon'$ (near-zero loss tangent) at normal operating frequencies so that we can count almost always on the inequality $G \ll \omega C$ in a transmission line. The 'almost' refers here to the possibility that there are regimes at operating frequencies much higher than usual, where the inequality no longer holds in certain dielectrics (e.g. water at 1–90 GHz). Thus

$$\varepsilon'' \ll \varepsilon' \qquad \longrightarrow \qquad G \ll \omega C \tag{20.35}$$

Equations (20.15) and (20.26) yield for the parallel-plate TL the ratio

$$\frac{R}{\omega L} = \sqrt{\frac{2\mu_\mathrm{m}}{\omega\sigma_\mathrm{m}\mu_0^2}}\,\frac{1}{d} \approx \sqrt{\frac{2}{\omega\sigma_\mathrm{m}\mu_0}}\,\frac{1}{d} = \frac{\delta}{d} \tag{20.36}$$

because $\mu_\mathrm{m} \approx \mu_0$. Dimensionally similar results arise for other transmission lines, possibly with coefficients of order unity multiplying the ratio δ/d (which is the ratio of the *skin depth* to the *plate separation distance*). This ratio, again, is usually very small at normal operating frequencies and it becomes smaller still at higher frequencies, so that

$$\delta \ll d \qquad \longrightarrow \qquad R \ll \omega L \tag{20.37}$$

Inequalities (20.35) and (20.37) are almost always valid and we shall explore their consequences.

20.2.1 Low-loss transmission line

In the extreme case of zero losses we have seen that $\beta = \omega\sqrt{LC}, \alpha = 0, Z_0 = \sqrt{L/C}$ and the phase velocity of the wave is $v_{ph} = \omega/\beta = 1/\sqrt{LC}$ is independent of frequency. The wave is *nondispersive* so that pulses and other waves of nonzero bandwidth retain their shape without distortion while they propagate through the transmission line. In the more realistic case that $G \ll \omega C$ and $R \ll \omega L$ we may approximate as follows:

$$\begin{aligned}\sqrt{G + j\omega C} &= \sqrt{j\omega C}\,(1 - jG/\omega C)^{1/2} \approx \sqrt{j\omega C}\,(1 - jG/2\omega C)\\ \sqrt{R + j\omega L} &= \sqrt{j\omega L}\,(1 - jR/\omega L)^{1/2} \approx \sqrt{j\omega L}\,(1 - jR/2\omega L)\end{aligned} \tag{20.38}$$

Multiplication of these two approximate equalities yields

$$\gamma \approx j\omega\sqrt{LC}\left[1 - \frac{j}{2\omega}\left(\frac{G}{C} + \frac{R}{L}\right) + \frac{RG}{4\omega^2 CL}\right] \tag{20.39}$$

The last term is very much smaller than the $1/\omega$ terms, which in turn are both very much less than unity in the low-loss TL. As $\gamma = \alpha + j\beta$ it follows that

$$\beta \approx \omega\sqrt{LC}\left(1 + \frac{RG}{4\omega^2 CL}\right) \qquad \alpha \approx \frac{1}{2}\sqrt{LC}\left(\frac{G}{C} + \frac{R}{L}\right) \tag{20.40}$$

Thus β has a very small term $\propto 1/\omega$, which indicates that the medium is weakly dispersive (small differences in the phase velocity $v_{ph} = \omega/\beta$ of different frequency components). The wave can travel relatively large distances into the TL before appreciable distortions of pulse shapes occur. The attenuation $\alpha \ll \beta$ is also relatively weak as the amplitudes of the fields are down by $1/e$ only after very many wavelengths. Division of the second of (20.38) by the first yields

$$Z_0 \approx \sqrt{\frac{L}{C}}\left[1 + \frac{j}{2\omega}\left(\frac{G}{C} - \frac{R}{L}\right) + \frac{RG}{4\omega^2 CL}\right] \tag{20.41}$$

so that the characteristic impedance has a reactive component much smaller than the resistive one.

20.2.2 Distortion-free approximation

Reconsider (20.33) and write it as

$$\gamma \equiv \alpha + j\beta = (R + j\omega L)\sqrt{\frac{G + j\omega C}{R + j\omega L}} \tag{20.42}$$

If the square root were a real number independent of frequency, say κ, then we would have $\beta = \omega\kappa L$ so that $v_{ph} = 1/\kappa L$ would be independent of frequency and no distortion of pulses would occur, other than attenuation due to the fact that $\alpha = \kappa R$. The condition for this to happen is thus

$$G + j\omega C = \kappa^2 (R + j\omega L) \qquad (20.43)$$

and this condition requires (because κ is real) that $G = \kappa^2 R$ and $C = \kappa^2 L$, from which we see that $G/R = C/L$ or

$$\frac{G}{C} = \frac{R}{L} \quad \text{and} \quad \gamma = \sqrt{\frac{C}{L}}(R + j\omega L) \qquad (20.44)$$

The second equality follows from the use of (20.43) and $\kappa^2 = C/L$ in (20.42). Thus we find that $\beta = \omega\sqrt{LC}$, so that $v_{\text{ph}} = 1/\sqrt{LC}$ is independent of frequency. Insertion of (20.43) into (20.34b) with $\kappa^2 = C/L$ shows that $Z_0 = \sqrt{L/C}$: the characteristic impedance is purely resistive. As almost always $R \ll \omega L$, we see from (20.44) that the attenuation $\alpha = R\sqrt{Z_0/L}$ is very small, although nonzero.

The lack of distortion is not perfect. From (20.30) we observe (at least for the parallel-plate TL) that $R \propto \omega^{-1/2}$, so that $R/L \propto \omega^{-1/2}$, whereas G/C is not a function of frequency. Consequently, if (20.44) holds for one frequency, it does not for other frequencies. In general, however, a pulsed signal will be almost distortion-free if its bandwidth is not large compared to a central (or carrier) frequency chosen so that (20.44) holds. Only for those materials and in those frequency ranges where G/C and R/L have the same frequency dependence can one expect the 'perfect' lack of distortion obtained when (20.44) holds.

20.2.3 High-frequency approximation

There are parameter and frequency regimes in which it is not true that $G \ll \omega C$, whereas the small ratio $R \ll \omega L$ diminishes rapidly. In the high-frequency approximation, we assume that $R \approx 0$ so that

$$\gamma = \sqrt{j\omega L(G + j\omega C)} = j\omega\sqrt{LC}\sqrt{(1 - jG/\omega C)} \qquad (20.45a)$$

With $G/C = \sigma/\omega\varepsilon'\varepsilon_0$, it follows that

$$\gamma = j\omega\sqrt{LC}\sqrt{(1 - j\sigma/\omega\varepsilon'\varepsilon_0)} \qquad (20.45b)$$

However, the analysis with TEM electromagnetic fields **E**, **H** has shown us that $\gamma = j\omega\sqrt{\mu_0\varepsilon}$ when $R = 0$, with $\varepsilon = \varepsilon_0(\varepsilon' - j\sigma/\varepsilon_0\omega)$, so that we obtain from that analysis

$$\gamma = j\omega\sqrt{\mu\varepsilon'\varepsilon_0}\sqrt{(1 - j\sigma/\omega\varepsilon'\varepsilon_0)} \qquad (20.46)$$

Comparison of (20.46) with (20.45a) yields the high-frequency approximation

$$LC \approx \mu_0\varepsilon'\varepsilon_0 \qquad (20.47)$$

While we cannot distill simplified formulas for β, α and v_{ph} in this case, unless $G \ll \omega C$, the high-frequency approximation does have one very useful consequence. The fact that

$$LC = \mu_0 \varepsilon' \varepsilon_0 \quad \text{and} \quad \frac{G}{C} = \frac{\sigma}{\varepsilon' \varepsilon_0} \tag{20.48}$$

implies that we can find all three unit-length quantities L, C, G if we know only one of them. If, for example, we have determined the capacitance per unit length C at high frequencies for a structure, then (20.48) yields both L and G from C.

20.3 Finite-length transmission lines

The situation described above is unrealistic in finite-length transmission lines after a transient time long enough for wave modes launched at one end of the TL to reach the other end and return. Modes are reflected at the ends – we will discuss the mechanism later – and there are wave modes in both $+\hat{\mathbf{z}}$ and $-\hat{\mathbf{z}}$ directions. We must expect solutions that are the sum of forward and backward components:

$$\begin{aligned} V(z) &= V_{\text{f}} e^{-\gamma z} + V_{\text{b}} e^{\gamma z} \\ I(z) &= I_{\text{f}} e^{-\gamma z} + I_{\text{b}} e^{\gamma z} \end{aligned} \tag{20.49}$$

instead of $V(z) = V_0 \exp(-\gamma z)$ and $I(z) = I_0 \exp(-\gamma z)$, the expressions we obtained for modes propagating only in one direction. If, again, we differentiate each term in (20.32), then use (20.32) – as it was before this differentiation – to eliminate the first derivatives, we obtain $d^2V/dz^2 - \gamma^2 V = 0$ and $d^2I/dz^2 - \gamma^2 I = 0$, with γ given by (20.33). Clearly (20.49) satisfies these equations with this definition of γ, and thus the complex propagation constant is unchanged from the one-way propagation case. However if (20.49) is inserted into the single-derivative equations (20.32) we obtain the more complicated relationships

$$\begin{aligned} V_{\text{f}} e^{-\gamma z} - V_{\text{b}} e^{\gamma z} &= Z_0 (I_{\text{f}} e^{-\gamma z} + I_{\text{b}} e^{\gamma z}) \\ V_{\text{f}} e^{-\gamma z} + V_{\text{b}} e^{\gamma z} &= Z_0 (I_{\text{f}} e^{-\gamma z} - I_{\text{b}} e^{\gamma z}) \end{aligned} \tag{20.50}$$

$$Z_0 \equiv \sqrt{(R + j\omega L)/(G + j\omega C)} \tag{20.51}$$

These relationships are satisfied by requiring

$$V_{\text{f}} = Z_0 I_{\text{f}} \quad \text{and} \quad V_{\text{b}} = -Z_0 I_{\text{b}} \tag{20.52}$$

and it is obvious that two more relationships are needed to be able to determine all four coefficients of (20.49).

20.3.1 Voltage, current, and impedance equations

In fact, relationships such as (20.52) are determined by the function that a finite transmission line plays in a circuit. If, for example, a TL of length l connects a time-harmonic voltage source V_s with internal impedance Z_s to a load characterized by a complex impedance Z_L, as in Fig. 20.7, then we do have a completely determined system, and we should be able to express all unknowns in terms of V_s, Z_s, and the TL's propagation constant γ, characteristic impedance Z_0, and length l, all of which are known.

Application of (20.49) to the load location at $z = l$, together with (20.52), yields

$$V_L = V_f e^{-\gamma l} + V_b e^{\gamma l} \qquad \text{and} \qquad Z_0 I_L = V_f e^{-\gamma l} - V_b e^{\gamma l} \tag{20.53}$$

By considering these to be two linear equations in the two unknowns V_f and V_b, we find

$$V_f = \tfrac{1}{2}(V_L + Z_0 I_L)e^{\gamma l} \qquad V_b = \tfrac{1}{2}(V_L - Z_0 I_L)e^{-\gamma l} \tag{20.54}$$

Thus, (20.52) and (20.54) determine the four coefficients in (20.49) in terms of the three TL quantities γ, Z_0, l and the load quantities V_L, I_L. The two load quantities are usually not specified explicitly, but they are determined by the source voltage and impedance and are connected to each other by the fact that $V_L = Z_L I_L$. The load impedance Z_L may be assumed to be specified in any situation depicted by Fig. 20.7. It is straightforward to insert (20.54) into (20.49) to obtain

$$\begin{aligned} V(z) &= \frac{I_L}{2}\left[(Z_L + Z_0)e^{\gamma(l-z)} + (Z_L - Z_0)e^{-\gamma(l-z)}\right] \\ I(z) &= \frac{I_L}{2Z_0}\left[(Z_L + Z_0)e^{\gamma(l-z)} - (Z_L - Z_0)e^{-\gamma(l-z)}\right] \end{aligned} \tag{20.55a}$$

These equations describe the voltage difference and current at location z (with $0 < z < l$) in terms of the load impedance and current. Several alternative forms

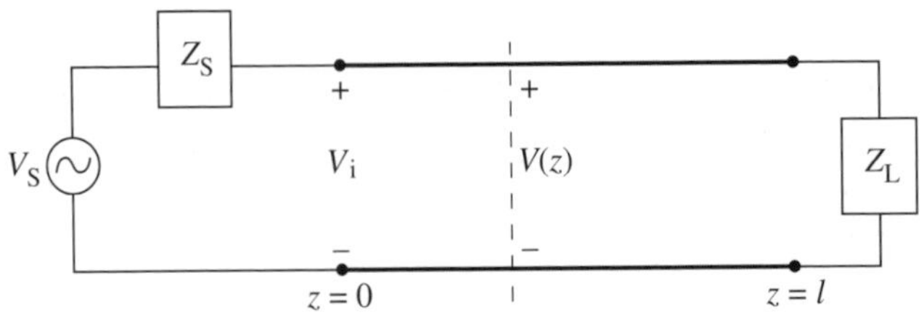

Fig. 20.7 A finite-length transmission line and its voltage-source elements.

are possible. For example, we can pull out a common factor $Z_L + Z_0$ to obtain

$$\begin{aligned} V(z) &= \frac{I_L}{2}(Z_L + Z_0)\left[e^{\gamma(l-z)} + \Gamma e^{-\gamma(l-z)}\right] \\ I(z) &= \frac{I_L}{2Z_0}(Z_L + Z_0)\left[e^{\gamma(l-z)} - \Gamma e^{-\gamma(l-z)}\right] \end{aligned} \qquad (20.55b)$$

with $\Gamma \equiv (Z_L - Z_0)/(Z_L + Z_0)$. This *reflection coefficient* plays the same role that $\mathcal{R}$ does in (18.27a), which in turn describes the *total* **E** and **H** fields on the incident side of a planar interface between free space and a dielectric. In fact the equations have a very similar form if we identify the voltage $I_L(Z_L + Z_0)$ with the electric field amplitude E_{i0}, and the impedances Z_L and Z_0 with η_0 and η_1 respectively. This identification makes quite plausible that $\exp[-\gamma(l - z)]$ represents a reflected wave, and we shall return later in more detail to this interpretation.

Now we use the hyperbolic functions by equating $\exp x = \frac{1}{2}(\cosh x + \sinh x)$ and $\exp(-x) = \frac{1}{2}(\cosh x - \sinh x)$ in (20.55a) to obtain

$$\begin{aligned} V(z) &= I_L\{Z_L \cosh[\gamma(l-z)] + Z_0 \sinh[\gamma(l-z)]\} \\ I(z) &= \frac{I_L}{Z_0}\{Z_L \sinh[\gamma(l-z)] + Z_0 \cosh[\gamma(l-z)]\} \end{aligned} \qquad (20.55c)$$

All three of these forms of (20.55) are equivalent. We will apply whichever form is most useful at any time. Very shortly, we will show how to express I_L in terms of the known circuit quantities. First, note that division of $V(z)$ by $I(z)$ yields the line impedance at location z, which is most usefully written as either one of the two following forms (note that $\tanh x = \sinh x/\cosh x$):

$$Z(z) = Z_0 \frac{Z_L + Z_0 \tanh[\gamma(l-z)]}{Z_0 + Z_L \tanh[\gamma(l-z)]} = Z_0 \frac{1 + \Gamma e^{-2\gamma(l-z)}}{1 - \Gamma e^{-2\gamma(l-z)}} \qquad (20.56)$$

In contrast to current or voltage difference, the line impedance at any location $0 < z < l$ can be expressed purely in the transmission-line quantities γ, Z_0, and l. Thus, the input impedance to the TL is obtained by setting $z = 0$ in (20.56):

$$Z_i = Z_0 \frac{Z_L + Z_0 \tanh \gamma l}{Z_0 + Z_L \tanh \gamma l} = Z_0 \frac{1 + \Gamma e^{-2\gamma l}}{1 - \Gamma e^{-2\gamma l}} \qquad (20.57)$$

Thus, Z_i can be considered as the impedance representing the transmission line plus load in an equivalent circuit containing the input voltage source. The Kirchhoff voltage law then expresses the input voltage and current in terms of the known circuit quantities: $V_i = V_s - Z_s I_i = V_s - V_i Z_s/Z_i$, so that

$$I_i = \frac{V_s}{Z_s + Z_i} \qquad V_i = \frac{Z_i}{Z_s + Z_i} V_s \qquad (20.58)$$

We are almost where we wish to be. We need only apply (20.55c), at, for example, $z = 0$ to obtain

$$\begin{aligned} V_i &= V_L \cosh \gamma l + Z_0 I_L \sinh \gamma l \\ I_i &= \frac{V_L}{Z_0} \sinh \gamma l + I_L \cosh \gamma l \end{aligned} \tag{20.59}$$

We then invert these equations (i.e. solve for output quantities in terms of input quantities) to obtain

$$\begin{aligned} V_L &= V_i \cosh \gamma l - Z_0 I_i \sinh \gamma l \\ I_L &= \frac{V_i}{Z_0} \sinh \gamma l + I_i \cosh \gamma l \end{aligned} \tag{20.60}$$

Now eliminate the input quantities with (20.58) to obtain

$$\begin{aligned} V_L &= \frac{V_s}{Z_s + Z_i} (Z_i \cosh \gamma l - Z_0 \sinh \gamma l) \\ I_L &= \frac{V_s/Z_0}{Z_s + Z_i} (-Z_i \sinh \gamma l + Z_0 \cosh \gamma l) \end{aligned} \tag{20.61}$$

We eliminate Z_i in favor of the specified load impedance Z_L with (20.57).

Let us recapitulate what has now been obtained. Equations (20.55) give the line voltage and current at z in terms of the transmission line quantities and the load current I_L. The latter is given by (20.61) in terms of the source quantities and input impedance Z_i. Finally, (20.57) eliminates the input impedance in terms of the load impedance.

The net result is that $V(z)$ and $I(z)$ are expressed solely in terms of γ, Z_0, l of the TL *and in terms of the source quantities V_s, Z_s.*

Equation (20.56) already gives $Z(z)$ in terms only of the TL quantities γ, Z_0, and l. Hence all three of the line quantities $V(z)$, $I(z)$, and $Z(z)$ can be calculated as functions of source and TL quantities as shown above.

A special situation arises when the load impedance Z_L is identical to the characteristic impedance Z_0 of the transmission line ($Z_L = Z_0$). Equation (20.56) then yields $Z_i = Z_0 = Z_L$; the input impedance is transferred to the load without change. In the customary parlance, one speaks of a TL that is *matched to its load.* Equations (20.55c) show, for $z = 0$ as well as for any z, that the input voltage and current then pass through the transmission line without change. In fact these equations lead to

$$\begin{aligned} V_i &= V_L e^{\gamma l} \quad \longrightarrow \quad V(z) = V_i e^{-\gamma z} \\ I_i &= I_L e^{\gamma l} \quad \longrightarrow \quad I(z) = I_i e^{-\gamma z} \end{aligned} \tag{20.62}$$

which is what was obtained in a TL of infinite length. There is no reflected wave mode.

20.3.2 Role as circuit elements

It should have become clear from subsection 20.3.1 that the transmission line plus load can be regarded as an equivalent impedance Z_i that is inserted in a circuit diagram of voltage source and impedance Z_S, e.g. as in Fig. 20.8; Z_i is given as in (20.57),

$$Z_\mathrm{i} = Z_0 \frac{Z_\mathrm{L} + Z_0 \tanh \gamma l}{Z_0 + Z_\mathrm{L} \tanh \gamma l} \tag{20.63a}$$

If the TL is essentially lossless, then we may replace Z_0 by a real quantity R_0 (a purely resistive characteristic impedance) and γ by $j\beta$. As $\tanh jx = j \tan x$, we obtain

$$Z_\mathrm{i} \approx R_0 \frac{Z_\mathrm{L} + jR_0 \tan \beta l}{R_0 + jZ_\mathrm{L} \tan \beta l} \tag{20.63b}$$

in the lossless approximation. The results of these last two equations applied to short-circuited loads ($Z_\mathrm{L} = 0$) and to open-circuit loads ($Z_\mathrm{L} = \infty$) gives a method of *measuring* the TL quantities γ and Z_0. For the short-circuited load we find $Z_\mathrm{i}^\mathrm{short} = Z_0 \tanh \gamma l$, whereas for the open-circuit load we obtain $Z_\mathrm{i}^\mathrm{open} = Z_0 \coth \gamma l$.

The product and ratio of these two input impedances, which we can measure, yield

$$Z_0 = \sqrt{Z_\mathrm{i}^\mathrm{short} Z_\mathrm{i}^\mathrm{open}} \qquad \gamma = \frac{1}{l} \tanh^{-1}\left(\sqrt{Z_\mathrm{i}^\mathrm{short} / Z_\mathrm{i}^\mathrm{open}}\right) \tag{20.64}$$

so that we indeed obtain these two TL quantities from direct impedance measurements at the input of the transmission line. The inverse function $y = \tanh^{-1} x$ is defined by the relationship $x = \tanh y$; the $\tanh^{-1} x$ function is shown in Fig. 20.9.

In the lossless approximation, for which $Z_\mathrm{i}^\mathrm{short} \approx jR_0 \tan \beta l$ and $Z_\mathrm{i}^\mathrm{open} \approx -jR_0 \cot \beta l$, the length l of the transmission line can be tailored to give purely reactive or purely capacitive short-circuit or open-circuit impedances. Certain special choices of length are also of importance, as follows.

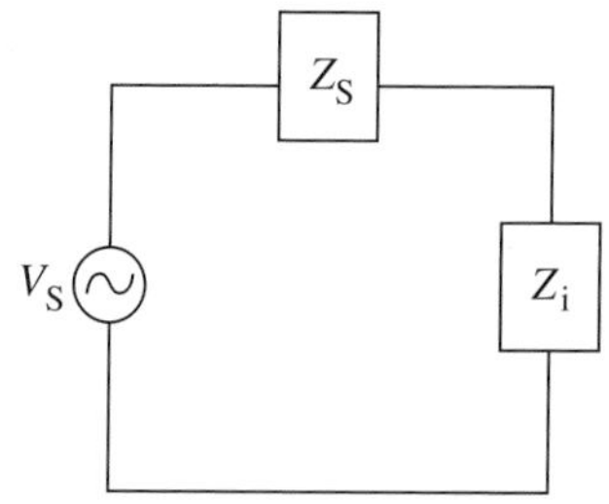

Fig. 20.8 The transmission line plus load as an impedance element Z_i in a circuit.

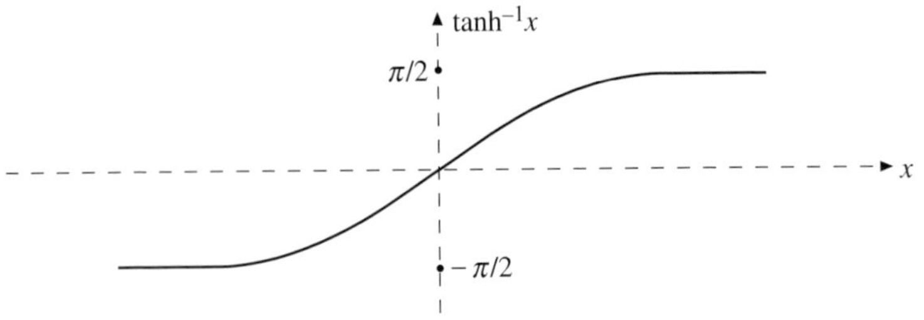

Fig. 20.9 Functional dependence of $\tanh^{-1} x$ upon its argument x.

Quarter-wave TL *section*
If l is chosen to be an odd multiple of a quarter wavelength, i.e. if l is chosen such that $\beta l = n\pi/2$, with $n = 1, 3, 5, \ldots$, then $\tan \beta l = \pm\infty$ and (20.63b) predicts that $Z_i Z_L = R_0^2$. Hence, Z_i is real for real load impedances Z_L and the TL is a resistive impedance in the circuit. It also follows that $Z_i = \infty$ implies a *short-circuited load* ($Z_L = 0$) whereas $Z_i = 0$ implies an *open-circuit load* ($Z_L = \infty$).

Half-wave TL *section*
If l is chosen to be a multiple of a half wavelength, i.e. if l is chosen such that $\beta l = n\pi$, with $n = 1, 2, 3, \ldots$, then $\tan \beta l = 0$ and we find $Z_i = Z_L$. Here we obtain a TL matched to the load impedance for any value of Z_L and Z_0, and thus without requiring that $Z_L = Z_0$. The *length* of the TL has been adjusted to obtain this result.

20.3.3 Reflection coefficients and standing-wave ratio

Equations (20.55b) showed an alternative way of expressing the voltage and current at z in terms of a *reflection coefficient* $\Gamma \equiv (Z_L - Z_0)/(Z_L + Z_0)$:

$$\begin{aligned} V(z) &= \frac{I_L}{2}(Z_L + Z_0)\left[e^{\gamma(l-z)} + \Gamma e^{-\gamma(l-z)}\right] \\ I(z) &= \frac{I_L}{2Z_0}(Z_L + Z_0)\left[e^{\gamma(l-z)} - \Gamma e^{-\gamma(l-z)}\right] \end{aligned} \tag{20.65a}$$

and the analogy was noted to Eq. (18.27a) for the total field on the free-space side of a vacuum–dielectric interface, which we repeat here for convenience:

$$\begin{aligned} \mathbf{E}(\mathbf{r}) &= -E_{i0}\hat{\mathbf{y}}(e^{-j\beta z} - \mathcal{R}e^{j\beta z}) \\ \mathbf{H}(\mathbf{r}) &= \frac{1}{\eta_0}E_{i0}\hat{\mathbf{x}}(e^{-j\beta z} + \mathcal{R}e^{j\beta z}) \end{aligned} \tag{20.65b}$$

$\mathcal{R} = (\eta_0 - \eta_1)/(\eta_0 + \eta_1)$ is comparable to $-\Gamma$ because $\eta_0 \leftrightarrow Z_0$ and $\eta_1 \leftrightarrow Z_L$. This comparison makes plausible why Γ is called a reflection coefficient, and one can interpret (20.65a) is a way comparable to that in which Eqs. (20.65b) are interpreted in the vacuum–dielectric interface situation. The ratio of V to I in (20.65a) is now

$$Z(z) = Z_0 \frac{1 + \Gamma e^{-2\gamma(l-z)}}{1 - \Gamma e^{-2\gamma(l-z)}} \tag{20.66}$$

A full propagation picture with reflections at $z = 0$ and $z = l$ can be obtained from the Kirchhoff voltage equation $V_s = V_i + Z_s I_i$ for the circuit in Fig. 20.8. We need to eliminate I_L from (20.65a) to get this picture. First apply (20.65a) for $z = 0$:

$$\begin{aligned} V_i &= \frac{I_L}{2}(Z_L + Z_0)e^{\gamma l}(1 + \Gamma e^{-2\gamma l}) \\ I_i &= \frac{I_L}{2Z_0}(Z_L + Z_0)e^{\gamma l}(1 - \Gamma e^{-2\gamma l}) \end{aligned} \tag{20.67}$$

Apply the Kirchhoff voltage equation to get

$$V_s = \frac{I_L}{2}(Z_L + Z_0)e^{\gamma l}\left[\left(1 + \frac{Z_s}{Z_0}\right) + \Gamma\left(1 - \frac{Z_s}{Z_0}\right)e^{-2\gamma l}\right]$$

which yields

$$\frac{I_L}{2}(Z_L + Z_0) = \frac{Z_0 V_s e^{-\gamma l}}{Z_s + Z_0}\left[1 - \Gamma\frac{Z_s - Z_0}{Z_s + Z_0)}e^{-2\gamma l}\right]^{-1} \tag{20.68}$$

We may interpret $(Z_s - Z_0)/(Z_s + Z_0) \equiv \Gamma_s$ as an *input reflection coefficient* of the transmission line, whereas $(Z_L - Z_0)/(Z_L + Z_0) \equiv \Gamma$ is the *output reflection coefficient*. Hence, with substitution of (20.68) into (20.65a) we find

$$V(z) = V_s \frac{Z_0 e^{-\gamma z}}{Z_s + Z_0}\left[\frac{1 + \Gamma e^{-2\gamma(l-z)}}{1 - \Gamma_s \Gamma e^{-2\gamma l}}\right] = \tilde{V}_s e^{-\gamma z}\frac{1 + \Gamma e^{-2\gamma(l-z)}}{1 - \Gamma_s \Gamma e^{-2\gamma l}} \tag{20.69}$$

This expresses $V(z)$ in terms of a modified input voltage and transmission-line quantities:

$$\tilde{V}_s \equiv Z_0 V_s/(Z_s + Z_0)$$

The line current $I(z)$ can be expressed similarly. If we expand the denominator by means of the expansion

$$\frac{1}{1 - x} = 1 + x + x^2 + x^3 + \cdots$$

which is valid for $|x| < 1$, we find $V(z) = V_1(z) + V_2(z) + Z_3(z) + \cdots$ with

$$
\begin{aligned}
V_1(z) &= \tilde{V}_s e^{-\gamma z} \\
V_2(z) &= \tilde{V}_s \Gamma e^{-\gamma(2l-z)} = \tilde{V}_s e^{-\gamma l} \Gamma e^{-\gamma(l-z)} \\
V_3(z) &= \tilde{V}_s \Gamma \Gamma_s e^{-2\gamma l} e^{-\gamma z} = \tilde{V}_s e^{-\gamma l} \Gamma e^{-\gamma l} \Gamma_s e^{-\gamma z} \\
V_4(z) &= \tilde{V}_s \Gamma^2 \Gamma_s e^{-2\gamma l} e^{-2\gamma(l-z)} = \tilde{V}_s e^{-\gamma l} \Gamma e^{-\gamma l} \Gamma_s e^{-\gamma l} \Gamma e^{-\gamma(l-z)}
\end{aligned}
\tag{20.70}
$$

etc.

Figure 20.10 illustrates the meaning of each of these terms. Consider as an example the second expression for $V_3(z)$ read from left to right: it indicates a change in $\tilde{V}_s$ by means of propagation over the entire length l (i.e. the factor $e^{-\gamma l}$), then reflection off the load, characterized by the factor Γ, then propagation all the way back to the source or *generator* (another factor $e^{-\gamma l}$) and reflection off the generator (the factor Γ_s), and finally forward propagation to z (the factor $e^{-\gamma z}$). In the figure, each of these factors is shown in the order in which they occur in the third line of (20.70). Equation (20.69) represents the *sum* of contributions to $V(z)$ produced at z from all possible paths (we speak of *multipaths*) with reflections from source and load. The analogy with the situation in Chapter 19 (plane waves reflected from interfaces) leaps out at us.

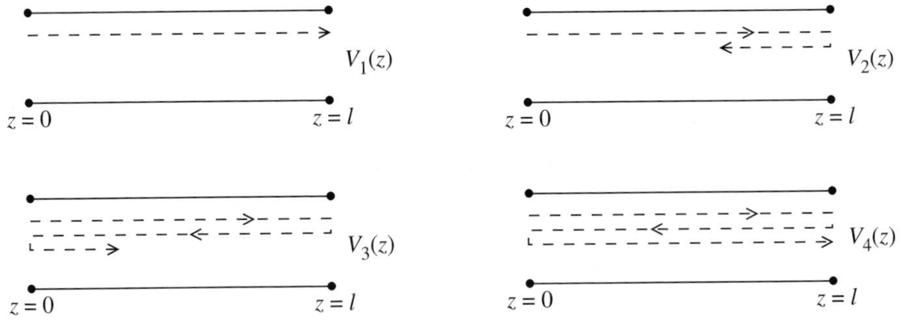

Fig. 20.10 Diagram of a wave mode inside a transmission line interpreted as successively reflecting from either end.

Returning to the expressions (20.65a) for voltage and for current, we find that we can write these as

$$
\begin{aligned}
V(z) &= \tilde{V} e^{\gamma(l-z)} \left[1 + \Gamma e^{-2\gamma(l-z)}\right] \quad \text{with} \quad \tilde{V} \equiv \frac{I_L}{2}(Z_L + Z_0) \\
I(z) &= \tilde{I} e^{\gamma(l-z)} \left[1 - \Gamma e^{-2\gamma(l-z)}\right] \quad \text{with} \quad \tilde{I} \equiv \tilde{V}/Z_0
\end{aligned}
\tag{20.71}
$$

We will now consider these expressions for transmission lines that are sufficiently short that losses along the line are negligible. Then γ may be replaced by $j\beta$. The reflection coefficient $\Gamma = (Z_L - Z_0)/(Z_L + Z_0)$ can be approximated by $\Gamma \approx (Z_L - R_0)/(Z_L + R_0)$, where R_0 is a real characteristic impedance. However, Z_L can be real or complex, so that even in this practically lossless case we must write $\Gamma = |\Gamma|e^{j\theta}$, where θ is the phase angle of an essentially complex reflection coefficient (when the load impedance is complex). Hence, in (20.71) we make the replacement

$$\Gamma e^{-2\gamma(l-z)} \quad \longrightarrow \quad |\Gamma|e^{j\varphi} \quad \text{with} \quad \varphi(z) = \theta - 2j\beta(l-z) \tag{20.72a}$$

and we have

$$|\Gamma| = \sqrt{\frac{(R_L - R_0)^2 + X_L^2}{(R_L + R_0)^2 + X_L^2}} \qquad \tan\theta = \frac{2R_0X_L}{(R_L + R_0)^2 + X_L^2} \tag{20.72b}$$

because $Z_L = R_L - jX_L$. The absolute values of voltage and current therefore can be shown, after using the trignometric identities $\cos\varphi = 2\cos^2(\frac{1}{2}\varphi) - 1 = 1 - 2\sin^2(\frac{1}{2}\varphi)$, to become

$$\begin{aligned} |V(z)| &= |\tilde{V}|\sqrt{(1-|\Gamma|)^2 + 4|\Gamma|\cos^2\left[\frac{1}{2}\varphi(z)\right]} \\ |I(z)| &= |\tilde{I}|\sqrt{(1-|\Gamma|)^2 + 4|\Gamma|\sin^2\left[\frac{1}{2}\varphi(z)\right]} \end{aligned} \tag{20.73}$$

As only φ is a function of z, and a linear function at that, it follows that the absolute values of voltage and current are *periodic functions* of distances z along the TL. There is no compelling reason to single out the voltage, other than that it can be easier to measure along a TL than the current, but if we do consider the voltage then it becomes clear that $|V(z)|$ has a maximum and a minimum value determined by the angle $\varphi(z)$. This in turn gives rise to the concept of *voltage standing-wave ratio* ($VSWR$) defined by the ratio of maximum to minimum absolute voltage,

$$VSWR \equiv \frac{|V(z)|_{\max}}{|V(z)|_{\min}} = \frac{1+|\Gamma|}{1-|\Gamma|} \tag{20.74}$$

20.3.4 Matched impedance

The usual concept of 'matched impedance' in a circuit derives from the wish to maximize the average power $P = \frac{1}{2}\text{Re}\{VI^*\}$, in the understanding that V and I are phasor representations. In a conventional circuit such as Fig. 20.8 one chooses the load impedance to have the relationship $Z_i = Z_s^*$ (the asterisk denotes the complex conjugate). It is shown in all introductory network-analysis texts that this maximizes the average power transferred to the load. However, in transmission-line parlance one speaks of a matched load when one chooses $Z_L = Z_0$. From the definition of the reflection coefficient, it then follows that $\Gamma = 0$, as a consequence of which it follows from (20.65a) that $V(z) = V_i e^{-\gamma z}$ and $I(z) = I_i e^{-\gamma z}$. In short it follows that Eqs. (20.9), (20.10) hold. The choice $Z_L = Z_0$ makes the finite-length TL behave as if it were infinite; no wave is reflected and the wave is transmitted beyond the end of the line without reflection. Do not confuse this concept with the choice of a half-wavelength section of transmission line (see subsection 20.3.2 above) in which it is found that $Z_i = Z_L$. The fortuitous combination $Z_i = Z_L = Z_0$ will allow one to match the input to the load impedance without loss of signal due to internal reflections in the TL. The term 'fortuitous' is used here because it may not always be possible to choose $Z_0 = Z_L$ for an arbitrary load.

20.4 Measurement of transmission-line parameters

The discussion in this section will focus on how measurements of voltages and distances in an essentially lossless transmission line can lead to determination of the wavelength and reflection-coefficient parameters of a transmission line with load impedance.

20.4.1 Measurement for transmission lines with purely resistive termination impedance

Let us first consider loads with real impedance, so that $|\Gamma| = |(R_L - R_0)/(R_L + R_0)|$ is either $+\Gamma$ or $-\Gamma$, depending upon whether R_L is either greater or less than R_0. Equations (20.73) then simplify to

$$\begin{aligned} |V(z)| &= |\tilde{V}|\sqrt{(1-\Gamma)^2 + 4\Gamma\cos^2[\beta(l-z)]} \\ |I(z)| &= |\tilde{I}|\sqrt{(1-\Gamma)^2 + 4\Gamma\sin^2[\beta(l-z)]} \end{aligned} \tag{20.75a}$$

and the extreme values depend upon whether $\Gamma > 0$ or $\Gamma < 0$. The reason why $\Gamma = |\Gamma|e^{j\theta}$ appears here instead of $|\Gamma|$ is that $\theta = 0$ for $R_L > R_0 (\Gamma > 0)$, whereas $\theta = \pm\pi$ for $R_L < R_0 (\Gamma < 0)$. These formulas also follow directly from (20.71). We see that

$$|V(l)| = |\tilde{V}|(1+\Gamma) \qquad |I(l)| = |\tilde{I}| = (1-\Gamma) \tag{20.75b}$$

so that $|V(l)|$ is a maximum and $|I(l)|$ is maximum when $\Gamma > 0$. However, $|V(l)|$ is a minimum and $|I(l)|$ is maximum when $\Gamma < 0$. So, while the absolute value of the termination impedance is at an extreme value for real load impedance, *that extreme*

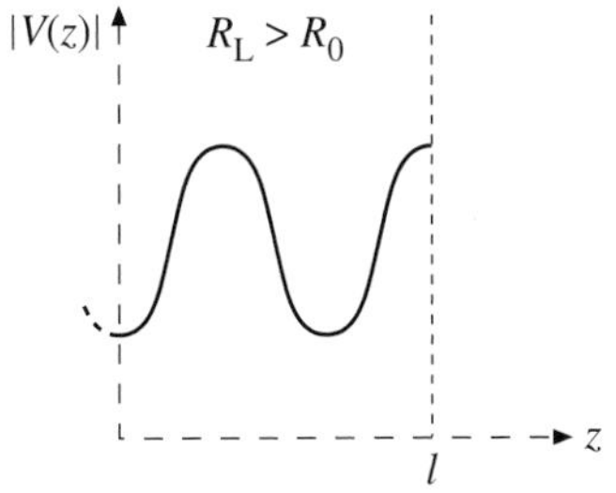

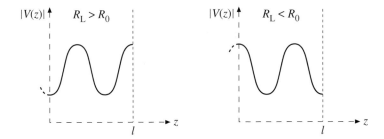

Fig. 20.11 Behavior of $|V(z)|$ vs. z near the termination of a transmission line.

value is a maximum only when $R_L > R_0$, *and it is minimum when* $R_L < R_0$ (see Fig. 20.11). The following special cases can be seen to follow from (20.65)–(20.74) without difficulty:

Short circuit

A short circuit indicates that $R_L = 0$ so that $\Gamma = -1$, but since $|\Gamma| = 1$ the $VSWR \to \infty$. From (20.65a) we find

$$\begin{aligned} V(z) &= jR_0 I_L \sin[\beta(l - z)] \\ I(z) &= I_L \cos[\beta(l - z)] \end{aligned} \tag{20.76}$$

i.e. a standing wave is set up in the transmission line. Note that $|V(l)| = |V|_{\min} \equiv 0$ and $|I(l)| = I_{\max} \equiv R_0 I_L$ at the termination $z = l$. Thus, both $V(z)$ and $I(z)$ reach extreme values at the termination.

Open circuit

An open circuit indicates that $R_L = \infty$ so that $\Gamma = 1$, and again $|\Gamma| = 1$ so that the $VSWR \to \infty$. From (20.65a) we find

$$\begin{aligned} V(z) &= V_L \cos[\beta(l - z)] \\ I(z) &= j(V_L/R_0) \sin[\beta(l - z)] \end{aligned} \tag{20.77}$$

i.e. a standing wave is set up in the transmission line in this case also, and now $|V(l)| = |V|_{\max} \equiv V_L$, whereas $|I(l)| = |I|_{\min} = 0$, at the termination. The extreme values are reversed from the short-circuit case.

Matched load

If $R_L = R_0$ then $\Gamma = 0$, so that the $VSWR = 1$ and (20.65a) with (20.67) yields

$$V(z) = V_L e^{j\beta(l-z)} = V_i e^{-j\beta z} \tag{20.78}$$

This, of course, corresponds to the previously discussed case in which the transmission line simply transmits the wave without change from beginning to end because the reflection coefficient is zero.

20.4.2 Measurement for transmission lines with complex termination impedance

If the load impedance Z_L is complex, then we can use measurements of $|V(l)|$ or $|I(l)|$ to obtain information about the phase constant θ of the reflection coefficient $\Gamma = |\Gamma|e^{j\theta}$ in the case of an essentially lossless transmission line. To see how, consider again the formula for the absolute value of voltage:

$$|V(z)| = |\tilde{V}|\sqrt{(1-|\Gamma|)^2 + 4|\Gamma|\cos^2\left[\frac{1}{2}\varphi(z)\right]}$$
$$\varphi(z) = \theta - 2\beta(l-z) \tag{20.79}$$

The use of $|I(z)|$ is also possible, as noted above, and it can be handled in a fashion similar to what follows here for $|V(z)|$. However, let us proceed with the absolute voltage. A plot of (20.79) as a function of z is shown in Fig. 20.12 for a hypothetical TL of length $l = 5$ cm, with a wavelength $\lambda = 2\pi/\beta = 3$ cm and with reflection coefficient such that $|\Gamma| = 0.8$ and $\theta = 106°$. Three features implicit to the expression (20.79) immediately come to the fore:

(1) The curve of $|V(z)|$ as a function of z is periodic.
(2) The minima of $|V(z)|$ are sharper than the maxima.
(3) The termination at $z = l$ is no longer an extreme value of $|V(z)|$.

As in the case of real termination, we observe that $|\Gamma|$ can be inferred from measuring the maximum and the minimum value of $|V(z)|$ and thus obtaining the

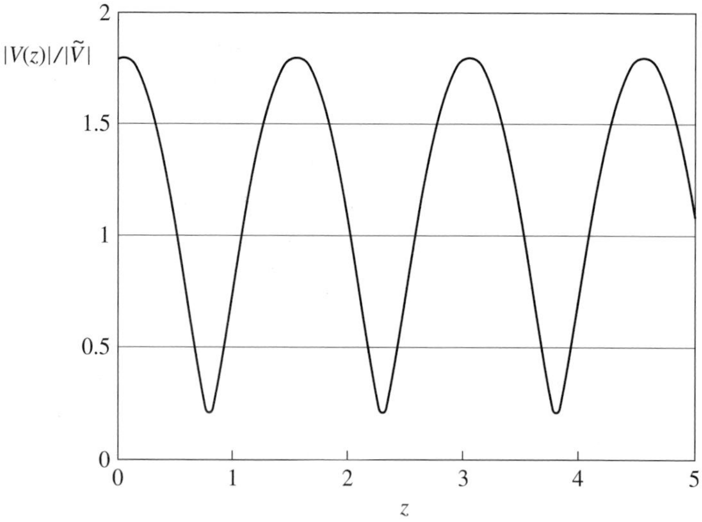

Fig. 20.12 $|V(z)|$ as a function of the depth of penetration into a transmission line with complex termination. $\lambda = 3$ cm, $|\Gamma| = 0.8$, $\theta = 106°$.

voltage standing-wave ratio $VSWR$:

$$|\Gamma| = \frac{1 - VSWR}{1 + VSWR} \tag{20.80}$$

The minimum value of $|V(z)|$ is obtained when $\varphi(z)/2 = \pi/2, 3\pi/2, 5\pi/2, \ldots$, so that the distance Δz between two successive minima is given by

$$\begin{aligned} &\tfrac{1}{2}[\varphi(z + \Delta z) - \varphi(z)] = \pi \qquad \text{and} \\ &\tfrac{1}{2}[\varphi(z + \Delta z) - \varphi(z)] = \beta\Delta z = 2\pi\Delta z/\lambda \end{aligned} \tag{20.81}$$

where the second line follows from (20.79). It follows that $\Delta z = \lambda/2$: *the distance between two successive minima of either absolute voltage or current is one half-wavelength*. Hence a measurement of Δz yields a value for β because $\beta = 2\pi/\lambda = \pi/\Delta z$. Of course, two successive maxima also could have been used, but two successive minima are preferred because these are sharper and hence their location can be determined more accurately.

The phase angle θ can be determined from a measurement of the location of the *last minimum* of $|V(z)|$ before termination. Call this location z_{lm}. At this value we know from the above that

$$\begin{aligned} &\frac{1}{2}\varphi(z_{\text{lm}}) \equiv \frac{1}{2}\theta - \beta(l - z_{\text{lm}}) = \left(n + \frac{1}{2}\right)\pi \qquad \text{or} \\ &z_{\text{lm}} = l - \frac{\lambda}{4}\left[\frac{\theta}{\pi} - (2n + 1)\right] \end{aligned} \tag{20.82}$$

We need to choose n from the values $0, \pm 1, \pm 2, \ldots$, i.e. from a positive or negative integer, such that z_{lm} corresponds to the last possible before $z = l$. Keeping in mind that $l - z_{\text{lm}}$ must not exceed $\lambda/2$ (otherwise z_{lm} is not the location of the last minimum), we can see that $n = -1$ is needed, so that

$$z_{\text{lm}} = l - \left(\frac{\theta}{\pi} + 1\right)\frac{\lambda}{4} \tag{20.83}$$

All complex values of Γ with constant $|\Gamma|$ are obtained by allowing θ to vary from $-\pi$ to $+\pi$. For θ exceeding the value $-\pi$ by a minuscule amount ε, we obtain $z_{\text{lm}} = l - \varepsilon$, i.e. the location of the last minimum lies very close to termination. As θ grows to the values of $\theta = 0$, we see that z_{lm} moves from l to $l - \lambda/4$. Now consider positive values of θ. For $\theta = 0$ we see that $z_{\text{lm}} = l - \lambda/4$ lies midway between two minima (i.e. at a maximum, in agreement with the open-circuit termination for real load impedance). When θ is moved from 0 to π, z_{lm} moves from a distance $\lambda/4$ from termination to a distance $\lambda/2$ from termination so that we then obtain the short-circuit real load-impedance result with minimum $|V(l)|$ value. Thus the entire distance between two successive minima is transversed by z_{lm} when θ is varied from $-\pi$ to $+\pi$. Therefore, if the convention is adopted that $-\pi < \theta < \pi$ (which allows for all possible complex values of Γ with constant $|\Gamma|$), we conclude that

$$\theta = \frac{4\pi}{\lambda}(l - z_{lm}) - \pi \tag{20.84}$$

As z_{lm} varies from l to $l - \lambda/2$, we again see that all values between $-\pi$ and $+\pi$ are found for θ. *The location of the last minimum of $|V(z)|$ determines the phase constant θ of the reflection coefficient.*

Summarizing what has been found, we see that we need *three* measurements of $|V(z)|$ to determine the wavelength and reflection coefficient of the TL, namely

(1) the distance Δz between two successive minima, to obtain λ
(2) the *VSWR*, to obtain $|\Gamma|$
(3) the location of the last minimum, to obtain θ

If R_0 has already been determined by an open- and short-circuit measurement of the TL's input impedance, as prescribed above in (20.64), then knowledge of Γ allows determination of the load impedance Z_L in a situation where it cannot be measured directly.

Suppose that it is not possible to measure the voltage and/or current inside the transmission line. Then one must resort to an *indirect method* to obtain the information, by adding to the transmission line an extra piece of length l' such that $|V(z)|$ is at minimum at the new termination $z = l + l'$. It is then certain that the termination impedance at $z = l + l'$ will be real; call it R'. We can determine R' from a measurement. Consider Fig. 20.13, and apply (20.63b) for the impedance representating the input to the extra piece of line (so that the new Z_i equals the old

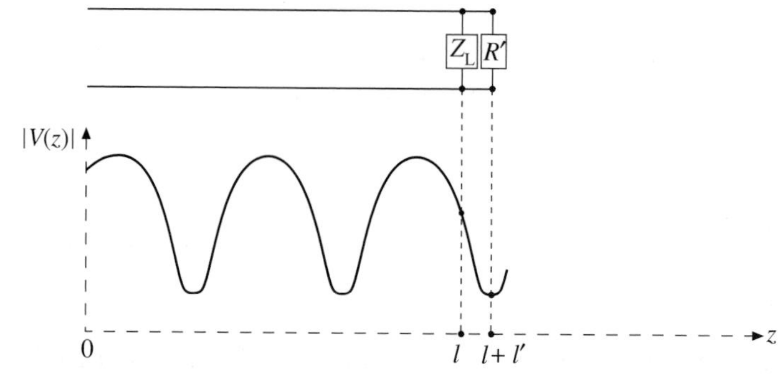

Fig. 20.13 Location of the last minimum of $|V(z)|$.

complex Z_L of the original TL, and the new final impedance is R'). This application yields

$$Z_L = R_0 \frac{R' + jR_0 \tan \beta l'}{R_0 + jR' \tan \beta l'} \tag{20.85a}$$

and here we know that $R' < R_0$ because we end at an absolute-voltage minimum (see the discussion immediately following Eqs. (20.75)). From (20.85a) we can form $\Gamma = (Z_L - R_0)/(Z_L + R_0)$ to obtain

$$\Gamma = -\frac{R' - R_0}{R' + R_0}\left[\frac{1 - j\tan \beta l'}{1 + j\tan \beta l'}\right] \equiv \frac{R_0 - R'}{R_0 + R'} e^{j(\pi + \theta')} \tag{20.85b}$$

The second bracketed (complex) factor has unit absolute value and its phase angle is $\theta' = -2\beta l$. To find this we use $\varphi = \beta l'$ in

$$\frac{1 - j\tan\varphi}{1 + j\tan\varphi} = \frac{(1 - \tan\varphi)^2}{1 - \tan^2\varphi} = \frac{(1 - \tan^2\varphi) - 2j\tan\varphi}{1 - \tan^2\varphi} \equiv e^{j\theta'} \qquad \text{with}$$

$$\tan\theta' = \frac{-2\tan\varphi}{1 - \tan^2\varphi}$$

in which one may recognize the trignometric formula for the tangent of the sum of two (equal) angles. Therefore $|\Gamma| = (R_0 - R')/(R_0 + R')$, so that $\Gamma = |\Gamma|e^{j\theta}$ with

$$|\Gamma| = \frac{R_0 - R'}{R_0 + R'} \qquad \theta = \pi - 2\beta l' \tag{20.86}$$

As a check with the more direct method given above, consider from this that $l' = (\pi - \theta)/2\beta = (\pi - \theta)\lambda/4\pi$. But the last minimum before termination must lie at $z_{\mathrm{lm}} = (l + l') - \lambda/2$ because $z = l + l'$ is the first minimum *beyond* termination. So we again find

$$z_{\mathrm{lm}} = l + \frac{\lambda}{4\pi}(\pi - \theta) - \frac{\lambda}{2} = l - \left(\frac{\theta}{\pi} + 1\right)\frac{\lambda}{4} \tag{20.87}$$

which is identical to (20.83). This indirect method, in which R' and l' are determined, also leads to the desired determination of Γ.

20.5 The Smith Chart

20.5.1 Essentials of a Smith Chart

The very fact that the mathematical expression for the reflection coefficient of a transmission line is of the form $\Gamma = (Z_L - Z_0)/(Z_L + Z_0)$ suggests the use of a graphical tool, developed by P.H. Smith (first published in the January 1939 issue

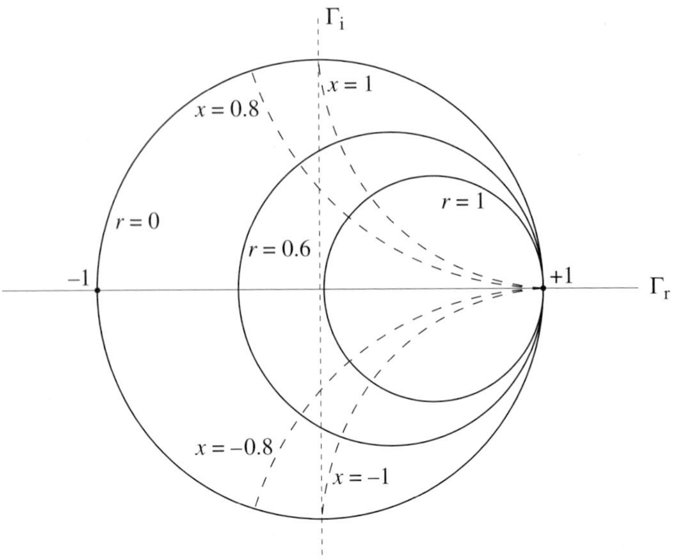

Fig. 20.14 The 'r' circles (solid lines) and 'x' circles (broken lines) of a Smith Chart.

of *Electronics*, vol. 12), for relating Z_L to Γ and vice versa. It is based upon the fact that we may write

$$\zeta \equiv \frac{Z_L}{Z_0} = \frac{1+\Gamma}{1-\Gamma} \tag{20.88}$$

Any complex number $\zeta = r + jx$ can be obtained from any other complex number $\Gamma = \Gamma_r + j\Gamma_i$ graphically by means of a *Smith Chart*. It is not necessary that ζ is a normalized load impedance, as in (20.88), or that Γ is a complex reflection coefficient. The Smith Chart plots two sets of curved lines (these are all parts of circles), as shown in Fig. 20.14. One set comprises lines on which r is constant, and other set comprises lines on which x is constant. The two sets intersect each other at right angles, i.e. each r line is perpendicular to each x line. The horizontal axis going through the center (read: *origin*) of the Smith Chart represents values of Γ_r between -1, the left-most point and $+1$, the right-most point. The vertical axis (not shown on a Smith Chart) through the origin represents Γ_i between the values of -1, the lowest point, and $+1$, the highest point. Figure 20.14 shows $r = r_0$ and $x = x_0$ curves for several constant values of $r_0 \leq 1$ and $|x_0| \leq 1$.

We note the following.

(1) The outermost circle represents $|\Gamma| = 1$, a circle with radius 1. All points inside that circle represent $|\Gamma| < 1$.

(2) The $r = r_0$ lines are full circles with center between $\Gamma_r = 0$ and $\Gamma_r = 1$. All these circles go through the '3 o'clock' point $\Gamma_r = 1$, $\Gamma_i = 0$.

(3) The left-most points of $r_0 > 1$ circles lie to the right of the origin; those for $r_0 < 1$ circles lie to the left of the origin. The value of an $r = r_0$ circle in a Smith Chart is noted along and just above the horizontal axis at the location of the left-most intersection.

(4) The $x = x_0$ lines are parts of circles with centers lying on the straight line $\Gamma_r = 1$. The $x_0 > 0$ lines lie above the horizontal axis; those with $x_0 < 0$ lie below it. The value of an $x = x_0$ circle arc is noted along and just inside the innermost circumference of the Smith Chart.

(5) Only those lines with $|x_0| < 1$ have points that lie to the left of a vertical axis through the origin. Those with $|x_0| > 1$ lie entirely in the $\Gamma_r > 0$ half of the Smith Chart.

Hence the intersection of an r and an x curve determines a point of which the Cartesian coordinates are (Γ_r, Γ_i). Likewise, any point inside the Smith Chart with Cartesian coordinates (Γ_r, Γ_i) is also the intersection of a unique r and a unique x curve (although one may have to interpolate values between two r and two x curves shown on the Smith Chart). Note that the discussion is restricted to $r > 0$ because resistances are positive. If $r < 0$ were relevant we would work with $-\zeta = (1+\tilde{\Gamma})/(1-\tilde{\Gamma})$ and (20.88) would show that $\Gamma = -1/\tilde{\Gamma}$.

The mathematics confirming the rules above is a result of (20.88), which is written more fully as

$$r + jx = \frac{(1+\Gamma_r) + j\Gamma_i}{(1-\Gamma_r) - j\Gamma_i} \tag{20.89}$$

The real and imaginary parts of this equation can be rearranged to yield

$$\begin{aligned} \left(\Gamma_r - \frac{r}{1+r}\right)^2 + \Gamma_i^2 &= \left(\frac{1}{1+r}\right)^2 \\ (\Gamma_r - 1)^2 + \left(\Gamma_i - \frac{1}{x}\right)^2 &= \left(\frac{1}{x}\right)^2 \end{aligned} \tag{20.90}$$

The first of these equations represents circles with centers at $(r/(1+r), 0)$, i.e. with centers on the horizontal axis and with $0 \leq \Gamma_r \leq 1$. The radius of each is $1/(1+r)$. Clearly all these circles contain the point $\Gamma_r = 1$, $\Gamma_i = 0$. For $r = 0$ the circle coincides with the perimeter of the Smith Chart. For $r = 1$, the radius is $1/2$ and the r circle is sandwiched precisely between the origin and the $\Gamma_r = 1$ point on the horizontal axis. For $r > 1$, the radius is less than $1/2$ and the r circles lie to the right of the origin.

The second of (20.90) also represents circles, but now with centers at $\Gamma_r = 1$, $\Gamma_i = 1/x$ and with radii also equal to $1/x$. These circles therefore also all are tangential to the horizontal axis at $\Gamma_r = 1$. As the centers lie at $\Gamma_i = 1/x$ it follows that positive values of x give circles that lie above and negative values of x give circles that lie below the horizontal axis. The circle arcs are small for large $|x|$ and they approach the horizontal axis as $|x| \to 0$. These circles are entirely in the right half of the Smith Chart of $|x| > 1$, but cross over into $\Gamma_r < 0$ when $|x| < 1$.

The Smith Chart also allows one to measure Γ in terms of polar coordinates, for which $\Gamma_r = |\Gamma|\cos\varphi$ and $\Gamma_i = |\Gamma|\sin\varphi$. The angle φ is shown just outside the innermost set of perimeter circles at $|\Gamma| = 1$. The value of $|\Gamma|$ must be obtained from a linear measurement along a line from the Smith Chart origin (where $|\Gamma| = 0$) to the innermost perimeter circle (where $|\Gamma| = 1$). A scale for doing so often is shown next to or below the chart, and it must be cut out or copied on the straight edge of another sheet of paper. The Smith Chart is a mathematical tool of limited accuracy, but at least it allows a pictorial representation of the (normalized) impedances that go with the reflection coefficients defined by (20.88).

Speaking of accuracy, inspection of the Smith Chart shows that one should expect low accuracies for ζ when $|\Gamma|$ lies close to unity and has a small imaginary part, i.e. when points lie close to three-o'clock position on the perimeter of the Smith Chart. In that case it is advantageous to consider not Γ but $\tilde{\Gamma} = -\Gamma = \Gamma e^{j\pi}$ (which lies in a high-accuracy part of the Smith Chart). The corresponding normalized impedance is $\eta = (1-\Gamma)/(1+\Gamma)$, according to (20.88). If $\zeta = |\zeta|e^{i\varphi}$, $\eta = |\eta|e^{i\varphi}$ we then obtain

$$
\begin{aligned}
|\zeta| &= \sqrt{\frac{1+|\Gamma|^2+2\Gamma_r}{1+|\Gamma|^2-2\Gamma_r}} \qquad & \tan\varphi &= \frac{2\Gamma_i}{1+|\Gamma|^2} \\
|\eta| &= \sqrt{\frac{1+|\Gamma|^2-2\Gamma_r}{1+|\Gamma|^2+2\Gamma_r}} \qquad & \tan\psi &= \frac{-2\Gamma_i}{1+|\Gamma|^2}
\end{aligned}
\tag{20.91}
$$

so that it is obvious that $\eta = 1/\zeta$. If ζ lies in the low-accuracy part of the Smith Chart, then $\eta = 1/\zeta$ lies in the high-accuracy part, and we can infer ζ from η. In this way we use a high-accuracy part of the chart to determine ζ from a reflection coefficient that lies in the low-accuracy part. The quantity $\eta = Z_0/Z_L$ is also equal to $Z_0 Y_L = Y_L/Y_0$, in terms of the *admittances* Y_0 and Y_L, which are the inverses of the impedances. To put it more briefly: if transformation of the impedance into reflection coefficient uses a low-accuracy part of the Smith Chart, consider working with the admittance because if $\zeta \to \Gamma$, then $\eta \to -\Gamma$.

20.5.2 Various uses of a Smith Chart

Example 20.1 Finding the reflection coefficient of a TL for given impedance $Z(z)$

If we consider the normalized impedance $\zeta(z) = Z(z)/Z_0$ at a location z of the transmission line, then (20.66) predicts for a lossless line that

$$\zeta(z) = \frac{1 + \Gamma e^{-2j\beta(l-z)}}{1 - \Gamma e^{-2j\beta(l-z)}} \equiv \frac{1+\tilde{\Gamma}}{1-\tilde{\Gamma}} \tag{20.92}$$

We use the Smith Chart to find first $\tilde{\Gamma}$ and then Γ. We shall see there are tools at the perimeter of the Smith Chart that enable one to do so with relative ease if not with great accuracy. Suppose that for a given normalized impedance $\zeta(z)$ we find a point P_1 in Fig. 20.15; we will then have found $\tilde{\Gamma} = |\tilde{\Gamma}|e^{j\varphi} = |\Gamma|e^{j\varphi}$ (the absolute value of $\tilde{\Gamma}$ is the same as that of Γ). We extend the line OP_1 to the perimeter, which consists of a number of concentric circles:

(1) an inner circle with numbers within it that give the x values of $\zeta = r + jx$. The numbers outside this circle (but inside the next circle) give the angle φ (in degrees) of any line OP_1 with respect to the positive horizontal axis. The angle φ is given as $0° < \varphi < 180°$ above the horizontal axis, and as $-180° < \varphi < 0°$ below that axis

(2) an outermost circle

(3) the remaining, next-to-last, circle measures the angle φ by interpreting it as $2\beta l_0/4\pi = l_0/\lambda$ where l_0 is a given length of transmission line. The numbers outside this circle depict values of l_0/λ from 0 at the $\Gamma = -1$ point (the nine-o'clock point) to 0.25 at the $\Gamma = 1$ point (the three-o'clock point) and,

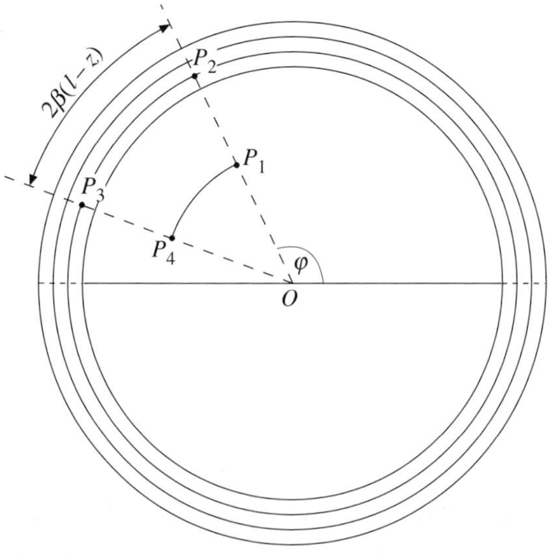

Fig. 20.15 Smith Chart calculation for Example 20.1.

continuing below the horizontal axis, to 0.50 back at the $\Gamma = -1$ point. Because the values of l_0/λ increase in a clockwise (negative-φ) direction, the chart states in words that the scale depicts *wavelengths toward the generator*. One entire clockwise circle indicates that l_0 has decreased by a half-wavelength, e.g. that a move has been made from one voltage minimum inside a transmission line to a previous one. The numbers inside this circle go in the reverse (positive-φ) direction. They measure *wavelengths towards load.* Note that the values shown on either side of this circle *are* of l_0/λ but they *specify* an angle $\varphi = 4\pi l_0/\lambda$, so the shown values must be multiplied by 4π to get the angle in radians.

At the perimeter, we define P_2 as the intersection of OP_1 either with the degree circle or with the l_0/λ circle. The point P_1 represents $\tilde{\Gamma}$, as we have already stated, and it also represents the load impedance ζ_L because the transformation (20.92) for $z = l$ holds at that point. The connection between Γ and $\tilde{\Gamma}$ is

$$\tilde{\Gamma} = |\Gamma|e^{j\varphi} \qquad \text{with} \qquad \varphi = \theta - 2\beta(l - z) \tag{20.93}$$

The line length OP_1 is equal to $|\Gamma|$. The angle φ is given by point P_2. To find θ we need to add the angle $2\beta(l - z)$ to it. This angle is specified by the value $(l - z)/\lambda$, which we assume to be known. If $z < l$ then we use the *wavelengths towards load* scale to measure an increase in φ due to the positive angle $4\pi(l - z)/\lambda$ to obtain point P_3. Then OP_3, by virtue of (20.93), makes an angle θ with the positive horizontal axis. Finally the point P_4 with $OP_4 = OP_1$ represents the reflection coefficient Γ.

Example 20.2 Find the $VSWR$ of a TL for given load impedance Z_L

The $VSWR$ is found without much difficulty in two steps. First, we use (20.92) for $z = l$, which yields $\zeta_L = (1 + \Gamma)/(1 - \Gamma)$. We use $\zeta_L = r_L + jx_L$ and find point P_1 in Fig. 20.16 as the usual intersection of r and x circles. Then, the Cartesian coordinates of P_1 represent the real and imaginary parts of Γ, and the length OP_1 is $|\Gamma|$. As $VSWR = (1 + |\Gamma|)/(1 - |\Gamma|)$ we may again use the Smith Chart to interpret $VSWR$ as a complex number $\zeta = r + jx$ for real positive 'Γ' $= |\Gamma|$ (in which case $x = 0$ and r lies on the right-hand half of the horizontal axis). To find r, we rotate point P_1 clockwise around the origin until point P_2 at the intersection with the horizontal axis between 0 and 1 has been found. The value of r at that intersection is the $VSWR$. Or, conversely, one may rotate counterclockwise to find point P_2' at the intersection with the horizontal axis between -1 and 0; the $VSWR$ gives $1/r$ at that point because it corresponds to $(1-|\Gamma|)/(1+|\Gamma|)$.

Example 20.3 Find the input impedance Z_i of a TL for a given load impedance Z_L

The first step is the same as in Example 20.2: find the location of Γ from the given ζ_L. This again gives point P_1, now shown in Fig. 20.17. Extend the line OP_1 to OP_1'.

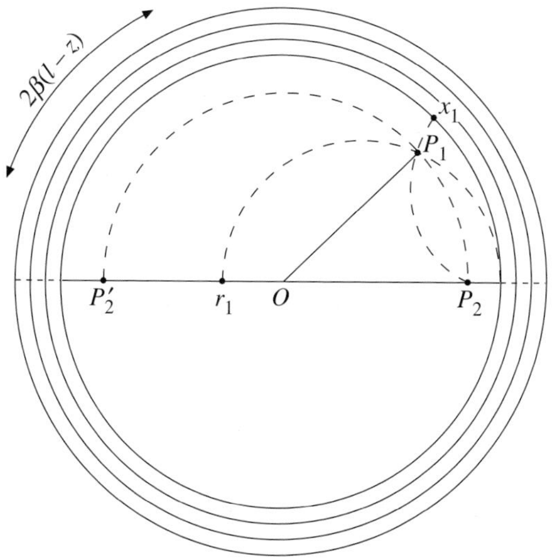

Fig. 20.16 Smith Chart calculation for Example 20.2.

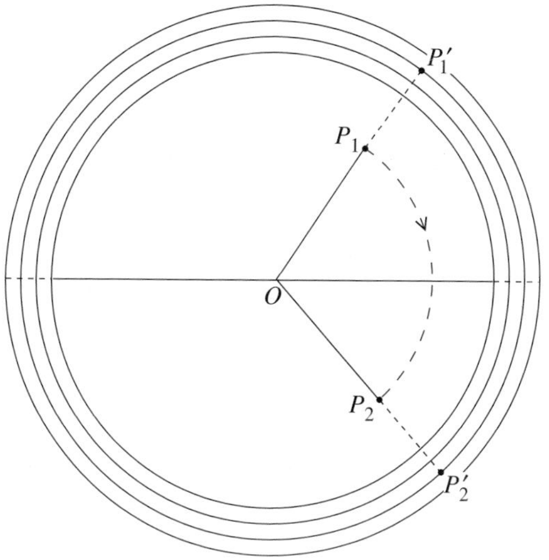

Fig. 20.17 Smith Chart calculation for Example 20.3.

From (20.92) we observe that

$$\zeta_i = \frac{1 + \Gamma e^{-2j\beta l}}{1 - \Gamma e^{-2j\beta l}} \equiv \frac{1 + \tilde{\Gamma}}{1 - \tilde{\Gamma}} \tag{20.94}$$

To obtain $\tilde{\Gamma}$ from Γ, we see that we must rotate over an angle $-2\beta l = -4\pi l/\lambda$. We can use the *wavelengths towards generator* scale to plot the value l/λ (with the nearest

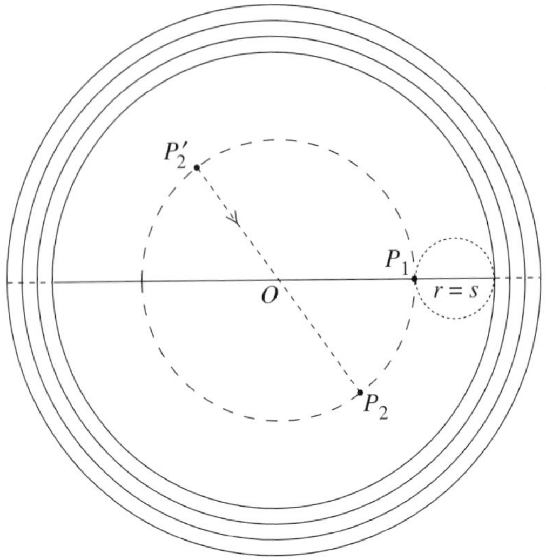

Fig. 20.18 Smith Chart calculation for Example 20.4.

lower half-integer subtracted because each 0.5 part of l/λ is an entire circle). This, for example, might yield points P_2' and P_2. The r and x values of P_2 give the desired input ζ_i.

Example 20.4 Find the reflection coefficient from the $VSWR$ and the location of the last voltage minimum of a TL

Let S be the value of the $VSWR$. It follows from (20.74) that $S = (1 + |\Gamma|)/(1 - |\Gamma|)$, so that the location of the real value of S on the horizontal axis as an r value gives a point P_1 in Fig. 20.18 such that $OP_1 = |\Gamma|$. From (20.84) we know that

$$\theta = 4\pi(l - z_{\mathrm{lm}})/\lambda - \pi = 4\pi[(l - z_{\mathrm{lm}})/\lambda - 0.25]$$

So we must rotate OP_1 around O by the angle θ. To do so, we rotate via the *wavelengths towards load* scale by $(l - z_{\mathrm{lm}})/\lambda$ units to P_2', and then in the opposite direction by a half-circle (or by projection through point O) to P_2. Point P_2 conforms to $OP_2 = |\Gamma|$ and to the correct phase angle θ of Γ, hence it corresponds to ζ_{L} and its Cartesian coordinates are those of $\Gamma_r + j\Gamma_i$.

Example 20.5 Find the impedance $Z(z)$ of a lossy TL for given load impedance Z_{L}

In this example, we need to use

$$\zeta(z) = \frac{1 + \Gamma e^{-2(\alpha+j\beta)(l-z)}}{1 - \Gamma e^{-2(\alpha+j\beta)(l-z)}} \equiv \frac{1 + \tilde{\Gamma} e^{-2j\beta(l-z)}}{1 - \tilde{\Gamma} e^{-2j\beta(l-z)}} \tag{20.95}$$

where use of $\tilde{\Gamma} = |\Gamma| \exp[-2\alpha(l - z)]$ reduces (20.95) to (20.92), with the only difference that $\tilde{\Gamma}$ plays the role here that Γ did in (20.92). If α were zero, the $\tilde{\Gamma}$ would

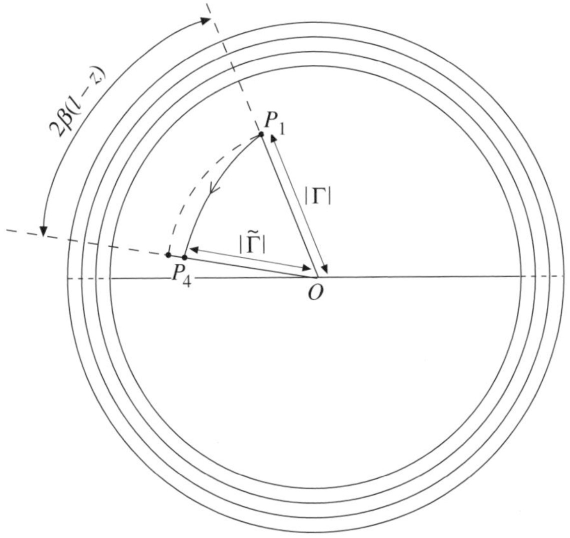

Fig. 20.19 Smith Chart calculation for Example 20.5.

be identical to Γ, and a rotation of $2\beta(l-z)$ would suffice, as in Fig. 20.15. However, $|\tilde{\Gamma}|$ changes in magnitude by the factor $\exp[-2\alpha(l-z)]$ as the rotation occurs (the rotation amounts to a decrease in z, hence a decrease in the exponential factor). The rotation from P_1, which represents the load impedance, to P_4 is therefore accompanied by an inward spiraling, so that the factor multiplying Γ changes from 1 to $\exp[-2\alpha(l-z)]$. Thus, the length OP_4 must be reduced by the factor $\exp[-2\alpha(l-z)]$, as shown in Fig. 20.19. The new point P_4 in Fig. 20.19 represents $\tilde{\Gamma}$ and the normalized impedance $\zeta(z)$.

20.6 Some impedance-matching techniques

20.6.1 Real load impedance

One application of transmission lines is the transport of current and power from a source to a load. Often, a coaxial cable or some other, essentially loss-free, transmission line of length l and characteristic impedance R_0 is used to transmit waves at frequency f with $\beta = 2\pi f/c = 2\pi/\lambda$ to a resistive load R_L. As the line is not matched to the load, there are reflections with coefficient $\Gamma \neq 0$, with concomitant inefficiencies in power transfer to the load. The question arises whether that inefficiency can be minimized. One attempt to answer that question involves the insertion of a short piece of a second transmission line with characteristic impedance R_c and length l' between the first line and the load. Application of (20.63b) for input impedance to this short piece indicates that

$$R(l) = R_{\rm c}\frac{R_{\rm L} + jR_{\rm c}\tan\beta' l'}{R_{\rm c} + jR_{\rm L}\tan\beta' l'} \tag{20.96}$$

In order that the main transmission line be matched, so that there are no reflections, we need to have $R(l) = R_0$. The right-hand side is real if

(1) $\tan\beta' l' = 0$, in which case $R_{\rm c} = R(l) \equiv R_0$, but then the short line is merely an extension of the main TL
(2) $\tan\beta' l' = \infty$, in which case $\beta' l' = \pi/2, 3\pi/2, \ldots$, and the minimum length is $l' = \lambda'/4$.

Also, $R(l) = R_{\rm c}^2/R_{\rm L}$ and, as $R(l) \equiv R_0$, we observe that such a short piece should have a quarter-wavelength and a characteristic impedance $\sqrt{R_0 R_{\rm L}}$. A more careful analysis is given below in an optional section; it shows the analogy with the quarter-wave optical filters discussed in Section 18.6.

Let us start this more careful method by returning to (20.49) and (20.52), which state that the voltage and current at the end of a TL of length l are

$$V(l) = V_{\rm f}e^{-\gamma l} + V_{\rm b}e^{\gamma l} \qquad I(l) = I_{\rm f}e^{-\gamma l} + I_{\rm b}e^{\gamma l} \tag{20.97a}$$

with $V_{\rm f} = Z_0 I_{\rm f}$, and $V_{\rm b} = -Z_0 I_{\rm b}$. If the line is lossless, so that $\gamma = j\beta$, then we have

$$V(l) = V_{\rm f}e^{-j\beta l} + V_{\rm b}e^{j\beta l} \qquad I(l) = \frac{1}{Z_0}(V_{\rm f}e^{-j\beta l} - V_{\rm b}e^{j\beta l}) \tag{20.97b}$$

Referring to Fig. (20.20), we could interpret $V_{\rm f} \equiv V_{\rm f1}^{\rm LHS}$ as the '*forward-propagating*' voltage at $z = 0$ and $V_{\rm b} = V_{\rm b1}^{\rm LHS}$ as the '*backward-propagating*' voltage at $z = 0$, so that $V_{\rm f1}^{\rm RHS} = V_{\rm f}e^{-j\beta l}$ is the forward-propagating voltage at $z = l$, and $V_{\rm b1}^{\rm RHS} = V_{\rm b}e^{j\beta l}$ is the backward-propagating voltage at $z = l$. Equation (20.97b)

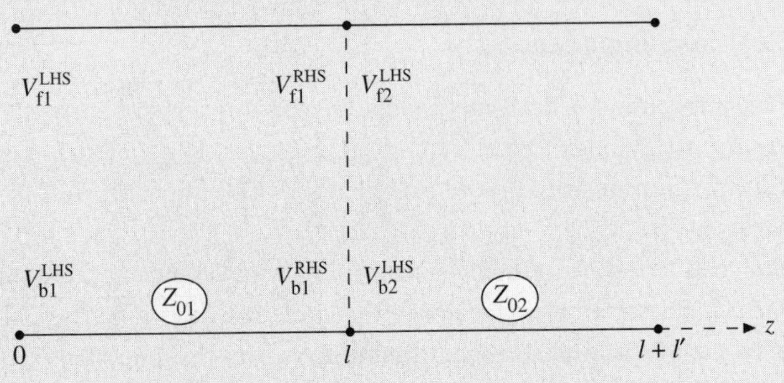

Fig. 20.20 Analogy between junction of two TLs and interface between two dielectrics.

then reads, in terms of these new names,

$$V(l) = V_{f1}^{RHS} + V_{b1}^{RHS} \qquad I(l) = \frac{1}{Z_0}(V_{f1}^{RHS} - V_{b1}^{RHS}) \tag{20.97c}$$

Now consider Fig. 20.20. At $z = l$ two transmission lines join each other. Because $V(l)$ and $I(l)$ are continuous, we must have

$$V^{RHS}{}_{f1} + V_{b1}^{RHS} = V_{f2}^{LHS} + V_{b2}^{LHS}, \; \frac{1}{Z_{01}}(V_{f1}^{RHS} - V_{b1}^{RHS}) = \frac{1}{Z_{02}}(V_{f2}^{LHS} - V_{b2}^{LHS}) \tag{20.98}$$

These equations are exactly the same as (18.57) for the electric fields on either side of an interface between two dielectrics; the intrinsic impedances there are replaced by the characteristic impedances here. The analogy therefore allows us to conclude, as we did in Chapter 18 or by solving (20.98) directly, that

$$\begin{bmatrix} V^{LHS}{}_{f2} \\ V^{LHS}{}_{b2} \end{bmatrix} = \frac{1}{2}\begin{bmatrix} 1 + \dfrac{Z_{02}}{Z_{01}} & 1 - \dfrac{Z_{02}}{Z_{01}} \\ 1 - \dfrac{Z_{02}}{Z_{01}} & 1 + \dfrac{Z_{02}}{Z_{01}} \end{bmatrix} \begin{bmatrix} V^{RHS}{}_{f1} \\ V^{RHS}{}_{b1} \end{bmatrix} \tag{20.99a}$$

and likewise that

$$\begin{bmatrix} V^{RHS}{}_{f2} \\ V^{RHS}{}_{b2} \end{bmatrix} = \begin{bmatrix} e^{-j\beta_2 l_2} & 0 \\ 0 & e^{j\beta_2 l_2} \end{bmatrix} \begin{bmatrix} V^{LHS}{}_{f2} \\ V^{LHS}{}_{b2} \end{bmatrix} \tag{20.99b}$$

So, by multiplying a number of two-by-two matrices together (alternating the two above types) we can relate the V_f and the V_b at either side of any junction between two of a set of connected transmission lines to the values either side of any other junction, exactly as was done in Section 18.6 for electric fields.

Returning, then, to the problem at hand of a lossless TL with length l and characteristic impedance R_0, we attach to it a short piece with length l' and characteristic impedance R_c. Figure 20.21 shows us how to express V_{f3}^{LHS}

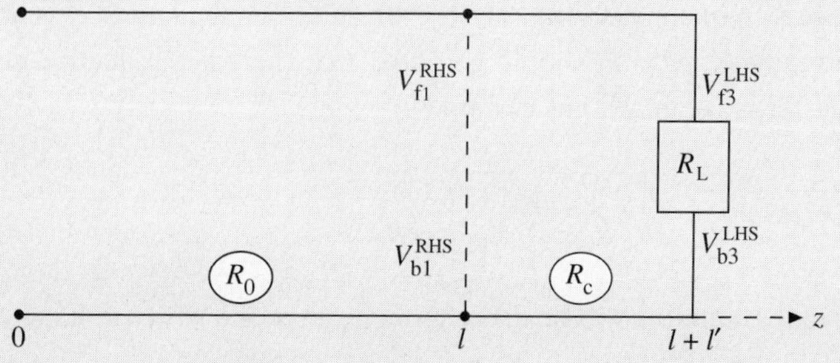

Fig. 20.21 Termination of a finite TL by a short extra piece of another TL of length l' and characteristic impedance R_c; the (real) load is R_L.

and V_{b3}^{LHS} in terms of V_{f1}^{RHS} and V_{b1}^{RHS}. The voltages just to the right of $z = l$ differ from those just to the left of $z = l + l'$. We need to look at

$$\begin{bmatrix} V^{LHS}{}_{f3} \\ V^{LHS}{}_{b3} \end{bmatrix} = \frac{1}{4} \begin{bmatrix} 1 + \dfrac{R}{R_c} & 1 - \dfrac{R}{R_c} \\ 1 - \dfrac{R}{R_c} & 1 + \dfrac{R}{R_c} \end{bmatrix} \begin{bmatrix} e^{-j\beta' l'} & 0 \\ 0 & e^{j\beta' l'} \end{bmatrix} \begin{bmatrix} 1 + \dfrac{R_c}{R_0} & 1 - \dfrac{R_c}{R_0} \\ 1 - \dfrac{R_c}{R_0} & 1 + \dfrac{R_c}{R_0} \end{bmatrix} \begin{bmatrix} V^{RHS}{}_{f1} \\ V^{RHS}{}_{b1} \end{bmatrix} \tag{20.100}$$

Here, the characteristic impedance of 'medium 3' (the location at the load) is the ratio of V to I at the load, denoted R_L.

The new piece of line does what it is designed to do, namely to match the original TL of length l to the load, if $V_b = 0$ everywhere. This specifically implies that V_{b1}^{RHS} and V_{b3}^{LHS} are zero in (20.100). This means that

$$\frac{1}{4}\left(1 - \frac{R}{R_c}\right)\left(1 + \frac{R_c}{R_0}\right)e^{-j\beta' l'} + \frac{1}{4}\left(1 + \frac{R}{R_c}\right)\left(1 - \frac{R_c}{R_0}\right)e^{j\beta' l'} = 0 \quad \text{or}$$
$$\left(1 - \frac{R}{R_0}\right)\cos\beta' l' + j\left(\frac{R}{R_c} - \frac{R_c}{R_0}\right)\sin\beta' l' = 0 \tag{20.101}$$

This equation can be satisfied only by the choice $\beta' l' = \pi/2$ (or values that differ from this by an integer times π), and the simultaneous choice $R_c = \sqrt{R_L R_0}$. A choice of $l' = \lambda'/4$ gives the shortest additional piece.

It should be emphasized that the quarter-wavelength piece of TL must have length $l' = \lambda'/4$ *in terms of the local wavelength* λ' *defined by* R_c (as $\beta' l' = \pi/2$ is the necessary condition). Remember, $R_c = \sqrt{L'/C'}$ for lossless lines, so that $\beta' = \omega\sqrt{L'C'}$ is not necessarily the same as the β for the TL of length l (where $R_0 = \sqrt{L/C}$ and $\beta = \sqrt{LC}$). By the same token, this matching is valid only at the chosen frequency, so that at other frequencies the length l' is no longer a quarter-wavelength. Also, it should be noted that there is a reflection coefficient Γ' for the quarter-wavelength piece, namely

$$\Gamma' = \frac{R_L - \sqrt{R_L R_0}}{R_L + \sqrt{R_L R_0}} = \frac{\sqrt{R_L} - \sqrt{R_0}}{\sqrt{R_L} + \sqrt{R_0}} \tag{20.102}$$

so that a standing wave exists on the short section with a voltage standing-wave ratio

$$VSWR = \frac{(1 + |\Gamma'|)}{(1 - |\Gamma'|)} = \sqrt{\frac{R_L}{R_0}}$$

20.6.2 Complex load impedance

If the load impedance Z_L is complex, then the above technique will not be effective. Various alternatives exist. One technique is to use a *stub* that is placed in parallel to the transmission line, starting at a distance l' *prior* to the load at distance l; see Fig. 20.22. The stub is basically a short piece of length l_{stub} of transmission line with the same R_0 as the main TL. Because inverse impedances of the components of two parallel branches add to form an effective inverse impedance, it is useful to work with admittances, which add linearly. Thus $Y_{line} = 1/Z_{line}$, and $\eta_{line} = 1/\zeta_{line}$ is the normalized admittance of the transmission line, found by multiplying Y_{line} by R_0. Likewise, η_{stub} is the normalized admittance of the parallel stub. A perfect match implies that

$$\eta_{line} + \eta_{stub} = 1 \tag{20.103}$$

at $z = l - l'$, because then there is no reflection coefficient at the end of the TL (up to the stub). One way to make this possible is to try to make $\eta_{line} = 1 + jx$ and $\eta_{stub} = -jx$. The Smith Chart manipulation illustrated in Fig. 20.23 shows how this can be achieved. Let P_L represent the normalized impedance ζ_L of the load at the

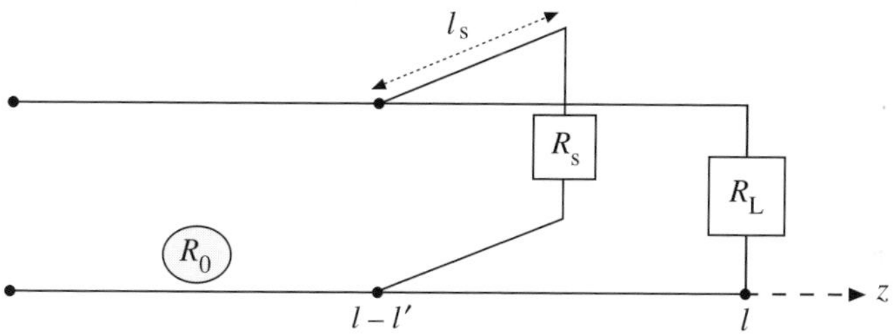

Fig. 20.22 Impedance matching by means of a single parallel 'stub'.

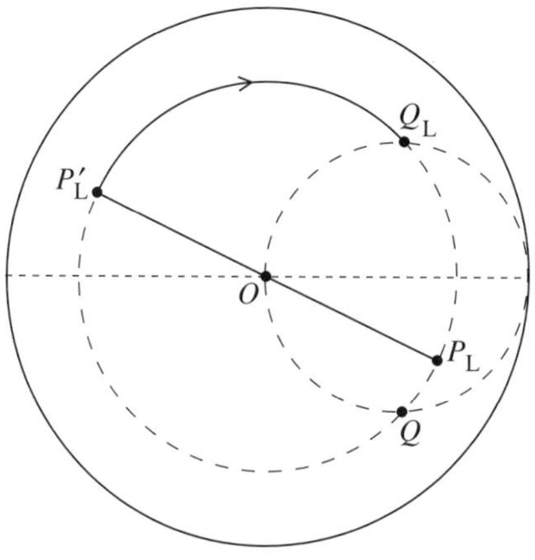

Fig. 20.23 Smith Chart calculation for the situation of Fig. 20.22.

location where the stub is joined. Then $P'_{\rm L}$, at $\lambda/4$ (half-circle) rotation from $P_{\rm L}$ around the origin O, must represent the normalized admittance $\eta_{\rm L} = 1/\zeta_{\rm L}$ (see the discussion above pertaining to (20.91)). To find the distance l', we rotate from $P'_{\rm L}$ towards the generator (clockwise) until we intersect the $r = 1$ circle in point $Q_{\rm L}$. This point obviously has $\eta_{\rm line} = 1 + jx$ and the distance l' corresponds to the arc $\varphi = 4\pi l'/\lambda$ from $P'_{\rm L}$ to $Q_{\rm L}$. The stub, therefore, must have an imaginary part equal to that of the admittance corresponding to point $Q_{\rm stub}$ (the complex conjugate of $Q_{\rm L}$) and zero real part. There are two possibilities for $\eta_{\rm stub} = -jx$:

(a) a short-circuited stub ($Z_{\rm stub} = 0$), which gives $\eta_{\rm stub} = -j\,{\rm cotan}\,\beta l_{\rm stub}$; this can be adjusted for sign (depending upon whether x is positive or negative) by making $l_{\rm stub}$ smaller or larger than $\lambda/4$;
(b) an open-circuit stub ($Z_{\rm stub} = \infty$) for which $\eta_{\rm stub} = j\tan\beta l_{\rm stub}$; here, too, $l_{\rm stub}$ is chosen so as to have the correct polarity for $\eta_{\rm stub} = -jx$.

The value of $l_{\rm stub}$ is not arbitrary (although the sign of $l_{\rm stub}$ can be + or −). The restriction to intersections $Q_{\rm L}$ or $Q_{\rm stub}$ indicates a value for $\eta_{\rm line} = 1 \pm jx$ that is fixed, hence x is given by the value of the x circle through $Q_{\rm L}$ or $Q_{\rm stub}$. Hence this value of x determines $l_{\rm stub}$, and the arc length from $P'_{\rm L}$ to $Q_{\rm L}$ (or $Q_{\rm stub}$) determines l'.

More precise solutions need to be obtained from the equations themselves. For example, we require ${\rm Re}\,\eta_{\rm line} = 1$. The admittance at $z = l - l'$ is

$$\eta_{\rm line} = \frac{1 + j\zeta_{\rm L}\tan\beta l'}{\zeta_{\rm L} + j\tan\beta l'} = \frac{1 - x_{\rm L}\tan\beta l' + jr_{\rm L}\tan\beta l'}{r_{\rm L} + j(x_{\rm L} + \tan\beta l')} \tag{20.104}$$

Setting the real part of this equal to unity delivers the equation

$$(1 - r_{\rm L})\tan^2\beta l' + 2x_{\rm L}\tan\beta l' + x_{\rm L}^2 + r_{\rm L}^2 - r_{\rm L} = 0 \tag{20.105}$$

This quadratic equation has one special solution, $\tan\beta l' = -\frac{1}{2}x_{\rm L}$, if $r_{\rm L} = 1$, but otherwise yields two possible solutions,

$$(1 - r_{\rm L})\tan\beta l' = -x_{\rm L} \pm \sqrt{r_{\rm L}[x_{\rm L}^2 + (1 - r_{\rm L})^2]} \tag{20.106}$$

From this, we extract a minimal positive distance l' by using the properties of the tangent function and the polarity (sign) of the expression for $\tan\beta l'$ that follows from (20.106). While the Smith Chart does essentially the same thing pictorially, (20.106) yields more accurate results. The imaginary part of $\eta_{\rm line}$ then can be found by inserting the above solution for $\tan\beta l'$ into the imaginary part of (20.104). Then $\eta_{\rm stub} = -{\rm Im}\,\eta_{\rm line}$. From these relationships, $l_{\rm stub}$ and l' are obtained directly.

Other techniques exist for matching transmission lines to loads. For example, a number of double-stub methods can be applied when one does not have the freedom to attach stubs to arbitrary locations along the line. The lengths of those stubs is then variable and this can be used to acquire the desired match. We shall not discuss such methods in detail; some of them can be found in other texts (see the Bibliography at the end of the book).

20.7 General lossless-transmission-line equations and transients

In this section, we develop equations for lossless transmission lines that hold for any signal that we can describe reasonably well mathematically. The development allows us to study how finite signals propagate along a finite transmission line, and thus reinforces the propagation picture discussed in an optional part of subsection 20.3.3. The disadvantage of the treatment there was that it holds for infinite plane waves, so that the sequence of reflections from both ends as a function of time could not be shown. A good starting point now will be Eqs. (20.20), which are repeated here:

$$L\frac{\partial i(z,t)}{\partial t} = -\frac{\partial v(z,t)}{\partial z} \quad \text{and} \quad C\frac{\partial v(z,t)}{\partial t} = -\frac{\partial i(z,t)}{\partial z} \tag{20.107}$$

Take the time derivative of either of these two equations and then use the other to turn the mixed double derivative in the second term into a spatial second derivative, to obtain

$$\frac{\partial^2 v}{\partial z^2} + \frac{1}{c^2}\frac{\partial^2 v}{\partial t^2} = 0 \quad \text{and} \quad \frac{\partial^2 i}{\partial z^2} + \frac{1}{c^2}\frac{\partial^2 i}{\partial t^2} = 0 \tag{20.108}$$

where $c = 1/\sqrt{LC} = 1/\sqrt{\mu_0\varepsilon_0}$ is the velocity of the signal in free space (see (20.16b) and the discussion preceding it for a reminder of this equality). We first encountered the prototype of these equations (in one spatial dimension) in (16.12). The crucial point is that both v and i may be *any* function of either $t - z/c$ or $t + z/c$. Any function of $t - z/c$ represents a signal propagating at velocity c in the $+\hat{\mathbf{z}}$ direction, whereas any function of $t + z/c$ represents a backward-propagating signal. Such functions automatically satisfy (20.108)! Therefore, we may assume very generally that

$$\begin{aligned} v(z,t) &= V_{\mathrm{f}}(t - z/c) + V_{\mathrm{b}}(t + z/c) \\ i(z,t) &= I_{\mathrm{f}}(t - z/c) + I_{\mathrm{b}}(t + z/c) \end{aligned} \tag{20.109}$$

From here on we shall develop these equations into a form that allows the analysis of transient effects. For those who wish to omit the details of the development, we refer to the result, Eq. (20.119):

From the low-loss equality $LC = \mu\varepsilon = 1/c^2$ it also follows that $cL = 1/cC = \sqrt{L/C} \equiv R_0$, the characteristic impedance of the lossless line; we can then obtain a useful pair of equations upon insertion of (20.109) in (20.107):

$$\begin{aligned} V_f'(t - z/c) - V_b'(t + z/c) &= R_0[I_f'(t - z/c) + I_b'(t + z/c)] \\ V_f'(t - z/c) + V_b'(t + z/c) &= R_0[I_f'(t - z/c) - I_b'(t + z/c)] \end{aligned} \tag{20.110}$$

The prime indicates a derivative with respect to time. Because V_f and V_b are independent solutions, it follows that these two equations require

$$\begin{aligned} V_f(t - z/c) &= R_0 I_f(t - z/c) \\ V_b(t + z/c) &= -R_0 I_b(t + z/c) \end{aligned} \tag{20.111}$$

We then multiply the second of Eqs. (20.109) by R_0, obtaining

$$\begin{aligned} v(z, t) &= V_f(t - z/c) + V_b(t + z/c) \\ R_0 i(z, t) &= V_f(t - z/c) - V_b(t + z/c) \end{aligned} \tag{20.112}$$

Apply these two equations to the termination $z = l$. By adding and subtracting these two equations from each other, and also using the fact that $v(l, t) = R_L i(l, t)$, it follows that

$$\begin{aligned} V_f(t - l/c) &= (R_L + R_0) i(l, t) \\ V_b(t + l/c) &= (R_L - R_0) i(l, t) \end{aligned} \tag{20.113}$$

from which we find immediately, upon division, that $V_b(t + l/c) = \Gamma_L V_f(t - l/c)$, $\Gamma_L \equiv (R_L - R_0)/(R_L + R_0)$ being the termination reflection coefficient. And, because t is an arbitrary time, it follows that we have found

$$V_b(t) = \Gamma_L V_f(t - 2l/c) \tag{20.114}$$

It was not necessary to restrict this derivation to real termination impedance, but we have done so only because that is often the case. We could, however, have replaced R_L by complex Z_L and thus made Γ_L complex too.

Consider now the input situation of Fig. 20.8 with real source (generator) impedance R_s and general time-dependent source voltage $v_s(t)$. The Kirchhoff voltage law then states that the input voltage obeys $v(0, t) = v_s(t) - R_s i(0, t)$.

Apply this to Eqs. (20.112) at time $t = 0$ to obtain

$$\begin{aligned} V_\mathrm{f}(t) + V_\mathrm{b}(t) &= v_\mathrm{s}(t) - R_\mathrm{s} i(0, t) \\ &= v_\mathrm{s}(t) - \frac{R_\mathrm{s}}{R_0}[V_\mathrm{f}(t) - V_\mathrm{b}(t)] \end{aligned} \tag{20.115a}$$

This can be rearranged without undue difficulty to yield

$$V_\mathrm{f}(t) - \Gamma_\mathrm{s} V_\mathrm{b}(t) = \tilde{v}_\mathrm{s}(t) \tag{20.115b}$$

in which expression the *input reflection coefficient* $\Gamma_\mathrm{s} = (R_\mathrm{s} - R_0)/(R_\mathrm{s} + R_0)$ appears naturally, as does the 'renormalized' source voltage

$$\tilde{v}_\mathrm{s}(t) \equiv \frac{R_0}{R_\mathrm{s} + R_0} v_\mathrm{s}(t) \tag{20.116}$$

Insertion of (20.114) into (20.116) produces the penultimate equation that is needed:

$$V_\mathrm{f}(t) = \tilde{v}_\mathrm{s}(t) + \Gamma_\mathrm{L}\Gamma_\mathrm{s} V_\mathrm{f}(t - 2l/c) \tag{20.117a}$$

This equation is solved by iteration, i.e. we take (20.117a) at $t - 2l/c$ and so obtain

$$V_\mathrm{f}(t - 2l/c) = \tilde{v}_\mathrm{s}(t - 2l/c) + \Gamma_\mathrm{L}\Gamma_\mathrm{s} V_\mathrm{f}(t - 4l/c)$$

We then substitute this into the last term of (20.117a) to obtain the first iteration:

$$V_\mathrm{f} = \tilde{v}_\mathrm{s}(t) + \Gamma_\mathrm{L}\Gamma_\mathrm{s}\Gamma_\mathrm{s}\tilde{v}_\mathrm{s}(t - 2l/c) + (\Gamma_\mathrm{L}\Gamma_\mathrm{s})^2 V_\mathrm{f}(t - 4l/c) \tag{20.117b}$$

Continuation of this process of iteration yields

$$V_\mathrm{f}(t) = \sum_{n=0}^{\infty} (\Gamma_\mathrm{L}\Gamma_\mathrm{s})^n \tilde{v}_\mathrm{s}(t - 2nl/c) \tag{20.117c}$$

With (20.114) we also obtain

$$V_\mathrm{b}(t) = \Gamma_\mathrm{L} \sum_{n=0}^{\infty} (\Gamma_\mathrm{L}\Gamma_\mathrm{s})^n \tilde{v}_\mathrm{s}[t - 2(n+1)l/c] \tag{20.118}$$

The last two equations can be used in (20.109) to obtain the ultimate equation needed for understanding transient effects. This is best written as

$$v(z, t) = \sum_{n=0}^{\infty} (\Gamma_\mathrm{L}\Gamma_\mathrm{s})^n \left\{ \tilde{v}_\mathrm{s}\left(t - \frac{2nl - z}{c}\right) + \Gamma_\mathrm{L}\tilde{v}_\mathrm{s}\left(t - \frac{2(n+1)l - z}{c}\right) \right\} \tag{20.119}$$

where the reader is reminded that $\tilde{v}_\mathrm{s}(t) \equiv R_0 v_\mathrm{s}(t)/(R_\mathrm{s} + R_0)$ and $\Gamma_\mathrm{s} = (R_\mathrm{s} - R_0)/(R_\mathrm{s} + R_0)$. The iterated terms, again, can be understood with the aid of Fig. 20.10.

Consider the $n = 0$ terms of (20.119):

$$[v(z,t)]_{n=0} = \tilde{v}_s\left(t - \frac{z}{c}\right) + \Gamma_L \tilde{v}_s\left(t - \frac{2l - z}{c}\right) \tag{20.120}$$

The first term represents a right-traveling signal at location z, given by the same signal at the origin at a time z/c earlier, i.e. representing propagation of the signal $\tilde{v}_s$ from 0 to z. The second term represents a left-traveling signal that has undergone a reflection by coefficient Γ_L at $z = l$, and is time-lagged by exactly the amount of time it takes a right-traveling signal to travel from 0 to l and then leftwards back to z after reflection. The two terms are represented by the top two diagrams in Fig. 20.10.

The $n = 1$ terms of (20.119) yield

$$[v(z,t)]_{n=0} = \Gamma_L \Gamma_s \tilde{v}_s\left(t - \frac{2l}{c} - \frac{z}{c}\right) + \Gamma_L \Gamma_s \Gamma_L \tilde{v}_s\left(t - \frac{2l}{c} - \frac{2l - z}{c}\right) \tag{20.121}$$

Inspection of these terms shows that they represent the bottom two sketches of Fig. 20.10. The extra factors $\Gamma_L \Gamma_s$ represent the two additional reflections in the time period $2l/c$. The higher-order n terms continue, pairwise, to add more reflections from the generator and terminator ends of the transmission line.

Let $\tilde{v}_s(t)$ represent a sharply edged pulse, created at time $t = 0$, of extent very much shorter than l/c in time. Thus $\tilde{v}_s(t) = 0$ for $t < 0$ and for $t > t_0$, where t_0 is a time very much less than the time l/c needed to propagate to the end of the transmission line. Let us choose t between the two times $(n-1)l/c$ and nl/c and such that the pulse edges do not overlap the ends of the transmission line. Then only one term in (20.119) is nonzero. If n is odd this term is

$$v(z,t) = (\Gamma_L \Gamma_s)^n \tilde{v}_s\left(t - \frac{2nl - z}{c}\right) \tag{20.122a}$$

If n is even, the term is

$$v(z,t) = (\Gamma_L \Gamma_s)^n \Gamma_L \tilde{v}_s\left(t - \frac{2(n+l)l - z}{c}\right) \tag{20.122b}$$

The (20.122a) term represents n sets of propagation from generator to terminator with reflection from the terminator and subsequent propagation back to the generator followed by reflection from the generator, with a final propagation from the generator to location z. The (20.122b) term differs only in that the last propagation goes all the way to the terminator, is followed by a reflection, and finally by backwards propagation to location z. The picture that this gives is entirely in accord with intuition: the narrow pulse travels back and forth between generator and terminator, in principle losing some energy with each reflection (represented by either Γ_L or Γ_s). Likewise, if a rectangular pulse that is long with respect to time l/c is initiated, then (20.119) will show a *transient* buildup of amplitude at location z. That buildup will maximize if a factor $(\Gamma_L\Gamma_s)^m$ reduces the signal to below the noise level so that terms with $n > m$ in (20.119) are negligible. The amplitude will decrease when the 'tail' of the signal passes z, and it will become negligible when that tail has made more than m pairs of reflections.

20.8 Reasons for considering TE, TM modes

Up to this point, only transverse electromagnetic (TEM) modes have been discussed, but there are reasons for utilizing TE, TM modes in waveguides that accommodate these, such as the rectangular waveguide discussed in Section 19.6 or the cylindrical ones of Section 19.7. One advantage of these two waveguides is the isolation of the interior fields from outside electromagnetic influences. This is not the case for strip lines, parallel-plate or two-wire transmission lines, all of which are open structures such that outside fields can penetrate into the interior regions where the modes are generated and contain most of their energy.

Another, more complicated, issue has to do with radiation losses into the walls of the guiding structures. The fact that metallic walls do not have infinite conductivity was shown in Section 20.2 to lead to a nonzero resistance parameter $\mathcal{R}$ as given in (20.26), which in turn leads to ohmic losses in the walls of the guide. If we may neglect the bulk conductivity of the dielectric inside the guide (for example, if it is considered to be essentially a vacuum), then the complex propagation constant γ given by (20.33) becomes

$$\gamma \approx j\omega\sqrt{LC}\sqrt{1 - j\mathcal{R}/\omega L} \approx j\omega\sqrt{LC} + \tfrac{1}{2}\mathcal{R}\sqrt{L/C} \tag{20.123}$$

The last form follows by using the fact that $\mathcal{R}/\omega L \ll 1$ under the square-root sign so that a binomial approximation can be applied. The attenuation coefficient α is then given approximately by

$$\alpha \approx \tfrac{1}{2}\mathcal{R}\sqrt{L/C} \tag{20.124}$$

If we insert values of the parameters for a parallel-plate TL, e.g. as given in (20.15) and (20.26), then straightforward substitution into (20.124) yields

$$\alpha \approx \frac{1}{\eta_0 d}\sqrt{\frac{\mu_0\omega}{2\sigma_m}} \tag{20.125}$$

with $\eta_0 = \sqrt{\mu_0 \varepsilon_0} \approx 120\pi$. For copper walls ($\sigma_m \approx 5.8 \times 10^7$ S/m), we obtain numerically

$$\alpha \text{ in m}^{-1} \approx 6.92 \times 10^{-7} \frac{\sqrt{f \text{ in MHz}}}{d \text{ in m}} \tag{20.126}$$

from which we see that $\alpha \approx 0.002$ per meter at $f = 1$GHz, with an increase by a factor $\sqrt{10}$ for every factor of 10 increase in frequency. Thus, the losses can be appreciable for long transmission lines at high frequencies if TEM modes are utilized.

So why would this loss be avoided by using TE or TM modes? The answer is given by the nature of the electromagnetic fields produced in TE or TM modes. For the two-dimensional planar TE_n modes given in (19.29b) it follows that the actual fields are

$$H_z(x, y, z, t) = H_0 \cos(\omega t - kz) \quad E_y(x, y, z, t) = E_0 \sin(\omega t - kz) \tag{20.127}$$

where the real coefficients E_0, H_0 do not need to be specified here. The issue at hand is that the *instantaneous* Poynting vector is

$$\mathbf{S} = E_y \hat{\mathbf{y}} \times H_z \hat{\mathbf{z}} = 2E_0 H_0 \hat{\mathbf{x}} \sin(2\omega t - 2kz) \tag{20.128}$$

and hence is purely oscillatory at twice the operating frequency with no ohmic losses at all, because the average power flux into the walls due to these fields is zero. Similar conclusions hold for other two-dimensional and three-dimensional TE and TM modes.

Problems

20.1. Use the TEM equations of subsection 19.2.1 to find the C, L parameters in a circularly cylindrical lossless coaxial cable with core radius a and sheath radius b.

20.2. A lossless transmission line (TL) carries a wave traveling at phase velocity $u = 2.1 \times 10^8$ m/s and has a characteristic resistance of $R = 45\,\Omega$. Find the L and C parameters.

20.3. Give the L, C, G, and R coefficients for a coaxial-cable transmission line. *Hint*: The R coefficient can be found by replacing $2/w$ by $1/c_c + 1/c_s$, where c_c, c_s are the circumferences of core and sheath.

20.4. The capacitance per unit length of a two-wire transmission line is $C = \pi\varepsilon/\cosh^{-1}(D/2a)$, given that a is the radius of each wire and D is the separation distance between the axes of the wires. Assume the high-frequency approximation ($R \approx 0$) and express the conductance G and inductance L (both per unit length) in terms of μ, ε, σ, D and a if σ is the effective conductivity of the medium in which the two wires lie.

20.5. A lossless transmission line has length $l = 5\,\text{cm}$ and is operated at frequency $f = 3.6\,\text{GHz}$.

(a) Calculate the propagation constant γ if it is air-filled (or with free space).

(b) Given that $C = 3\varepsilon_0\,\text{F/m}$ calculate the ratio of width w to separation distance d if this were a parallel-plate transmission line.

(c) Calculate L in H/m without using the parallel-plate TL formula.

20.6. A parallel-plate TL along the z direction has width $w = 25\,\text{mm}$ and plate separation distance $d = 2.5\,\text{mm}$. The plate conductivity is $\sigma_\text{m} = 1.5 \times 10^7\,\text{S/m}$, and the dielectric filling has $\varepsilon' = 3.5$ and $\sigma = 0.001\,\text{S/m}$. The operating frequency is 750 MHz, and it has free-space permeability everywhere.

(a) Calculate the L, C, G, R parameters of the TL. Also calculate $R/\omega L$ and $G/\omega C$.

(b) Calculate the ratio $|E_z(z)/E_y(z)|$, given that the main **E** field component is in the $\hat{\mathbf{y}}$ direction.

(c) Calculate the complex γ and Z_0 parameters of the TL.

20.7. Explain the difference between the low-loss and the high-frequency approximations for the TL propagation and impedance parameters.

20.8. A parallel-plate transmission line has separation distance $d = 3\,\text{cm}$ and width $w = 7.5\,\text{cm}$. It is made of copper ($\sigma_\text{m} = 5.8 \times 10^7\,\text{S/m}$), has free-space permeability everywhere, and is filled with a slightly lossy gas ($\varepsilon_\text{r} = 1$, $\sigma = 10^{-5}\,\text{S/m}$). At what frequency f should it be operated so that a narrow-band signal around f can propagate with minimal distortion?

20.9. An $l = 3.5\,\text{m}$ transmission line with characteristic resistance $R_0 = 50\,\Omega$ and propagation constant $\gamma = 20j\,\text{m}^{-1}$ is terminated with a $75\,\Omega$ load. It is fed by an ac source $V_\text{s}(t) = 15e^{j\omega t}$ with an internal resistance $R_\text{s} = 150\,\Omega$.

(a) Calculate the reflection coefficient Γ.

(b) Calculate the final voltage $V(l)$ and current $I(l)$.

(c) Calculate the initial voltage $V(0)$ and current $I(0)$.

20.10. A transmission line of known length l and characteristics γ, Z_0 is fed by an ac source with internal resistance. Can you predict the outcome of measurements of the input impedance for open-circuit and for short-circuit termination of the TL?

20.11. It is desired to measure the impedance Z of an immovable object by means of a circuit that cannot be brought near to that object. How can we use a transmission line to enable us to make this measurement on the circuit?

20.12. Suppose that you are given a coaxial TL with characteristic resistance R_0 and $\beta = 2\pi/\lambda$ at an available channel frequency with which to connect your TV set to the cable-TV box outside (losses in the TL may be ignored). The TV load impedance is $75\,\Omega$.

(a) How should l/λ be chosen so that this impedance is transferred without change to the cable-TV box?

(b) If, after having chosen l ideally as in part (a) we wish also to minimize internal reflection losses in the TL, how can this be done?

20.13. In measuring the absolute value of the voltage, $|V(z)|$, of a hollow coaxial transmission line with $R_0 = 50\,\Omega$, we observe that the $z = l$ value is a minimum. The $VSWR$ is measured to be 3.

(a) What is the load resistance?

(b) The last maximum occurs at a distance of $z = l - 20$ cm. At what frequency is it operated?

20.14. Given a load resistance $Z_L = 120 - 150j\,\Omega$ terminating a TL with characteristic impedance of $30\,\Omega$, use a Smith Chart to find the reflection coefficient Γ.

20.15. It is found from a $VSWR$ measurement that $\Gamma = 0.5e^{-j0.78\pi}$. Use this to find the load resistance on a Smith Chart, given that $Z_0 = 100\,\Omega$.

20.16. A TL with characteristic resistance $R_0 = 60\,\Omega$ is terminated with a $Z_L = 180 + 120j\,\Omega$ load. Use a Smith Chart to find the $VWSR$.

20.17. In a Smith Chart, why do the left intersections of the r and $1/r$ circles with the horizontal axis lie symmetrically on opposite sides of the origin?

20.18. Find the impedance $Z(\frac{1}{2}l)$ for a TL of length $l = 7.2\lambda$ terminated by a normalized load impedance $\zeta_L = 0.5(1 + j)$. How does the answer change if, in addition, it is given that there is a power loss of 0.004 dB per wavelength? The characteristic resistance is $R_0 = 100\,\Omega$.

20.19. A parallel-plate TL is operated at 275 MHz and terminated by a normalized $\zeta_L = 3.5 - 1.75j\,\Omega$ load. A stub is placed across the TL at a distance l' from the termination. Use the Smith Chart to find the shortest distance l' which will amount to either an inductive or capacitative stub and which will also effect a maximal transfer of power across the line. Is the stub capacitative or inductive?

21

Selected topics in radiation and antennas

In many electrical-engineering curricula, there is a separate course at the upper-half or early-graduate level on radiating systems and antennas (which can radiate as well as receive electromagnetic radiation). The purpose of this chapter is to present some of the most salient features of radiation and antennas, leaving the details to the more specialized courses. We start by introducing the concept of infinitesimal-source radiation, and proceed from there to a short discussion of some of the most elementary electric and magnetic dipole radiators and of several canonical antenna situations. Then we proceed to discuss two complementary but different aspects of high-frequency radiation in materials: the propagation of waves through smoothly varying dielectric media and the diffraction of waves by discontinuities in the refractive index. The former leads to the concept of rays (or wavefronts) that progress with slow bending through such a medium; this description is known as *geometrical optics*. An example is a slant path between a ground-based telescope and a star in the night sky; the apparent position of the star is slightly displaced in angle from where it really is because the light rays from the start to the telescope undergo bending in the earth's atmosphere. The placing of an object of contrasting refractive index in the path of a beam leads to *diffraction* of the beam around the object and so to interference phenomena beyond the object. An example is the often-seen phenomenon of sun rays seeming to bend around the edges of an opening in the clouds.

21.1 Time-harmonic radiation by an infinitesimal source

In Chapter 16, we discussed the most general form of the electromagnetic potentials $V(\mathbf{r}, t)$ and $\mathbf{A}(\mathbf{r}, t)$, which would arise due to current and charge distributions that are time dependent. The expressions we found were (16.18a, b), and we write those down again here as a starting point for a discussion on radiation:

$$V(\mathbf{r},t) = \int dv' \frac{\varrho_v(\mathbf{r}', t - |\mathbf{r}-\mathbf{r}'|/c)}{4\pi\varepsilon|\mathbf{r}-\mathbf{r}'|}$$
$$A(\mathbf{r},t) = \int dv' \frac{\mu\mathbf{J}(\mathbf{r}', t - |\mathbf{r}-\mathbf{r}'|/c)}{4\pi|\mathbf{r}-\mathbf{r}'|} \tag{21.1}$$

An important feature is that in the integrands the *retarded* time $t' = t - |\mathbf{r}-\mathbf{r}'|/c$ at which a current or charge source point at $\mathbf{r}'$ contributes to the potentials at $\mathbf{r}$ at time t represents a time that is earlier by an amount $|\mathbf{r}-\mathbf{r}'|/c$, i.e. by the time any electromagnetic wavefront needs to travel the distance $|\mathbf{r}-\mathbf{r}'|$. This is equivalent to saying, as we pointed out earlier, that point $\mathbf{r}$ cannot experience the electromagnetic influence of point $\mathbf{r}'$ in a time less than it would take for a wavefront to travel at a velocity c from $\mathbf{r}'$ to $\mathbf{r}$. The relativistic principle of *finite travel time of signals* holds under all circumstances.

The potentials of (21.1) are not independent because these expressions result from the partial differential equations (16.9), which in turn require the *Lorentz condition*

$$\nabla \cdot \mathbf{A}(\mathbf{r},t) + \mu\varepsilon \frac{\partial}{\partial t} V(\mathbf{r},t) = 0 \tag{21.2}$$

If there is a time-harmonic current-source distribution $\mathbf{J}(\mathbf{r},t) = \mathbf{J}(\mathbf{r})e^{j\omega t}$ and/or a time-harmonic charge-source distribution $\varrho_v(\mathbf{r},t) = \varrho_v(\mathbf{r})e^{j\omega t}$, and if we set $V(\mathbf{r},t) = V(\mathbf{r})e^{j\omega t}$ and $\mathbf{A}(\mathbf{r},t) = \mathbf{A}(\mathbf{r})e^{j\omega t}$ to define phasor quantities, then the time-harmonic forms ensue for those phasors:

$$V(\mathbf{r}) = \int dv' \frac{\varrho_v(\mathbf{r}')e^{j(\omega t - k|\mathbf{r}-\mathbf{r}'|)}}{4\pi\varepsilon|\mathbf{r}-\mathbf{r}'|}$$
$$A(\mathbf{r}) = \int dv' \frac{\mu\mathbf{J}(\mathbf{r}')e^{j(\omega t - k|\mathbf{r}-\mathbf{r}'|)}}{4\pi|\mathbf{r}-\mathbf{r}'|} \tag{21.3a}$$

where $k = \omega/c = \omega\sqrt{\mu\varepsilon}$, with the Lorentz condition

$$\nabla \cdot \mathbf{A}(\mathbf{r}) + j\mu\varepsilon\omega V(\mathbf{r}) = 0$$

If we have an *infinitesimal* charge and current distribution at $\mathbf{r}'$ then we can write

$$V(\mathbf{r}) = [\varepsilon^{-1}\varrho_v(\mathbf{r}')\,dv'] \frac{e^{j(\omega t - kR)}}{4\pi R}$$
$$A(\mathbf{r}) = [\mu\mathbf{J}(\mathbf{r}')\,dv'] \frac{e^{j(\omega t - kR)}}{4\pi R} \tag{21.3b}$$

where $R \equiv |\mathbf{r} - \mathbf{r}'|$. To find the electromagnetic fields, we invoke again the definitions (15.32) for time-harmonic phasor potentials and fields:

$$\mathbf{E}(\mathbf{r}) = -\nabla V(\mathbf{r}) - j\omega\mathbf{A}(\mathbf{r}) \qquad H(\mathbf{r}) = \frac{1}{\mu}\nabla \times \mathbf{A}(\mathbf{r}) \tag{21.4a}$$

The first of these expressions can be transformed as follows by substituting for ∇V, from the above Lorentz condition:

$$\begin{aligned}-\nabla V - j\omega\mathbf{A} &= \frac{1}{j\mu\varepsilon\omega}\nabla(\nabla\cdot\mathbf{A}) - j\omega\mathbf{A} = \frac{1}{j\mu\varepsilon\omega}[\nabla\times(\nabla\times\mathbf{A}) + \Delta\mathbf{A}] - j\omega\mathbf{A}\\ &= \frac{1}{j\mu\varepsilon\omega}[\nabla\times(\nabla\times\mathbf{A})] + \frac{1}{j\mu\varepsilon\omega}(\Delta\mathbf{A} + k^2\mathbf{A})\end{aligned}$$

Equations (16.9) for time-harmonic fields show that $\Delta\mathbf{A} + k^2\mathbf{A}$ is nonzero only at locations where a current density exists. If $\mathbf{r}$ is not a location where there is a source, then (21.4a) becomes

$$\mathbf{E}(\mathbf{r}) = \frac{1}{j\mu\varepsilon\omega}\nabla\times[\nabla\times\mathbf{A}(\mathbf{r})] \qquad H(\mathbf{r}) = \frac{1}{\mu}\nabla\times\mathbf{A}(\mathbf{r}) \tag{21.4b}$$

Consequently, only the vector potential field in (21.3) needs to be considered in determining the **E** and **H** fields for an oscillatory infinitesimal source (which, in turn, can be considered as a time-harmonic *current* source). So, in replacing the $\mathbf{J}(\mathbf{r}')\,dv'$ element by an infinitesimal length of (infinitesimally thin) wire $I\,d\mathbf{l}'$ at $\mathbf{r}'$, we find

$$\mathbf{A}(\mathbf{r}) = \mu I\,d\mathbf{l}'\frac{e^{-jkR}}{4\pi R} \tag{21.5}$$

Thus, the *mathematical* structure of the **E** and **H** fields is determined entirely by ∇ operators working on the factor $e^{-jkR}/(4\pi R)$. To obtain the expressions for the fields, consider $d\mathbf{l}' = \hat{\mathbf{z}}\,dl'$ and apply the derivative operator in spherical coordinates. We then observe from (21.4b) and (21.5), and from the vector identity $\nabla\times[\hat{\mathbf{z}}f(\mathbf{r})] = -\hat{\mathbf{z}}\times\nabla f(\mathbf{r})$, that

$$\begin{aligned}\mathbf{H}(\mathbf{r}) &= (Idl')\nabla\times\left(\hat{\mathbf{z}}\frac{e^{-jkR}}{4\pi R}\right) = -(Idl')\hat{\mathbf{z}}\times\nabla\left(\frac{e^{-jkR}}{4\pi R}\right)\\ &= (Idl')\hat{\mathbf{z}}\times\hat{\mathbf{R}}\frac{e^{-jkR}}{4\pi R}\left(jk + \frac{1}{R}\right) = \hat{\boldsymbol{\varphi}}(Idl'\sin\theta)\frac{e^{-jkR}}{4\pi R}\left(jk + \frac{1}{R}\right)\end{aligned} \tag{21.6a}$$

because $\hat{\mathbf{z}}\times\hat{\mathbf{R}} = \hat{\boldsymbol{\varphi}}\sin\theta$ in spherical coordinates. To continue from here, we observe from (21.4b) that $\mathbf{E} = (\nabla\times\mathbf{H})/(j\varepsilon\omega)$. Because **H** is in the spherical $\hat{\boldsymbol{\varphi}}$ direction, it follows that there are only two spherical-coordinate components of **E**:

$$E(\mathbf{r}) \equiv E_\theta\hat{\boldsymbol{\theta}} + E_R\hat{\mathbf{R}} = \frac{1}{j\varepsilon\omega}\left[\mathbf{R}\frac{1}{R\sin\theta}\frac{\partial}{\partial\theta}(H_\varphi\sin\theta) - \hat{\boldsymbol{\theta}}\frac{1}{R}\frac{\partial}{\partial R}(RH_\varphi)\right] \tag{21.6b}$$

From the last form in (21.6a) we then find

$$\begin{aligned} E_\theta &= j\eta(Idl')k\sin\theta\,\frac{e^{-jkR}}{4\pi R}\left(1 - \frac{j}{kR} - \frac{1}{k^2R^2}\right) \\ E_R &= 2j\eta(Idl')k\cos\theta\,\frac{e^{-jkR}}{4\pi R^2}\left(1 - \frac{j}{kR}\right) \\ H_\varphi &= j(Idl')k\sin\theta\,\frac{e^{-jkR}}{4\pi R}\left(1 - \frac{j}{kR}\right) \end{aligned} \tag{21.6c}$$

As before, $\eta = \sqrt{\mu/\varepsilon}$ is the intrinsic impedance of the medium, measured in Ω. For free space, $\eta_0 = \sqrt{\mu_0/\varepsilon_0} \approx 120\pi\,\Omega$. While these expressions are useful if we choose dl' to be in the $\hat{\mathbf{z}}$ direction, the following forms express the fields in coordinate-free notation so that they can be adapted to radiators built up out of many pieces $I\mathbf{dl}'$ in diverse directions:

$$\begin{aligned} \mathbf{E}_\perp &= j\eta Ik\frac{e^{-jkR}}{4\pi R}\left(1 - \frac{j}{kR} - \frac{1}{k^2R^2}\right)\hat{\mathbf{R}} \times (\hat{\mathbf{R}} \times \mathbf{dl}') \\ \mathbf{E}_\parallel &= 2j\eta Ik\frac{e^{-jkR}}{4\pi R^2}\left(1 - \frac{j}{kR}\right)(\hat{\mathbf{R}} \cdot \mathbf{dl}')\hat{\mathbf{R}} \\ \mathbf{H} &= -jIk\frac{e^{-jkR}}{4\pi R}\left(1 - \frac{j}{kR}\right)\hat{\mathbf{R}} \times \mathbf{dl}' \end{aligned} \tag{21.7}$$

Note that **H** is always in a direction $\perp$ the direction of propagation $\hat{\mathbf{R}}$, whereas **E** has components both $\|$ and $\perp$ to $\hat{\mathbf{R}}$. So, to recapitulate: (21.6a, b, c) represent the received **E** and **H** fields at **r** due to an infinitesimal current source I of length $\mathbf{dl}'$ at $\mathbf{r}'$. The various terms of **E** and **H** have differing dependences upon inverse powers of kR; these play a differing role in the transmitted power depending upon whether this parameter combination is large or small compared to unity.

The average transmitted Poynting vector $\frac{1}{2}\mathrm{Re}\{\mathbf{E} \times \mathbf{H}^*\}$ in a lossless medium (k is real) is easily calculated from (21.6). The parallel component of **E** yields a Poynting-vector component S' perpendicular to the direction of propagation:

$$S'(\theta) = \frac{\eta}{2}\frac{(Idl')^2}{(4\pi)^2R^3}k^2\sin\theta\left(1 + \frac{1}{k^2R^2}\right)\hat{\boldsymbol{\varphi}} \tag{21.8a}$$

Note that this component decreases in strength as R^{-3} with increasing distance R. The other component of **E** yields a Poynting-vector component

$$S(\theta) = \frac{\eta}{2}\frac{(Idl')^2}{(4\pi R)^2}(k\sin\theta)^2\hat{\mathbf{R}} \tag{21.8b}$$

parallel to the direction of propagation. It decreases less rapidly with distance, as R^{-2}. The total power radiated through *any* sphere with $\mathbf{r}'$ as origin is obtained from the integral of $R^2 \mathbb{S}(\theta, \varphi) \sin\theta \, d\theta \, d\varphi$ over all possible solid angles defined by θ and φ. Thus the transmitted power is easily found to be

$$P_t = R^2 \int_0^{2\pi} d\varphi \int_0^{\pi} d\theta \sin\theta \, S(\theta) = \eta k^2 \frac{(Idl')^2}{12\pi} \tag{21.8c}$$

and we see that P_t is constant in a lossless medium. It depends on at least the second power of frequency, and can be proportional to higher powers for certain types of current elements such as oscillating electric dipoles (see below). The S' component does not contribute to P_t. It represents a weaker tangential radiation that dies out more rapidly with distance than does S.

It is conventional to define the region $kR \ll 1$ as the *near-field zone* and the region $kR \gg 1$ as the *far-field zone*, at least for infinitesimal radiation sources. (It should be noted that a similar terminology is used when comparing the distance kD^2 to R for a source with diameter D.) In the near-field approximation, the electric field is

$$\begin{aligned} E_\theta &= -j\eta(Idl') \sin\theta \frac{e^{-jkR}}{4\pi k R^3} \\ E_R &= 2j\eta(Idl')k \cos\theta \frac{e^{-jkR}}{4\pi R^2} \\ H_\varphi &= -(Idl') \sin\theta \frac{e^{-jkR}}{4\pi R^2} \end{aligned} \tag{21.9}$$

whereas in the far-field approximation we may neglect the longitudinal E_R component and find

$$\begin{aligned} E_\theta &= j\eta(Idl')k \sin\theta \frac{e^{-jkR}}{4\pi R} \\ H_\varphi &= j(Idl')k \sin\theta \frac{e^{-jkR}}{4\pi R} = \frac{1}{\eta} E_\theta \end{aligned} \tag{21.10}$$

In the far-field expressions both the **E** and the **H** fields are perpendicular to the radial direction of propagation (as shown in Fig. 21.1), and therefore the Poynting vector is essentially radial. The electromagnetic energy propagates in the form of spherical wavefronts moving radially away from the point source in the far-field $kR \gg 1$.

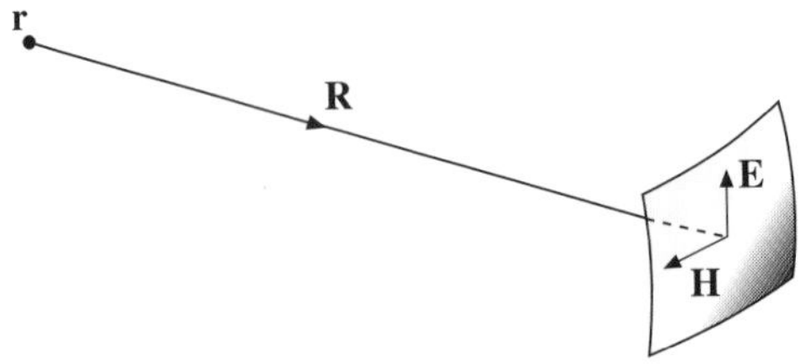

Fig. 21.1 Far-field **E** and **H** fields.

21.2 The time-harmonic infinitesimal electric-dipole source

The above discussion of radiation from an infinitesimal length of current $I d\mathbf{l}'$ at $\mathbf{r}$ gives rise to several elemental forms of antenna. The infinitesimal electric dipole antenna is obtained from the discussion in the previous section by carefully considering the nature of an oscillating infinitesimal current source. Consider a negative point source $-q$ anchored at $\mathbf{r}'$ and a positive point source $+q$ oscillating around it; $+q$ is at the time-dependent location $\mathbf{r}' + \Delta\mathbf{r}'(t)$, see Fig. 21.2. The charge density infinitesimally close to $\mathbf{r}'$ is given by

$$\varrho_{\mathrm{v}}(t) = q\{\delta[\mathbf{r}' + \Delta\mathbf{r}'(t)] - \delta(\mathbf{r}')\} \tag{21.11}$$

The function $\delta(\mathbf{r}')$ is the Dirac delta function, also known as the impulse response function when the variable $\mathbf{r}'$ is a scalar. $\delta(\mathbf{r}')$ is zero whenever $\mathbf{r}' \neq 0$ and infinite when $\mathbf{r}' = 0$ in such a fashion that the volume integral $\int dv' f(\mathbf{r}')\delta(\mathbf{r}')$ equals $f(0)$ whenever the integration volume encompasses 0, no matter how small that volume is.

We may write $\Delta\mathbf{r}'(t) = \hat{\mathbf{z}}\Delta z e^{j\omega t}$ (in phasor notation) so that careful evaluation of the time derivative yields

$$\frac{\partial \varrho_{\mathrm{v}}}{\partial t} = j\omega[q\Delta\mathbf{r}'(t)] \cdot \nabla\delta(\mathbf{r}') = j\omega\mathbf{p} \cdot \nabla\delta(\mathbf{r}') \tag{21.12}$$

where $\mathbf{p} = (q\Delta z)\hat{\mathbf{z}}$. The conservation-of-charge equation, $\partial\varrho_{\mathrm{v}}/\partial t + \nabla \cdot \mathbf{J} = 0$ then predicts $\mathbf{J} = -j\omega\mathbf{p}\,\delta(\mathbf{r}')$, so that the infinitesimal current element $\mathbf{J}d\mathbf{v}' = \mathbf{I}dl'$ must be $-j\omega\mathbf{p}$. Consequently, for the infinitesimal dipole $\mathbf{p}$, it follows that the small source element in the radiation formulas is

$$I d\mathbf{l}' = -j\omega\mathbf{p}$$

From (21.8c), one important consequence for the radiation is that the total radiated power through any sphere around the source is

$$P_{\mathrm{t}} = \frac{\mu\omega^4 p^2}{12\pi c} \tag{21.13}$$

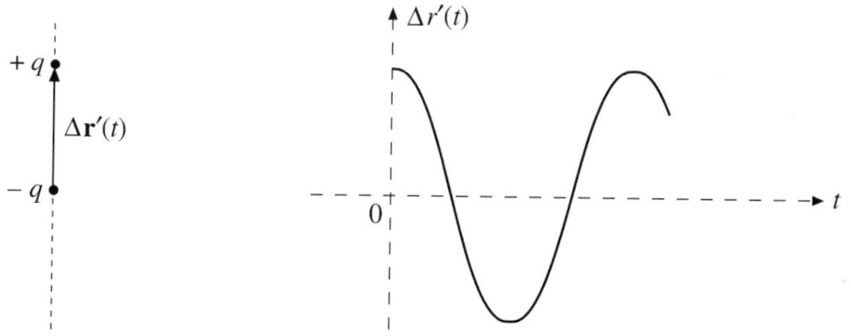

Fig. 21.2 The infinitesimal electric dipole as a radiator.

thus indicating that *the power radiated from an electric dipole is proportional to the fourth power of the frequency*. This is one reason why it is favorable to transmit at high frequencies, but it is not the only one; the other has to do with the length of an antenna with respect to the wavelength and this will be dealt with when nonzero-length antennas are discussed below.

21.3 The time-harmonic infinitesimal magnetic-dipole source

The results for an infinitesimal magnetic dipole are obtained less directly because the current element Idl needs to be integrated around a vanishingly small circle. It is easiest to work with the expression for the vector potential $\mathbf{A}(\mathbf{r})$. Reconsider an integral of (21.5) for an infinitesimal current loop in the xy plane around the z axis, see Fig. 21.3; at a point $\mathbf{r}$ in the yz plane, $\mathbf{A}(\mathbf{r})$ is given by

$$\mathbf{A}(\mathbf{r}) = \mu I \oint \mathbf{d}\boldsymbol{l}\, \frac{e^{-jkR}}{4\pi R} \tag{21.14}$$

Here, we have $\mathbf{R} = \mathbf{r} - \mathbf{r}'$ and $\mathbf{d}\boldsymbol{l} = (a\, d\varphi)\hat{\boldsymbol{\varphi}}$. The idea is to approximate R, both in the exponent and in the factor $1/R$, by the first two terms in the binomial expansion of $R = \sqrt{r^2 - 2\mathbf{r}\cdot\mathbf{r}' + (\mathbf{r}')^2}$. This yields

$$\begin{aligned} &R \approx r - \hat{\mathbf{r}}\cdot\mathbf{r}' = r - a\sin\theta\sin\varphi \\ &\frac{1}{R} \approx \frac{1}{r}\left(1 + \frac{\hat{\mathbf{r}}\cdot\mathbf{r}'}{r}\right) \qquad e^{-jkR} \approx e^{-jkr}(1 + jk\hat{\mathbf{r}}\cdot\mathbf{r}') \end{aligned} \tag{21.15}$$

for infinitesimal r'. The choice of coordinates dictates that $\hat{\mathbf{r}}\cdot\mathbf{r}' = a\sin\theta\sin\varphi$. As a consequence,

$$\frac{e^{-jkR}}{4\pi R} = \frac{e^{-jkr}}{4\pi R}\left[1 + (1 + jkr)\frac{a\sin\theta\sin\varphi}{r}\right] \tag{21.16}$$

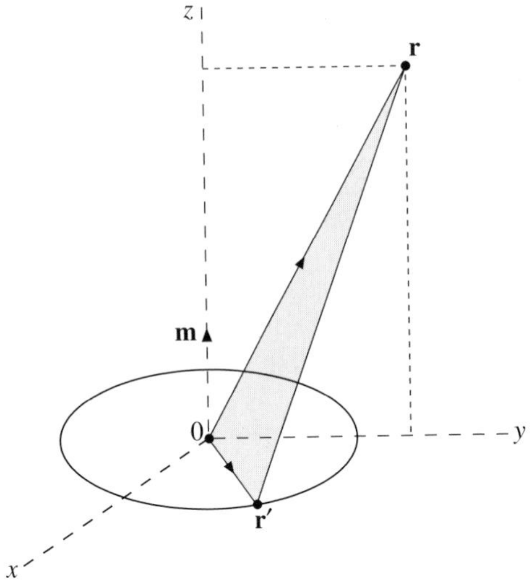

Fig. 21.3 The infinitesimal magnetic dipole as a radiator.

and if we apply Cartesian coordinates to $d\boldsymbol{l}$ we observe that $d\boldsymbol{l} = a\,d\varphi[-\hat{\mathbf{x}}\sin\varphi + \hat{\mathbf{y}}\cos\varphi]$. The insertion of this and (21.16) into (21.14) yields a nonzero term only for the $a\sin\theta\sin\varphi$ term in (21.16). We obtain

$$\mathbf{A}(\mathbf{r}) = -\mu I\hat{\mathbf{x}}\,\frac{e^{-jkr}}{4\pi r}\int_0^{2\pi} d\varphi(1+jkr)\frac{a^2\sin\theta\sin^2\varphi}{r} \tag{21.17}$$

and the $d\varphi$ integration delivers a factor π which, using the dipole-moment notation $m \equiv I\pi a^2$, yields

$$\begin{aligned}\mathbf{A}(\mathbf{r}) &= -\hat{\mathbf{x}}\frac{\mu m\sin\theta e^{-jkr}}{4\pi r^2}(1+jkr)\\ &= \frac{\mu\mathbf{r}\times\mathbf{m}}{4\pi}\left(\frac{e^{-jkr}}{r^2}\right)(1+jkr)\end{aligned} \tag{21.18}$$

The second form of (21.18) can be applied to any situation in which an infinitesimal magnetic dipole is placed somewhere in space. Note that $\mathbf{r}\times\mathbf{m} \propto -\hat{\boldsymbol{\varphi}}$ when $\mathbf{r}$ lies in an arbitrary vertical plane. In this case, the first form in (21.18) can be used with $\hat{\mathbf{x}}$ replaced by $\hat{\boldsymbol{\varphi}}$, and curl $\mathbf{A}$ is found simply, because it has only two components that involve the $\hat{\boldsymbol{\varphi}}$ component:

$$\nabla\times\mathbf{A} = \hat{\mathbf{r}}\frac{1}{r\sin\theta}\frac{\partial}{\partial\theta}(A_\varphi\sin\theta) - \hat{\boldsymbol{\theta}}\frac{1}{r}\frac{\partial}{\partial r}(rA_\varphi) \tag{21.19}$$

With relatively minor calculational effort, it is found that

$$\begin{aligned}\frac{\partial}{\partial\theta}(A_\varphi \sin\theta) &= \frac{2\mu m \sin\theta\cos\theta\, e^{-jkr}}{4\pi r^2}(1+jkr)\\ \frac{\partial}{\partial r}(rA_\varphi) &= -\frac{\mu m \sin\theta\, e^{-jkr}}{4\pi r^2}(1+jkr-k^2r^2)\end{aligned} \tag{21.20}$$

Now apply the second of (21.4b) to obtain

$$\begin{aligned}H_\theta &= \frac{m\sin\theta\, e^{-jkr}}{4\pi r^3}(1+jkr-k^2r^2)\\ H_r &= \frac{m\cos\theta\, e^{-jkr}}{2\pi r^3}(1+jkr)\end{aligned} \tag{21.21a}$$

We obtain the electric field from $\mathbf{E} = (\nabla\times\mathbf{H})/j\varepsilon\omega$. Only the two $\hat{\boldsymbol{\varphi}}$ terms of $\nabla\times\mathbf{H}$ in spherical coordinates are nonzero, and we find from (21.21a) that

$$E_\varphi = -j\eta\frac{km\sin\theta\, e^{-jkr}}{4\pi r^2}(1+jkr) \tag{21.21b}$$

In terms of coordinate-free expressions we obtain

$$\begin{aligned}\mathbf{H}_\perp &= \frac{e^{-jkr}}{4\pi r^3}(1+jkr-k^2r^2)\hat{\mathbf{r}}\times(\hat{\mathbf{r}}\times\mathbf{m})\\ \mathbf{H}_\parallel &= \frac{e^{-jkr}}{2\pi r^3}(1+jkr)(\hat{\mathbf{r}}\cdot\mathbf{m})\hat{\mathbf{r}}\\ \mathbf{E} &= j\eta\,\frac{ke^{-jkr}}{4\pi r^2}\hat{\mathbf{r}}\times\mathbf{m}\end{aligned} \tag{21.22}$$

The far-field expressions obtained from (21.21a) are

$$\begin{aligned}H_\theta &= -\frac{k^2 m\sin\theta\, e^{-jkr}}{4\pi r}\\ E_\varphi &= \eta\,\frac{k^2 m\sin\theta\, e^{-jkr}}{4\pi r} = -\eta H_\theta\end{aligned} \tag{21.23}$$

The time-averaged Poynting vector is $S = \frac{1}{2}\eta|E_\varphi|^2\hat{\mathbf{r}}$, with magnitude

$$\mathbb{S} = \frac{\eta k^4(m\sin\theta)^2}{32\pi^2 r^2} \tag{21.24a}$$

The total power transmitted by radiation through a sphere surrounding the dipole, obtained similarly to (21.13), is

$$P_t = \eta\frac{m^2k^4}{12\pi} \tag{21.24b}$$

As in the electric-dipole case, the total power radiated through a sphere is proportional to the fourth power of the frequency.

21.4 A linear current source of nonzero length; antenna parameters

A nonzero-length linear antenna radiator can be build up from the infinitesimal current-source results established above in Eqs. (21.6)–(21.10). As shown in Fig. 21.4, such an antenna may be considered as the sum (integral) of a linear superposition of infinitesimal current sources Idl' at locations $\mathbf{r}' = (0, 0, z')$ with $-L/2 < z' < L/2$. Our infinitesimal-current results can be adjusted by replacing R by $|\mathbf{r} - \mathbf{r}'|$, which is equal to $\sqrt{r^2 - 2\mathbf{r}\cdot\mathbf{r}' + r'^2} = \sqrt{r^2 - 2rz'\cos\theta + z'^2}$.

The most useful results are obtained for a parameter regime defined roughly by $\lambda \leq L \ll r$, i.e. for antennas that are comparable in length to at least an appreciable fraction of a wavelength, or are several wavelengths long but with the radiation observed at a distance r that is very large compared to L. In this case, it is an excellent approximation to replace all inverse powers of R by r, but we must use $|\mathbf{r} - \mathbf{r}'| \approx r - z'\cos\theta$ in the exponential $\exp(-jkR)$ because we do not assume that kL is small compared to π. However, neglect of the next-highest order terms in the exponential implies that we are assuming that $kL^2 \ll r$. This, then, is the crucial approximation to be considered in this section; it is known in optics as the *Fraunhofer* or *radiation zone* (see Section 21.8).

The results of (21.6)–(21.10) are repeated, except that $(Idl')\sin\theta$ is replaced by

$$\int_{-L/2}^{L/2} dz\, I(z)e^{jkz\cos\theta} \tag{21.25}$$

The next issue is that of an appropriate expression for $I(z)$. A distribution of current in a conductor is a difficult problem to solve rigorously because it is one in which fields and sources must satisfy Maxwell's equations at every point, and in particular at the interfaces, so that a complicated boundary-value problem must be solved to obtain $I(\mathbf{r})$ for a non-infinitesimally thin wire before taking the limit of

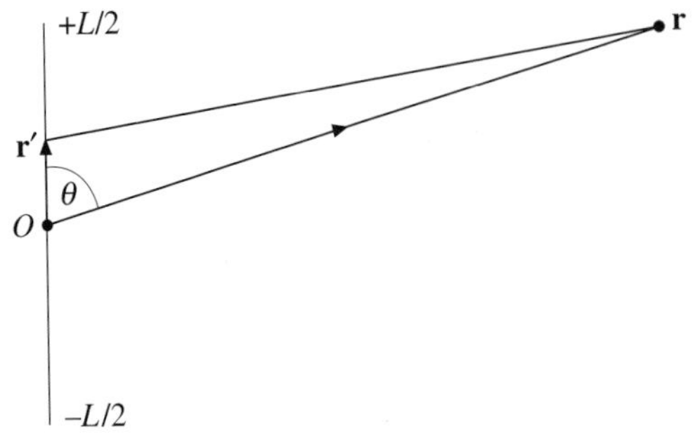

Fig. 21.4 A linear current-source as an antenna.

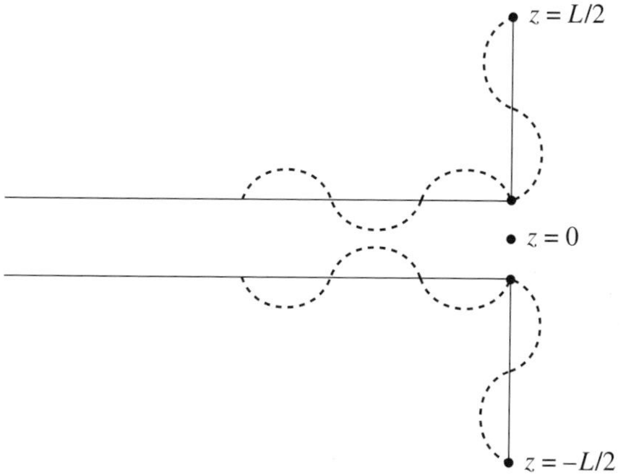

Fig. 21.5 A transmission line terminating in a linear antenna.

infinitesimal thickness. It is not feasible to introduce $I(z)$ in that way here, although approximate treatments can be found in texts specializing in antenna theory. One simplified way to approach a possible radiation current is to take a thin two-wire transmission line and to bend by 90° a length preceding each termination, in order to obtain a radiator, as shown in Fig. 21.5.

On the stretch of transmission line that is horizontal in the figure, the upper wire has a current distribution of opposite sign to that of the lower wire. Therefore, if the spacing is small compared to the wavelength then the radiation from this part will be negligible (and zero in the limit of zero spacing).

On the upper and lower vertical pieces the current pattern is symmetric and sinusoidal. Hence a possible form for $I(z)$ might be

$$I(z) = I_0 \sin(\varphi - k|z|) \tag{21.26a}$$

Transmission lines with open-circuit termination have zero current at termination; see (20.77) for a reminder. It follows that $\varphi = kL/2$, and thus the current distribution is

$$I(z) = I_0 \sin[k(L/2 - |z|)] \tag{21.26b}$$

Hence $Idl' \sin\theta$ must be replaced, see (21.25), as follows:

$$Idl' \sin\theta \longrightarrow I_0 \sin\theta \int_{-L/2}^{L/2} dz\, e^{jkz\cos\theta} \sin[k(L/2 - |z|)]$$

$$= 2I_0 \sin\theta \int_0^{L/2} dz\, \cos(kz\cos\theta) \sin[k(L/2 - |z|)] \tag{21.27a}$$

The calculation of the integral is straightforward, although somewhat lengthy; the result is simply

$$\begin{aligned} Idl' \sin\theta &\longrightarrow 2\left(\frac{I_0}{k}\right)F(\theta) \\ F(\theta) &= \frac{\cos(\frac{1}{2}kL\cos\theta) - \cos(\frac{1}{2}kL)}{\sin\theta} \end{aligned} \tag{21.27b}$$

As a consequence, we obtain in the Fraunhofer regime of radiation from a linear antenna

$$\begin{aligned} E_\theta &= j\eta I_0 F(\theta)\frac{e^{-jkr}}{2\pi r}\left(1 - \frac{j}{kr} - \frac{1}{k^2r^2}\right) \\ E_r &= 2j\eta I_0 F(\theta)\frac{e^{-jkr}}{2\pi r^2}\left(1 - \frac{j}{kr}\right) \\ H_\varphi &= jI_0 F(\theta)\frac{e^{-jkr}}{2\pi r}\left(1 - \frac{j}{kr}\right) \end{aligned} \tag{21.28}$$

and in the far field this reduces to

$$\begin{aligned} E_\theta &= j\eta I_0 F(\theta)\frac{e^{-jkr}}{2\pi r} \\ H_\varphi &= jI_0 F(\theta)\frac{e^{-jkr}}{2\pi r} = \frac{1}{\eta}E_\theta \end{aligned} \tag{21.29}$$

The antenna pattern function $F(\theta)$ defines the dependence of each field component – whether it is in the far-field or not – upon the polar angle of direction. Note that $F(\theta)$ is not a function of azimuthal angle φ; there is rotational symmetry around the linear current source. It is easily seen that $F(\theta) \to \sin\theta$ as $L \to 0$. As L is increased the antenna power pattern $|F(\theta)|^2$ shows a shape with an increasing number of 'lobes'. A lobe is defined by angular region $\theta_1 < \theta < \theta_2$ such that $|F(\theta)|^2$ is zero at $\theta = \theta_1$ and θ_2 and reaches some maximal value in between. The main direction of radiated power defines a *main lobe*. Other lobes are known as *side lobes*. The effective width of the main lobe can also be defined by the half-power points, or by some other criterion such as the -10 dB power points. The full main-lobe width, as defined above, is 180° when $L/\lambda = 1.2$ (there are side lobes for the linear antenna under consideration above when $L/\lambda > 1.2$). For example, the -10 dB width of the main lobe is defined by θ_m as given by

$$|F(\theta_m)|^2 = 0.1|F(\theta_0)|^2 \tag{21.30}$$

if the maximum is at $\theta = \theta_0$. The linear antenna described above does not emit maximum power at $\theta = \pi/2$ when $l > 1.2\lambda$; the maximum power is then sent in a lobe the direction of which deviates from $\theta = \pi/2$. For those antennas for which

$F(\theta)$ differs for $E(\mathbf{r})$ and $H(\mathbf{r})$ it is necessary to define separate antenna patterns for the electric and magnetic fields.

The average Poynting vector S in the far-field zone represents the power flux through 1 m^2 at a distance r from the center of the source, and we have observed that it is proportional to $1/r^2$ because (21.29) predicts

$$S = \eta \frac{I_0^2 |F(\theta)|^2}{8\pi^2 r^2} \hat{\mathbf{r}} \tag{21.31}$$

This quantity gives rise to a related scalar quantity used in antenna theory: the *radiation intensity* $U(\theta, \varphi) = r^2|S|$, which thus represents the *power radiated per unit solid angle* in the direction indicated by the angles θ and φ. It is useful to compare the radiation intensity to the *average radiation intensity* $\bar{U}$, which is equal to the average power radiated per unit solid angle: $\bar{U} = P_t/4\pi$. The ratio is known as the *gain* in the direction θ, φ, and it is defined by

$$G_t(\theta, \varphi) = U(\theta, \varphi)/\bar{U}(\theta, \varphi) = 4\pi U(\theta, \varphi)/P_t \tag{21.32}$$

The radiated flux (average Poynting vector) in any direction can then be expressed as

$$S(\theta, \varphi) = \frac{P_t G_t(\theta, \varphi)}{4\pi r^2} \hat{\mathbf{r}} \tag{21.33}$$

The gain $G_t(\theta, \varphi)$ varies from direction to direction as given by the antenna pattern, whereas the remaining factors in this equation do not. An *isotropic* antenna has no preferred direction of radiation, hence an isotropic antenna has unit gain; $G_t = 1$.

It is easily seen, from the flux and power calculations derived from (21.10), that the infinitesimal (Hertzian) electric dipole antenna has a gain $G_t(\theta, \varphi) = 1.5 \sin^2 \theta$; the same result holds for the infinitesimal magnetic dipole, as can be verified from (21.23). These gains are relatively small deviations from isotropy. The more the gain of an antenna can deviate from unity, the more that antenna deviates from an isotropic radiator. Many types of radar, for example, have antennas with high gain so that distant small targets can be observed, provided the radar is pointed in the right direction. Radio transmitters, by contrast, are designed so that they have an antenna pattern that is broad in horizontal directions, so that there is a wide-area coverage of ground receivers (in homes, etc.). There is no point in wasting power in directions other than those parallel to the surface of the earth, hence the vertical pattern is designed to be narrow.

The effectiveness of an antenna is often measured by the *radiation resistance* $\mathcal{R}$, as follows. The average power dissipated in a resistor $\mathcal{R}$, fed by current $I(t) = I_0 \cos \omega t$, is $P = \frac{1}{2}\mathcal{R} I_0^2$. Likewise, the current on the antenna gives rise to radiated power, which is 'dissipated' into space. The radiation resistance therefore is defined in an antenna by

$$\mathcal{R} = 2P_t/I_0^2 \tag{21.34}$$

and is measured in ohms (Ω). Thus we find for an antenna with pattern function $F(\theta)$ and with

$$P_t = \eta\frac{I_0^2}{4}\int_0^{\pi} d\theta \sin\theta |F(\theta)|^2 \tag{21.35a}$$

(see (21.31)) that the radiation resistance is

$$\mathcal{R} = \frac{\eta}{2}\int_0^{\pi} d\theta \sin\theta |F(\theta)|^2 \tag{21.35b}$$

21.5 An array of linear current sources

It is clear after consideration of (21.27b), second equation, that the linear antenna formulas (21.28) do not give rise to narrow antenna patterns with high gain; the $F(\theta)$ antenna pattern is too broad or too multilobed at any value of L/λ. One way to obtain a much narrower pattern with higher gain is to construct an *array* of antennas. If, for example, we consider linear current sources, as described in the previous section, and put N of these in a linear *array* with spacing l in a direction indicated by the unit vector $\hat{\mathbf{r}}'$ (see Fig. 21.6), with source n centered at vector $\mathbf{r}_n = nl\hat{\mathbf{r}}'$, then we must replace $I_0 e^{jkr}$ in Eqs. (21.28) by a sum:

$$I_0 \sum_{n=0}^{N-1} e^{-jk|\mathbf{r}-\mathbf{r}_n|} \tag{21.36}$$

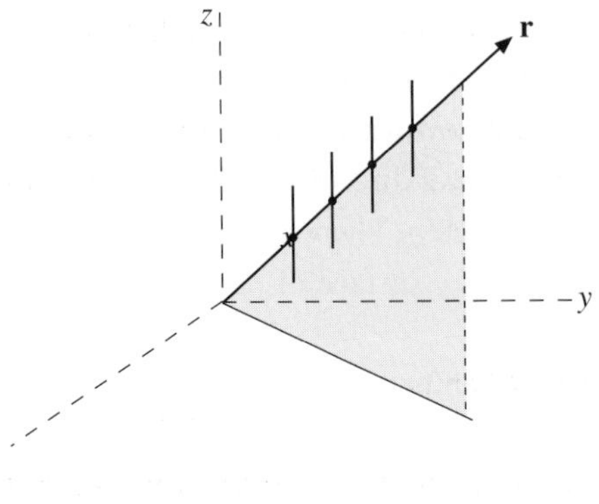

Fig. 21.6 An array of linear current sources.

The point of observation remains $\mathbf{r}$ and is still assumed to be in the far-field zone of all N linear sources. Hence the other factors in (21.28) remain unchanged. The above sum can be evaluated by noting that

$$|\mathbf{r}-\mathbf{r}_n| = \sqrt{r^2 - 2\mathbf{r}\cdot\mathbf{r}_n + r_n^2} \approx r - \hat{\mathbf{r}}\cdot\mathbf{r}_n$$
$$= r - nl\hat{\mathbf{r}}\cdot\hat{\mathbf{r}}' \equiv r - nl\cos\psi \tag{21.37}$$

The angle ψ is between the direction of observation $\hat{\mathbf{r}}$ and the line of sources in direction $\hat{\mathbf{r}}'$. Thus, we replace $I_0 e^{jkr}$ by

$$I_0 e^{jkr}\sum_{n=0}^{N-1} e^{-jnkl\cos\psi} \tag{21.38}$$

The sum is easily evaluated, as it is a simple power series in the exponential $e^{-j\beta}$, with $\beta \equiv kl\cos\psi$: it is

$$\frac{e^{-jN\beta}-1}{e^{-j\beta}-1} = \frac{e^{-\frac{1}{2}jN\beta}}{e^{-\frac{1}{2}j\beta}}\frac{\sin(\frac{1}{2}N\beta)}{\sin(\frac{1}{2}\beta)}$$

so that the power antenna pattern becomes

$$|F(\theta)|^2\left|\frac{\sin(\frac{1}{2}Nkl\cos\psi)}{\sin(\frac{1}{2}kl\cos\psi)}\right|^2 \tag{21.39a}$$

This pattern is much more sharply peaked than that for a single source; this can be seen by writing (21.39a) as

$$|F(\theta)|^2 N^2\left|\frac{\sin(\frac{1}{2}Nkl\cos\psi)}{\frac{1}{2}Nkl\cos\psi}\right|^2\left|\frac{\sin(\frac{1}{2}kl\cos\psi)}{\frac{1}{2}kl\cos\psi}\right|^2 \tag{21.39b}$$

The peak value is N^2, at $\cos\psi = 0$, and this value falls off rapidly to a first zero as $\frac{1}{2}Nkl\cos\psi$ approaches $\pi/2$. If $Nkl \gg 1$, as is likely to be the case, then this criterion tells us that a maximal value occurs at $\psi = \pi/2$ and that the lobe falls off to zero at $\psi \approx \pi/2 - 2/Nkl$. At this value, the last factor (29.39b) is still very close to unity if N is large. So, a main lobe stretches out *broadside* with a half width $\sim 2/Nkl$, which is certainly much narrower than the lobe $F(\theta)$ of (21.27b) if l and L are of the same order of magnitude.

21.6 Receiving antennas and radar equation

Here, we only discuss a few points about antennas as used for receiving radiation (from other antennas). There is much *reciprocity* between transmitting and receiving antennas. Without going into details here, one might guess, correctly, that a transmitting antenna would work as a receiving antenna if time were reversed: when the radiated *EM* fields flowed back into the antenna they would produce the very same current I that gave rise to those fields when time progressed in the usual positive fashion. If, however, a receiving antenna were positioned in a beam emanating from a transmitting antenna some distance away, then the received power could be expressed as

$$P_\mathrm{r} = A_\mathrm{r}\mathbb{S}_\mathrm{r}(\theta, \varphi) \tag{21.40}$$

where A_r is a characteristic reception area for the receiving antenna and $\mathbb{S}_\mathrm{r}(\theta, \varphi)$ is the power flux received from the transmitting antenna. In terms of the transmitting-antenna gain $G_\mathrm{t}(\theta, \varphi)$ and the distance R from the receiving antenna, one can write (21.40) as

$$P_\mathrm{r} = \frac{A_\mathrm{r}G_\mathrm{t}(\theta, \varphi)P_\mathrm{t}}{4\pi R^2} \tag{21.41}$$

Reciprocity and other arguments can be shown (in a lengthy derivation that goes beyond the material to be covered in this text) to lead to a linear relationship between the characteristic area A_r and the gain factor of the receiving antenna $G_\mathrm{r}(\theta', \varphi')$, where the angles θ' and φ' pertain to the orientation of the receiving antenna with respect to the incoming radiation:

$$A_\mathrm{r} = \frac{\lambda^2}{4\pi}G_\mathrm{r}(\theta', \varphi') \tag{21.42}$$

Equation (21.41) then becomes

$$P_\mathrm{r} = \frac{\lambda^2 G_\mathrm{t}(\theta, \varphi)G_\mathrm{r}(\theta', \varphi')}{16\pi^2 R^2}P_\mathrm{t} \tag{21.43}$$

The received power is thus expressed in terms of two gain factors. One of these involves the angles θ, φ that express the direction of radiation with respect to the main axis of the transmitting antenna; the other is a function of the angles θ', φ' between the radiation and the main axis of the receiving antenna. Equations (21.42)–(21.43) require the solid angle A_r/R^2 to be quite small, otherwise the gain factors are not well defined and angle integrals will be needed to account for variations in the gain factors over this solid angle (as would be the case for distances between the antennas that are not large compared to antenna dimensions).

Consider now the situation of Fig. 21.7, in which a transmitting radar (T) illuminates an object of dimensions small enough (or at a sufficiently distant location) that the gain factor is constant over the effective width of the illuminated object. Owing to the mismatch in dielectric permittivities of the air and of the object, the incident radiation will scatter in many directions, and some of it will propagate towards the receiving radar (R).

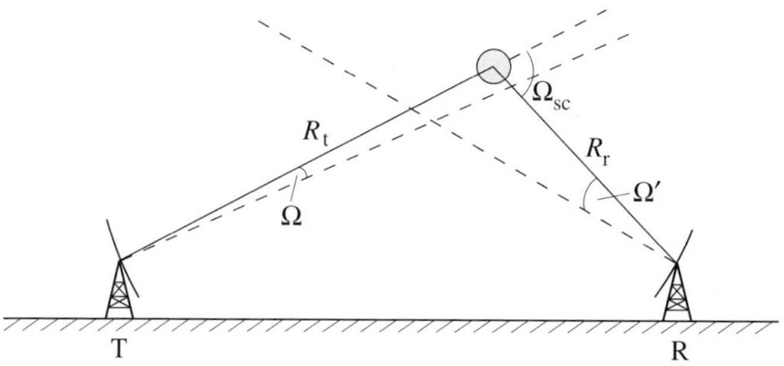

Fig. 21.7 Two ground radars with parameters for the radar equation. The direction of the scatterer with respect to the main axis of the transmitting radar is given by θ, φ (abbreviated to Ω); the direction of the scattered radiation with respect to the same axis is given by θ_{sc}, $\varphi_{sc}(\equiv \Omega_{sc})$; the direction of the scattered radiation with respect to the main axis of the receiving radar is given by $\theta'\varphi'(\equiv \Omega')$.

The power flux incident upon the scatterer is

$$\mathbb{S}_t(\Omega) \equiv \frac{1}{2}\eta|E_{inc}|^2 = \frac{P_t G_t(\Omega)}{4\pi R_t^2} \tag{21.44}$$

given that Ω is an abbreviation for the angles θ, φ that the line of sight R_t makes (see Fig. 21.7) with the main axis of the transmitting radar. It may be presumed that both $\theta = 0$ and $\varphi = 0$ if the radar is aimed optimally at the scattering object. The first expression in (21.44) is merely a definition of E_{inc} and η is the intrinsic impedance of air.

At a location sufficiently distant from the illuminated object, in any direction the scattered fields will appear to be emanating from a point source, and hence are spherical waves. Consequently, we may write for the electric field scattered at a large distance from the object, in a direction defined by angles θ_{sc}, φ_{sc} $(\equiv \Omega_{sc})$, again with respect to the main axis of the transmitting radar,

$$E_{sc} = f(\Omega_{sc})E_{inc}\frac{e^{-jkR_r}}{4\pi R_r} \tag{21.45a}$$

Here $f(\Omega_{sc})$ represents a 'strength factor' (or *scattering amplitude*), which gives the fraction of the incident electric field that has been scattered into the direction Ω_{sc}. Hence the power flux incident upon the receiving radar must be

$$\mathbb{S}_r(\Omega') = \frac{|f_{sc}(\Omega_{sc})|^2}{(4\pi R_r)^2}\mathbb{S}_t(\Omega) = \frac{\sigma_{sc}(\Omega_{sc})}{(4\pi R_r)^2}\mathbb{S}_t(\Omega) \tag{21.45b}$$

where $\sigma(\Omega_{sc}) \equiv |f(\Omega_{sc})|^2$ is known as the *differential cross section* for scattering by the object from the incident direction into the direction under consideration. The dimension of $\sigma(\Omega_{sc})$ is that of an *area* (hence the appellation 'cross section').

The integral of $\sigma(\Omega_{sc})$ over all possible directions is known as the *total cross section* σ, which is a function only of frequency and of the intrinsic dielectric properties and shape of the scatterer.

The quantity $\sigma_b = \sigma(\theta_{sc} = \pi)$ is known as the *backscatter cross section*, and it is a measure of the backscattered radiation of a *monostatic* radar (receiver and transmitter at the same location). The differential cross section $\sigma(\Omega_{sc})$ is also a function of the orientation as well as of the shape of the object. The prediction of $\sigma(\Omega_{sc})$, σ_b, or σ is a subject that has been widely investigated, and it is one for which analytical solutions exist only in a very restricted number of cases. For example, the scattering from dielectric and/or conducting spheres was investigated by G. Mie early in the twentieth century, and the so-called Mie scattering cross sections are widely used as approximations for rain drops and other hydrometeors that can occur in the atmosphere. These Mie cross sections have simple analytical forms when the diameters D are either small or large with respect to a wavelength λ, but must be expressed as complicated power series in the parameter D/λ in the intermediate regime where wavelength and particle diameter are of the same order of magnitude.

We now consider the scattered radiation from the point of view of the receiving radar. Again we use Ω' to indicate the angles θ', φ' which give the scattered direction with respect to the main axis of the receiving radar, so that we also may write for the power that is received by R from that direction

$$P_r = A_r \mathbb{S}_r(\Omega') = \frac{\lambda^2 G_r(\Omega')}{4\pi} \frac{\sigma(\Omega_{sc})}{(4\pi R_r)^2} \mathbb{S}_t(\Omega) \tag{21.45c}$$

The final *bistatic radar equation* then becomes

$$P_r = \frac{\lambda^2 \sigma(\Omega_{sc}) G_t(\Omega) G_r(\Omega')}{(4\pi)^3 R_t^2 R_r^2} P_t \tag{21.46}$$

This equation clearly demonstrates for a *monostatic* radar (receiver and transmitter at the same location so that $R_r = R_t$) that the power backscattered from an illuminated object is proportional to the inverse fourth power of the distance, which in turn helps explain why quite powerful radars are needed to obtain echoes from small object. Consider for example a small airplane, for which one may estimate that $\sigma(\Omega_{sc}) \approx 1.5\,\text{m}^2$. At 3 GHz frequency, the transmitted wavelength λ is 0.1 m. At 10 km distance we would find, for the power received per unit solid angle from the direction Ω',

$$P_r = 0.76 \times 10^{-13} G_t(\Omega) G_r(\Omega') P_t \text{ in W} \tag{21.47a}$$

If we assume a 10 000 W transmitter with a gain of 20 dB in the optimal direction (which we may assume to be the case when illuminating the object) and a receiver with a gain of 10 dB in the main lobe (assuming also that the receiving radar is pointing optimally), then another factor of 10^7 is gained, and

$$P_r \approx 0.76 \times 10^{-6}\,\mathrm{W} \tag{21.47b}$$

Thus a sensitive receiver is clearly needed to sense the airplane at that distance because the received signal (for this representative example) is in the order of only 1 μW.

21.7 Geometrical optics and refraction

By 'high frequencies' we usually imply that the wavelength of the radiation in question is smaller than any characteristic physical length of the medium through which it is propagating. This is the case, for example, if the relative dielectric permittivity ε_r (or refractive index n with $\varepsilon_r = n^2$) varies very little over one wavelength in a medium through which monochromatic (or almost monochromatic) electromagnetic waves propagate. Under these conditions, very useful approximations can be made to the electromagnetic fields. The purpose of this section is to write down such an approximation for the electric field phasor after showing the considerations that lead to it. The starting point is the wave equation for time-harmonic waves, in the absence of sources, which can be found from (16.23) to be

$$\nabla \times (\nabla \times \mathbf{E}) - k^2 \varepsilon_r \mathbf{E} = 0 \tag{21.48a}$$

in terms of the relative dielectric permittivity ε_r; this then can be rewritten as

$$\Delta \mathbf{E} + k^2 \varepsilon_r \mathbf{E} = \nabla(\nabla \cdot \mathbf{E}) \tag{21.48b}$$

The Maxwell divergence equation, in the absence of sources, states that $\nabla \cdot (\varepsilon_r \mathbf{E}) = 0$, so that some rearrangement of this yields $\nabla \cdot \mathbf{E} = -\mathbf{E} \cdot \nabla(\ln \varepsilon_r)$, which in turn leads to

$$\Delta \mathbf{E} + k^2 \varepsilon_r \mathbf{E} = -\nabla(\mathbf{E} \cdot \nabla \ln \varepsilon_r) \tag{21.48c}$$

A careful analysis, which goes beyond the scope of this treatment, shows that the right-hand side of this equation is negligible when $\lambda|\nabla(\ln \varepsilon_r)| \ll 1$. A plausibility argument confirming this goes as follows: the field $\mathbf{E}$ varies roughly (see below) as e^{jknz}, if the z axis represents the local direction of propagation of a wavefront that is essentially plane in a uniform-n medium, and therefore each derivative of $\mathbf{E}$ supplies a factor $kn = 2\pi n/\lambda$. Thus the two left-hand-side terms are of order E/λ^2, whereas the right-hand-side term is of order $E|\nabla(\ln \varepsilon_r)|/\lambda$. Hence we have the condition $\lambda|\nabla(\ln \varepsilon_r)| \ll 1$ for neglecting the right-hand side. An approximate scalar equation for each component E of $\mathbf{E}$ follows:

$$\Delta E + k^2 n^2(\mathbf{r}) E = 0 \tag{21.49}$$

This equation then describes propagation through a medium with slowly varying refractive index.

The next step is to exploit the idea of slow change and to write the electric field component as

$$E = Ae^{-j\phi} \tag{21.50}$$

The quantity ϕ is known as the *eikonal*. By taking the divergence, ∇E, one finds

$$\Delta E = \left[-2j\frac{\nabla A}{A} \cdot \nabla\phi - (\nabla\phi)^2 - j\Delta\phi + \frac{\Delta A}{A} \right] E$$

which gives rise to two equations after substitution into (21.49), one for the real and one for the imaginary part:

$$\begin{aligned} \frac{\Delta A}{A} + k^2 n^2 - (\nabla\phi)^2 &= 0 \\ \nabla \cdot (A^2 \nabla\phi) &= 0 \end{aligned} \tag{21.51}$$

If $\Delta A/A$ is negligible, then the first of these equations becomes the defining equation of geometrical optics, namely

$$\nabla\phi = kn\hat{\mathbf{s}} \tag{21.52}$$

where $\hat{\mathbf{s}}$ is a unit vector perpendicular to the surface $\phi =$ constant passing through the location in question. We shall apply this equation first to a tube of field lines of $\nabla\phi$ (or $\hat{\mathbf{s}}$) that links two such surfaces. Let us suppose that the tube has a cross-sectional area S, which is a function of a position $\mathbf{s}$ along the field line. The conditions for validity of (21.52) are listed below in Eq. (21.55) and a derivation of those conditions follows here in an optional-reading section.

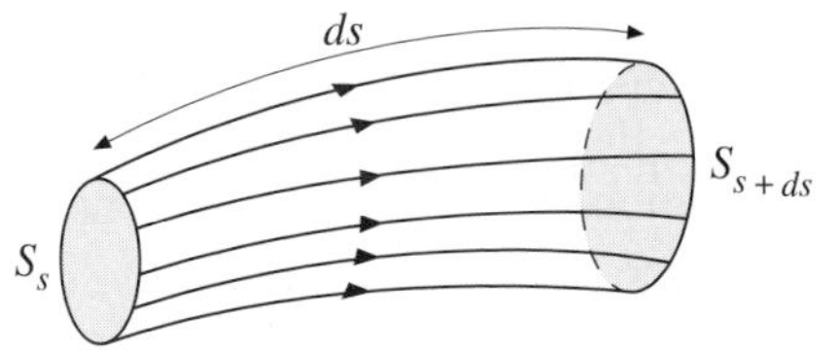

Fig. 21.8 A tube of geometrical-optics rays. The two ends are parts of constant-ϕ surfaces. The left-hand end has area $S(s)$, the right-hand end has area $S(s + ds)$.

If (21.52) holds, then the second of (21.51) becomes

$$\nabla \cdot (A^2 n\hat{\mathbf{s}}) = 0 \tag{21.53a}$$

and a volume integral of this equation over a tube of ∇S (or $\hat{\mathbf{s}}$) lines yields, using Gauss's divergence theorem,

$$\int dv[\nabla \cdot (A^2 n\hat{\mathbf{s}})] = \oint dS\, A^2 n = (A^2 nS)_{s+ds} - (A^2 nS)_s = 0 \tag{21.53b}$$

Note carefully that S here is a surface area perpendicular to unit vector $\hat{\mathbf{s}}$ in the tube. Thus (21.53b) states that the quantity $A^2(s)n(s)S(s)$ is constant along a field line defined by $\hat{\mathbf{s}}$:

$$A^2(s) = \frac{\text{constant}}{n(s)S(s)} \tag{21.53c}$$

From this we find that

$$\nabla A = -\frac{A}{2}\left[\frac{1}{n}\nabla n + \frac{1}{S}\nabla S\right] \tag{21.54a}$$

where both gradient operators are directed along direction $\hat{\mathbf{s}}$. Upon further differentiation,

$$\Delta A = \frac{A}{4}\left[\frac{1}{n}\nabla n + \frac{1}{S}\nabla S\right]^2 - \frac{A}{2}\left[\left(\frac{1}{n}\nabla n\right)^2 - \frac{1}{n}\Delta n + \left(\frac{1}{S}\nabla S\right)^2 - \frac{1}{S}\Delta S\right] \tag{21.54b}$$

If we neglect the curvature (the second derivatives) of n and S, then (21.45b) states that $|\Delta A/A|$ is determined entirely by the *squares* of the fractional derivatives, $|\nabla n/n|$ of the refractive index and $|\nabla S/S|$ of the surface areas perpendicular to the tube of field lines. Because $|\nabla n/n| = |\nabla(\ln n)| = \frac{1}{2}|\nabla(\ln \varepsilon_r)| \ll \lambda^{-1}$, by virtue of the neglect of the right-hand side of (21.48c), we need only require that $|\nabla S/S| \ll \lambda^{-1}$ in order that $\Delta A/A \ll \lambda^{-2}$. This in turn is enough to make the first term of (21.51) negligible, which is what we assumed in the first place to derive this. Neglect of the curvature means that we have also assumed that $|\Delta n/n| \ll \lambda^{-2}$ and $|\Delta S/S| \ll \lambda^{-2}$.

The assumption that $\Delta A/A$ is negligible, and therefore that (21.52) is a reasonable approximation, is guaranteed by the inequalities

$$\lambda\left|\frac{1}{n}\nabla n\right| \ll 1, \quad \lambda\left|\frac{1}{S}\nabla S\right| \ll 1, \quad \lambda^2\left|\frac{1}{n}\Delta n\right| \ll 1, \quad \lambda^2\left|\frac{1}{S}\Delta S\right| \ll 1 \tag{21.55}$$

The second inequality merely states that the area S does not change appreciably in a wavelength, i.e. that the tube of field lines of $\hat{\mathbf{s}}$, or of $\nabla\phi$, must not change appreciably in diameter in a wavelength. These are the necessary and sufficient conditions for what is known as geometrical optics.

Integration of (21.52) then yields

$$\phi(s) = k \int_0^s ds' \, n(s') \tag{21.56}$$

so that the eikonal function $\phi(s)$ is seen to be k times the *optical pathlength* and is therefore equal to the phase of the electric field phasor. Together with (21.53c), this result predicts an electric field phasor

$$E(s) = \frac{\text{constant}}{\sqrt{S(s)n(s)}} \exp\left\{ -jk \int_0^s ds' \, n(s') \right\} \tag{21.57}$$

in the understanding that $E(s)$ is and remains perpendicular to $\hat{\mathbf{s}}$ to the order of approximation involved in obtaining (21.57) and that $S(s)$ is the cross-sectional area of a tube of field lines of $\hat{\mathbf{s}}$ over which $E(s)$ is essentially constant.

Thus the major effects in geometrical optics are that the phase change over a path of length s is

$$\delta\phi = k \int_0^s ds' [n(s') - 1] \tag{21.58}$$

and that some 'bending' or refraction of the wavefront can occur. The refraction can be seen from Eq. (21.52): because the left-hand side of that equation is a gradient, it follows that

$$\nabla \times n(s)\hat{\mathbf{s}} = 0 \tag{21.59a}$$

Consider Fig. 21.9, in which an optical path $\hat{\mathbf{s}}$, which indicates the direction of $\nabla\phi$, travels through a medium in which the refractive index $n(z)$ is a function of only one direction (at least locally). Then the surface integral over the rectangle shown

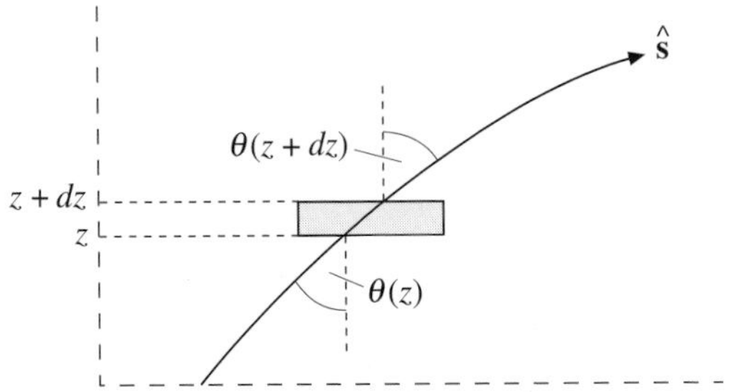

Fig. 21.9 Refraction of a geometrical-optics ray.

yields, via Stokes's law,

$$0 = \int d\mathbf{S} \cdot [\nabla \times n(s)\hat{\mathbf{s}}] = \oint d\boldsymbol{l} \cdot n(s)\hat{\mathbf{s}} \tag{21.59b}$$

with $d\boldsymbol{l} \cdot \hat{\mathbf{s}} = dl \sin\theta$; θ is the angle between the $\hat{\mathbf{z}}$ axis and $\hat{\mathbf{s}}$. Thus, the geometry of the infinitesimally thin rectangle in Fig. 21.9 indicates that (21.59b) yields

$$n(z) \sin\theta(z) = \text{constant} \tag{21.60}$$

which is the continuous form of Snell's law. Equations (21.56) and (21.60) express the major consequences of the geometrical-optics approximation: the wavefronts move along trajectories or 'rays' $\hat{\mathbf{s}}$ with the phase changes as given by (21.56), and these trajectories undergo the slow refractive bending as given by (21.60). The major assumption is that $n(\mathbf{r})$ changes slowly in a wavelength.

21.8 Diffraction

The scalar wave equation (21.49), which governs high-frequency wave propagation in a medium in which the refractive index $n(\mathbf{r})$ is slowly varying, leads to a rather different phenomenon when sharp discontinuities in the refractive index are present. Consider a finite volume v in free space ($n = 1$) as indicated in Fig. 21.10, with an observation point P in its interior. Exclude from the volume v an infinitesimal sphere of radius ϵ around P. Thus, the volume v has an exterior surface S and an interior surface S_0. Gauss's divergence theorem can verify that, if we define $G \equiv e^{-jkr}/R$ with $R = |\mathbf{r} - \mathbf{r}'|$, given that $\mathbf{r}'$ is any point between or on the surfaces S and S_0 and that $\mathbf{r}$ is the coordinate of the observation point P, then

$$\int_v dv'(E\Delta G - G\Delta E) = \oint_{S+S_0} d\mathbf{S}' \cdot (E\nabla' G - G\nabla' E) \tag{21.61}$$

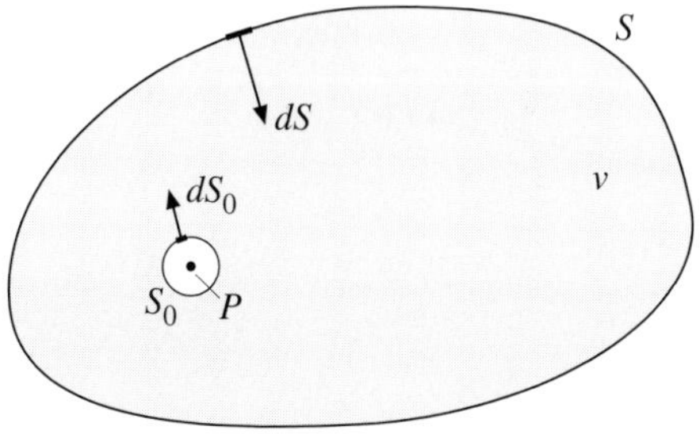

Fig. 21.10 The volume with exclusion of an infinitesimal section around the point of observation for Green's theorem (21.61).

This is one of a number of vector-integral identities known as *Green's theorems*; G is a Green's function. The surface integral over S_0 contains the integrand

$$\epsilon^2\, d\Omega \left[\left(-jk - \frac{1}{\epsilon}\right) E \frac{e^{-jk\epsilon}}{\epsilon} - \frac{e^{-jk\epsilon}}{\epsilon} \hat{\mathbf{n}} \cdot \nabla E \right] \tag{21.62}$$

which $\to -4\pi E$ as $\epsilon \to 0$. For any point $\mathbf{r}$ *inside* volume v it can be shown that $\Delta G = -k^2 G$ (this would no longer be true if $\mathbf{r}$ were allowed to be on the surface), whereas (21.49) in free space predicts $\Delta E = -k^2 E$, as a consequence of which the volume integral vanishes. The result for (21.61) is

$$E(\mathbf{r}) = \frac{1}{4\pi} \oint_S dS' \cdot [E\nabla' G - G\nabla' E] \tag{21.63}$$

Now we apply this to the situation depicted in Fig. 21.11, in which a point source O radiates fields through an aperture in a metallic plate, which in turn are observed at a point $P(\mathbf{r})$ beyond the plate. The surface S here consists of all points $\mathbf{r}'$ on the plate, on the surface surrounding the source-and-plate configuration. With respect to the infinite surface, it is safe to assume that the field reaching any point on it will be proportional to $e^{-jkr'}/r'$ and therefore that $\nabla' E$ is proportional to $-(jk + 1/r')Er' \approx -jkEr'$ on the infinite surface. The reason for this is that any finite source appears as a point to an observer who is essentially infinitely far away, and the point source plus aperture plate can be considered together as a finite source with dimensions negligible compared to r'. The definition of G indicates that $\nabla' G \approx jkG'\hat{\mathbf{R}} \approx -jkG\hat{\mathbf{r}}$, from which it then follows that $(E\nabla' G - G\nabla' E)dS'$ is zero on the infinite surface. Some care must be taken in establishing this as $dS' \to \infty$, even for very small solid angles, but it does follow.

On the aperture plate, more specifically on the surface facing point P, it is reasonable to assume that $E = 0$ and that $\hat{\mathbf{n}} \cdot \nabla E = 0$. This assumption is probably less valid very close to the rim of the aperture, and the results must therefore be

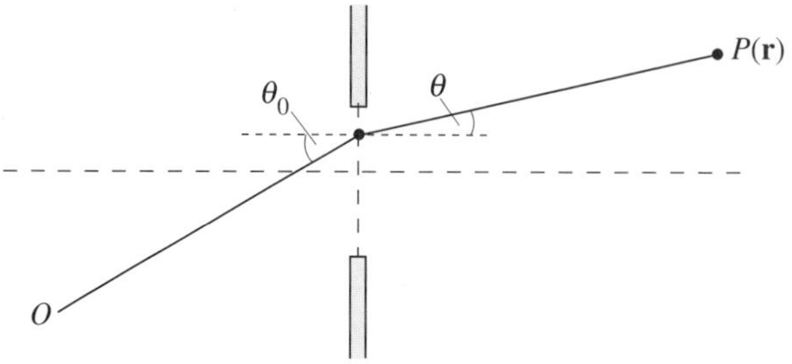

Fig. 21.11 Diffraction through an aperture.

regarded only as approximate, but the zero-field and normal-derivative approximation on the aperture surface hold much more closely as boundary conditions when the locations are at least several wavelengths away from the rim of the aperture; the surface then appears to be part of an infinite surface.

So then only the surface across the aperture is left; it can be taken as a planar surface, on which one may pose the *Kirchhoff boundary conditions*

$$E \approx \frac{Ae^{-jkr'}}{r'} \qquad \nabla' E \approx \left(jk + \frac{1}{r'} \right) E\hat{\mathbf{r}}' \tag{21.64}$$

In the far-field approximation ($k \gg 1/r'$, $k \gg 1/R$) we obtain

$$E \approx \frac{jkA}{4\pi} \int_S dS'(\cos\theta_0 + \cos\theta) \frac{e^{-jk(r'+R)}}{r'R} \tag{21.65}$$

This is a classic result in simple scalar diffraction theory known as the *Fresnel–Kirchhoff diffraction integral.* A number of variations of this expression exist. If the aperture is sufficiently small, then r' will not vary appreciably around the center of the aperture at $\mathbf{r}_0 = (0, 0, r_0)$. If we also take the origin O to lie on the central axis through the aperture, which is perpendicular to the aperture plane, then $\cos\theta_0 = 0$ and

$$E \approx \frac{jkA}{4\pi} \frac{e^{-jkr_0}}{r_0} \int_S dS'(1 + \cos\theta) \frac{e^{-jk|r-r'|}}{|r - r'|} \tag{21.66}$$

and $\cos\theta$ is essentially constant if the aperture diameter is very small compared to distance R. We also obtain this approximate result if the surface in the aperture is replaced by a part of a sphere with its center at O (which need not be on the central axis), provided that the surface differs from planar by no more than a small fraction of a wavelength.

A complementary situation is sketched in Fig. 21.12; the plate is removed and the aperture is replaced by a metallic disk of the same size and shape. *Babinet's principle*

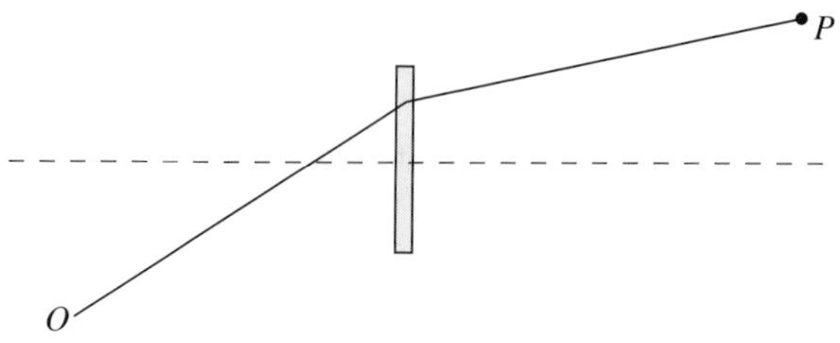

Fig. 21.12 The Babinet's principle complement to Fig. 21.11.

states that we can deduce the field at P for this situation from a superposition of the two situations. In the superposition, the sum of the two situations implies that we have a continuous metallic plate so that the field at P is just zero. Then, subtraction of the situation of Fig. 21.11 from this basic situation yields the situation of Fig. 21.12, which describes diffraction around an intervening object. Thus, zero minus the field of (21.66) yields the electric field for the situation of Fig. 21.12: the field for the complementary metal disc is *minus* that for the aperture.

Now let us try to calculate (21.66) for points near the axis; here $\cos\theta \approx 1$, so that

$$E \approx \frac{jkA}{2\pi}\frac{e^{-jkr_0}}{r_0}\int_S dS'\frac{e^{-jk|r-r'|}}{|r-r'|} \tag{21.67}$$

Referring to Fig. 21.12, we note that

$$R \equiv |r-r'| = \sqrt{L^2+(\rho-\rho')^2} \approx L+(\rho-\rho')^2/2L$$

so that the integral becomes

$$E \approx \frac{jkA}{2\pi}\frac{e^{-jkr_0}e^{-jk(L+\rho^2/2L)}}{r_0 L}\int_S d^2\rho' e^{-jk\rho\cdot\rho/2L}e^{-jk\rho'^2/2L} \tag{21.68}$$

For large distances L such that $k\rho'^2 \ll 2L$ (as mentioned earlier, this is known as the *Fraunhofer* or radiation-zone approximation), the second exponential factor in the integrand is essentially unity, and the integration variable ρ' can be replaced by $\xi' = \rho'\sqrt{k/L}$ to yield

$$E(\xi) \approx \frac{jA}{2\pi}\frac{e^{-jk(L+r_0)}e^{-jk\rho^2/2L}}{r_0}\int_S d^2\xi' e^{-j\xi\cdot\xi'} \tag{21.69}$$

given that $\xi = \rho\sqrt{k/L}$. If the aperture is circular with radius a, then

$$\int_S d^2\xi' e^{-j\xi\cdot\xi'} = 2\pi\int_0^{\xi_0} d\xi'\xi' J_0(\xi\xi') \tag{21.70}$$

with $\xi_0 = a\sqrt{k/L}$; $J_0(\xi\xi')$ is a Bessel function of the first kind and of zero order. If the aperture is not many wavelengths in diameter then $\xi_0^2 = ka^2/L \ll 1$, and the integral (21.70) reduces to the factor $\pi\xi_0^2 = \pi ka^2/L$ because the Bessel function's argument is much less that unity for all ξ' (which makes $J_0(\xi\xi') \approx 1$). Let $E(r_0) = A\exp(-jkr_0)/r_0$. We then obtain at $r = (x, y, L)$ with $\rho = \sqrt{x^2+y^2}$

$$\frac{E(r)}{E(r_0)} \approx \frac{j}{2L}e^{-jk(L+\rho^2/2L)} \tag{21.71}$$

So there are circular fringes at distance L of width $\sim \sqrt{\lambda L}$ in which the E field is either relatively strong (reinforcement of the rim contributions) or relatively weak (destructive interference of the rim contributions). This is typical of what is known as a *diffraction pattern*. The calculation becomes more difficult in the *Fresnel zone*, where $k\rho'^2$ is not much less than $2L$, so that the last factor in the integrand of (21.68) must be taken into account (but then we do need to have $k\rho'^3 \ll 2L^2$ if (21.68) is to be valid).

In contrast to the refraction of a wave, discussed in the previous section, diffraction is due to the shape of an object and comes about by scattering of the electric field at 'edges', i.e. locations where the refractive index undergoes sharp discontinuities. Refraction and diffraction can be considered as different aspects of the scattering of high-frequency waves; nevertheless these fascinating phenomena are not always completely separable.

Appendices

Many engineering students taking upper-level courses in undergraduate electromagnetism find that they need to draw on mathematical skills obtained in introductory calculus, linear algebra, and even late high-school mathematics courses – which, by the time they take a course in electromagnetism, can lie one or more years behind them. The mathematical concepts may have become somewhat hazy in memory. The purpose of the appendix is to go over some of the key concepts as a refresher. The material treated here will not suffice to replace the full mathematical course material but will serve merely to remind the reader of the key concepts and how these relate to more familiar material. Nor are proofs given with the rigor mathematicians require, but it is hoped that enough is provided below for the reader to be able to understand the given mathematics, and therefore also to be able to apply that meaningfully.

A.1 Some results from calculus

A.1.1 Taylor series

It is often useful to estimate the value of a function $f(x)$ of a single variable x at a location that is close to a location x_0 at which $f(x_0)$ has a known value. The mathematical statement of that estimate is a Taylor series:

$$\begin{aligned} f(x) &= f(x_0) + (x - x_0)f'(x_0) + \frac{1}{2!}(x - x_0)^2 f''(x_0) + \frac{1}{3!}(x - x_0)f^{(3)}(x_0) + \cdots \\ &= \left[e^{(x-x_0)d/d\xi} f(\xi)\right]_{\xi=x_0} \end{aligned} \tag{A.1a}$$

In this formula, $f^{(n)}(x_0)$ indicates the nth derivative of the function $f(x)$ evaluated at the location x_0. This is indicated more clearly by the second *symbolic* form in (A.1a) in which the exponential is handled as if $d/d\xi$ were an ordinary variable yielding $d^n/d\xi^n$ when the exponential is expanded into its power series. This symbolic form should be considered only as an aid to memory; among other things, the order of the exponential and $f(\xi)$ cannot be interchanged! The dummy variable ξ is used here to distinguish a variable quantity from the fixed values x and x_0. First we take derivatives with respect to ξ, only afterwards do we set $\xi = x_0$. An alternative form is

$$f(x+\Delta x) = f(x) + (\Delta x)f'(x) + \frac{1}{2!}(\Delta x)^2 f''(x) + \frac{1}{3!}(\Delta x)^3 f^{(3)}(x) + \cdots$$
$$= e^{(\Delta x)d/dx} f(x) \tag{A.1b}$$

where Δx is a fixed interval along the x axis.

A number of conditions must hold for the validity of (A.1). First, the function $f(\xi)$ needs to be continuous in the entire interval $x_0 \le \xi \le x$, and it must be differentiable infinitely often at $\xi = x_0$. A two-dimensional analog is found from the symbolic form

$$f(x+\Delta x, y+\Delta y) = e^{(\Delta x\partial/\partial x + \Delta y\partial/\partial y)} f(x,y) \tag{A.2a}$$

Expansion of the symbolic exponential into its power series (and treatment of x, y as independent variables) will yield the term-by-term two-dimensional Taylor series. The first few terms are

$$f(x+\Delta x, y+\Delta y) = f(x,y) + \Delta x\frac{\partial f}{\partial x} + \Delta y\frac{\partial f}{\partial y}$$
$$+ \frac{1}{2!}\left[(\Delta x)^2\frac{\partial^2 f}{\partial x^2} + (\Delta y)^2\frac{\partial^2 f}{\partial y^2} + 2(\Delta x)(\Delta y)\frac{\partial^2 f}{\partial x\partial y}\right] + \cdots \tag{A.2b}$$

The symbolic form (A.2a) is easily generalized to three dimensions: just add $\Delta z\partial/\partial z$ to the exponential, and replace $f(x,y)$ by $f(x,y,z)$. Equations (A.1) state something interesting, which is often not stressed: if you know all the derivatives of $f(x)$ then you can predict the entire function for all values of $x > x_0$. This of course is not practical, but it *is* practical to make use of the related fact that knowledge of the first N derivatives allows us to approximate $f(x)$ for some limited range Δx of values larger than x_0. In general, the more derivatives we know the larger Δx is. This is a property of which much use will be made in this course.

The proofs of these theorems usually derive from the mean-value theorem of calculus, which states that somewhere between x_0 and x there is a point x_1 such that the derivative at x_1 has the value $[f(x) - f(x_0)]/(x - x_0)$.

A.1.2 Multiple integrals

Figure A.1 illustrates two possible ways of doing a two-dimensional integral of $f(x,y)$ over the area shown in the figure. In Fig. A.1(a) the double integral can be

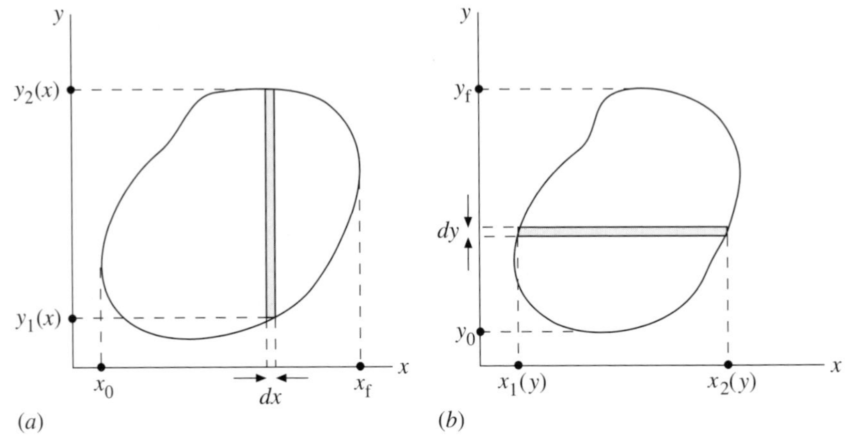

written as

$$I = \int_{x_0}^{x_f} dx \int_{y_1(x)}^{y_2(x)} dy\, f(x, y) \tag{A.3}$$

We perform this integral in two parts. First, we keep x at some fixed value between x_0 and x_f and we do the dy integration. This yields

$$F(x) = \int_{y_1(x)}^{y_2(x)} dy\, f(x, y) \tag{A.4a}$$

In the second part, we simply calculate

$$I = \int_{x_0}^{x_f} dx\, F(x) \tag{A.4b}$$

In Fig. A.1(b), the integral is simply done in the reverse order (the diagram shows the same procedure as that for Fig. A.1(a), except that x and y are reversed).

$$I = \int_{y_0}^{y_f} dy \int_{x_1(y)}^{x_2(y)} dx\, f(x, y) \tag{A.4c}$$

We strongly advise use of the way in which the multiple integrals are written in (A.4): the infinitesimal intervals dx and dy are placed immediately beyond their respective integral signs. This avoids the ambiguity that is implicit in, say,

$$\int_0^{2\pi}\int_0^{\pi} d\theta\, d\varphi f(\theta, \varphi)$$

where it might be unclear which variable is integrated between 0 and 2π, and which between 0 and π. Mathematicians use one, engineers another convention. Some people associate the second infinitesimal $(d\varphi)$ with the left-most integral sign, others with the right-most one. Equations (A.4) avoid these ambiguities!

Extension to more than two variables is straightforward when the above conventions are adopted. As an example, here is an integral of a function $g(x, y, z)$ over the volume between two surfaces:

$$I = \int_{x_0}^{x_f} dx \int_{y_1(x)}^{y_2(x)} dy \int_{z_1(x,y)}^{z_2(x,y)} dz\, g(x, y, z) \equiv \int_{x_0}^{x_f} dx \int_{y_1(x)}^{y_2(x)} dy\, f(x, y) \tag{A.5}$$

First an area element $dx\,dy$ is kept constant, and $g(x, y, z)$ is integrated over z, from the intersection of the infinitesimal $dx\,dy$ column with one bounding surface to its intersection with the other, to obtain $f(x, y)$. This reduces the problem to the above two-dimensional case.

A.1.3 Logarithms

The natural logarithm of a product of two functions is

$$\ln[f(x)y(x)] = \ln[f(x)] + \ln[g(x)] \tag{A.6a}$$

and that of a power is therefore

$$\ln a^p = p \ln a \tag{A.6b}$$

These rules are useful, for example, in reducing $\ln\sqrt{x^2 + \varepsilon^2}$ when $|\varepsilon| \ll |x|$, i.e. when ε is very small in magnitude compared to x. We then can use (A.6) to obtain:

$$\begin{aligned}\ln\sqrt{x^2 + \varepsilon^2} &= \ln[(x^2 + \varepsilon^2)^{1/2}] = \frac{1}{2}\ln(x^2 + \varepsilon^2) \\ &= \frac{1}{2}\ln x^2 + \frac{1}{2}\ln\left(\frac{x^2 + \varepsilon^2}{x^2}\right) = \ln x + \frac{1}{2}\ln\left(1 + \frac{\varepsilon^2}{x^2}\right)\end{aligned} \tag{A.7}$$

The last term can be reduced by using the power-series expansion (for $|\epsilon| \ll 1$)

$$\ln(1+\epsilon) = \epsilon - \frac{\epsilon^2}{2} + \frac{\epsilon^3}{3} - \frac{\epsilon^4}{4} + \frac{\epsilon^5}{5} - \cdots \tag{A.8}$$

Another useful item is the conversion from base a to base b. Let $f \equiv \ln_a x$. That implies that $a^f = x$. Therefore $\ln_b a^f = \ln_b x$, and thus one obtains $f \ln_b a = \ln_b x$ so that

$$\ln_a x = \frac{\ln_b x}{\ln_b a} \tag{A.9}$$

A form we often use is the conversion from base 10 to base e: $\log x = \ln x / \ln 10 \approx 0.4343 \ln x$. See below for the definition of the logarithm of a complex number.

A.1.4 Binomial theorem

It is sometimes useful to expand $(1+x)^p$ into a power series in x (when x and p are not large) by means of the binomial theorem:

$$\begin{aligned}(1+x)^p &= 1 + px + \frac{1}{2}p(p-1)x^2 + \frac{1}{3!}p(p-1)(p-2)x^3 + \cdots \\ &= 1 + \sum_{n=1}^{\infty} \binom{p}{n} x^n \quad \text{with} \quad \binom{p}{n} \equiv \frac{p!}{n!(p-n)!}\end{aligned} \tag{A.10}$$

Validity of this expansion is ensured when $|x| < 1$ and $|px| < 1$. This theorem is used often in the main text.

A.2 Some elementary facts about complex numbers

The need for complex numbers can be shown to arise in many ways. One that is perhaps familiar to the reader is arrived at through solving quadratic equations. Readers of these lines should be familiar with the solution of the equation $ax^2 + bx + c = 0$, which is

$$x = -\frac{b}{2a} \pm \frac{1}{2a}\sqrt{b^2 - 4ac} \tag{A.11a}$$

It is quite possible that $b^2 < 2ac$, in which case

$$x = -\frac{b}{2a} \pm \frac{1}{2a}\sqrt{4ac - b^2}\,\sqrt{-1} \tag{A.11b}$$

which is of the form $A \pm B\sqrt{-1}$, where A and B both are real quantities. If we abbreviate $\sqrt{-1}$ as j, as is the custom in engineering (an 'i' is used in physics and mathematics), then we arrive at what is known as a *complex number* $A + Bj$.

Both the sum and the product of two complex numbers are complex numbers. To get the inverse of a complex number $a + bj$ we use the following procedure:

$$\frac{1}{a+bj} = \frac{a-bj}{(a+bj)(a-bj)} = \frac{a-bj}{a^2+b^2} \tag{A.12}$$

The numerator $a - bj$ is known as the *complex conjugate* and the denominator $a^2 + b^2$ is the square of the *modulus* or *absolute value* of the original complex number. If $\zeta = a + bj$, then the notation $\zeta^* \equiv a - bj$ is used for the complex conjugate, and the notation $|\zeta|$ for the modulus $\sqrt{a^2 + b^2}$. Equation (A.12) allows for the division of two complex numbers, and this also leads to a new complex number:

$$\frac{c+dj}{a+bj} = \frac{(c+dj)(a-bj)}{(a+bj)(a-bj)} = \frac{(ac+bd)+(ad-bc)j}{a^2+b^2} \tag{A.13}$$

which is of the form $A + Bj$. The addition of two complex numbers reminds one of the addition of vectors:

$$\begin{aligned}(a+bj)+(c+dj) &= (a+c)+(b+d)j \\ (a\hat{\mathbf{x}}+b\hat{\mathbf{y}})+(c\hat{\mathbf{x}}+d\hat{\mathbf{y}}) &= (a+c)\hat{\mathbf{x}}+(b+d)\hat{\mathbf{y}}\end{aligned} \tag{A.14}$$

This suggests that – at least for addition and subtraction – we can represent a complex number $\zeta = a + bj$ by a location on a two-dimensional graph with Cartesian axes. The horizontal axis is labeled Re (for 'real') and the vertical one Im (for 'imaginary'), as shown in Fig. A.2. From the above, we see that $|\zeta|$ is equivalent (in Fig. A.2) to the length of a vector from the origin to the point characterized by (a, b). The figure also shows that

$$a = |\zeta|\cos\varphi \quad b = |\zeta|\sin\varphi \qquad \rightarrow \qquad \zeta = |\zeta|(\cos\varphi + j\sin\varphi) \tag{A.15}$$

so that $\cos\varphi + j\sin\varphi$ is a complex number with unit modulus. Also, a fixed value of $|\zeta|$ represents the radius of a circle around the origin that is the locus of all complex

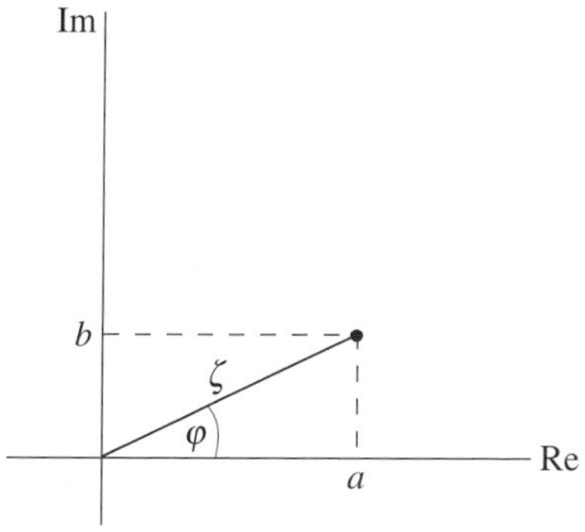

Fig. A.2 An Armand diagram showing a complex number as a vector.

numbers with modulus $|\zeta|$. The *Euler relationship* establishes that

$$e^{j\varphi} = \cos\varphi + j\sin\varphi \tag{A.16}$$

Now using the expansions of $\cos(\varphi_1 + \varphi_2)$ and $\sin(\varphi_1 + \varphi_2)$, we can show that

$$\cos(\varphi_1 + \varphi_2) + j\sin(\varphi_1 + \varphi_2) = (\cos\varphi_1 + j\sin\varphi_1)(\cos\varphi_2 + j\sin\varphi_2)$$

If we then use (A.16) on both sides of this equation, we obtain

$$e^{j(\varphi_1+\varphi_2)} = e^{j\varphi_1}e^{j\varphi_2} \tag{A.17}$$

from which all the other properties of $e^{j\varphi}$ follow. As a consequence, we also can represent a complex number as

$$\zeta \equiv a + bj = |\zeta|(\cos\varphi + j\sin\varphi) = |\zeta|e^{j\varphi} \tag{A.18}$$

This latter form is known as the *phasor* representation of a complex number. It is now possible to use Fig. A.2 to compute products of two complex numbers, e.g. $\zeta_1 = a_1 + b_1 j$ and $\zeta_2 = a_2 + b_2 j$. Write them in phasor representation to obtain

$$\zeta_1\zeta_2 = |\zeta_1||\zeta_2|e^{j(\varphi_1+\varphi_2)} \tag{A.19}$$

Thus, the moduli are multiplied together to produce a new circle radius, and this new radius is at the sum angle $\varphi_1 + \varphi_2$ to the horizontal axis. These phasor representations are most useful for applications. The relationships of *de Moivre* follow rapidly from (A.16):

$$\cos(n\varphi) + j\sin n\varphi = e^{jn\varphi} = (e^{j\varphi})^n = (\cos\varphi + j\sin\varphi)^n \tag{A.20}$$

Figure A.3 illustrates the additions (A.14), and the use of phasors to do so. Through use of the ordinary plane-geometry relationship between the sides of a triangle, we find for the sum of $\zeta_1 = a_1 + jb_1$ and $\zeta_2 = a_2 + jb_2$, with (A.15) holding

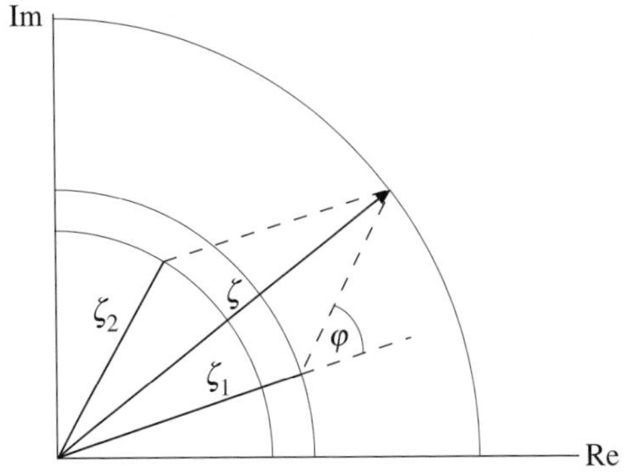

Fig. A.3 Addition of two complex number ζ_1 and ζ_2. The phase angles φ_1 and φ_2 are

for each of the a's and b's,

$$\zeta = \sqrt{\zeta_1^2 + \zeta_2^2 + 2\zeta_1\zeta_2 \cos\varphi}, \quad \varphi = \varphi_2 - \varphi_1 \tag{A.21}$$

As the figure shows, this is identical to a vector sum.

The logarithm of a complex number now can be worked out easily. Use $\zeta = a + bj = |\zeta|e^{j\varphi}$ with $\zeta = \sqrt{a^2+b^2}$ and $\tan\varphi = b/a$:

$$\begin{aligned}\ln(a+bj) &= \ln(|\zeta|e^{j\varphi}) = \ln|\zeta| + \ln(e^{j\varphi}) = \ln|\zeta| + j\varphi \\ &= \tfrac{1}{2}\ln(a^2+b^2) + j\arctan(b/a)\end{aligned} \tag{A.22}$$

Here, it is essential to realize that the exponential and the natural logarithm are inverse functions of each other, i.e. $e^{\ln x} = x$ and $\ln e^x = x$. Or, in other words, if $y = e^x$ then $x = \ln y$.

A.3 Vectors

A.3.1 The principal vector operators

In this appendix, V is any scalar field, $\mathbf{A}$ is any vector field.

Cartesian (rectangular) coordinates

$$\nabla V = \frac{\partial V}{\partial x}\hat{\mathbf{x}} + \frac{\partial V}{\partial y}\hat{\mathbf{y}} + \frac{\partial V}{\partial z}\hat{\mathbf{z}} \qquad \text{grad } V$$

$$\nabla\cdot\mathbf{A} = \frac{\partial A_x}{\partial x} + \frac{\partial A_y}{\partial y} + \frac{\partial A_z}{\partial z} \qquad \text{div } \mathbf{A}$$

$$\nabla\times\mathbf{A} = \left(\frac{\partial A_z}{\partial y} - \frac{\partial A_y}{\partial z}\right)\hat{\mathbf{x}} + \left(\frac{\partial A_x}{\partial z} - \frac{\partial A_z}{\partial x}\right)\hat{\mathbf{y}} + \left(\frac{\partial A_y}{\partial x} - \frac{\partial A_x}{\partial y}\right)\hat{\mathbf{z}} \qquad \text{curl } \mathbf{A}$$

$$\Delta V \equiv \nabla^2 V = \frac{\partial^2 V}{\partial x^2} + \frac{\partial^2 V}{\partial y^2} + \frac{\partial^2 V}{\partial z^2} \qquad \text{Laplacian of } V$$

Cylindrical coordinates

$$\nabla V = \frac{\partial V}{\partial \rho}\hat{\boldsymbol{\rho}} + \frac{1}{\rho}\frac{\partial V}{\partial \varphi}\hat{\boldsymbol{\varphi}} + \frac{\partial V}{\partial z}\hat{\mathbf{z}} \qquad \text{grad } V$$

$$\nabla \cdot \mathbf{A} = \frac{1}{\rho}\frac{\partial}{\partial \rho}(\rho A_\rho) + \frac{1}{\rho}\frac{\partial A_\varphi}{\partial \varphi} + \frac{\partial A_z}{\partial z} \qquad \text{div}\,\mathbf{A}$$

$$\nabla \times \mathbf{A} = \left[\frac{1}{\rho}\frac{\partial A_z}{\partial \varphi} - \frac{\partial A_\varphi}{\partial z}\right]\hat{\boldsymbol{\rho}} + \left[\frac{\partial A_\rho}{\partial z} - \frac{\partial A_z}{\partial \rho}\right]\hat{\boldsymbol{\varphi}} + \frac{1}{\rho}\left[\frac{\partial}{\partial \rho}(\rho A_\varphi) - \frac{\partial A_\rho}{\partial \varphi}\right]\hat{\mathbf{z}} \qquad \text{curl}\,\mathbf{A}$$

$$\Delta V \equiv \nabla^2 V = \frac{1}{\rho}\frac{\partial}{\partial \rho}\left(\rho\frac{\partial V}{\partial \rho}\right) + \frac{1}{\rho^2}\frac{\partial^2 V}{\partial \varphi^2} + \frac{\partial^2 V}{\partial z^2} \qquad \text{Laplacian of } V$$

Spherical coordinates

$$\nabla V = \frac{\partial V}{\partial r}\hat{\mathbf{r}} + \frac{1}{r}\frac{\partial V}{\partial \theta}\hat{\boldsymbol{\theta}} + \frac{1}{r\sin\theta}\frac{\partial V}{\partial \varphi}\hat{\boldsymbol{\varphi}} \qquad \text{grad}\,V$$

$$\nabla \cdot \mathbf{A} = \frac{1}{r^2}\frac{\partial}{\partial r}(r^2 A_r) + \frac{1}{r\sin\theta}\frac{\partial}{\partial \theta}(A_\theta \sin\theta) + \frac{1}{r\sin\theta}\frac{\partial A_\varphi}{\partial \varphi} \qquad \text{div}\,\mathbf{A}$$

$$\nabla \times \mathbf{A} = \frac{1}{r\sin\theta}\left[\frac{\partial}{\partial \theta}(A_\varphi \sin\theta) - \frac{\partial A_\theta}{\partial \varphi}\right]\hat{\mathbf{r}} + \frac{1}{r}\left[\frac{1}{\sin\theta}\frac{\partial A_r}{\partial \varphi} - \frac{\partial}{\partial r}(rA_\varphi)\right]\hat{\boldsymbol{\theta}}$$
$$+ \frac{1}{r}\left[\frac{\partial}{\partial r}(rA_\theta) - \frac{\partial A_r}{\partial \theta}\right]\hat{\boldsymbol{\varphi}} \qquad \text{curl}\,\mathbf{A}$$

$$\Delta V \equiv \nabla^2 V = \frac{1}{r^2}\frac{\partial}{\partial r}\left(r^2\frac{\partial V}{\partial r}\right) + \frac{1}{r^2\sin\theta}\left[\frac{\partial}{\partial \theta}\left(\sin\theta\frac{\partial V}{\partial \theta}\right) + \frac{1}{\sin\theta}\frac{\partial^2 V}{\partial \varphi^2}\right]$$

Laplacian of V

A.3.2 Useful vector identities

In this appendix and the next, $u, v, w, \ldots$ are scalars and $\mathbf{A}, \mathbf{B}, \mathbf{C}, \ldots$ are vectors.

(1) $\mathbf{A}\cdot\mathbf{B} = \mathbf{B}\cdot\mathbf{A}$ — Commutative law for dot product
(2) $\mathbf{A}\times\mathbf{B} = -\mathbf{B}\times\mathbf{A}$ — Anticommutative law for cross product
(3) $\mathbf{A}\cdot(\mathbf{B}\times\mathbf{C}) = \mathbf{B}\cdot(\mathbf{C}\times\mathbf{A}) = \mathbf{C}\cdot(\mathbf{A}\times\mathbf{B})$ — Scalar triple product
(4) $\mathbf{A}\times(\mathbf{B}\times\mathbf{C}) = (\mathbf{A}\cdot\mathbf{C})\mathbf{B} - (\mathbf{A}\cdot\mathbf{B})\mathbf{C}$ — Vector triple product
(5) $(\mathbf{A}\times\mathbf{B})\cdot(\mathbf{C}\times\mathbf{D}) = (\mathbf{A}\cdot\mathbf{C})(\mathbf{B}\cdot\mathbf{D}) - (\mathbf{A}\cdot\mathbf{D})(\mathbf{B}\cdot\mathbf{C})$
(6) $(\mathbf{A}\times\mathbf{B})\times(\mathbf{C}\times\mathbf{D}) = [(\mathbf{A}\times\mathbf{B})\cdot\mathbf{D})]\mathbf{C} - [(\mathbf{A}\times\mathbf{B})\cdot\mathbf{C})]\mathbf{D}$

A.3.3 Useful vector-operator identities

(1) $\nabla(uv) = u\nabla v + v\nabla u$ — 1st chain rule
(2) $\nabla\cdot(u\mathbf{A}) = u\nabla\cdot\mathbf{A} + \mathbf{A}\cdot\nabla u$ — 2nd chain rule

(3) $\nabla \times (u\mathbf{A}) = u\nabla \times \mathbf{A} + (\nabla u) \times \mathbf{A}$ 3rd chain rule

(4) $\nabla \cdot (\mathbf{A} \times \mathbf{B}) = \mathbf{B} \cdot (\nabla \times \mathbf{A}) - \mathbf{A} \cdot (\nabla \times \mathbf{B})$ 4th chain rule

(5) $\nabla(\mathbf{A} \cdot \mathbf{B}) = \mathbf{A} \times (\nabla \times \mathbf{B}) + \mathbf{B} \times (\nabla \times \mathbf{A}) + (\mathbf{A} \cdot \nabla)\mathbf{B} + (\mathbf{B} \cdot \nabla)\mathbf{A}$

(6) $\nabla \times (\mathbf{A} \times \mathbf{B}) = \mathbf{A}(\nabla \cdot \mathbf{B}) - \mathbf{B}(\nabla \cdot \mathbf{A}) + (\mathbf{B} \cdot \nabla)\mathbf{A} - (\mathbf{A} \cdot \nabla)\mathbf{B}$

(7) $\mathbf{A} \times (\nabla \times \mathbf{B}) = (\nabla\mathbf{B}) \cdot \mathbf{A} - (\mathbf{A} \cdot \nabla)\mathbf{B}$ $(\nabla\mathbf{B})$ is a dyadic

(8) $\nabla \times (\nabla V) = 0$ curl grad $V = 0$

(9) $\nabla \cdot (\nabla \times \mathbf{A}) = 0$ div curl $\mathbf{A} = 0$

(10) $\Delta\mathbf{A} \equiv \nabla \cdot \nabla\mathbf{A} \equiv \nabla^2\mathbf{A} = \nabla(\nabla \cdot \mathbf{A}) - \nabla \times (\nabla \times \mathbf{A})$ Laplacian of a vector

A.4 Some mathematical and physical constants*

Quantity	Symbol and value	Unit
pi	$\pi = 3.14159265358979$	
base of natural logarithms	$e = 2.71828182845905$	
speed of light in free space	$c = 2.99792458 \times 10^8$	m/s
permittivity of free space	$\varepsilon_0 = 8.8541878 \times 10^{-12}$	F/m or N m^2/C^2
Planck's constant	$h = 6.6260755 \times 10^{-34}$	J s
permeability of free space	$\mu_0 = 4\pi \times 10^{-7}$	H/m
charge of an electron	$e = 1.60217733 \times 10^{-19}$	C
mass of an electron	$m_e = 9.1093897 \times 10^{-31}$	kg
mass of a proton	$m_p = 1.6726231 \times 10^{-27}$	kg
charge-to-mass ratio of electron	$e/m = 1.75881962 \times 10^{11}$	C/kg
proton-to-electron mass ratio	$m_p/m_e = 1836.15$	
electron magnetic moment	$m_s = 9.2847701 \times 10^{-24}$	kg m^2 or J/T
classical radius of electron	$r_e = 2.818 \times 10^{-15}$	m
Avogadro's constant	$N = 6.0221367 \times 10^{23}$	mol^{-1}
Boltzmann's constant	$k = 1.380658 \times 10^{-23}$	J/K

*Values from *A Physicist's Desk Reference*, 2nd edition (ed. H.L. Anderson), American Institute of Physics, New York, 1989.

A.5 Various material constants

Dielectric permittivities of selected materials

Material	Permittivity[†]
air (0 °C)	1.00059
paraffin	2.0–2.5
lucite, plexiglass	2.9–3.5
rubber	2.9–6.6
sugar	3.3
quartz (fused)	3.8
paper	2.0–4.0
glass, Vycor	3.8–3.9
glass, Pyrex	4.0–6.0
glass, Corning	6.0–6.8
porcelain	6.0–8.0
salt	6.1
mica	7.5
barium chloride	9.4
ferrous oxide (15 °C)	14.2
lead sulfide (15 °C)	17.9
acetone	20.7
ethanol (25 °C)	24.3
lead nitrate	37.7
thallium chloride	46.9
water (25 °C)	78.5
teflon	100.0

[†]Values from *CRC Handbook of Chemistry and Physics*, 64th edition, CRC Press, Boca Raton FL, 1983. All values of $\varepsilon_r = \varepsilon/\varepsilon_0$ are those measured at room temperature and pressure and for low frequency, unless indicated differently.

Conductivities of selected materials

Material	Conductivity*
copper	5.9×10^7
gold	4.1×10^7
aluminum	3.82×10^7
nickel	1.45×10^7
iron (99.99%)	1.05×10^7
tin	8.77×10^6
gallium (hard-wire)	5.8×10^6
bismuth	9.4×10^5
graphite	10^5
seawater	4.0
water	10^{-4}–10^{-3}
window glass	2×10^{-5}
nylon	10^{-13}–10^{-12}
mica	10^{-15}–10^{-11}
polystyrene	10^{-19}–10^{-15}

*Values from *CRC Handbook of Chemistry and Physics*, 64th edition, CRC Press, Boca Raton FL, 1983. All values are in S/m and are as measured at room temperature and pressure and at low frequency.

Dielectric breakdown fields of selected materials

Material	Breakdown field**
mica	39.4–78.7
polystyrenes	11.8–25.6
hard rubber	13.5–25.6
nylon	16.5–19.7
air	~ 3
electrical ceramics	2.2–11.8

**Values from *Engineering Materials and Their Applications*, 4th edition, R.A. Flin and P.K. Trojan, Houghton Mifflin, Boston MA, 1990. All values are in MV/m and are as measured at room temperature and pressure.

Bibliography: Selected texts on electromagnetics

INTRODUCTORY LEVEL

D.K. Cheng, *Field and Wave Electromagnetics*, 2nd edition, Addison-Wesley, Reading MA, 1989

D.K. Cheng, *Fundamentals of Engineering Electromagnetics*, Addison-Wesley, Reading MA, 1993

K.R. Demarest, *Engineering Electromagnetics*, Prentice-Hall, Upper Saddle River NJ, 1998

B.S. Guru and H.R. Hiziroglu, *Electromagnetic Field Theory Fundamentals*, PWS Publishing, Boston, 1998

H.A. Haus and J.R. Melcher, *Electromagnetic Fields and Energy*, Prentice Hall, Englewood Cliffs NJ, 1989

W.H. Hayt, Jr., *Engineering Electromagnetics*, 5th edition, McGraw-Hill, New York, 1989

S.R.H. Hoole and P.R.P. Hoole, *A Modern Short Course in Engineering Electromagnetics*, Oxford University Press, Oxford UK, 1996

U.S. Inan and A.S. Inan, *Engineering Electromagnetics*, Addison-Wesley, Menlo Park CA, 1999

C.T.A. Johnk, *Engineering Electromagnetic Fields and Waves*, J. Wiley & Sons, New York, 1988

R.W.P. King and S. Prasad, *Fundamental Electromagnetic Theory and Applications*, Prentice Hall, Englewood Cliffs NJ, 1986

J.D. Kraus and D.A. Fleisch, *Electromagnetics*, 5th edition, McGraw-Hill, New York, 1999

S.V. Marshall, R.E. DuBroff and G.G. Skitek, *Electromagnetic Concepts and Applications*, Prentice Hall, Englewood Cliffs NJ, 1996

G.F. Miner, *Lines and Electromagnetic Fields for Engineers*, Oxford University Press, New York, 1996

M.N. Nayfeh and M.K. Brussel, *Electricity and Magnetism*, J. Wiley & Sons, New York, 1985

H.P. Neff, Jr., *Basic Electromagnetic Fields*, 2nd edition, Harper & Row, New York, 1987

H.P. Neff, Jr., *Introductory Electromagnetics*, J. Wiley & Sons, New York, 1991

D.T. Paris and F.K. Hurd, *Basic Electromagnetic Theory*, McGraw-Hill, New York, 1969

C.R. Paul, K.W. Whites and S.A. Nasar, *Introduction to Electromagnetic Fields*, 3rd edition, McGraw-Hill, 1998

N.N. Rao, *Elements of Engineering Electromagnetics*, 4th edition, Prentice Hall, Englewood Cliffs NJ, 1994

M.A. Plonus, *Applied Electromagnetics*, McGraw-Hill, New York, 1978
M.N.O. Sadiku, *Elements of Electromagnetics*, HRW Saunders, New York, 1989
S.E. Schwarz, *Electromagnetics for Engineers*, HRW Saunders, New York, 1990
F.T. Ulaby, *Fundamentals of Applied Electromagnetics*, Prentice-Hall, Upper Saddle River NJ, 1996
P.B. Visscher, *Fields and Electrodynamics*, J. Wiley & Sons, New York, 1988

INTERMEDIATE LEVEL

R.S. Elliott, *Electromagnetics – History, Theory, and Applications*, IEEE Press Series on Electromagnetic Waves. Institute of Electrical and Electronics Engineers, New York, 1993
R.P. Feynman, R.B. Leighton and M. Sands, *The Feynman Lectures on Physics, Vol. II*, Addison-Wesley, Reading MA, 1964
S. Liao, *Engineering Applications of Electromagnetic Theory*, West Publishing Co., St. Paul MN, 1988
S. Ramo, J.R. Whinnery and T. Van Duzer, *Fields and Waves in Communication Electronics*, 3rd edition, J. Wiley & Sons, New York, 1994
L.C. Shen and J.A. Kong, *Applied Electromagnetism*, 3rd edition, PWS Publishing, Boston MA, 1987
D.H. Staelin, A.W. Morgenthaler and J.A. Kong, *Electromagnetic Waves*, Prentice-Hall, Englewood Cliffs NJ, 1994

ADVANCED OR GRADUATE LEVEL

C.A. Balanis, *Advanced Engineering Electromagnetics*, J. Wiley & Sons, New York, 1989
R.F. Harrington, *Time-harmonic Electromagnetic Fields*, McGraw-Hill, New York, 1961
J.D. Jackson, *Classical Electrodynamics*, 3rd edition, J. Wiley & Sons, New York, 1999
J.A. Stratton, *Electromagnetic Theory*, McGraw-Hill, New York, 1941
J. Van Bladel, *Electromagnetic Theory* (*Revised printing*), Hemisphere Publishing, New York, 1985
J.R. Wait, *Electromagnetic Wave Theory*, Harper & Row, New York, 1985

Index